2nd International Conference on Mechanical Engineering Research

(ICMER 2013)

IOP Conference Series: Materials Science and Engineering Volume 50

Pahang, Malaysia
1 – 4 July 2013

ISBN: 978-1-63439-861-9
ISSN: 1757-8981

Printed from e-media with permission by:

Curran Associates, Inc.
57 Morehouse Lane
Red Hook, NY 12571

Some format issues inherent in the e-media version may also appear in this print version.

Printed by Curran Associates, Inc. (2015)

Institute of Physics
Dirac House, Temple Back
Bristol BS1 6BE UK

Phone: 44 1 17 929 7481
Fax: 44 1 17 920 0979

techtracking@iop.org

Additional copies of this publication are available from:

Curran Associates, Inc.
57 Morehouse Lane
Red Hook, NY 12571 USA
Phone: 845-758-0400
Fax: 845-758-2634
Email: curran@proceedings.com
Web: www.proceedings.com

TABLE OF CONTENTS

AUTOMOTIVE

APPLIED MECHANICS

COMPUTATIONAL ANALYSIS

ENERGY

MANUFACTURING

MATERIAL

IOP Conf. Series: Materials Science and Engineering **50** (2013) 011001 doi:10.1088/1757-899X/50/1/011001

In the name of Allah the Most Beneficent, Most Merciful

It is with great pleasure that I welcome the participants of the International Conference of Mechanical Engineering Research 2013. The Prophet Muhammad (peace be upon him) said "acquire knowledge and impart it to the people" (Al Tirmidhi). The quest for knowledge has been from the beginning of time but knowledge only becomes valuable when it is disseminated and applied to benefit humankind. It is hoped that ICMER 2013 will be a platform to gather and disseminate the latest knowledge in mechanical engineering. Academicians, Scientist, Researchers and practitioners of mechanical engineering will be able to share and discuss new findings and applications of mechanical engineering. It is envisaged that the intellectual discourse will result in future collaborations between universities, research institutions and industry both locally and internationally. In particular, it is expected that focus will be given to issues on environmental energy sustainability. Researchers in the mechanical engineering faculty at UMP have a keen interest in technology to harness energy from the ocean. Lowering vehicle emission has been a primary goal of researchers in the mechanical engineering faculty, and the automotive engineering centre, including developing vehicles using alternatives fuels such as biodiesel and renewable sources (such as solar).

Finally I would like to congratulate the organising committee for their tremendous efforts in organizing the conference. I pray to Allah S.W.T that the conference will be a successful event.

PROF. DR ZAHARI TAHA

Chairman

Honorary Chairman:

Profesor Dato' Dr. Daing Nasir Ibrahim

Chairman:

Profesor Dr. Hj. Zahari Taha

Secretary:

Dr.Rizalman Mamat

Mohamad Faizal Mohamed Zahri

Main Committee:

Profesor Ir. Dr. Yusoff Ali

Profesor Ir. Dr. Hassan Ibrahim

Dr. Muhamad Arifpin Mansor

Dr. Kumaran Kadirgama

Dr. Mohd Fairusham Ghazali

Dr. Syarifah Nur Aqida Syed Ahmad

Dr. Mahadzir Ishak@Muhammad

Ir. Dr. Hj. Nik Mohd Zuki Nik Mohamed

Publicity & Website

Prof. Madya Dr. Tuan Mohammad Yusoff Shah Tuan Ya (Chief)

Dr. Mohamad Firdaus Basrawi

Idris bin Mat Sahat

Mohammad Khalid Wahid

Azim Mohd Arshad

Mohd Padzly Radzi

Mohd Tarmizy Che Kar

Sponsorship

Mr. Lee Giok Chui (Chief)

Prof. Ir. Dr. Yusoff Ali

Luqman Hakim Ahmad Shah

Gan Leong Ming

Rosmazi Rosli

Mohd Sazali Salleh

Keynote Speaker

Dr. Kumaran Kadirgama (Chief)

Mohd Fadhlan Mohd Yusof

Technical

Dr. Mohd Fairusham Ghazali (Chief)

Nasrul Hadi Johari

Hazami Che Hussain

Mahdhir Mohd Yusof

Muhammad Adib Shaharun

Mohd Adib Mohd Amin

IOP Conf. Series: Materials Science and Engineering **50** (2013) 011001 doi:10.1088/1757-899X/50/1/011001

Proceedings & Journals Publication

Dr. Mahadzir Ishak@Muhammad (Chief)

Prof. Madya Dr. Md. Mustafizur Rahman

Dr.Siti Rabiatull Aisha

Amir Aziz

Che Ku Eddy Nizwan Che Ku Husin

Zulkifli Ahmad@Manap

Secretariat

Ir. Dr. Hj. NikMohdZuki Nik Mohamed (Chief)

Prof. Ir. Dr. Hassan Ibrahim

Dr. Syarifah Nur Aqida Syed Ahmad

Dr. Kumaran Kadirgama

Azim Mohd Arshad

Mohamad Zairi Baharom

Nur Azhani Abd Razak

Norhaida Ab. Razak

Nurazima Ismail

At-Tasneem Mohd Amin

Mohd Fadhlan Mohd Yusof

Mohammad Khalid Wahid

Wan Anuar Wan Hassan

Treasurer

Dr. Siti Rabiatull Aisha Idris

Rahimah Che Ramli

IOP Conf. Series: Materials Science and Engineering **50** (2013) 011002 doi:10.1088/1757-899X/50/1/011002

Salam 1Malaysia!

First and foremost, I want to thank all 2nd International Conference on Mechanical Engineering Research (ICMER 2013) organisers for inviting me to address and officiate this conference. This 2nd ICMER 2013 provide a platform to bring together not only researchers, postgraduate students but also industrial people. With this platform, ICMER will embarks whole process of making new discoveries and then translating them into products and services for the marketplace is only made possible with people like all of you.

I would like to congratulate ICMER 2013 organisers for the achievement of collecting 208 papers for this conference. Submission received from 17 local universities, 7 industrial companies and 9 different countries is a great achievement for UMP. I am very happy to welcome all of you from Malaysia, Iran, Turkey, Japan, India, Australia, Ireland, Saudi Arabia and Bangladesh to this conference. As one of Malaysia's Public Universities, UMP's main challenge is to remain competitive and relevant by offering high quality technical academic programmes and research activities, focusing on its niche areas. New knowledge and findings cannot be generated without any research and development (R&D) activities. These efforts will undoubtedly generate lots of interesting results and new knowledge either bring further commercialisation activities. Malaysian government has invested a huge sum of Ringgits in R&D over the years. Therefore, UMP must produce more quality researchers and graduates to ensure Malaysia reaps the returns from these investments vice versa of progressive economic growth for the country. UMP's 2011-2015 Strategic Plans determines to strengthen and sustained its financial support by allocating a research grants and industry collaboration and consultations through its business and commercialise unit. On behalf of UMP, I would like to express my appreciation to all committee members of ICMER 2013 from Faculty of Mechanical Engineering for their hardwork and relentless effort. Without their commitment and contributions, this event would not been possible and successfully delivered right this time.

I wish you all a fruitful discourse. Thank you.

PROFESSOR DATO' DR. DAING NASIR IBRAHIM

Vice-Chancellor, Universiti Malaysia Pahang

Development of formula varsity race car chassis

M A Abdullah[a], M R Mansur, N Tamaldin and K Thanaraj
Centre for Advanced Research in Energy, Faculty of Mechanical Engineering, Universiti Teknikal Malaysia Melaka, Hang Tuah Jaya, 76100 Durian Tunggal, Melaka, Malaysia

E-mail: [a]mohdazman@utem.edu.my

Abstract. Three chassis designs have been developed using commercial computer aided design (CAD) software. The design is based on the specifications of UTeM Formula Varsity™ 2012 (FV2012). The selection of the design is derived from weighted matrix which consists of reliability, cost, time consumption and weight. The score of the matrix is formulated based on relative weighted factor among the selections. All three designs are then fabricated using selected materials available. The actual cost, time consumption and weight of the chassis's are compared with the theoretical weighted scores. Standard processes of cuttings, fittings and welding are performed in chassis mock up and fabrication. The chassis is later assembled together with suspension systems, steering linkages, brake systems, engine system, and drive shaft systems. Once the chassis is assembled, the studies of driver's ergonomic and part accessibility are performed. The completion in final fittings and assembly of the race car and its reliability demonstrate an outstanding design for manufacturing (DFM) practices of the chassis.

1. Introduction

Based on the events of Formula SAE [1-2] and Formula Student [3], Universiti Teknikal Malaysia Melaka (UTeM) has started Formula Varsity in 2006 [4]. This biennial student level race competition is initiated to test the undergraduate students' analytical and technical skills in designing, analysing and developing a single-seater race car [5-7]. In 2012, a team of second year Bachelor of Mechanical Engineering in Automotive (BMCA) students has developed the race car named UTeM Racing Team (URT). This race car is an improved design from the previous cars [8-11] and components [12-19]. A conceptual design based on hybrid composite [20] and electric [21] have also been developed. In previous report, 2 concepts have been designed, analyzed and fabricated [22-23]. However, the chassis's were fabricated to meet the specification [23] and based on available materials. This has led to material waste when the fabricated chassis was not used in actual race event due to its low reliability. In this extended report of chassis development, 3 chassis concept designs are compared. The selected design is undergone with detail static analysis, acceleration analysis, braking analysis, cornering analysis and stiffness analyses.

2. Methodology

The CATIA V5 R20 CAR software is used in the development of chassis designs. Figure 1 – 3 show the concept designs. Table 1 tabulated the summary of each design. Table 2 shows the selection procedure based on the advantages and disadvantages of each design. Table 3 shows the actual components masses that included for the later stress analysis.

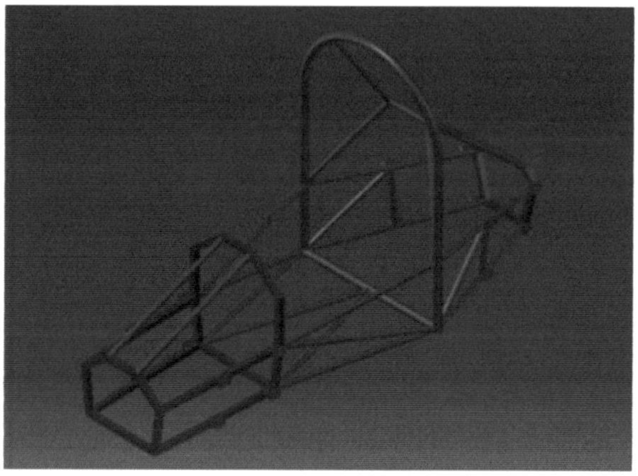

Figure 1. Concept design No. 1.

Figure 2. Concept design No. 2.

Figure 3. Concept design No. 3.

Table 1. Chassis concept designs.

	Concept design No. 1	Concept design No. 2	Concept design No. 3
Chassis volume	0.004 m^3	0.004 m^3	0.003 m^3
Material	ASTM A36	ASTM A500	ASTM A500
Material density	7860 kg/m^3	7850 kg/m^3	7850 kg/m^3
Yield strength	250 MPa	200 MPa	290 MPa
Theoretical Weight	31.761 kg	31.241 kg	21.857 kg
Poisson's ratio	0.266	0.300	0.300
Material composition			
i) Fe (%)	99.00 %	99.00 %	99.00 %
ii) C (%)	00.26 %	00.30 %	00.30 %

Table 2. Design selection.

	Concept design No. 1	Concept design No. 2	Concept design No. 3
Advantages	Able to withstand very strong impact. Meets safety regulations.	Easy to fabricate. Fulfil component layout. Able to withstand very strong impact.	Lightweight. Fulfil component layout. Meets safety regulations.
Disadvantages	Does not fulfil component layout. Heavyweight	Heavyweight. Does not meet safety regulations.	Difficult to fabricate. Unable to withstand very strong impact.
Selection	Not selected.	Not selected.	Selected.

3. Analysis

The total mass of the car with its components and driver (70 kg) is about 270.3 kg. This value is used in the Von Mises stress analysis. Figure 4 to 9 show the location of the forces and clamps for static, acceleration, braking, cornering, front stiffness and rear stiffness analyses respectively. As shown in Figure 4, for static case analysis, only the major components that were placed inside the chassis were considered. These weight components were the weight of the engine and the weight of the driver. The mass of the engine was 20 kg while the mass of the driver was approximately 70 kg which were about 200 N and 700 N respectively. The clamps were placed at all the parts of the chassis where the brackets were attached. The weight of the driver was distributed to the seat mountings while the weight of the engine was distributed to the engine mountings. The loads were directed downwards. In Figure 5, for acceleration analysis, only the major weight components of the car inside the chassis which were the engine and driver were considered. The loads and clamps were placed at the same parts of the chassis as in the static analysis. However, the loads were directed backwards, since, during acceleration, the engine and driver experience inertia whereby the mass of the driver and engine tends to stay where it was (pushing backwards) when the car was moving forward. In Figure 6, for braking analysis, the loads and clamps were placed at the same parts of the chassis as in the acceleration analysis. However, the loads were directed forward as during braking, the engine and driver experience inertia whereby the mass of the driver and engine tends to continue moving forward when the car is being stopped from moving forward.

Table 3. Component's mass.

Component	Mass (kg)
Front A-arm plus ball joint	4.12
Rear A-arm plus ball joint	4.28
Front knuckle plus brake set	14.89
Rear knuckle plus brake set	10.40
All four wheels (tire & rim)	38.48
Front absorbers	2.29
Rear absorbers	3.02
Steering rack	2.16
Differential	4.8
Radiator	0.15
Brake pumps plus pedals	1.88
Drive shaft	4.00
Engine	30.00

In Figure 7, the same magnitude of loads was distributed to the same mountings in the cornering analysis. However, the loads were directed to the side, as during cornering, the driver and engine experience inertia whereby the mass of the driver and the mass of the engine tends to stay where it was (pushing to the side way) as the car turns the other way. The front stiffness analysis was performed to test the stiffness at the front of the chassis as the chassis experiences torsion during cornering. As shown in Figure 8, the clamps were placed at parts of the chassis where the brackets at the back were attached. The load was divided at ratio of 40-60 whereby for front torsion, 40 % of the estimated total mass of the car was used (100 kg). This mass of 100 kg which is approximately 1000 N, placed at parts of the chassis where the brackets were attached at the front. The load was divided equally on both sides, one directing upwards while one directing downwards. The rear stiffness analysis was performed to test the stiffness at the back of the chassis as the chassis experiences torsion during cornering (Figure 9). As shown in Figure 9, the clamps were placed at parts of the chassis where the brackets at the front were attached. The load was divided at ratio of 40-60 whereby for rear torsion, 60 % of the estimated total mass of the car was used (150 kg). This mass of 150 kg which is approximately 1500 N, placed at parts of the chassis where the brackets were attached at the back. The load was divided equally on both sides, one directing upwards while one directing downwards.

4. Results and Discussion

Figure 10 – 15 show the stress distribution for each analysis. Figure 10 shows the Von Misses stress obtained from the analysis. The maximum stress obtained, which was 99.3 MPa, is the allowable stress while the yield stress (ultimate stress) of the material is 290 MPa. These values were used to calculate the factor of safety. The factor of safety was obtained by dividing the ultimate stress with the allowable stress which produced a value of 2.92. Figure 11 shows that for acceleration analysis, the maximum Von Misses stress value obtained was 52.8 MPa. The same method of calculation for factor of safety yielded a value of 5.49.

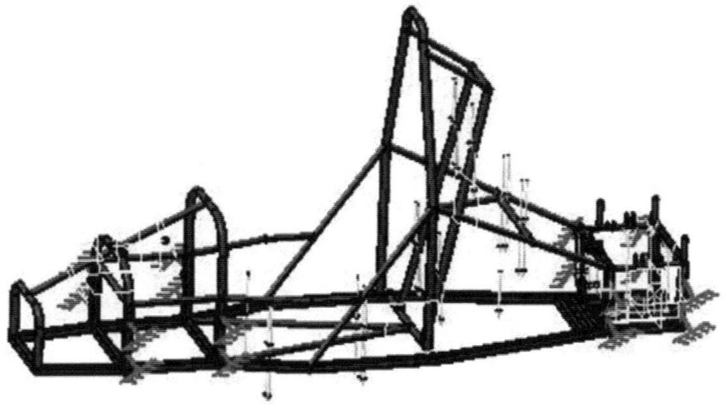

Figure 4. Static analysis.

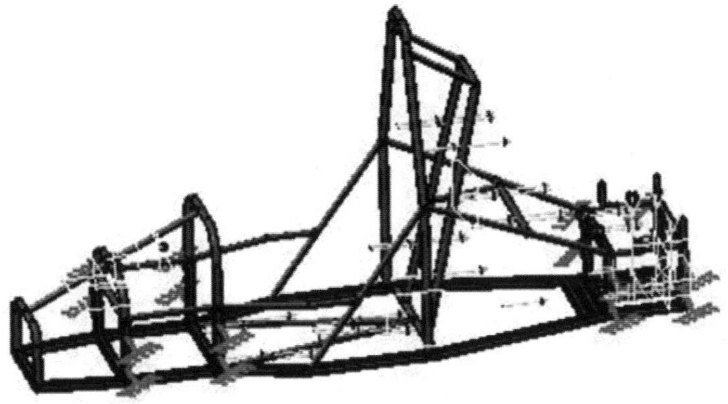

Figure 5. Acceleration analysis.

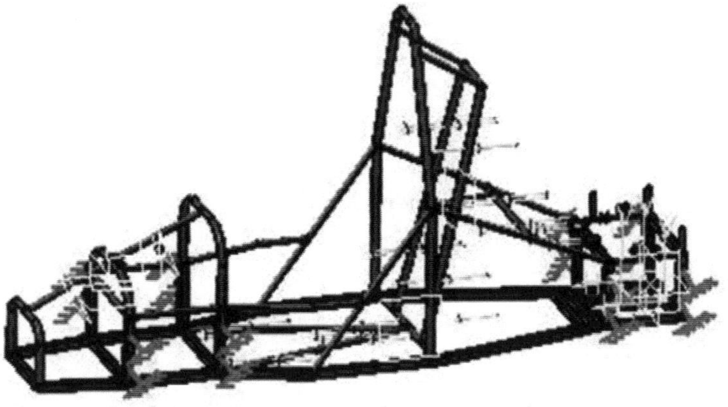

Figure 6. Braking analysis.

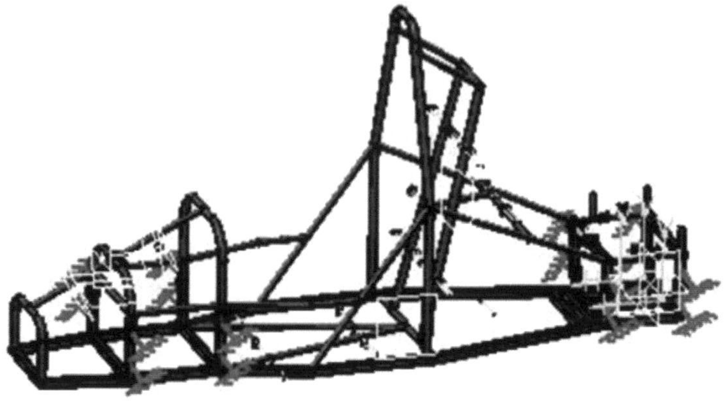

Figure 7. Cornering analysis.

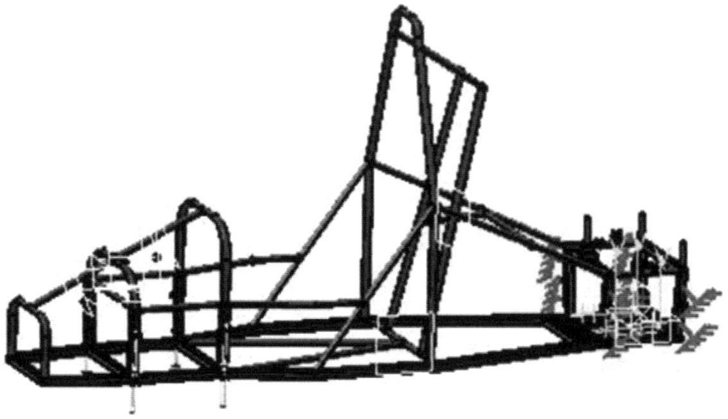

Figure 8. Front stiffness analysis.

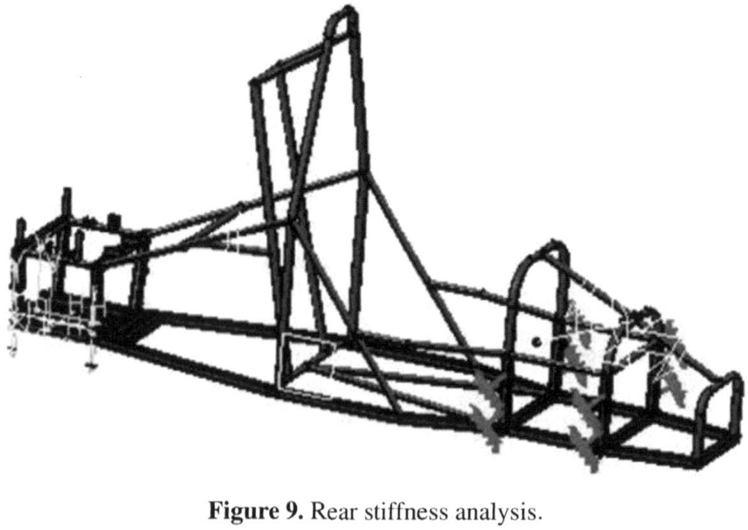

Figure 9. Rear stiffness analysis.

IOP Conf. Series: Materials Science and Engineering **50** (2013) 012001 doi:10.1088/1757-899X/50/1/012001

Figure 12 shows that for braking analysis, the maximum Von Misses stress obtained was same as in the acceleration analysis which was 52.8 MPa thus it yielded the same factor of safety, 5.49. Figure 13 shows that for cornering, the maximum value of Von Misses stress produced was 68.4 MPa. The corresponding calculation for factor of safety produced a value of 4.24. Figure 14 shows the maximum Von Misses stress value obtained from this analysis which was 133 MPa. When the yield stress of the material was divided by this value, it produced a factor of safety of 2.18. Figure 15 shows the maximum Von Misses stress value obtained from this analysis which was 254 MPa. When the yield stress of the material, 290 MPa was divided by this value, it produced a factor of safety of 1.14.

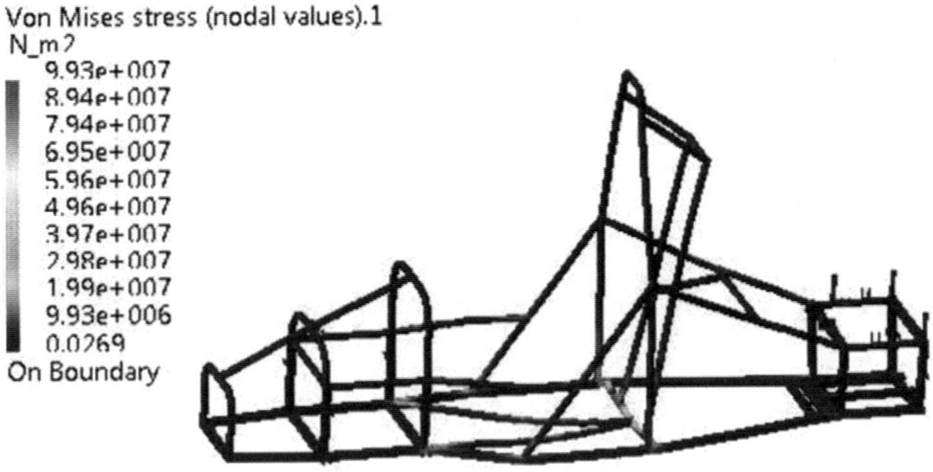

Figure 10. Stress distribution for static analysis.

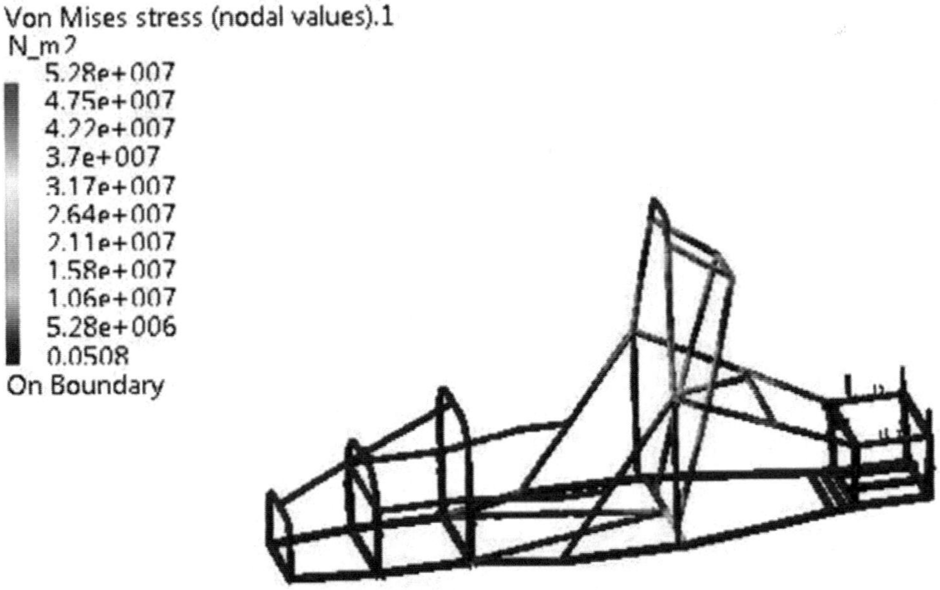

Figure 11. Stress distribution for acceleration analysis.

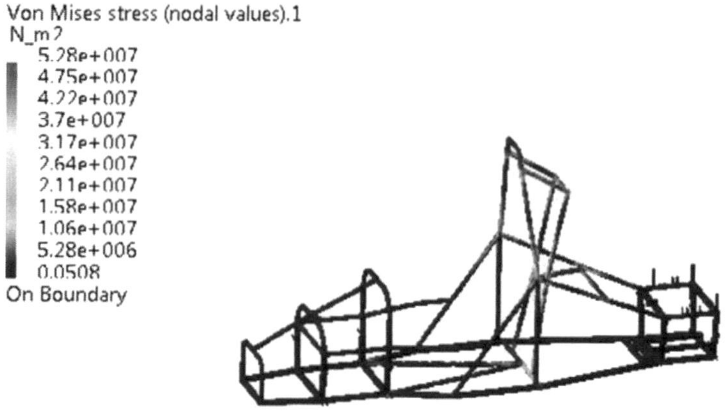

Figure 12. Stress distribution for braking analysis.

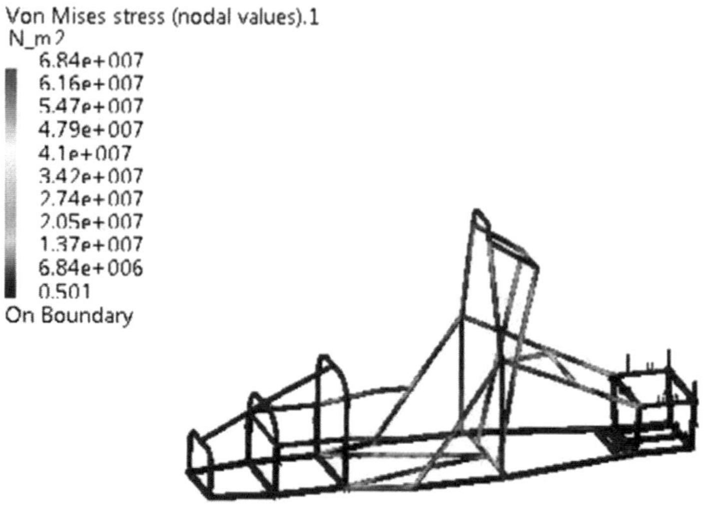

Figure 13. Stress distribution for cornering analysis.

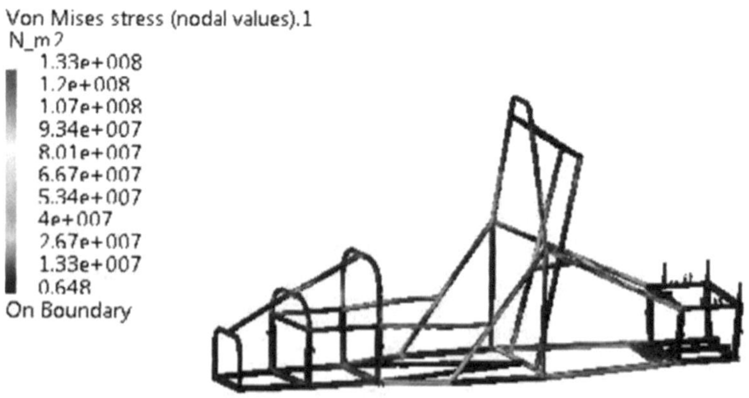

Figure 14. Stress distribution for front stiffness analysis.

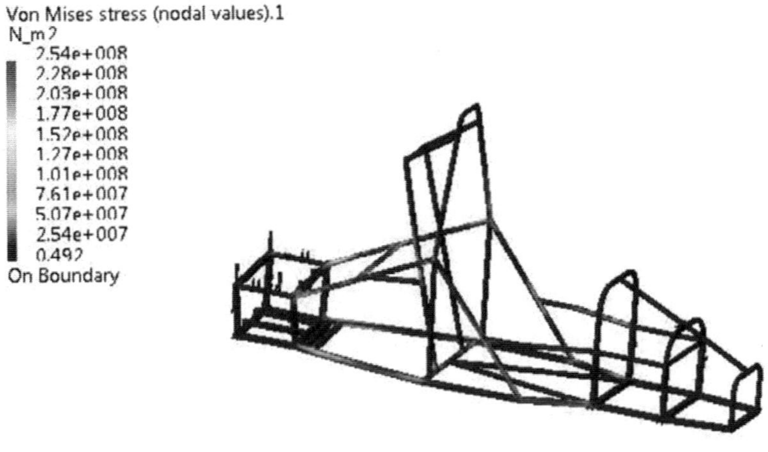

Figure 15. Stress distribution for rear stiffness analysis.

5. Conclusion

The simulation analysis results have established approving outcome in term of factor of safety at different loading tests. In overall, from all the analysis, conclusion can be made that the chassis was safe to be used. The factor of safety ranged from 1.14 to 5.49 which were acceptable. When the factor of safety is lower than 1, a structure is deemed to be unsafe. In industry a factor of safety ranging from 2-5 is preferred depending on the situation. In our case on the other hand, a factor of safety of 1 is also acceptable as this chassis was made for a one-off race event and not for durability. Thus, having a factor of safety equalling to 1 in certain situations is considered to be safe as the structure is strong enough to support the load while certain factors such as fatigue was not considered. If the factor of safety is too high, this would cause an unnecessary increase in cost and the mass of the chassis.

6. Acknowledgments

The authors gratefully acknowledged the financial support from Universiti Teknikal Malaysia Melaka and The Ministry of Higher Education, Malaysia (MoHE) under Exploratory Research Grant Scheme (ERGS), grant no.: ERGS/1/2012/TK08/UTEM/02/1/E00007.

7. References

[1] Milliken, William F and Milliken, Douglas L 1997 Race Car Vehicle Dynamics *Society of Automotive Engineers (SAE).*

[2] Society of Automotive Engineers (SAE) 2012 Formula SAE Competition Rules and Regulations *SAE International.*

[3] Institution of Mechanical Engineers (IMechE) 2012 Formula Student Competition Rules and Regulations *IMechE.*

[4] Universiti Teknikal Malaysia Melaka (UTeM) *2012 Formula Varsity 2012 Competition Rules and Regulations, UTeM.*

[5] Sariman M F 2012 Merekabentuk dan Fabrikasi Casis Kerangka Ruang Kereta Lumba Elektrik Formula Varsiti UTeM (In Malay) *Final Year Project Report* Universiti Teknikal Malaysia Melaka (UTeM), Melaka, Malaysia.

[6] Ashari M H 2011 Design and Fabrication of Space Frame Chassis for UTeM Formula Style Race Car *Final Year Project Report* Universiti Teknikal Malaysia Melaka (UTeM), Melaka, Malaysia.

[7] Indral F 2010 Conceptual of a Single Seater Formula Racing Car *Final Year Project Report* Universiti Teknikal Malaysia Melaka (UTeM), Melaka, Malaysia.

[8] Rahmat M S 2011 Design and Fabrication of Engine Mounting for UTeM Formula Style Race Car *Final Year Project Report* Universiti Teknikal Malaysia Melaka (UTeM), Melaka, Malaysia.

[9] Mohd Hashim M S 2010 Design and Development of Lightweight Chassis for UTeM Formula Style Race Car *Final Year Project Report* Universiti Teknikal Malaysia Melaka (UTeM), Melaka, Malaysia.

[10] Shaharuzaman M A, Abdul Munir F, Hassan M Z, Mohamad Zin M R, Boejang H, Muhammad Farhan F H and Abdul Jamal A H 2010 Development of a Single Seat Race Vehicle for Educational Motorsport Engineering *3rd International Conference on Science & Technology: Applications in Industry & Education*, 16-17 December 2010, Universiti Teknologi MARA, Pulau Pinang, MALAYSIA.

[11] Nasruddin F 2009 Design And Analysis Of The Drivetrain System For Formula SAE Race Car *Final Year Project Report* Universiti Teknikal Malaysia Melaka (UTeM), Melaka, Malaysia.

[12] Abdul Munir F, Munir, Shaharuzaman M A, Hassan M Z, Mohamad Zin M R, Mohd Azmi M I and Mat Nuri N R 2011 Components Design of a Single Seat Educational Race Vehicle *Malaysian Technical Universities International Conference on Engineering & Technology (MUiCET 2011)*, 13-15 November 2011, UTHM, Batu Pahat, Johor.

[13] Mustaffa Albakri A A 2011 Design And Fabrication Of A Semi Automatic Gear Shifting Mechanism For UTeM Formula Style Race Car *Final Year Project Report*, Universiti Teknikal Malaysia Melaka (UTeM), Melaka, Malaysia.

[14] Suhaimi K 2011 Design And Fabrication Of An Upright With Brake Caliper Mounting For Formula Varsity Race Car *Final Year Project Report* Universiti Teknikal Malaysia Melaka (UTeM), Melaka, Malaysia.

[15] Daem M F 2011 Ergonomics Study Of Cockpit Arrangement For UTeM Formula Varsity Race Car *Final Year Project Report* Universiti Teknikal Malaysia Melaka (UTeM), Melaka, Malaysia.

[16] Kamal M, Rahman M M, Rahman A G A 2013 *Inter J Automot Mech Eng* **7** 912-923.

[17] Hamzah M A 2011 An Experimental Study Of A Steering Wheel Vibration For UTeM Formula Varsity Race Car *Final Year Project Report* Universiti Teknikal Malaysia Melaka (UTeM), Melaka, Malaysia.

[18] Kamal M, Rahman M M, Rahman A G A 2012 *J Mech Eng Sci* **3** 291-300.

[19] Ridha M, Fonna S, Huzni S, Supardi J, Ariffin A K 2013 *Inter J Automot Mech Eng* **7** 1015-1023.

[20] Zamri M F 2009 Design And Development Of Hybrid Composite Chassis For Formula Student Race Car *Final Year Project Report* Universiti Teknikal Malaysia Melaka (UTeM), Melaka, Malaysia.

[21] Othman M N 2012 Rekabentuk dan Fabrikasi Sistem Stering Kereta Lumba Elektrik Formula Varsiti UTeM (In Malay) *Final Year Project Report* Universiti Teknikal Malaysia Melaka (UTeM), Melaka, Malaysia.

[22] Abdullah M A, Mansor M R, Mohd Tahir M, Abdul Kudus S I, Hassan M Z and Ngadiman M N 2013 Analysis and Fabrication of Chassis Frame for UTeM Formula Varsity™ Race Car *International Journal of Mining, Metallurgy & Mechanical Engineering (IJMMME)* Volume 1 Issue 1 ISSN 2320–4060 (Online) 75-77.

[23] Abdullah M A, Mansor M R , Mohd Tahir M, Abdul Kudus S I, Hassan M Z and Ngadiman M N 2012 Analysis and Fabrication of Chassis Frame for UTeM Formula Varsity™ Race Car *Proceeding for International Conference on Manufacturing and Industrial Engineering (ICMIE'12)* International Scientific Academy of Engineering & Technology (ISAET), Kuala Lumpur, Malaysia, (December 29th to 30th, 2012).

Lubrication performance of double fraction palm olein using pin-on-disk tribotester

S Syahrullail[a] and M S J Ismail
Faculty of Mechanical Engineering,
Universiti Teknologi Malaysia, 81310 UTM Skudai, Johor, Malaysia.

E-mail: [a]syahruls@mail.fkm.utm.my

Abstract. Friction, wear and fatigue are the three most commonly encountered industrial problems leading to the replacement of components and assemblies in an engineering system. The present paper experimentally investigated the effect of sliding speed on friction and wear behaviour of an aluminium pin sliding against a SKD11 disc with a constant normal load of 4 kg, lubricated with double fraction palm olein (DFPO) as a test lubricant and commercial hydraulic oil (HO) as a reference lubricant. Experiments were carried out at various sliding speeds from 900rpm to 1800rpm. The experimental works were carried out at room temperature. After the testing was completed, the friction coefficient and wear rates were calculated. The morphology of the worn surface was observed using a charge couple device (CCD) camera. Results showed that the friction coefficient for both lubricants decreased with increments in sliding speed.

1. Introduction

Sliding between mating components in operations of extended duration will produce high levels of friction and wear, as well as ultimately failure of the materials if there is no effective lubrication between the contact surfaces. These high frictional values are usually caused by an inherent roughness of the materials in contact and the large pressures that develop between them. In order to minimise the undesirable effects of wear, lubricants are generally applied along the interface of materials in contact [1, 2].

Lubricants are thus used to control friction and minimise wear [3-5]. When lubricants are applied between the contact interfaces of two sliding surfaces, the material loss due to wear and the energy consumption due to friction are reduced by several orders of magnitude [6-7]. To understand the wear behaviour of materials, wear tests are often carried out with suitable techniques that can provide more analysis of the behaviour of wear. The pin-on-disc test is usually used to conduct wear experiments.

In the pin-on-disk test, the wear volume can be determined either by the changed geometry of the specimen or by reduction in the mass of the specimen. In most studies on wear, the measurements commonly taken are of wear, volume loss, weight loss and wear coefficient. Such results also provide an indication of the anti-friction, anti-wear and anti-scuff properties of oil or an additive in oil. The operating conditions can be made to simulate, as far as possible, those in a real industrial machine. The continuous supply of lubricant conducted in this test causes boundary lubrication, where metal-to-metal contact occurs, and the chemistry of the metals and the oil control the friction and wear. In this study, a pure aluminium pin was tested on a SKD11 disc and lubricated with double fraction palm olein (DFPO) as a test lubricant. Similar tests were conducted using commercial hydraulic oil (HO), as a benchmark. Normal load was set to 4 kg and rotational speed was set between 900rpm and 1800rpm. After the testing was completed, the friction coefficient and wear rates were calculated. The morphology of the worn surface was observed using a charge couple device (CCD) camera. Results showed that the friction coefficient for both lubricants decreased with increments in the sliding speed.

2. Material and method

Figure 1 is a schematic sketch of a pin-on-disk tribometer. The disc was made of SKD11 and the material for the pin was pure aluminium A1100. The average hardness of the pin before undergoing heat treatment was 52HV and after heat treatment it was 36HV. The hardness of the specimens was measured using a Vickers hardness tester. Rotational speed was set between 800rpm and 1800rpm. Normal load applied on the aluminium pin was set to 4 kg.

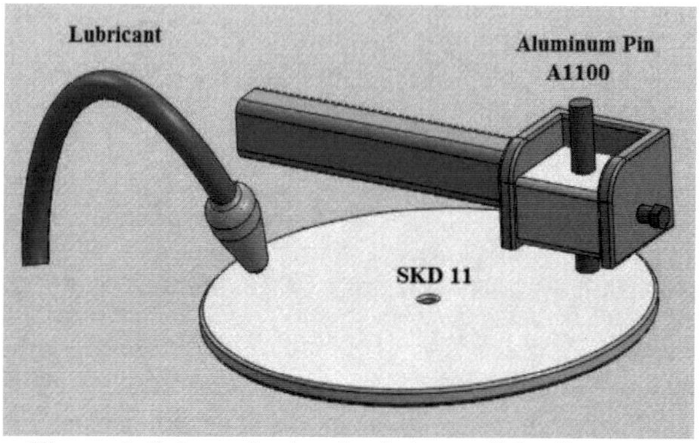

Figure 1. Schematic sketch of pin-on-disk tribotester.

3. Results and discussion

The coefficient of friction can be defined as a measure of the amount of resistance that a surface exerts on substances moving over it, equal to the ratio between the maximal frictional force that the surface exerts and the force pushing the object towards the surface. The coefficient of friction is not always the same for objects that are motionless and objects that are in motion; motionless objects often experience more friction than moving ones, requiring more force to put them in motion than to sustain them in motion.

Figure 2 shows the comparison of the average friction coefficient for an aluminium pin sliding on the rotating SKD11 disc, lubricated with double fraction palm olein (DFPO) and hydraulic oil (HO). Generally, both lubricants showed the same characteristic, namely that the friction coefficient decreased when the rotating speed was increased from 900rpm to 1800rpm. This indicates that the lubrication conditions were favourable at high speeds, probably in or close to the hydrodynamic region, while at low velocities the conditions obviously corresponded to the mixed or boundary regime. Velkavrh et al. [8] obtained the same results: a decrease in friction with the increase of velocity in the low-speed regime and an increase in friction with the increase of speed at high velocity. This is typical behaviour, and the regime transition is found along the Stribeck curve [9].

Moreover, the decrease in friction coefficient with increasing sliding speed may be due to the change in the shear rate, which can influence the mechanical properties of the mating materials. These findings are in agreement with those of Chowdhury et al. [10]. Similar trends were observed for the friction coefficient with variation in the duration of rubbing; the results are presented in Figure 1, which shows that friction decreased with increasing sliding speed within the observed range. Also, increased surface roughening and a large quantity of wear debris are believed to be a major reason for the decrease in the friction coefficient with increasing sliding speed.

The level of the coefficient of friction was also affected by the surface condition. The high temperature generated during the rubbing process between the two oxide surfaces, so that the spherical aluminium pin had a tendency to reduce the friction. Furthermore, the layer formed on the surface contributed to the friction. Tiong et al. [11] and Syahrullail et al. [12] similarly found that, when the layer substantially covers the sliding surfaces, sliding occurs within the layer and friction is characteristic of that process. When the rate of removal of the surface layer is much faster than that of its formation, the surface is always partially covered and a single-valued friction coefficient results. When the rate of formation is higher than that of removal, there is always a well-formed layer, and a low, single-valued friction coefficient can be anticipated.

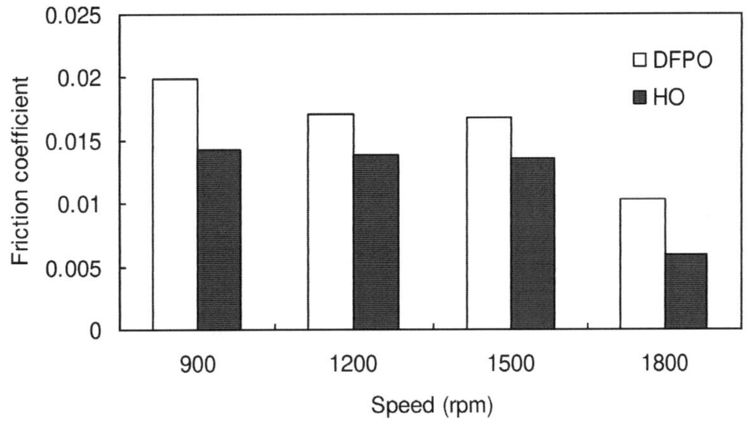

Figure 2. Friction coefficient distribution.

When two surfaces are in loaded sliding contact, stresses are imposed on both solids as a result of normal and tangential forces that arise from sliding. Frictional heat is also generated at the sliding-contact interface. The imposed stresses and frictional heating at the contact interface are the principal factors in causing wear at the sliding-contact interface. Consequently, the wear rates and wear mechanisms are determined in large part by the magnitude of these factors. Figure 3 represents the correlation between wear rates of a spherical aluminium pin with respect to differences in rotational speed using two different types of lubricant. As we can see from the data plotted, the wear rate clearly declines with increasing rotational speed in the case of both lubricants. However, in the range of rotational speed of 900rpm to 1200rpm, the wear rate seems to exhibit a higher difference according to the lubricant, whereas in the range of 1500rpm to 1800rpm, the decrements in the case of both lubricants are comparable with each other.

The decrease of the average wear rate with increasing sliding speed may be due to direct asperity collisions in steel/steel contacts under a constant contact pressure without using any additives. This leads to adhesive wear, as also seen at low velocities, which implies that such asperity contacts occur very seldom on steel surfaces at high velocities, even at the lowest viscosity.

During the rubbing process between two surfaces, a high temperature oxide of aluminium (Al_2O_3) is formed, which subsequently covers both the pin and the disc track surfaces in a layer and reduces friction and wear. This is mainly due to higher temperatures generated at the interface. As this layer offers a certain amount of protection from wear, wear rates of both the pin and the disc are lower at higher speeds. With increasing speed, the amount of layer formation increases, owing to the higher temperatures generated. This may be why the wear rate decreases with increasing speed, as also observed by Ravikiran et al. [13].

The wear rate decrease with increasing sliding can be attributed to several further factors. The first factor influencing the tribological behaviour may be the easier occurrence of plastic flow caused by the decrease of the shear strength in a small volume next to the coatings contact area, or by local melting, owing to the high flash temperatures reached. This means that, at high rotational speeds, the time between passes of the pin over the same point on the disk is reduced, so that the time available for heat dissipation is shortened and the lubricated surface temperature and flash temperatures, especially in the contact area between the spherical aluminium pin and the disk, can rise significantly. The results found by Betancourt-Dougherty [14] were similar to those of this study.

Rotational speed is one of the factors that may allow the reduction of wear. Because of low speeds, the debris generated is not ejected from the wear track, and will continue to contribute to the track wear process in the form of abrasive wear. Thus, low speeds may contribute to a high wear rate. In contrast, at high rotational speeds, most of the debris generated is ejected out of the wear track, such that debris abrasive wear is minimised.

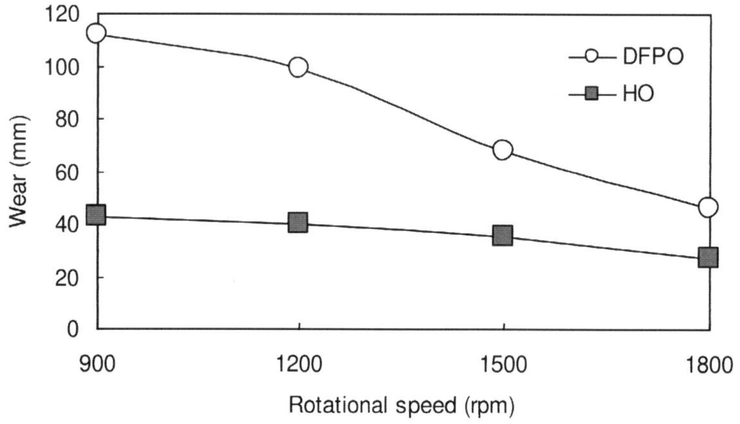

Figure 3. Wear vs rotation speed.

Figure 4 shows the worn surface characteristics of the aluminium pin for both types of lubricant. The cluster of debris or particulates produced around the surface for 900rpm using DFPO was significantly greater compared to HO. This indicates that the cluster of debris formed during the experiment decreased with increasing sliding speed. The worn characteristic seen at 1800rpm, which is the wear debris, was greatly reduced around the spherical aluminium surface.

The reduction in debris generated during the experiment may be due to the low speeds, which means that the debris generated is not ejected from the wear track and will continue to contribute to the track wear process in the form of abrasive wear. In contrast, most of the debris generated is ejected out of the wear track at the high rotational speed, thus minimising the debris abrasive wear. This is why a lot of debris results from slower speeds and the amount of debris is reduced when the speed is increased. However, the wear debris generated when using HO seems to be less than when using DFPO. This is because HO has additives that can significantly reduce the friction. In contrast, the DFPO is pure palm oil, free of additives. Thus, we can conclude that the additives in a lubricant can influence the worn characteristic of the direct contact between two sliding surfaces.

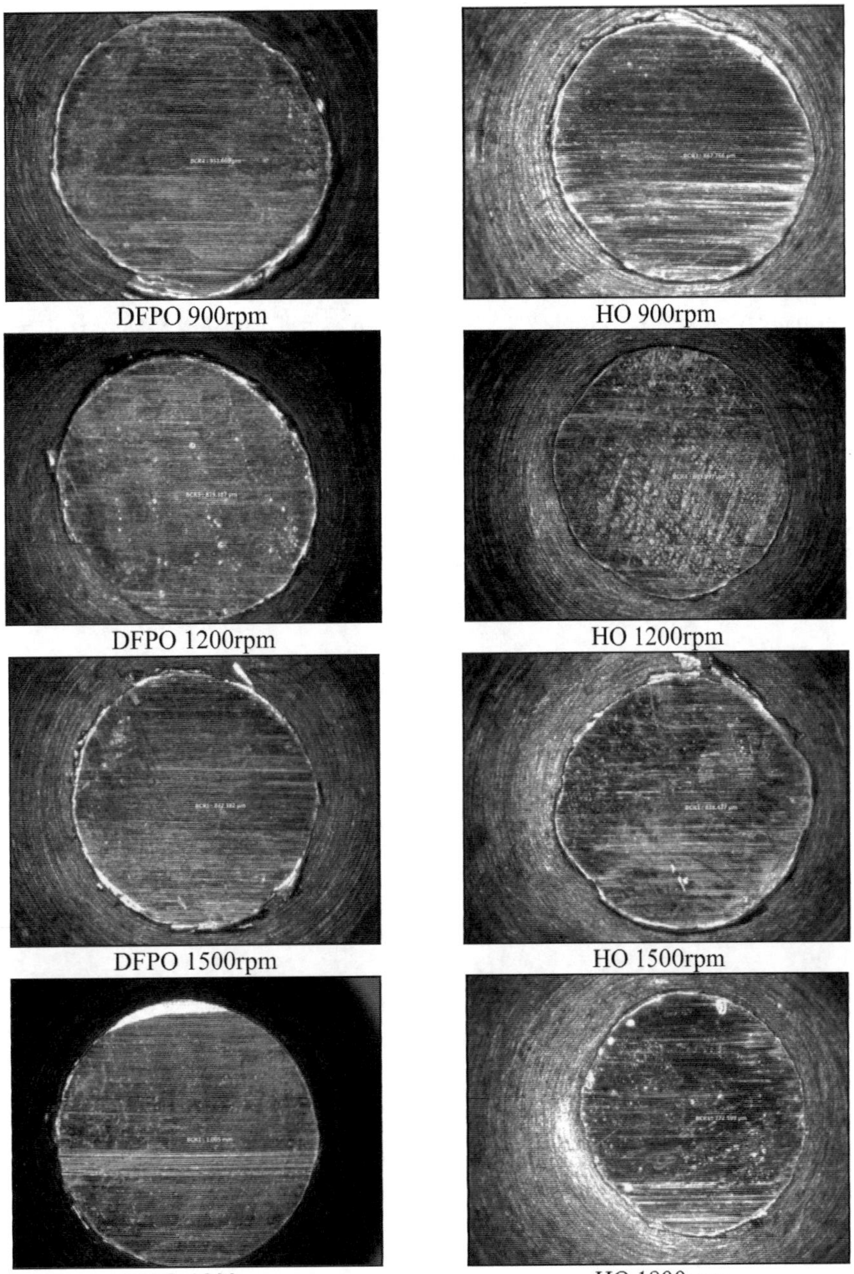

DFPO 900rpm HO 900rpm

DFPO 1200rpm HO 1200rpm

DFPO 1500rpm HO 1500rpm

DFPO 1800rpm HO 1800rpm

Figure 4. Worn surface observation.

4. Conclusion

The tribological performance of double fraction palm olein was tested using a pin-on-disk tribotester. Commercial hydraulic oil has a lower friction coefficient and wear rate than double fraction palm olein. However, the difference in value is acceptable and commercial hydraulic oil with additives is judged to give superb performance in terms of anti-friction and anti-wear.

Acknowledgement
The authors wish to thank the Faculty of Mechanical Engineering at the Universiti Teknologi Malaysia for their support and cooperation during this study. The authors also wish to thank Research Management Centre (RMC) for the Research University Grant (GUP-03H58) from the Universiti Teknologi Malaysia, Fundamental Research Grant Scheme (FRGS) from the Ministry of Higher Education (MOHE) and E-Science Grant and ERGS from the Ministry of Science, Technology and Innovation (MOSTI) of Malaysia for their financial support.

References
[1] Syahrullail S Nakanishi K and Kamitani S 2005 *J. Jap. Soc. Trib.* **50** 877-85
[2] Syahrullail S Zubil B M Azwadi C S N and Ridzuan M J M 2011 *Int. J. Mech. Sci.* **53** 549-55
[3] Najiha M S Rahman M M Yusoff A and Kadirgama K 2012 *Inter. J. Automot. Mech. Eng.* **6** 766-74
[4] Ghobadian B Najafi G and Nayebi M 2013 *J. Mech. Eng. Sci.* **4** 373-82
[5] Najiha M S Rahman M M Kamal M Yusoff A R and Kadirgama K 2012 *J. Mech. Eng. Sci.* **3** 340-45
[6] Wan N W B Maleque M A Ani F N and Masjuki H H 2007 *Ind. Lub. Trib.* **59** 200-08
[7] Shukla S and Deheri G 2013 *J. Mech. Eng. Sci.* **4** 532-47
[7] Velkavrh I Kalin M and Vizintin J 2009 *Trib. Int.* **42** 1752-57
[8] Wan Nik W B Ani F B Masjuki H H and Eng Giap S G 2005 *Ind. Crops and Products* **22** 249 55
[9] Tiong C I Azli Y Rafiq A K M and Syahrullail S 2012 *J. Zhejiang Univ. Sci. A.* **13** 633-40
[10] Chowdhury M A Khalil M K Nuruzzaman D M and Rahaman M L 2011 *Int. J. Mech. And Mechatronics Eng.* **11**
[11] Syahrullail S Tiong C I Rafiq A K M and Azli Y 2012 *Trib. Trans.* **55** 539-48
[12] Ravikir A and Surappa M K 1997 *Wear* **206** 33–38
[13] Betancourt-Dougherty L C and Smith R W 1998 *Wear* **217** 147-54

Soot particle trajectories of a Di diesel engine at 18° ATDC crankshaft angle

M H M Hafidzal[1,a], W M F W Mahmood[2] , M Z A Manaf [1], M S Zakaria[1], M N A Saadun[1] and M N A Nordin[1]
[1]Faculty of Mechanical Engineering, Universiti Teknikal Malaysia Melaka, Hang Tuah Jaya, 76100, Durian Tunggal, Melaka, Malaysia
[2] Department of Mechanical and Materials Engineering, Faculty of Engineering and Built Engineering, Universiti Kebangsaan Malaysia, Malaysia

E-mail: [a]hafidzal@utem.edu.my

Abstract. Among the major pollutants of diesel engine is soot. Soot is formed as an unwelcome product in combustion systems. Soot emission to the atmosphere leads to global air warming and health problems. Furthermore, deposition of soot particles on cylinder walls contaminates lubricant oil hence increases its viscosity. This reduces durability of lubricant oil, causing pumpability problems and increasing wear. Therefore, it is necessary to study soot formation and its movement in diesel engines. This study focuses on soot particle trajectories in diesel engines by considering the diameter of soot particles that were formed at 18° ATDC crankshaft angle. These soot particle movements are under the influence of drag force with different radial, axial and angular settings and simulated by using MATLAB routine. The mathematical algorithm which was used in the MATLAB routine is trilinear interpolation and 4th order of Runge Kutta. Simulation was carried out for a combustion system of 4 valves DI diesel engine from inlet valve closing (IVC) to exhaust valve opening (EVO). The results show that small diameter of soot particles were transferred near the cylinder wall while bigger soot particle mostly moved in inner radius of the combustion chamber.

1. Introduction

Diesel engine has its drawbacks especially from an environmental perspective. Combustion in a diesel engine will release particulate matters (PM), nitrogen oxides (NOx), sulfur oxides (SOx) and noise emissions [1-6]. Particulate matter contains solid soot particles which will contaminate the lubricant oil, thus viscosity of the lubricant oil increases. Subsequently the walls of the combustion chamber will not be adequately coated which will increase its wear [7]. Besides that, when the viscosity of the lubricant oil increases, the interval between oil changes will also decrease [8].

The wear mechanism due to the soot is still not fully understood, therefore fundamental study about this area is needed. Previous researchers investigated on soot distribution in a diesel engine by using the CFD software, Kiva-3v. Kiva-3v is reliable software to simulate the soot formation in a diesel engine combustion chamber [9]. Kiva-3v implements a two stage of Hiroyasu soot model in the simulation. The two stage soot models are for soot formation and soot oxidation [10]. Research made by [8] carried out a soot particle size prediction along its path using combination of Hiroyasu's soot formation [6] and Nagle and Strickland-Constable soot oxidation [12] rate expression. Meanwhile, [13] also developed a model for soot formation process in diesel engine but the result of the model was different to the experimental result. Other mechanisms that occur in the cylinders that causes soot deposition are thermophoresis, inertial deposition, electrophoresis and gravitational sedimentation. Among those mechanisms, thermophoresis was shown to be the most dominant [14]. According to another study, it is essential to understand how force acts on the soot particles in order to determine the soot movements [15]. The forces that act on the soot particles are drag, electrostatic, gravitational, acoustic, diffusiophoretic and thermophoretic[16].

The objective of this study is to analyze soot movement at 18° crank angle ATDC in a cylinder of a diesel engine under the influence of drag force with different radial, axial and angular settings to predict the possibility of soot particles to stick to the wall of the cylinders.

2. Kiva-3v

Kiva-3v was used in previous investigation to obtain data of in-cylinder soot formation. The data that were obtained from the Kiva-3v simulation were velocity vectors of the soot, fuel, temperature, pressure and others [8]. In order to get the soot particle trajectories, the data from Kiva-3v was implemented by MATLAB software.

3. Specifications of engine

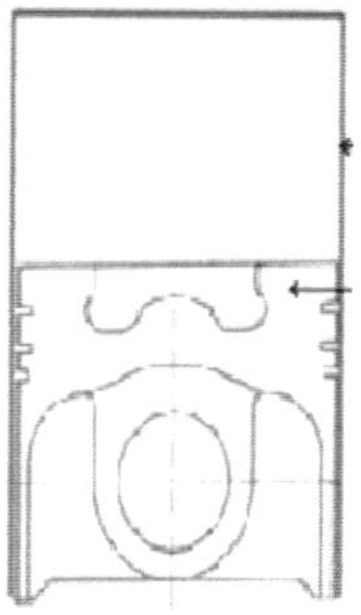

Figure 1. Crown piston configuration.

The engine has a bowl-in-piston and flat cylinder head face configuration, with a seven-hole injector installed vertically and centrally. The schematic diagram for crown piston configuration is shown in figure 1. The simulations were carried out for the combustion system of a 4 valve DI diesel engine from inlet valve closing (IVC) to exhaust valve opening (EVO). The specifications of the engine are bore × stroke: 86.0 × 86.0 mm, squish height: 1.3 mm, compression ratio: 18.2:1 and displacement: 500 cm3, speed: 2500rpm. The details of the specification of engine and mesh specification are shown in table 1 and table 2 [8].

Table 1. Specifications of the engine

Parameters	Specifications
Engine Type	4 valve DI diesel
Bore × Stroke	86.0 × 86.0 mm
Squish height	1.3 mm
Compression Ratio	18.2:1
Displacement	500 cm^3
Piston Geometry	Bowl-in-piston

Table 2. Specifications of the mesh configuration.

Parameters		Mesh configuration
Total number of cells		201,900
Number of cells	Azimuthal	150
	Radial	37 (20 in bowl region)
	Axial	39 (15 in bowl region)
Resolution	Azimuthal (°)	2.4°
	Radial (mm)	0.83 – 1.15
	Axial (mm)	0.99 – 3.63

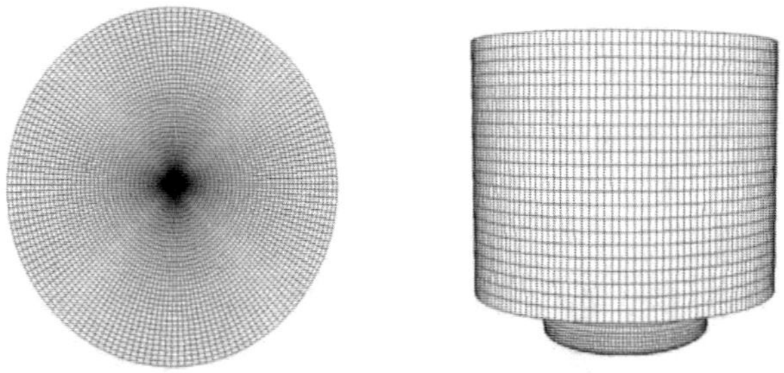

Top view Side view

Figure 2. Mesh configuration of piston bowl-in.

Figure 2 shows the top and side view of the mesh configuration of the piston. The data from experimental result by [4] was used for the simulation and was used for model validation.

4. Soot particle tracking

By using MATLAB, the data for soot concentration that was used by [8] was employed to investigate the soot movement in the combustion chamber. There are two methods to determine the formation of soot in a combustion chamber, which is individual tracking and cumulative tracking. This studies focusing on the individual tracking. The starting location of soot can be determined by assuming its velocity is the same as that particular flow velocity.

The soot movement starting from its formation which is after combustion process until the exhaust valve is open can be predicted by using numerical method. Euler method or also known as first stage of Runge-Kutta method is an easy way to calculate and determine the soot location. This method takes velocity at a certain location as soot velocity, and considered distance traveled by the soot. In order to get an accurate value, higher stage of the Runge - Kutta method can be used. And as for this study 4th Stage of Runge-Kutta is used to calculate the soot's location. On top of that, to investigate soot velocity at every condition trilinear interpolation method is used. In this study, the drag force equation is included in the simulation to determine soot movement. The drag force equation is

$$F_d = \frac{1}{2}\rho v^2 C_d A \qquad (1)$$

F_d is the force of drag, ρ is the density of the fluid, v is the velocity of the object relative to the fluid, A is the area and C_d is the drag coefficient. To analyze the movement of the soot particles, it is necessary to identify the initial position of the soot particle. This can be found based on the zone of soot distribution with certain crank angle ATDC [8]. After selecting the crank angle and zone, the starting coordinate in radial (rho), angular (θ) and axial direction (z) were also decided as below:

a) Rho: 2.0-2.6 cm, θ=10° and z=10 cm
b) Rho= 2.3 cm, θ =15°-30°, z=10 cm
c) Rho= 2.3 cm, θ =10°, z=9.5-10.2 cm

Soot particle movement paths have been studied by considering reasonable crank angle .In present paper the reasonable crank angle that is investigated in is at 18° ATDC. 18° ATDC is the point where injection is already completed and soot formation process is dominating. A comparison of the soot particle influenced by different radial (rho), angular (theta) and axial direction (z) were carried out from the results of the simulation.

5. Results and discussions

By considering the radial sensitivity, rho: 2.0-2.6 cm, angular (θ): 10° and axial direction (z): 10 cm, the soot movement path from 18° ATDC to 120° ATDC was shown in figure 3. The results shows that when the piston moves downward, almost all the soot particles move downwards in anti-clock wise motion and moves towards middle of the cylinder wall. The combination of swirl, squish and reverse squish were the reason the soot particle moves towards the cylinder wall. It can be observed that, at rho=2cm the length path of soot movement is shorter compared to the other paths. In addition, it can be seen that an increase of radial, increases possibility of soot particle to move down towards middle of cylinder wall. The movement of soot near to the cylinder wall will cause contamination to the lubricant oil.

a) Rho: 2.0-2.6 cm, θ=10° and z=10 cm

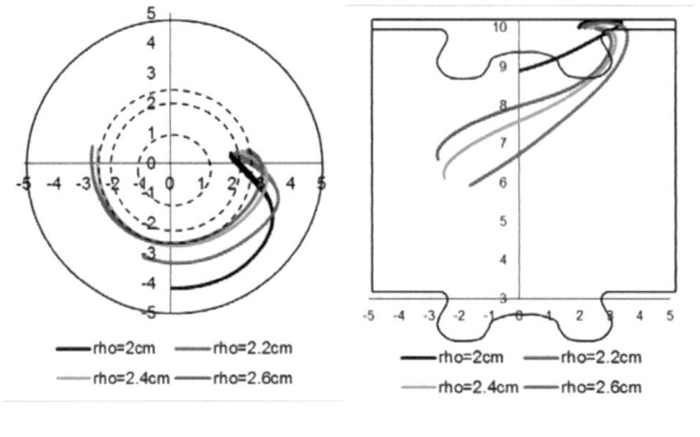

Figure 3. Radial sensitivity.

Analysis of the soot particle movement by considering the sensitivity of angular was also carried out. Figure 4 shows the Angular sensitivity: radial distance= 2.3 cm, θ =15°-30°, z=10 cm. Almost of the soot particles move downwards in and anti-clockwise motion. It can be seen that, in this zone, the angular sensitivity has less influence on the soot particle movement as the comparison between movements of soot particle showed very similar result.

b) Rho= 2.3 cm, θ =15°-30°, z=10 cm

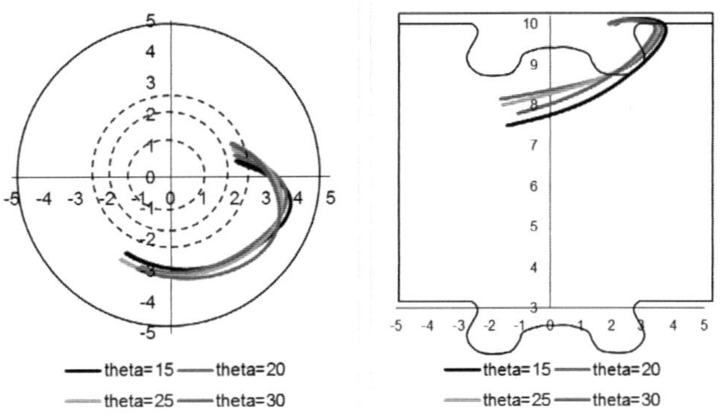

Figure 4. Angular sensitivity.

Figure 5 demonstrates the soot particle movement between axial distance= 2.3 cm, θ =10°, z=9.5-10.2 cm: of soot particle movement from 18° ATDC to 120° ATDC. At z=9.5 cm, the soot particle paths are not available in the graph. It is because soot particles follow the velocity of air flow in the combustion chamber.

Besides that, it can be seen that, the soot particle movement paths become shorter as axial direction increases. On the other hand, as z decreases, soot particle has high tendency to go near to cylinder wall. Therefore, it is suggested to improve design of piston or fuel spray angle based on the results of the study.

c) rho= 2.3 cm, θ =10°, z=9.5-10.2 cm

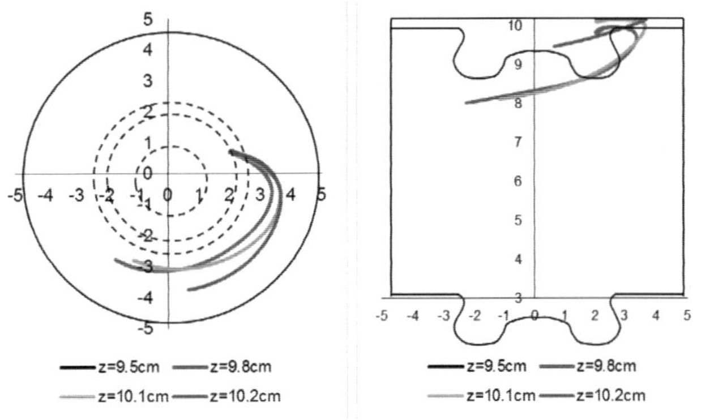

Figure 5. Axial sensitivity.

6. Conclusions

In order to reduce lubricant oil contamination is to minimize the transfer of soot particles to the cylinder wall. By understanding the movement path of soot particle in diesel engine, measures to reduce soot particle movement towards the cylinder wall can be identified.

Soot particle movement in a diesel engine depends on the starting point in radial, angular and axial. Most of the soot particles at 18° ATDC at certain locations move downwards, anti-clockwise motion and not approaching the cylinder wall. Hence, the soot particles from these points are likely to give a low risk to lubricant oil contamination problem. In addition the modification of the bowl, piston shape and spray angle is expected to reduce the possibilities of the soot particle moving towards the cylinder wall. Nevertheless, this will be investigated further in the future.

Acknowledgement
The authors would like to acknowledge Universiti Teknikal Malaysia Melaka (UTeM) PJP/2012/FKM(8A)/S01082 and Universiti Kebangsaan Malaysia (UKM) through GGPM-2011-055 Research Grant for supporting and funding these research activities.

References

[1] Shah A N, Yun-shan G and Jian-wei T 2009 Carbonyls emission comparison of a turbocharged diesel engine fuelled with diesel, biodiesel, and biodieselediesel blend, *Jordan Journal of Mechanical and Industrial Engineering* **3** 111-118
[2] Mat Yasin M H, Mamat R, Sharma K V and Yusop A F 2012 *J Mech Eng Sci* **3** 331-339
[3] Ghobadian B, Najafi G and Nayebi M 2013 *J Mech Eng Sci* **4** 373-382
[4] Azad A K, Ameer Uddin S M and Alam M M 2012 *Inter J Automot Mech Eng* **5** 576-586
[5] Rahim R, Mamat R, Taib M Y and Abdullah A A 2012 *J Mech Eng Sci* **2** 226-236
[6] Sundar Raj C and Sendilvelan S 2010 *Inter J Automot Mech Eng* **2** 144-156
[7] George S, Balla S and Gautam M 2007 Effect of diesel soot contaminated oil on engine wear, Wear **262** 1113–22
[8] Mahmood W M F W 2011 Computational Studies of Soot Paths to Cylinder Wall Layers of a Direct Injection Diesel Engine. Phd Tesis Department of Mechanical, Materials and Manufacturing Engineering, The University of Nottingham
[9] Bonatesta F, Rocca A L, Shayler P J and Wahab E 2007 The Influence of Swirl Ratio on Soot Quantity and Distribution in the Cylinder of a Diesel Engine, *Third European Combustion Meeting ECM*
[10] Boussouara K and Kadja M 2009 Numerical investigation of soot formation in diesel jet flame with KIVA-3V *Revue des Energies Renouvelables* **12** 55-62
[11] Hiroyasu H and Nishida K 1989 Simplified Three-dimensional Modeling of Mixture Formation and Combustion in a D.I. Diesel Engine. *SAE Papers* 890269
[12] Nagle J and Strickland-Constable 1962 R.F. Oxidation of Carbon Between 1000-2000°C. in *Proc of the Fifth Conf. On Carbon* (Pennsylvalnia State University, University Park) (Pennsylvania: Pergamon Press).
[13] Hong S, Wooldridge M S, Im H G, Assanis D N and Pitsch H 2005 Development and application of a comprehensive soot model for 3D CFD reacting flow studies in a diesel engine. Combustion and Flame **143** 11-26.
[14] Suhre B R and Foster D E 1992 In-cylinder soot deposition rates due to thermophoresis in a direct injection diesel engine *SAE Papers* **921629**.
[15] Eduardo M, Thiago P and Fernando F 2009 An extended analysis for the dynamic of soot particle in droplet combustion, *20th International Congress of Mechanical Engineering*
[16] Rosner D E, Mackowski D W and Garcia-Ybarra P 1991 Size and Structure Insensitivity of the Thermophoretic Transport of Aggregated Soot Particles in Gases *Combustion Science and Technology* **80** 87-101

Review of the investigation of mixture formation and combustion process using rapid compression machine and direct visualization system

M Jaat[1,a], Amir Khalid[2,b], B Manshoor[2], Him Ramsy[1]
[1]Centre for Energy and Industrial Environment Studies (CEIES), Universiti Tun Hussein Onn Malaysia, Parit Raja, Batu Pahat, 86400, Johor, Malaysia
[2]Universiti Tun Hussein Onn Malaysia, Parit Raja, Batu Pahat, 86400, Johor, Malaysia.

E-mail: [a]norrizam@uthm.edu.my , [b]amirk@uthm.edu.my

Abstract. This paper reviews of some applications of optical visualization systems to compute the fuel-air mixing process during early stage of mixture formation in Diesel Combustion Engines. A number of studies have contributed to the understanding of fuel air mixing in DI diesel engine. This review has shown that the mixture formation process affects initial flame development. The review also found that injection pressure has a great effect on the mixture formation then the flame development and combustion characteristics. The method of the simulation of real phenomenon of diesel combustion with optical access rapid compression machine is also reviewed and experimental results are presented. The application of these methods to the investigation of diesel sprays highlights mechanisms which govern propagation and distribution of the formation of a combustible fuel-air mixture. A summary of the implementation of constant volume chamber and optical visualization system are shown in the accompanying tables and figures. The visualization of the formation process of diesel spray and its combustion in the diesel combustion chamber of diesel engine has been recognized as one of the best ways to understand the characteristics of the mixture formation.

1. Introduction

Nowadays diesel engines are the most important power source in the heavy transport sector and off-road machinery, and they are increasing their presence in the light and passenger transport too. In a diesel engine, combustion and emission characteristics are significantly influenced by spray atomization and fuel-air mixing processes in the combustion chamber [1-5]. The study on the mixture formation and combustion process was performed in Rapid Compression Machine (RCM) [2, 6-8]. RCM is an instrument designed to stimulate a single compression event of an engine cycle of internal combustion engine.

In diesel engine, ignition delay is a major factor determining the rapid pressure rise in the initial burning stage and the subsequent combustion stages. The ignition delay, which is defined as the time between the start of fuel injection and the start of the combustion, also influences the combustion process strongly. The ignition delay in a diesel engine is generally seen as consisting of two different which is the physical and the chemical ignition delay [9-11]. Physical ignition delay corresponds to the mixture formation, and the chemical delay to the time necessary to get an exponential increase in the chemical reaction rate. The physical processes involved such as atomization, penetration, entrainment and vaporization and the chemical processes such as fuel decomposition and accumulation and oxidation [12-14].

Direct injection (DI) diesel engines are widely used worldwide because of their high efficiency [15]. However, the main problems associated with diesel engines are the high level of NO_x and particulate emissions. According to Breda Kegl [16], NO_x emissions are significantly affected by injection timing and injection pressure. High pressure fuel injection induces a higher level of dispersion of the fuel with better atomization and fuel-air mixing. The aim of the fuel injection process in a diesel engine is the preparation of a fuel-air mixing to achieve a clean and efficient combustion process [17, 18]. A comparative analysis was made by N.A.Heinan between the effects of increasing the fuel injection pressure and swirl ratio on the auto ignition reactions, cool flames and premixed combustion fractions[19]. Significant reductions in particulate matter (PM) and smoke emissions from DI diesel engines with high pressure injection have been reported [20, 21]. High pressure injection was also expected to result in improvements in exhaust odor.

The visualization of the formation process of diesel spray and its combustion in the combustion chamber of diesel engine has been recognized as one of the best ways to understand in-cylinder injection performance, spray characteristics and combustion for controlling emissions [22-24]. Flame development distribution in DI diesel engine which is an important factor of NO_x and soot formation was measured by the image analysis photograph based on two color method [25].

2. Experimental Setup

The different specifications and operating conditions of the experimental are given in table 1 and table 2. A Rapid compression Machine (RCM) was used to simulate diesel combustion in a constant volume and this RCM is the same as that used in the previous report [2]. A schematic of the machine is shown in figure 1 together with outline of the fuel injection system. The main body or called cylinder liner consists of several cylinder connected in series each having an inner diameter of 50 mm and total length of 1550 mm. This machine can generate a high temperature and pressure by a method in which a piston is shot by compressed air and is rammed in to a stopper installed in front of the combustion chamber. The combustion chamber was disc type with a diameter of 60 mm and a width of 20 mm. One of the base surfaces of the chamber was composed of pyrex glass to observe spray and flame developments, and the other side surface had a injector holder.

Table 1. Experimental Specifications [2]

Base swirl velocity, r_s	19 m/s
Ambient temperature,	850 K
Ambient pressure, p_i	4 Mpa (p_c = 100 kPa)
	6 Mpa (p_c = 150 kPa)
	8 Mpa (p_c = 200 kPa)
Ambient density, ρ	16.6 kg/m^3 (p_c=100 kPa)
	25.0 kg/m^3 (p_c=150 kPa)
	33.3 kg/m^3 (p_c 200 kPa)
Nozzle diameter, d_n	0.129 mm
Injection system type	Common rail,
	Electronic control
Injection pressure	100 -160 MPa

Table 2. Experimental Specifications [1]

Ambient temperature,	≤ 900 K
Ambient pressure, p_i	4.2 Mpa
	8 Mpa
	12 Mpa
Nozzle diameter, d_n	0.25 mm
Injection system type	Electronic control
Injection pressure	30 -150 MPa

The temperature and pressure at the end of the compression can be varied over a side range by changing the starting position of the piston. To control the temperature, a heater is installed around the cylinder liner. A swirler with a 14mm x 16mm-size connecting port installed at inlet of the chamber created swirl flow inside the chamber. The port inclination angle controlled swirl velocity. Swirl velocity was defined as the velocity at 2/3- location of radial direction from the chamber center. Ambient pressure pi was changes by changing initial charging pressure pc before compression by the piston. The increasing ambient pressure simulated high boost pressure condition. Injection commencement was measured from the needle lift detected by a hole sensor installed in the injector. Gas pressure inside the combustion chamber was measured by a piezoelectric pressure transducer (Kistler, 601A). NOx concentration was measured by a chemiluminescence analyzer (Yanako, ECL-77A). The heat release rate dQ/dt was calculated from the combustion pressure.

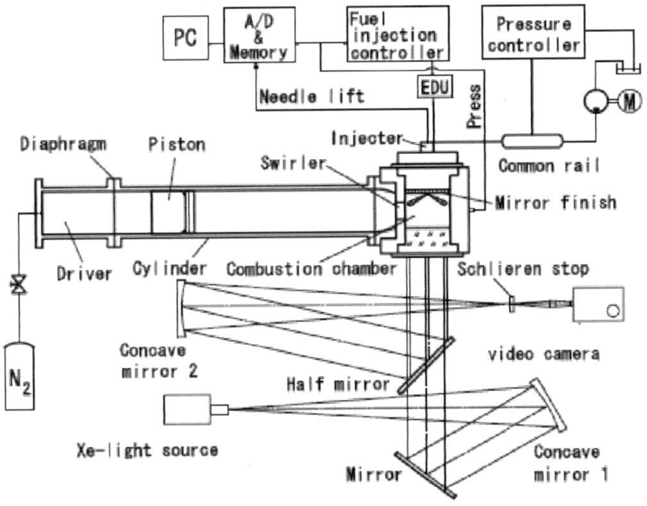

Figure 1. Schematic diagram of experimental set up [2].

3. Experimental Results

3.1 Combustion Characteristics in Diesel Combustion

figure 2 shows the effect of ambient pressure on ignition delay τ. According to the figure, increasing ambient gas pressure will results in shortens ignition delay because of the better mixing achievable at higher injection pressures. Beside that by increasing the fuel injection pressure, the fuel droplet size becomes finer, the entrainment of the air increases, and the injected spray becomes more uniform. As a result, the ignition delay and combustion periods become shorter, resulting combustion improvement. The ignition delay τ, decreases rapidly with pressure and finally reaches a constant value above a certain pressure [8, 21].

Yamaguchi et al. [4] studied the relation between the injection pressure and heat release rate and found that the reduction in heat release results in a lower maximum pressure rise rate. Figure 3 shows the heat release obtained from the injection pressure analysis. The increased injection pressure caused an advance in the ignition timing and an increased peak rate of heat release [5]. The advance in ignition timing was due to the good atomization brought about by the high pressure injection and the short energizing duration for a given injection quantity. The high pressure fuel injection system provided a significant reduction in injection duration, a reduction in combustion delay, an increase in peak heat release rates and a reduction in combustion duration [26, 27]. At higher injection pressure, the better mixing formation is achievable. M. Ikegami [8] reported that net heat release due to chemical reactions might be obtained by comparing the pressure history at normal firing with the corresponding non-fired one. The mixture is heated to a considerably high temperature before the beginning of large scale heat release [28, 29]. Only a limited amount of mixture reaches the stage of net heat release and the other mixtures are still in the physical process which is vaporizing and cracking fuel. Most important for a good mixing and utilization of the complete combustion space is an intense interaction of the injected fuel jets with the combustion chamber walls and the in-cylinder flow as well as inner spray interactions [30].

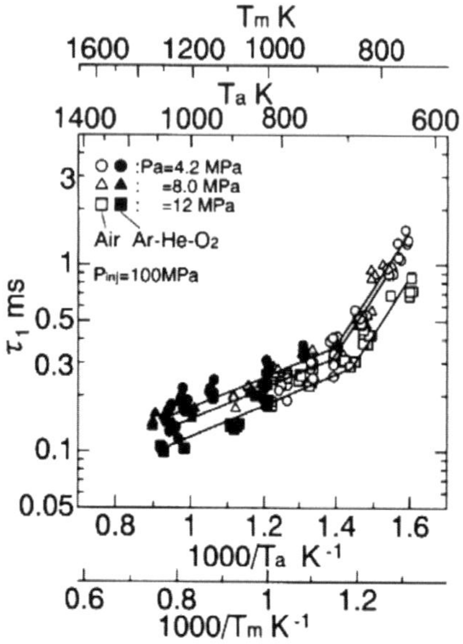

Figure 2. Effect of ambient gas pressure on ignition delay [1].

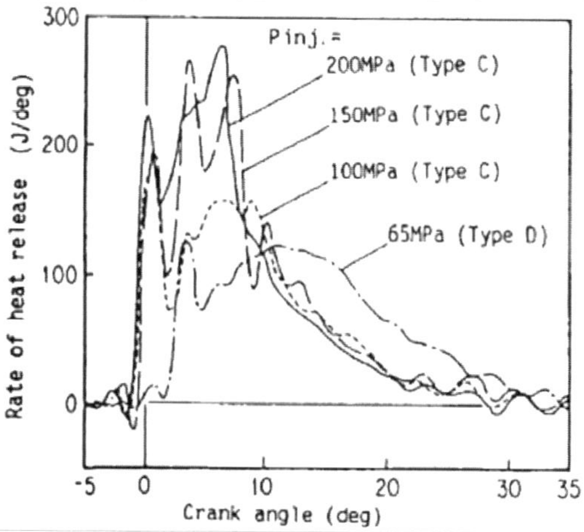

Figure 3. Heat release rate at different injection pressures [4].

3.2 Mixture Formation Effects

Effect of Mixture Formation on diesel combustion will be discussed in this section. Figure 4 shows the images of the effect of ambient density on mixture formation. At p_c=100 kPa, the mixture covers wider area of combustion chamber than the other higher ambient pressure conditions. In this chamber, the condition of p_c=100 kPa seems to produce better distribution of the mixture than the cases of p_c=150 kPa and 200 kPa. The dark flame is a kind of signal that tells the position where well-mixed mixture is prepared before ignition. Combustion photograph with different injection pressures are compared in figure 5. When comparing photographs at the early stage of combustion, the first explosion and ignited flames are observed at the spray swirl downstream near the nozzle tip. But in the middle and late stage of combustion, the portions of unburned fresh air region become smaller in the case of higher injection pressure. Jiang. H. F et al. [31] observed that the initial ignition positions appear at the center of the combustion chamber. Flame extends along all injection directions of the spray and it is a typical diesel combustion process.

IOP Conf. Series: Materials Science and Engineering **50** (2013) 012004 doi:10.1088/1757-899X/50/1/012004

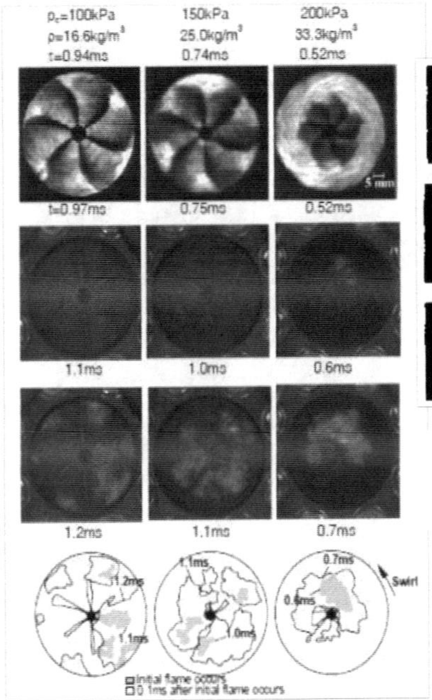

Figure 4. Effect of ambient density on mixture formation and initial flame development [2] .

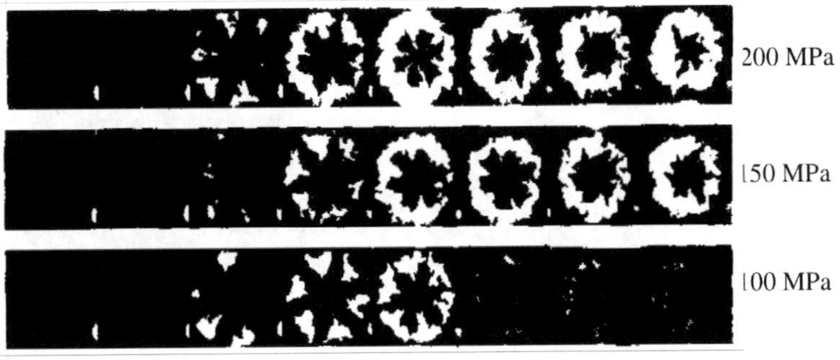

Figure 5. Combustion photographs at different injection pressure [4].

3.3 The Influences of Combustion Process on Emission Production

The formation of NO_x cannot be prevented in the combustion of fuel with air, as nitrogen is the main component of ambient air, unless gas temperatures are significantly reduced [32]. The increase of ignition delay period would influence the progress of combustion and the formation of emissions. Moreover, the reduction in oxygen flow rate associated with the application of EGR reduces the combustion temperature and pressure which, in turn, leads to lower NO_x formation as well as lower soot oxidation [33, 34]. The measured Particulate Matter PM and NO_x specific emissions are separately shown in figure 6, as a function of Start of Ignition (SOI). Three different injection pressure levels were tested to see the effect of SOI on NO_x emission and PM. By comparing the emission data for a fixed SOI value, an average decrease was recorded for NO_x emissions as the injection pressure was increase. On the other hand, an increase in PM emissions was detected when reducing the injection pressure [35].

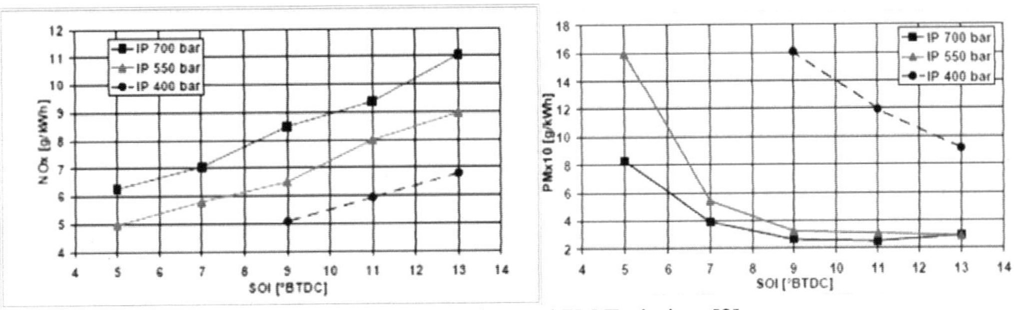

Figure 6. Effect of SOI on NO_x Emission and PM Emission [3].

Figure 7 illustrates exhaust emissions characteristics, such as soot, nitrogen dioxide NO_x, hydrocarbon HC, and carbon monoxide CO. Soot emissions showed a somewhat increasing trend with advanced injection timings. The increased injection pressure resulted in low soot emissions due to the good droplet atomization and long premixed duration [36, 37]. NO_x emissions increased with the advancing injection timing and it is indicated an active combustion reaction. CO and HC emissions were generally similar regardless of injection timing and injection pressure. CO and HC emissions increased were due to the incomplete combustion. Yoshio et al. [38] was compared between the 5-hole injection nozzle and 3-hole injection nozzle regarding HC, CO and NOx emissions. With the 5-hole injection nozzle HC and CO emissions are sufficiently low when a small quantity of fuel is injected, but increase as the injection quantity increases. On the contrary, with the 3-hole injection nozzle HC emissions are high overall but decrease as the injection quantity increases.

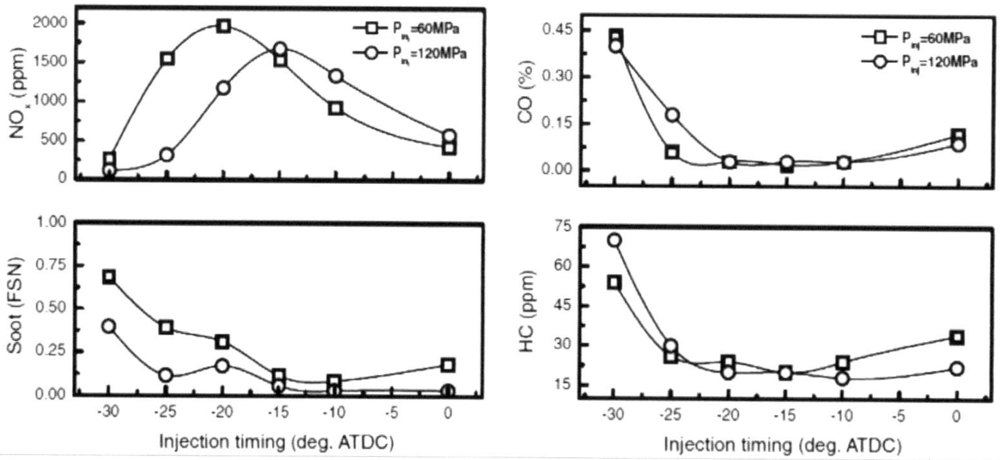

Figure 7. Effect of injection pressure and timing on the exhaust emissions characteristics [5].

4. Conclusion

The results obtained in this study are summarized as follows:

1. The rapid compression machine was used to study the fundamental of diesel combustion. Although the mechanism is looks simple, the performance and the reliability is verified to be sufficient.
2. The increased injection pressure caused an advance in ignition timing and this is due to the good atomization brought about by the high pressure injection and the short energizing duration for a given injection quantity. Also by increasing the fuel injection pressure, the fuel droplet size become finer, the entrainment of the air increases and the injected spray becomes more uniform and finally, the ignition delay and combustion periods become shorter.
3. Comparison of multi-hole injection nozzle, it's found that HC and CO emissions are sufficiently low when a small quantity of fuel is injected, but increase due to the poor air introduction to the spray at a large quantity of fuel injection.
4. The physical and the chemical process are strongly linked with each other and their part on the ignition delay can't be seen as independent from each other.
5. The higher injection pressure, higher oxygen content, results in lower smoke and CO emissions and in a slightly higher HC emission. The conditions in the cylinder during the first part of injection and combustion process influence to great extent the NOx formation.

Acknowledgements

The authors also would like to thank the Universiti Tun Hussein Onn Malaysia and Ministry of Higher Education, Malaysia for supporting this research under the Fundamental Research Grant Scheme VOT.1064.

References

[1] Kamil M, Rahman M M and Bakar R A 2011 *J. Mech. Eng. Sci.* **1** 87
[2] Ghobadian B, Najafi G and Nayebi M 2013 *J. Mech. Eng. Sci.* **4** 373
[3] Aziz A R A, Firmansyah and Shahzad R 2010 *Inter. J. Automot. Mech. Eng.* **2** 157
[4] Mohanamurugan S and Sendilvelan S 2011 *Inter. J. Automot. Mech. Eng.* **3** 279
[5] Wandel A P, Noor M M and Yusaf T F 2012 *Inter. J. Automot. Mech. Eng.* **6** 731
[6] Oki M 2006 *SAE Paper* **01-0274** 1
[7] Miwa K, Ohmija T and Nishitani T 1988 *JSME Int. J Ser II* **31**166
[8] Ikegami M, Miwa K and Inada M 1981 *Bulletin of the JSME* **24** 1608-1615.
[9] Rosseel E and R Sierens 1996 *SAE Paper* **961123** 1
[10] Espey C, Pinson J A and Litzinger T A 1990 *SAE Paper* **902076**
[11] Shundoh S 1990 *International Symposium Comodia* 607-612
[12] Singal S K, Pundir B P and Mehta P S 1993 *SAE Paper* **930604** 1-8.
[13] Corcione E F, Vaglieco B M and Valentino G 1998 *SAE Paper* **980502**
[14] Nakakita K 1990 *SAE Paper* **902081**
[15] Takahashi H 1991 *SAE Paper* **910226** 1
[16] Kegl B 2011 *Applied Energy* **88** 1803
[17] Arregle J, Pastor J V and Ruiz S 1999 *SAE Paper* **01-0200** 1
[18] Fang T 2010 *SAE Paper* **01-0339**
[19] Henein NA 2006 *SAE Paper* **01-0076**
[20] Su T F 1995 *SAE Paper* **952360**
[21] Roy M M and Tsunemoto H 2002 *SAE Paper* **01-2874**
[22] Shiozaki T 1999 *SAE Paper* **01-3496** 1
[23] Wang T C 1999 *SAE Paper* **01-3496** 1
[24] Oda Y, Kajitani S and Suzuki S 2003 *SAE Paper* **01-1925**
[25] Khalid A and Manshoor B 2013 *Applied Mechanics and Materials* **315** 293
[26] Kobori S, Kamimoto T and Aradi A A 1999 *Int J Engine Research* **1** 29
[27] Khalid A 2009 *SAE Paper* **32-0018** 1
[28] Woods M., Kamo R and Bryzik W 2000 *SAE Paper* **01-1186**
[29] Riva G E, Reggiori A and Daminelli A G 1993 *SAE Paper* **930926**
[30] Ishiyama T, Miwa K and Horikoshi O 1995 *JSME Inter. J.* **33** 483
[31] Kee S S 2003 *SAE Paper* **01-1792** 1
[32] Jiang H F, Wang J X and Shuai S J 2005 *SAE Paper* **01-0136**
[33] Arcoumanis C and Schindler K P 1997 *SAE Paper* **972681** 1
[34] Ladommatos N, Abdelhalim S M and Zhao Z H H 1998 *SAE Paper* **980184** 1
[35] Miyamoto T 1999 *SAE Paper* **01-0229**
[36] Khalid A 2013 *Applied Mechanics and Materials* **315** 517
[37] Senda J 2004 *SAE Paper* **01-0084**
[38] Pickett L M and Siebers D L 2003 *SAE Paper* **01-3080**
[39] Sato Y, Noda A and Jun L 2001 *SAE Paper* **01-3634** 1

Development and modification of a single overhead camshaft 4-valve 4-stroke 135 cc formula varsity race car engine

M A Abdullah[a], N Tamaldin, H Rusnandi, T Manoharan and M A Samsir
Centre for Advanced Research in Energy, Faculty of Mechanical Engineering, Universiti Teknikal Malaysia Melaka, Hang Tuah Jaya, 76100, Durian Tunggal, Melaka, Malaysia

E-mail: [a]mohdazman@utem.edu.my

Abstract. The engine that was chosen to be developed and modified is Yamaha LC 135 Single Overhead Camshaft (SOHC) 4-valve 4-stroke 135cc liquid-cooled engine. The engine selection is based on the specification, rule and regulation in UTeM Formula Varsity 2012 (FV 2012). The engine performance is determined by engine operating characteristics. The engine air flow affects the filtration, intake and exhaust systems. The heat from the engine rejected to the surrounding through the active cooling system which has radiator and fan. The selection of the engine is based on weighted decision matrix which consists of reliability, operating and maintenance cost, fuel consumption and weight. The score of the matrix is formulated based on relative weighted factor among the selections. It been compared between Yamaha LC 135 Single Overhead Camshaft (SOHC) 4-valve 4-stroke 135cc liquid-cooled engine, Honda Wave 125 X Air Cooled, 4 Cycle Engine Overhead Camshaft (OHC) and Suzuki Shogun RR 4 stroke air cooled Single Overhead Camshaft (SOHC). The modification is applied to the engine through the simulation and tuning of Capacitor Discharge Ignition (CDI).

1. Introduction

In the previous researches, the development and modification of engine part were not presented [1][2]. Original and standard engine was used to drive the race car. In this research, the chassis is designed and developed [3-4] and in addition the engine is modified to increase its capability. The development and modification of Formula Varsity (FV) race car are performed in proper manner so that it will not harm the driver. The development and modification are performed with consideration of rake angle, gear ratios, carburettor type, vehicle weight, the number of cylinders and the brake and suspension systems [5-9]. Motorcycle that manufactured by manufacturer produced different power output based on displacement volume and engine setting whereby it is not always produce optimum performance due to improper design and manufactures error. However, in order to produce optimum engine performance, the engine needed to be analyzed via simulation. Lotus Single Engine simulation software is suitable for the engine performance analysis. Investigation on which parameters will produce optimum engine performance is carried out by modification of the engine specifications. Comparison is also performed for engine performance between actual and simulation by using parameters from manufacturer sources. The specifications of the engine are shown by table 1 [10].

Table 1. Yamaha LC 135 engine specifications.

	Specification
Engine type	Liquid-cooled 4-stroke, SOHC, 4-valve
Cylinder arrangement	Single cylinder
Displacement	134.4 cm^3
Bore x stroke	54.0 mm x 58.7 mm
Connecting rod length	125 mm
Intake valve size	2 x 23.19 mm
Exhaust valve size	2 x 20.13 mm
Compression ratio	10.9:1
Maximum output	8.93 kW @ 8500 rpm
Maximum torque	11.79 Nm @ 5500 rpm
Starter	Electric starter and kick starter
Lubrication	Wet sump
Engine oil capacity	0.9 l
Fuel tank volume	4.0 l
Carburetion	Mikuni BS25 x 1
Ignition	DC-CDI

2. Methodology

Lotus Single Engine Simulation is used to analyze by changing parameter such as bore size of piston to get the result which can sustain the engine for 30 Laps in Melaka International Motorsport Circuit (MIMC) for 48 km. The analyses of the engine include engine fuel consumption, engine heat rejection, and factor of safety. Through this theoretical proof modification is made on the engine precisely and accurately. The modifications are:

i. Substitution of current Ignition Capacitor Discharge Ignition (CDI) to Capacitor Discharge Ignition (CDI) of Racing Rextor and unlimited Rotation per Minute (RPM).
ii. Engine overhaul and increment of bore size of piston to 57 mm with new connecting rod.
iii. Improvement of Carburetor with Throttle Position Sensor (TPS) and accelerator pump to UMA Racing Carburetor with Throttle Position Sensor (TPS) and accelerator pump (Main Jet size 142 - Pilot Jet size 45).

The modification follows the specification, rule and regulation that been fixed in UTeM Formula Varsity 2012 [9]. The engine is shown in figure 1. Figure 2 shows the software program (capacitor discharge ignition (CDI) of Racing Rextor Tuning Program) used to in the development and modification of the engine for Formula Varsity race car.

Figure 1. Yamaha LC135 engine [10].

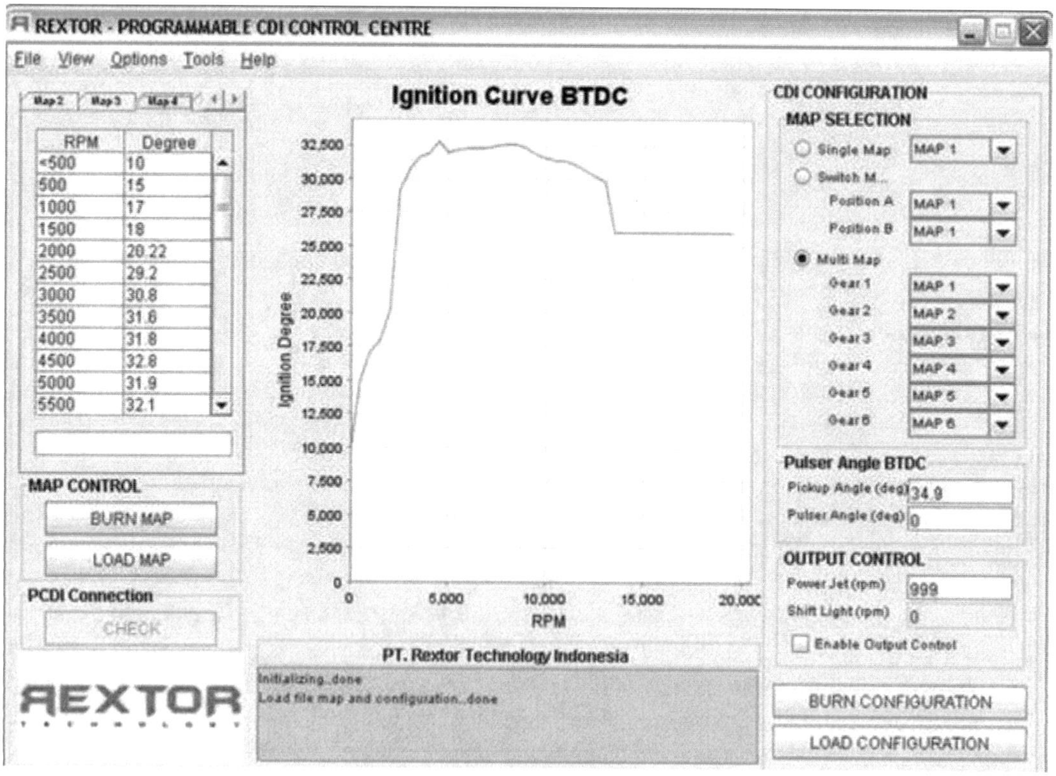

Figure 2. Concept design No. 2.

3. Analysis

The Lotus Single Engine simulation program is used to simulate the performance of the engine by 3 basic steps:

i. Construction of simulation model and engine specifications.
ii. Data and test conditions are defined for the cycle simulation.
iii. Calculation results for cycle-averaged data (volumetric efficiencies, brake specific fuel consumption (BSFC), torque, power), and intra-cycle data (pressure, temperature, mass flow rate) [11-13] in the form of report quality summary sheet and through quick-to-use graph plotting.

The engine performance is calculated theoretically with the new modified specifications from the simulation results (table 2) using Rextor tuning program.

Table 2. Yamaha LC 135 engine modifications.

	Specification
Bore, B x stroke, S	65.0 mm x 58.7 mm
Compression ratio, r_c	12.2:1
Maximum output power, P	8.45 kW @ 8500 rpm
Maximum torque, T	11.65 Nm @ 5500 rpm

The volume displacement is[13],

$$V_d = \frac{\pi B^2 S}{4} \quad (1)$$

The engine torque is calculated at max power [14].

$$P = \frac{2\pi NT}{60} \quad (2)$$

The average piston speed at max power is [15],

$$U_p = N_S \quad (3)$$

The volumetric efficiency is [16],

$$\eta_{ve} = \frac{ma}{pa} \quad (4)$$

The thermal efficiency is calculated by [14],

$$\eta_{thermal} = 1 - \left[\frac{T_4 - T_1}{T_3 - T_2}\right] \quad (5)$$

The results of the calculation and data gathering from engine trial are tabulated in Table 3. Throughout the modification process, the engine components are machined and amended according to the standard procedures of workshop and machining practices [17-21]. The engine trial is performed on the Melaka International Motorsport Circuit (MIMC).

Table 3. Results and data of the modified engine.

Engine Specification	Value
Engine Capacity	194.78 cc
Engine Torque at maximum Power	29.8235 N.m
Average piston speed@ max power	16.63 m/s
Bore-Stroke ratio (B/S)	1.1073
Volumetric Efficiency	0.878 @ 87.8%
Engine Power at maximum Torque	26.55 kW @ 35.6 hp
Thermal Efficiency	0.567 @ 56.7%

Fuel flow rate	8.0952 ml/s
Fuel per millage	0.2125 ml/m

4. Results and Discussion

Through the simulation results obtained for original engine condition, the maximum torque is 10.23 Nm @ 5500 rpm and the maximum power is 8.80 kW @ 8,500 rpm. The compression ratio is 10.9 : 1. The comparison of performance summary between the original and modified engine is shown in figure 4. The simulation analysis result for the modification engine has established approving outcome in term of 194.78 cc of engine capacity, compression ratio 12.2:1, bore size of piston 65 mm, engine power at maximum torque 26.55kW @ 35.6hp, engine torque at maximum power 29.8235 Nm, and thermal efficiency of 56.7%. It is validated that the theoretical data calculated and the results are precise. With the modification of the engine, the performance is improved for endurance and racing.

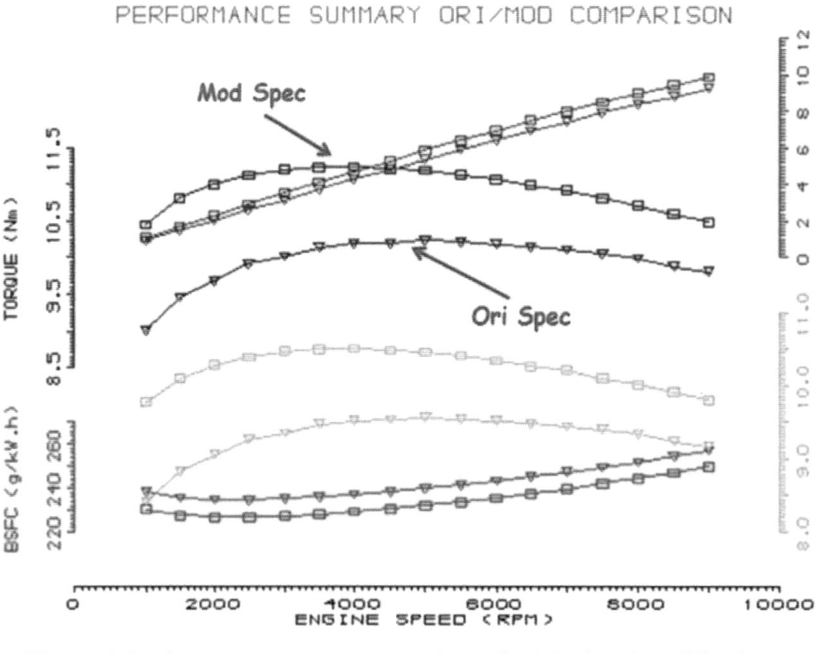

Figure 4. Performance summary comparison of original and modification.

5. Conclusion

Through the research, development and modification of FV race car engine is successfully performed. The parameters such as bore size, stroke length, compression ratio, valve timing give significant impact to the engine performance when modified. The simulation analysis result has established approving outcome in term of engine capacity, compression ratio, engine power at maximum torque, engine torque at maximum power and thermal efficiency. The engine is later assembled with power train and chassis for the FV racing competition. The validation of the results between simulation and theoretical calculation shows the availability and practicality of engine modification.

Acknowledgments

The authors gratefully acknowledged the financial support from Universiti Teknikal Malaysia Melaka and The Ministry of Higher Education, Malaysia (MoHE) under Exploratory Research Grant Scheme (ERGS), grant no.: ERGS/1/2012/TK08/UTEM/02/1/E00007.

References

[1] Shaharuzaman, M A, Abdul Munir F, Hassan M Z, Mohamad Zin M R, Boejang H, Muhammad Farhan, F H and Abdul Jamal A H 2010 Development of a Single Seat Race Vehicle for Educational Motorsport Engineering, *3rd International Conference on Science & Technology: Applications in Industry & Education*, 16-17 December 2010, Universiti Teknologi MARA, Pulau Pinang Malaysia

[2] Abdul Munir, Munir F, Shaharuzaman M A, Hassan M Z, Mohamad Zin M R, Mohd Azmi, M I and Mat Nuri N R 2011 Components Design of a Single Seat Educational Race Vehicle, *Malaysian Technical Universities International Conference on Engineering & Technology (MUiCET 2011)* 13-15 November 2011 UTHM Batu Pahat Johor

[3] Abdullah M A, Mansor M R, Mohd Tahir M, Abdul Kudus S I, Hassan M Z and Ngadiman M N 2013 Analysis and Fabrication of Chassis Frame for UTeM Formula Varsity™ Race Car *International Journal of Mining, Metallurgy & Mechanical Engineering (IJMMME)* **1** 75-77

[4] Abdullah M A, Mansor M R, Mohd Tahir M, Abdul Kudus S I, Hassan, M Z and Ngadiman, M. N 2012 Analysis and Fabrication of Chassis Frame for UTeM Formula Varsity™ Race Car, *Proceeding for International Conference on Manufacturing and Industrial Engineering (ICMIE'12)*, International Scientific Academy of Engineering & Technology (ISAET), Kuala Lumpur, Malaysia, (December 29th to 30th, 2012)

[5] SAE, 2012 Formula SAE Competition Rules and Regulations, Society of Automotive Engineers, USA

[6] Kamal M, Rahman M M and Rahman A G A 2012 *J Mech Eng Sci* **3** 291

[7] Nagarkar M P, Vikhe G J, Borole K R and Nandedkar V M 2011 *Inter J Automot Mech Eng* **3** 364

[8] Kamal M, Rahman M M and Rahman A G A 2013 *Inter J Automot Mech Eng* **7** 912.

[9] UTeM FV2012 2012 Formula Varsity Competition Rules and Regulations, Universiti Teknikal Malaysia Melaka, Melaka, Malaysia

[10] Motomalaya 2012 Original Yamaha 135LC Big Valve Head. Retrieved 13 December, 2012, from http://motomalaya.net/blog/2010/09/17/original-yamaha-135lc-big-valve-head-intake-23-19mm-exhaust-20-13mm/yamaha-135lc-big-valve-head-ori-9/

[11] Nice K, 2012 How Camshafts Work Retrieved 12 December, 2012, from http://www.howstuffworks.com/camshaft.htm.

[12] Auto-Ware 2012 Valve Timing & Performance. Retrieved 11 December, 2012, from http://www.auto-ware.com/combust_bytes/valvetiming.html.

[13] Auto-Ware 2012 Cam Specs & Effects Retrieved 12 December, 2012, from http://www.auto-ware.com/combust_bytes/camspecs.html.

[14] Pulkrabek WW 1997 *Engineering Fundamental of Internal Combustion Engine* (Prentice Hall: New Jersey)

[15] Heywood J B 1998 *Internal Combustion Engine Fundamental* (McGraw Hill: New York)

[16] Challengers 101, 2012. Exhaust and Intake Tuning. Retrieved 12 December, 2012, from http://www.challengers101.com/IntakeTuning.html.

[17] Serope Kalpakjian, Steven Schmid 2006 *Manufacturing Engineering and Technology* (Canada: Prentice Hall)

[18] Miller R And Richard M 2004 *Machine Shop Basics All new* (Wiley Publishing, Inc)

[19] W. A. J. Chapman. 2001. *Workshop Technology: Part 1*, 5(United Kingdom. Edward Arnold Publishers Limited)

[20] Khurmi R S and Gupta J K 2008 *Theory of Machines (*14th ed.) (Eurasia Publishing House) p 371

[21] Budynas R and Nisbett K 2006 *Shigley's Mechanical Engineering Design* (8th ed.) (McGraw Hill) p 504

Performance of a fully mechanical parking brake system for passenger cars

A H Rozaini[a], M R Ishak, A R Abu Bakar and M Z Mohd Zain

Faculty of Mechanical Engineering, Universiti Teknologi Malaysia, 81310, Johor, Malaysia

E-mail: [a]humaizi@fkm.utm.my

Abstract. In order to ensure that a vehicle remains stationary when it is parked at a certain road slope, the driver has to apply sufficient pulling force on the handbrake lever. Otherwise, the vehicle will start to rollaway where the torque generated by the parking brake system is lower that the torque required by the vehicle to remain stationary. This poses a danger situation not only to the vehicle's occupants but also to the people surrounding it. Thus, this paper aims to investigate performance of a typical parking brake system used in passenger cars. A theoretical model of drum-type parking brake system is derived and later being validated by test data that measured from the parking brake test bench. A good agreement is achieved between calculated and test results. Results from the model show that the parking brake system used in this work can hold the vehicle stationary at 11 degree slope less than 200 N of the applied force and thus it meets the regulation requirement, and also the vehicle will not rollaway even though there are four adult passengers inside it.

1. Introduction

In a ground vehicle, mechanical parking brake unit is a mechanism that used to hold the vehicle stationary either on the even or slope road. It consists of a cable (usually adjustable for length) directly connected to the brake mechanism on one end and to a mechanism that can be actuated by the driver on the other end. This actuating mechanism is often a hand-operated lever on the floor on either side of the driver, or a pull handle located below and near the steering wheel column or a (foot-operated) located far apart from the other pedals [1].

European Economic Community (EEC) regulation [2] specified that the handbrake system of a laden vehicle in class M1 (passenger cars comprising no more than eight seats in addition to the driver's seat) must be able to hold vehicle in 20% gradient. FMVSS 135[3], on the other hand, stated that the vehicle should able to stay stationary on 20% grade road for 5 minutes in both forward and reverse direction with 400N or less force applied at hand control mechanism.

To date, there is wealth information available in the open literature to evaluate braking performance for the service brakes [4-8], but very limited studies have been done on the parking brakes. For instance, McKinlay et al [9, 10] investigated conventional parking brake performance in term of vehicle rollaway for the disc brakes through laboratory tests and theoretical approaches. In the theoretical approach a one-dimensional (1-D) model of disc-type parking brake system using simple linear spring element had been proposed and analyzed to predict clamping force. A laboratory parking brakes test rig had been developed to measure clamping force of the parking brakes system which is used to validate results from the 1-D model. In the studies, they considered thermal effects on the braking torque and subsequently vehicle rollaway.

To authors' knowledge, there is no information available in the public domain to investigate performance of on drum-type of parking brakes system through theoretical and laboratory test approaches. Thus, this paper attempts to establish mathematical model of parking brake and develop reliable laboratory parking brake test bench in order investigate and assess parking brake performance.

2. Theoretical Model

In order to study performance of the parking brake system, a mathematical model is derived based on diagrams given in figure 1(a) and 1(b). Thus, it can be deduced that the torque produced by the drum-type of parking brake in the forward and reverse directions are given in Eqs. (1) and (2), respectively.

$$T_{forward} = \frac{\mu * r_d * F_{hb} * 8.8333}{l_c}\left[\left(\frac{l_d * l_f}{l_e + \mu * l_h}\right) + \left(\frac{l_m * l_j}{l_e - \mu * l_h}\right)\right] \quad (1)$$

$$T_{reverse} = \frac{\mu * r_d * F_{hb} * 8.8333}{l_c}\left[\left(\frac{l_d * l_f}{l_e - \mu * l_h}\right) + \left(\frac{l_m * l_j}{l_e + \mu * l_h}\right)\right] \quad (2)$$

where μ is coefficient of friction, Fhb is handbrake force and rd is radius of the drum.

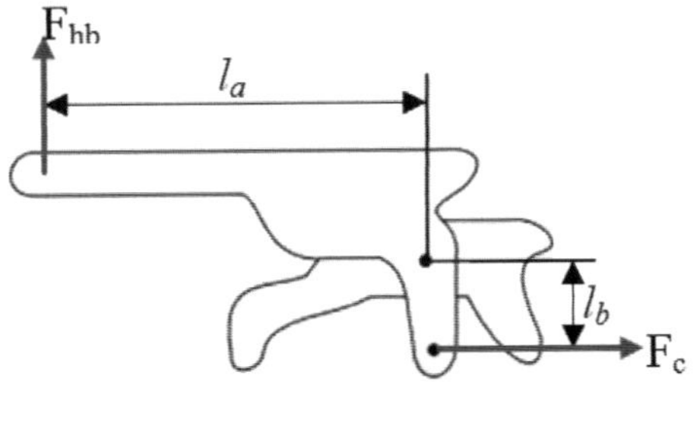

(a) Hand lever mechanism

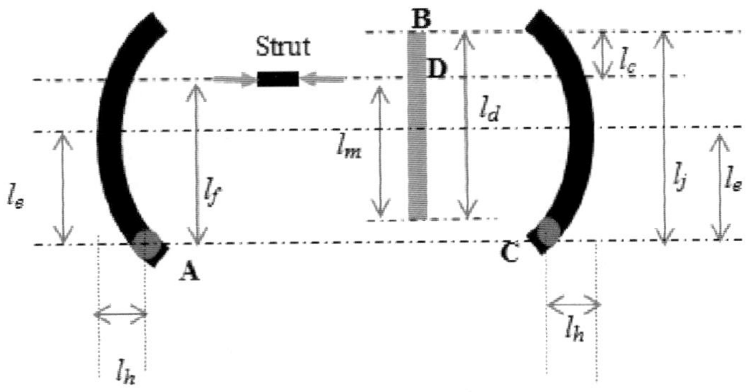

(b) Drum brake

Figure 1.One-dimensional parking brake model.

The torque required to hold a passenger car stationary on a slope road can be estimated based on the free body diagram as shown in figure 2. Thus, the required torque to hold the vehicle can be given as follows;

$$T_{req} = \mu_r * F_r * r_{wheel} = mg\sin\theta * r_{wheel} \tag{3}$$

where m is the vehicle weight, θ is the road slope, g is the gravity and r_{wheel} is radius of the wheel

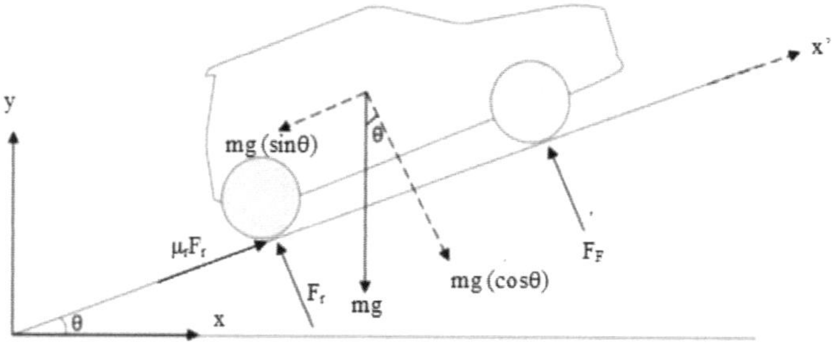

Figure 2. Free body diagram of a vehicle on the slope

3. Experimental Work

The experiment is conducted using parking brake test bench as shown in figure 3 and all data are captured and recorded by Dewetron data logger[11]. The bench consists of a set of parking brake mechanism connected to a pair of rear-drum-brake unit, two hydraulic pumps, an actuator, a flywheel, a 5.5 kW motor with speed controller, and one load cell. The parking brake unit is taken from one of the national passenger cars. One of the hydraulic pumps is used to supply brake pressure to the drum brake and to heat up the brake unit, whilst the other one is connected to the actuator which is used to pull a cable that connects to the flywheel and the load cell. This will measure brake torque produced by the drum brake unit when the hand lever is being pulled. In order to engage the parking brake, force on the hand lever is applied using deadweight by connecting them through a cable.

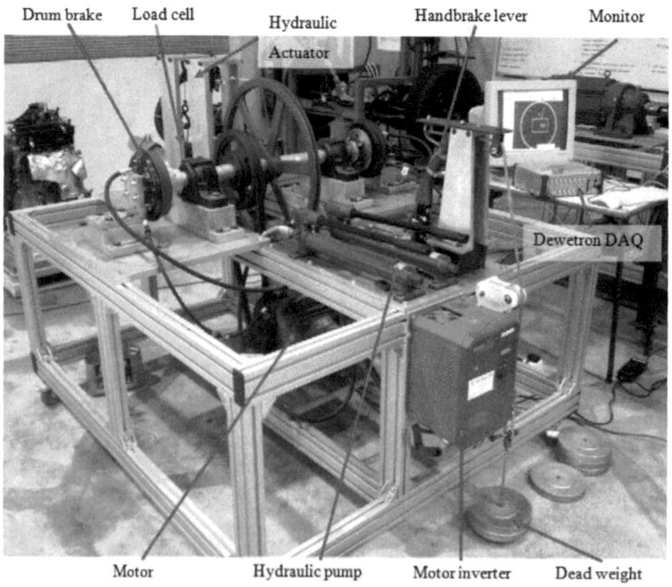

Figure 3. Parking brake test bench.

4. Results And Discussion

It can be seen in figure 4(a) and 4(b) that there is good correlation achieved between model and test results within 200 N of the applied force. It is also shown that the parking brake torque is increased with the increase of applied force. It is noticed that the torque in the forward direction is slightly lower than the torque in the reverse direction. This is due to design of the leading-trailing type of a drum brake unit [1]. Table 1 shows value of the parameters used in this work.

Table 1. Value of the parameters used in the model and experiment

Model		Experiment	
Parameter	Value	Parameter	Value
μ	0.25-0.28	M	1150 kg
r_d	0.091 m	r_{wheel}	0.289 m
l_c	0.022 m	G	9.81 m/s^2
l_d	0.113 m		
l_e	0.068 m		
l_f	0.112 m		
l_h	0.070 m		
l_j	0.135 m		

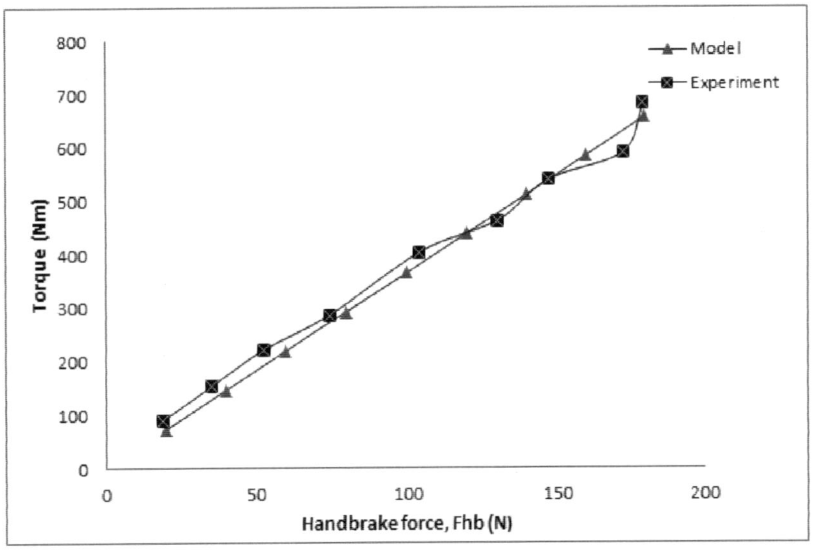

(a) Forward direction

Figure 4. Correlation of the parking brake torque between model and experiment.

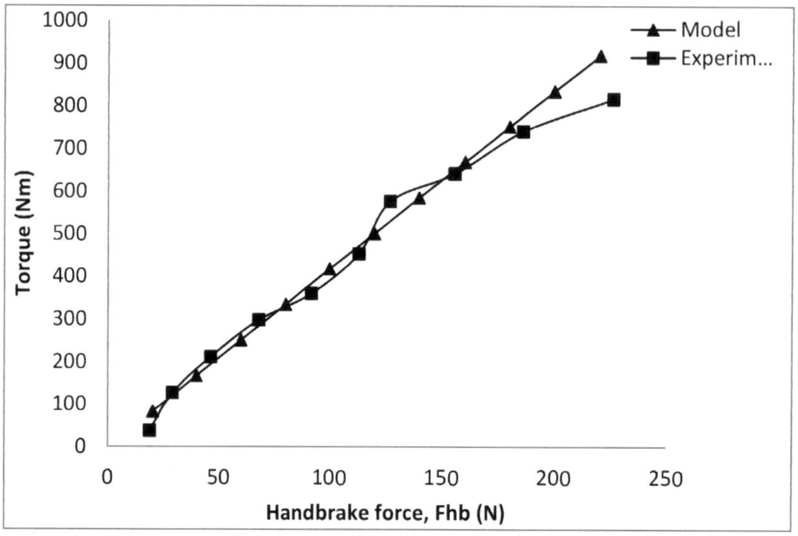

(b) Reverse direction

Figure 4 (continued). Correlation of the parking brake torque between model and experiment

4.1. Effect of road slope

It is interesting to see performance of the parking brake with various road slopes. FMVSS 135 [3] stated that a vehicle must be stayed stationary for five minutes at 20% slope (or 11 degree slope) with 400 N applied force. Figure 5(a) and 5(b) show that the vehicle is not rollaway at 400N because the torque at 11 degree is very much lower than the torque produced by the parking brake (PB Torque). It is also shown that the minimum forces required to hold the vehicle stationary are 60N (4 degree), 120N (8 degree) and 180N (11 degree) for forward direction and 58N (4 degree), 110N (8 degree) and 160N (11 degree) for reverse direction.

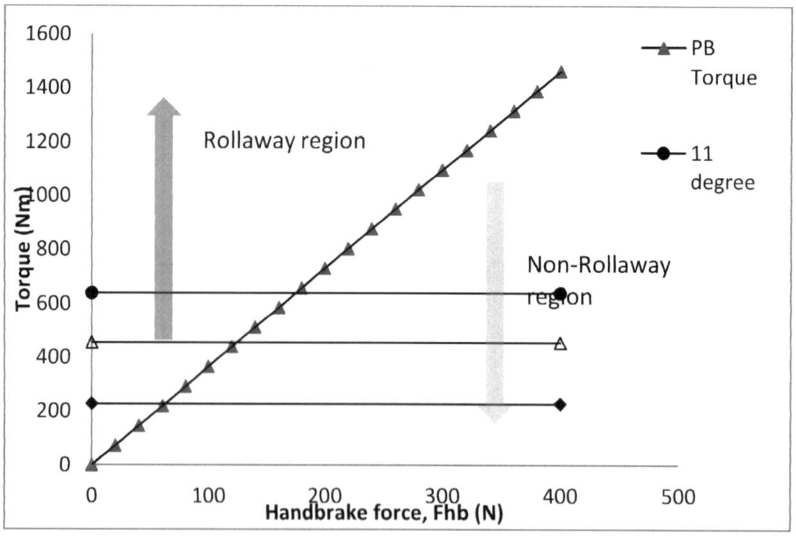

(a) Forward direction

Figure 5. Parking brake performance at different road slopes.

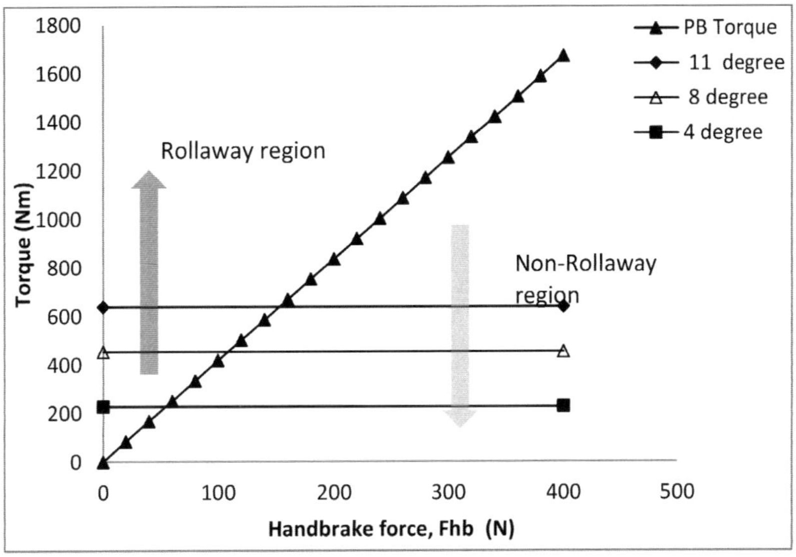

(b) Reverse direction

Figure 5 (continued). Parking brake performance at different road slopes.

4.2. Effect of passenger weight

The effect of passenger weight is also studied up to four adult passengers with 70kg each. It can be seen in figure 6(a) and 6(b) that the vehicle remains stationary up to 4 passengers when the force applied by the driver is 400N. The minimum force needs to be applied with 4 passengers is 220N in the forward direction and 200N in the reverse direction.

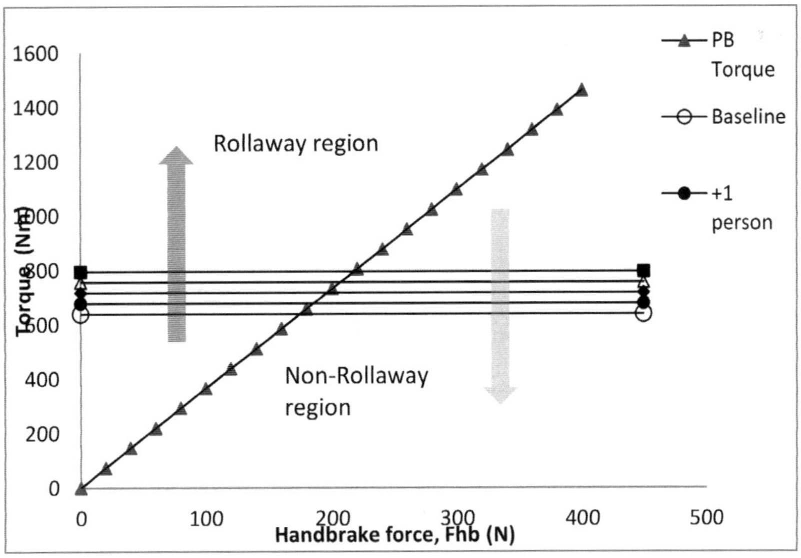

(a) Forward direction

Figure 6. Parking brake performance with various number of passengers.

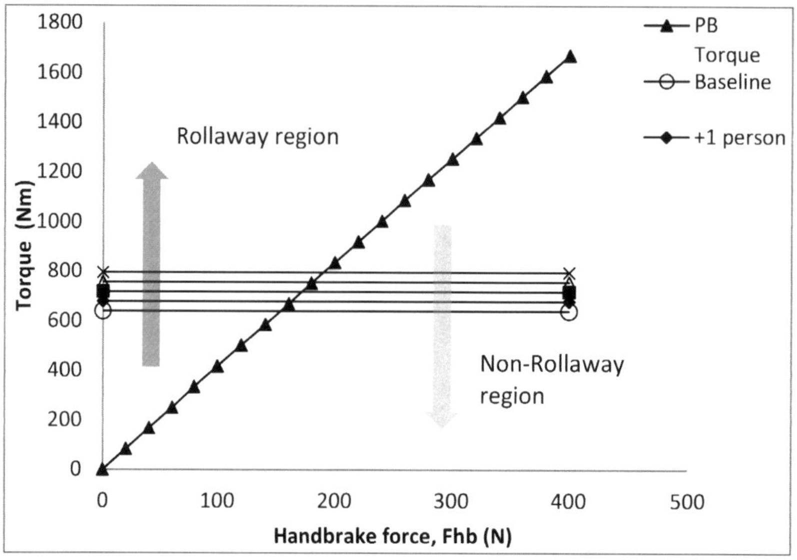

(b) Reverse direction

Figure 6 (Continued). Parking brake performance with various number of passengers.

5. Conclusion

From the study, it can be deduced that:

- The parking brake model is valid where good correlation is achieved between calculated and measured results,
- Due to design of the leading-trailing type of drum brake unit, the torque produced in the forward and reverse direction is slightly different,
- The parking brake system is complied with FMVSS 135 regulation,
- The vehicle will not rollaway at 11 degree slope and with 4 passengers inside it.

Acknowledgements

The authors would like to thank the Faculty of Mechanical Engineering in Universiti Teknologi Malaysia and Ministry of Higher Education for financial support under FRGS Grant Vot. No. 4F164.

References

[1] Limpert R 1999 Brake Design and Safety 2nd ed. *(Warrendale, Pa: Society of Automotive Engineers, Inc).* 525.
[2] U.N. Inland Transport Division 2008 UNECE Regulations *(United Nations:United Nations Economic Commission)*
[3] U.S. National Highway Traffic Safety Administration 2005 Light Vehicle Brake Systems FMVSS 135 *(United State:National Highway Traffic Safety Administration)*
[4] Xiaoyan Z, Q Peixin, and L Xiaojing 2012 Braking performance study for automobile with numerical simulation. *Applied Mechanics and Materials* **155-156** 1159-63
[5] Tirovic M, et al. 2011 Structural analysis of a commercial vehicle disc brake caliper *Journal of Automobile Engineering* **226** 613
[6] Hwang P, X Wu, and Y B Jeon 2009 Thermal--mechanical coupled simulation of a solid brake disc in repeated braking cycles *Journal of Engineering Tribology* **223** 1041
[7] Koylu H and A Cinar 2012 Experimental design of control strategy based on brake pressure changes on wet and slippery surfaces of rough road for variable damper setting during braking with activated anti-lock brake system *Journal Automobile Engineering* **226(10)** 1303–24.

[8] Zhang Z, J Chen, and BofuWu 2012 The control strategy of optimal brake energy recovery for a parallel hydraulic hybrid vehicle *Journal of Automobile Engineering* **226(11)** 1445–53.
[9] McKinlay A J, et al. 2004 The mystery of vehicle rollaway. in *Braking 2004: Vehicle Braking and Chassis Control* (Leeds, United Kingdoms: IMechE)
[10] McKinlay A J, P C Brooks, and D C Barton. 2006 A Study of Vehicle Handbrake Rollaway: Thereotical, Numerical and Experimental Assessment. in *Vehicle Braking Technology* IMechE).
[11] *Dewetron Data Acquisition Instrument*, http://www.dewetron.com, Retrieved 23 July 2013

2nd International Conference on Mechanical Engineering Research (ICMER 2013) IOP Publishing
IOP Conf. Series: Materials Science and Engineering **50** (2013) 012007 doi:10.1088/1757-899X/50/1/012007

A comparison of cylindrical and row trenched cooling holes with alignment angle of 0 degree near the combustor endwall

E Kianpour, C S Nor Azwadi[a] and I Golshokouh
Department of Thermo Fluid, Faculty of Mechanical Engineering, UTM
81300, Skudai, Johor, Malaysia

E-mail: [a]azwadi@fkm.utm.my

Abstract. We studied the effects of cylindrical and row trenched cooling holes with alignment angle of 0° at BR=3.18 on the film cooling performance near the endwall surface of a combustor simulator. In this research, a three-dimensional presentation of gas turbine engine was simulated and analyzed with a commercial finite volume package FLUENT 6.2.26 to gain fundamental data. The current study has been performed with Reynolds-averaged Navier-Stokes turbulence model (RANS) on internal cooling passages. This combustor simulator combined the interaction of two rows of dilution jets, which were staggered in the stream wise direction and aligned in the span wise direction. The entire findings of the study declared that with using the row trenched holes near the enwall surface; film cooling effectiveness is doubled compared to the cooling performance of baseline case.

1. Introduction

Gas turbine industries try for higher engine efficiencies [1]. Bryton cycle is a key to this objective [2]. According to this cycle, the turbine inlet temperature should increase to gain more efficiency. However increasing the turbine inlet temperature creates harsh environment for downstream components. So, it is needed to design a cooling technique in this area. Film cooling is the way which is used. In this system, a thin thermal boundary layer such as buffer zone is formed and attached on the protected surface. Cylindrical and trenched cooling holes are two layouts of film cooling. With trenching the cooling holes, the injected coolant is suddenly spread before exiting the cooling holes and entering the main flow and as a result the coolant attached better on the protected surface [3-4]. According to the importance of this research, a broad literature search was conducted to collect the information. Stitzel and Thole [5] indicated that with no dilution, the exit profile was relatively uniform however, the high temperature and low total pressure mainstream flow was found. Kianpour et al.[6-7] simulated the combustor endwall cooling holes with two different layouts and exit section area. The results declared that for both configurations, the temperatures adjacent the wall and between the jets was a while cooler with less cooling holes. From Yiping et al.[8] researches, it is found that the narrow trenched holes were more effective than other trenched cases and baseline. Also they mentioned that row trenches have dominant effect on film cooling performance compared to the individual trenches. Sundaram and Thole [9] and Lawson and Thole [10] studied the effects of trenched depth and width on film cooling performance at the vaneendwall. The results showed that the maximum cooling effectiveness is obtained at the trench depth of 0.80D. According to the importance of this issue, more studies are required. There are questions should be answered: Which type of trenched cooling holes with different alignment angles has more effect on film cooling performance?

In order to measure the validity of the results, a comparison between the data gained from this study and Vakil and Thole[1] project was made.

2. Methods and Materials

As seen in figure 1, the combustor was a container with a width and height of 111.8cm and 99.1cm. The length of the combustor was 156.9cm and the contraction angle was 15.8°. While the inlet cross-sectional area was 1.11m2, the exit cross-sectional area was 0.62m^2. The contraction angle began at X=79.8cm. The combustor simulator involved four stream wise series of film-cooling panels. The starting point of these panels was at 1.6m upstream of the turbine vanes. The length of the first and second panels was 39centimeter and 41cm respectively. However the third and fourth panels were 37 and 43 centimetre in length. The thickness of the panels was 1.27cm. The thermal conductivity (k= 0.037 W/mk) was such light that adiabatic surface temperature measurement was possible. There were two rows of dilution holes within the second and third panels of cooling panels. These dilution holes were located at 0.67m and 0.90m downstream of the beginning of the combustor liner panels. The diameter of the first row and second row of dilution holes was 8.5cm and 11.9cm respectively.

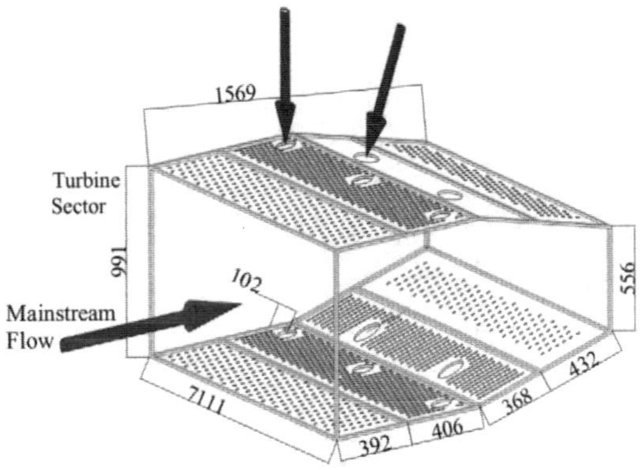

Figure 1. The 3-D view of the combustor simulator.

The centreline of the second row was staggered with respect to the first row of dilution holes. To verify the purpose of this study, a three-dimensional representation of a Pratt and Whitney engine was simulated. The present combustor simulator included two configurations of cooling holes. The first arrangement (baseline or case 1) was designed similar to the Vakil and Thole [2]. In both cases, the film cooling holes were placed in equilateral triangles. The diameter of the film cooling holes was 0.76cm and drilled at an angle of 30° from the horizontal surface. The length of film cooling holes in the baseline case was 2.5cm. For the second case (case 2) the cooling holes placed within a row trench with alignment angle of 0 degree which is shown in figure 2. Furthermore, the trench depth and width was 0.75D and 1.0D respectively. A global coordinate system (X, Y and Z) was also selected.

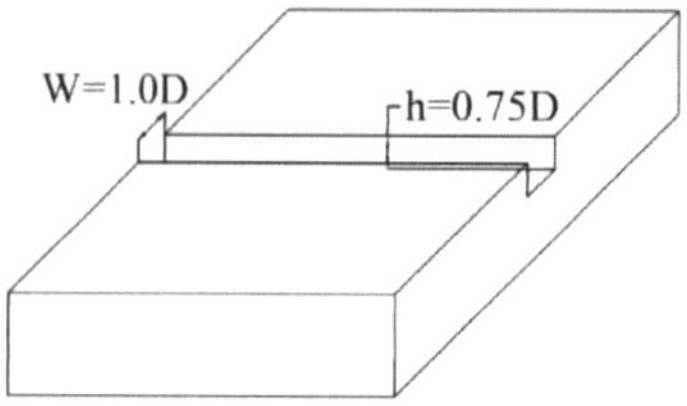

Figure 2. The schematic of row trenched cooling holes with alignment angle of 0°.

The thermal distribution inside a combustor simulator was measured along the specific measurement planes. These measurement planes are shown in figure 3.

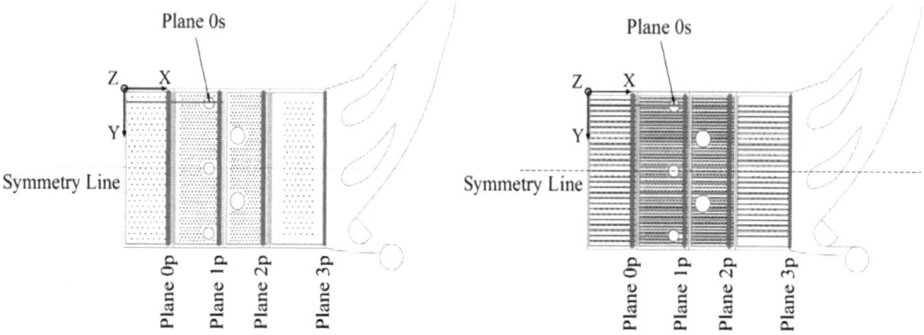

Figure 3. Location of the measurement planes (a) Case 1 (b) Case 2.

In order to get more accurate data and reasonable time consumption, about 8×10^6 tetrahedral meshes were used and this is in concurred with Stitzel and Thole [5] study. According to the specific flow ratio at the inlet of volume control, inlet mass flow boundary condition was defined. Wall boundary condition and slip less boundary condition were applied to limit the interaction zone between fluid and solid layer. Also at the end of volume control the pressure outlet boundary condition was used. In addition, both cases were completely symmetric along the X-Y and X-Z planes. According to this issue, symmetry boundary condition $\partial/\partial n=0$was applied. In addition theses equations were used as well.

Continuity equation:

$$\frac{\partial \rho}{\partial t}+\frac{\partial}{\partial x}\frac{dx}{dt}+\frac{\partial \rho}{\partial y}\frac{dy}{dt}+\frac{\partial \rho}{\partial z}\frac{dz}{dt}=-\rho(\nabla.V) \tag{1}$$

Momentum equation:

$$\frac{\partial}{\partial t}(\rho u_i)+\frac{\partial}{\partial x_j}=(\rho u_i u_j)=-\frac{\partial P}{\partial x_i}+\frac{\partial \tau_{ij}}{\partial x_i}+\rho g_i+\vec{F_i} \tag{2}$$

Energy equation:

$$\frac{\partial}{\partial t}(\rho E) + \frac{\partial}{\partial x_i}\big(u_i(\rho E + P)\big) = \frac{\partial}{\partial x_i}\left(K_{eff}\frac{\partial T}{\partial x_i} - \sum_j h_j J_j + u_j(\tau_{ij})_{erf}\right) + S_h \tag{3}$$

and RNG K-ε equation:

$$\frac{\partial}{\partial t}(\rho k) + \frac{\partial}{\partial x_i}(\rho k u_i) = \frac{\partial}{\partial x_j}\left[\left(\mu + \frac{\mu_t}{\sigma_k}\right)\frac{\partial k}{\partial x_j}\right] + P_k - \rho\epsilon \tag{4}$$

$$\frac{\partial}{\partial t}(\rho \epsilon) + \frac{\partial}{\partial x_i}(\rho \epsilon u_i) = \frac{\partial}{\partial x_j}\left[\left(\mu + \frac{\mu_t}{\sigma_\epsilon}\right)\frac{\partial \epsilon}{\partial x_j}\right] + C_{1\epsilon}\frac{\epsilon}{k}P_k - C_{2\epsilon}^{*}\rho\frac{\epsilon^2}{k} \tag{5}$$

To understand the thermal field results, the quantities should be defined. Film cooling effectiveness is defined as below:

$$\eta = \frac{T - T_\infty}{T_C - T_\infty} \tag{6}$$

3. Findings and Discussion

The comparison has been made for baseline case among the numerical and experimental results by Stitzel and Thole[5] and Vakil and Thole [2], and the current study. Figure 4 presents the comparison of film cooling effectiveness for plane 1p and 2p at Y/W=0.4. The deviation was calculated as follows:

$$\%Diff = \frac{\sum_{i=1}^{n}\frac{x_i - x_{i,benchmark}}{x_{i,benchmark}}}{n} \times 100 \tag{7}$$

According to this formula, the deviation was equal to 8.34% and 9.76% compared to findings for plane 1p and equal to 11.96% and 13.36% compared to numerical and experimental data for plane 2p.

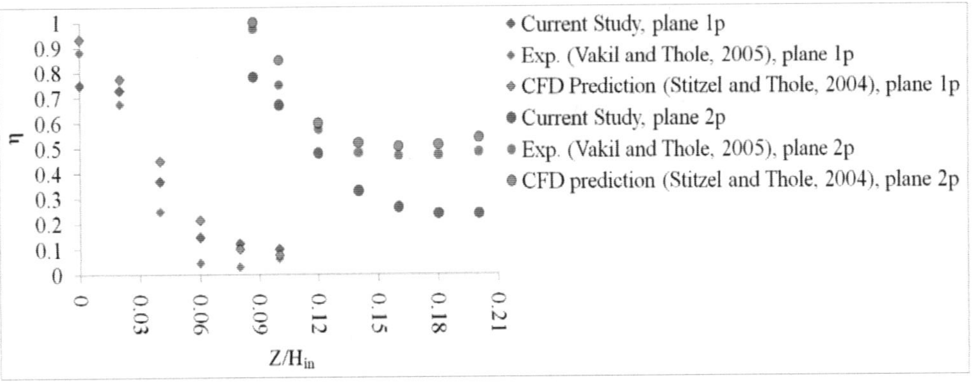

Figure 4. Film cooling effectiveness comparison for plane 1p and 2p along Y/W=0.4.

The distribution of film cooling effectiveness of plane 1p is shown in figure 5a (case 1) and 5b (case 2). Note that, at the right side of figure 5a and 5b (50cm<Y<54cm) thermal field contours show that film cooling is being entrained by upward motion of dilution jet. The v and w velocity vectors of plane 1p are shown as well. From the middle of the temperature distribution contour, a significant movement of vortexes toward the left and right sides is found. This is the effect of dilution injection on the thermal behaviour of flow.

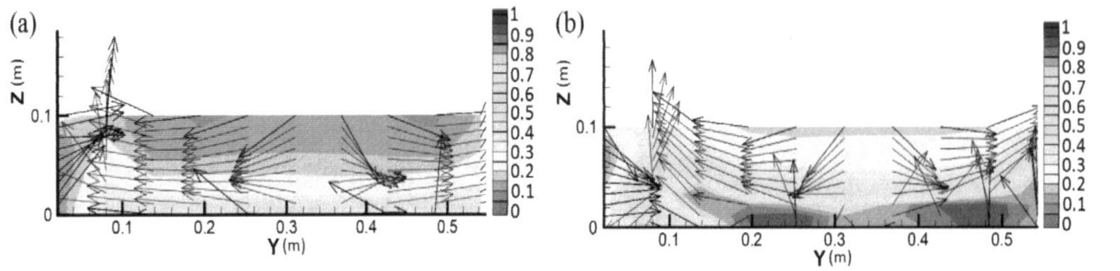

Figure 5. The vectors of v and w with film cooling effectiveness contours for plane 1p (a) case 1 (b) case 2.

Figure 6a (baseline) and 6b (case 2) show the film cooling distribution for plane 2p. It is declared that the rotating flow is seen on the left side of figure and it is entrained along the span wise direction. However, this rotating area is weaker for the trenched case. Also, at the right side of figure 6a the hot gases covered more extended area in comparison with trenched case. Lastly, these figures show the v and w velocity vectors superimposed on the thermal field contours of this measurement plane. The sweeping of the coolant toward the second row of dilution jet is visible.

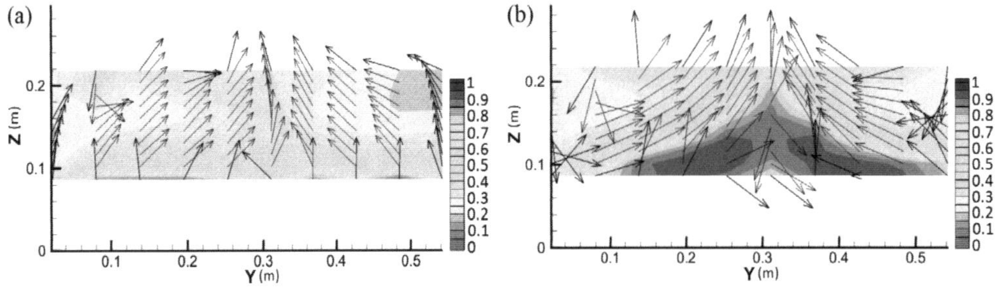

Figure 6. The vectors of v and w with film cooling effectiveness contours for plane 2p (a) baseline (b) case 2.

The variation of film cooling effectiveness for different measurement planes at Y=30cm and along Z axis is shown in figure 7.

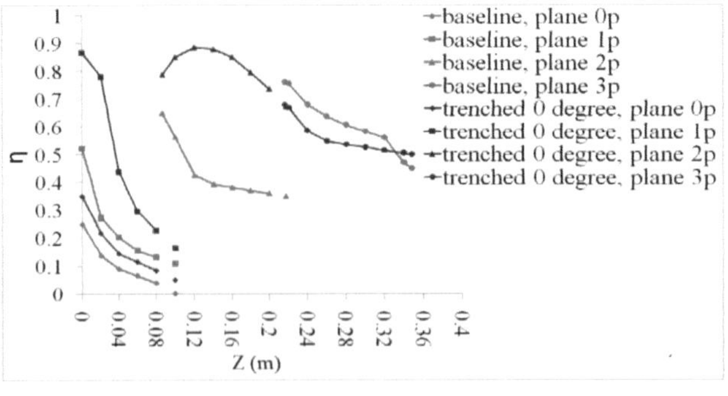

Figure 7. The variation of film cooling effectiveness for different measurement planes.

It is declared that for the three first measurement planes, case 2 performed better. Also, the film cooling effectiveness difference between the baseline and trenched hole is growing step by step from plane op to plane 2p. While, at the end of first cooling panel, film cooling effectiveness is increasing 40% by trenching cooling holes, at the end of second cooling panel, this ratio is equal to 66%. Also, it is seen that film cooling effectiveness increased intensively (about 92%) at further distance compared to baseline at the end of third cooling panel, although, at the first of this panel, this ratio increased only 21%.

4. Conclusion and Recommendation

In this study a three-dimensional representation of a Pratt and Whitney engine was analyzed. To sum up, the usage of trenched cooling holes significantly to development of the film cooling layer. Also, the central part of the plane 2p showed the intense penetration of the coolant and a thick film cooling layer creation in the trenched case. However, the temperature adjacent the wall and between the jets was cooler with trenching the cooling holes because by trenching the cooling holes, coolant spread better in this area. Initially, the results declared that at the end of three first measurement cooling panels, trenching cooling holes has intense effect on film cooling. Based on the results, there are several recommendations to consider. In future research within this area, different configurations of trenched cooling holes and baseline case should be considered for different cooling panels. The effects of different blowing ratios on the film cooling performance should notice as well.

Acknowledgements

The authors would like to thanks the Universiti Teknologi Malaysia and Ministry of Higher Education, Malaysia for sponsoring this research. This research is supported by research grant Vote 4F114 and 4L074.

References

[1] Ibrahim T K and Rahman M M 2013 *J Mech Eng Sci* **4** 383
[2] Vakil S S and Thole K A 2005 *J. Eng. Gas Turb. Power* **127** 257
[3] Smajevic I 2010 *Inter J Automot Mech Eng* **1** 1.
[4] Wandel A P, Noor M M and Yusaf T F 2012 *Inter J Automot Mech Eng* **6** 731
[5] Stitzel S and Thole K A 2004 *J. Turbomach.* **126** 122.
[6] Kianpour E, Azwadi C S N and Agha S M B M 1992 *Jurnal Teknologi* **58** 5
[7] Kianpour E, Azwadi C S N and Agha S M B M *Int. J. Heat Mass Tran.* **61** 389
[8] Yiping L, Dhungel A, Ekkad S V and Bunker R S 2009 *J. Turbomach.* **131** 1
[9] Sundaram N and Thole K A 2008 *J. Turbomach.* **130** 1
[10] Lawson S A and Thole K A 2012 *J. Turbomach.* **134** 1

IOP Conf. Series: Materials Science and Engineering **50** (2013) 012008 doi:10.1088/1757-899X/50/1/012008

Effects of pilot injection timing and EGR on a modern V6 common rail direct injection diesel engine

Nik Rosli Abdullah[1,a], Rizalman Mamat[2], Miroslaw L Wyszynski[3], Anthanasios Tsolakis[3] and Hongming Xu[3]

[1]Faculty of Mechanical Engineering, Universiti Teknologi MARA, 40450, Shah Alam, Selangor, Malaysia

[2]Faculty of Mechanical Engineering, Universiti Malaysia Pahang, 26600, Pekan, Pahang, Malaysia

[3]School of Mechanical Engineering, University of Birmingham, Birmingham B15 2TT, United Kingdom

E-mail: [a]nikrosli@salam.uitm.edu.my

Abstract. Nitric oxide and smoke emissions in diesel engine can be controlled by optimising the air/fuel mixture. Early injection produces premixed charge resulted in simultaneous NO_x and smoke emissions reduction. However, there could be an increase in hydrocarbons and CO emissions due to fuel impinged to the cylinder wall. The focus of the present work is to investigate the effects of a variation of pilot injection timing with EGR to NO_x and smoke level on a modern V6 common rail direct injection. This study is carried out at two different engine load conditions of 30 Nm and 55 Nm, at constant engine speed of 2000 rpm. The results show that the early pilot injection timing contributed to the lower smoke level and higher NO_x emissions. The higher level of NO_x is due to higher combustion temperatures resulting from the complete combustion. Meanwhile, the lower smoke level is due to complete fuel combustion and soot oxidation. The early pilot injection timing produces an intermediate main ignition delay which also contributed to complete combustion. The formation of smoke is higher at a high engine load compared with low engine load due to the higher amount of fuel being injected.

1. Introduction

In direct injection diesel and compression ignition engines, the start of fuel injection (SOI) influences the fuel-air reaction rates and combustion behaviour [1-2]. A number of studies have been carried out aiming to promote premixed lean charge and minimise diesel engine emissions [3-7]. This is due to the improved premixed combustion leading to low combustion temperatures and condition favourable for low NO_x and PM emissions. Several studies have revealed that the pilot injection strategy produced a shorter ignition delay on the main injection thus contributing to the lower emissions [8-12] and cleaner combustion process [13-14].

In diesel and compression ignition engines, ignition delay and combustion phasing are amongst the main parameters influencing emissions formation and engine performance. Long ignition delay promotes premixed uncontrolled combustion which can lead to knocking and high combustion noise [15-19]. Murari Mohon Roy and Hideyuki Tsunemoto studied the effects of injection pressure (200 - 1200 bar) and split injection on emissions and engine noise using a common rail four-stroke multi cylinder 7.7 L direct injection diesel engine. The results showed that the proper post-pilot injection strategy has a significant improvement in the emissions and engine noise [20]. Keiichi Okude et al. investigated the effect of multiple injections of engine performance and emissions using a single-cylinder engine 0.744 L. The results showed that the multiple pilot injection substantially reduced HC and CO emissions [9]. A. Vanegas et al. used a 4-cylinder common-rail diesel engine to study the effect of multiple injection on emissions. The overall results showed that the NO_x emissions from the split injection strategy were significantly lower compared to the single injection strategy [14]. D. T. Hountalas et al. demonstrated that the cooled EGR is favourable for NO_x reduction [21]. Siddappa et al. utilised a single cylinder direct injection engine to study the performance and emissions at different injection pressures and injection timing fuelled with diesel and biodiesel. According to their experimental results the higher injection pressure produces improved engine performance and emissions in terms of smoke, HC and CO but NO_x emissions were slightly increased [19].

From the literature review, it was concluded that the effects of variation of pilot injection timing on engine performance and emissions of a diesel engine can be further investigated. Moreover, the influence of the pilot injection timing on the main ignition delay has not been extensively discussed. A number of research programs have been carried out at the University of Birmingham to investigate diesel engine performance and emissions [22-31]. In present study, the variations of pilot injection timing were performed in order to investigate their effects on the emissions of NO_x and smoke level by using a common rail direct injection V6 diesel engine equipped with variable turbine geometry (VTG) turbochargers and cooled EGR. The pilot injection has a potential to control the pressure rise during the main combustion through a shorter main ignition delay. This experiment run with a split injection consists of 10 % pilot injection and 90 % main injection fuel quantity. The use of 10 % pilot fuel quantity is due to the large amount of pilot injection tends to increase the NO_x emissions due to higher rate of heat release [32].

2. Experimental Setup

The experiments were carried out on a fully instrumented multi-cylinder V6 diesel engine, common rail multiple direct fuel injection system, twin water-cooled variable geometry turbochargers and cooled EGR. The engine specification is shown in table 1. An eddy-current dynamometer type Schenck W230 and an engine starter motor, are used to load and start the engine respectively. The Schenck series 2000 controller is used to control the dynamometer.

The exhaust gas of the engine is passed through the gas analyser via a sample line and NO_x and smokes were measured. The exhaust sample acquisition time is approximately 10 seconds at an operating temperature of 28 to 30°C and the relative humidity is approximately 40-50 percent. The emissions presented are an average value from 10 reading samples. The smoke emissions were measured by using an AVL 415S smoke meter which provides results directly as a Filter Smoke Number (FSN) unit. Main injection quantity, timing, pressure and EGR rate were held constant throughout the experiment. The amount of fuel injected for both pilot and main injection was at a constant ratio of 10:90 for pilot and main respectively. All the tests were performed with ultra low sulphur diesel (ULSD) and the fuel properties are given in table 2. The effects of pilot injection timing on the emissions of NO_x and smoke level were examined. These experiments consisted of three different regions of pilot injection timing (early, middle and late) at two different engine loads (30 Nm, 55 Nm) operating with a constant speed 2000 rpm. The details about the experimental conditions have been summarized in table 3.

Table 1. Engine Specifications.

Bore	81.0 mm
Stroke	88.0 mm
Displacement Volume	2720 cm^3
Maximum Torque	435 Nm @ 1900 rpm
Maximum Power	152 kW @ 4000 rpm
Compression Ratio	17.3 : 1
Connecting Rod Length	160.0 mm

Table 2. Fuel Properties.

Properties	ULSD
Cetane Number	53.9
Density at 15 °C (kg/m^3)	827.1
Viscosity at 40 °C (cst)	2.5
50% Distillation (°C)	264
90% Distillation (°C)	329
LCV (MJ/kg)	42.6
Sulphur (mg/kg)	46
Mano-aromatics (% wt.)	21
Di-aromatics (% wt.)	3.1
Molecular Mass (eq.)	209
C (wt.)	86.5
H (wt.)	13.5
O (wt.)	-

Table 3. Summary of Test Conditions.

Test	Injection Timing CAD Pilot Injection	Engine Load (Nm)	Notes
1	9 bTDC	30	
2	12 bTDC	Fixed	Late Injection
3	15 bTDC	Fixed	
4	19 bTDC	Fixed	Middle Injection
5	21 bTDC	Fixed	
6	24 bTDC	Fixed	
7	27 bTDC	Fixed	Early Injection
8	30 bTDC	Fixed	
9	9 bTDC	55	
10	12 bTDC	Fixed	Late Injection
11	15 bTDC	Fixed	
12	19 bTDC	Fixed	Middle Injection
13	21 bTDC	Fixed	
14	24 bTDC	Fixed	
15	27 bTDC	Fixed	Early Injection
16	30 bTDC	Fixed	

Constant Engine Speed: 2000 rpm
Constant Main Injection Timing: 1.4 CAD aTDC
Constant EGR Rate: 40%

3. Results and Discussion

The NO_x emissions are mainly influenced by the peak in-cylinder pressure and high combustion temperatures. It can clearly be seen from figure 1 that the NO_x concentration increases monotonically with early and late pilot injection timing. The early pilot injection timing tends to produce intermediate main ignition delay resulting in a complete combustion process due to long residence time for the reaction of fuel-air (Guntram A. Lechner, Timothy J. Jacobs et al. 2005-01-0167, 2005). As a result, the early pilot injection timing produces the higher in-cylinder pressures leading to the higher temperatures and NO_x emissions (Cenk Sayin and Canakci 2009; Cenk Sayin, Murat Ilhan et al. 2009). The middle pilot injection produces longest main ignition delay that leads to the slower combustion and low NO_x emissions (K.Verbiezen, A. J. Donkerbroek et al. 2007).

Therefore, in general the reduction in NO_x can be controlled through the reduction of peak in-cylinder pressure by reducing the main ignition delay, the air temperature and combustion duration (Mohammad Ghaffarpour and Noorpoor 2007). The overall results show that the NO_x emissions increased with the early and late of pilot injection but decreased with the middle pilot injection timing in both engine loads.

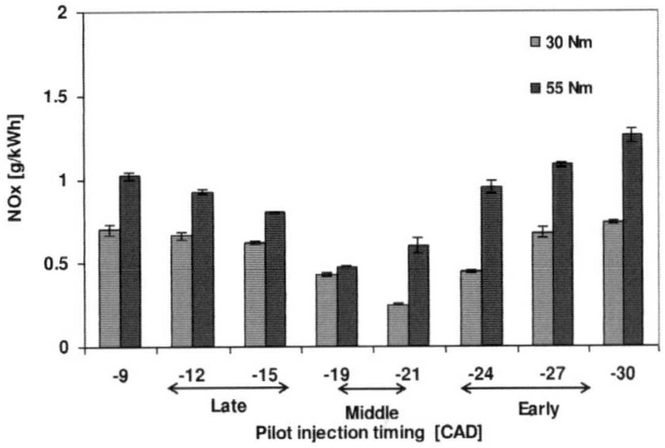

Figure 1. NOx emissions as a function of pilot fuel injection timings (-9 to -30 CAD) at two different engine loads (30 Nm and 55 Nm) and constant engine speed 2000 rpm, error bars represent 95 percent confidence.

The effect of pilot injection timing on the smoke number (FSN) is shown in figure 2 with the engine operating at two different loads at a constant engine speed of 2000 rpm. The amount of smoke formation is strongly related to the fraction of diffusive combustion phase (Ren, Huang et al. 2006). Smoke formation occurs at extreme local air deficiency. It increases as the air/fuel ratio decreases. The higher fraction of diffusive combustion phase will result in an increase in smoke formation due to insufficient of oxygen. The presented results are in a good agreement with Keiichi Okude et al. which showed that the smoke decrease as the pilot injection timing is advanced (Keiichi Okude, Kazutoshi Mori et al. 2007). It is strongly believed due to improvement in mixture formation with an early pilot injection. This in turn resulted in complete combustion due to better mixing process. The early pilot injection timing produces an intermediate main ignition delay which also contributed to complete combustion. The formation of smoke is higher at a high engine load due to the higher amount of fuel being injected, resulting in higher smoke formation. Complete combustion also resulted in higher combustion temperature leading to the higher soot oxidation during expansion and exhaust stroke (Nicolas Dronniou, Marc Lejeune et al. 2005). The most obvious finding to emerge from this study is that the early pilot injection timing is better for smoke reduction.

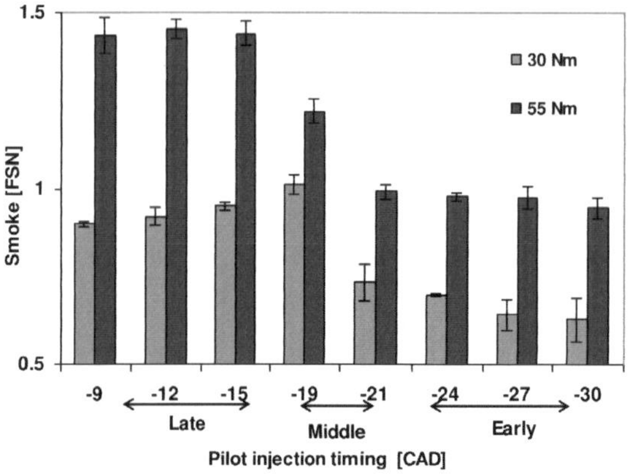

ke level (FSN) as a function of pilot fuel injection timings (-9 to -30 CAD) at two different engine loads (30 Nm and 55 Nm) and constant engine 2000 rpm, error bars represent 95 percent confidence.

4. Conclusion

In the present work, the emissions of NO_x and smoke level were measured for various pilot injection timings, operating with two different engine loads at a constant engine speed. As accepted, the pilot injection favoured for the main combustion behaviour. It has been shown smoke level were increased with retarded pilot injection and then decreased after an injection timing of 19 CAD onwards. Both engine loads produced similar patterns. However, the NO_x emissions were decreased with retarded pilot injection then increased after 19 CAD onwards in both engine loads. Therefore, it has been proven that advancing pilot injection timing of 24, 27 and 30 CAD are better for smoke. Conversely, the injection timings of 9, 12, 15, 19 and 21 are worse for this particular emission. In terms of NO_x emissions, the injection timings of 19 and 21 CAD are the best. In summary, advanced pilot injection has promising emissions reduction associated with it in terms of smoke but not for NO_x.

5. References

[1] Cenk S and Canakci M 2009 *Energy Conversion and Management* **50** 203
[2] Cenk S 2009 *Renewable Energy* **34** 1261
[3] Ghobadian B, Najafi G and Nayebi M 2013 *J Mech Eng Sci* **4** 373
[4] Mat Yasin M H, Mamat R, Sharma K V and Yusop A F 2012 *J Mech Eng Sci* **3** 331
[5] Ghobadian B, Najafi G and Nayebi M 2013 *J Mech Eng Sci* **4** 373
[3] Montgomery D T and Reitz R D 2001 *SAE Paper* **01-0195**
[4] Dec J E 2009 *Proc. of the Combustion Institute* vol 32
[5] Zhang L 1999 *SAE Paper* **01-3493**
[6] Keiichi O 2007 *SAE Paper* **01-4178**
[7] Gavin D 2008 *SAE Paper* **01-0941**
[8] Vanegas A 2008 *SAE Paper* **01-1191**
[9] Nwafor O M I 2000 *Renewable Energy* **21** 495
[10] Shuji K 2001 *SAE Paper* **01-0200**
[11] Sundar R C and Sendilvelan S 2010 *Inter J Automot Mech Eng* **2** 144
[12] Aziz A R A, Firmansyah and Shahzad R 2010 *Inter J Automot Mech Eng* **2** 157
[13] Mohanamurugan S and Sendilvelan S 2011 *Inter J Automot Mech Eng* **3** 279
[14] Timothy J J, Lean and Rich 2005 *SAE Paper* **01-0166**
[15] Nwafor I, and O M 2007 *Renewable Energy* **32** 2361

[16] Siddappa S B, Babu M K G and Subrahmanyam J P 2007 *SAE Paper* **01-0613**
[17] Murari M R and Tsunemoto H 2002 *SAE Paper* **01-2874**
[18] Hountalas D T, Mavropoulos G C and Binder K B 2008 *Energy* **33** 272
[19] Abdullah N R 2009 *SAE Paper* **24-0049**
[20] Tsolakis A, Megaritis A and Wyszynski M L 2004 *Fuel* **83** 1837
[21] Tsolakis A and Megaritis A 2004 *Biomass and Bioenergy* **27** 493
[22] Tsolakis A and Megaritis A 2005 *Inter. J. of Hydrogen Energy* **30** 731
[23] Abu-Jrai A, Tsolakis A and Megaritis A 2007 *Inter. J. of Hydrogen Energy* **32** 3565
[24] Tsolakis A 2007 *Energy* **32** 2072
[24] Watanabe H 2007 *Conversion and Management* **48** 2077
[26] Chuepeng S 2007 *SAE Paper* **01-0072**
[27] Tsolakis A, Megaritis A and Yap D 2008 *Energy* **33** 462
[28] Chuepeng S 2008 *SAE Paper* 01-0074
[29] Ghaffarpour M and Noorpoor A R 2007 *Int. J. Automotive Technology and Management* **7** 15

6. Acknowledgements

Sincere thanks to Dr K. Thennoi and Dr G. Tian for providing valuable comments and suggestions. The authors are grateful to the Future Power Group members of Birmingham University, Jaguar Car Ltd and Shell Global Solutions (UK) for their support during this experiment.

IOP Conf. Series: Materials Science and Engineering **50** (2013) 012009 doi:10.1088/1757-899X/50/1/012009

Investigation of the effects of inlet system configuration on the airflow characteristics

Shafiq Suhaimi[1,a], **Rizalman Mamat**[1], **Nik Rosli Abdullah**[2]
[1]Faculty of Mechanical Engineering, Universiti Malaysia Pahang, 26600, Pekan, Pahang
[2]Faculty of Mechanical Engineering, Universiti Teknologi MARA, 40450, Shah Alam, Selangor

E-mail: [a]shafiq.suhaimi@gmail.com

Abstract. The effort towards the development of an engine that is efficient and green has become one of the biggest efforts of the major part of the automotive industry. The contribution of air quality to the engine plays an important role to increase power while minimizing the exhaust emission. The main purpose of this article is to investigate the characteristic of the air intake system of a naturally aspirated engine on a steady flow bench. The experiment was conducted using an industrial flow bench apparatus where the pressure different test pressures can be set and, at every test pressure, the valve is lifted at an increment of 1mm until the maximum possible valve lift where the flow rate is obtained in terms of cubic foot per minute or CFM. The experiment process was then repeated at different configuration. The experimental result shows that the flow rate increases with each valve lift in a linear pattern until at a certain point where the value becomes constant. The result also indicated that the flow rate increases as the different pressure increases.

1. Introduction

The basic works of an inlet system is simple and that is to provide sufficient air into the combustion chamber of an IC engine for the combustion process. While the main purpose of system is simple, the science that is involved with the system is quite complicated and the effort of improving the system as a whole has been a focus of the automobile industry as the inlet system plays a vital role in improving the performance and efficiency of the engine, plus it plays a vital role in the engine's emissions [1-5]. When it comes to emissions and the efficiency of an engine, the fuel to air ratio plays a major importance where engines that operate on lean mixtures (fuel to air ratio more than 1.0) results in a cleaner and more efficient engines. This is due to several reasons, some of which are; high compression ratios, low throttling losses at part load, and favourable thermodynamic properties in burned gasses. A major part in reducing the particulate matter emissions is the design of the engine itself. One of which is the optimized air motion through the combustion chamber and the intake air turbulence that provides adequate mixing of the air and fuel at low speeds and avoid over rapid mixing at high speeds that leads to hydrocarbons, nitro oxides and increased fuel consumption (A.Faiz et al., 1996).

The role that a good inlet system plays in the performance, efficiency and emission of an IC engine are the main motivations of this study, which, hopefully, leads to the production of an optimized and improved inlet system.

IOP Conf. Series: Materials Science and Engineering 50 (2013) 012009 doi:10.1088/1757-899X/50/1/012009

The secret to make a good inlet system design lies in making the inlet system to be able to provide the most amount of air velocity as possible into the combustion process so that the combustion process inside the engine (P.C. Vorum, 1976). For this condition to happen, the efforts that have been done are either by increasing the volumetric efficiency of the engine or through supercharging the air itself.

In an ideal intake process, the pressure of at the intake valve should be equal of that in the intake manifold. This, however, does not happen. The ideal conditions are not met due to several factors (J.M. Desantes et al., 2010), some of which are:

- The pulsations of pressure where the pressure in the intake manifold is strongly affected by the wave effects that is created by the pulsating engine operation.
- The residue that is left from the combustion process as the exhaust is imperfect. This happens when the intake valve opens before the exhaust valve closes, causes the exhaust gas mixes with the intake gas.
- The existence of different surfaces in the different part of the intake system, namely the intake manifold, the inlet valve, the engine runners, and the cylinder wall itself affected both the heat transfer of the engine and the pressure and flow operation during the intake process. The existence of different part of an engine and how it affected the airflow during the intake process will be the main focus of this study.

The main purpose of this study is to investigate in the different factors that affect the mass flow rate of an engine's intake flow. To be more specific, this article investigates how the flow rate changes with changing valve lifts, changing pressure and different configurations.

2. Experimental Set Up

The experiment was conducted using the SuperFlow SF-1200 flow bench, which is an apparatus that generates a steady flow and at the same time allows for the measurements of the global flow characteristics such as; mass flow rate, velocity and pressure (SuperFlow SF-2010 Instructional Manual). Table 1 shows the specifications of the flow bench that was used during the course of the study. The specimen that was used was the cylinder head of a Honda GX120 single piston two-stroke 4 horsepower engine. Other characteristics of the engine are shown in Table 2.

Table 1. Specification of the SF-1020SB Flow Bench

SF-1020 Specification	**Values**
Calibration Test Pressure	25" of water
Range	0-1, 100 CFM
Exhaust Capacity	1,000 CFM @ 25" H_2O
Power Phase	240VAC, 75A, Single Phase
Weight	480lbs (281kg)
Dimensions	48 x 33 x 43 in.

The test apparatus required the engine head to be attached to the flow bench by a cylinder head adaptor that have the same diameter as the bore diameter of the tested engine as shown in figure 2. The engine head, head adaptor and flow bench set up is held together using a type of sealant and foam tape to ensure that the leakage is as little as possible. Any small leakage that is still present is keyed into the flow bench so that it the flow bench apparatus can adjust to the leakage present when it is measuring the intake flow rate.

Table 2. Specification of the Honda GX120 engine

Dimensions and Weight	Values
Length X Width X Height	300 x 345 x 320 mm (11.8 x 13.6 x 12.6 in)
Dry Weight	12.0 kg (26.5 lb)
Engine type	4-stroke, overhead valve, single cylinder
Displacement (Bore X Stroke)	118cc (7.3 cu in) [60 x 42 mm (2.4 x 1.7 in)]
Max. Output	4.0 HP / 3,600 rpm.
Max. Torque	0.75 kg (5.4 ft-lb)/ 2,500 rpm
Fuel Consumption	230 g/PS h
Cooling system	Forced air
Ignition system	Transistorized magneto
PTO shaft rotation	Counter-clockwise

The cylinder head was tested at the pressure of 20.095 inH_2O, 40.19 inH_2O, and 60.285 inH_2O. At each test pressure, the valve was lifted from 0 mm at an increment of 5 mm until the valve is lifted at 11.5 cm, which is the maximum valve lift for this engine model, and at each of those valve lift, the flow rate was measured.

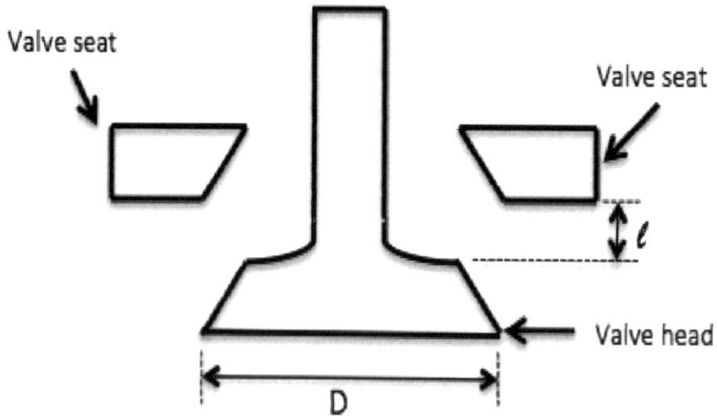

Figure 1. Diagram of the valve lift where, *l* is the valve lift and D, is the valve diameter.

The valve of the engine was opened, as shown in figure 1, where the valve head is lifted from the valve seat to create an opening that allows for the air to flow into the system as it was during an intake process. The valve was lifted using a manual valve actuator, which is a custom made plate that is fixed above the valve tip where the rocker arm should be. The valve is lifted by a turn of a blot that is attached to the plate, which pushes the valve open. A micrometre is fixed above the bolt to measure the actual valve lift.

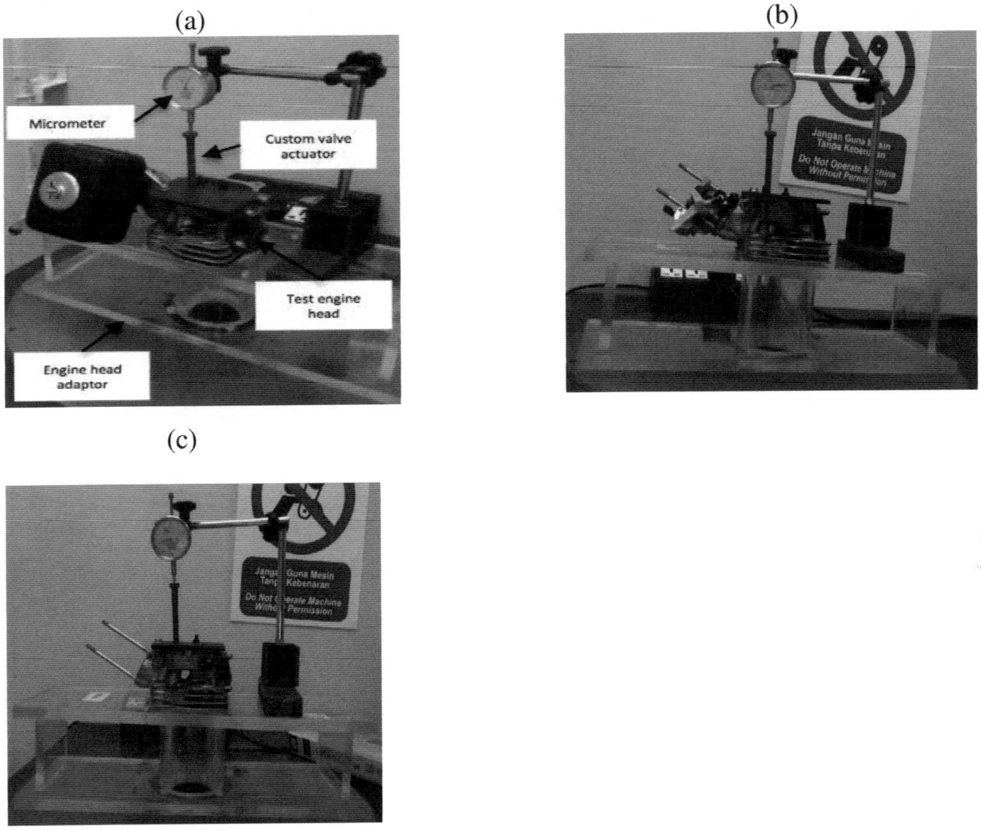

Figure 2. Experimental setup (a) at full configuration; (b) removed air filter; (c) port only configuration.

The test continues with different configurations for the intake system. The first configuration was the complete configuration as shown in figure 2(a). Second configuration was the port and the carburettor (the air filter removed) as shown in figure 2(b) and third configuration was the intake port only configuration (without anything attached to the intake port) as shown in figure 2(c). Each configuration was tested and the mass flow rate was recorded.

3. Result and Discussion

From the result that was obtained from the experiment shows that as the flow rate changes with valve lift in a polynomial pattern where the flow rate changes in a linear pattern at low valve lifts. The gradient of the flow rate to valve lift curve then changes substantially at higher valve lifts and when the valve lift reached at a certain point, the flow rate begun to show low changes and at later valve lifts, there were no changes to the flow rate, even as the valve lift increases. This pattern is true for most of the different conditions that were run during the course of this study. This pattern is due to the increase in area of the airflow gap.

The study also made a point of testing the engine at the valve lift of 11.5 mm because it is the maximum amount of valve lift allowed by the engine's current design's valve spring. While it is possible to test the engine at higher valve lift, the data shows that the flow rate will not have any significant changes beyond a certain point. Plus, the valve lift range that is used in this study is already beyond the normal operating conditions of the engine as the cam shaft design will not open the valve beyond 6mm and if the valve is opened anymore, the design will suffer a grate reduction in efficiency and increase in cost for the benefit of a very little improvement in flow intake.

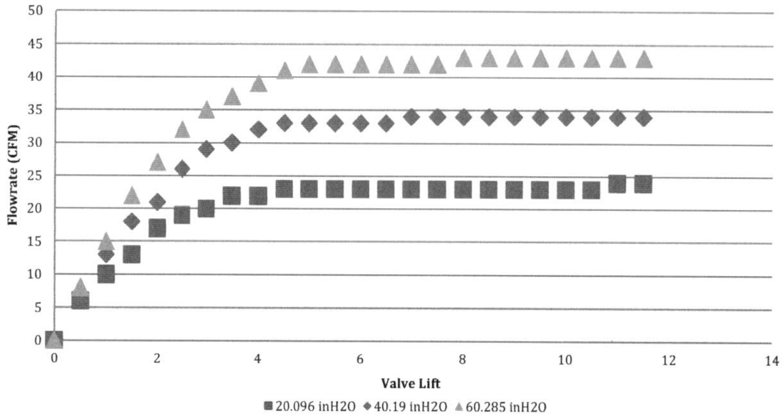

Figure 3. Mass flow rate of the tested port at full configuration at different pressure.

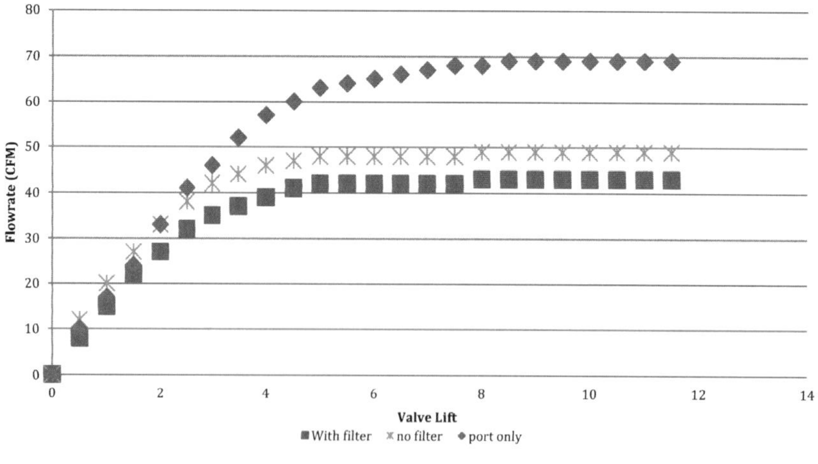

Figure 4. Mass flow rate of the tested cylinder head at 60.285 inH2O at different configuration.

Figure 3 shows how the flow rate changes with different valve lift at different pressure. As expected, with increasing pressure, the flow rate that was produced becomes greater. The different pressure does not changes the pattern in which the flow rate changes with valve lift. This means that the different pressure curves do not intersect with each other at any valve lift other than the initial valve lift of 0 mm. The increment of pressure from 20.095 inH2O to 40.19 inH2O leads to an average of 45% increase in flow rate, while the increment of 40.19 inH2O to 60.285 inH2O leads to an average of 25% increase in flow rate. The increase in flow rate is attributed to the higher-pressure difference inside the inlet system and the outside atmosphere and the fact that there is a direct correlation between the pressure and flow rate for every rise in pressure, shows that no special phenomenon will occurs during the intake process.

Figure 4 shows the flow rate changes with valve lift at different engine configuration. The result shows that when the engine is at its complete configuration, the flow rate is less produced. When the air filter is taken off and the engine is left at only its port and the its carburettor, the flow rate exhibit an average increase of 18% and when the carburettor is removed, leaving only the engine port, the flow rate exhibit an average increase of 25%. The difference in flow rate is due to the losses that occur in the internal flow due to the changing geometry and the different surfaces that the air flows through when it is at full or port and carburettor only configuration. It is not until the air is only flow directly through the port of the engine did the flow rate exhibits a significant increase.

For both the engine at full configuration and for the engine with the air filter removed, the curve flow rate against valve lift curve seems to pattern the same way as before with the initial valve lift gives a linear correlation between the flow rate and the valve lift and the gradient of the curve decreases as the valve lift increases as to a point where there is no apparent change in flow rate. As for the case of where there is only the engine port is tested, the flow rate against valve lift curves seems to have a lower gradient at low valve lift and the gradient increases before the gradient lowers again at higher valve lifts and later on the flow rate barely changes with increasing valve lift.

It is also important to note that at lower valve lift, the port only configuration seems to have lower flow rate when compared to the port and carburettor configuration. It is not until at a higher valve lift where the port only configuration has a higher flow rate when compared to the port and carburettor configuration, which the two curves intersect at valve lift of 2 mm. This phenomenon can be attributed to the geometry of the port only configuration that creates turbulence and significant losses to the airflow. This also shows the importance of creating a laminar flow before the air enters the engine.

It is also worthy to note that the curve pattern between the full configuration and, the port and carburettor configuration seems to have similar shapes and did not show any signs of turbulence or any other unexpected losses to the internal flow. This shows that the creation of laminar flow for the intake process is not attributed only to the air filter of the engine but also to the carburettor. It is also can be argued that, from a volumetric efficiency stand point, the carburettor is enough to create a laminar flow for the intake process and the current design of full configuration engine does exhibit a problem of creating unnecessary losses.

4. Conclusion

In this paper, it is found that the flow rate will increase with the increase of the valve lift up to a certain point before the flow rate decrease its rate and later on remains constant and as expected the flow rate increases with increasing test pressure. The full configuration will have the lowest overall flow rate followed by the port and carburetor configuration and the port only configuration exhibits the highest overall flow rate, which demonstrates the effects and losses caused by the different engine parts. At low valve lift, the port only configuration seems to have lower flow rate compared to the port and carburetor configuration due to the turbulence formed when the engine is at this configuration.

The results of the study serves as a very good baseline for further investigations especially when it comes to computer simulations. Plus the phenomenon observed at low valve lift at port only configurations begs for further investigations in the turbulence formation during the intake process of the same test engine.

5. References

[1] Al-Falahi A A, Yusoff M Z and Ahmed D I 2012 *Inter J Automot Mech Eng* **6** 702.
[2] Sulaiman S A, Murad S H M, Ibrahim I and Abdul K Z A 2010 *Inter J Automot Mech Eng* **1** 91.
[3] Satya N P V, Ramireddy G and Venkataramana S 2011 *Inter J Automot Mech Eng* **3** 350.
[4] Najiha M S, Rahman M M, Yusoff A R and Kadirgama K 2012 *Inter J Automot Mech Eng* **6** 766
[5] Naga S S, Sita R R A V, Kalyani R K and Syam S L 2012 *Inter J Automot Mech Eng* **6** 795.
[6] Asif F, Christopher S. W, Michael P W 1996 Air Pollution from Motor Vehicles: Standards and Technologies for Controlling Emissions *The International Bank for Reconstruction and Development*
[7] Vorum P C 1976 Short Pipe Manifold Design for Four Stroke Engines *ASME Paper* **76-WA/DGP-4**
[8] Desantes J M, Galindo J, Guardiola C and Dolz V 2010 *Experimental Thermal and Fluid Sciences* **34** 37
[9] *SuperFlow SF1200 Instruction Manual*
[10] Morgan P, Kach R, Hill C, Demitroff D, Monroe B, Dettloff T, McCoy W and Battermann D 1998 Investigation of a Ford 2.0L Duratec for Touring Car Racing *SAE Paper* **98303**
[11] Weclas M, Melling A and Durst F 1994 Flow asymmetry in geometrically symmetric engine head configurations *Inter. J. in Heat and Fluid Flow* **5**.
[12] Pavkin-Zylenkov V P, Ponomarev E G, Ryabikov O B and Samofalov V B 1990 The Effect of Inlet Port Geometry on In-Cylinder Flow Structure *Inter Symposium Comodia*
[13] Ugur K 2005 Study on the design of inlet and exhaust system of a stationary internal combustion engine *J. in Energy Conservation and Management*
[14] Hans S 1979 A mathematical model for simulation of processes in an internal combustion engine *Acta Astronautica* **6**
[15] Mont H Dynamic Computer Simulation of An Engine Intake and Carburation System
[16] P. Li A Two-Dimensional Intake Manifold Flow Simulation Scientific Research Laboratory, Ford Motor Company 20000 Rotunda, Dearborn, Michigan 48121
[17] Maria D *CFD-procedures for Air Induction Systems* Master of Science in Engineering Technology Space Engineering
[18] Ramin G Engine Simulation Model for a Formula SAE Race Car *Applied Design, Development, Correlation and Optimization*
[19] Ceviz M A and Akın M 2010 Design of a new SI engine intake manifold with variable length plenum *Journal for Energy Conservation and Management*
[20] Ceviz M A 2007 Intake plenum volume and its influence on the engine performance, cyclic variability and emissions *Energy Conversion and Management* **48** 961
[21] Makgata K W 2005 Computational Analysis And Optimisation Of The Inlet System Of A High-Performance Rally Engine (University of Pretoria)
[22] Sellnau M, Kunz T, Sinnamon J and Burkhard J 2006 2-step Variable Valve Actuation: System Optimization and Integration on an SI Engine *SAE Paper* **01-0040**
[23] Jemni M A 2011 Influence of intake manifold design on in-cylinder flow and engine performances in a bus diesel engine converted to LPG gas fuelled, using CFD analyses and experimental investigations *Journal in Energy* **2701**
[24] Cho S K and Korivi V M 2004 Port Design Optimization Using CFD Analysis *Journal of Advanced Manufacturing Systems* **3** 21
[25] Willar W P 2004 *Engineering Fundamentals of the Internal Combustion Engine* (Pearson Prentice-Hall)

IOP Conf. Series: Materials Science and Engineering **50** (2013) 012010 doi:10.1088/1757-899X/50/1/012010

Analysis on oscillating actuator frequency influence of the fluid flow characterization for 2D contractile water jet thruster

M F Shaari[1,2,a], H Abu Bakar[2], N Nordin[1], S K Saw[2], Z Samad[2]
[1]Faculty of Mechanical and Manufacturing Engineering,
Universiti Tun Hussein Onn Malaysia, Parit Raja, 86400 Batu Pahat, Johor, Malaysia
[2]School of Mechanical Engineering, Universiti Sains Malaysia,
Engineering campus, Seri Ampangan, 14300 Nibong Tebal, Pulau Pinang, Malaysia

E-mail: [a]mdfarid@uthm.edu.my

Abstract. Contractile body is an alternative mechanism instead of rotating blade propeller to generate water jet for locomotion. The oscillating motion of the actuator at different frequencies varies the pressure and volume of the pressure chamber in time to draw in and jet out the water at a certain mass flow rate. The aim of this research was to analyze the influence of the actuating frequency of the fluid flow in the pressure chamber of the thruster during this inflation-deflation process. A 70mm x 70mm x 18mm (L x W x T) 2D water jet thruster was fabricated for this purpose. The contractile function was driven using two lateral pneumatic actuators where the fluid flow analysis was focused on the X-Y plane vector. Observation was carried out using a video camera and Matlab image measurement technique to determine the volume of the flowing mass. The result demonstrated that the greater actuating frequency decreases the fluid flow rate and the Reynolds number. This observation shows that the higher frequency would give a higher mass flow rate during water jet generation.

1. Introduction

Contractile water jet thruster (CWT) is an alternative propulsion method for underwater robot. CWT had been utilized either as a main propulsion unit [1-4] or underwater vehicle's direction guiding unit [5]. Basically, this CWT applies the pressurized water jet to accelerate its body for locomotion. Compared to motor driven rotary blade propeller, this CWT uses pressure chamber volume reduction to generate this pressurized water jet (Figure 1). The pressure chamber volume reduction or contraction occurred by applying an amount of force to certain parts of its body. During actuation, the actuator contracts the pressure chamber body to expel the water from the chamber via nozzle or orifice. Recurring contraction is required for continuous locomotion, thus exhibiting the function of actuating frequency. Though intensive research had been carried out to characterize the CWT mechanism, few detail studies focused on the actuating frequency influence on the internal fluid flow of the CWT.

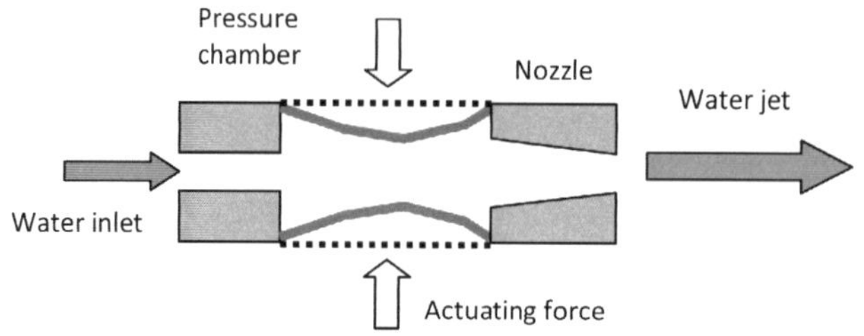

Figure 1. Fundamental concept of contractile water jet mechanism

A maximum mass flow rate of optimum velocity during the jetting period would create ultimate propulsion efficiency for a certain CWT specification. In the CWT context, the maximum mass flow rate could be only achieved by the right actuating frequency. Higher frequency would entrain less water and thus it would decrease the mass flow rate magnitude. Otherwise too low frequency reduces the velocity of the jetted fluid. In this research, the oscillating actuator frequency was varied at low frequency and the fluid flow behaviour was observed during the inflation and deflation process. At this moment, the studied focus on the 2D fluid flow characterization. In real application, smart actuator will be used. However in this research, the smart actuator was substituted with pneumatic cylinder for easy control purpose. A 70mm x 70mm x 3mm (L x W x T) 2D water jet thruster was fabricated and attached to two pneumatic cylinders. Maximum static pressure for cylinder actuation was set on the compressor regulator to obtain the required actuating force. Oscillating motion and frequency of the actuators was controlled using Arduino UNO microcontroller. The fluid flow during inflation and contraction of the CWT was observed through the motion of the red colour, supplied at the inlet and nozzle using transfusion needle. In order to gain the fluid velocity, a 30 frame-per-second video camera was used to catch the motion of the red colour. The results shows that the increased actuating frequency decreases the volume flow rate and Re number.

2. Research Overview

Oscillating actuation frequency depends on the type of the actuator. Each of the potential actuator has its own character and hence must match the design optimization of the CWT for ultimate thrust force generation [6]. For instance, piezo based actuator has high actuating frequency which would be up to 8MHz [7] could form a small and compact CWT because of the properties of the piezo material. IPMC actuator which acts by bending mechanism has lower actuating frequency compared to the piezo actuator but larger bending radius allows more mass flow rate [8-9]. Generally CWT has three stages of action during its operation which are the rest, inflation and deflation stage. These three stages mimic the nature of water jet locomotion aquatic animals such as squid, jellyfish and nautilus [10-11]. The rest stage would be the initial condition or recovery stage where there is neither kinematic change of the body nor actuation occurred. The second stage is the inflation stage where the volume of the pressure chamber increase dramatically depends on the speed of the actuator (Figure 2a). Based on Boyle's Law, the lower pressure entrained surrounding water to fill in the cavity. The third stage is the deflation process where again the actuator contract the CWT body and pressure increment enforce the water out via a nozzle (Figure 2b).

(a) (b)

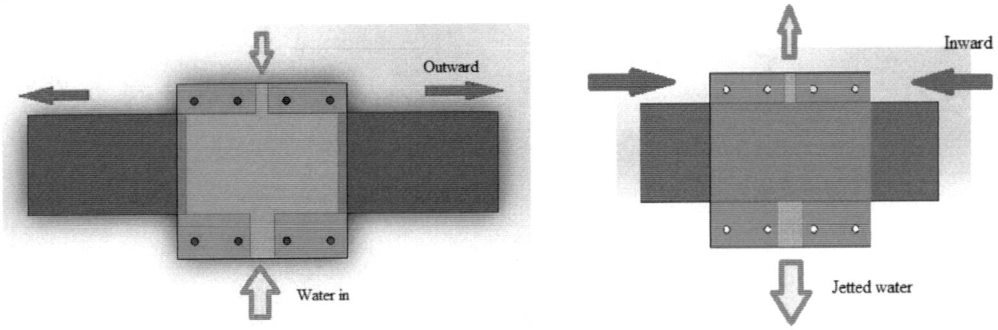

Inflation Deflation

Figure 2. Two different states of the CWT: (a) Inflation state; (b) Deflation/contraction state

The continuous inflation-deflation process imparts the pressure chamber volume differentiation, which is depending on the actuation frequency. A typical relation between the pressure chamber volume and the actuating frequency could be determined based on Eq. (1) below [12];

$$V_{pc} = V_x \sin(2\pi f_{act} t) \tag{1}$$

where V_{pc} is the volume of the pressure chamber, V_x is the volume variation amplitude, f_{act} is the actuator frequency and t is time. Due to the above relation, the sine function gives the positive and negative V_{pc}. If $t = \pi$ or 2π, then $V_{pc} = V_x = 0$. This mean at total deflation, V_x is not equal to zero. Thus, shifted square function had been used to model the contractile pressure chamber volume as shown in Eq. (2) below;

$$V_{pc} = V_x square(2\pi f_{act} t - \pi) + V_x \tag{2}$$

A simulated graph was generated using Matlab software to observe the V_{pc} variation in time (Figure 3). Using a square function has an advantage in controlling the frequency using microcontroller. In the designed system, the total inflation would be the maximum V_{pc} and total deflation would make the V_{pc} is equal to zero. It is understood that V_x has nothing to do with the f_{act} but the flow rate of the digested fluid could be influenced by the fact and nozzle area function. Eq. (3) demonstrates the function of V_{pc} and the fluid flow.

$$Q(x,t) = \frac{V_{pc}}{dt} = A_i \frac{dv}{dt} \tag{3}$$

Where Q is the fluid flow rate, A_i is the inlet cross sectional area and dv/dt is the velocity of the fluid. Substituting Eq.(2) into Eq.(3) then we get:

$$\frac{dv}{dt} = \frac{V_x}{A_i}\left[square(2\pi ft - \pi) + 1\right] \tag{4}$$

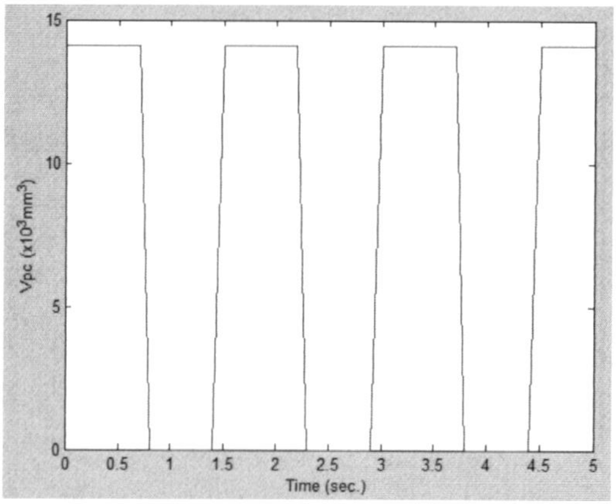

Figure 3. Simulation result to show the V_{pc} changes in time

3. Methodology

3.1. CWT Fabrication

The 2D CWT was developed based on stacked acrylic layers. As mentioned in Chapter 2, the term 2D refers to length-to-thickness ratio which was 11.67. Translucent acrylic panel with 3mm thickness was cut into two 70mm x 70mm (L x W) panels. These panels acted as cover panels. 6mm thick acrylic panels with several shapes were placed between the cover panels to form a nozzle, inlet and pressure chamber (Figure 4). In this research, the volume for CWT inlet and nozzle were designed based on the real squid mantle-funnel ratio where the internal mantle volume refers to pressure chamber and funnel refers to the nozzle. In order to gain the volume ratio, five squid samples (*Loligo spp*) with different dorsal mantle length (DML) was taken and their internal volume of the mantle and funnel were measured. The volume ratio result was plotted on a graph as shown in Figure 5. Based on this ratio, the volume of the CWT pressure chamber and nozzle was decided. In this research the nozzle volume was $0.6cm^3$. Another two acrylic panels with 40mm width were slotted into the pressure chamber. Two pneumatic cylinders (model SMC CJ2B10-75) were attached to these panels. The position of these pneumatic cylinders was adjusted to ensure that the maximum pressure chamber volume was 15 cm^3 as decided previously.

(a) (b)

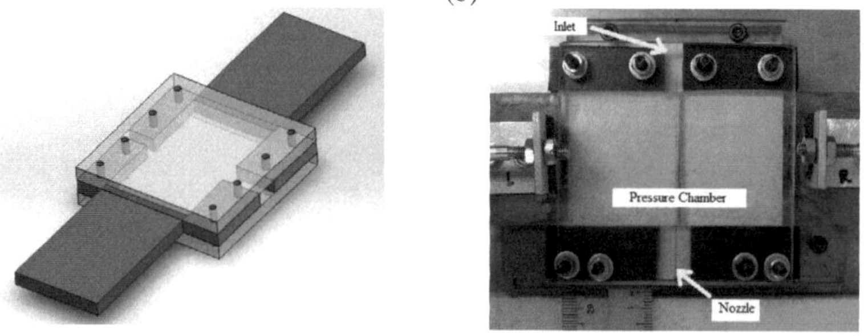

Isometric view Top view

Figure 4. Different view of the CAD model and fabricated 2D CWT

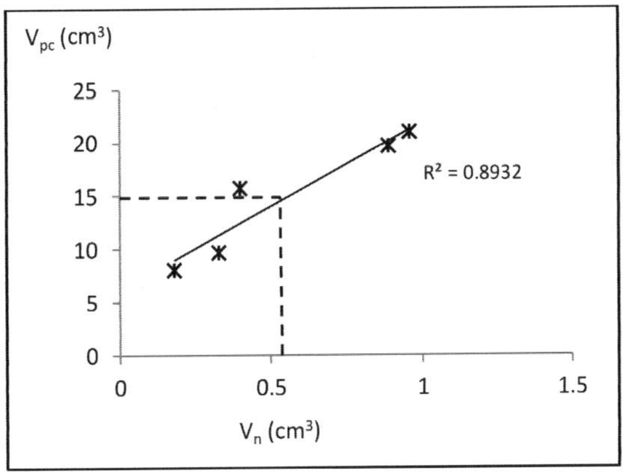

Figure 5. Pressure chamber-Nozzle volume ratio

3.2. Experimental Setup and Procedure

The 2D CWT was fixed a base to make it rigid and this base was allocated in a 485mm x 365mm x 75mm (L x W x T) water container. The depth of the CWT was set at 10mm under from the surface of the water. Red cooking colour was attached to the inlet and nozzle using drip infusion set. A video camera was set at the top of the CWT to catch video picture of the fluid flow from top view. The image acquisition was set at 35 frames per second (fps). The Matlab image processing tool was utilized to measure the volume of the entrained water in the pressure chamber. These cylinders were supplied with compressed air at different pressure value to provide different actuating force (Figure 6). The calculation of actuating force could be determined by Eq. (5):

$$F_a = A_B P_c \tag{5}$$

Where F_a was the actuating force, A_B was the internal bore area of the cylinder and P_c was the supplied pressure which was controlled by the regulator. In this case, A_B was 7.86E^{-5}m^2. F_a was set to 5N by adjusting the pressure regulator. The oscillating actuator frequency, f_{act} was set to 0.91Hz, 0.77Hz, 0.67Hz, 0.59Hz and 0.53Hz. These frequencies were gained by adjusting the time interval in the microcontroller programme. Arduino UNO was utilized as the microcontroller.

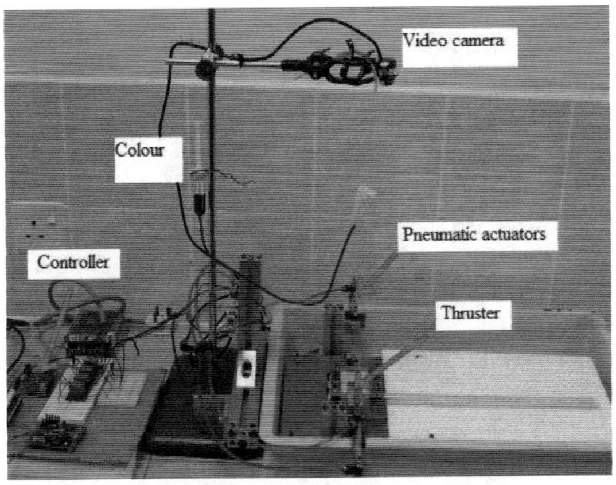

Figure 6. Experimental setup.

4. Result and Discussion

This section discusses the gained results of the experimental works. Basically there was no significant fluid flow rate at the initial stage of the inflation process. However, a notable fluid mass motion into the pressure chamber was observed at a different actuating frequency at the maximum inflation stage which took around 0.09sec (Figure 7). The higher frequency decreases the ingoing fluid volume and thus reducing the fluid flow rate. Full result is shown in table 1. The volume measurement was carried out by detecting the red region in Matlab image processing tool and converted as fluid volume. Figure 7 also depicts the ingoing fluid tended to be attracted at the lateral actuator surface. This phenomenon could be defined using the Navier-Stokes equation previously where the $\partial P/\partial y$ value is greater than the $\partial P/\partial x$. The pressure in x-axis direction became lower because of rapid motion of the actuator in line with the x-axis. Higher frequency enabled less water being entrained into the cavity. This condition occurred as the consequence of the high displacement velocity of the actuator compared to the lower actuating frequency. Higher actuator displacement velocity would not allow much time for the water to fill in the cavity and hence it reduces the number of the entrained water. The relation between the actuating frequency and the fluid flow rate is also shown in Figure 8. Regression analysis was carried out using Minitab software. The result exhibits that the P-value of the relation was 0.014 and the R^2 was 89.9%. The R^2 has answered that there is a strong relation between the frequency and the flow rate. The P-value result shows that the relation of these two parameters falls under normal distribution.

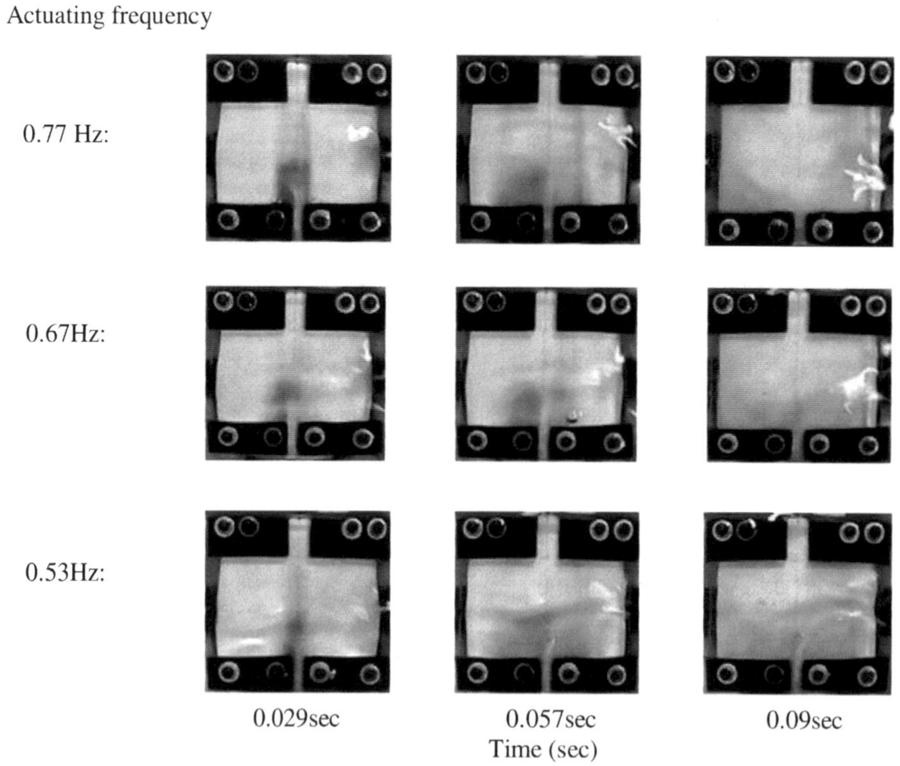

Figure 7. Fluid flow trend at different actuating frequency in inflation process.

Table 1. Flowrate result in different actuating frequency.

Actuating Frequency (Hz)	Fluid volume (mm^3)	Flowrate (mm^3/s)
0.91	7938.11	9.26E+04
0.77	8074.08	9.42E+04
0.67	11035.69	1.29E+05
0.59	13008.14	1.52E+05
0.53	13679.57	1.60E+05

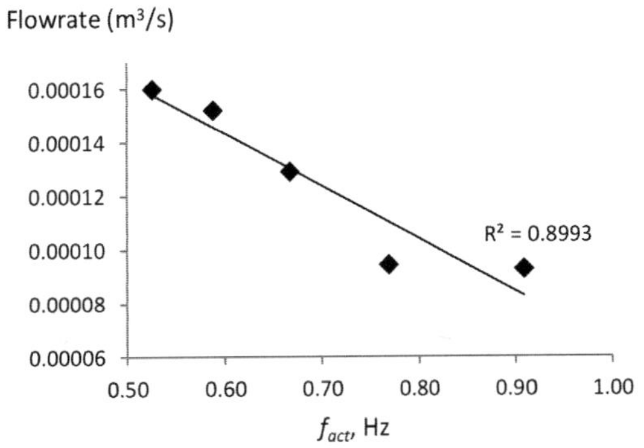

Figure 8. Relation between actuating frequency and fluid flow rate

The Reynolds number was calculated for each of the predetermined actuating frequency. The relation between the Reynolds number and the frequency is presented in Figure 9. Generally the Reynolds number decreases at a constant rate as the frequency increase as a result of the decreasing fluid flow rate. The observed Reynolds number was ranged between 22900 and 13300 by regarding to the actuating frequency. The high turbulence of the fluid flow during inflation could be clearly exhibited from Figure 7. The Reynolds number dropped around 9600 at 0.38Hz increments which means turbulence would change at almost 25000 per Hz.

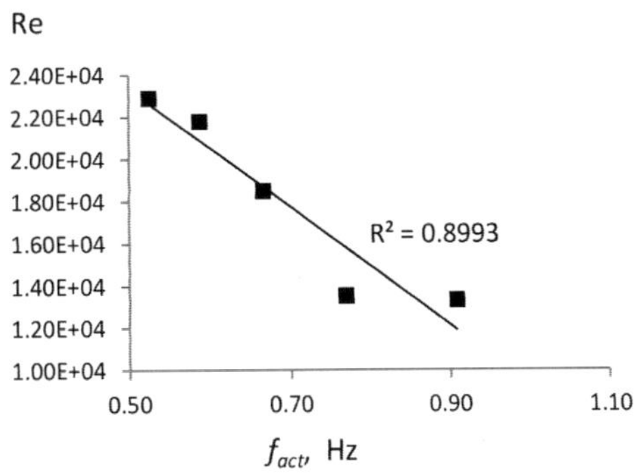

Figure 9. Reynolds number over actuating frequency.

5. Conclusion

In general, the actuating frequency has strong influence to the fluid characterization in the pressure chamber of the CWT during the inflation process. Higher frequency would decrease the fluid flow rate as well as the Reynolds number. At 0.91Hz frequency, the Reynolds number would reach 22900. This behaviour occurred because of the higher actuator velocity could not provide ample time for the water to fill in the cavity of the pressure chamber and thus decreasing the entrained water volume. In designing the CWT, an optimum actuating frequency must be obtained to achieve a maximum mass flow rate during the deflation period.

Acknowledgment

The authors would like to acknowledge compliment to the Ministry of Higher Education (MOHE) for funding this project via ERGS grant, University Tun Hussein Onn Malaysia for sponsoring the research and Universiti Sains Malaysia for providing research facilities and technical assistance.

References

[1] Yeom S W and Oh I K 2009 *Smart Material Structure* **18** 1
[2] Barber A Najem J Donald L and Blotman J 2011 Design and Development of Bio-inspired Underwater Jellyfish like Robot Using Ionic Polymer Metal Composite (IPMC) Actuators. *Proc. of SPIE* **7976** 24-11
[3] Jamil N Yusoff A R Mansor M H 2012 *J. Mech. Eng. Sci.* **3** 311-319
[4] Shi L W Guo S and Asaka K 2011 *Proc. of the IEEE Int. Conf. of Mechatronics and Automation* 853
[5] Thomas A P Milano M G Shell M G Fischer K and Burdick J 2005 *Proceedings of the IEEE Int. Conf. on Robotics and Automation* 181
[6] Shaari M F Samad Z Abu Bakar M E and Jaafar M 2012 *Advanced Materials Research II* **463-464** 1583-88
[7] Tan A C H and Hover F S 2010 *OCEANS 2010* 1
[8] Shahinpoor M and Kim K J 2005 *Smart Material Structure* **14** 197
[9] Muthucumaraswamy R Valliammal V 2010 *Inter. J. Automot Mech Eng.* **2** 231-38
[10] Gladfelter W B 1972 *Helgoländer wiss. Meeresunters* **23** 38
[11] Bartol I K Patterson M R and Mann R 2001 *The Journal of Experimental Biology* **204** 3655
[12] Stemme E and Stemme G 1993 *Sensors and Actuators A* **39** 159

A two pressure-velocity approach for immersed boundary methods in three dimensional incompressible flows

O Sabir[1,a], T M Y S Tuan Ya[2], Norhafizan Ahmad[1] and Y Nukman[1]
[1]Faculty of Engineering, University of Malaya, 50603 Kuala Lumpur, Malaysia
[2]Faculty of Mechanical Engineering, Universiti Malaysia Pahang, 26600, Pekan, Pahang, Malaysia

E-mail: [a]osama_sabir@siswa.um.edu.my

Abstract. This paper describes an innovative method for computing fluid solid interaction using Immersed boundary methods with two stage pressure-velocity corrections. The algorithm calculates the interactions between incompressible viscous flows and a solid shape in three-dimensional domain. The fractional step method is used to solve the Navier-Stokes equations in finite difference schemes. Most of IBMs are concern about exchange of the momentum between the Eulerian variables (fluid) and the Lagrangian nodes (solid). To address that concern, a new algorithm to correct the pressure and the velocity using Simplified Marker and Cell method is added. This scheme is applied on staggered grid to simulate the flow past a circular cylinder and study the effect of the new stage on calculations cost. To evaluate the accuracy of the computations the results are compared with the previous software results. The paper confirms the capacity of new algorithm for accurate and robust simulation of Fluid Solid Interaction with respect to pressure field.

1. Introduction

In many fluid-solid engineering the structure is in general flexible and interacts with the surrounding fluid. Commonly a body-fitted meshes techniques is used to model such systems since its consideration to both fluid and solid dynamics like The arbitrary Lagrangian–Eulerian method[1]. The Immersed boundary methods (IBM) is a nonconforming methods to calculate the fluid solid interaction. Charles S, Peskin invented the methods[2] and since then it is become popular among the researcher because of simplicity and geometric flexibility. IBMs cancel the need of mesh regenerations and handle well the boundary deformations.

IBMs represent the immersed boundary by a set of moving Lagrangian points with a fixed Eulerian grid for the fluid. The group of material points linked with fibres (springs) shows the solid interface mode. This technique simplifies treating immersed boundaries of nearly any arbitrary shapes, sizes and configurations, without use of ALE descriptions or adaptive meshing[3] . An efficient way to transform between the Eulerian fluid motion and the Lagrangian structure motion is to use the Dirac delta function. For numerical view the big advantage of IBM is the ability to employ a simple finite difference or finite volume scheme on a Cartesian mesh. Apply elastics boundary algorithm to simulate moving boundaries problems is easy, and it is work in three dimension simulation. One of the major weakness that it cannot work efficiency when simulate a flow in high Reynolds numbers[4].
[5] developed a method that extracts the forcing direct from the numerical solution for which a preliminary estimate can be determined[6]. The discrete approach is better suited for higher Reynolds

numbers, because of imposing the velocity boundary conditions at the immersed boundary, without introducing or computing any forcing term. The governing equations are most of the time discretized as follows. A second-order Adams-Bash forth scheme is employed for the convective terms, while the diffusion terms are discretized using an implicit Crank-Nicolson scheme. This cancels the viscous stability constraint, which can be severe in simulation of viscous flows. The (spectral) method of Mohd-Yusof settle the force term by the variance between the interpolated velocities in the boundary points and the anticipated (physical) boundary velocities. This method reduces the errors between the calculated velocities and the desired velocity profile on the body surface. If the boundary is stationary, the force can be determined by pairing the velocity at the internal point to the velocity at the external point with a weighted linear interpolation [7].

The obstacle of using a grid does not adapt to the body border is including the boundary conditions in the governing equations will involve enormous adjustments nearby the boundaries. Comparing with the traditional methods, it is challenging and had a bad impact on accuracy and conservation properties when treat the boundaries in IBM. In the boundary conforming grid methods, the control of grid resolutions on the body surrounding areas is much better. IBM has effects on grid resolutions to achieve the certain amount of control which leads to increase of the grid size with increasing Reynolds numbers. The alignment between the body surface and the grid lines places some grid points to be inside the body boundary where solving fluid flow equations is not essential [8].

To enforce Dirichlet conditions at the boundary, interpolations are performed between Lagrangian velocities at the solid surface and Eulerian velocities at neighbouring grid points. These interpolations have proved to be un-robust, which is undoubtedly as obstacle for this method to become widespread for practicable applications. In [9] body force method, the force between the solid and fluid is modelled by a volume fraction function of the solid volumetric fraction and the relative velocities of the two phases. This method ensures no momentum leakage between the phases as both fluid and solid share the rectilinear Eulerian grid and, therefore, making interpolation unnecessary. However, continuity is affected because of changes in velocity field enforced by volume averaging of the local fluid velocity and local solid velocity. Using Kajishima's method, the fluid-solid interaction is coupled by the body force. To use pressure as means of coupling of new two-way fluid-solid interaction, additional correction of pressure and velocity is needed.

In this paper we develop a model for three dimensions arbitrary elastic solid in a viscos incompressible domain based on two stage pressure-velocity correction immersed boundary methods. Following the steps of [8] Method, we develop a code for three dimensions domain.

2. Mathematical Model

The equations governing the flow around the immersed boundaries are the continuity and momentum equations for viscous incompressible flow past a body is describe by the non-dimensional Navier-Stokes equations, including the body force term are given by

$$\frac{\partial \vec{u}}{\partial t} + \vec{u} \cdot \nabla \vec{u} + \frac{1}{\rho} \nabla p - \frac{1}{Re} \Delta \vec{u} = \vec{f}, \tag{1}$$

$$\nabla \cdot \vec{u} = 0 \quad in\ \Omega_f \tag{2}$$

$$\vec{u} = \vec{u}_\Gamma \quad on\ \Gamma_b \tag{3}$$

where $\vec{u}(\vec{x},t)$ represent the velocity field of the flow, $p(\vec{x},t)$ is the pressure and $\vec{f}$ represents any external source terms. The quantities ρ and μ are constant fluid density and viscosity, respectively. The solid body occupies the domain Ω_b, with boundary showed by Γ_b, and the surrounding fluid domain noted by Ω_f. The governing equations were transformed into dimensionless forms on incorporating the following nondimensional variables.

This set of equations is solved with the immersed boundary condition, which constrains the velocity of the fluid to the velocity of the local IB, exactly at the IB. The immersed boundary condition (IBC) can be implemented in different ways that lead to different methods, which can be

both of first- or second-order accurate. A second-order accurate method employs a piecewise linear representation of Eq. (3); that is, linear interpolation is employed to relate the IB to the surrounding points. In a first order accurate method, the IB is approximated to the nearest discrete velocity point.
For cells partially occupied by the solid structure, Kajishima et al [9] proposed an immersed boundary method that solves the momentum exchange at the fluid structure boundary. This method is briefly described by Yuki et al[10] as follows. A velocity field $\vec{u}$ is introduced by calculating the volume fraction and finding the local fluid velocity $\vec{u}_{FLUID}$ and the local solid phase velocity $\vec{u}_{SOLID}$ in each cell by:

$$\vec{u} = (1-\alpha)\vec{u}_{FLUID} + \alpha\,\vec{u}_{SOLID} \tag{4}$$

where $(0 \leq \alpha \leq 1)$ is the local solid volumetric fraction in a cell. The fluid-structure interaction then can be solved at the interface by assuming the velocity field $\vec{u}$ follows the adjusted Navier-Stokes equations:

$$\frac{\vec{u}^{n+1} - \vec{u}^{n}}{\Delta t} = -\vec{u}\cdot\nabla\vec{u} - \frac{1}{\rho}\nabla p + \frac{1}{Re}\Delta\vec{u} + \vec{f}, \tag{5}$$

$$\frac{\vec{u}^{n+1} - \vec{u}^{n}}{\Delta t} = -\frac{1}{\rho}\nabla p + \overrightarrow{RHS} + \vec{f}, \tag{6}$$

and by sing projection method

$$\vec{u}_I = \vec{u}^n + \Delta t\left(\frac{3}{2}\overrightarrow{RHS}^n - \frac{1}{2}\overrightarrow{RHS}^{n-1}\right) \tag{7}$$

where superscripts n represent time, Δt is the time increment and $\vec{u}_I$ is intermediate velocity. Time advancement is then carried by second-order Adam-Bashforth method and Fractional Step Method as a single continuum. Then we solve the pressure A Poisson equation is then solved with diverging the intermediate velocity $\vec{u}_I$ as a source term to provide a pressure p^{n+1},

$$\nabla^2 p^{n+1} = \frac{\nabla.\vec{u}_I}{\Delta t} \tag{8}$$

this is then used to correct the intermediate velocity, providing a divergence free velocity $\vec{u}^{n+1}$ and integration continuing to the next time step.

$$\vec{u}^{n+1} = \vec{u}_I - \Delta t\,\nabla p^{n+1} \tag{9}$$

That finish the first correction stage.
The second correction for the velocity field $\vec{u}_{2nd}^{n+1}$ is derived by the volume averaging the local fluid velocity $\vec{u}^{n+1}{}_{FLUID}$ and the local solid phase velocity $\vec{u}^{n+1}{}_{SOLID}$ in each fluid cell that is partially or fully occupied by solid cell,

$$\vec{u}_{2nd}^{n+1} = (1-\alpha)\vec{u}^{n+1}{}_{FLUID} + \alpha\,\vec{u}^{n+1}{}_{SOLID} \tag{10}$$

Since there is significant change of the velocity field, the value of pressure should also reflect these changes. Therefore developing the second-stage correction. This procedure is also applied to fluid cells that are partially or fully occupied by solid cells bounded by a region called small domain. First, the SMAC method is used to find the scalar potential, Φ^{n+1}. Φ^{n+1} then used as the velocity and pressure correction factor to get $\vec{u}_{3rd}^{n+1}$ and p_{2nd}^{n+1}

$$\nabla^2\phi^{n+1} = \frac{\nabla.\vec{u}_{2nd}^{n+1}}{\Delta t} \tag{11}$$

$$\vec{u}_{3rd}^{n+1} = \vec{u}_{2nd}^{n+1} - \Delta t\,\nabla\phi^{n+1} \tag{12}$$

$$p_{2nd}^{n+1} = p^{n+1} + \phi^{n+1} \tag{13}$$

where $\vec{u}_{3rd}^{n+1}$ and p_{2nd}^{n+1} are velocity and pressure that satisfies continuity equation. The velocity and pressure fields are used for the next time step.

2.1. Discretization of the governing equations

The continuity and momentum equations are discretized on a staggered grid[11, 12], which is more natural than a collocated variable arrangement for Cartesian problems, as there is no artificial boundary conditions are needed and false modes are prevented. On a staggered grid, the cell faces contain the normal velocity components. The remaining variables, such as density and pressure, are stored at cell centres. A pressure projection method is employed to obtain the correct pressure after an initial guess of the velocity field.

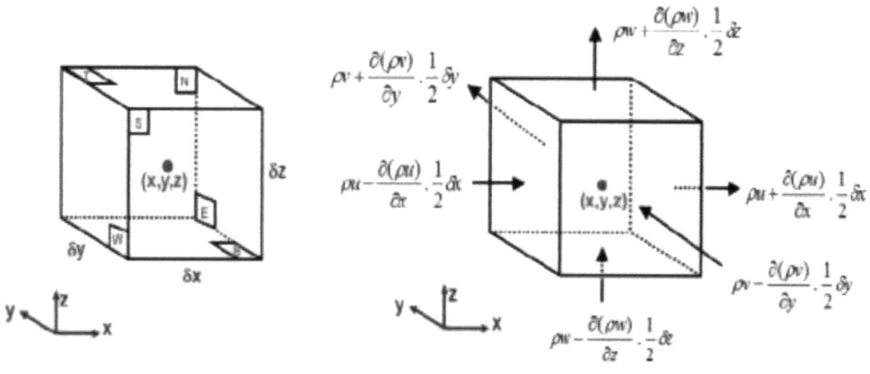

Figure 1. (a) Fluid element for conservation laws; (b) mass flows of fluid element

In figure 1(a), the six faces are labelled N, S, E, W, T and B, which stands for North, South, East, West, Top and Bottom. The pressure in the centre of the element is found at position (x,y, z). Deriving the mass conservation equation is to write down a mass balance for the element lead to:

$$\begin{aligned}\frac{\partial u}{\partial t}+u\frac{\partial u}{\partial x}+v\frac{\partial u}{\partial y}+w\frac{\partial u}{\partial z}&=-\frac{\partial p}{\partial x}+\frac{1}{Re}\left(\frac{\partial^2 u}{\partial x^2}+\frac{\partial^2 u}{\partial y^2}+\frac{\partial^2 u}{\partial z^2}\right)+f_x\\ \frac{\partial v}{\partial t}+u\frac{\partial v}{\partial x}+v\frac{\partial v}{\partial y}+w\frac{\partial v}{\partial z}&=-\frac{\partial p}{\partial y}+\frac{1}{Re}\left(\frac{\partial^2 v}{\partial x^2}+\frac{\partial^2 v}{\partial y^2}+\frac{\partial^2 v}{\partial z^2}\right)+f_y\\ \frac{\partial w}{\partial t}+u\frac{\partial w}{\partial x}+v\frac{\partial w}{\partial y}+w\frac{\partial w}{\partial z}&=-\frac{\partial p}{\partial z}+\frac{1}{Re}\left(\frac{\partial^2 w}{\partial x^2}+\frac{\partial^2 w}{\partial y^2}+\frac{\partial^2 w}{\partial z^2}\right)+f_z\end{aligned} \tag{14}$$

$$\frac{\partial u}{\partial x}+\frac{\partial v}{\partial y}+\frac{\partial w}{\partial z}=0\ in\ \Omega_f \tag{15}$$

This equation will discretize using finite difference schemes with second order. By using the same local volumetric fraction it maintain second order accurate interpolation.

This algorithm impose as:

- Solve the Navier-Stokes equations using projection methods
- Use second-order Adam-Bashforth method and Fractional Step Method to get the intermediate velocity component
- Apply Successive of relaxation to calculate the pressure using the Poisson equation.
- Correct the velocity.
- Calculate the local solid volumetric fraction and update the velocity.
- Use SMAC method to find the scalar potential ϕ then solve Poisson equation using SOR.
- Update the velocity and pressure.
- Go to step (1).

3. Validation

Flow past a circular cylinder follow the shape of the body provided the velocity of the flow is slow; this is known as laminar flow. Validate our algorithm by Recreating the same bench mark test which suggested by Schäfer, Michael, et al. [13]. For the 3D test cases the flows around a cylinder with circular cross–sections are considered.

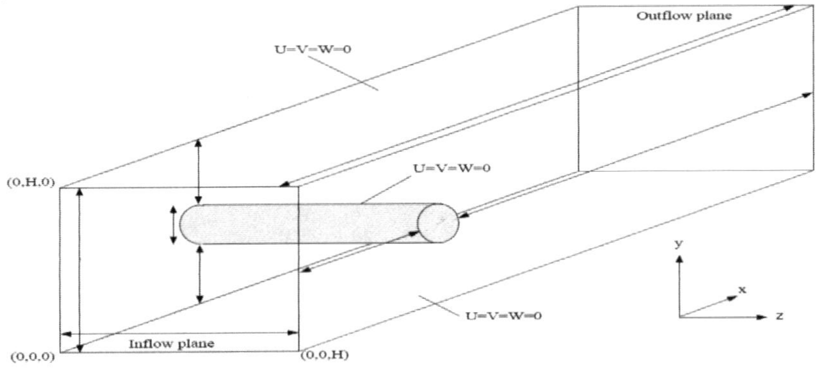

Figure 2. Configuration for flow around a cylinder [13].

4. Results

Animated movies were created to foresee the results data from the model. The representation of the simulation of viscous Newtonian fluid flow around solid cylinder at definite time.

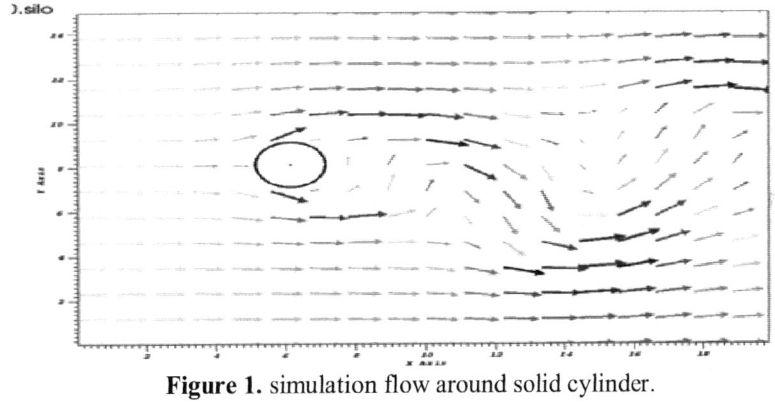

Figure 1. simulation flow around solid cylinder.

The time step and grid size in the simulations are chosen with consideration to the stability coefficient. We calculate the drag and lift coefficients and make comparison with the bench mark data. In chart (4, 5) the simulated drag coefficient Cd and simulated lift coefficient Cl are plotted against the result obtained from the bench mark tests. It can be seen that the simulated drag coefficients lie very close to the established relations. The drag coefficient differs slightly from the test data because the domain size starts to affect the flow and the known drag relationships are determined for cylinder immersed in an infinite fluid.

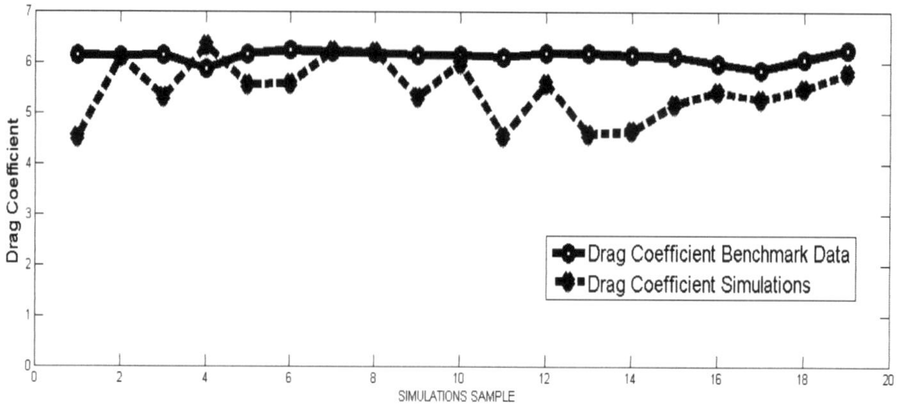

Figure 2. Drag Coefficient.

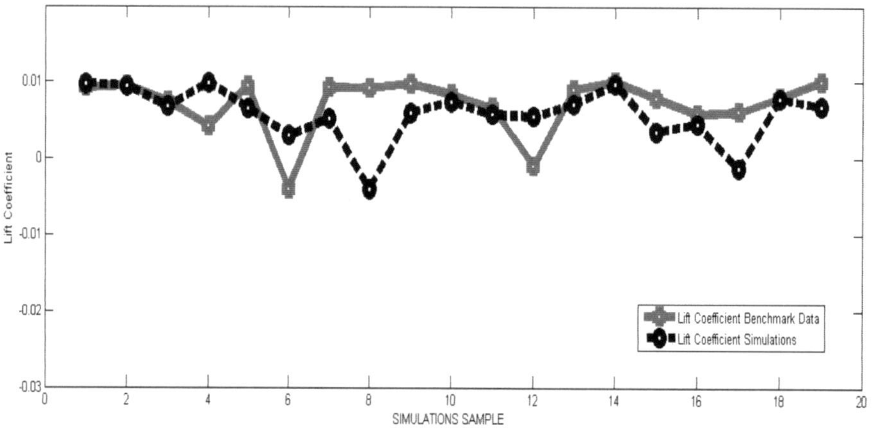

Figure 3. Lift Coefficient.

We assume in or hypothesis that the cylinder is very rigid so there is no need to calculate the solid movement by using finite element methods. That leads to the different in the pressure as figure 6 shows. Although, it is not severely violate the standard deviation limits.

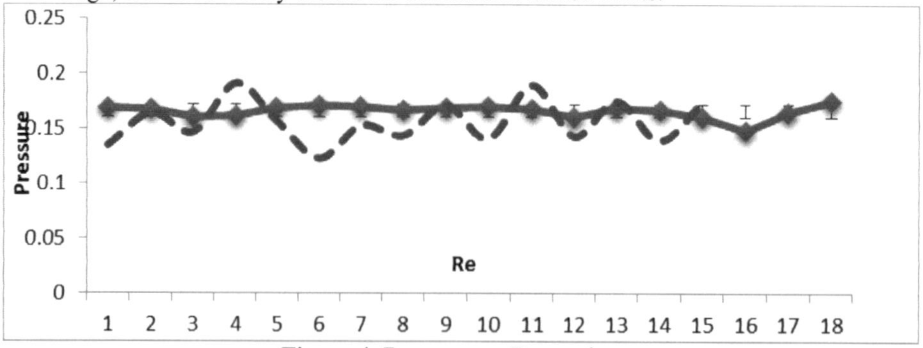

Figure 4. Pressure vs. Re numbers.

5. Conclusions

In this paper, a novel implicit second-order accurate immersed boundary methods with two stage pressure-velocity corrections is developed, implemented and validated. The method is successfully simulating the fluid solid interaction phenomena in three dimensions. Those results encourage us to pursue our quest to upgrade the algorithm to deal with any arbitrary shape. For future work we intend to try new methods for calculating the local volume fraction and optimize the code for better computation cost.

6. Acknowledgements

The authors would like to acknowledge Ministry of Science, Technology and Innovation, Malaysia and University of Malaya for their financial support under: FRGS grant - code: FP064/2010B - Fundamental development and experimental validation of immersed boundary and finite element methods for interaction of a 3D elastic body and fluid by two-stage correction of velocity and pressure.

7. References

[1] Donea J, Giuliani S, and Halleux J P 1982 An arbitrary lagrangian-eulerian finite element method for transient dynamic fluid-structure interactions *Computer Methods in Applied Mechanics and Engineering* 1982 **33** 689-723

[2] Charles S 1977 Numerical analysis of blood flow in the heart *Journal of Computational Physics* **25** 220-252

[3] Wang X S 2010 *Immersed Boundary/Continuum Methods Computational Modeling in Biomechanics* Springer Netherlands 3-48

[4] Peskin C S 2002 The immersed boundary method *Acta Numerica* **11** 479-517

[5] Mohd-Yusof J 1997 Combined immersed-boundary/B-spline methods for simulations of fow in complex geometries in *Center for Turbulence Research Annual Research Briefs*

[6] Mohd-Yusof J 1998 Development of immersed boundary methods for complex geometries *Center for Turbulence Research Annual Research* Briefs 325-336

[7] Zhang N and Zheng Z C 2007 An improved direct-forcing immersed-boundary method for finite difference applications *Journal of Computational Physics* **221** 250-268

[8] Ya T M Y S T, Takeuchi S, and Kajishima T 2007 Immersed Boundary and Finite Element Methods Approach for Interaction of an Elastic Body and Fluid by Two-Stage Correction of Velocity and Pressure (ASME)

[9] Kajishima T et al. 2001 Turbulence Structure of Particle-Laden Flow in a Vertical Plane Channel Due to Vortex Shedding *JSME International Journal Series B Fluids and Thermal Engineering* **44** 526-535

[10] Yuki Y, Takeuchi S, and Kajishima T 2007 Efficient Immersed Boundary Method for Strong Interaction Problem of Arbitrary Shape Object with the Self-Induced Flow *Journal of Fluid Science and Technology* **2** 1-11

[11] Harlow F H and Welch J E 1965 Numerical Calculation of Time-Dependent Viscous Incompressible Flow of Fluid with Free Surface *Physics of Fluids* **8** 2182

[12] Patankar S 1980 Numerical heat transfer and fluid flow (Taylor & Francis)

[13] Schäfer M, et al. 1996 *Benchmark computations of laminar flow around a cylinder Notes on numerical fluid mechanics* **52** 547-566

IOP Conf. Series: Materials Science and Engineering **50** (2013) 012012 doi:10.1088/1757-899X/50/1/012012

Numerical prediction of air flow within street canyons based on different two-equation *k*-ε models

Afiq Witri Muhammad Yazid[1], Nor Azwadi Che Sidik[1,a], Salim Mohamed Salim[2] and Nur Hamizah Mohamad Yusoff[1]

[1]Faculty of Mechanical Engineering,
Universiti Teknologi Malaysia, 81310, Skudai, Johor Bahru, Malaysia
[2]School of Engineering, Taylor's University, 47500, Subang Jaya, Selangor, Malaysia

E-mail: [a]azwadi@fkm.utm.my

Abstract. Numerical simulations on airflow within street canyons were performed to investigate the effect of the street aspect ratio and wind speed on velocity profiles inside a street canyon. Three-dimensional Standard, Renormalization Group (RNG) and Realizable k-ε turbulence model are employed using the commercial CFD code FLUENT to solve the Reynolds-averaged Navier-Stokes (RANS) equations. A comparison of the results from the presently adopted models with those previously published demonstrated that the k-ε model is most reliable when simulating wind flow. The model is then employed to predict the flow structures in a street canyon for a range of aspect ratios (building height to street width ratio) between 0.5 - 2 at Reynolds number of 9000, 19200 and 30700 corresponding to the ambient wind speeds of 0.68m/s, 1.46m/s and 2.32m/s respectively. It is observed that the flow structure in the street canyon is influenced by the buildings aspect ratios and prevailing wind speeds. As the street aspect ratio increases, the air ventilation within the canyon reduces.

1. Introduction

Street canyons are one of the most prominent features found in city environments where the flow is dominated by the interaction of prevailing wind and building layout. Rapid urbanization demands the constructions of high rise buildings resulting in low air ventilation hence affecting inhabitants comfort. Besides buildings configuration/layout, the local meteorological conditions and trees plantings are amongst the many factors affecting wind flow within street canyons. The formation of a centrally located vortex is one of the characteristic features of street canyon flow when prevailing winds are perpendicular to the street canyon. While oblique or parallel approach flow provides the best wind flow condition for the purpose of ventilation, perpendicular approach flow is the worst configuration due to existence of vortex that has a very low wind speed at pedestrian level[1]. Parameterization of buildings layout to represent canyon geometry has been widely used in terms of the ratio of the building height (H) to the width (W) between buildings. The flow regimes of isolated roughness flow, wake interference flow and skimming flow at certain range of buildings aspect ratio has been well described [2-4].

The introduction of two-equation turbulence closure schemes more than two decades ago has benefited most Computational Fluid Dynamics (CFD) researchers as it is computationally efficient [5-8]. The wind engineering community has long adopted the two-equation turbulence models, specifically the Standard k-ε turbulence model, to represent successfully the wind flow field [5,9-11], although it is not so accurate for flows involving impingement and separation regions [12] which may lead to the poor prediction of flow structures at the wake of buildings. Therefore, another alternative turbulence model should be adopted to ensure reliable results. Up till now, the authors are only aware of one numerical study available on the assessment of flow in street canyons using two-equation turbulence models by Chan [13]. They found out that the RNG k-ε turbulence model provide the best results on the pollutant concentration distribution in a street canyon when compared to Standard k-ε and Realizable k-ε model. However, there is missing information regarding the wind flow field within the street canyon. Furthermore, there are only a few studies on the flow induced in a street canyon for different ambient wind speeds by Baik and Kim [14] at H/W = 2.0 and by Nazridoust and Ahmadi [15] at H/W = 0.5 and 1.0.

Therefore, the objective of this study is to analyze the performance of k-ε turbulence models and its variants to simulate three-dimensional (3-D) flow inside a street canyon. This is important as there is no study that has been carried out to assess the performance of various k-ε turbulence models (Standard, RNG and Realizable k-ε) on the flow structures inside a 3-D street canyon. The comparison of a wide range of ambient wind speeds against experimental data also helps to increase confidence in our results. The most accurate k-ε turbulent model is then adopted to simulate different buildings aspect ratio (H/W) coupled with a wide range of ambient wind speeds by analyzing the flow structure and air ventilation inside the street canyon. This enhances our understanding on the capacity of buildings aspect ratio parameter towards air ventilation in urban street canyon.

2. Numerical Model

In the current CFD model, 1:200 scaled model of a street canyon aligned perpendicular to wind direction is constructed in a reduced 3-D spatial domain formed by two buildings of equal height H and separated by a street width W based on the experimental setup by Allegrini [16]. The distance of the street canyon from the inlet and top boundary is based on the provided boundary conditions of experiment. The outlet distance is at a distance H from the cavity and outflow condition is imposed. Only the central 1/3rd of the total lateral distance from the actual experiment was modelled while imposing symmetry boundary condition at both sides as illustrated in figure 1.

Incompressible, 3-D steady turbulence flows are solved using two-equation turbulence models of RANS equations. The models considered for the purpose of validation are Standard k-ε, RNG k-ε and Realizable k-ε combined with enhanced wall treatment for the near wall treatment [17]. Similar inlet profiles of velocity and turbulent kinetic energy from the experiment were imposed at the inlet of the numerical domain ranging from Re = 9000, 19200 and 30700 which corresponds to 0.68, 1.46 and 2.32 m/s of freestream velocity. The dissipation rate can be approximated by ε = (Cμ3/4k3/2)/l. The Reynolds number is based on the free stream velocity, U at buildings height, H. The governing equations are discretized using second order upwind for all the advection term and second order for pressure interpolation while the SIMPLE algorithm is used to solve the equations. The convergence criteria are set at 1.0e-6 to ensure the convergence of the results. The numerical simulation is performed using the commercial CFD software Fluent 6.3.

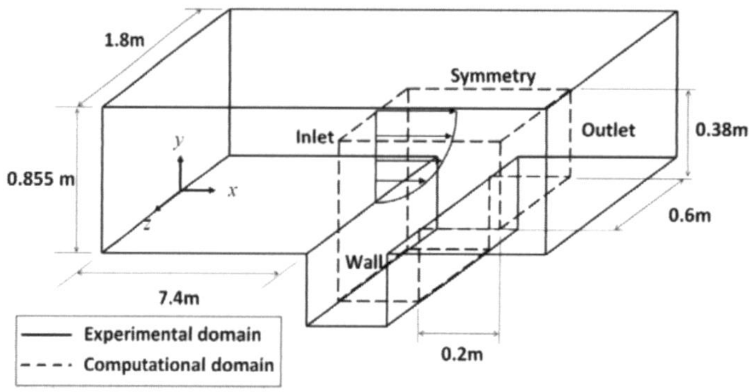

Figure 1. Experimental and computational domain (not to scale).

3. Model Validation

The accuracy of the CFD model is evaluated using extensive wind-tunnel experimental results of Allegrini [4]. The data consists of isothermal and thermal case for buildings aspect ratio H/W = 1 at a wide range of Reynolds number. For the purpose of validation, only the benchmark of isothermal case is compared. Mesh analysis is done through grid independence study with different mesh configuration of a structured hexahedral mesh. Since the enhanced wall treatment is used, the dimensionless parameter of wall y^+ is kept below 5 to ensure the variable gradient near the wall is adequately resolved [18]. In addition to that, cell stretching away from the buildings wall and ground floor with 1.2 is used to reduce the computational time while maintaining its accuracy.

After an appropriate mesh is identified for each turbulence models at each Reynolds number, the performances of the turbulence models are assessed using mesh independence. The mesh independence study of each flow case and turbulence models were analyzed based on the vertical profile of wind x-velocity generated inside the street canyon as illustrated in figure 2 while the percentage of average error between the numerical and experimental results are then calculated based on the vertical x-velocity profile inside street canyon with vertical interval of 0.01m as tabulated in table 1.

Table 1. Percentage of average error.

Turbulence models	Percentage of average error (%)		
	Re = 9000	Re = 19200	Re = 30700
Standard k-ε	28.18	51.67	59.58
RNG k-ε	22.02	41.43	53.85
Realizable k-ε	19.94	41.62	49.99

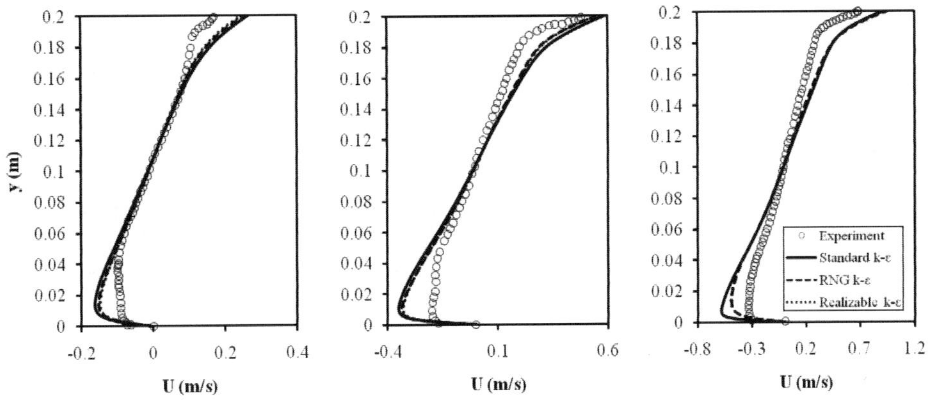

(a) (b) (c)

Figure 2. Vertical x-velocity profile: (a) Re = 9000, (b) Re = 19200, (c) Re = 30700.

It can be seen that the results of all two-equation turbulence models have over predicted the vertical profile particularly at 0 m < y < 0.02 m and y > 0.18 m. Nevertheless, the Realizable k-ε yields the least error for Re = 9000 and 30700 whilst for Re = 19200, the error is slightly higher than RNG k-ε where the deviation is less than 0.2%. With the good flow profile achieved by the Realizable k-ε, the model is then chosen for the investigation of the remaining objectives of this paper.

4. Results And Discussion

This section presents the numerical solutions of turbulence flow at different buildings aspect ratio and ambient wind speeds. The height of the building is fixed at H while only the width of the canyon is varied to get different buildings aspect ratio. This allows the Reynolds numbers to be maintained throughout the investigations. Figure 3 shows the flow patterns in the vertical cross-section for different Reynolds numbers. In the case of low wind speed (Re = 9000) with H/W = 0.5, the ambient flow at roof level penetrates into the canyon between buildings creating 'wake interference flow' described by Oke [2]. In the case of higher H/W (1.0, 2.0), the flow above roof is separated from the canyon flow and thus is categorized as 'skimming flow'. The secondary vortex located at the bottom of leeward wall with H/W = 0.5 and at the bottom of street canyon with H/W = 2.0 is disappeared or reduced in size as the wind speed aloft increased to Re = 30700. Looking at H/W = 2.0, the strong ambient speed aloft (Re = 30700) enhanced the momentum transfer deep into the lower region of the street canyon and in turn elongated the primary vortex further down and at the same time reduced the size of secondary vortex.

Figure 4 shows the x-velocity profile along the middle height of the street canyon at different buildings aspect ratio and Reynolds number. Comparison between different H/W on the vertical velocity shows that the air ventilation inside street canyon is reduced by the increment of H/W. This is most probably due to low air penetration from the above causing reduction of momentum transfer at deep street canyon.

IOP Conf. Series: Materials Science and Engineering **50** (2013) 012012 doi:10.1088/1757-899X/50/1/012012

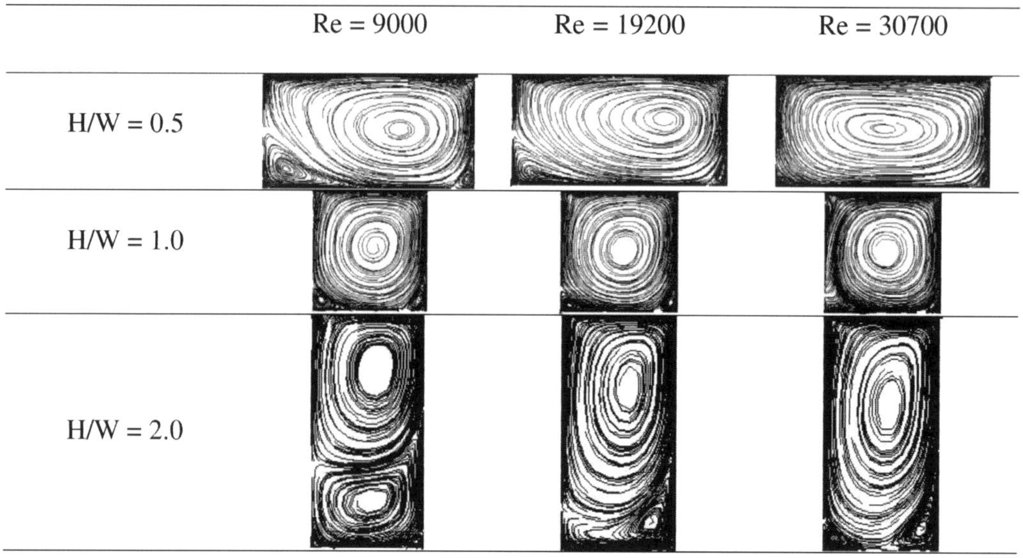

Figure 3. Streamlines inside street canyon (model scale) with different Reynolds numbers and buildings aspect ratios.

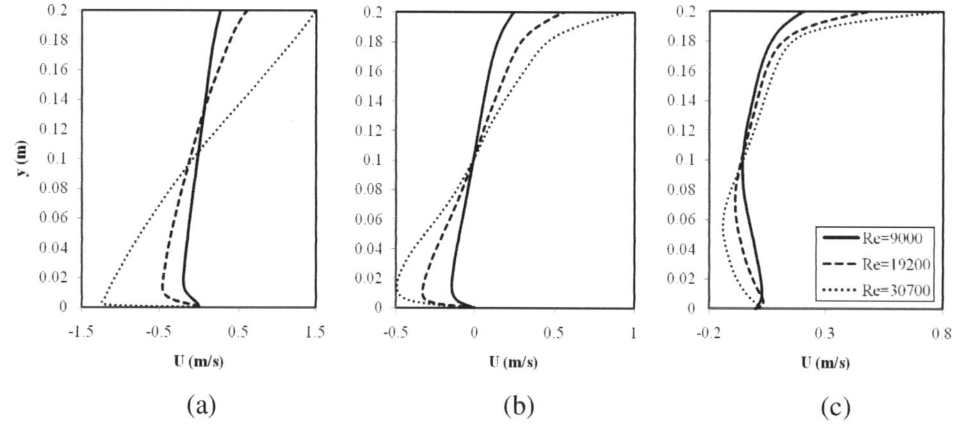

Figure 4. Vertical x-velocity profile: (a) H/W = 0.5, (b) H/W = 1.0, (c) H/W = 2.0.

5. Conclusions

The numerical results of the vertical velocity profile and streamline on air flow are investigated. It is observed that:

- The Realizable k-ε is the most accurate amongst other *k*-ε turbulence models to predict the wind flow profile inside street canyon.
- A secondary vortex is observed only at low Reynolds number.
- The air ventilation inside street canyon reduces with the increase of H/W.

6. Acknowledgements
The authors would like to thanks the Universiti Teknologi Malaysia and Ministry of Higher Education, Malaysia for sponsoring this research. This research is supported by research grant Vot. 4F114 and 4L074.

References
[1] Soulhac L, Perkins R J and Salizzoni P 2008 *Bound-Lay Meteorol.* **126** 365.
[2] Oke T R 1998 *Energ. Buildings* **11** 103.
[3] Kovar P A, Louka P, Sini J F, Savory E, Czech M, Abdelqari A, Mestayer P G and Toy N 2002 *Water Air Soil Poll.* **2** 237.
[4] Afiq W M Y, Azwadi C S N and Saqr K M 2012 *International Journal of Mechanical and Materials Engineering* **7** 158.
[5] Mochida A, Tominaga Y, Murakami S, Yoshi R, Ishihara T and Ooka R 2002 *Wind Struct.* **5** 227.
[6] Tahseen T A, Ishak M and Rahman M M 2013 *J Mech Eng Sci* **4** 418-430.
[7] Tahseen T A, Ishak M and Rahman M M 2012 *Inter J Automot Mech Eng* **6** 753-765.
[8] Najiha M S, Rahman M M, Yusoff A R and Kadirgama K 2012 *Inter J Automot Mech Eng* **6** 766-774.
[9] Vardoulakis S, Dimitrova R, Richards K, Hamlyn D, Camilleri G, Weeks M, Sini J F, Britter R, Borrego C, Schatzmann M and Moussiopoulos N 2011 *Environ Model Assess.* **16** 169.
[10] Idris M S, Irwan M A M and Ammar N MM 2012 *J Mech Eng Sci* **2** 206-216.
[11] Mohd Adib M A H, Mohd Hasni N H, Osman K, Maskon O and Kadirgama K 2012. *J Mech Eng Sci* **2** 217-225.
[12] Meroney R N, Leitl B M, Rafailidis S and Schatzmann M 1999 *J. Wind Eng. Ind. Aerod.* **81** 333.
[13] Chan T L, Dong G, Leung C W, Cheung C S and Hung W T 2002 *Atmos. Environ.* **36** 861.
[14] Baik J J and Kim J 1999 *Electron. J. Appl. Meteorol.* **38** 1576.
[15] Nazridoust K and Ahmadi G 2006 *J. Wind Eng. Ind. Aerod.* **94** 491.
[16] Allegrini J, Dorer V and van Carmeliet J 2013 *Build Environ.* **59** 315.
[17] FLUENT Documentation: User guide, ANSYS Inc, 2005
[18] Ariff M, Salim S M and Cheah S C 2009 7th International Conference on CFD in the Minerals and Process Industries CSIRO, Melbourne, Australia. December

Hydrodynamical analysis of the effect of fish fins morphology

Nor Azwadi Che Sidik[a] **and Tey Wah Yen**
Faculty of Mechanical Engineering, UniversitiTeknologi Malaysia
81310, Skudai, Johor, Malaysia

E-mail: [a]azwadi@fkm.utm.my

Abstract. The previous works on the biomechanics of fishes focuses on the locomotion effect of the fish bodies. However, there is quite a insufficiency in unveiling the respective function of fins when the fishes pose statics and exposed to fluid flow. Accordingly, this paper's focus is to investigate the hydrodynamic effect of the fins configuration to the fluid flow of shark-shaped-inspired structure. The drag and lift coefficient is computed for different cases of fish fins addition and configuration. The k-epsilon turbulence model is deployed using finite volume method with the aid of commercial software ANSYS CFX. The finding will demystify some of the functions of the fish's fins in term of their contribution to the hydrodynamic flow around the fishes.

1. Introduction

The high manoeuvre and swimming efficiency of the fishes have inspired human to design the unmanned underwater vehicles such as by manipulating the pectoral fins [1].This is needed as a powerful resolution for the deep sea exploration and survey of offshore industry [2]. In the real nature of fish fins configuration is a consummate association of both types of fins to ensure their ability in swimming due to the evolution of billions of years [3]. Although a variety of researches have been done on the undulation effects and fluid structure interaction of swimming fish, the respective functions of fish fins during static state are remained unclear.

Fins play instrumental roles in the propulsion, stabilization, and manoeuvre of the fishes. In general, the fins could be classified into ceratotrichia and lepidotrichia. Ceratotrichia fins are fairly stiff, unbranched, and unsegmented meanwhile the lepidotrichiafins are flexible, branched and segmented [4]. Specifically in this paper, only the ceratotrichiafins are considered in order to have a clearer insight into the functions of the fins.

From figure 1 as follows, it could be observed that the main fins existed in most of the fishes are caudal fin, dorsal fin and pectoral fins. The pelvic fins and adipose fin or second dorsal fin are the minor fins existed. They play their own roles contributing to the swimming ability of the fishes.

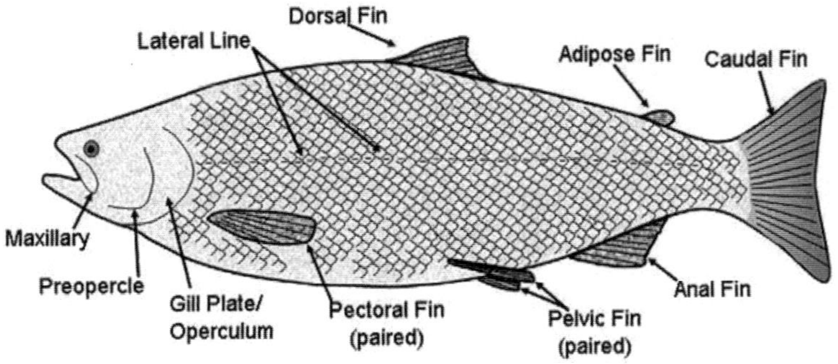

Figure 1. The general morphological structure of a fish. (Retrieved from: http://www.dfw.state.or.us/mrp/fishid/FishIDGlossary.asp).

Caudal fin determines the normal swimming speed of the fishes. Caudal fins can be categorized into homocercal tail, or lunate shaped tail, and heterocercal tail. For homocercal tail, the upper lobe has the similar length with the lower lobe of caudal fin, and it normally exists in the structure of fast-swimmer. Theheterocercal tail has a longer upper lobe, acting to lift the posterior body of the sharks [4].

Dorsal fin is able to provide the stability to the fishes. Upon folding of the fin by some of the fishes, the surface area of the fished which is exerted by the normal force will be dwindled, and hence reducing the drag coefficient. Yet the method of drag force reduction is not available in sharks [4].

Pectoral fins, which are located symmetrically at the left and right side of the fishes, operate as the stabilizer, and the "diving planes". Thus when the fishes are about to lift, the fins are adjusted to some angle of attack. For fast swimming fins, the pectoral fins are long and pointed while for slow swimmers, the fins are rounded [4].

For the minor fins, pelvic fins act as the steering and braking fins. However the function of adipose fin or second dorsal fin, a tiny fleshy fin, is still a remained mystery [1].

2. Mathematical Model

The simulation is solved by deploying the Reynolds-Averaged Navier-Stokes (RANS) Model, or more specifically the eddy-viscosity models. First and foremost the general Navier-Stokes equations (momentum and continuity equation) for both laminar and turbulent flow is [5-10]:

$$\rho\left(\frac{\partial u_i}{\partial t} + u_j\frac{\partial u_i}{\partial x_j}\right) = -\frac{\partial P}{\partial x_i} + \mu\frac{\partial^2 u_i}{\partial x_j{}^2} \quad (1)$$

$$\frac{\partial u_j}{\partial x_j} = 0 \quad (2)$$

Taking Reynolds averaging in order to decompose the flow fields into the averaged and fluctuated components such that:

$$u = \bar{u} + u', \qquad P = \bar{P} + P' \quad (3)$$

Upon some modification, the Reynolds stress transport equation can be derived in the following form:

$$\frac{\partial \overline{u'_i u'_j}}{\partial t} + \bar{u}_k \frac{\partial \overline{u'_i u'_j}}{\partial k} = -\frac{1}{\rho}\left(\overline{u'_j}\frac{\partial \bar{p}}{\partial x_i} + \overline{u'_i}\frac{\partial \bar{p}}{\partial x_j}\right) - 2\upsilon \frac{\partial \overline{u'_i}}{\partial k}\frac{\partial \overline{u'_j}}{\partial k} - \frac{\partial \overline{u'_i u'_j u'_k}}{\partial k} - u'_j u'_k \frac{\partial \overline{u_i}}{\partial k} + \upsilon \nabla^2 k \tag{4}$$

The term $k = 0.5\overline{u'_i u'_j}$ is the kinetic energy per unit mass. In equation (4), the first term on the right hand side represents the redistribution of the flow. Meanwhile the second and third term denotes dissipation and turbulent transport respectively. The fourth and fifth terms are the production term [11].

To close the equation (4), the eddy viscosity concept is applied to develop the $k - \epsilon$ model. The $k - \epsilon$ model yielded after some mathematical derivation would be:

$$\frac{\partial(\rho k)}{\partial t} + \frac{\partial(\rho \bar{u}_j k)}{\partial x_j} = \frac{\partial}{\partial x_j}\left[(\mu + \frac{\mu_t}{\sigma_k})\frac{\partial k}{\partial x_j}\right] + P_k - \rho\epsilon + P_{kb} \tag{5}$$

$$\frac{\partial(\rho \epsilon)}{\partial t} + \frac{\partial(\rho \bar{u}_j \epsilon)}{\partial x_j} = \frac{\partial}{\partial x_j}\left[(\mu + \frac{\mu_t}{\sigma_\epsilon})\frac{\partial \epsilon}{\partial x_j}\right] + \frac{\epsilon}{k}(C_{\epsilon 1}P_\epsilon - C_{\epsilon 2}\rho\epsilon + C_{\epsilon 3}P_{\epsilon b}) \tag{6}$$

and the turbulent viscosity, μ_t can be linked to the turbulence kinetic energy, k, and dissipation, ϵ with the relationship [12] :

$$\mu_t = C_\mu \rho \frac{k^2}{\epsilon} \tag{7}$$

whereby $C_{\epsilon 1}$,$C_{\epsilon 2}$,C_μ, σ_ϵ, and $\sigma_\epsilon$are constant.

All the above equations are solved using software ANSYS CFX. First, the domains are set with eight cases with different configuration of fish morphology. The meshing is then made using tetrahedron patch independent meshing with high smoothing and fine relevance centre. The outer wall of the cylinder enclosure is set to be in free slip condition while the structure wall is set to be in non-slip boundary condition. The inlet of the flow will be at 5m/s. The convergence criterion is 1×10^{-6}.

To examine the effect of fish morphology to the hydrodynamic features of the water flow, the data generated from the CFX solver will be applied to calculate the drag and lift coefficients of the fishes with the equations [13]:

$$C_D = \frac{F_x}{\frac{1}{2}\rho S U^2} \tag{8}$$

$$C_L = \frac{F_y}{\frac{1}{2}\rho S U^2} \tag{9}$$

where F_x and F_y are the force exerted on the examined structure in X- and Y- direction respectively, ρ is the water density with value 1000kg/m3, S is length of the overall fish, and U is the characteristics velocity.

The Caudal Fin Ratio, CFR can be defined as the ration of the length of lower lobe to the upper lobe of caudal fin, for instance, CFR is less than 1 if its tail is *heterocercal.*

$$CFR = \frac{\text{Horizontal length of lower caudin fin lobe}}{\text{Horizontal length of upper caudin fin lobe}} \tag{10}$$

3. Results and Discussions

From the simulation done upon eight cases, the corresponding drag and lift coefficients of the shark-inspired fish structure is shown in Table 1and figure 2 and figure 3.

Table 1. The drag and lift coefficient with the different morphological structure.

Case	Description	Drag Coefficient	Lift Coefficient
1	Simple ellipse	2.44641×10^{-07}	2.10128×10^{-07}
2	Addition of caudal fin with CFR 1.0	0.000898034	-6.60256×10^{-07}
3	Addition of dorsal fin with CFR 1.0	0.00098859	1.52949×10^{-05}
4	Addition of pectoral fin with CFR 1.0	0.001786795	-1.8391×10^{-05}
5	Addition of pelvic fin with CFR 1.0	0.002003269	-1.12607×10^{-05}
6	Addition of adipose fin with CFR 1.0	0.001901752	-1.93141×10^{-05}
7	Deployment of caudal fin with CFR 0.5	0.001741282	-1.1967×10^{-05}
8	Deployment of caudal fin with CFR 0.25	0.001608365	-8.88942×10^{-05}

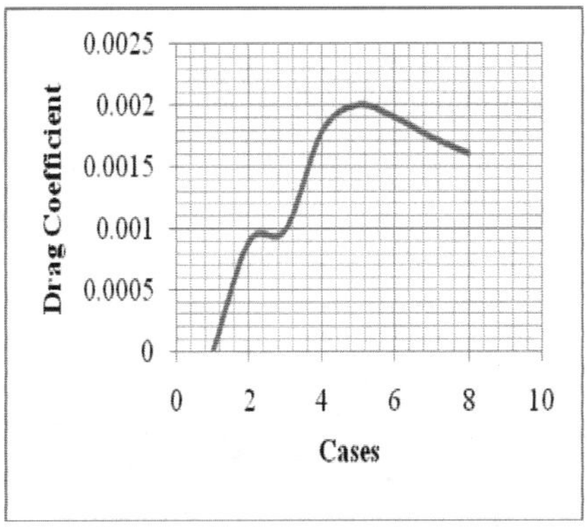

Figure 2. The corresponding plot of drag coefficient with different morphological structures.

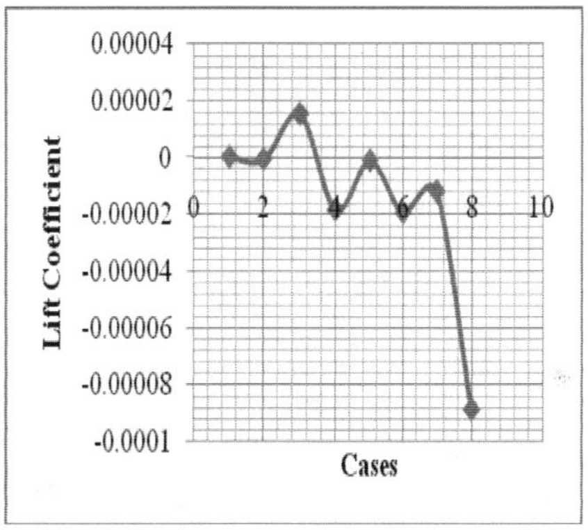

Figure 3. The corresponding plot of lift coefficient with different morphological structures.

The velocity streamlines profiles of all the eight cases above have been illustrated in figure 4. From the simple elipse, upon addition of caudal fin, the drag coefficients increases and lift coefficient decreases. This addition is a burden for the structure to thrust and lift, and thus in the nature the undulation of caudal fin is required to counter the effect and generate forces for propulsion and lifting.This could explain the phenomenon that a reverse Bernord von Karman Vortex sheet is required to be produced in order to produce the thrust and significantly reduce the drag force [14].

Pectoral fins increase the drag coefficient and decrease lift coefficient to drag down and slow down the structure. Pectoral fins require further morphological adaptation to decrease the drag [15]. In a case of normal pectoral fin without undulation, Emily and Timothy suggest that it would generate a significant negative lift for the purpose of gripping [16] and this has met with the negative lift generation as shown in Table 1.

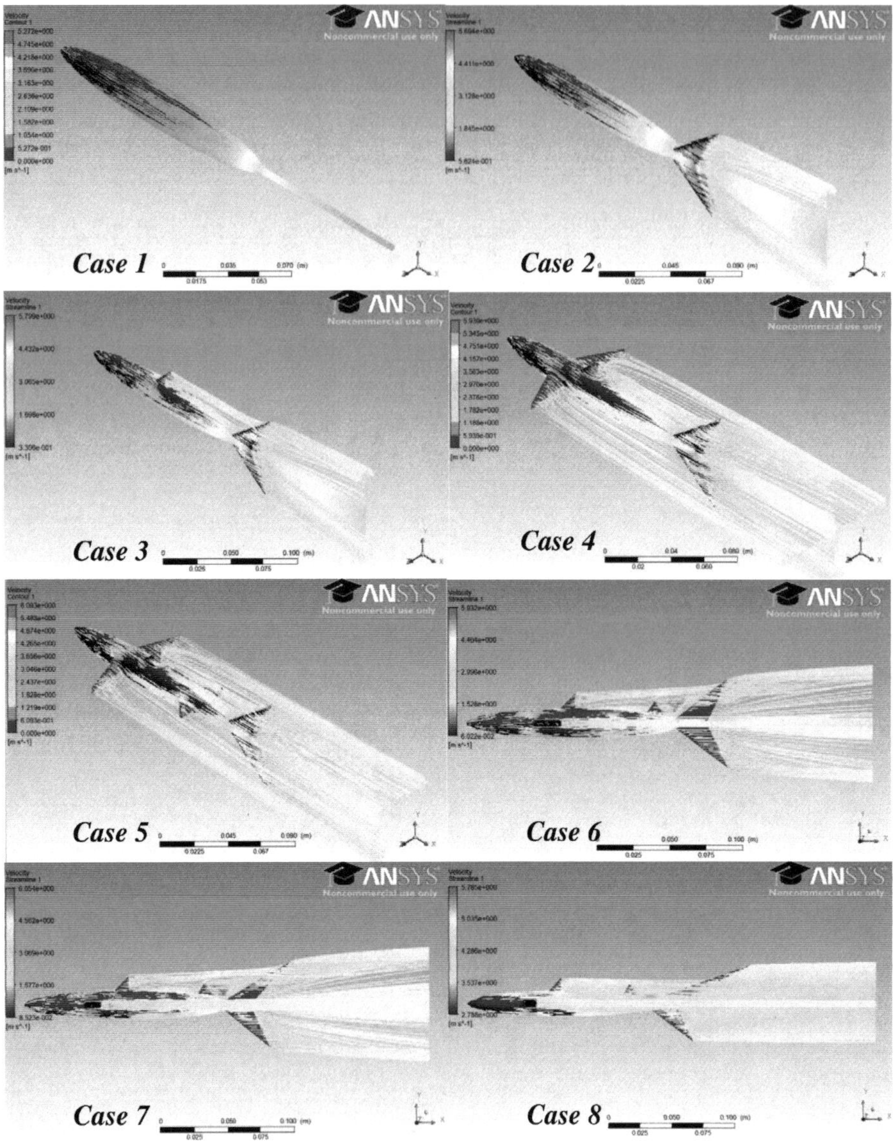

Figure 4. Velocity streamline plot for case (1) to case (8), starting from the upper left to the lower right.

The addition of dorsal fin increases the drag coefficient but increases the lift coefficient as well. This finding is in coherence with the natural function of dorsal fin to stabilize the flow and provide lift for fishes[1]. This reaches an agreement as well with the Anabela and Cheryl's perception that the first dorsal fin contributes to stability [17].

When the minor fins are configured, it could be observed that pelvic fins are able to further stabilize the structure but they could enhance its lifting ability. The adipose fins function which was remains unknown before, when it is analysed by static hydrodynamic point of view, it decreases the lift and drag coefficient at the same time. The reduction of CFR shows unstable lift coefficient

fluctuation, and it keeps decreasing the drag coefficient of the structure. In can be seen the extending upper lobe of the caudal fin would enhance the thrusting ability of the structure.
Anyhow, from a general view point, when the structure remained static without any locomotion and adjustment of angle of attack of the fins, it tends to dive and slow down. Fish locomotion and swimming kinematics are required to be connoted to ensure an effective trusting and lifting.

4. Conclusion

As a conclusion, during static and stiff condition, the dorsal fin and pelvic fins provides lift to the structure while the other tends to dive down the structure. The entire caudal fin, except the adipose fin, the dorsal fin, pectoral fins and pelvic fins tend to drag and slow down the structure. When the Caudal Fin Ratio (CFR) keeps decreasing, it shows a drag reduction which would enhance the thrusting ability, but it shall be highlighted that it has unstable performance on the lift coefficient. More morphological configuration studies are recommended for deeper understanding of fins functions in the future.

5. Acknowledgements

The authors would like to thanks the Universiti Teknologi Malaysia and Ministry of Higher Education, Malaysia for sponsoring this research. This research is supported by research grant Vot. 4F114 and 4L074.

References

[1] Muljowidodo J S D, Budiyono A and Nugroho S A 2007 Recent progress on design development and testing of underwater vehicles at ITB[C] Workshop on Underwater System Technology, Putrajaya, Malaysia
[2] Lauder G V and Drucker E G 2004 *Journal of Oceanic Engineering* **29** 566.
[3] Chong X H, Zhang L P and Xin He 2012 *Comput. & Fluids* **68** 54.
[4] Peter B M and Jospeh J C 2004 Fishes: An Introduction to Ichthyology San Francisco United States: Pearson Benjamin Cummings
[5] Afiq W, Azwadi C S N and Saqr K M 2012 *International Journal of Mechanical and Materials Engineering* **7** 158.
[6] Tahseen T A, Ishak M and Rahman M M 2012 *Inter J Automot Mech Eng* **6** 753.
[7] Najiha M S, Rahman M M, Yusoff A R and Kadirgama K 2012 *Inter J Automot Mech Eng* **6** 766.
[8] Tahseen T A, Ishak M and Rahman M M 2013 *J Mech Eng Sci* **4** 418.
[9] Lam S Y, Shuaib N H, Hasini H and Shuaib N A 2012 *Inter J Automot Mech Eng* **6** 783.
[10] Syam Sundar L and Sharma K V 2011 *Inter J Automot Mech Eng* **3** 265.
[11] Durbin P A and Petterson R B A 2011 Statistical Theory and Modeling for Turbulent Flows United Kingdom: John Wiley and Sons Ltd
[12] ANSYS Incorporation 2011 ANSYS CFX-SolverTheory Guide USA: ANSYS Incorporation
[13] Xu Y G and Wan D C 2012 *J. Hydrodyn.* **24** 263.
[14] Tansley Triantafyllou M S, Triantafyllou G S and Gopalkrishnan R 1991 *Phys. Fluids A* **3** 2835.
[15] Koehl M A R 1984 *Am. Zool.* **24** 57.
[16] Emily A K and Timothy E H 2012 *Zoology* **115** 223.
[17] Wilga C D and Anabela M 2007 Functional morphology of dorsal fins in two shark species Kingston, RI, USA, University of Rhode Island

IOP Conf. Series: Materials Science and Engineering **50** (2013) 012014 doi:10.1088/1757-899X/50/1/012014

Experimental study of circle grid fractal pattern on turbulent intensity in pipe flow

B Manshoor[a], I Zaman, M F Othman and Amir Khalid
Centre for Energy and Industrial Environment Studies (CEIES),
Faculty of Mechanical and Manufacturing Engineering, Universiti Tun Hussein Onn Malaysia, 86400, Pt Raja, Batu Pahat, Johor, Malaysia

E-mail: [a]bukhari@uthm.edu.my

Abstract. Fractal turbulence is deemed much more efficient than grid turbulence in terms of a turbulence generation. In this paper, the hotwire experimental results for the circle grids fractal pattern as a turbulent generator will be presented. The self-similar edge characteristic of the circle grid fractal pattern is thought to play a vital role in the enhancement of turbulent intensity. Three different beta ratios of perforated plates based on circle grids fractal pattern were used in the experimental work and each paired with standard circle grids with similar porosity. The objectives were to study the fractal scaling influence on the flow and also to explore the potential of the circle grids fractal pattern in enhancing the turbulent intensity. The results provided an excellent insight of the fractal generated turbulence and the fractal flow physics. Across the circle grids fractal pattern, the pressure drop was lower but the turbulent intensity was higher than those across the paired standard circle grids.

1. Introduction

Since turbulent flows were discriminated from laminar flows by Reynolds [1] more than a century ago, they have been experimentally generated mostly using simple shapes. There have been attempts at understanding turbulence using fractal formalism [2] but until today, most of the grids used to generate turbulence rely mostly on classical geometry. However, recently a new class of turbulent flows is attracting interest [3-9]. These flows are based on geometrically complex, multiscale shapes, often following a fractal pattern. In parallel to these new experimental approaches there have been attempts at understanding how particular interactions would generate particular energy spectra but the researchers are still far from any complete theory relating fractal and spectral aspects of turbulence. The study here is related to this new trend of turbulent flows but the interest though still theoretical focuses more towards applications. The purpose of this study is to investigate the effect of a circle grids fractal pattern on the generation of turbulent flows in a pipe.

An orifice flowmeters is the most common form of differential pressure flowmeters used in industry [10]. The orifice flowmeters have become the most popular form of flowmeter due to its simple construction, low maintenance costs, simplicity and ease of installation, and its wide range of applications to different fluids, from liquid to gas. In addition, there is a great weight of experience to confirm its operations and this method have been documented in more detail in BS 1042 [11] and ISO 5167 [12]. Though the orifice plate as a differential flowmeter has gained an overwhelming popularity because of its low maintenance cost [13] compared to many other existing flowmeters, the inefficient mixing which it generates results in a larger than required pressure drop. To sustain the flow rate, the necessity to overcome the accumulating pressure drop caused by many orifice plates over the entire pipe network may become expensive. Ideally we would like the flow to recover as fast as possible in order to lower the energy cost associated to the pressure drop necessary for flow metering.

One way to speed up the flow recovery is to fill the pipe as fast as possible by improving the flow mixing, perhaps by a forced energy cascade. This is where practical interest meets the theoretical research on fractal-generated turbulence which can provide interesting alternatives for forcing turbulence. To tackle this problem from the engineering point of view, we developed fractal orifice plates, and studied their performance as turbulent generators.

2. Experimental work

2.1. Circle grids fractal

The work described here is based on a novel family of fractal space filling grids with circles. Figure 1 show the generation of the fractal space filling circle grids from the initiator (first iteration) to higher order iterations up to the fourth level of fractal pattern.

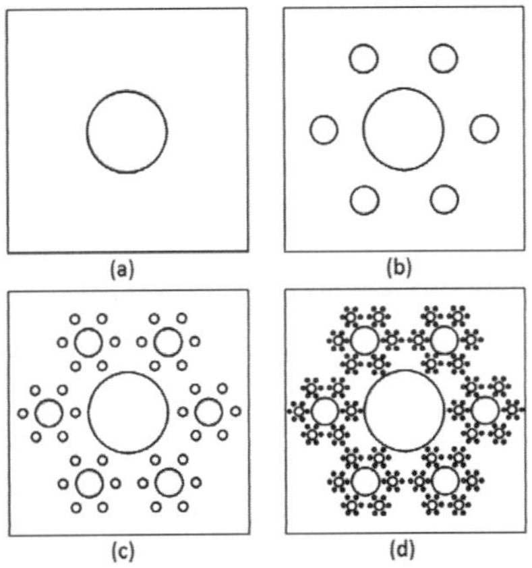

Figure 1. Fractal space filling circle grids.

As this pattern will be used as turbulent generators in a pipe flow, a small modification was made to the design to fit with the size and shape of the pipe. Starting from third iteration, the circle grids fractal orifices used for this work is shown in figure 2. The first orifice is equivalent to 14.45% porosity, followed by the second orifice with 25% porosity and the last is the orifice with 51.85% porosity.

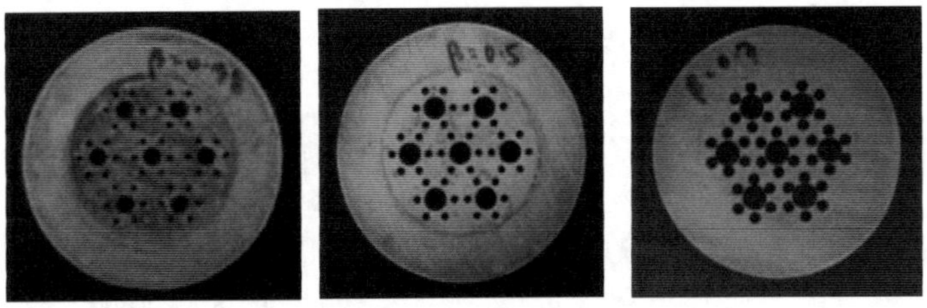

Figure 2. Circle grids fractal pattern.

2.2. Hot wire anemometry

Hot-wire anemometry is the most common method used to measure instantaneous fluid velocity. The technique depends on the convective heat loss to the surrounding fluid from an electrically heated sensing element or probe. A hot-wire anemometry measures fluid velocity by sensing changes in heat transfer from a small, electrically-heated sensor (hot-wire) exposed to the flow. Hot-wire anemometry is a well-established technique for fluid flow [14]. It works based on convective heat transfer from a heated sensor to the surrounding fluid, the heat transfer being primarily related to the fluid velocity. By using very fine wire sensors placed in the fluid and electronics with servo-loop technique, it is possible to measure velocity fluctuations of fine scale and of high frequencies.

When the wire is placed close to a solid wall, heat will be conducted through the flowing medium to the wall. If not corrected for this will cause the velocity to be measured too high. The wall influence starts at:

$$y+\leq \frac{yU\tau}{v} = 3.5 \tag{1}$$

Where, y is distance to the wall, $U\tau$ is friction velocity and **v** is kinematic viscosity. The critical wall distance is typically 0.1 to 0.2mm depending on free streaming [14]. The advantages of the CTA over other flow measuring principles are ease-of-use; the output is an analogue voltage, which means that no information is lost, and very high temporal resolution, which makes the CTA ideal for measuring spectra.

2.3. Experiment set-up

An experimental setup was established to measure the pressure drop after the circle grids fractal orifices as a function of the distance apart from the orifice position. The airflow in the test section was supplied through a open circuit tunnel with a cross section area of 0.00196 m^2 and a total length of 2.75 m. The inlet velocity was measured by a pitot tube located 15D downstream of the pipe entrance. The measurements were taken every 25 mm starting from 1.5D to 5D downstream of the circle grids fractal orifices, then in 50 mm increments to 8D and finally at the last location at 9.5D to give total of 12 measurement positions. A schematic drawing of the experimental set-up is shown in figure 3.

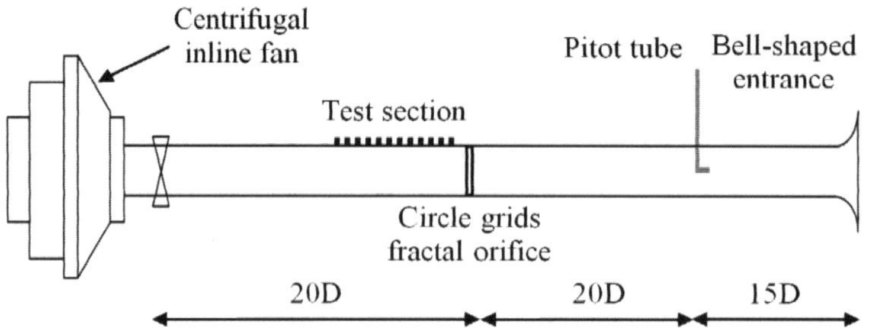

Figure 3. Schematic drawing of the experiment.

The flow was driven by a centrifugal inline fan with a speed controller. A bell mouth was used at the inlet of the pipe to ensure that the flow was fully developed in a short distance. Enough length has been allocated before the test section from the pipe entrance to ensure the flow is fully developed turbulent flow. To avoid the influence of the fluctuating flow through the measuring test section, more than 20 reading have been recorded per measuring station, and then the averaged was recorded.

The pressure holes were taped at fixed locations in the order 1/4 of the diameter starting from the orifice position and downstream of the flow in x-direction. The pressure taps were connected to a digital differential pressure manometer using plastic tubes of 4.5 mm in diameter. The digital differential pressure manometer is a microprocessor-based precision measuring instrument for low range differential pressures. It contains a highly sensitive differential pressure transducer and it displays the measured pressure in 11 different measuring units selected from the menu.

All the runs have been done at the same flow rate, in practice at the same control valve position. All circle grids fractal orifices are placed at the same position and several measured pressure drops, Δp, at each tape have been recorded. For each measured orifice, the orifice and the tap-hoses connection have been investigated in order to reduce any leakage. The flow is initiated in the pipe by operating the fan and then left to an appreciated time, transient filling time, to ensure that the pipe is filled completely by the fluid. Next, the measurement is taking using the digital micro-manometer and adjusted to ensure the flow rate is constant for all different circle grids fractal orifices.

3. Results and discussion

The test was divided in two sections and the first is measuring turbulent intensity while using the same porosity of plate but different Reynolds number. The second is measuring turbulent intensity while using the same Reynolds number but different porosity of plate.

3.1. Turbulent intensity with different Reynolds number

As in figure 4 to 6, with the increase of the position downstream the flow the turbulent intensity is reduced for all range of the Reynolds number. This may be explained as the circle grids fractal orifices have significant effect on the flow downstream the flow and enhance the mixing properties leading to this turbulent intensity reduction. From the nearest location downstream of the circle grids fractal pattern, the turbulence intensity decreases dramatically and then decreases slowly until it achieved constant value. The highest turbulence intensity generated is 0.95 by plate with $\beta = 0.38$ at the lowest Reynolds number, while the lowest turbulence intensity was generated is 0.52 by plate with $\beta = 0.7$ at the highest Reynolds number.

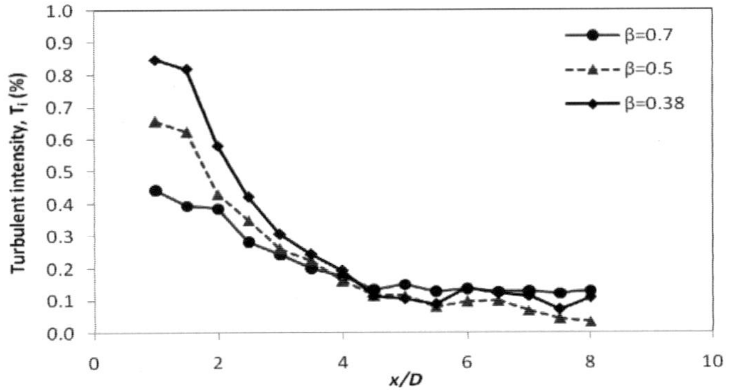

Figure 4. Turbulent intensity for Reynolds number 11528.

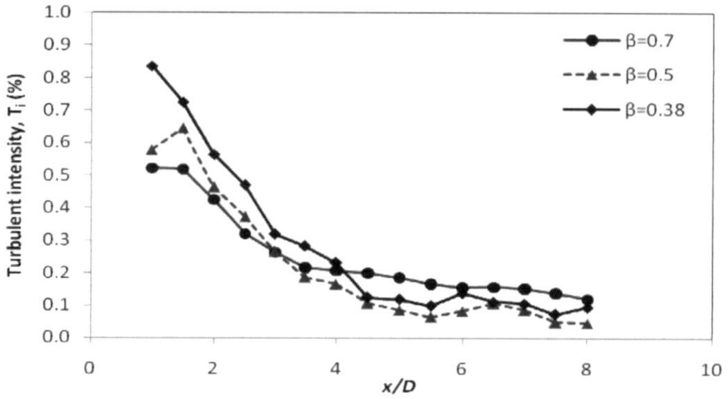

Figure 5. Turbulent intensity for Reynolds number 9722.

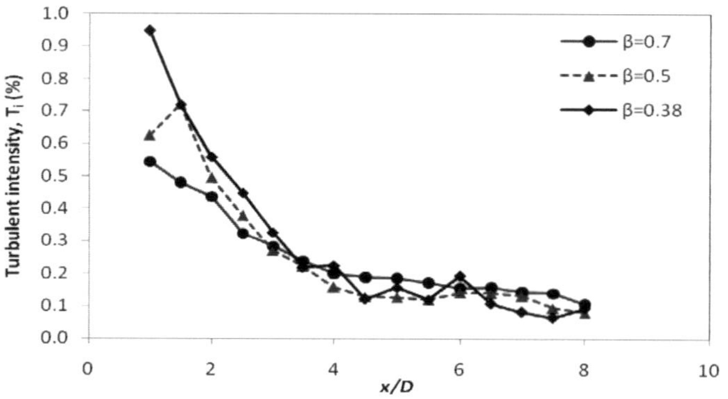

Figure 6. Turbulent intensity for Reynolds number 7917.

3.2. Turbulent intensity with different plate porosity

Another interesting result to discuss here are the relation between the turbulent intensity and the plate porosity. From the figure 7 to 9, the results show that with the increase of the porosity of the fractal plate, the turbulent intensity also increases. From 1D to 4D downstream of the plate, the turbulence intensity decreases dramatically and after the 4D downstream the decreasing of turbulent intensity starts to become constant.

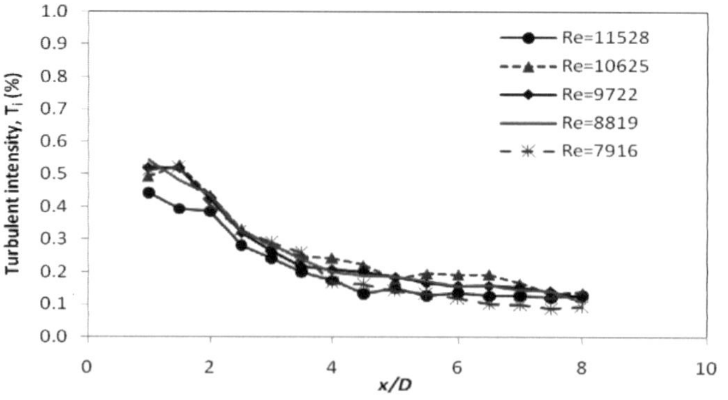

Figure 7. Turbulent intensity for plate porosity 14.15%.

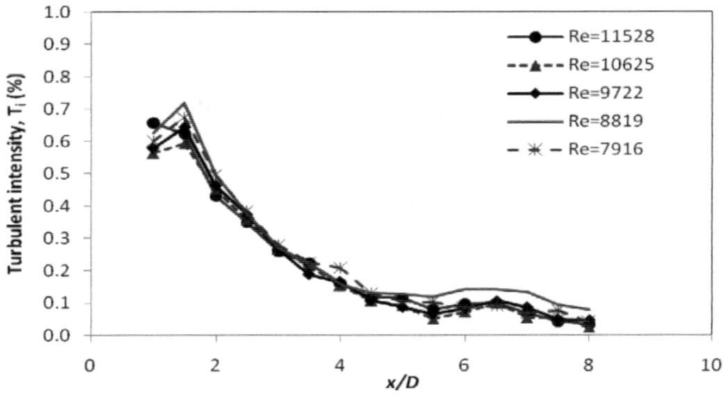

Figure 8. Turbulent intensity for plate porosity 25%.

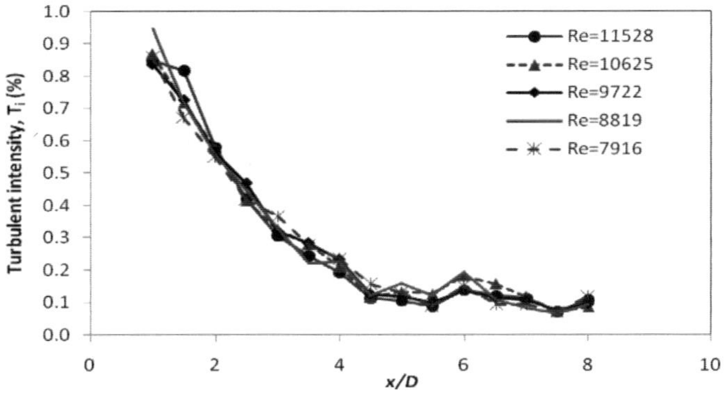

Figure 9. Turbulent intensity for plate porosity 51.85%.

4. Conclusion

For conclusion, all of the fractal patterns used in this research was found to be effective to generate turbulence flow. The turbulence flow can be measured by measuring the velocity of the air flow after through the fractal plate at multiple-point. But, the value of turbulence flow is different following the porosity of plate. Three of the plates were compared and it was obtained that the plate with higher porosity (51.85%) generated low turbulence intensity at nearly downstream. Otherwise, in range 1D to 4.5D the plate with low porosity (14.45%) generated higher turbulence intensity than other plate. But, the turbulence intensity was decreased dramatically at that range. In contrast with the plate with higher porosity, although it generated low turbulence in short distances downstream it decreases slowly until a long distance downstream.

Acknowledgments

The authors would like to thank to the Ministry of Higher Education, Malaysia (MOHE) and Universiti Tun Hussein Onn Malaysia (UTHM) for financial support under Fundamental Research Grant Scheme (FRGS) Vote 1056.

References

[1] Reynolds O 1883 An experimental investigation of the circumstances which determine whether the motion of water shall be direct or sinuous, and the law of resistance in parallel channels *Phil. Trans. R. Soc. London* **174** 935-982
[2] Meneveau C and Sreenivasan K R 1990 Interface dimension in intermittent turbulence *Physics review* 2246-2248
[3] Mohd Adib M A H, Mohd Hasni N N T N H and Osman K 2013 *J Mech Eng Sci* **4** 496.
[4] Vijaya Lakshmi B, Subrahmanyam T, Dharma Rao V and Sharma K V 2011 *Inter J Automot Mech Eng* **4** 455.
[5] Idris M S, Irwan M A M and Ammar N M M 2012 *J Mech Eng Sci* **2** 206.
[6] Naga Sarada S, Sita Rama Raju A V, Kalyani Radha K and Syam Sunder L 2012 *Inter J Automot Mech Eng* **6** 795.
[7] Tahseen T A, Ishak M and Rahman M M 2013 *J Mech Eng Sci* **4** 418.
[8] Tahseen T A, Ishak M and Rahman M M 2012 *Inter J Automot Mech Eng* **6** 753.
[9] Syam Sundar L and Sharma K V 2011 *Inter J Automot Mech Eng* **3** 265.
[10] Johnson D 2005 Differential flowmeters: Simple can be better Retrieved July, 2010 from http://www.manufacturing.net
[11] British Standard 1042 (1992). Measurement of fluid flow in closed conduits.
[12] ISO 5167 (1991). Measurement of fluid flow by means plates, nozzles and venturi tube inserted in circular cross section conduits running full.
[13] Miller R W 1989 Flow measuring engineering handbook London, McGraw Hill
[14] Bruun H H 1995 Hot-wire anemometry, Principles and signal analysis Oxford Science Publication.

IOP Conf. Series: Materials Science and Engineering **50** (2013) 012015 doi:10.1088/1757-899X/50/1/012015

The effect of different aspect ratio and bottom heat flux towards contaminant removal using numerical analysis

M N A Saadun[1,a], C S Nor Azwadi[2] Z A A Malek[2], M Z A Manaf[1], M S Zakaria[1] and M H M Hafidzal[1]

[1]Faculty of Mechanical Engineering, Universiti Teknikal Malaysia Melaka, Hang Tuah Jaya, 76100 Durian Tunggal,Melaka, Malaysia
[2]Faculti of Mechanical Engineering, Universiti Teknologi Malaysia, 81310 Skudai, Johor, Malaysia

E-mail: [a]asril@utem.edu.my

Abstract. Cubic Interpolated Pseudo-particle (CIP) numerical simulation scheme has been anticipated to predict the interaction involving fluids and solid particles in an open channel with rectangular shaped cavity flow. The rectangular shaped cavity is looking by different aspect ratio in modelling the real pipeline joints that are in a range of sizes. Various inlet velocities are also being applied in predicting various fluid flow characteristics. In this paper, the constant heat flux is introduced at the bottom wall, showing the buoyancy effects towards the contaminant's removal rate. In order to characterize the fluid flow, the numerical scheme alone is initially tested and validated in a lid driven cavity with a single particle. The study of buoyancy effects and different aspect ratio of rectangular geometry were carried out using a MATLAB govern by Navier-Stokes equation. CIP is used as a model for a numerical scheme solver for fluid solid particles interaction. The result shows that the higher aspect ratio coupled with heated bottom wall give higher percentage of contaminant's removal rate. Comparing with the benchmark results has demonstrated the applicability of the method to reproduce fluid structure which is complex in the system. Despite a slight deviation of the formations of vortices from some of the literature results, the general pattern is considered to be in close agreement with those published in the literature.

1. Introduction

Despite having various numbers of analytical methods of solving the Navier-Stokes Equation, it is still not sufficient enough to predict all flow models. The complexity of the flow in real engineering applications, such as in shaped pipe or combustor requires the CFD modelling to be used [1-6]. In studying the contaminant removal from rectangular shaped pipeline joints, there are also a lot of interactions between particle to particle and particle to fluid that attract the usage of CFD in solving these models. Therefore the study is focused on studying the real engineering problem of interest, which is the contaminant and deposits that accumulated in the pipeline joints. The methods in representing various sizes of gap at the joints is being modelled and represented by various aspect ratio of the cavity.

Previously, there were methods of cleaning the accumulated deposits. One of them is by using solvents. Nowadays, many researchers pointed out that the solvents used may be harmful and will cause corrosion and erosion to the pipeline material itself. The solvents used also might cause the end product to differ or contain alienated composition. Therefore the study is focused on replacing the solvents by using the natural flow to clear the deposits. Fluid flow with particle motion understanding is very important in natural and industrial applications. It is crucial for the environmental management and design and operation of such engineering equipment to comprehend the major mean particle-fluid flow phenomena, such as fluid-solid interaction and particles removal or mixing. Formation of recirculating vortices in cavities or steps can either enhance or prevent the removal of contaminant and depend on various flow parameters and characteristic of the contaminant itself. Tsorng [7] reported details experimental results on the behaviour of solid particles in lid-driven cavity flow from micro to macro size of particles. There are other researches done experimentally too [8-12]. These experimental efforts have added substantial knowledge to the state of the art of understanding on fluid-solid interaction. However, high accuracy of laser equipment together with high-speed digital image capture and data interpretation system is required to obtain reliable experimental data. Therefore as an alternative approach, many researchers considered fully computational scheme in their investigations [13-15]. From the behaviour of one particle in a lid-driven cavity flow to thousands of particles in expansion horizontal pipe has been studied in their research works sheds new hope in understanding this problem. Kosinski [16] applied the combination of continuum Navier-Stokes equations to predict fluid flow and second Newton's law for solid particle. Fang previously managed to predict the contaminant removal rate from a rectangular shaped cavity without heat effects [17]. Therefore this paper focused on the additional effects by the addition of heat, towards the removal rate with different aspect ratio of the cavity.

2. Physical model

Each of the specification and size of the model in figure 1 have been determined based on the previous modeling of the past study [18]. The length of the channel plate was A = 800mm to provide the laminar flow with the low Reynolds number of boundary layer. The thickness of the channel plate has eliminated because of the studies will done in two dimensional (2-D) analysis. For the height of the channel between the upper plate and downstream plate was B = 50mm the size is small for the height was to make sure that the flow created was the incompressible flow. Figure 1 shows that at the bottom of the cavity plate will have heat to give some heat convection effect on fluid flow in the cavity channel with interest geometry on horizontal plate.

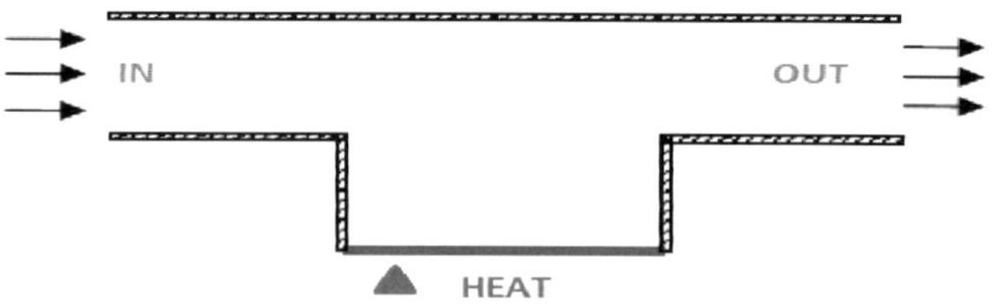

Figure 1. Schematic for square cavities with flow direction and haet effect at the bottom.

The size of the opening was based on the aspect ratio (AR) of the cavities used. The cavity aspect ratio is the ratio of the length over height (L/H) of the cavity dimension size as shown in figure 2. Four different aspect ratio of cavity have been selected to be the parameter for aspect ratio studies. The four different aspect ratio for this rectangular shape of cavity are AR = 1, AR = 2, AR = 3 and AR = 4. For the aspect ratio of cavity AR = 2, the opening will be L = 100mm. While for the aspect ratio of cavity AR = 3 and 4, the opening will be L = 150mm and 200mm respectively.

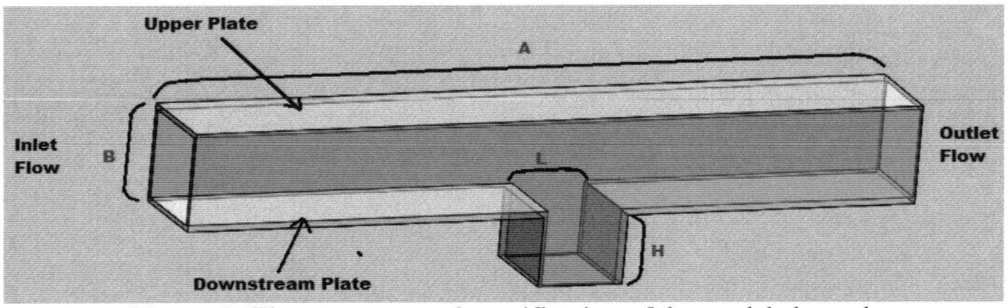

Figure 2. The geometry and specification of the model channel.

3. Mathematical formulation

In predicting the dynamics of solid particles in shear driven cavity flow, the equation of motion for solid particle is being expressed as

$$m_p \frac{dv_p}{dt} = f_p \quad (1)$$

where m_p, v_p and f_p are the mass of particle, its velocity and drag force acting on particle due to surrounding fluid respectively. According to Kosinski [16] et al, the drag force can be written as follows:

$$f_p = C_D A_p \rho \frac{|u - v_p|(u - v_p)}{2} \quad (2)$$

where A_p is the projected area of solid particle and C_D is the drag coefficient which is given as

$$C_D = \frac{24}{\mathrm{Re}_p} \quad (3)$$

The particle's Reynolds number in the above equation is calculated as follows

$$\mathrm{Re}_p = \frac{d_p |u - v_p|}{2} \quad (4)$$

where d_p is the diameter of solid particle.

In the computational technique, the new value of particles' velocity, v_p^{n+1} can be determined since the pre-calculated value of v_p^n is known at previous time step as follows

$$m_p \frac{v_p^{n+1} - v_p^n}{\Delta t} = f_p \quad (5)$$

Then the new position of solid particle can be determined as follows

$$x_p^{n+1} = v_p^{n+1} \Delta t + x_p^n \quad (6)$$

4. Numerical simulations

There are three main governing equations that govern the flow in the channel. The three equations are namely continuity, momentum and energy equations.

a) Continuity Equation

$$\frac{\partial u}{\partial x}+\frac{\partial v}{\partial y}=0 \tag{7}$$

b) Momentum Equations

$$\frac{\partial u}{\partial t}+u\frac{\partial u}{\partial x}+v\frac{\partial u}{\partial y}=-\frac{1}{\rho}\frac{\partial p}{\partial x}+\upsilon\left(\frac{\partial^2 u}{\partial x^2}+\frac{\partial^2 u}{\partial y^2}\right) \tag{8}$$

$$\frac{\partial v}{\partial t}+u\frac{\partial v}{\partial x}+v\frac{\partial v}{\partial y}=-\frac{1}{\rho}\frac{\partial p}{\partial y}+\upsilon\left(\frac{\partial^2 v}{\partial x^2}+\frac{\partial^2 v}{\partial y^2}\right)+g\beta(T-T_c) \tag{9}$$

c) Energy Equation

$$\frac{\partial T}{\partial t}+u\frac{\partial T}{\partial x}+v\frac{\partial T}{\partial y}=\frac{k}{\rho C_p}\left(\frac{\partial^2 T}{\partial x^2}+\frac{\partial^2 T}{\partial y^2}\right) \tag{10}$$

Vorticity-stream function approach is much easier since of the velocity variable were reduced. Vorticity and vorticity-stream equations

$$\omega=-\left(\frac{\partial^2\psi}{\partial x^2}+\frac{\partial^2\psi}{\partial y^2}\right) \tag{11}$$

$$\frac{\partial\omega}{\partial t}+u\frac{\partial\omega}{\partial x}+v\frac{\partial\omega}{\partial y}=\upsilon\left(\frac{\partial^2\omega}{\partial x^2}+\frac{\partial^2\omega}{\partial y^2}\right)+\frac{Gr}{Re^2}\frac{\partial\theta}{\partial x} \tag{12}$$

The most advantages reason for using dimensionless variables is the value is independent of dimension used, either metric or SI units. It is also to reduce the number of parameters pertaining to the physical situation and yet completely describes the behaviour of solution. Furthermore, many research papers present the variable in this way. From now on, any mentioned variable is considered in the dimensionless form. The followings list down all the dimensionless variables used in this study, and its derivation. Dimensionless variables used:

$$\Psi=\frac{\psi}{u_\infty D} \qquad \Omega=\frac{\omega D}{v} \qquad X=\frac{x}{D}$$

$$U=\frac{u}{u_\infty} \qquad V=\frac{v}{u_\infty} \qquad Y=\frac{y}{D}$$

$$\tau=\frac{tu_\infty}{D} \qquad \theta=\frac{T-T_C}{T_H-T_C}$$

For solving two dimensional Navier-Stokes equations using CIP method, it is important to give particular attention to both advection and non-advection parts of the equation. In this approach, the non-advection part is solved separately using Finite Difference while the advection part is solved using

CIP approach. In this section, CIP will be implemented base on the following advection equation hence, the advection part of NSE is:

$$\frac{\partial \Omega}{\partial \tau} + U\frac{\partial \Omega}{\partial X} + V\frac{\partial \Omega}{\partial Y} = 0 \tag{13}$$

$$\frac{\partial \theta}{\partial \tau} + U\frac{\partial \theta}{\partial X} + V\frac{\partial \theta}{\partial Y} = 0 \tag{14}$$

for the non-advection part solution the Finite Difference method should be used. It is then discretized by using Central Finite Difference. The process starts with the dimensionless momentum equation given as follows:

$$\frac{\partial \Omega}{\partial T} = g = \frac{1}{Re}\left(\frac{\partial^2 \Omega}{\partial X^2} + \frac{\partial^2 \Omega}{\partial Y^2}\right) + \frac{Gr}{Re^2}\frac{\partial \theta}{\partial X} \tag{15}$$

The corresponding $\partial_X g$ and $\partial_Y g$ for equation (15) are

$$\frac{\partial_X \Omega}{\partial T} = \frac{1}{Re}\left(\frac{\partial^3 \Omega}{\partial X^3} + \frac{\partial^3 \Omega}{\partial X \partial Y^2}\right) + \frac{Gr}{Re^2}\frac{\partial^2 \theta}{\partial X^2} - \frac{\partial U}{\partial X}\frac{\partial \Omega}{\partial X} - \frac{\partial V}{\partial X}\frac{\partial \Omega}{\partial Y}$$

$$\frac{\partial_Y \Omega}{\partial T} = \frac{1}{Re}\left(\frac{\partial^3 \Omega}{\partial X^2 \partial Y} + \frac{\partial^3 \Omega}{\partial Y^3}\right) + \frac{Gr}{Re^2}\frac{\partial^2 \theta}{\partial X^2} - \frac{\partial U}{\partial Y}\frac{\partial \Omega}{\partial X} - \frac{\partial V}{\partial Y}\frac{\partial \Omega}{\partial Y}$$

The cubic polynomial of square grid that is used to interpolate $\Omega(X,Y)$ within a grid cell $(i. j)$ $-$ $(i. j + 1)$ $-$ $(i + 1, j + 1)$ $-$ $(i + 1, j)$ is given by

$$\begin{aligned}\Omega_{i,j}(X,Y) &= [(A1_{i,j}\hat{X} + A2_{i,j}\hat{Y} + A3_{i,j})\hat{X} + A4_{i,j}\hat{Y} + \partial_X \Omega_{i,j}]\hat{X} \\ &+ [(A5_{i,j}\hat{Y} + A6_{i,j}\hat{X} + A7_{i,j})\hat{Y} + \partial_Y \Omega_{i,j}]\hat{Y} + \Omega_{i,j}\end{aligned} \tag{16}$$

Where $\hat{X} = X - X_{i,j}$ and $\hat{Y} = Y - Y_{i,j}$

The following parameters is incorporated in CIP method in determining the value of (X,Y)

$$\begin{aligned}\Omega_{i,j}(X,Y) &= [(A1_{i,j}\hat{X} + A2_{i,j}\hat{Y} + A3_{i,j})\hat{X} + A4_{i,j}\hat{Y} + \partial_X \Omega_{i,j}]\hat{X} \\ &+ [(A5_{i,j}\hat{Y} + A6_{i,j}\hat{X} + A7_{i,j})\hat{Y} + \partial_Y \Omega_{i,j}]\hat{Y} +\end{aligned}$$

$$A8_{i,j} = \Omega_{i,j} - \Omega_{i+1,j} - \Omega_{i,j+1} + \Omega_{i+1,j+1}$$
$$A1_{i,j} = \left[-2d_i + \partial_X(\Omega_{i+1,j} + \Omega_{i,j})\Delta X\right]/\Delta X^3$$
$$A2_{i,j} = \left[A8_{i,j} - \partial_X d_j \Delta X\right]/(\Delta X^2 \Delta Y)$$
$$A3_{i,j} = \left[3d_i - \partial_X(\Omega_{i+1,j} + 2\Omega_{i,j})\Delta X\right]/\Delta X^2$$
$$A4_{i,j} = \left[-A8_{i,j} + \partial_X d_j \Delta X + \partial_Y d_i \Delta Y\right]/(\Delta X \Delta Y)$$
$$A5_{i,j} = \left[-2d_j + \partial_Y(\Omega_{i,j+1} + \Omega_{i,j})\Delta Y\right]/(\Delta Y^3)$$
$$A6_{i,j} = \left[A8_{i,j} - \partial_Y d_i \Delta Y\right]/(\Delta X \Delta Y^2)$$
$$A7_{i,j} = \left[3d_i - \partial_Y(\Omega_{i,j+1} + 2\Omega_{i,j})\Delta Y\right]/\Delta Y^2$$

The solution for Ω, $\partial_X \Omega$ and $\partial_Y \Omega$ after period of ΔT can be estimated and after one time step of ΔT where $\Omega_{i,j}^{n+1} = \bar{\omega}_{i,j}(X_{i,j} - U\Delta T, Y_{i,j} - V\Delta T)$, $\partial_X \Omega_{i,j}^{n+1} = \partial_X \bar{\omega}_{i,j}$ and $\partial_Y \Omega_{i,j}^{n+1} = \partial_Y \bar{\omega}_{i,j}$ can be explicitly written as

$$\begin{aligned}\Omega_{i,j}^{n+1} &= [(A1_{i,j}\xi + A2_{i,j}\eta + A3_{i,j})\xi + A4_{i,j}\eta + \partial_X \Omega_{i,j}^*]\xi + \\ &+ \left[(A5_{i,j}\eta + A6_{i,j}\xi + A7_{i,j})\eta + \partial_Y \Omega_{i,j}^*\right]\eta + \Omega_{i,j}^*\end{aligned} \tag{17}$$

$$\partial_X \Omega_{i,j}^{n+1} = (3A1_{i,j}\xi + 2A2_{i,j}\eta + 2A3_{i,j})\xi + (A4_{i,j} + A6_{i,j}\eta)\eta + \partial_X \Omega_{i,j}^* \tag{18}$$

$$\partial_Y \Omega_{i,j}^{n+1} = (3A5_{i,j}\eta + 2A6_{i,j}\xi + 2A7_{i,j})\eta + (A4_{i,j} + A2_{i,j}\xi)\xi + \partial_Y \Omega_{i,j}^* \tag{19}$$

Where

$\xi = -U\Delta T, \eta = -V\Delta T$.

The following equations are used to determine the value of stream function, Ψ, U and V velocity.

$$\Psi_{i,j}^{n+1} = \frac{\dfrac{\Psi_{i+1,j}^n + \Psi_{i-1,j}^n}{(\Delta X)^2} + \dfrac{\Psi_{i,j+1}^n + \Psi_{i,j-1}^n}{(\Delta Y)^2} + \Omega_{i,j}^{n+1}}{\left(\dfrac{2}{(\Delta X)^2} + \dfrac{2}{(\Delta Y)^2}\right)} \tag{20}$$

$$U_{i,j}^{n+1} = \frac{\Psi_{i,j+1}^{n+1} + \Psi_{i,j-1}^{n+1}}{2\Delta Y} \tag{21}$$

$$V_{i,j}^{n+1} = \frac{\Psi_{i-1,j}^{n+1} + \Psi_{i+1,j}^{n+1}}{2\Delta X} \tag{22}$$

In summary, the evolution of the CIP scheme consists of three steps. The initial value of $\Omega^n_{i,j}$, $\Omega^n_{Xi,j}$ and $\Omega^n_{Yi,j}$ specified at each grid points and evolves by:

1. Since the pre-advected value of $\Omega^n_{i,j}$, $\Omega^n_{Xi,j}$ and $\Omega^n_{Yi,j}$ are known on each grid, the constrained interpolation process can be completed according to Eqs. (17), (18) and (19).
2. After the interpolation, advection takes place, and $\Omega^{n*}_{i,j}$, $\Omega^{n*}_{Xi,j}$ and $\Omega^{n*}_{Yi,j}$ are obtained.
3. The values of $\Omega^{n+1}_{i,j}$, $\Omega^{n+1}_{Xi,j}$ and $\Omega^{n+1}_{Yi,j}$ on the mesh grid are then computed from the newly advected values in step 2 by solving the non-advection phase of the governing equation. Then the interpolation and the advection processes are repeated.

5. Results and discussion

This section presents the numerical solutions of laminar forced convection heat transfer for a rectangular type of cavity with heat convection at the bottom wall. Figure 3 shows that the cavity with aspect ratio=1 produces lowest removal percentage at 8.85% followed by AR = 2 (33.6%), AR = 3 (54.25%) and AR = 4 produces highest removal rate (74.1%). For AR = 1 there is small vortex formed at top right corner, and moving towards to the centre. Since there is high number of particles packed in a small cavity, it caused the strength of the vortex weaken compared with the three others. For AR = 4 the particles waved out initially due to increased cavity width. This all is done by fixing the Reynolds number to 50 and Grashof number = 100.

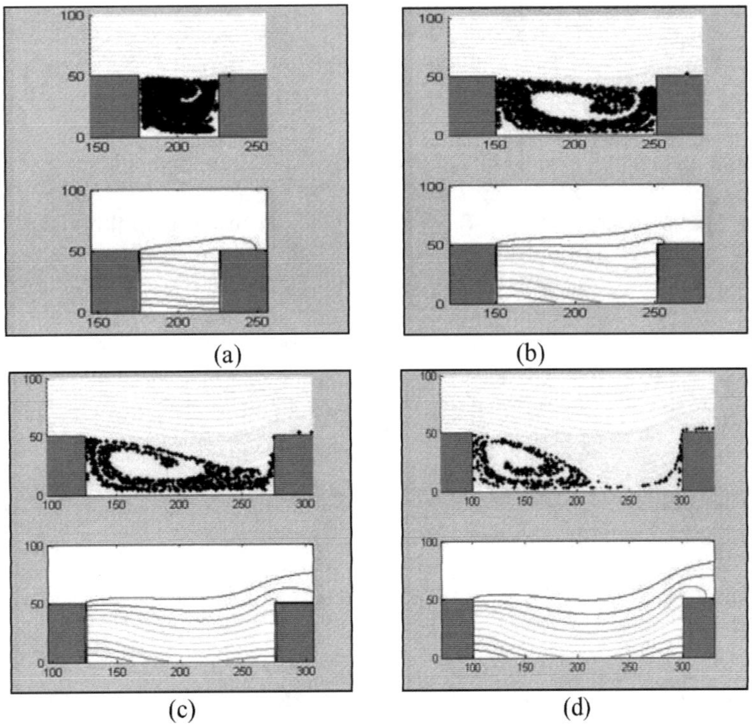

Figure 3. Streamline and isotherm pattern for Re=50 to Gr=100
(a) AR = 1, (b) AR = 2, (c) AR = 3 and (d) AR = 4

Figure 3 also shows the isotherm pattern that indicates the buoyancy effect in the cavity. It can be seen that when the aspect ratio change the pattern of isotherm also will change but not a big difference. The pattern becomes more fluctuation or waving when the aspect ratio is higher. It have been shows in the figure 3 (c) and 3 (d). This pattern directly will influence the movement of particle in the cavity and then produce less time for particle to flush out from the cavity. The time consume for each aspect ratio are different where until it reach at some period which is called steady state. At this time (steady state), the particle totally cannot be removed from the cavity due to the weaknesses of the vortex occur. This phenomenon indicates that the vortex formation is strong enough at the early stage of flow stream.

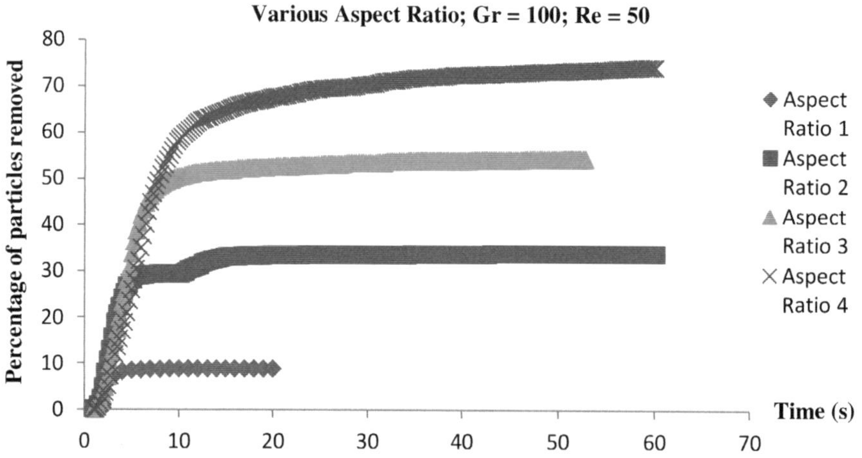

Figure 4. Comparison of removal rate AR1 to AR4

Figure 4 shows that the removal rate of higher AR generally is higher compared to lower AR and in this study. This is shown that the Reynolds number is affecting the initial unsteady state removal. It can be seen that the unsteady removal rate is much higher and it is in this region of time where the primary and secondary vortices will be formed. After the primary vortex stabilizes, then the effect of removal is due to the buoyancy force due to the temperature effect from the bottom. Generally, the higher the buoyancy force, removal rate will be higher, but specifically for higher Reynolds number, the buoyancy force and the counter-rotating vortices produced, bring the particles back to the bottom of the cavity, and this caused the removal rate at higher Reynolds number is not always higher compared to lower one at certain AR. Figure 5 shows the comparison of various Grashof number which is make the buoyancy effect towards contaminant's removal rate.

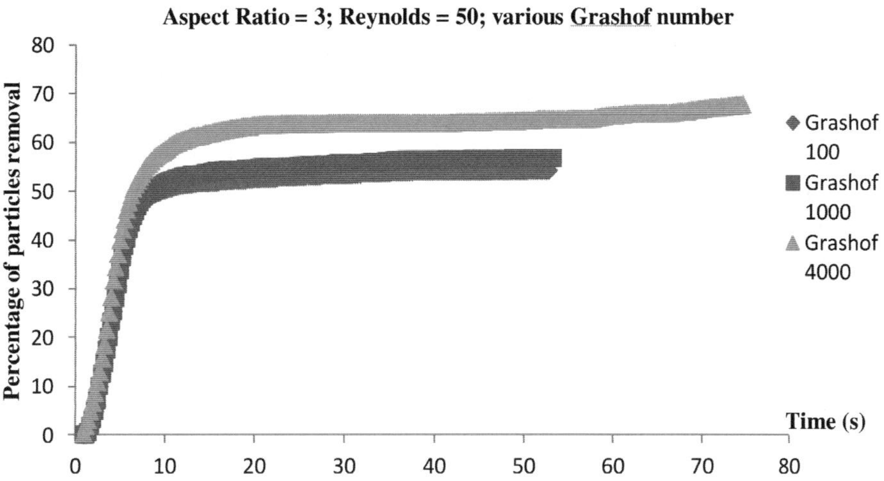

Figure 5. Comparison of various Grashof number towards particle's removal.

6. Conclusions

The simulated numerical result of the natural convection heat transfer in a square cavity at various aspect ratios shows the effectiveness of CIP-NSE method as one of the appropriate model to be applied in studying the fluid behaviour. The comparison results between the previous methods are made to validate the accuracy of the CIP-NSE method throughout this study. In this case, the CIP method is sufficiently able to solve the advection phase equation based on Navier-Stokes equations. CIP provides an alternative method in obtaining to solution for advection phase equation, however, the non-advection phase only can be solved through finite difference method. The implementation of CIP in finite difference method might not be available, thus it causing an inaccuracy for the numerical simulation. Therefore, it can be concluded that the CIP-NSE is a particular useful method to be applied in the convection heat transfer problem.

There are various variable in this study with the main one is the cavity aspect ratio, followed by different values of Reynolds and Grashof numbers. The percentage of removal is mainly affected by variation of aspect ratios, followed by Reynolds and Grashof number. The study also managed to show results with addition of temperature effects towards the rate of removal and it can be seen that the percentage of removal is increased with addition of heat for most of the cases.

Acknowledgements

The authors wish to thank to Universiti Teknikal Malaysia Melaka (UTeM) and Center of research and Innovation Management (CRIM) through research grant PJP/2012/FKM(9A)/S01084 for supporting and funding this research activities. The authors also thanks to Kementerian Pengajian Tinggi (KPT) and Universiti Teknologi Malaysia (UTM) for sponsorship study and facilities provided.

References

[1] Tahseen T A, Ishak M and Rahman M M 2013 *J. Mech. Eng. Sci.* **4** 418
[2] Mohd Adib M A H, Mohd Hasni N N T N H and Osman K 2013 *J. Mech. Eng. Sci.* **4** 496
[3] Tahseen T A, Ishak M and Rahman M M 2012 *Inter. J. Automot. Mech. Eng.* **6** 753
[4] Najiha M S, Rahman M M, Kamal M, Yusoff A R and Kadirgama K 2012 *J. Mech. Eng. Sci.* **3** 340
[5] Wandel A P, Noor M M and Yusaf T F 2012 *Inter. J. Automot. Mech. Eng.* **6** 731

[6] Najiha M S, Rahman M M, Yusoff A R and Kadirgama K 2012 *Inter. J. Automot. Mech. Eng.* **6** 766
[7] Tsorng S J, Capart H, Lo D C, Lai J S and Young D L 2008 Behaviour of macroscopic rigid spheres in lid-driven cavity flow *Intl. J. Multiphase Flow* **34** 76-101
[8] Ilea C G, Losinski P, Hoffmann A C 2008 Three-Dimensional simulation of a dust lifting process with varying parameters *Intl. J. Multiphase Flow* **34** 869-878
[9] Azwadi C S and Takahiko T 2006 Simplified Thermal Lattice Boltzmann in Incompressible Limit *Intl. J. Mod. Phys.* B **20** 2437-49
[10] Martys N S and Chen H 1996 Simulation of multicomponent fluids in complex three dimensional geometries by the lattice Boltzmann method *Phys. Rev.* A **53** 743-750
[11] Frish U, Hasslacher B and Pomeau Y 1986 Lattice Gas Automata for the Navier-Stokes equation *Phys. Rev. Lett.* **56** 1505-09
[12] McNamara G and Alder B 1993 Analysis of the lattice Boltzmann treatment of hydrodynamic *Phys.* A **194** 218-228
[13] Ghia U, Ghia K Nand Shin C Y1982 High-Re solutions for Incompressible Flow using the Navier-Stokes Equations and a Multigrid Method *J. Comp.Phys.* **48** 387-411
[14] Nor Azwadi C S , Attarzadeh S R 2011 An accurate numerical prediction of solid particle fluid flow in a lid-driven cavity *Intl. J. Mech.* **5** 123-128
[15] Fang L C et. al. 1999 Transient removal of a contaminated Fluid from a cavity, *Inter. J. of Heat and Fluid Flow* **20** 605-613
[16] Kosinski P, Kosinska A and Hoffmann A C 2009 Simulation of solid particles behaviour in a driven cavity flow, *Powder Tech.* **191** 327-339
[17] Lih-Chuan Fang 2002 Effect of mixed convection on transient hydrodynamic removal of a contaminant from a cavity *Inter. J. of Heat and Mass Transfer* **46** 2039–49
[18] Faure T M, Adrianos P and Lusseyran F O and Pastur L 2006 Visualizations of The Flow Inside an Open Cavity at Medium Range Reynolds Numbers

IOP Conf. Series: Materials Science and Engineering **50** (2013) 012016 doi:10.1088/1757-899X/50/1/012016

Slip flow coefficient analysis in water hydraulics gear pump for environmental friendly application

A A Yusof[a], F Wasbari, M S Zakaria and M Q Ibrahim
Faculty of Mechanical Engineering, [2]Faculty of Engineering Technology, Universiti Teknikal Malaysia Melaka, Hang Tuah Jaya, 76100 Durian Tunggal, Melaka, Malaysia.

E-mail: [a]anas@utem.edu.my

Abstract. Water hydraulics is the sustainable option in developing fluid power systems with environmental friendly approach. Therefore, an investigation on water-based external gear pump application is being conducted, as a low cost solution in the shifting effort of using water, instead of traditional oil hydraulics in fluid power application. As the gear pump is affected by fluid viscosity, an evaluation has been conducted on the slip flow coefficient, in order to understand to what extent the spur gear pump can be used with water-based hydraulic fluid. In this paper, the results of a simulated study of variable-speed fixed displacement gear pump are presented. The slip flow coefficient varies from rotational speed of 250 RPM to 3500 RPM, and provides volumetric efficiency ranges from 9 % to 97% accordingly.

1. Introduction

The use of water hydraulics promotes possibility in designing fluid power systems with sustainable and environmental friendly elements [1]. In this context, water hydraulics can be understood as a technology, which is using tap water or water-based fluids for transmission of energy and power [2]. The aim of using water as the hydraulic medium is to transfer the energy, power and the resource sustainably, at a rate that does not compromise the natural environment. The use of non-renewable resources, such as fossils fuel, natural gas and coal has to be decreased, in promoting green technology for a better future. This includes the use of hydraulic oil, which is acquired from petroleum-based mineral oil. The use of water is vital in developing sustainable fluid power system. Simultaneously, water characteristics involving hygiene, safety and low maintenance cost provide fascinating perspectives for design engineers due to concerns over hydraulic fluid disposal, contamination, costly maintenance and flammability. The impressive uniqueness of energy and power transmission has also brought the technology to wider scopes of applications. The challenging problems of the low lubricity and highly corrosive properties of water have been overcome with the recent development of new materials, particularly ceramics and the machining of parts to greater tolerances. All these factors have contributed to increase hardware costs when using water [3]. However, water is significantly different from oil, which at the same time, in many ways, can provide advantages in one aspect, while producing disadvantages in other aspects. Fortunately, these technical difficulties have been solved effectively by using special materials, coatings and designs [4]. Water hydraulics can offer a design for hygiene solution in various industries, as demonstrated by the development of water hydraulics driven burger machine, beef cutter and ice filled machine in the food processing industry, and the use of environment-friendly waste packer lorry [5]. Water hydraulics can also be applied in industrial water cleaning application, die-castings, robotics, humidification equipment's and fire protection and fighting systems [6]-[8]. Therefore, based on the above reality, this paper investigates the effect of using water on a modified external gear pump system. Previously, a test was conducted at Centre of Advanced Research on Energy, Universiti Teknikal Malaysia Melaka. The test concentrated on the operation of a hydraulic scissor lift, by replacing the hydraulic oil with reverse osmosis water. It was noted that the hydraulic pump driven scissor lift managed to lift up to 400 kg of load, with workable water temperature of 41.4°C [9]. Unfortunately, no simulation was conducted in order to understand the effectiveness of the system in a wider range of flow rates. Thus, improved experimental and simulation study is being conducted in order to determine the effect of water hydraulics system on typical spur gear pump, especially the slippage under the influence of various loading pressure and slip flow coefficient.

2. Mathematical Modeling and Analysis

In general, external gear pumps produce flow by carrying fluid in between two meshing gears. One gear is driven by the drive shaft and turns the idler gear. Pumping chambers are formed between the gear teeth, which are enclosed by the pump housing and side plates, as shown in figure 1. The clearances between gear faces, gear tooth crests and the housing create an almost constant loss or slippage in any pumped volume at a fixed pressure. Volumetric efficiencies of gear pumps run as high as 90%. At low speed, the volumetric efficiency is poor, thus for an optimum volumetric efficiencies, the gear pumps should be run close to their optimum rotational speeds.

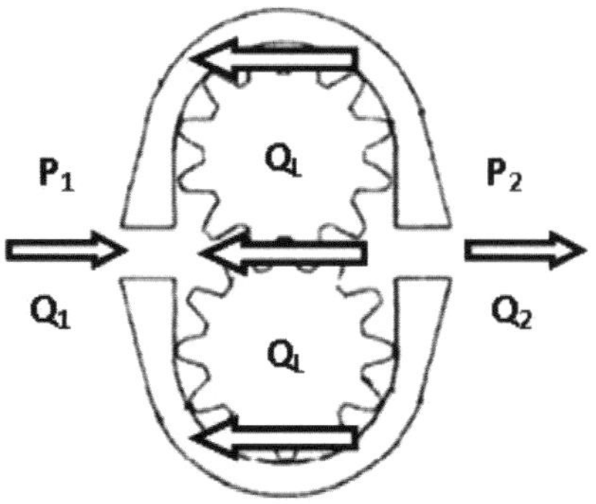

Figure 1. The flow and pressure of a gear pump.

The delivery or flow rate of an external gear pump can only be calculated by identifying the volumetric displacement of the pump. In the gear pump, one simple method of knowing the pump displacement is by multiplying the area of the chamber that displaced the hydraulic fluid, with the gear teeth width of the pump. Thus, the volumetric displacement of a gear pump, V_D can be represented by,

$$V_D = \frac{\pi}{4}\left(D_o^2 - D_i^2\right)L \tag{1}$$

where D_o, D_i, and L is referred to outside diameter, inside diameter and the width of the gear teeth. The ideal flow rate of the gear pump, Q_I, can be represented by,

$$Q_1 = V_D N \tag{2}$$

where N is the rotational speed of the prime mover. Table 1 shows the parameter for the simulated pump.

Table 1. Parameter for Simulated Pump

Parameter	Dimension
Outside Diameter	4.0 cm
Inside Diameter	3.6 cm
Gear Width	1.4 cm
Displacement	3.34 cm^3/rev
Flow Rate[a]	3340 cm^3/min

[a] Flow Rate at 1000 RPM

For any positive displacement pump, there is an inlet chamber that is increasing in volume and a discharge chamber that is decreasing. There may be leakage directly from the high pressure discharge chamber to the low pressure inlet chamber. There may also be discharge from both chambers to the drain. The volumetric efficiency of a pump can be influence by the leakage, with specific operating conditions. Any slippage or the leakage through the running clearances can increase with pressure, but at constant amount, as rotational speed and flow rate changes [11]. Because the clearances are so small, leakage flow is treated as laminar, thus,

$$Q \propto \Delta P \tag{3}$$

therefore, the outlet to inlet leakage or slip flow, Q_L, is shown as,

$$Q_L = SF\,(P_2 - P_1) \tag{4}$$

where SF is the slip flow coefficient, and P_2 and P_1 are the pump's outlet and inlet pressure. The changes of the pressure can be represented by a variable load pressure, P_L.

$$P_L = (P_2 - P_1) \tag{5}$$

By applying the principle of continuity to the inlet and outlet pipe, and taking account the slip flow, the actual flow rate, Q_2, can be represented by,

$$Q_2 = Q_1 - Q_L \tag{6}$$

where Q_1 is the ideal flow rate. Therefore,

$$Q_2 = V_D N - SF(P_L) \tag{7}$$

and the expression for volumetric efficiency, η_V, can be stated as,

$$\eta_V = 1 - \frac{SF(P_L)}{V_D N} \tag{8}$$

3. Results and Discussion

The simulated value of volumetric efficiency, with respect to slip flow coefficient, SF, is presented in figure 2 to figure 7. The efficiency is represented with various pressures, ranging from 30 bar, 60 bar, 90 bar, 120 bar, 160 bar, and finally 260 bar. The data is also represented with various rotational speeds, ranging from 0 to 4000 RPM. All figures show the trend of the volumetric efficiency, with slip flow coefficient of —◆— 0.005, —■—0.025, —▲—0.05 and —■— 0.1. Basically, the slip flow coefficient is increased from 5, 10 and 20 times of the original SF = 0.005. The simulated data is recorded for an external, spur gear pump, having theoretical volumetric displacement of 3.34 cm^3/rev. Figure 2 shows that the value of simulated volumetric efficiency at loading pressure P_L of 30 bar, for slip flow coefficient, SF = 0.005. It is noted that the efficiency is around 80% to 98%, from 250 RPM up to 3500 RPM. The efficiency for the pump decreases when the slip flow coefficient is increased 5 times the original value, into SF = 0.025. At the same loading pressure, the efficiency is around 9% to 93%, ranging from 250 RPM up to 3500 RPM. At SF = 0.05, or 10 times the original value, the efficiency declines from 9% to 87%, which is recorded from 500 RPM to 3500 RPM. At 20 times the original value (SF = 0.1), the volumetric efficiency is around 9% to 74%, and it is only measurable from 1000 RPM to 3500 RPM.

In figure 3, the value of simulated volumetric efficiency changes significantly when the loading pressure increases to 60 bar. It is noted that for slip flow coefficient, SF=0.005, the efficiency is around 63% to 97%, from 250 RPM up to 3500 RPM. The efficiency for the pump decreases when the slip flow coefficient is increased to SF = 0.025. At the same loading pressure, the efficiency is around 9% to 87%, ranging from 500 RPM up to 3500 RPM. At SF = 0.05, the efficiency decreases from 9% to 74%, which is recorded from 1000 RPM to 3500 RPM. At SF = 0.1 (20 times the original value), the volumetric efficiency is around 9% to 48%, and it is only measurable from 2000 RPM to 3500 RPM. From this figure, it shows that at a typical rotational speed of 1750 RPM, no efficiency can be recorded with the use of pumps having slip flow coefficient of 0.1. Therefore, if the slip flow coefficient represents the use of water hydraulics in gear pumps, it would be acceptable if a higher RPM is used. And even if the highest RPM is possible, for example around 3500 RPM, it will only provide volumetric efficiency around 48%.

In figure 4, the changes can be seen more clearly when the loading pressure increases to 90 bar. For slip flow coefficient, SF = 0.005 and SF = 0.025, the efficiency is around 45% to 96%, and around 9% to 80%, from 250 RPM up to 3500 RPM and from 750 RPM to 3500 RPM respectively. The efficiency for the pump decreases when the slip flow coefficient is increased to SF = 0.05 and SF = 0.1. At the same loading pressure, the efficiency is ranging from the efficiency is around 9% to 61%, and around 9% to 22%, from 1500 RPM up to 3500 RPM and from 3000 RPM to 3500 RPM respectively. In figure 5 and figure 6, where the pressure is set at 120 bar and 160 bar, the volumetric efficiency for slip flow coefficient SF = 0.01 is out of range. This means that, if the slip flow coefficient represents the use of water hydraulics in gear pumps, the system cannot be used for loading pressure of 120 bar onwards. For slip flow coefficient of SF = 0.005 and SF = 0.025, with the same pressure of 120 bar and 160 bar, and within selected rotational speed that ranges from 2000 RPM to 3000 RPM, the volumetric efficiency are around 90% to 93%, and 54% to 69%, respectively. In figure 7, where the maximum pressure of 260 bar is set, the volumetric efficiency for slip flow coefficient SF = 0.05 and SF = 0.1 is out of range. For the same selected rotational speed of 2000 RPM to 3000 RPM, the volumetric efficiency are around 80% to 86%, and 2% to 34%, for slip flow coefficient of SF = 0.005 and SF = 0.025, respectively.

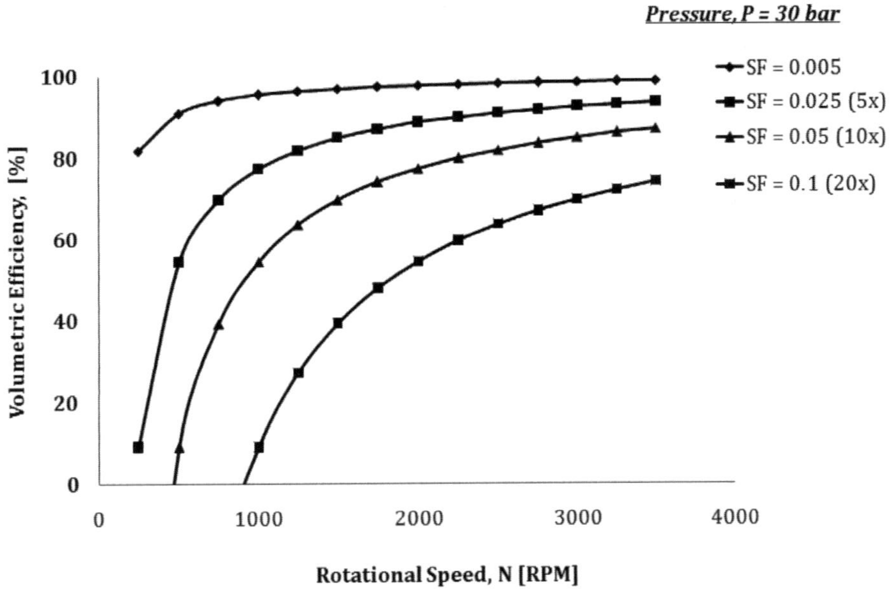

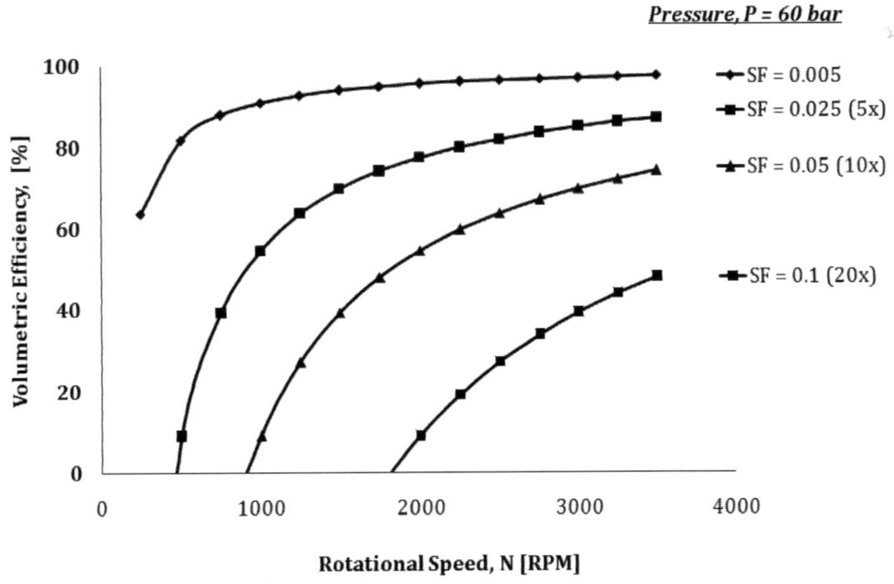

Figure 3. Volumetric efficiency at loading pressure of 60 bar.

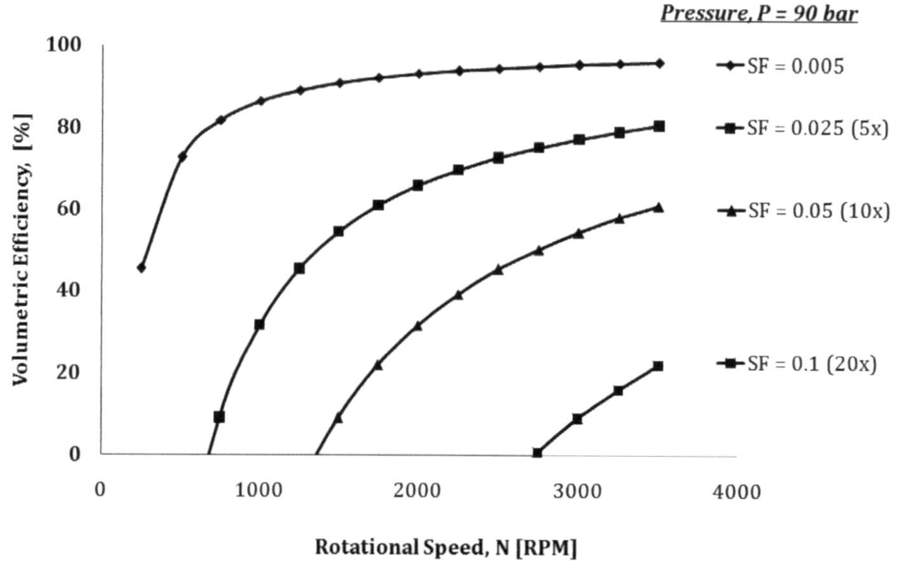

Figure 4. Volumetric efficiency at loading pressure of 90 bar.

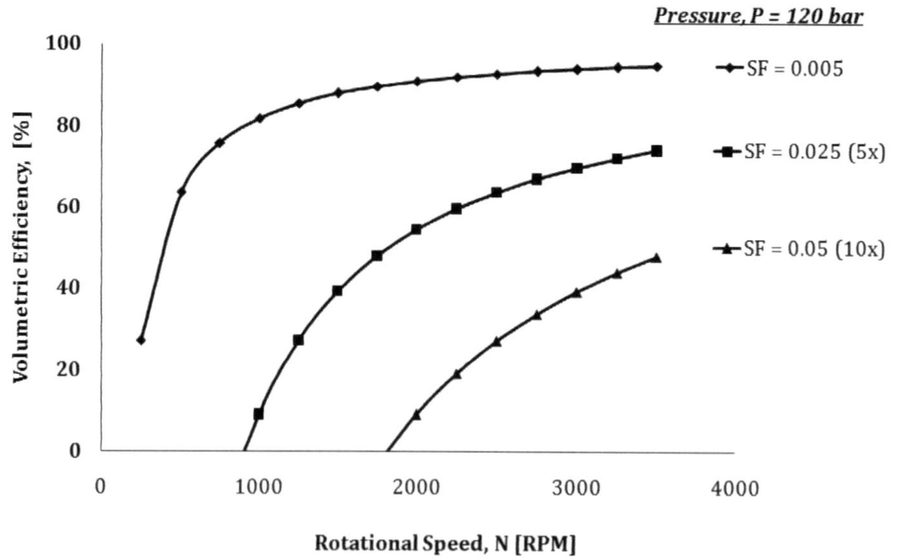

Figure 5. Volumetric efficiency at loading pressure of 120 bar.

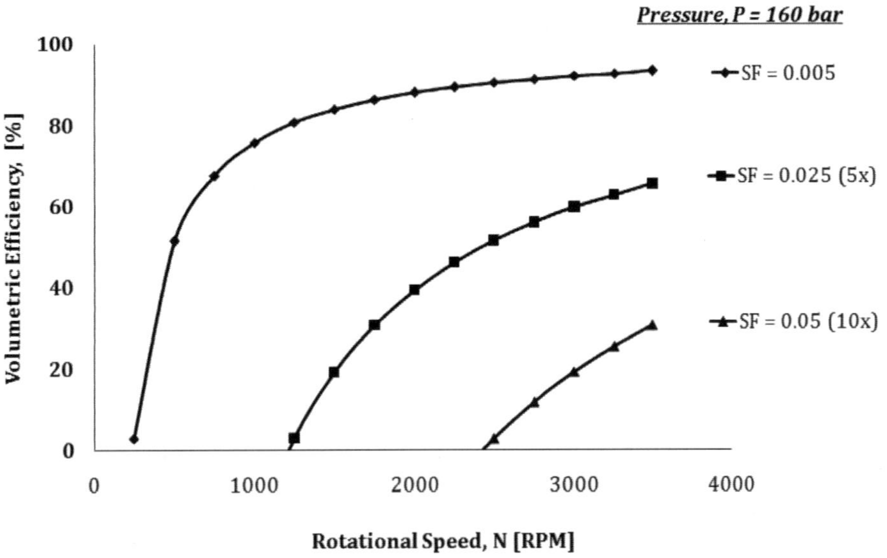

Figure 6. Volumetric efficiency at loading pressure of 160 bar.

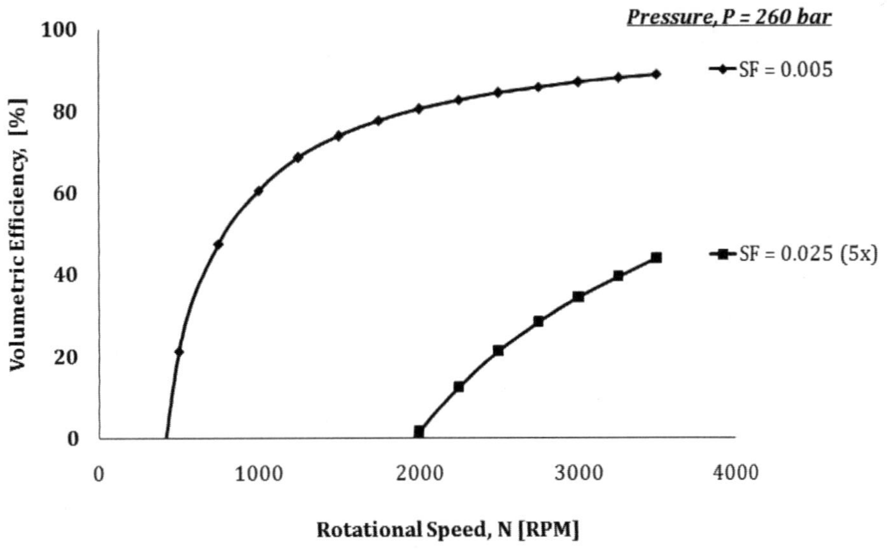

Figure 7. Volumetric efficiency at loading pressure of 260 bar.

4. Conclusion

The study on rotary gear pump for water hydraulics application is being conducted at the Centre of Advanced Research on Energy, University Teknikal Malaysia Melaka. The objective of the study is to provide variety information regarding to the use of low viscosity, sustainable water hydraulics in external, spur gear pump. In this study, the focus is limited to the slippage in between the outlet and inlet of the pump. The result suggests that the use of water is possible, if the slip flow coefficient is at constant value 0.005, with system pressure of 30 to 260 bar. The effectiveness varies from rotational speed of 250 to 3500 RPM, and provides volumetric efficiency ranges from 9 % to 97% accordingly. The result suggests that water can be applied in typical traditional oil hydraulic gear pump, and this is true if an acceptable slip flow coefficient value can be achieved.

Acknowledgement

The research work is supported under the short term project (PJP/2012/FKM(15A)/S01090). The authors wish to thank Universiti Teknikal Malaysia Melaka for their financial support.

References

[1] Koskinen K T, Leino T Riipinen H 2008 Sustainable development with water hydraulics – Possibilities and challenges *Proc.of the 7th JFPS Int. Symp. on Fluid Power* Toyama Japan 11 – 18
[2] Trostmann E, Frolund B, Olesen B H and Hilbrecht B 2001 Tap Water as a hydraulic pressure medium (New York: Marcel Dekker)
[3] Backe W 1999 Water or oil hydraulics in the future *Proc. of the 6th Scandinavian Int. Conf. on Fluid Power* Tampere Finland 51 – 65
[4] Hilbrecht B 2000 Water as a pressure medium in water hydraulics *Proc. of the 48th National Conf. on Fluid Power* Chicago Illinois USA 555 – 559
[5] Conrad F 2005 Trends in design of water hydraulics – Motion control and open-ended solutions *Proc. of the 6th JFPS Inter. Symp. on Fluid Power* Tsukuba Japan 420 – 430
[6] Higgins M 1996 Water hydraulics – The real world *Industrial robot: An Int. J.* **23** 13 – 18
[7] Ikeo S, Nakashima H and Ito K 2008 Water hydraulics system for high speed cylinder drive *Proc. of the 7th JFPS Int. Symp. on Fluid Power* Toyama Japan 95 – 100
[8] Lim G H, Chua P S K and He Y B 2003 Modern water hydraulics – The new energy transmission technology in fluid power *Applied Energy* **76** 239 – 246
[9] Yusof A A, Mat S and Din AT 2013 Promoting sustainability through water hydraulics technology – The effect of water hydraulics in industrial scissor lift *Applied Mechanics and Materials* **315** 488 – 492
[10] Esposito A 2003 *Fluid power with application* (New York: Pearson)
[11] Merritt H 1956 *Hydraulic control systems* (New York: McGraw Hill)

The influence of gouge defects on failure pressure of steel pipes

N A Alang[a], N A Razak and M R Zulfadli
Corrosion and Fracture Focus Group, Faculty of Mechanical Engineering, Universiti Malaysia Pahang, 26600, Pekan, Pahang.

E-mail: [b]azuan@ump.edu.my

Abstract. Failure pressure of API X42 steel pipes with gouge defects was estimated through a nonlinear finite element (FE) analysis. The effect of gouge length on failure pressure of different pipe diameters was investigated. Stress modified critical strain (SMCS) model was applied as in predicting the failure of the pipe. The model uses strain based criteria to predict the failure. For validation of the model, the FE results were compared to experimental data in literature showing overall good agreement. The results show that the gouge length has significant influence on failure pressure. A smaller pipe diameter gives highest value of failure pressure.

1. Introduction

Since few years ago, pipelines become the most preferred medium for oil and gas transportation. After period of time in services, number of cases of pipe failure has been reported [1-3]. The failure of the pipes is due to the reduction of wall thickness that may cause by corrosion phenomenon and gouge defect. The gouge on pipe surface normally occurred during installation of the pipes in which the collision between the pipes was occurred. Study [4] has shown that the defect on the pipes give significant effect on the failure pressure. In this respect, huge number of mathematical models has been developed to assess the remaining strength of pipelines containing defect [5-8]. Most of the models use stress based failure criterion to predict the failure of the pipes. However, these models sometimes were underestimate or too conservative [9]. Another models use in predicting the failure is based on local strain criterion. Oh et al [10] uses strain based criterion which is SMCS model to predict the failure pressure of API X65 steel pipes. The results published by Oh et al [8] show that the SMCS model is capable to predict the failure pressure for both pipes with corrosion and gouge defects. Oh et al [10] has determined the failure pressure of API X65 steel pipes with constant diameter. Mathematically, the SMCS model is evaluated by equation (1) that is express by:

$$\varepsilon_f = A\exp\left(-\frac{3\sigma_m}{2\sigma_e}\right) \tag{1}$$

where ε_f is fracture strain, σ_m / σ_e is stress triaxiality and A is the material constant that can be found through experiment.

In this work, the effects of gouge length on failure pressure of API X42 steel pipes with different pipe diameters were investigated. The pipe with gouge defect was modeled and analyzed using MSC PATRAN/MARC 2008r1 software. The SMCS model was applied to estimate the failure pressure of the steel pipes.

2. Material

The material used in this study was API X42 steel. The chemical compositions of the material have been identified using spectrometer machine. The uniaxial tensile test was performed according to ASTM E08-2008 [11]. The chemical compositions and mechanical properties of the material are tabulated in table 1 and table 2, respectively. Figure 1 shows the true plastic stress-strain curve obtained from tensile test.

Table 1. Chemical composition of API X42 steel (%wt).

C	P	Mn	S	Si	Fe	Ceq
0.03	0.01	0.98	0.003	0.19	98.6	0.21

Table 2. Mechanical properties of API X42 steel at room temperature.

Young Modulus, E (GPa)	Poisson Ratio, v	Yield Strength, σ_y (MPa)	Tensile Strength, σ_u (MPa)
207	0.3	284.7	464.4

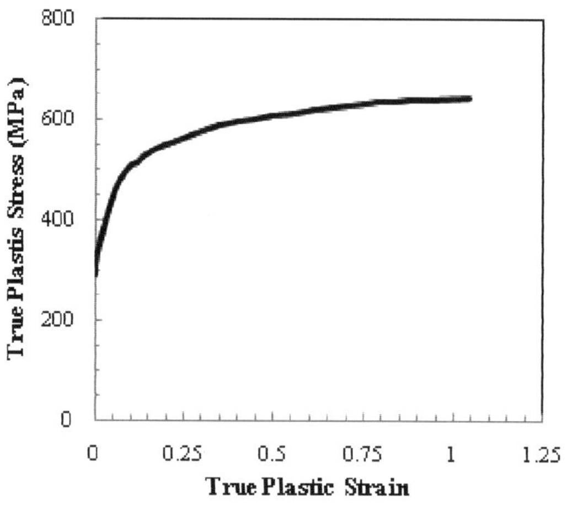

Figure 1. True plastic stress-strain data employed in FE analysis.

3. Finite Element Analysis

A 3-D nonlinear FE analysis was performed. The pipe with gouge length, l and outer diameter, D was modelled using eight node solid elements, isotropic elastic-plastic material with large deformation using MSC PATRAN/MARC 2008r1 software. Reduced integration scheme is applied in all simulation works. The simulation was performed with fully utilized the symmetrical condition for computational efficiency. Therefore, only a quarter of the pipe was simulated. The internal pressure was applied at the inner surface of the pipe. The schematic illustration of the pipe with gouge is shown in figure 2. The total lengths of the pipe, L gouge depth, d and wall thickness, t were kept constant to be 2300 mm, 8.75 mm and 17.5 mm, respectively. The gouge is characterized by the 45 degree V-notch with the radius of 2 mm. Figure 3 shows the detail FE mesh applied on the gouge defect. Since the failure is assumed will be occurred at gouge defect, the FE mesh is applied sufficiently small at that particular region.

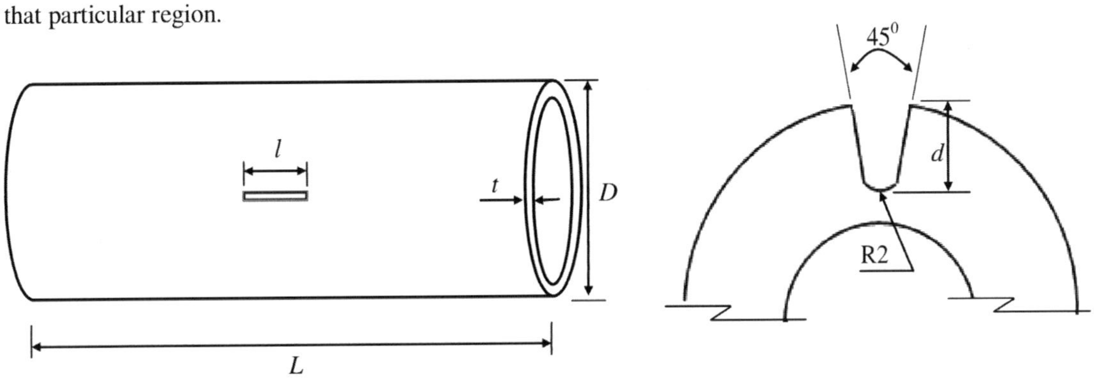

Figure 2. Schematic of pipe with gouge.

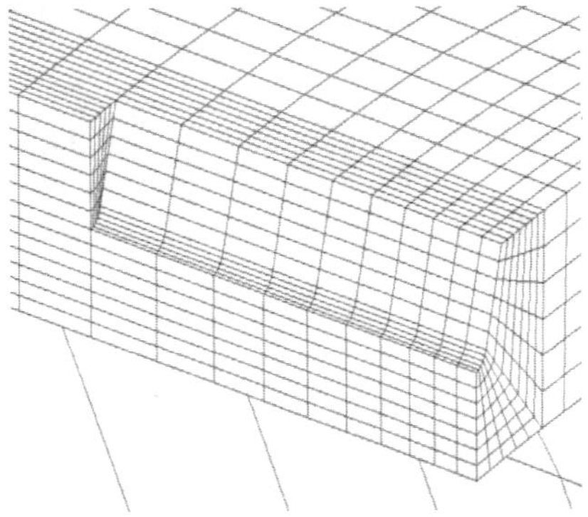

Figure 3. Detail FE mesh on gouge defect.

In order to determine the failure pressure of the defective pipe, the local stress strain results was collected. The stress triaxiality and equivalent strain were calculated for the entire loading history. The failure is assumed to occur when the equivalent strain is equal or greater than fracture strain of the pipe material. Recently, the author [9] has developed the SMCS model for API X42 steel pipes. In mathematics, the model can be written as:

$$\varepsilon_f = 1.732\exp\left(-1.5\frac{\sigma_m}{\sigma_e}\right) \tag{2}$$

4. Validation of Model Parameter

The model has been validated by comparing the FE results coupled with SMCS model with experimental data from literature. The small scale burst pressure test had been performed on pipes with rectangular artificial defect and been summarized in table 3. The pipes were pressurized by hydraulic oil and failure pressures were experimentally determined. Detail on the experimental works can be found in literature [12]. Figure 4 shows the failure pipe after the test. The maximum error between these two methods is 9%.

Table 3. Comparison of failure pressure between FE and experiment.

Pipe No.	Defect Length, *l* (mm)	Experiment (MPa)	FEA (MPa)	Deviation (%)
A1	50	54	57.62	6.7%
A2	70	46	50	8.7%

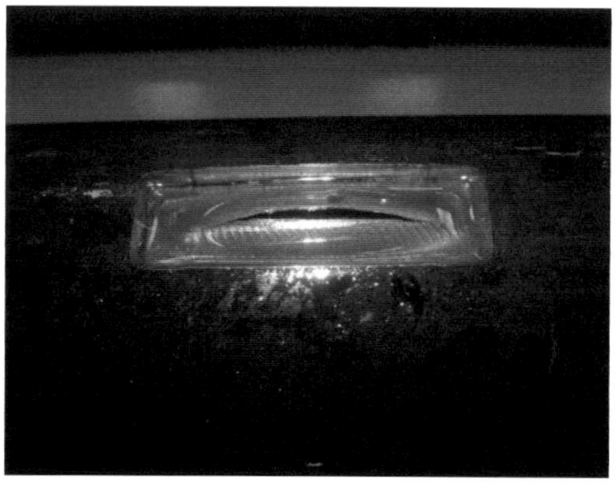

Figure 4. Failure pipe after the test.

5. Result and Discussion

A total 9 cases of pipe parameters were simulated that represent by PN1 to PN9. Three different gouge lengths were analyzed for each pipe outer diameter. The outer diameter, *D* that has been selected in this study is 508 mm, 762 mm and 1016 mm. Table 4 summarizes the parameters for each case including the predicted failure pressure results. Figure 5a presents the effect of gouge length on failure pressure of the API X42 steel pipes. Figure 5a clearly shows that the failure pressure decreases as the gouge length increases. There is no significant change on failure pressure value for the pipe with outer diameter of 762 mm and 1016 mm. However, when the diameter of the pipe changes from 508 mm to 762 mm, the failure pressure significantly decreases. It is due to the ratio of wall thickness to the pipe diameter. For the pipe with outer diameter of 508 mm and 762 mm, the ratio of wall thickness to diameter is 3.44% and 2.29%, respectively. Meanwhile, for pipe with outer diameter of 1016 mm, the wall thickness to pipe diameter ratio is 1.72%. The graph of failure pressure as a function of this ratio was plotted and shown in figure 5b. The relationship between these two parameters can be represented by linear regression demonstrate a strong correlation. Figure 6 shows the von Mises stress distribution on gouge defect area for pipe PN4. This figure also shows that the bulging phenomenon was occurred at the onset of pipe bursting.

Table 4. Summary of the failure pressure results.

Case	Pipe Diameter, (mm)	*t/D* (%)	Defect Dimension, (mm)		Failure Pressure (MPa)
			Length, *l*	Depth, *d*	
PN1	508	3.44	100	8.75	28.8
PN2			200		25.2
PN3			300		23.6
PN4	762	2.29	100		19.2
PN5			200		18.0
PN6			300		17.0
PN7	1016	1.72	100		16.8
PN8			200		15.0
PN9			300		14.2

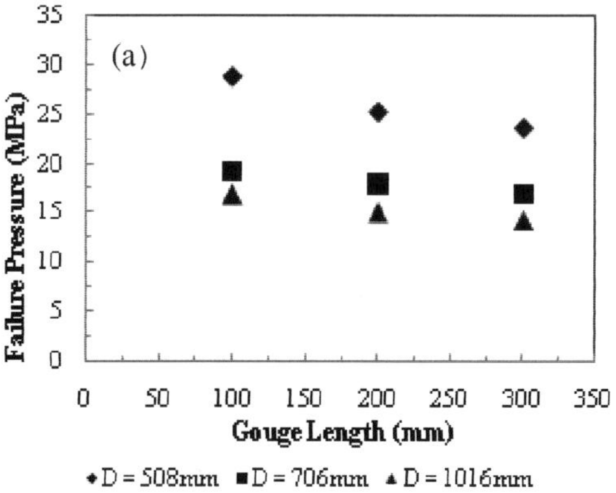

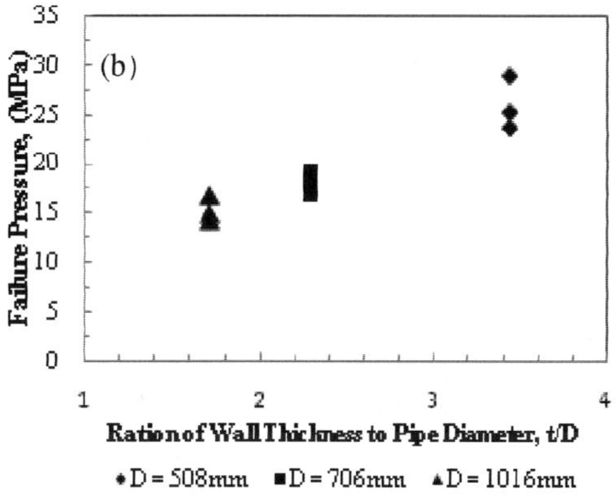

Figure 5. Failure pressure for different pipe diameter: (a) effect of gouge length, (b) effect of wall thickness to pipe diameter ratio.

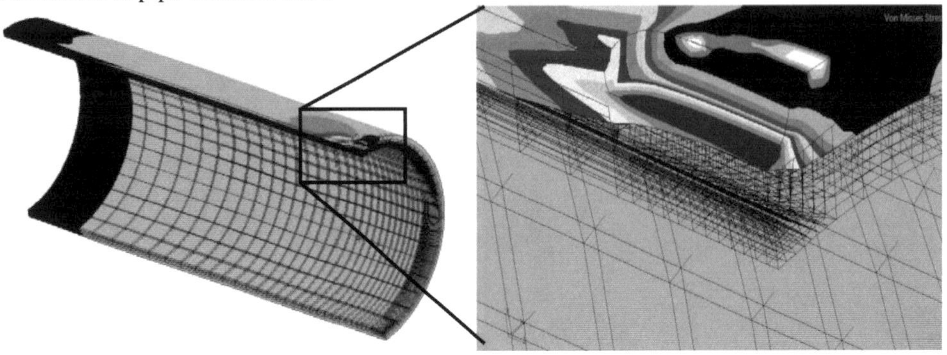

Figure 6. von Mises stress distribution and enlargement on gouge defect.

6. Conclusion

This paper has presented the effect of gouge length on failure pressure of API X42 steel pipes. The conclusions of the study are as follow:

1) The failure pressure of API X42 pipe was influenced by the length of the gouge defects.
2) The failure pressure dropped significantly when the pipe diameter reduced from 508 mm to 762 mm. However, predicted failure pressure for pipe diameter of 762 mm and 1016 mm is slightly different. It is due to highest deviation of wall thickness to pipe diameter ratio between pipe diameter 508 mm and 762 mm.

Acknowledgement

The authors would like to express appreciation to all members of Corrosion and Fracture Focus Group (CFRAC) for technical support. Not forgotten to the sponsors of the research, University Malaysia Pahang, Malaysia (RDU100312).

References

[1] MacDonald K A, Cosham A, Alexander C R and Hopkins P 2007 Assessing mechanical damage in offshore pipelines – Two case studies *Engineering Failure Analysis* **14** 1667–79

[2] Liu P F, Zheng J Y, Zhang B J and Shi P 2010 Failure analysis of natural gas buried X65 steel pipeline under deflection load using finite element method *Materials and Design* **31** 1384–91

[3] Hossam A and Kishaway 2010 Review of pipelines integrity management practices International *Journal of Pressure Vessel and Piping* **87** 373-380

[4] Netto T A, Ferraz U S and Estefen S F 2005 The effect of corrosion defects on the burst pressure of pipelines *Journal of Constructional Steel Research* **61** 1185–1204

[5] DNV Recommended Practice RP-F101 2010 Corroded pipelines *Det Norske Veritas*

[6] American Petroleum Institute 2000 API RP579 – Recommended practice for fitness-for-service. American Petroleum Institute

[7] Kamal M, Rahman MM and Rahman A G A 2012 *J Mech Eng Sci* **3** 291

[8] Kamal M, Rahman M M and Rahman A G A 2013 *Inter J Automot Mech Eng* **7** 912.

[9] Alang N A, Razak N A, Safie K A and Sulaiman A 2012 Finite element analysis on burst pressure of defective steel pipes *European Conference on Fracture* Kazan Russia

[10] Oh C K, Kim Y J, Baek J H and Kim W S 2007 Ductile failure analysis of API X65 pipes with notch-type defects using a local fracture criterion *International Journal of Pressure Vessels and Piping* **84** 512–525

[11] ASTM Standard E8-08 2008 Standard test method for tension testing of metallic material American and Society for Testing and Materials

[12] Alang N A, Razak N A, Safie K A and Sulaiman A 2013 Finite element analysis on burst pressure of steel pipes with corrosion defects *International Conference on Fracture* Beijing China

2nd International Conference on Mechanical Engineering Research (ICMER 2013) IOP Publishing
IOP Conf. Series: Materials Science and Engineering **50** (2013) 012018 doi:10.1088/1757-899X/50/1/012018

The effects of bolted joints on dynamic response of structures

I Zaman[1,a], A Khalid[1], B Manshoor[1], S Araby[2] and M I Ghazali[1]
[1]Faculty of Mechanical and Manufacturing Engineering, Universiti Tun Hussein Onn Malaysia, 86400 Batu Pahat, Johor, Malaysia
[2]School of Advanced Manufacturing & Mechanical Engineering, University of South Australia, SA 5095, Australia

E-mail: [a]izzuddin@uthm.edu.my

Abstract. Joint is an universal fastening technology for structural members; in particular bolted joints are extensively used in mechanical structures due to their simple maintenance and low cost. However, the components of bolted joints are imperative because failure could be catastrophic and endanger lives. Hence, in this study, the effects of bolted joints on vibrating structures are investigated by determining the structural dynamic properties, such as mode shapes, damping ratios and natural frequencies, and these are compared with the monolithic structures (welding). Two approaches of experimental rigs are developed: a beam and a frame where both are subjected to dynamic loading. The analysis reveals the importance of bolted joints in increasing the damping properties and minimizing the vibration magnitude of structures, this indicates the significant influence of bolted joints on the dynamic behaviour of assembled structures. The outcome of this study provides a good model for predicting the experimental variable response in different types of structural joints.

1. Introduction

Joints, in most engineering structures are considered vital part due to its functionality to unite and assemble the structural members. This result in using a choice of fastening techniques such as bolting, welding and riveting [1,2]. Although joint is known to introduce flexibility to the structure, its behaviour when subjected to a dynamic loading is still not fully understood. In most cases, vibration is the most critical problem occurred in the mechanical structures which concerns to joint failure. The structural vibration causes the components to move in rapid linear motion, as a consequent produces a harmful effect mainly to the torque of joints. This awareness makes most people in particular structural engineers to realize the importance of joint especially in a giant construction where a failure instigates catastrophic and people death.

Until today, much knowledge and research about joints has been discovered and performed. For example, Ahmadian and Jalali [3] identified the parameters of bolted joints in assembled structures through experiment and analytical solution. In their study, the lateral stiffness and torsional stiffness of assembled structures were identified using the first three natural frequencies whilst the damping coefficient was identified using a nonlinear response function [4-5]. Similar study was performed by Ma et al. [6] who compared the dynamics structure of beam with and without a joint through experimental approach. Three levels of joints tightening were distinguished to investigate non-linear effects and loosening effects occurred from the damping of joint. The experimental result indicated that this approach was found applicable to every bolted structure to know their parameters.

In order to accurately model the properties of joints, researchers calculated the frequency response function (FRF) data using the finite element analysis [7-11]. The assembly structure was divided into three systems: (i) first was substructure system, (ii) second was joint system and (iii) the third was the whole assembly system. Regardless it is difficult to get the accurate calculation because FRF data were contaminated by measurement errors. Kim et al. [12] introduced four finite element models of bolted joints, such as a solid bolt model, a coupled bolt model, a spider bolt model and a non-bolt model in order to investigate the accurateness of these modelling techniques. The models were analysed using ANSYS.

Although there are many research in the past have been devoted in the structural dynamics system, until today there are still lacking of studies in getting the accurate parameters of the bolted joints effects in particular. Aforementioned it is important to know and understand the behaviours of joint ant its effect on dynamic environments. Therefore, this study aims to evaluate and describe the effects of joints using the experimental rig testing. The experimental is carried out using modal testing—where impact hammer is excited through beams and frames structure in order to get the structural dynamic characteristics.

2. Research Methodology

2.1. Sample Preparation

There are two assembly structures were analysed in the study, which are a beam and frame structures. Both beam and frame were having two types; monolithic structure, where components are merged using welding method, and jointed structure, where components are merged using bolt and nut (See figure 1). The structures were weighted equivalently to avoid errors. In the bolted structure, 12 mm diameter hexagon bolts were employed. Figure 2a shows the dimension of beam structure which given as 5.2×2.5×0.635 cm while figure 2b shows the dimension of frame structure at 26.7×25.4×0.635 cm.

Figure 1. Monolithic structure and bolted-joint structure.

2.2. Design of Experiment

For the modal testing, the structure was suspended by 1 m length of elastic rope in order to stimulate a free-free boundary condition as shown in figure 3. An impact hammer was used to excite the structure at all points (11 points for beam and 30 points for frame) and an accelerometer was fixed at point 1 to accept acceleration response from the structures. The force and acceleration data were collected through data analyser which later converting to FRF using Dynamic Signal Analyzer software.

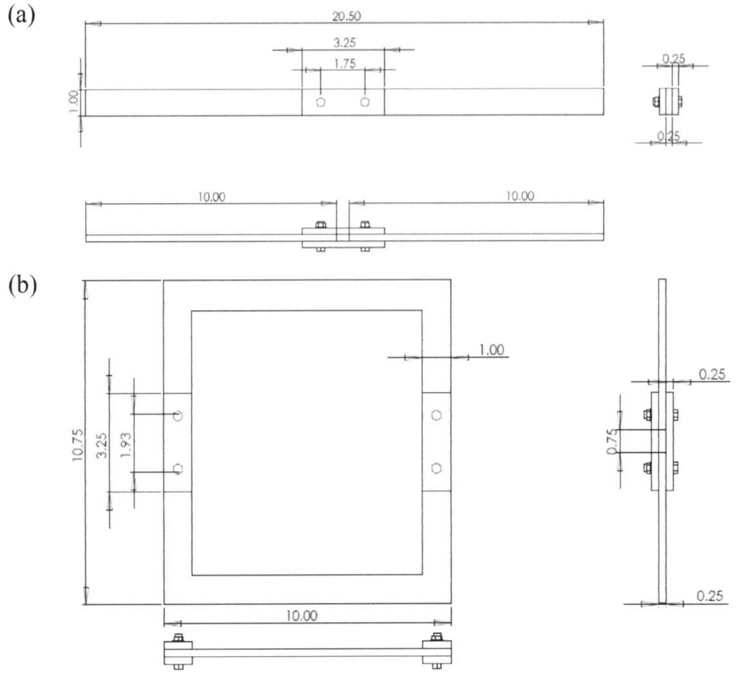

Figure 2. Dimensions of (a) beam structure (b) frame structure from front, top and side view.

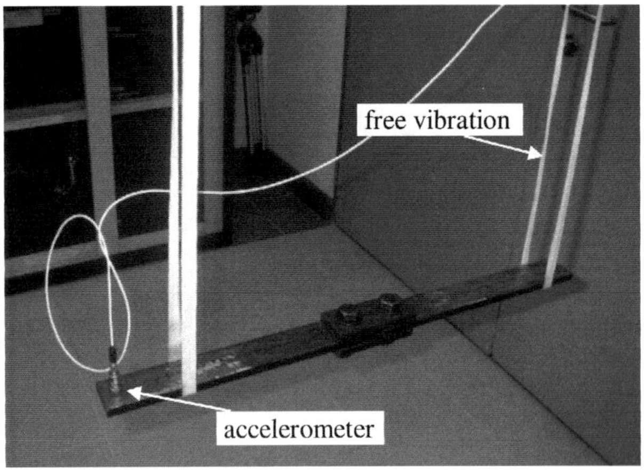

Figure 3. Typical experimental set-up for impact hammer test.

3. Results and discussion

3.1. Analysis of Beam

Table 1 tabulates the results of natural frequencies and damping ratios of each type of beams that obtained from experimental modal testing. It can be seen that the natural frequencies increase from each mode within the range of 0–2000 Hz. The bolt-jointed beam, however, shows reduction of natural frequencies about 1–15% compares to monolithic structure. By contrast, the damping ratios increased significantly in jointed beam, where the highest increment recorded was ~440% at mode three.

Table 1. Structural dynamic properties of beams.

Mode	Monolithic beam		Jointed beam	
	Nat. freq. (Hz)	Damp. ratios	Nat. Freq (Hz)	Damp. ratios
1	189	1.77	159	1.75
2	444	0.224	437	0.166
3	1010	0.306	868	1.66
4	1420	0.222	1320	1.15

These results were further analysed using FRF where the graph log magnitudes were plotted against frequencies (see figure 4a). It can be seen that the maximum peak in the third mode (1010 Hz) is shifted about 14%, and 7% for mode four (1420 Hz). Since the mass of both structures are similar, the only parameter that could affect natural frequency is stiffness. The stiffness of bolt-jointed beam was found smaller than monolithic beam due to the gaps at the lap joints. The magnitude of jointed beam was also found smaller where it decays about 77% and 85% for mode three and four, respectively. This result revealed that the bolted joints can reduce vibration in assembly structure and the hypothesis is supported by higher damping factor as shown in table 1. Similar result was observed by (Ma et al., 2001) when they compared monolithic beam and jointed beam in their experiment. The corresponding mode shape of the beam at mode one is illustrated in figure 5. It was found that the highest residue magnitude occurred at both end sides of the beam, where the lowest is observed in bolt-jointed beam.

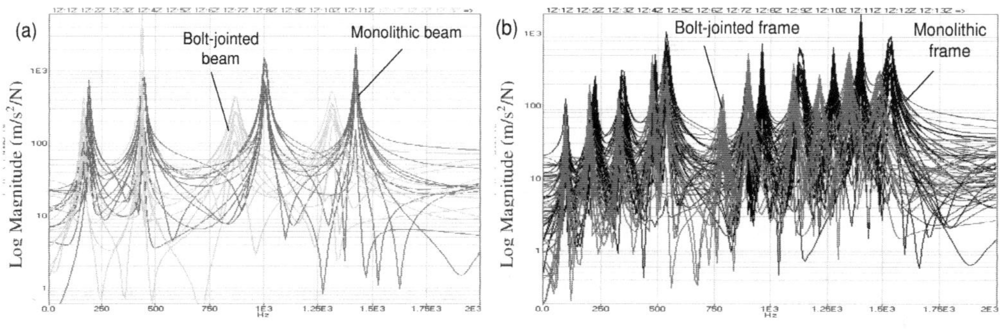

Figure 4. Graph log magnitude vs. frequency for (a) beam and (b) frame.

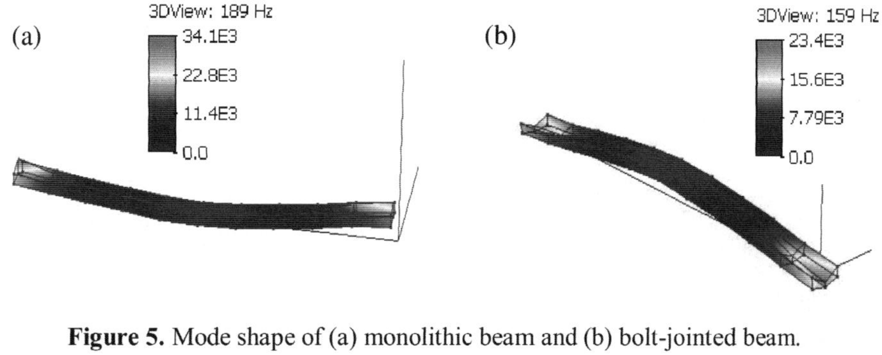

Figure 5. Mode shape of (a) monolithic beam and (b) bolt-jointed beam.

3.2. Analysis of Frame

Figure 4b shows the FRF result of frames against frequencies. Similarly, bolt-jointed frame was compared to monolithic frame. There are ten mode were collected in range of 0 to 2000 Hz of frequency. The peaks are shifted for several modes where the joints are located. The first four peaks show almost similar phases, while the rest shows significantly different. Table 2 tabulates the dynamic properties of frames for the first four modes. It can be seen that bolt-jointed frame produces lower natural frequencies and higher damping ratios than monolithic frame. The average percentage increase of damping ratio recorded was ~290%. This can be further verified in figure 4b, where the FRF of jointed frame is more stable and has lower amplitude from peak to peak. Figure 6 illustrates the mode shape of frames at first mode which observed as mode torsion.

Table2. Structural dynamic properties of frames.

Mode	Monolithic frame		Jointed frame	
	Nat. freq. (Hz)	Damp. ratios	Nat. Freq (Hz)	Damp. ratios
1	105	0.231	98.4	0.819
2	221	0.255	199	1.000
3	343	0.254	328	0.906
4	486	0.17	476	0.476

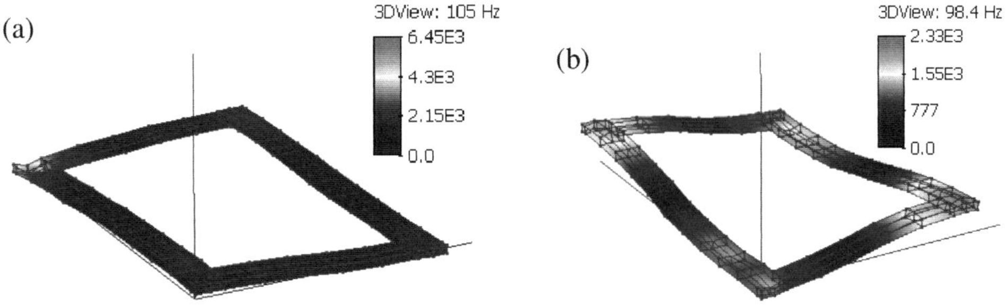

Figure 6. Mode shape of (a) monolithic frame and (b) bolt-jointed frame.

4. Conclusion

This study investigated the effects of joints on dynamic response of assembly structures. The importance conclusions of this study are summarized as follows:

- The addition of bolted joints decreases the structure' natural frequencies by adding additional mass to the structure.
- The assembled structure using bolted joint significantly increases the damping ratio, thus this will immensely affect a complex structure.
- The monolithic structures have higher stiffness compare to jointed structures.

Acknowledgements

The authors thank Universiti Tun Hussein Onn Malaysia for the support under Short Term Grant, vote 0441.

References

[1] Blakke A 1985 *Design of Mechanical Joints* (New York: Marcell Dekker)
[2] Bickford J H and Nassar S 1998 *Handbook Of Bolts and Bolted Joints* (New York: Marcel Dekker)
[3] Ahmadian H and Jalali H 2007 Identification of Bolted Lap Joints Parameters in Assembled Structures *Journal of Mechanical Systems and Signal Processing* **21** 1041-50

[4] Kamal M, Rahman M M and Rahman A G A 2012 J Mech Eng Sci **3** 291
[5] Kamal M, Rahman M M and Rahman A G A 2013 Inter J Automot Mech Eng **7** 912
[6] Ma X, Bergman L and Vakakis A 2001 Identification of Bolted Joints Through Laser Vibrometry *Journal of Sound and Vibration* **246** 441-60
[7] Guo T, Li L, Cai L and Zhao Y 2012 Alternative Method for Identification of the Dynamic Properties of Bolted Joints *Journal of Mechanical Science and Technology* **26** 3017-27
[8] Ren Y and Beards C F 1994 Identification of Joint Properties of a Structure Using FRF Data *Journal of Sound and Vibration* **186** 567-87
[9] Reid J D and Hiser N R 2004 Detailed Modeling of Bolted Joints With Slippage *Journal of Finite Elements in Analysis and Design* **41** 547-62
[10] Saxena M, Tripathi K and Kanchwala H 2012 Investigation of Dynamic Response of Bolted Joint Test Rig Under Free Vibrations using FEA *International Journal of Applied Engineering Research* **7** 1359
[11] Kadirgama K, Rahman M M, Ismail A R and Bakar R A 2011 *J Mech Eng Sci* **1** 37
[12] Kim J, Yoon, J C and Kang B S 2006 Finite Element Analysis and Modeling of Structure with Bolted Joints *Journal of Applied Mathematical Modelling* **31** 895-911

Reducing Mechanical Formation Damage by Minimizing Interfacial Tension and Capillary Pressure in Tight Gas

Arshad Ahmed[1], Muhannad Talib Shuker[1], Khalil Rehman[1], Hassan Bahrami[2] and Muhammad Khan Memon[1]
[2]Petroleum Engineering Department, Curtin University Australia

E-mail: arshad.lashari@live.com

Abstract. Tight gas reservoirs incur problems and significant damage caused by low permeability during drilling, completion, stimulation and production. They require advanced improvement techniques to achieve flow gas at optimum rates. Water blocking damage (phase Trapping/retention of fluids) is a form of mechanical formation damage mechanism, which is caused by filtrate invasion in drilling operations mostly in fracturing. Water blocking has a noticeable impact on formation damage in gas reservoirs which tends to decrease relative permeability near the wellbore. Proper evaluation of damage and the factors which influence its severity is essential to optimize well productivity. Reliable data regarding interfacial tension between gas and water is required in order to minimize mechanical formation damage potential and to optimize gas production. This study was based on the laboratory experiments of interfacial tension by rising drop method between gas-brine, gas-condensate and gas-brine. The results showed gas condensate has low interfacial tension value 6 - 11 dynes/cm when compared to gas-brine and gas- diesel which were 44 - 58 dynes/cm and 14 - 19 dynes/cm respectively. In this way, the capillary pressure of brine-gas system was estimated as 0.488 psi, therefore diesel-gas system was noticed about 0.164 psi and 0.098 psi for condensate-gas system. A forecast model was used by using IFT values to predict the phase trapping which shows less severe phase trapping damage in case of condensate than diesel and brine. A reservoir simulation study was also carried out in order to better understand the effect of hysteresis on well productivity and flow efficiency affected due to water blocking damage in tight gas reservoirs.

1. Introduction Introduction

1.1. *Water Blocking /Water Phase Trapping*

Howard Johnson estimated that 10% of the world's proven reserves of hydrocarbons some 200 x 10^9 BOF in low permeability reservoirs. Even in successfully fractured completions, the formations susceptible for water blocking (phase trapping) which one of the major mechanism of reduced productivity near wellbore[1].

A considerable effort required to diagnosing and evaluating its effects[2]. Due to filtrate invasion, the introduction of an additional immiscible or an increase existing phase saturation with porous media

[1] Petroleum Engineering Department, Universiti Teknologi PETRONAS Malaysia
[2] Petroleum Engineering Department, Curtin University Australia

can cause deleterious relative permeability effects which can substantially impact the permeability and relative permeability of oil and gas[3].
According to the simulation and modelling result performed in the study of tight sand reservoir concluded that the importance of water blocking (liquid phase trapping) is one of the major damage due to relative permeability and capillary pressure effects[4].

1.2. *Mechanism of Water Blocking*
Figure 2 shows a schematic of the establishing water blocking within a low permeability gas reservoir application. For the better understanding, this mechanism was divided into three stages[5].

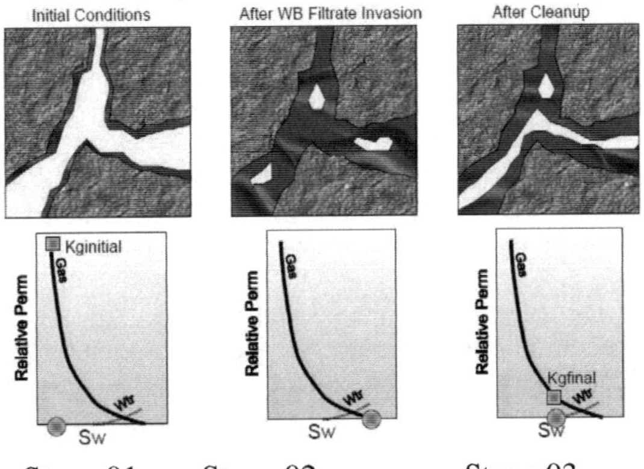

Figure 1. Mechanism of Water Blocking[5]

Table 1. Stages of Water Blocking Damage Mechanism

Stage 1(Initial condition)	Stage 2 (After Water Based Filtrate Invasion)	Stage 3 (After clean up)
It can be observed in the initial desiccated condition, the pre-existing Swi existing in porous media. Majority of the cross sectional area is available for gas flow which tends to high initial relative permeability	After flushing with water based filtrate (i-e drilling mud filtrate, completion fluid, kill fluids etc) results in the establishment of a high water saturation tends to be establishment of a high saturation of water	Subsequently the drawdown of the reservoir results in the affected zone reverting to the irreducible water saturation dictated by the capillary mechanics of the system rather than back to the potentially very low initial water saturation

The irreducible saturation of water is greater than initial water saturation(Sirr > Swi) , this results a notably restriction in the cross sectional area available for fluid flow observed in Figure 1 and causes reduction in relative permeability to gas. Figure 2 shows a schematic of these mechanisms.

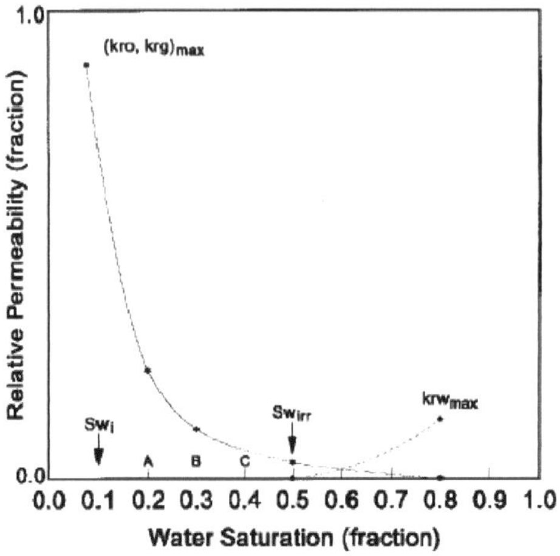

Figure 2. Water blocking-relative permeability relations[3]

1.3. *Factors Which Affects Severity of Water Phase Trapping*
The factors were highly influenced on the water blocking damage such as interfacial tension, capillary pressure. rock wettability, fluid type and fluid composition[6, 7].

2. Methodology

2.1. *Collection of Samples*

2.1.1. Fluid Samples
The fluid samples were collected from the Melaka refinery PETRONAS Penapis Sdn Bhd Sungai Udangan Melak. Bintulu. Three samples was undergoes for the experimental study of interfacial tension and its effect on phase trapping in tight gas reservoirs.

2.1.2. Gas Sample
The nitrogen was used to imitate the produced gas after fluid invaded near wellbore in the tight gas reservoirs.

2.2. *Interfacial Tension Measurement*
The Vinci Interfacial Tension Meter (IFT-700) was used for the IFT measurement by using rising drop method in different temperature and pressure of following samples. The density of fluids at high temperature was required as input. Anton Par Density meter model DMA4500 was used to measure the density of these fluids at high temperature, shown in Table 2. Three Gas-liquid systems were used in the interfacial tension measurement at high temperature and pressure conditions.

- Brine-Nitrogen
- Condensate-Nitrogen
- Diesel-Nitrogen

Table 2. Density of liquids at 70 °C

Fluid	Density
Brine	0.9915
Diesel	0.8198
Condensate	0.7012

3. Capillary Pressure Influences on Phase Trapping Damage

Interfacial tension a dominant factor of capillary pressure and they were also directly proportional to each other. The Interfacial tension between the damage fluid and produced fluid is high; it leads high capillary pressure which is caused the more serious phase trapping damage. In order to control the phase trapping, it is required to reduce capillary pressure this will helpful to displace the trapped fluid towards the wellbore by the reservoir drawdown. The relationship of capillary pressure and interfacial tension are given in the form of following equation[8],

$$P_c = \frac{2\sigma \cos\theta}{r^2} A \qquad (1)$$

Where,

Pc = Capillary pressure, psi

σ= interfacial tension, dynes/cm

θ = contact angle, degree

r = pore throat radius, microns

A= 145 x 10-3 (constant, to convert in psi)

4. Correlations to Predict the Potential Severity of Aqueous Phase Trapping

In various stages of development of well, an evaluation of the potential severity of aqueous phase trapping into the formation is essential. The following tool provides a set of equation that is easy to use and helpful to evaluate the damage severity. The potential sensitivity of reservoir to the development of phase trap can also be predicted[3].

Many aqueous phase trap tests which have been carried out to over a wide range of permeabilities, gas reservoir and sub irreducible oil saturation applications. The regression analysis is a base of these equations generally for the phase trap tests. Aqueous phase trapping index (APTi) uses by focusing initial formulation and establish a new methodology to diagnose the problem related to aqueous phase trapping. The permeability and initial water saturation values are the key factors of this basic equation which have two strongly characteristics for oil and gas reservoirs[3]. The APTi formulation generally said to be a conservative estimate, which was given by Equation 2.

$$APT_i = 0.25\,[\log 10\,(k_a)] + 2.2\,(S_{wi}) \tag{2}$$

where:

APTi = aqueous phase trap index

ka = uncorrected average formation air permeability (mD)

Swi = initial water saturation (fraction)

The above equation is valid for the ranges of absolute permeability and intial water saturation values, shown in Table 3.

Table 3. Range of Validity[3]

Parameter	Value
Ka	5000mD
Swi	0 - 1.00

The prediction criteria depend on the APTi value and refer the Table 4 for evaluation of damage potential.

Table 4. Interpretation of APTi[3]

Parameter	Value
APTi > 1.00	Formation unlikely to exhibit significant permanent sensitivity to aqueous phase
0.80 < APTi < 1.00	Formation may exhibit sensitivity to aqueous phase trapping
APTi < 0.80	Formation will likely exhibit significant sensitivity to aqueous phase trapping

5. Phase Trapping co efficient – A Forecast Method

The coefficient of phase trapping was introduced to evaluate potential damage of the water phase trapping by considering influences of initial water saturation, reservoir pressure, interfacial tension between fluids, liquid properties and pore structure on water phase trapping[9].

$$PTC = e^{\sqrt{\frac{k}{\emptyset}}\frac{\Delta p}{\sigma\cos\theta\cdot\mu_m}\frac{S_{wi}}{S_{wirr}}} \quad (3)$$

Where,

k = permeability, μm2

ø = porosity, fraction

Δp = the biggest difference pressure for the driving fluid, kPa

μm = the viscosity ratio to the trapping phase and oil/gas,

Swi= initial water saturation

Swirr= irreducible water saturation (not movable)

The outcomes of the interfacial tension and capillary pressure was observed low in condensate-gas system, so it leads to low severity of phase trapping in the tight gas reservoirs. However, it required to use forecast model in order to predict the potential of damage. The prediction of phase trapping was taken place by applying phase trapping co-efficient (PTC) dimensionless factor, table 05 presented the criteria of prediction of phase trapping damage.

Table 5. Prediction criteria of phase trapping[9]

PTC	Damage Severity
PTC < 0.05	None
0.05≤PTC<0.3	Weakly
0.3≤PTC<0.5	Weakly to medium
0.5≤PTC<0.7	Medium to intensely
PTC≥0.7	Intensely

6. Result and Discussion

The measurement of interfacial tension between the damage fluid and produced fluid at different pressure and temperature was examined. The result shows that the interfacial tension of gas-condensate was low as compared the diesel and brine system. The values of IFT were measured 6 to 11 dynes/cm, mentioned in the Figure 5.

However the highest values of IFT was found in the brine-gas system stated from 44 to 59 dynes/cm, shown Figure 3. The diesel-gas system had moderate values of IFT that was noted as 14 – 19 dynes/cm varies with temperature and pressure, see Figure 4.

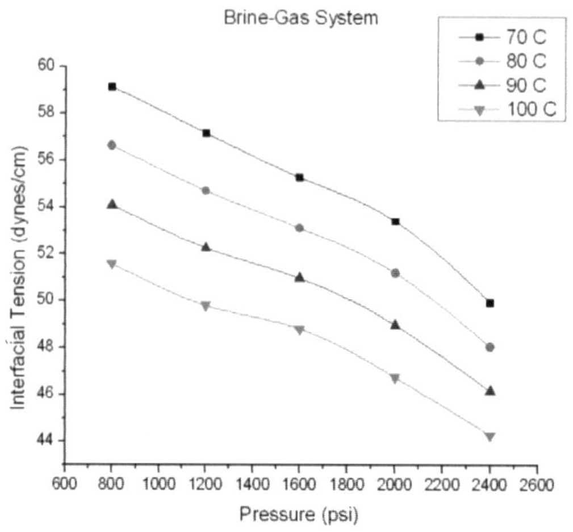

Figure 3. Interfacial tension of brine-gas system at different temperature and pressure

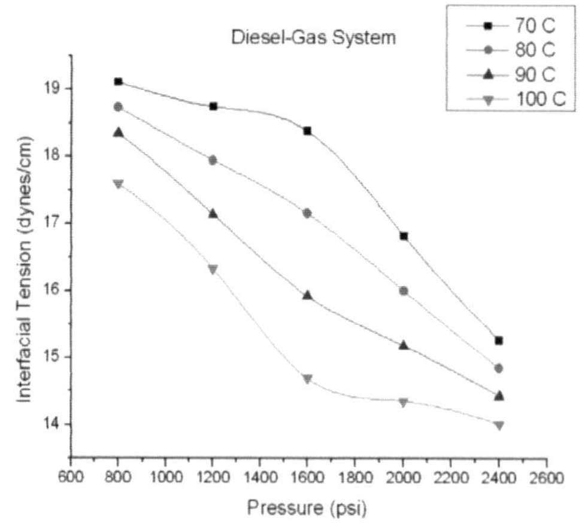

Figure 4. Interfacial tension of diesel-gas system at different temperature and pressure

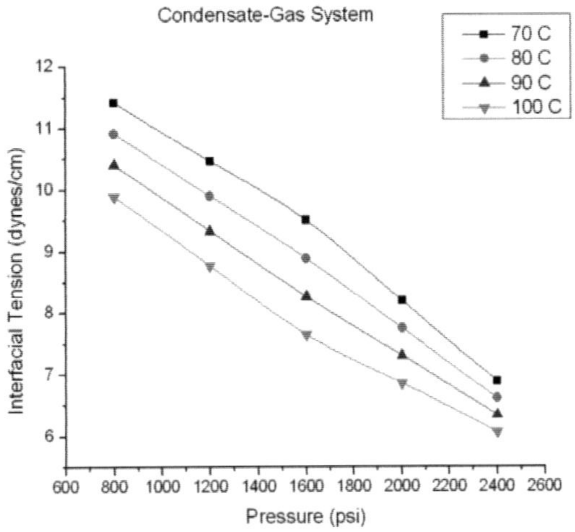

Figure 5. Interfacial tension of condensate-gas system at different temperature and pressure

The capillary pressure was estimated with the help of Equation 1 by substituting the value of interfacial tension of each fluid. As the capillary pressure is directly proportional to the interfacial tension so the high value of Pc exist in the case of brine-gas system which is visualized by the Figure 6. However the moderate and low values of capillary pressure are presented in the Figure 7 and 8 respectively.

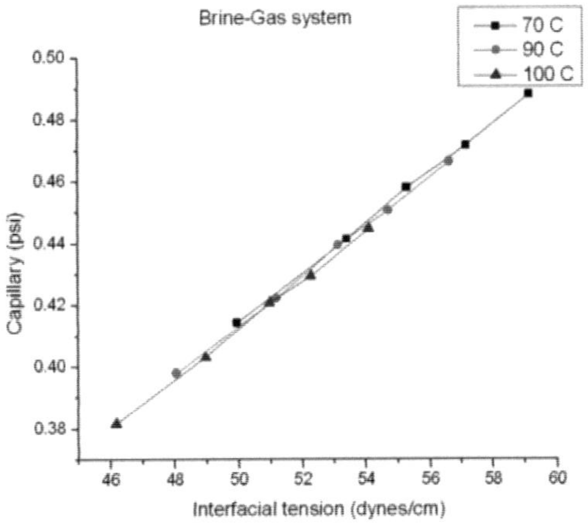

Figure 6. Capillary pressure vs Interfacial tension of brine-gas system at different temperature conditions

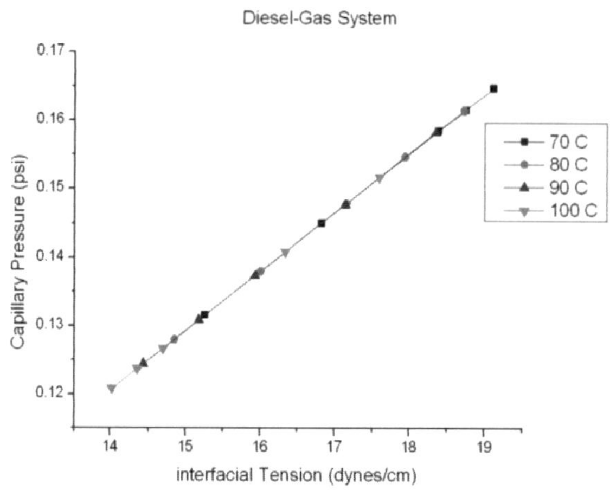

Figure 7. Capillary pressure vs Interfacial tension of diesel-gas system at different temperature conditions

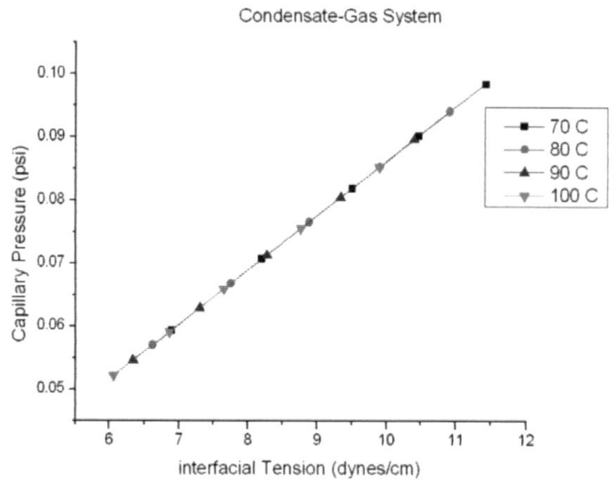

Figure 8. Capillary pressure vs Interfacial tension of condensate-gas system at different temperature conditions

With the help of Equation 3, PTC was estimated by considering interfacial tension of three gas-liquid systems, shown in Table 6. In this way, it is concluded that condensate-gas system has low potential to phase trapping damage and displaced efficiently by gas produced which results improve in well productivity.

Table 6. Predication results of phase trapping

Fluid	PTC
Brine	0.66
Diesel	0.48
Condensate	0.42

7. SCAL Reservoir Simulation

Furthermore, simulation approach used to simulate the core flooding in the tight core samples in order to evaluate the permeability damage caused by phase trapping. Sendra (Weatherford) software used for SCAL reservoir simulation which made for revealing relative permeability and capillary pressure from two-phase and multi-phase flow experiments performed in the SCAL laboratory. The following results based on the simulation by using tool Sendra Software. The Figure 9 shows the relationship between the relative permeability Kr, water saturation Sw and capillary pressure Pc in tight permeability core sample by using Corey's & Burdin Correlation. The Krg and Pc were decreasing with increment in Sw. Corey & LET correlation was used in the Figure 10 in Sendra Simulation[10] in which Krg was declined gradually with Sw value raises and Pc was also reduced during middle region of curves.

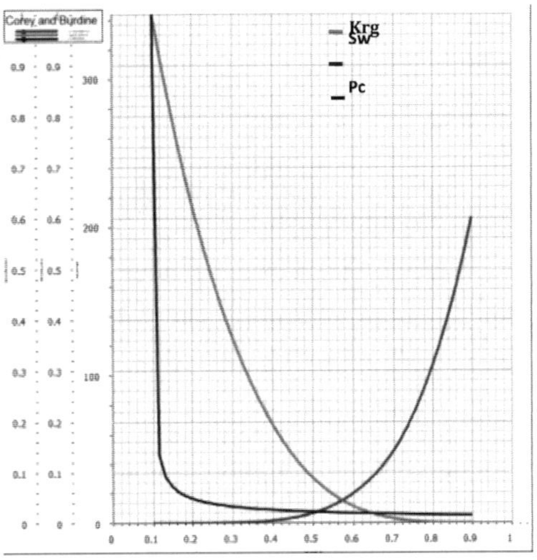

Figure 9. Relationship between Kr, Sw and Pc using Corey & Burdine Correlation[10]

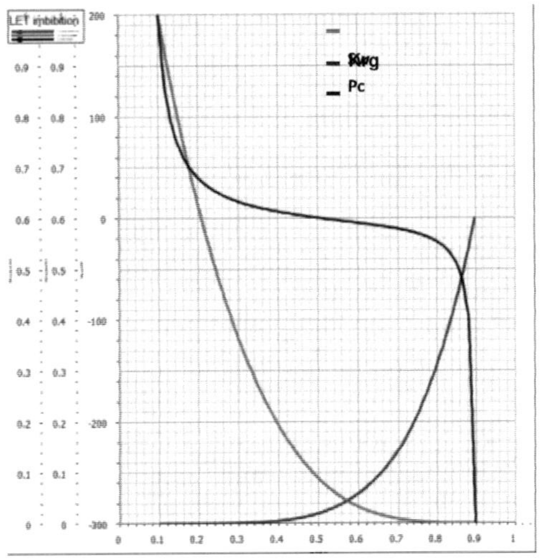

Figure 10. Relationship between Kr, Sw and Pc using LET Correlation[10]

8. Conclusion

The interfacial tension less with low capillary pressure which has low tendencies to block mechanism and required core flooding to validate its effect. On the basis of IFT results, condensate has less severity to phase trapping in the low permeability and tight core samples as compared diesel and brine. Core flooding is required to evaluate the damage potential by injecting these fluids, estimate irreducible saturation which caused trapping of gas production and reduced well productivity. Proper evaluation and diagnosis of water blocking damage by laboratory experiments tends to be an effective prevention for the reduction of formation damage and optimized productivity of gas in low permeability reservoirs

ACKNOWLEDGEMENTS

This research work was supported by Universiti Teknologi PETRONAS, Malaysia. Authors wish to express their sincere thanks to Mr.Riduan (Lab Technician UTP) for generous help in the experiments.

REFERENCES

[1] H. Johnson 2001 Low permeability reservoirs: introduction *Petroleum Geoscience*, **7** 2

[2] D. Bennion, R. Bietz, F. Thomas, and M. Cimolai 1994 Reductions in the productivity of oil and low permeability gas reservoirs due to aqueous phase trapping *Journal of Canadian Petroleum Technology* **33**

[3] D. Bennion, F. Thomas, R. Bietz, and D. Bennion 1996 Water and hydrocarbon phase trapping in porous media-diagnosis, prevention and treatment *Journal of Canadian Petroleum Technology* **35**

[4] J. Amaerule, D. Kersey, D. Norman, and P. Shannon 1988 Advances in formation damage assessment and control strategies *Annual Technical Meeting*

[5] D. B. Bennion and F. B. Thomas 2005 Formation damage issues impacting the productivity of low permeability, low initial water saturation gas producing formations *Journal of energy resources technology* **127** 240-247

[6] D. Bennion, F. Thomas, D. Imer, and T. Ma 2000 Low permeability gas reservoirs and formation damage-tricks and traps *SPE/CERI Gas Technology Symposium*

[7] D. Bennion, F. Thomas, and T. Ma 2000 Recent advances in laboratory test protocols to evaluate optimum drilling, completion and stimulation practices for low permeability gas reservoirs *SPE Rocky Mountain Regional/Low-Permeability Reservoirs Symposium and Exhibition*

[8] V. Joekar-Niasar, S. Hassanizadeh, and A. Leijnse 2008 Insights into the relationships among capillary pressure, saturation, interfacial area and relative permeability using pore-network modeling *Transport in Porous Media* **74** 201-219

[9] L. YOU and Y. Kang 2009 Integrated evaluation of water phase trapping damage potential in tight gas reservoirs *8th European Formation Damage Conference*

[10] "Sendra," 2011 ed: Weatherford, 2011.

IOP Conf. Series: Materials Science and Engineering **50** (2013) 012020 doi:10.1088/1757-899X/50/1/012020

Stress intensity factors of three parallel edge cracks under bending moments

A E Ismail[a]
Faculty of Mechanical and Manufacturing Engineering, Universiti Tun Hussein Onn Malaysia, Batu Pahat, 86400 Johor, Malaysia

E-mail: [a]emran@uthm.edu.my

Abstract. This paper reports the study of stress intensity factors (SIF) of three edge cracks in a finite plate under bending moments. The goal of this paper was to analyze the three edge crack interactions under such loading. Several studies can be found in literature discussing on mode I SIF. However, most of these studies obtained the SIFs using tensile force. Lack of SIF reported discussing on the SIFs obtained under bending moments. ANSYS finite element program was used to develop the finite element model where singular elements were used to model the cracks. Different crack geometries and parameters were utilized in order to characterize the SIFs. According to the present results, crack geometries played a significant role in determining the SIFs and consequently induced the crack interaction mechanisms.

1. Introduction

Parallel edge cracks are one of the most common flaws especially in industrial failures [1]. Accurate stress analysis of these components in the presence of cracks and evaluation of fracture design parameters are needed for reliable estimation of the crack growth rates and residual fracture strength [2]. This paper discusses three parallel cracks in a sheet with finite width where the sheets are subjected to the bending moment. The crack will be appeared on the surface component which can reduce the component durability [3-8]. According to the literature survey [9], most of the work conducted to analyze the crack with the assumption that only a single crack can be used to represent such problems. In many cases the multiple cracks also can be carried out and must be studied to increase the safety factor of the component structure [10]. This paper discusses three parallel cracks in a sheet with finite width. This sheet is subjected to bending moment in the longitudinal direction remote from the cracks. Then, crack interactions are analyzed and discussed.

2. Displacement Extrapolation Method

The finite element method is an appropriate approach to calculate the stress intensity factor (SIF) for linear elastic fracture mechanics problems. In order to determine the SIFs, a displacement extrapolation method [11] is used in this study. Several other works have implemented a similar method are also available [12-13]. After obtaining the elastic finite element solution of the particular problem, nodal displacements between two crack faces are determined and used to compute the SIFs as follows

$$K_I = \frac{2G\sqrt{2\pi}}{1+\kappa}\frac{|v_b - v_d|}{\sqrt{r}} = \frac{2G\sqrt{2\pi}}{1+\kappa}\frac{|\Delta v|}{\sqrt{r}} \quad (1)$$

$$K_{II} = \frac{2G\sqrt{2\pi}}{1+\kappa}\frac{|u_b - u_d|}{\sqrt{r}} = \frac{2G\sqrt{2\pi}}{1+\kappa}\frac{|\Delta u|}{\sqrt{r}} \quad (2)$$

$$K_{III} = 2G\sqrt{2\pi}\frac{|w_b - w_d|}{\sqrt{r}} = 2G\sqrt{2\pi}\frac{|\Delta w|}{\sqrt{r}} \quad (3)$$

where, K_I, K_{II} and K_{III} are the respective mode I, II and III SIFs, Δv, Δu and Δw are the relative nodal displacements between two crack faces in the direction of y-axis, x-axis and z-axis, respectively, and G is the modulus of rigidity. For plain strain condition, $\kappa = 3\text{-}4\upsilon$, where, υ is the Poisson's ratio. All the SIFs obtained from the analysis are converted into normalized values in order to ensure the generality of the results. A normalized SIF, F, can be defined as follows [14-15]

$$F_{I,a} = \frac{K_{I,a}}{\sigma_a\sqrt{\pi a}} \quad (4)$$

$$F_{I,b} = \frac{K_{I,b}}{\sigma_b\sqrt{\pi a}} \quad (5)$$

$$F_{II} = \frac{K_{II}}{\tau_{xy}\sqrt{\pi a}} \quad (6)$$

$$F_{III} = \frac{K_{III}}{\tau_{xy}\sqrt{\pi a}} \quad (7)$$

where, σ_a, σ_b and τ_{xy} are the axial, bending and shear stresses, respectively and a is a crack depth.

3. Methodology and Model Validation

ANSYS commercial finite element (FE) analysis software is used to model the parallel edge cracks through the use of ANSYS Parametric Design Language (APDL) as in figure 1(a). Due to the symmetrical problem, half finite element model is sufficiently used to analyze the problem as in figure 1(b) where singular finite element is used around the crack tips as revealed in the enlarged area of Figure 1(b). Two important parameters are used in this work such as b/a and a/W. Both parameters are in the range of 0.25 to 2.00 and 0.1 to 0.5, respectively. The calculation of stress intensity factor is based on the theory of displacement extrapolation method [10]. In order to validate the results, the present model must be compared with the previous results as produced by Jiang et al. [3, 4]. Unfortunately, there are unavailable stress intensity factors under bending moment, especially for multiple parallel edge cracks. Therefore, the SIFs obtained under tensile force are used for validation purposes. Figure 2(a) shows the meshed edge crack model subjected to axial stress and Figure 2(b) represents the stress distribution due to the crack interactions. Figure 5 shows the model validations where it is showed that the present model is well in agreement with the previous model. Therefore, it is assumed that the present model can be validated significantly using the results obtained from Jiang et al. [3, 4]. For the present analysis, it is assumed that plain strain finite element model is used where PLANE83 element type in ANSYS library is utilized. In order to bend the plate remotely, an independent node is created on the top of the plate. Then, all nodes situated on the top edge of plate are connected to this newly created node. Bending moment is applied to this node and then consequently creating mode I crack mechanism as shown in figure 2(c). Due to symmetrical problem, stress intensity factor (SIF) for crack 3 is not accounted since it is similar to crack 1 (outer crack). Middle crack in this work is called crack 2. All the SIFs are normalized according to Eqs. (4)-(7). The normalized SIF is also called geometrical correction factor. It is important to normalize the SIFs for the sake of analysis generalization.

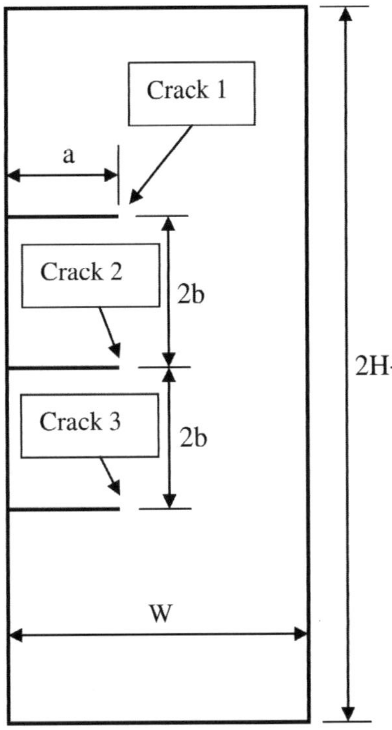

(a)

Figure 1(a). Parallel edge cracks in a finite plate.

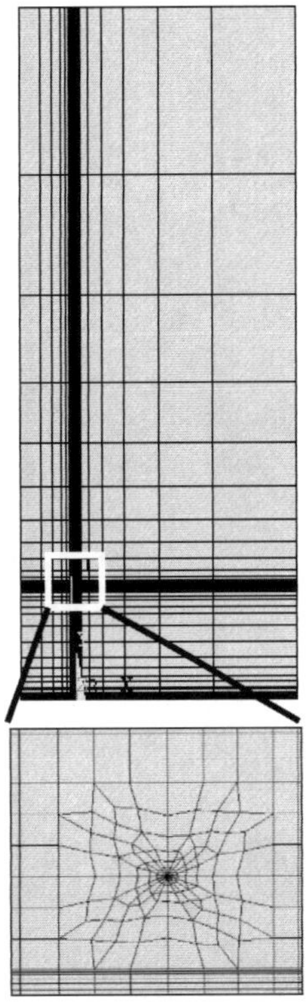

(b)

Figure 1(b). Half symmetrical finite element model of parallel edge cracks with enlarged crack tip area.

IOP Conf. Series: Materials Science and Engineering **50** (2013) 012020 doi:10.1088/1757-899X/50/1/012020

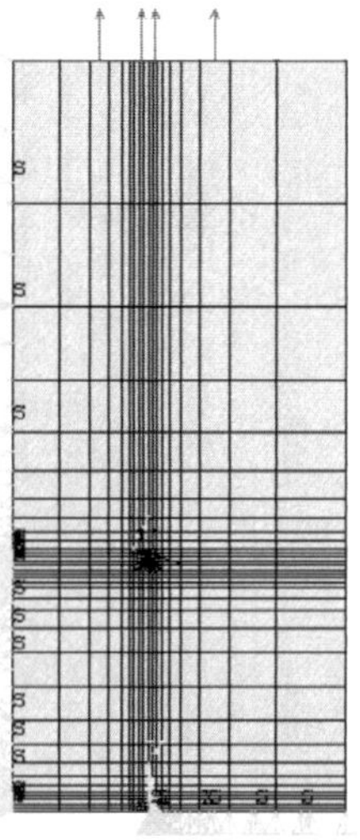

(a)

Figure 2(a). Finite element model under axial stress.

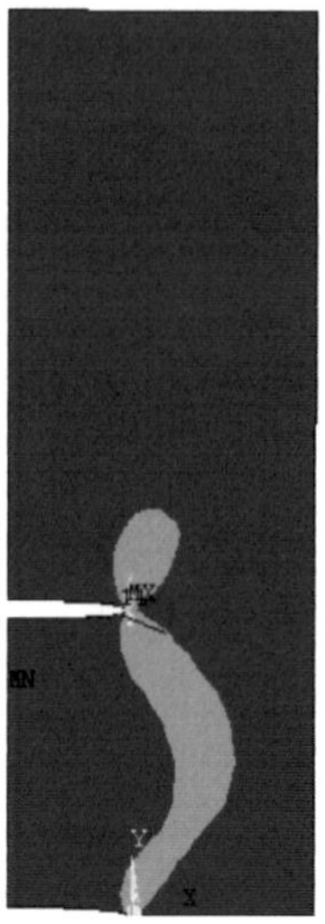

(b)

Figure 2(b). The stress distribution and the crack interaction of parallel edge cracks

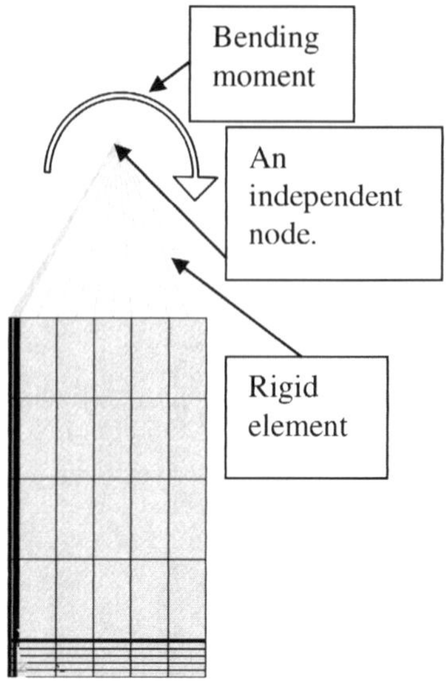

(c)

Figure 2(c). Remotely applied bending moment.

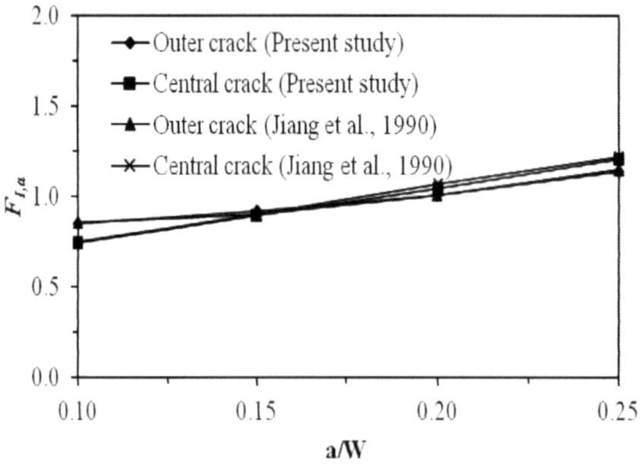

Figure 3. Validations of the present finite element models with the previous model developed [3, 4].

4. Results and Discussion

Tables 1 and 2 list the geometrical correction factors for outer and central multiple edge cracks under bending moments, respectively. Figure 4 and 5 show the plot of SIF against *a/W* when *b/a* is varied for cracks 1 and 2, respectively when compared with non-interacting cracks. It is showed that the SIF increased when the cracks become deeper. These characteristics are widely reported in literature [3, 4]. Figure 4 shows the stress intensity factors (SIFs) of single edge crack under bending moment. It is found that the effects of crack eccentricities are played an important role in increasing the SIFs. However, the increment of SIFs occurred for the cracks situated within $b/a \leq 0.50$. For the cracks with $b/a \geq 2.0$, it is seemed that the SIFs decreased as a function of *a/W* compared with central crack (b/a = 0.0). The decrement of SIFs is due to the fact when the location of crack is away from the central line of finite plate, bending moment or stress decreased.

Table 1. Geometrical correction factor, $F_{I,b,o}$ for outer edge multiple cracks.

b/a	a/W								
	0.10	0.15	0.20	0.25	0.30	0.35	0.40	0.45	0.50
0.25	0.7980	0.8110	0.8360	0.8770	0.9400	1.0130	1.1130	1.2410	1.4060
0.50	0.8400	0.8760	0.9150	0.9690	1.0410	1.1300	1.2420	1.3810	1.5560
1.00	0.9440	0.9780	1.0260	1.0880	1.1630	1.2530	1.3580	1.4940	1.6470
2.00	1.0750	1.1050	1.1430	1.1860	1.2390	1.3040	1.3910	1.4980	1.6520

Table 2. Geometrical correction factor, $F_{I,b,c}$ for central edge multiple cracks.

b/a	a/W								
	0.10	0.15	0.20	0.25	0.30	0.35	0.40	0.45	0.50
0.25	0.5290	0.5290	0.5410	0.5680	0.6000	0.6780	0.7710	0.8990	1.0750
0.50	0.6000	0.6280	0.6620	0.7180	0.7990	0.9080	1.0490	1.2310	1.4420
1.00	0.7600	0.8010	0.8640	0.9620	1.0720	1.1930	1.3260	1.4780	1.6440
2.00	0.9950	1.0540	1.1180	1.1790	1.2390	1.3060	1.3930	1.4980	1.6540

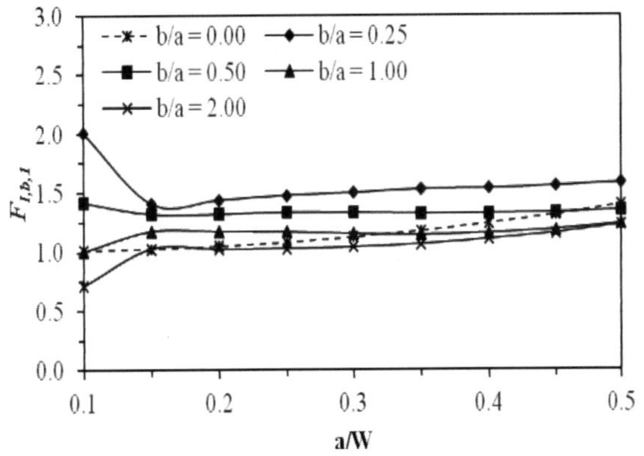

Figure 4. Stress intensity factors (SIFs) of single edge cracks under bending moment for crack 1.

Figure 5 reveals the SIFs distribution for multiple edge cracks subjected to bending moment for cracks 1 and 2, respectively. Figure 5(a) shows the SIFs of multiple edge cracks for crack 1. It is found that the SIF increased in the similar pattern due to the interactions and stress relaxations among the cracks. As the results, the SIFs for multiple cracks are lower than the offset single cracks. This is due to the stress field redistribution when the present of cracks. As expected, the SIFs of multiple cracks reduced when the cracks interacted to each others as in figure 5(b). It is also found that greater amount of interactions occurred when closer crack intervals are used. However, such interactions are diminished when the neighbouring crack is situated greater than $b/a \geq 2.0$.

5. Conclusion

This paper present the calculation of stress intensity factors of three parallel edge cracks in a finite plate subjected to remote bending moment. Several main conclusions can be drawn from this work are as follows:

(a) The results show that due to the presence of multiple crack edge cracks, the stress distribution is relaxed and therefore, the stress intensity factors for all cracks decreased.

(b) When the distance between the cracks is increased, interaction is seemed to be diminished and it is can be neglected especially for swallow crack.

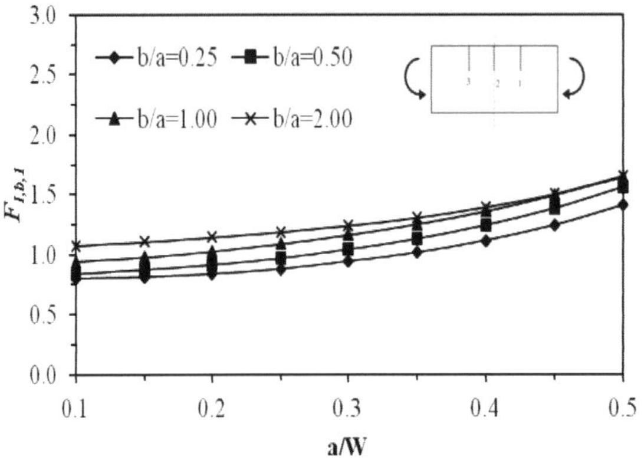

(a)

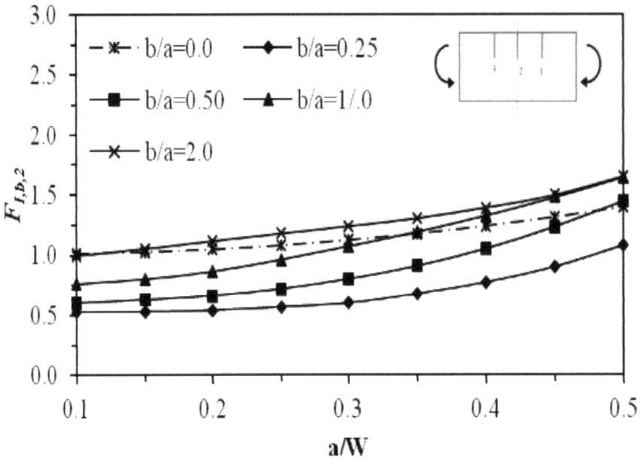

(b)

Figure 5. Stress intensity factors (SIFs) of multiple edge cracks under bending moment for (a) crack 1 and (b) crack 2.

Acknowledgement

Author acknowledges Universiti Tun Hussein Onn Malaysia (UTHM) and Office for Research, Innovation, Commercialization and Consultancy Management (ORICC), UTHM for sponsoring the research work (vot. 1001) and participating in International Conference on Mechanical Engineering Research (ICMER2013), Gambang, Kuantan, Pahang.

References

[1] Daud R, Ariffin A K, Abdullah S, Ismail A E 2012 *Applied Mechanics and Materials* **52-54** 1326-31

[2] Ismail A E, Ariffin A K, Abdullah S, Ghazali M J, Abdulrazzaq M, Daud R 2012 *Journal of Zhejiang University – Science A.* **13**(1) 1-5

[3] Jiang Z D, Petit J, Benzine G 1991 *Engineering Fracture Mechanics* **40**(2) 345-54

[4] Jiang Z D, Zeghloul A, Benzine G, Petit J 1990 *Engineering Fracture Mechanics* **35**(6) 1073-79

[5] Kamal M, Rahman M M and Rahman A G A 2012 *J. Mech. Eng. Sci.* **3** 291

[6] Kadirgama K, Rahman M M, Ismail A R and Bakar R A 2011 *J. Mech. Eng. Sci.* **1** 37
[7] Kamal M, Rahman M M and Rahman A G A 2013 *Inter. J. Automot .Mech. Eng.* **7** 912
[8] Daud R, Ariffin A K and Abdullah S 2013 *Inter. J. Automot. Mech. Eng.* **7** 995
[9] Ismail A E, Ariffin A K, Abdullah S, Ghazali M J. 2012 *Meccanica* **47** 1141-56
[10] Lam K Y, Phua S P 1991 *Engineering Fracture Mechanics* **40**(3) 585-92
[11] Gustavo V G, Jaime P, Manuel E 2000 *Engineering Fracture Mechanics* **66** 243-55
[12] Aslantas K, Ergun E, Tasgetiren, S 2006 *International Journal of Mechanics of Materials and* Design **3**(2) 201-08
[13] Aslantas K 2003 *Int. J. of Solids and Structures.* **40**(26) 7475-81.
[14] Toribio J, Matos J C, Gonzalez B, Escuadra J. 2009. *Engineering Failure Analysis* **16**(2) 618-30
[15] Ismail A E, Ariffin A K, Abdullah S, Ghazali M J, Daud R. *Advanced Materials Research* **214**:192-96

2nd International Conference on Mechanical Engineering Research (ICMER 2013) IOP Publishing
IOP Conf. Series: Materials Science and Engineering **50** (2013) 012021 doi:10.1088/1757-899X/50/1/012021

Effect of plenum chamber depth in a swirling fluidized bed

M A Hafiz, M F M Batcha[a] and N Asmuin
Faculty of Mechanical and Manufacturing Engineering,
Universiti Tun Hussein Onn Malaysia, 86400 Parit Raja, Batu Pahat, Johor, Malaysia

E-mail: [a]mfaizal@uthm.edu.my

Abstract. This paper presents the numerical investigation via Computational Fluid Dynamic (CFD) to study the effect of plenum chamber depth on air flow a distribution in a swirling fluidized bed (SFB). A total of 9 simulations were conducted for 3 plenum chamber depths of 175 mm, 350 mm and 525 mm (below the distributor) for 3 different inlets: single, double and triple inlets. Air flow distribution was analyzed based on the tangential velocity distribution ad pressure drop at the distributor outlet. Statistical parameters used in characterizing the air flow distribution were standard deviation, skewness and kurtosis together with system pressure drop. An optimum plenum chamber depth has low statistical values, implying a uniform velocity distribution inside the bed while low pressure drops are necessary to reduce energy loss in the system. The findings yield that plenum chamber with 175 mm depth with via triple inlets suffices both criteria of high uniformity and low pressure drops.

1. Introduction

Fluidization is a process by which solid particles are made to behave like a fluid, by being suspended in a gas or liquid. One of the recent developments in providing a variant in fluidized bed operation is the swirling fluidized bed (SFB), which provides swirling motion inside the bed apart from fluidization. In contrast with conventional fluidization, in a SFB the fluidizing gas enters the bed at an inclination to the horizontal directed thus by a suitable design of distributor which as an array of blades with centre body, which forms annular opening as shown in figure 1 [1].

The swirling fluidized bed is an outcome of studies carried out to overcome disadvantages a conventional fluidized bed, particularly the inadequate lateral mixing. It also possesses the ability to well fluidize the Geldart type-D particles which are large in size and usually difficult to fluidize [2]. Although many aspects of the bed have been studied over the years, less attention have been given on the plenum chamber design. Among related works on the plenum chamber design were by [3,4] whom have reported their CFD investigation and experimental works. The present work is an extension to the previous study by [5] whom have investigated the aerodynamics of a SFB by taking into the effect of distributor design on velocity distribution and pressure drop inside the bed. As for this study, particular attention is given on the plenum chamber design by varying its depth and number of inlet and their effect on aerodynamic characteristics of the bed. The findings are important to improve the plenum chamber design in the attempt to increase overall performance of the system.

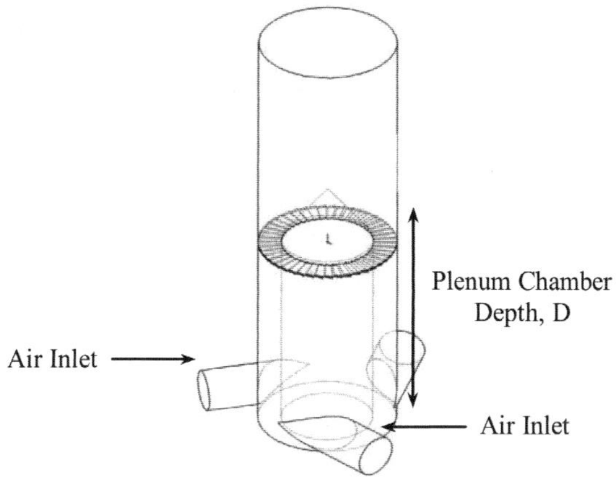

Figure 1. Configuration of a Swirling Fluidized Bed (SFB).

2. Methodology

Investigation of the air flow distribution in a SFB was conducted using commercial CFD software – FLUENT 6.3. The computation domain and grid generation was developed via GAMBIT 2.4.6. The 60 blades with 15° horizontal inclination has been selected based on previous studies by [6]. Two parameters were varied to observe the relation between the plenum chamber depth and number of inlets as shown in table 1.

Table 1. Configuration of plenum chamber in a swirling fluidized bed

Case	Plenum Chamber Depth (mm)	Number of Inlet
1		Single
2	175	Double
3		Triple
4		Single
5	350	Double
6		Triple
7		Single
8	525	Double
9		Triple

The air inlet was modeled as velocity inlet boundary condition of 2.5 m/s. The velocity and other parameters were based on the actual SFB system in the Faculty of Mechanical and Manufacturing Engineering, UTHM, which is currently used for drying of biomass. The air outlet was modeled as a pressure outlet of 1.01325 bar (atmospheric pressure). Meshing of the SFB system was generated in such a way that it may capture airflow taking place through narrow opening of the distributor blades. Grid sensitivity was done prior to actual investigation to ensure the independence of grid size on the flow and to reduce possible numerical errors. The suitable grid size is chosen in such a way that they total number of elements are less (to reduce compunational costs) without compromising simulation results.

Hence, the Tri:Pave Meshing Scheme was applied to the surface and it allowed GAMBIT 2.4.6 to create a face mesh consisting of irregular triangular mesh elements. The Tet/Hybrid parameter type that specifies tetrahedral, hexahedral, pyramidal and wedge element were defined to the meshing algorithm. The mesh elements in the computation domain as well as the plenum chamber depth is presented in figure 2.

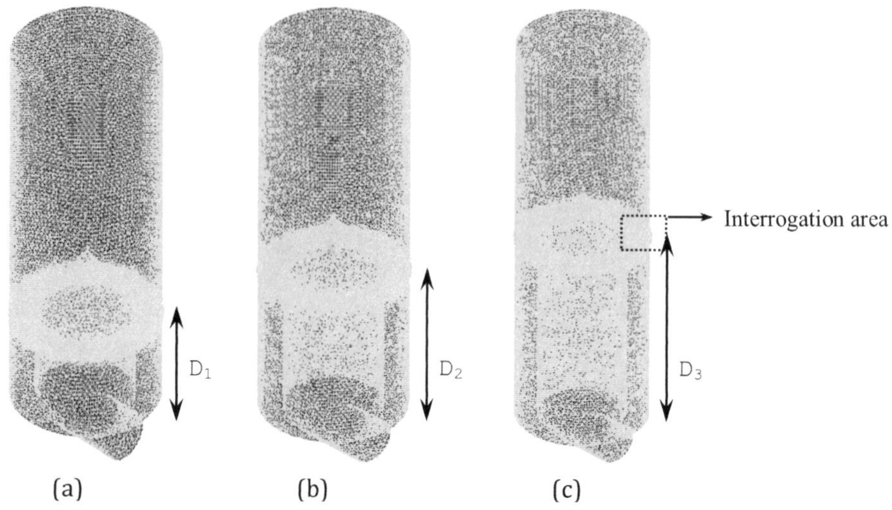

Figure 2. Computational domain in CFD for plenum chamber with various depths: (a) $D_1 = 175$ mm; (b) $D_2 = 350$ mm; (c) $D_3 = 525$ mm.

The mesh quality was be evaluated using the EquiAngle Skew (Q_{EAS}) criterion, which are lower or equal to 0.2 for more than 95 % of the control volumes. From this evaluation the mesh quality could be considered satisfactory. In FLUENT environment, the Reynolds Averaged Navier Stokes (RANS) turbulence equation of the (Re-Normalization Group) RNG methods based on model transport equations for the turbulence kinetic energy (k) and its dissipation rate (ε) which is RNG k-ε model has been selected [7]. This turbulence model is similar to the semi-empirical model namely standard k-ε model but has additional term in its dissipation rate (ε) equation that significantly improves the accuracy for rapidly strained flows [7]. Apart from this, it also provides an analytical formula for turbulence Prandtl numbers and also the effect of swirl on turbulence [8]. A second-order upwind scheme was selected for the discreatisation of the momentum equations which is suitable to moderate swirl and SIMPLE algorithm have be applied to solve the pressure-velocity coupling algorithms in the simulation.

The analyze the flow distribution inside the SFB due to the effect of varying plenum chamber depth, three statistical parameters were applied in this study; the standard deviation, σ, skewness, S and kurtosis, K. Standard deviation denotes the variation to the mean value of the probability distribution while skewness is a measure of asymmetry of the probability distribution. As for the kurtosis, it is a measure whether the data are peaked or flat, relative to normal distribution. In the present study, a combination of low standard deviation, skewness and kurtosis is desired as it implies high uniformity of velocity distribution inside the bed. The equations for above mentioned statistical parameters are as follows [9]:

- *Standard deviation:* $$\sigma = \left[\frac{1}{N-1} \sum_{1}^{N} (V_i - V_{mean}) \right]^{0.5} \quad (1)$$
- *Skewness:* $$S = \sum_{1}^{N} \frac{(V_i - V_{mean})^3}{(N-1)\sigma^3} \quad (2)$$
- *Kurtosis:* $$K = \sum_{1}^{N} \frac{(V_i - V_{mean})^4}{(N-1)\sigma^4} \quad (3)$$

where N is the number of data points, σ is the standard deviation, S is the skewness, K is the kurtosis, V_i is the univariate data of velocity distribution and V_{mean} is the mean velocity of velocity distribution.

3. Result and Discussion

Upon completion of simulation, data was extracted on a horizontal plane, 10 mm above the distributor as in figure 2 since this is the area where gas-solid contact starts in an actual bed. Discussion in this paper however is limited to tangential velocity alone because it has the highest magnitude and responsible for swirling in the annular region of the bed. The velocity profile for various plenum chamber depths and number of inlets were shown in figure 3.

Upon completion of simulation, data was extracted on a horizontal plane, 10 mm above the distributor (as in figure 2) since this is the area where gas-solid contact starts in an actual bed. Discussion in this paper however is limited to tangential velocity alone because it has the highest magnitude and responsible for swirling in the annular region of the bed. The velocity profile for various plenum chamber depths and number of inlets were shown in figure 3.

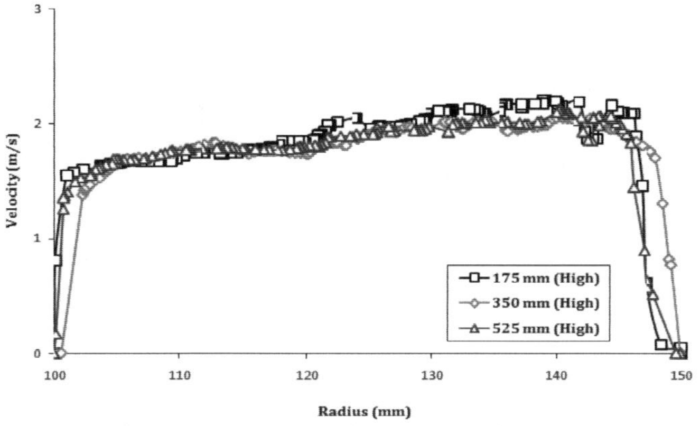

(a)

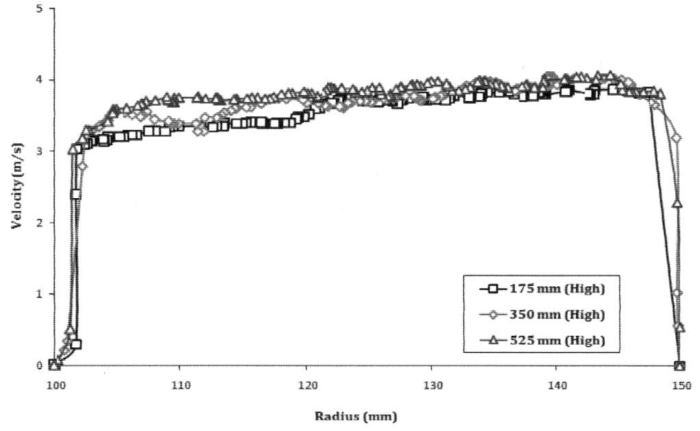

(b)

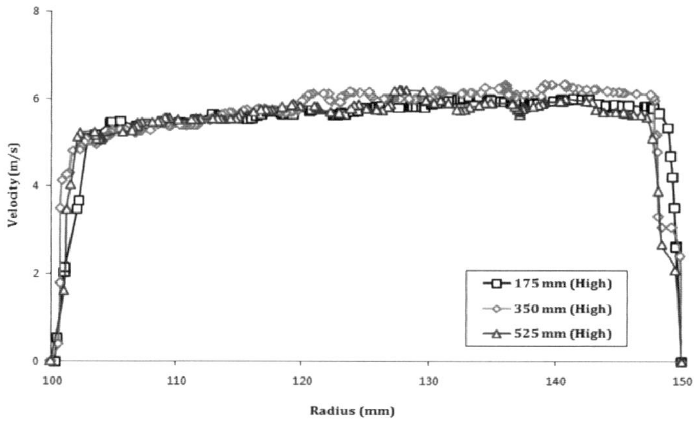

(c)

Figure 3. Tangential velocity distribution for various plenum chamber depths and different number of inlet: (a) Single; (b) Double; and (c) Triple.

Generally, the velocity profile increases along the radius towards the bed wall as a result of swirling motion which generates centrifugal force. This centrifugal force pushes the air to mass at outer periphery towards the bed wall, hence higher momentum is present at this location. The effect is more pronounced at higher inlet velocities as reported by Faizal et al, 2012. However, the velocity of air is zero at the wall itself due to no-slip condition as a result of shear force at wall.

It was evident that more inlets result in higher velocity magnitude due to the increase in overall mass flowrate. For single inlet as in figure 3 (a), lower plenum chamber depth (175 mm) result in noticeable skewed profile but the velocity distribution becomes almost identical as the number of inlet increases (figure 3 (c)). The tangential velocity magnitude for various plenum chamber is summarized in table 2 below.

Table 2. Tangential velocity magnitude for various plenum chamber configuration

Case	Number of Inlet	Plenum Chamber Depth (mm)	Tangential Velocity (m/s)
1		175	1.843
2	Single	350	1.787
3		525	1.776
4		175	3.479
5	Double	350	3.510
6		525	3.645
7		175	5.428
8	Triple	350	5.559
9		525	5.456

Apart from velocity distribution, pressure drop of the system were also extracted from the simulation to gain a better understanding of the system behavior. The pressure drops were distributor pressure drop, which was extracted at equal distances of 10 mm below and above the distributor blades. The findings were shown in figure 4.

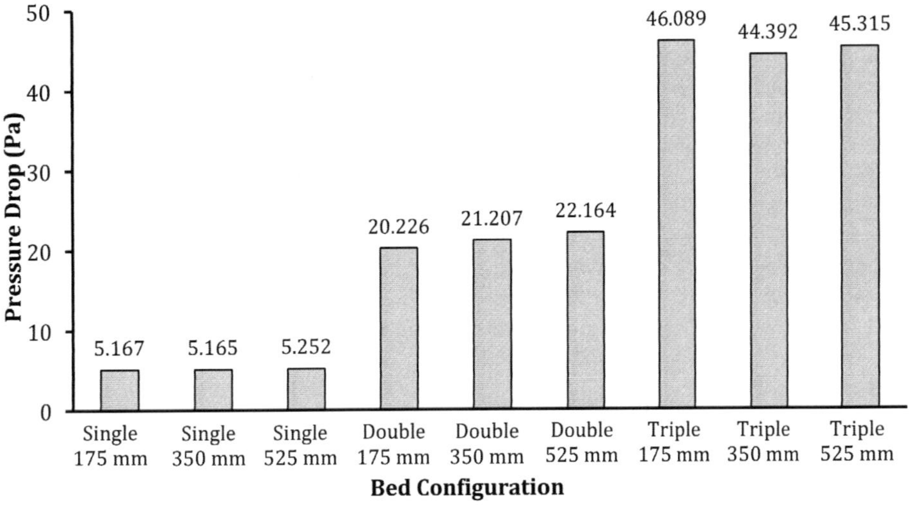

Figure 4. Pressure drop for various plenum chamber configuration.

Naturally the higher number of inlets have higher pressure drop due to increasing amount of air flowrate. Interestingly, only small variation in pressure drop found for different plenum chamber. This shows the strong dependence of pressure drop on air flowrate alone, while plenum chamber depth has negligible effect. From both velocity distribution and pressure drop values obtained, a statistical analysis was conducted to arrive at the optimum design of the plenum chamber from all 9 configurations. The calculated statistical parameters are as in able 3.

Table 3. Statistical parameters

Case	Number of Tangential Entry	Plenum Chamber Depth (mm)	Standard Deviation of Velocity (σ)	Skewness (*S*)	Kurtosis (*K*)
1	Single	175	0.39401	-2.28685	7.5284
2	Single	350	0.34553	-2.81901	1.23656
3	Single	525	0.38584	-2.42711	6.27859
4	Double	175	0.59668	-2.72354	12.2002
5	Double	350	0.77327	-2.82005	8.7888
6	Double	525	0.72372	-3.15304	11.7699
7	Triple	175	1.08481	-2.50476	7.4591
8	Triple	350	1.02326	-2.39202	8.2536
9	Triple	525	0.96792	-2.53916	9.0162

From the statistical analysis summary in table 2 above, it was found that case 2 has better velocity distribution in relative to others. This was indicated by the low standard deviation and kurtosis, implying small variation in tangential velocity as well as less local velocity peaks (flatter velocity profile). However, the air flow in the studied location was skewed to the left towards the centre body. This skew was anticipated due to the centrifugal force from the swirling motion and hence evident for all other configurations.

4. Conclusion

CFD simulation is an effective method to understand the complex phenomena of velocity distribution such as in the swirling fluidized bed. These aerodynamic characteristics are imperative in optimizing the plenum chamber design towards increasing the overall efficiency of the system. It can be concluded that number of inlet has stronger influence on velocity distribution and distributor pressure drop in comparison to plenum chamber depth. From the statistical point of view, case 2: single inlet with 350 mm depth can be considered the best configuration in the preset study.

Acknowledgments

The authors wish to convey their sincere gratitude to the Ministry of Higher Education, Malaysia for the financial support to this study via Fundamental Research Grant Scheme (FRGS). A sincere thanks also goes to Prof. Dr. Vijay R. Raghavan for his comments in preparing this paper.

References

[1] Batcha M F M, Nawi M A M, Sulaiman S A and Raghavan V R 2013 Numerical Investigation of Airflow in a Swirling Fluidized Bed *Asian Journal of Scientific Research* **6** 157 – 66

[2] Batcha M F M, Sreenivasan B, Sulaiman, S A and Raghavan V R 2012 Heat Transfer in a Swirling Fluidized Bed with Geldart Type-D Particles *Korean J. Chem. Eng.* **29 (7)** 862-67

[3] Othman S, Wahab A A and Raghavan V R 2009 Validation by PIV of the numerical study of flow in the plenum chamber of a swirling fluidized bed *CFD Letters* **2 (2)** 78-86

[4] Depypere F, Pieters J G and Dewettinck K 2004 CFD Analysis of Air Distribution in Fluidised Bed Equipment *Powder Tech.* **145** 176 – 89

[5] Faizal M, Seri S M, Al-Hafiz M and Raghavan V R 2012 CFD Studies on Velocity Distribution of Air in a Swirling Fluidized Bed *Advanced Materials Research* **468-471** 25-29

[6] Mohideen M F, Seri S M and Raghavan V R 2012 Fluidization of Geldart Type-D Particles in a Swirling Fluidized Bed *Applied Mechanics and Materials.* **110-116** 3720 – 27

[7] Versteeg H K and Malalasekera W 2007 An introduction to Computational Fluid Dynamics. 2nd ed. Essex: Pearson Education Limited

[8] Safiah O, Wahab A A and V R Raghavan 2008 Numerical Study of The Plenum Chamber of a Swirling Fluidized Bed *Proc. of Int. Conf. on Mechanical & Manufacturing Eng.,* Johor Bahru, Malaysia

[9] Joanest D N and Gill C A 1998 Comparing measures of sample Skewness and Kurtosis *The Statistician*, **47 (1)** 183-89

IOP Conf. Series: Materials Science and Engineering **50** (2013) 012022 doi:10.1088/1757-899X/50/1/012022

Pulverization of sweet potato vine at different mower speeds

Amer N N Kakahy[1,a], D Ahmad[1], M D Akhir[2], S Sulaiman[1] and Ishak A[1]
[1]Faculty of Engineering, University Putra Malaysia, 43400 UPM, Serdang, Selangor, Malaysia
[2]Mechanization and Automation Research Center, MARDI, Serdang, Selangor, Malaysia

Email: [a]amer_kakahy@yahoo.com.

Abstract. The effects of different mower speeds (2358, 2440 and 2553 rpm) and different knife angles (30°, 40° and 50°) on sweet potato vine pulverization were studied. The results indicated that all the treatments were significant at $p < 0.05$ and $p < 0.01$ significance level for the pulverized percentage of sweet potato vine. The 30° knife angle gave the best result with highest vine pulverized percentage of 54.60 %, and a mower speed of 2553 rpm had the finest vine pulverized percentage of 46.99 %. The best performance for interaction effect between knife angle and speed of mower was achieved by the 30° knife angle and a mower speed of 2440 rpm resulting in an average percentage of 61.27% of pulverized vine.

1. Introduction

A review was conducted on different types of mowers, knives typically used to cut the vines and leaves and concluded that the flail type and vertical type mowers sweeping is most effective in the removal of vegetation [1]. Cutting and grass cutting has many variables that have an impact on crushing and some of the important variables are the cutting speed and the percentage of moisture from the vine, and there is a need to establish a model of cutting mechanics and to relate knife parameters to forage material properties [1, 2].

The results indicate that optimisation of cutting speed and blade oblique angle will result in significant energy savings and increased efficiency of miscanthus harvesting machinery [3]. Justin et al. [4] indicated from earlier studies that cutting speed and blade configurations play a significant character in crop harvesting. Variety, knife bevel angle, moisture content and cutting speed are the main factors affecting shearing strength and energy of the wheat stem [5].

The information available is shown to be diverse and largely empirical but probably sufficient for conventional cutting heads to have been developed to near optimal form. Research shows that blades used in forage chopping should have a blade angle in the range 30°- 40° with a rake angle of 10°- 20°, cutting at speeds up to 30 m/s. The optimum radius of the cutting edge is approximately 0.05 mm. A rotary mower has shown minimum power requirements at blade angles between 25° and 30°. A similar result suggests minimum energy requirements for a 25° angle, even though they found slight difference over a range 25°-50° [1], and some have studied the effect of rake angle and found that an angle of 30° gave least energy.

O'Dogherty [1], reported from other studies that a continuous reduction of specific energy with increasing rake angle for maize stalks which indicated that a value of at least 30° was required with an optimum at approximately 50°. Most of the investigations on the effect of blade cutting speed on single stems have been concerned with impact cutting, and found a relatively slow linear fall of about 25 % in specific energy for assemblages of stems as velocity was increased from 20 to 60 m/s. In both laboratory experiments on single stems and field experiments on mowers, the evidence is that a high impact velocity is required. Typical velocities employed by disc and rotary mowers are in the range of 71- 84 m/s [1].

The results for many studies [6, 7] indicate that all the treatments were significant at 99 % significance level for the grass leave area and the length of stem. The best results obtained for the mower speeds at 2500 rpm and 2700 rpm had the best average value of 9.47 and 6.19 cm2 for the leave area and 15.83 and 17.82 cm for length of stem, respectively.

Another study for [8] showed that the best results were at a mower speed of 2500 rpm and 20.37% of grass moisture content which resulted in average values of 81.03 % for the percentage of the leave area and 82.08 % for the percentage of the length of the stem. Kakahy et al. [9], results indicated that all the treatments had significant effects at 99 % significance level for the percentage of grass leave area and the percentage of length of stem.

Study by Jorge et al.[10] found that the power required for pineapple fields shredding increases with the feeding speed and the cutting apparatus angular velocities. Impact cutting of stems required critical speeds of 15 and 25 m/s for the sharp and blunt blades, respectively, and the results for 5 and 20 m/s cutting speed show that there was a reduction in the percentage of uncut stems as rake angle was increased over the range -30 ° to +45 ° [11].

The minimum cutting speed increased from 12.9 to 18.0 m/s for a knife rake angle range of 20-60° as the knife bevel angle was increased from 30 to 70°, and a blade speed of 25 ms-1 was essential to slash most forage material without complexity, also the flail-type devices tended to show a minor reduction in cutting energy requirements as the knife speed increased [12, 13].

Further study of the concept of blade usage is required before it may be usefully applied to the design and operation of rotary mowers [14]. There are requirements for additional, cutting mechanics and for further studies of the cutting action of blades, and there is a need to create a model of cutting mechanics and to relate knife parameters to forage material properties [15, 16].

2. Methods

The study was conducted at the Department of Biological and Agricultural Engineering Laboratory, Faculty of Engineering, University Putra Malaysia, to investigate the effects of three different knife angles (30°, 40° and 50°) and three different speeds (2358, 2440 and 2553 rpm) of a mower on sweet potato vine slashing (pulverizing), at 36.15 % moisture content, wet base (w.b%).

Data were analyzed statistically using ANOVA (analysis of variance) and the least significant difference LSD calculated at $p < 0.05$ and $p < 0.01$ (probability of obtaining a test statistic) to estimate the differences between the averages.

3. Results and Discussion

Table 1 and 2 and figure 1, 2, 3 and 4 indicate that all the treatments had significant effects on the percentage of vine passing through the sieve (< $28 mm^2$) at $p < 0.05$ and $p < 0.01$. The 30° knife angle gave the highest pulverized vine of 54.60% passing through the sieve while the lowest percentage of 36.84% was by the 50° knife angle. The 2553 rpm mower speed gave the highest percentage of vine passing through the sieve at 46.99%. It is well known that speeding up the cutterhead provides a shorter cut for the vine, the results are have a good agreement with conducted results by [6, 7].

On interaction effect between the knife angles and mower speeds, the best result was by the 30° knife angle with 2440 rpm mower speed giving 61.27 % of the percentage of vine; in the meantime the lowest one at 35.71 % was by the 50° knife angle at 2440 rpm mower speed. The results agree with the founding of [1, 6, 7, 16].

Table 1. Analysis of variance (ANOVA)

Source of variation (S.O.V)	Degree of freedom (d.f)	Percentage of vine passing through the sieve (<28mm^2)%
Duplicates	2	
Transactions	8	
Angles of knife (a)	2	723.91[a]
Speeds of mower (v)	2	2.2409[b]
Interaction between (a, v)	4	116.01[a]
Experimental error	16	0.5697
Total	26	
		L.S.D1%=1.5919, L.S.D5%=1.076

[a]significant at $p < 0.01$ level.
[b]significant at $p < 0.05$ level.

Table 2. Factors influencing the percentage of vine passing through the sieve (%).

A	2358	2440	2553	Mean[a]
30°	53.16	61.27	49.37	54.60
40°	49.57	41.01	53.21	47.93
50°	36.44	35.71	38.38	36.84
Mean[b]	46.39	46.00	46.99	

[a]Knife angles.
[b]Mower speeds.

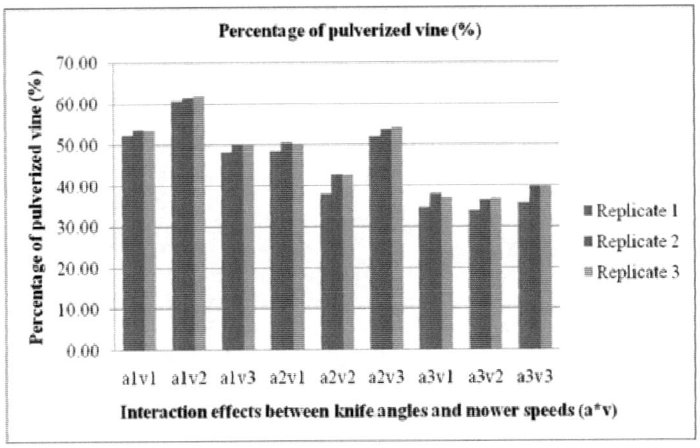

Figure 1. Effects of the treatments on the percentage of vine passing through the sieve (< 28mm^2).

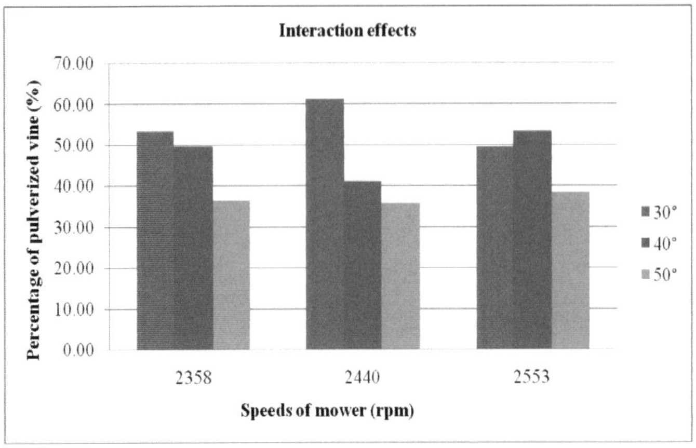

Figure 2. Effects of the interaction on the percentage of vine passing through the sieve (< 28mm^2).

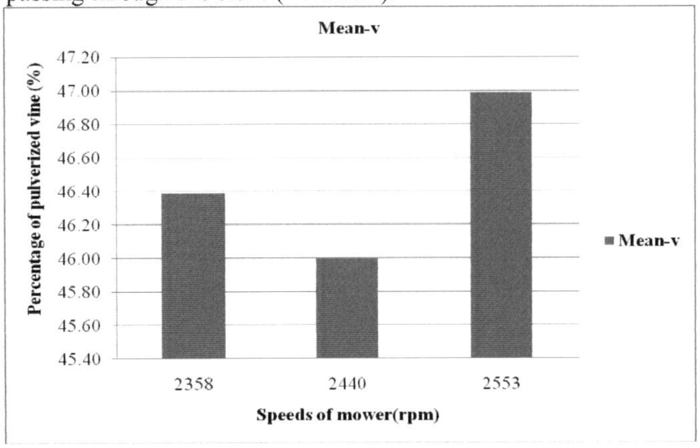

Figure 3. Effects of the speeds of mower on the percentage of vine passing through the sieve (< 28mm^2)

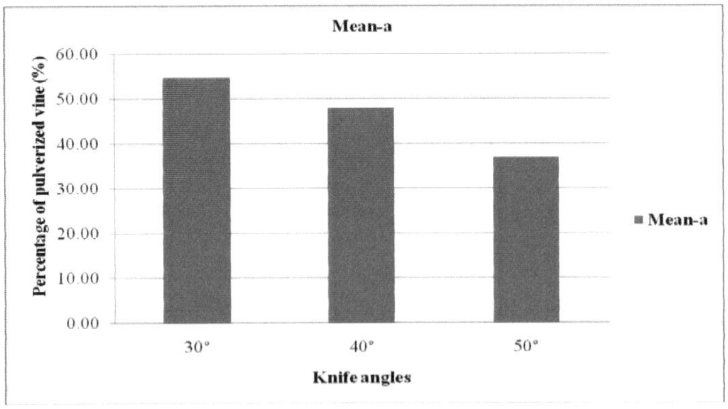

Figure 4. Effects of knife angles on the percentage of vine passing through the sieve (< 28mm^2).

4. Conclusion

The study indicated that the best result on pulverization was achieved by the 30o knife angle at 2440 rpm mower speed, and there was significant difference of all the studied characters. Other mower speeds and knifes angle will be deliberated in the future.

Acknowledgements

Thanks are due to the staff of Agricultural Mechanization & Automation Center, MARDI especially to Mr. Othman Omer Jena , Senior Technician, and Research Assistance M.V. for their generous help and advice during the preparation of this paper. Special thanks to Mr. Hairul Anuar b. Abd Mubin (Technician) and Zainal Abidin b. Abdul Ghani (Assistant Engineer), in the Department of Biological and Agricultural Engineering, Faculty of Engineering, University Putra Malaysia.

References

[1] M J O'Dogherty 1982 A review of research on forage chopping *Journal of Agricultural Engineering Research* **27** 267-89

[2] A N N Kakahy, D Ahmad, M Akhir, A Ishak, and S Sulaiman 2011 Design and development of an integrated slasher (pulverizer) for sweet potato harvester: A Review *1st International Conference in Mechanical Engineering Research (ICMER 2011)*

[3] P C Johnson, C L Clementson, S K Mathanker, T E Grift, and A C Hansen 2012 Cutting energy characteristics of Miscanthus x giganteus stems with varying oblique angle and cutting speed *Biosystems engineering* **112** 42-48

[4] D M Justin, K M Sunil, M F Brian, E G Tony, and C H Alan 2012 Impact of Cutting Speed, Blade Type, and Blade Angle on Miscanthus Harvesting Energy Requirement *American Society of Agricultural and Biological Engineers (ASABE)* **12** 1338182

[5] B Hoseinzadeh, A Esehaghbeygi, and N Raghami 2009 Effect of Moisture Content, Bevel Angle and Cutting Speed on Shearing Energy of Three Wheat Varieties *World Applied Sciences Journal* **7** 1120-23

[6] A N N Kakahy, D Ahmad, M Akhir, A Ishak, and S Sulaiman 2012 Effects of different cutting speeds on Grass pulverizing *International Conference on Green Technology & Ecosystems for Global Sustainable Development 2012 (ICGTEC2012)*, University of Tuzla, Bosnia and Herzegovina

[7] A N N Kakahy, D Ahmad, M Akhir, S Sulaiman, and A Ishak 2012 Effects of Cutting Speeds and Moisture Content on Grass Chopping(Pulverizing) *The International Conference on Agricultural, Food and Biological Engineering (ICAFBE 2012,* China, Guangzhou

[8] A N N Kakahy, D Ahmad, M Akhir, S Sulaiman, and A Ishak 2012 Effects of Types of Slashers and Speeds on Grass Chopping (Pulverizing) *2nd Malaysian International Conference on Trends in Bioprocess Engineering (MICOTriBE 2012)*, Langkawi, Malaysia.

[9] A N N Kakahy, D Ahmad, M Akhir, S Sulaiman, and A Ishak and H Ali 2012 Effects of Types of Slashers and Cutting Speeds on the Quality of Grass Chopping (Pulverizing) at High Moisture Content *International Conference on Mechanical Engineering and Advanced Technology (ICMEAT2012)*, Iran, Isfahan

[10] S P d C F Jorge, G P Francesco, C I Coronel, F G Reina, and P Spugnoli 2009 Power demand of a flail mower during pineapple field demolition Cien *Cien. Inv. Agr.* **36** 59-68

[11] C R Tuck, M J O'Dogherty, D E Baker, and G E Gale 1991 Laboratory studies of the performance characteristics of mowing mechanisms *Journal of Agricultural Engineering Research* **50** 61-80

[12] P S Chattopadhyay and K P Pandey 1999 Effect of Knife and Operational Parameters on Energy Requirement in Flail Forage Harvesting *Journal of Agricultural Engineering Research* **73** 3-12

[13] P S Chattopadhyay and K P Pandey 2001 PM—Power and Machinery: Impact Cutting Behaviour of Sorghum Stalk using a Flail-Cutter—a Mathematical Model and its Experimental Verification *Journal of Agricultural Engineering Research* **78** 369-76

[14] D M McRandal and P. B. McNulty 1978 Impact cutting behaviour of forage crops II. Field tests *Journal of Agricultural Engineering Research* **23** 329-38

[15] M J O'Dogherty 1984 Chop length distributions from forage harvesters and a simulation model of chopping *Journal of Agricultural Engineering Research* **30** 165-73

[16] M J O'Dogherty and G E Gale 1986 Laboratory studies of the cutting of grass stems *Journal of Agricultural Engineering Research* **35** 115-29

Predicting brain acceleration during heading of soccer ball

Zahari Taha, Mohd Hasnun Arif Hassan[a], Mohd Azri Aris and Zulfika Anuar
Centre for Sports Engineering, Faculty of Mechanical Engineering,
Universiti Malaysia Pahang, 26600 Pekan, Pahang, Malaysia.

E-mail: [a]hasnun@ump.edu.my

Abstract. There has been a long debate whether purposeful heading could cause harm to the brain. Studies have shown that repetitive heading could lead to degeneration of brain cells, which is similarly found in patients with mild traumatic brain injury. A two-degree of freedom linear mathematical model was developed to study the impact of soccer ball to the brain during ball-to-head impact in soccer. From the model, the acceleration of the brain upon impact can be obtained. The model is a mass-spring-damper system, in which the skull is modelled as a mass and the neck is modelled as a spring-damper system. The brain is a mass with suspension characteristics that are also defined by a spring and a damper. The model was validated by experiment, in which a ball was dropped from different heights onto an instrumented dummy skull. The validation shows that the results obtained from the model are in a good agreement with the brain acceleration measured from the experiment. This findings show that a simple linear mathematical model can be useful in giving a preliminary insight on what human brain endures during a ball-to-head impact.

1. Introduction

Soccer is the most popular sport in the world played by millions of people. There are currently 265 million players worldwide that are actively involved in the game of soccer [1]. The uniqueness of this game is that the players are allowed to use their heads to direct the ball to the teammates or even trying to score goal. Thus, soccer players are susceptible to head injuries and concussions. Studies have shown that more than 70% of concussed soccer players did not realize that they had suffered concussion [2].

Most of the concussions in soccer occur due to head-to-head contact. Other factors include the impact between the head and elbow, knee, ground and also the impact with the ball itself that occurs during purposeful heading [3]. In a single game of soccer, a player can be subjected to six to twelve heading occasions [3–5]. A professional soccer player experiences a median of 800 heading occasions every season excluding the headings that take place during training sessions [3]. There has been a long debate among researchers of whether purposeful heading can lead to brain injury. Several studies have shown some results that support the argument that intentionally heading the ball in soccer can cause mild traumatic brain injury (TBI).

Tysvaer and Lochen have examined 37 former soccer players from Norway through a comprehensive series of psychological tests [6]. The tests have shown that 81% of the former soccer players exhibited mild to severe impairments regarding attention, concentration, memory and judgement. Moreover, ten regular headers showed a higher degree of severe to gross neuropsychological impairment compared to non-headers. They have concluded that the impact of heading the ball to the head shows convincing evidence of brain damage similar to that found in patients who have sustained minor head injuries.

Another study was conducted to determine whether amateur soccer players exhibit the sign of chronic TBI [5]. Thirty-three amateur soccer players were studied alongside 27 amateur swimming and track athletes that served as control group. They have found that amateur soccer players showed lower performance on tests of planning and memory compared to the control group. Furthermore, they have discovered that the number of headings is inversely related to the neuropsychological performance; hence suggest that heading in soccer can lead to chronic TBI.

Matser et al. have also examined a group of 84 professional soccer players through neuropsychological tests to determine the relation between the number of headings in a season and the number of soccer-related concussions with the cognitive function [7]. It was found that increasing number of headings has resulted in poorer performance in tests regarding focused attention and memory, whereas the number of soccer-related concussions was proportional to poorer results on sustained attention and visuoperceptual processing evaluation.

A recent study has revealed that repetitive heading occasions experienced by soccer players could lead to degeneration of brain cells [8]. In this study, 37 amateur soccer players were examined using diffusion tensor imaging (DTI), an advanced magnetic resonance (MR) technique. They have measured the movement of water molecules along nerve fibres called axons. This measurement is known as fractional anisotropy (FA), in which in a healthy brain, the FA values are high. They have found that frequent headers had notably lower FA values in five brain regions that are responsible for attention, memory, executive functioning and higher-order visual functions. They have also discovered a threshold level of approximately 885 to 1,550 headings per year, in which exceeding this threshold will cause the FA values to decrease significantly.

Those studies have shown that there is a need to explore what actually happens to the brain during heading occasion. The acceleration of the brain during ball-to-head impact is the cause of concussion that can lead to traumatic brain injury. Measuring the acceleration of a human brain during an impact is almost impossible. Previous studies have only measured the head acceleration during heading, but none has attempted to specifically measure the brain acceleration. This study proposes a linear mathematical model that measures the acceleration of the brain during an impact with a ball. From the simulation, not only brain acceleration can be measured, but also the duration of the brain vibration.

2. Proposed mathematical model

A two-degree-of-freedom linear mathematical model is developed as shown in figure 1. It is assumed that a ball that travels with a defined velocity impacts the skull that is represented by a mass m_1. It is attached to a spring (stiffness, k_1) and a dashpot (damping coefficient, c_1) that represent the neck. Another mass m_2, which symbolizes the brain, is attached to the skull by a spring-damper system (stiffness, k_2 and damping coefficient, c_2) that represents the suspension characteristics of the brain.

The mass of the skull is taken as $m_1 = 3.5$ kg [9], whereas the mass of the brain is taken as $m_2 = 1.3$ kg [10]. The stiffness and damping coefficient of the neck are taken as 1800 kNm^{-1} and 450 Nsm^{-1} respectively [11], whereas for the brain, the stiffness is defined as 156 kNm^{-1} and the damping coefficient is taken as 340 Nsm^{-1} [12]. The acceleration of the skull and the brain are given by following equations:

$$m_1\ddot{x}_1 = u + m_1 g - k_1 x_1 - c_1\dot{x}_1 + k_2\left(x_2 - x_1\right) + c_2\left(\dot{x}_2 - \dot{x}_1\right) \quad (1)$$

$$m_2\ddot{x}_2 = m_2 g - k_2\left(x_2 - x_1\right) - c_2\left(\dot{x}_2 - \dot{x}_1\right) \quad (2)$$

Equation (1) and equation (2) are second-order differential equations, which are numerically solved using Euler method with step size of 0.0001.

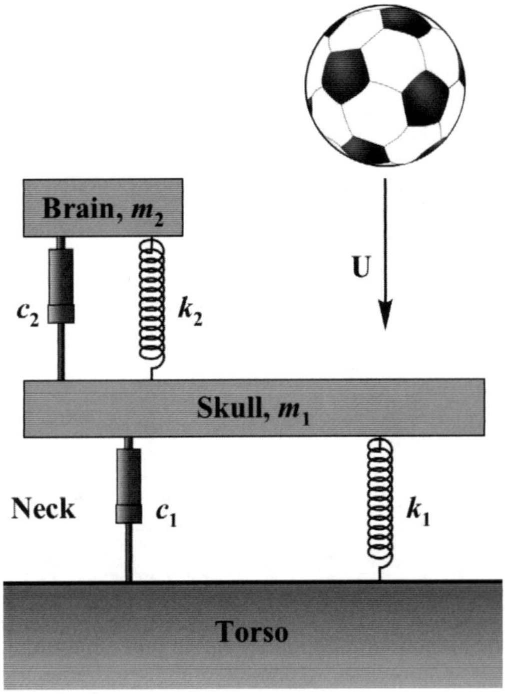

Figure 1. Proposed mathematical model.

3. Model validation

A dropped-ball experiment was conducted to validate the results obtained from the proposed mathematical model. A ball was dropped from different heights (0.5, 1.0, 1.5 and 2.0 m) onto an instrumented dummy skull, thus creating different ball velocities. The dummy skull (figure 2) is made of ABS and fabricated through rapid prototyping technique. Its inner cavity was filled with ultrasound gel (type EcoGel 200, Eco-Med Pharmaceutical Inc.) that represents the brain, and an accelerometer (model 356A67, PCB Piezoelectronics) was placed inside the gel (approximately at the centre of gravity of the skull) to measure the acceleration of the gel upon impact. The acceleration of the gel is assumed to represent the brain's acceleration during ball impact. Figure 3 illustrates the experimental setup.

A high-speed camera (model SV643C, EPIX Inc.) was used to record the motion of the ball. The camera is placed on flat surface parallel to the dropping position of the ball. The camera was set to record 1,000 frames per second. From the high-speed videos, the duration of impact as well as the velocities of the ball before and after impact were measured. The impact duration measured was 11 ms and the value was defined in the model. Further, the coefficient of restitution calculated from the ball velocities is equal to 0.88.

For each height, the experiment was repeated three times and the average peak brain accelerations were recorded. The tri-axial accelerometer was connected to a data acquisition device (model NI cDAQ-9171, National Instruments) and the accelerations were recorded using commercial data acquisition software. The data was then imported into spread sheet software for further analysis.

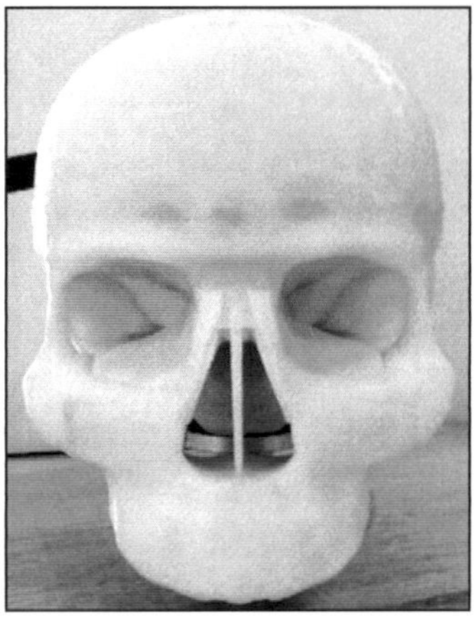

Figure 2. Instrumented dummy skull.

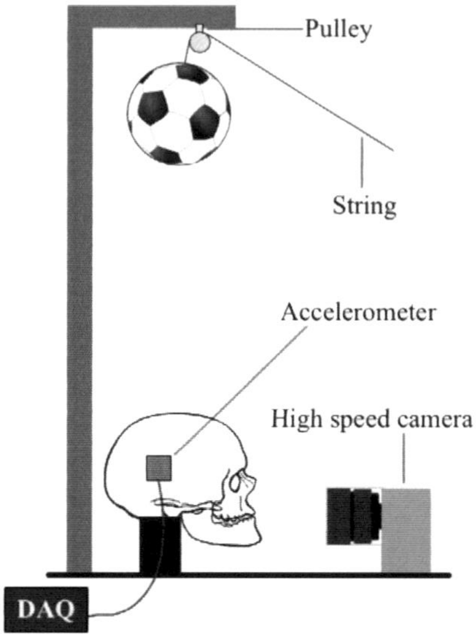

Figure 3. Dropped-ball test setup.

4. Results and discussions

Table 1 shows the comparison between peak brain acceleration obtained from the experiment and the mathematical model. It is shown that peak brain accelerations obtained from the mathematical model are in a good agreement with the experimental data. Differences of less than 10% suggest that the values are acceptable.

Table 1. Comparison between peak brain accelerations from experiment and model.

Dropped-height (m) / Ball velocity (ms^{-1})	Peak brain acceleration (g)		Difference
	Experiment	*Model*	
0.5 / 2.93	3.88	3.98	3%
1.0 / 3.93	5.32	5.32	0%
1.5 / 4.83	6.25	6.53	4%
2.0 / 5.55	8.18	7.49	8%

Aside of the peak acceleration value, the acceleration curve obtained from both experiment and proposed model was also compared as shown in figure 4. Figure 4(a) shows the gel acceleration profiles for different dropped-heights, whereas figure 4(b) depicts the brain acceleration profiles generated by the proposed model. Apart from the similarity in pattern, it is observed from figure 4(a) that the duration of the gel acceleration is approximately 50 ms, which concurs with the duration of brain acceleration predicted by the model as shown in figure 4(b). Further, it took approximately 24 ms for the gel to reach peak acceleration, whilst the model predicted that the time taken to reach peak acceleration is about 16 ms. The difference is only 8 ms and therefore it is still considered as acceptable.

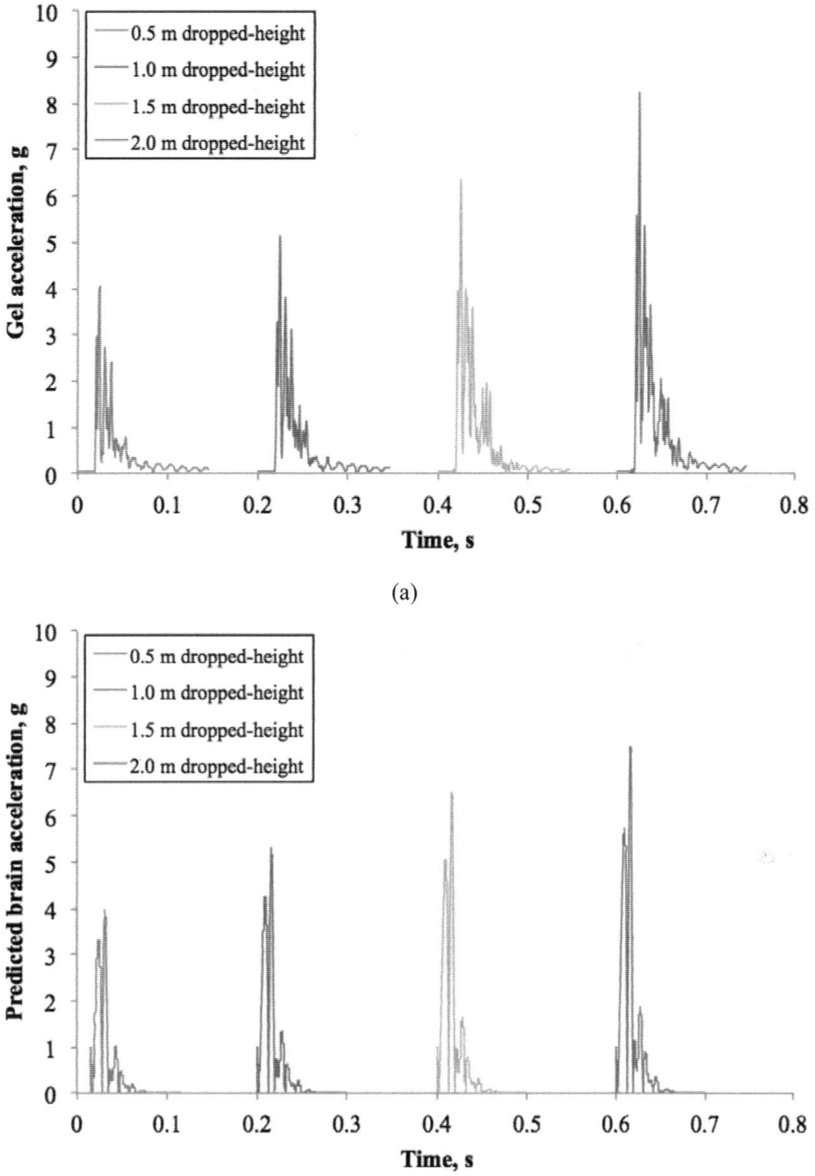

(b)

Figure 4. Comparison of acceleration profiles. (a) Gel accelerations obtained from dropped-ball test. (b) Brain accelerations obtained from the model.

5. Conclusion

A two-degree-of-freedom linear mathematical model was proposed to estimate the linear acceleration endured by the brain during ball-to-head impact in soccer. The model was validated through dropped-ball test on a dummy skull. It is shown that the data obtained from both experiment and model are in a good agreement. This proves that a simple mathematical model is useful in giving an estimation of the acceleration endured by the brain during a ball-to-head impact in soccer. Using mathematical model is beneficial since it offers the flexibility in varying the parameters involved during the impact.

Acknowledgments

The authors would like to thank the Universiti Malaysia Pahang and the Ministry of Higher Education, Malaysia for giving the funding for this research under the grant RDU110207. Sincere thanks are expressed to the Centre for Product Design and Manufacturing (CPDM), University of Malaya for providing the dummy skull for the experiment.

6. References

[1] FIFA.com 2006 Big Count - FIFA.com

[2] Delaney J, Lacroix V, Leclerc S and Johnston K 2002 Concussions among university football and soccer players *Clinical Journal of Sport Medicine* **12** 331–38

[3] Levy M L, Kasasbeh A S, Baird L C, Amene C, Skeen J and Marshall L 2012 Concussions in soccer: a current understanding *World Neurosurgery* **78** 535–44

[4] Spiotta A M, Bartsch A J and Benzel E C 2012 Heading in Soccer: Dangerous Play? *Neurosurgery* **70** 1–11

[5] Matser E, Kessels A, Lezak M, Jordan B and Troost J 1999 Neuropsychological Impairment in Amateur Soccer Players *JAMA: The Journal of the American Medical Association* **282** 971–73

[6] Tysvaer A and Løchen E 1991 Soccer injuries to the brain: A neuropsychologic study of former soccer players *The American Journal of Sports Medicine* **19** 56–60

[7] Matser J T, Kessels a G, Lezak M D and Troost J 2001 A dose-response relation of headers and concussions with cognitive impairment in professional soccer players *Journal of Clinical and Experimental Neuropsychology* **23** 770–74

[8] Lipton M, Kim N and Zimmerman M 2013 Soccer Heading Is Associated with White Matter Microstructural and Cognitive Abnormalities *Radiology*

[9] Gekhman D and Elert G 2006 Mass of a Human Head

[10] Scott R and Elert G 2003 Mass of a Human Brain

[11] Dirisala V, Karami G and Ziejewski M 2012 Effects of neck damping properties on brain response underimpact loading *International Journal for Numerical Methods in Biomedical Engineering* 472–94

[12] Baghaei S, Sadegh A and Rajaai S 2009 A Mathematical Head/Brain Model for Investigation of Damping Characteristics of SAS in Low Velocity Head Impacts *25th Southern Biomedical Engineering Conference 2009, 15 – 17 May 2009, Miami, Florida, USA* **24** 283–86

2nd International Conference on Mechanical Engineering Research (ICMER 2013) IOP Publishing
IOP Conf. Series: Materials Science and Engineering **50** (2013) 012024 doi:10.1088/1757-899X/50/1/012024

Determination of air side heat transfer coefficient in a mini-channel heat exchanger using Wilson Plot method

K K Thoo[1,a], W M Chin[2] and M R Heikal[3]
[1]OYL Research & Development Centre Sdn Bhd, Lot 60334 Persiaran Bukit Rahman Putra 3, Taman Perindustrian Bukit Rahman Putra, 47000 Sungai Buloh, Selangor Darul Ehsan, Malaysia
[2]OYL Research & Development Centre Sdn Bhd, Lot 60334 Persiaran Bukit Rahman Putra 3, Taman Perindustrian Bukit Rahman Putra, 47000 Sungai Buloh, Selangor Darul Ehsan, Malaysia
[3]University Teknologi Petronas, Bandar Seri Iskandar, 31750, Tronoh, Perak Darul Ridzuan, Malaysia

Email: [a]thookk@oyl.com.my

Abstract. In this study, the air side heat transfer coefficient of an aluminium mini-channel heat exchanger was investigated for single-phase flow in the mini-channel, with water in the tubes and air on the outside. Research methods included hydraulic tests on a single mini-channel tube, Wilson Plot experiments and experiment validation. Results obtained from the hydraulic test showed that turbulent flow occurred in the tube at a Reynolds number of 830. Wilson Plot experiments were conducted to determine air side heat transfer coefficient of the heat exchanger. The tube side Reynolds number was maintained above 1000 to ensure turbulent flow and tube side heat transfer coefficient was calculated using Gnielinski equation for turbulent flow. The air side heat transfer coefficients obtained from the Wilson Plot experiments were in good agreement with known correlations. The outcome of this study is to use the air side heat transfer coefficient to calculate the performance of refrigerant condensers for different tube pass ratios and flow pass configurations.

1. Introduction

Past research on mini-channel heat exchangers had focused on condensation heat transfer, pressure drop, two-phase flow patterns and system configuration while evaporator mini-channels research has gained much popularity in recent years. Due to the high price of copper, aluminium Mini-Channel Heat Exchangers, or in short MCHX, are gaining popularity to replace the conventional copper-aluminium air-conditioning heat exchangers. MCHX are made from extruded aluminium tubes brazed to aluminium louvered fins in a temperature controlled environment using a process such as the Nocolok brazing. The tubes are brazed at both ends to aluminium manifolds or headers. Thus the entire construction is from aluminium.

Air is drawn by a blower and passes through the matrix of the fins and tubes of the MCHX while the refrigerant passes through the tubes. The advantages of MCHX are high heat transfer surface area per unit volume (A/V), light weight and resistance to galvanic corrosion.

As in the conventional heat exchangers, the MCHX is also subject to air and refrigerant side flow maldistribution. An example of air side maldistribution in conventional finned-heat exchangers had been studied by Chin and Raghavan [1]. Various factors that contribute to refrigerant flow maldistribution include the heat exchanger geometry, operating conditions, multiphase flow and fouling [2]. Maldistribution causes a degradation in the heat exchanger performance which would mean less efficient heat exchange as well as more energy required to overcome the pumping losses.

Other problems that would bring about a high pressure drop and performance degradation to the mini-channel heat exchanger are the tube pass ratio and number of passes.

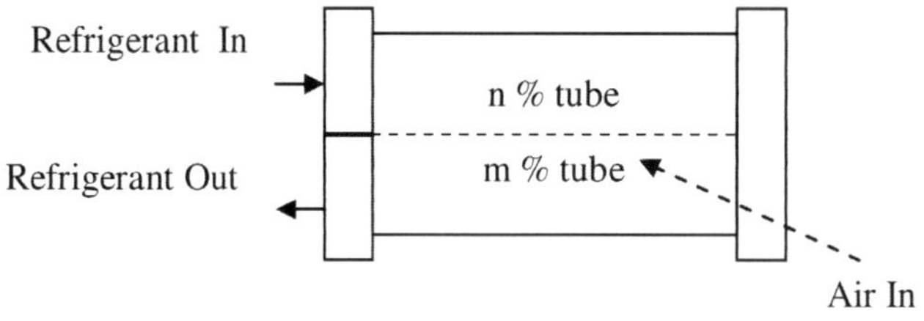

Figure 1. Division of the Mini-Channel according to Tube Pass Ratio.

Table 1. Type of Mini-Channel used in this Study according to Different Tube Pass Ratio.

Type	1	2	3	4	5
n	50	60	70	80	90
m	50	40	30	20	10

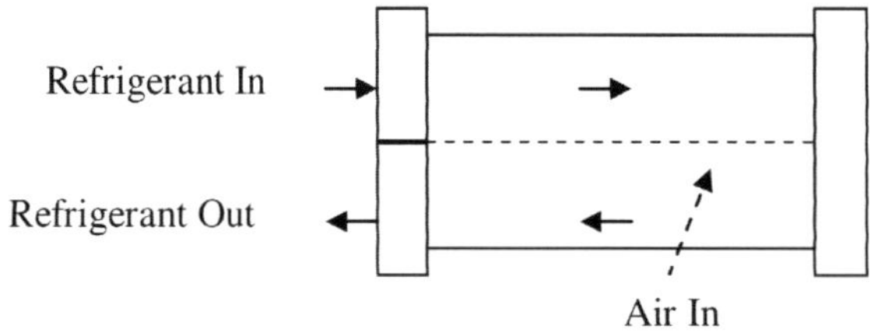

(a)

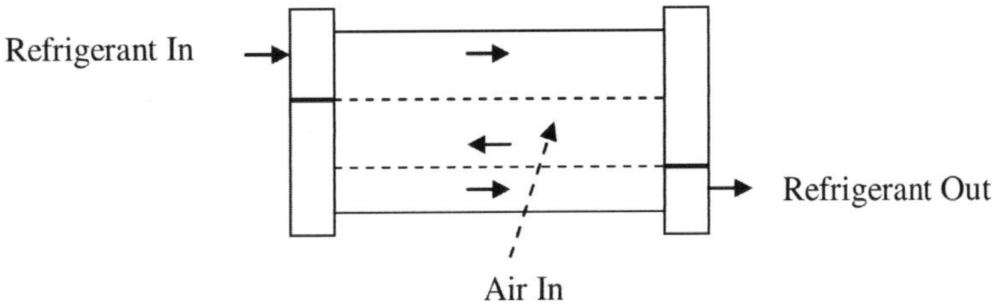

(b)

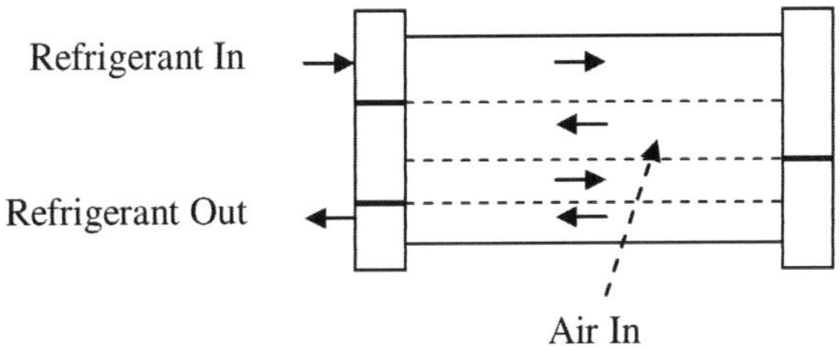

(c)

Figure 2. (a) Two Pass Flow Configuration, (b) Three Pass Flow Configuration, (c) Four Pass Flow Configuration.

With equal flow passes in all configurations shown above, the heat transfer and pressure drop would be sub-optimal. At the inlet the volume flow rate would be high in view of the high vapor fraction; the velocities would be high leading to a good thermal performance. Due to condensation of vapor, the velocities will be lower in the second pass giving rise to low pressure drop but also poor heat transfer. The aim of good design would be to balance the number of tubes in each pass as well as the number of passes in such a way as to get the maximum heat transfer at minimum pressure drop. In turn this would mean a higher flow area in the first pass.

The overall scope of this work is to investigate different tube pass ratio which includes 50/50 as a baseline case, 60/40, 70/30, 80/20 and 90/10 and different number of flow configuration which includes 2 as a baseline case, 3 and 4 pass.Water is used as the tube-side working medium in the hydraulic test and Wilson Plot experiments. For the Wilson Plot experiments, the water and air side velocities range between 1.0-3.0 m/s.

2. Literature Review

A literature survey was undertaken to review the research on various design considerations for heat exchangers that cause degradation of their performance, with focus given to mini-channel heat exchangers. The outcome of this survey showed that the proposed work has not been carried earlier out by any researcher.

A study of flow maldistribution was conducted by Bassiouny and Martin [3,4] in plate heat exchangers (PHE). The study used a theoretical calculation method to determine flow distribution and pressure drop in plate heat exchangers for the U-type, where both the inlet and outlet ports were on the same side and also the Z-type, where the inlet and outlet ports were on opposite sides.

Another study conducted by Rao et al [5] looked into the effect of port-to-channel flow maldistribution on the thermal behavior of multipass plate heat exchangers (PHE). The variation of the heat transfer coefficient due to flow variation from channel to channel was also considered. It was found that multipass heat exchangers were generally found to be less affected by flow maldistribution compared to single pass heat exchangers.

Hwang et al [6] studied the refrigerant inlet location in the manifold and compared the behavior of side-inlet and end-inlet locations. The side-inlet location showed better liquid refrigerant flow distribution than the end-inlet location by more effectively mixing the liquid and vapor refrigerant from the inlet. They found that by improving refrigerant maldistribution through the manifold inlet, the mini-channel evaporator performance could be improved.

Kandlikar et al [7] stated that the pressure drop penalty needs to be carefully evaluated when selecting the channel size. The pressure gradient (pressure drop per unit length) increases drastically with channel size reduction [8] and therefore mini-channel heat exchangers need to be suitably designed to provide short flow lengths to limit the overall pressure drop.

Cavallini et al. [9] studied condensation heat transfer and pressure drop in a horizontal smooth tube using a new HFC refrigerant. The effects of vapour quality, mass velocity, saturation temperature, driving temperature difference and reduced pressure were discussed. It was discovered that low-pressure and mid-pressure refrigerants such as R22 perform better than high pressure refrigerants such as R410A and that at the same mass velocity and vapor quality, low pressure fluids such as R22, on the other hand, show a higher pressure drop penalty than R410A.

Del Col et al. [10] studied the effect of the cross sectional shape on the condensation behaviour in a single square mini-channel and compared it to the circular mini-channel. It was discovered that at lower mass fluxes the square channel yielded higher heat transfer coefficients than that of a circular channel. This was thought to be due to the effect of surface tension pulling the liquid towards the corners and reducing the average thermal resistance.

Kandlikar and Schmitt [11] studied the effect of surface roughness on the pressure drop in single phase flow in a mini-channel pipe and found that the average relative roughness value ε / D_h was greater than the threshold value of 0.05 used in the Moody diagram. They introduced three surface characterization parameters together with equivalent roughness and constricted hydraulic diameter. Brackbill and Kandlikar [12] proved that using these parameters, the relationship between the critical Reynolds value and $\varepsilon / D_{h,cf}$ agree theoretically and experimentally.

Using known correlations and computer software, Subramaniam and Garimella [13] studied the parameters that influence the performance of R410A micro-channel condensers and suggested an optimum condenser design from a baseline design.

3. Methodology

The effects of the different tube pass ratios and number of flow passes of the mini-channel are to be compared in terms of thermal and hydraulic performance. In order to do this, an experiment is conducted using the Wilson Plot Method to determine the air side heat transfer coefficient, h_o and overall heat transfer coefficient, U_o. As the flow of refrigerant inside the tube must be turbulent flow for the Wilson Plot experiment, the turbulent flow Reynolds number, Re_t is first determined from flow experiments using a single tube.

For the single tube test, an experimental rig (figure 3) is set-up and water is used as the medium flowing inside a single mini-channel. Temperature and pressure drop on the water side across the tube and water flow rate are measured using resistance temperature detectors (RTD), differential pressure transmitter and magnetic flow meter respectively. Data collected from this experiment are used to calculate Re and Fanning friction factor, f_F and using (1) and (2):

$$\mathrm{Re} = \frac{\rho v d_h}{\mu} \tag{1}$$

where ρ is the density of the liquid, v is the velocity of the fluid, d_h is the hydraulic diameter of the mini-channels and μ is the viscosity of the liquid

$$f_F = dP\left(\frac{d_h}{L}\right)\left(\frac{1}{2\rho v^2}\right) \tag{2}$$

where f_F is the Fanning friction factor, L is the tube length, v is the flow velocity and dP is the pressure drop across the tube.

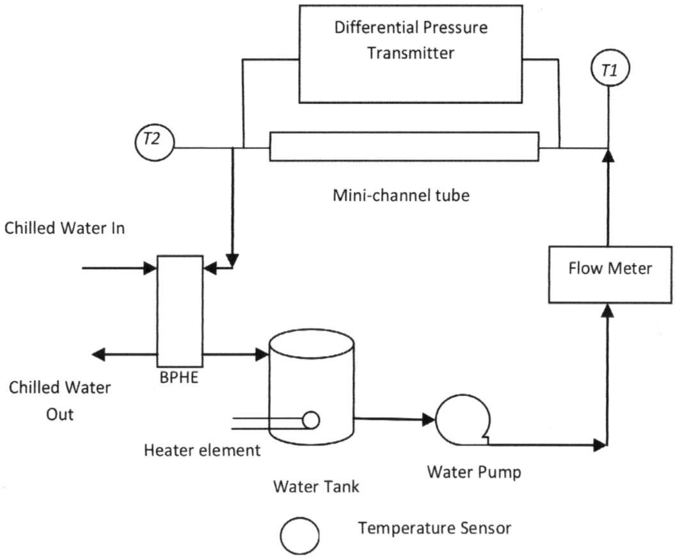

Figure 3. Schematic of Experimental Rig to determine Re_t.

The f_F and Re values are then used to plot a f_F vs. Re chart from which Re_t is determined.

Water is used as the medium inside the mini-channel and the water side heat transfer coefficient h_i is calculated using the Gnielinski equation.

$$h_i = \frac{(\mathrm{Re}-1000)\,\mathrm{Pr}(f/2)(k/d_h)}{(1+12.7(\mathrm{Pr}^{2/3}-1)(f/2)^{0.5})} \tag{3}$$

where Re is the Reynolds number, Pr is the Prandtl No, f is Fanning the friction factor, k is the thermal conductivity and d_h is the hydraulic diameter of the mini-channel.

From the Wilson Plot experiment data, h_o and U_oA_o can be determined.

$$\frac{1}{U_o} = \frac{1}{n_o h_o} + R_w + \frac{A_o}{h_i A_i} \tag{4}$$

where U_o is the overall heat transfer coefficient, h_o is the air side heat transfer coefficient, h_i is the water side heat transfer coefficient, η_o is the surface effectiveness, R_w is the tube wall thermal resistance, A_o is the is the total heat transfer area of external fin and tube and A_i is the internal tube heat transfer area.

$$\eta_o = 1 - \frac{A_f}{A}\left(1 - \eta_f\right) \tag{5}$$

$$\eta_f = \frac{\tanh(mL)}{mL} \tag{6}$$

$$m = \left(\frac{2h}{kt}\right)^{1/2} \tag{7}$$

where A_f is the entire fin surface area, A is the total surface area (fin plus exposed base), n_f is the efficiency of a single fin, h is the convective coefficient of the fin, k is the thermal conductivity of the fin and t is the thickness of the fin.

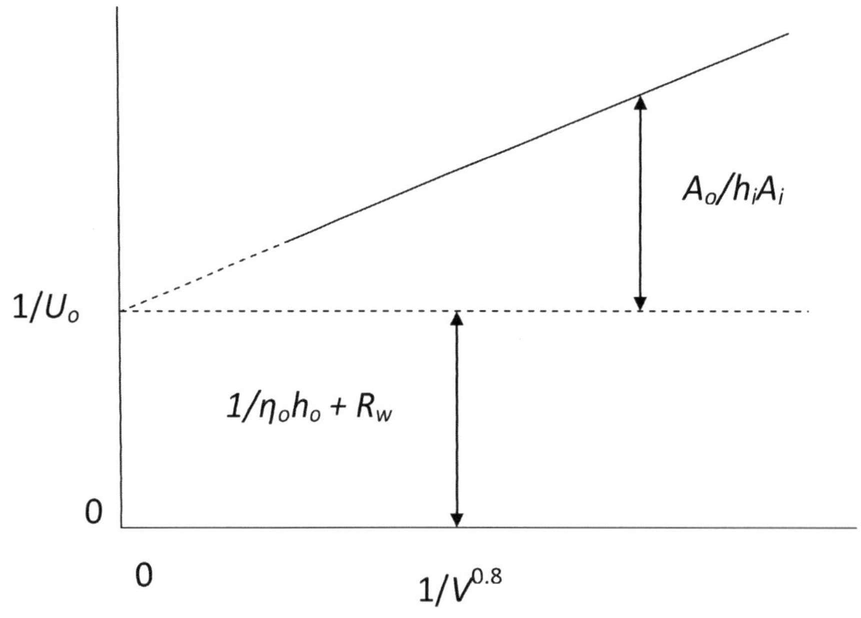

Figure 4. $1/U_o$ versus $1/V^{0.8}$ determined from Wilson Plot Experiment.

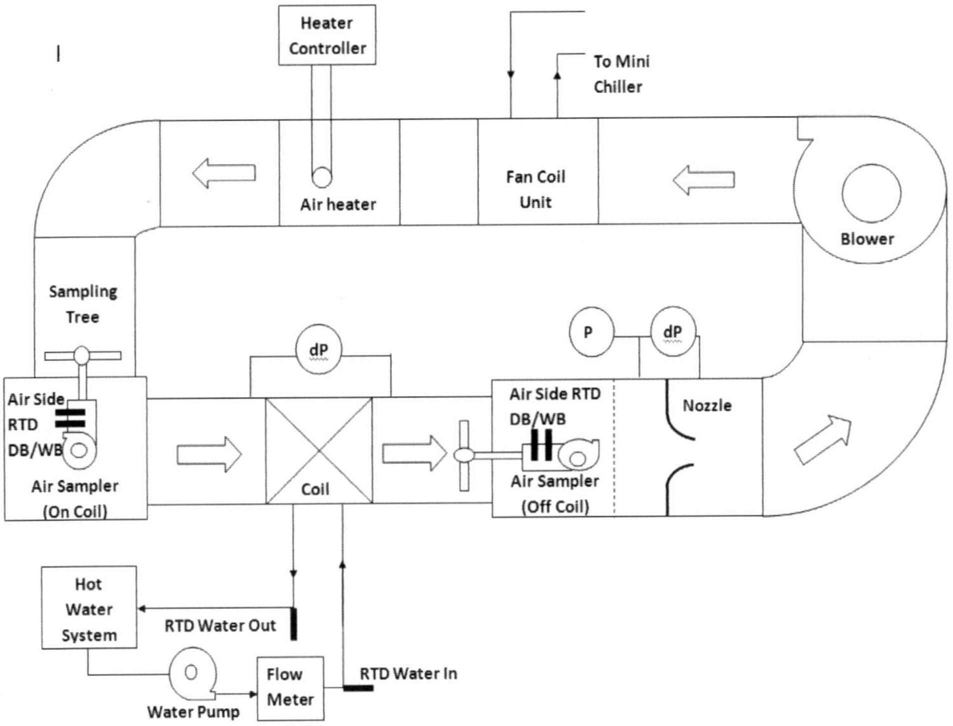

Figure 5. Wilson Plot Test Rig Set-Up.

The Wilson Plot test set-up is shown in figure 5. The air side and water side temperatures are measured using resistance temperature detectors (RTD). The air and water flow rate are measured using air nozzle and flow meter respectively. Air temperature is controlled at the air side inlet to the mini channel heat exchanger (coil) using chilled water fan coil unit and air heaters while water temperature is controlled at the water side inlet to the heat exchanger using water heaters. A frequency inverter is used to control the water pump in order to maintain the flow rate of the water.

All h_o values determined from the Wilson Plot are used to calculate the performance of the mini-channel heat exchanger. A discrete element analysis is used to calculate the performance of the mini-channel heat exchanger for different tube pass ratios and flow pass configurations. This is done by dividing the heat exchanger into a number of discrete elements and using the *ε-NTU* method to evaluate the local heat transfer rate q_i of each element. The sum of q_i values yields the total heat transfer rate Q, of the mini-channel heat exchanger. The pressure drop across the mini-channel heat exchanger is calculated using Bassiouny and Martin analysis method [3,4]

4. Results and Discussion

The Reynolds numbers for turbulent flow to initiate in the mini-channel tube is 830. See figure 6. The Wilson Plot Test are run at *Re* above 1000. The Wilson Plot experiment results are shown in table 2.

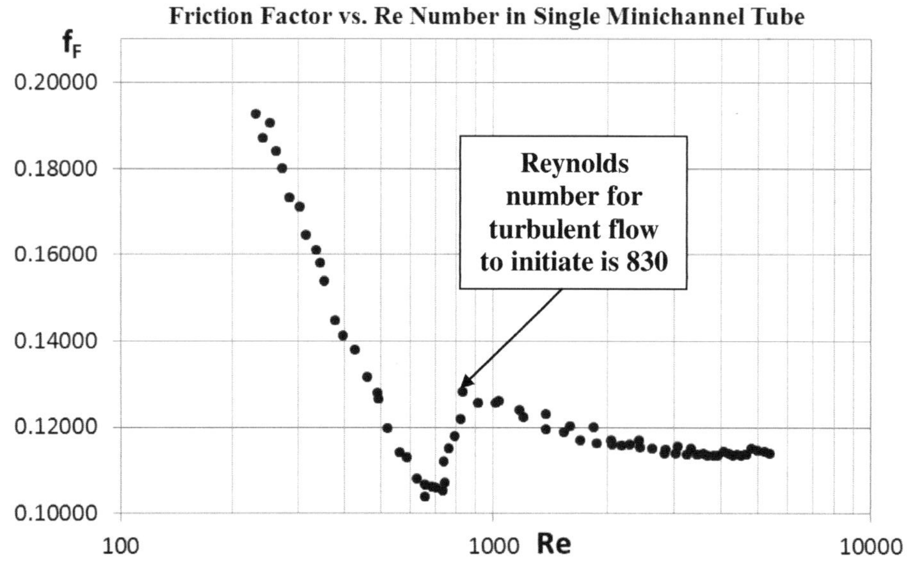

Figure 6. Friction factor vs. Reynolds Number in Single Mini-Channel Tube.

Table 2. Air Side Heat Transfer Coefficient vs. Frontal Air Velocity.

$n_o h_o$, ($W/m^2.K$)	243.38	225.11	199.58	185.82	178.37	121.64
Frontal Air velocity (m/s)	8.20	7.29	6.39	5.47	3.64	1.83

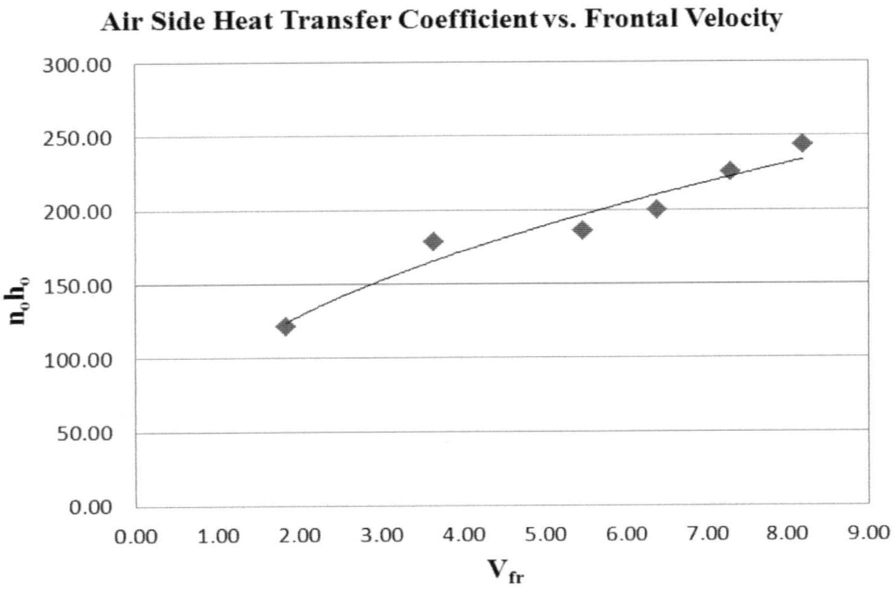

Figure 7. Air Side Heat Transfer Coefficient vs. Frontal Air Velocity.

A graph is plotted as shown in figure 7. The fitted equation of the curve is $y = 95.6x^{0.42}$. The exponent is 0.42. This agrees well with the general equation for flow over cylinders shown in (8) [14]. The small deviation is due to the flattened tube over which air flows in the MCHX.

$$h = 2.755V^{0.471} / D^{0.529} \quad (8)$$

The value of h_o thus determined is to be used to calculate the overall heat transfer coefficient U of the mini-channel heat exchanger. In order to arrive at the best arrangement, the outlet state of the refrigerant is to be computed for different tube pass ratios and flow pass configurations, by utilizing several correlations for MCHX passages.

5. Conclusions

The Reynolds number at which turbulent flow initiates in the mini-channel tube is 830 as determined from the hydraulic test. Results of the Wilson Plot experiment show a close agreement with known correlations for the air-side. Therefore the h_o values determined from the experiment are suitable to be used to calculate the performance of the mini-channel heat exchanger for different tube pass ratios and flow pass configurations which is a sequel to this study.

Acknowledgements

The first author would like to thank Professor Vijay R. Raghavan for his invaluable advice and OYL Research & Development Centre Sdn Bhd who has financially supported the experiment rig and test equipment.

References

[1] Chin W M and Raghavan V R 2011 On the adverse influence of higher statistical moments of flow maldistribution on the performance of a heat exchanger *Int'l J. Thermal Sci.*, **50**(4), 581-91

[2] Rao B P, Kumar P K and Das S K 2002 Effect of flow distribution to the channels on the thermal performance of a plate heat exchanger *Chem. Eng. and Processing*, **41**(1), 49-58

[3] Bassiouny M K and Martin H 1984 Flow distribution and pressure drop in plate heat exchangers – I , *Chem. Eng. Sci.*, **39**(4), 693-00

[4] Bassiouny M K and Martin H 1984 Flow distribution and pressure drop in plate heat exchangers – II , *Chem. Eng. Sci.*, **39**(4), 701-04

[5] Rao B P and Das S K 2004 Effect of Flow Distribution to the channels on the thermal performance of the multipass plate heat exchangers *Heat Transfer Eng.* **25**(8) 48- 59

[6] Hwang Y, Jin D H and Radermacher R 2007 Refrigerant distribution in mini-channel evaporator manifolds *ASHRAE HVAC&R* **13**(4) 543-55

[7] Kandlikar S G 2007 A roadmap for implementing mini-channel in refrigeration and air-conditioning systems – current status and future directions *Heat Transfer Eng.* **28**(12), 979-85

[8] Kandlikar S G and Grande W J 2003 Evolution of microchannel flow passages-thermohydraulic performance and fabrication technology *Heat Transfer Eng.* **24**(1), 3-17

[9] Cavallini A, Censi G, Del Col D and Dorettti L 2001 Experimental investigation on condensation heat transfer and pressure drop of new HFC refrigerant, *Int'l J. of Refrigeration* **24**(1) 73-87

[10] Del Col D, Bortolin S, Cavallini A and Matkovic M 2011 Effect of cross sectional shape during condensation in a single square mini channel *Int'l J. Heat and Mass Transfer* **54** 3909-20

[11] Kandlikar S G, Schmitt D, Carrano A L and Taylor J B 2005 Characterization of surface roughness effects on pressure drop in single-phase flow in mini channels *Phys. of Fluids* **17** 100606

[12] Brackbill T P and Kandlikar S G 2007 Effect of sawrooth roughness on pressure drop and turbulent transition in micro channels *Heat Transfer Eng.* **28**(8-9) 662-69

[13] Subramaniam V and Garimella S 2005 Design of air-cooled R-410A microchanel condensers *ASHRAE Trans.* **111**(1) 471-86

[14] 2009 Fundamentals SI Edition *ASHRAE Handbook* 4.19

IOP Conf. Series: Materials Science and Engineering **50** (2013) 012025 doi:10.1088/1757-899X/50/1/012025

Rotating cylinder design as a lifting generator

Azharrudin Asrokin[a], Mohammad Rizal Ramly[b] and Abdul Halim Ahmad
[1]Universiti Kuala Lumpur, Malaysia Institute Of Aviation Technology, Lot 2891, Jalan Jenderam Hulu, Jenderam Hulu, 43800 Dengkil, Selangor Darul Ehsan, Malaysia

E-mail: [a]azharrudin@miat.unikl.edu.my, [b]mdrizal@miat.unikl.edu.my

Abstract. The airfoil shape of a wing has always been the design to generate lift. But few realized that a simple rotating cylinder can also create lift. However, the explanation and study of how a rotating cylinder creates lift are still complex. In remote area where it is difficult for air vehicle to access, the exploration and discovery of different configuration for design concept is rather important. Due to this reason, there is a need to think of a lift generator that can produce better lift (few fold better than conventional airfoil) at lower speed to take off in a short distance of time. This paper will explain the conditions and the design of such a wing using the rotating cylinder concept that will take off in a short time and requires little takeoff and landing strip. Spokes will be attached to the cylinder to force the surrounding air to rotate along with the cylinder. This will create a vortex that hastens the speed of the air on top of the cylinder and at the same time retarding the speed of air below the cylinder. From the results, the rougher surface cylinder produces more lift when rotating and also, higher speed rotation of the cylinder greatly changes the speed of the surrounding air, thus better lift.

1. Introduction

According to research done by Anton Flettner year 1924, the design of a rotorship in figure 1 has already been experimented and found to increase sailing ship efficiency by 50 percent. The sphere or cylinder spinning in an airstream develop a force of the moving air created a vacuum ahead of the cylinder area and high pressure behind, results in forward movement. Based on the discovery by Flettner, the rotating cylinder from the ship could extract up to 15 times energy from the wind. The rotorship has two hollow cylinders 120 meter high and 3 meters in diameter and rotated at 120rpm using diesel engine have successfully crossed the Atlantic. These studies have shown interest in the aviation field to look into the possibility of using rotating cylinder to replace wing.

Figure 1. The Flettner rotor ship [1]

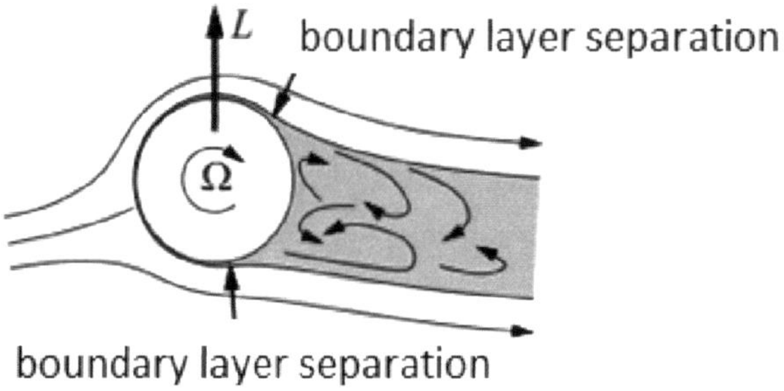

Figure 2. The direction of Lift due to rotating cylinder and boundary layer separation

According to Magnus effect, named after a 19th century German engineer, and is related to the circulation around an object in a flow field, the phenomenon whereby a spinning object flying in a fluid creates a whirlpool of fluid around itself, and experiences a force perpendicular to the line of motion [8]. Rayleigh studied the lift of a rotating cylinder for an inviscid fluid, and related lift to the circulation of a rotating cylinder. The overall behaviour is similar to that around an airfoil with a circulation which is generated by the mechanical rotation, rather than by aerofoil action. The pressure gradient can be explained simply by Bernoulli's principle, in which pressure and velocity are inversely proportional. In this research, the two criteria are explored in detail regarding the rotating surface that can make them viable for flying contraption uses. They are the roughness of the surface, and the rotating speed (vortex strength) of the cylinder in moving laminar airflow.

2. Literature Review

Flow over rotating cylinder is important in a wide number of applications, as such research have been done for shafts and axles, spinning projectiles, and other engineering application. In aviation, the important consideration is in lift generation properties. As a brief explanation, lift is the force created by an object that is perpendicular to the laminar flow direction. The different pressure created between the upper surface and lower surface of some shape will produce the force. This can be produced by airfoil shape, plate shape and spinning of the cylinder and ball. Airflow is moving from left to right, and the cylinder is rotating clockwise. The friction from the cylinder is pushing the motion of the air on top of the cylinder to go faster and deflecting air more towards the top. Meanwhile the friction slows down the air movement below the cylinder and thus creating a higher static pressure there than on top of the cylinder. Flow about a rotating cylinder is equivalent to the combination of flow past a cylinder and a vortex. As such in addition to superimposed uniform flow and a doublet, a vortex is thrown at the doublet centre which will simulate a rotating cylinder in uniform stream. The pressure distribution will result in a force, a component of which will culminate in lift force .The phenomenon of generation of lift by a rotating object placed in a stream is known as Magnus effect. Considering the interior of the circle (on which $\psi = 0$) to be a solid cylinder, the outer streamline pattern is shown in Figure 3.

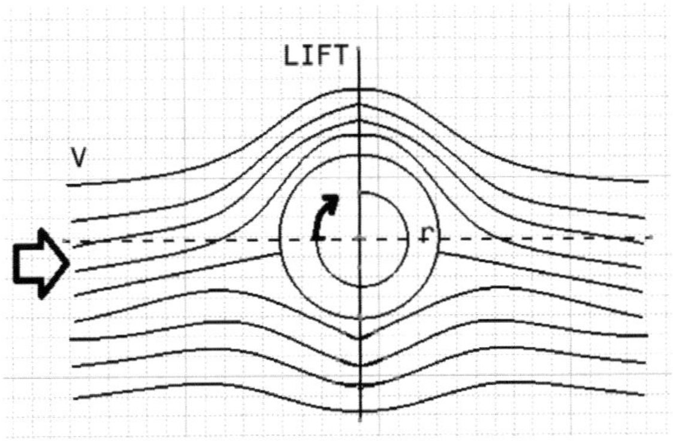

Figure 3. How lift is generated by rotating cylinder.

The studies regarding rotating cylinders and spheres have started centuries ago. Basically, a rotating cylinder moving in a uniform stream experiences a force normal to the direction of the stream. Goldstein (1938) and Magnus (1853) refer to several historical papers on both rotating spheres and cylinders, with the first laboratory experiments examining the lift on a rotating cylinder.Reid (1924), Prandtl (1925), and Thom (1926, 1931) performed experiments on a circular cylinder rotating about its axis in a uniform flow and found that the mean lift of a cylinder was a function of its rotation rate. Experiments and simulations done by Badr & Dennis (1989), Ingham & Tang (1990), and Tang & Ingham (1991) for the steady flow past a rotating cylinder, at low and higher Reynold's numbers show that there is no periodic vortex shedding from a cylinder that is rotating with a surface velocity greater than two or three times the free-stream velocity. [7][8]

According to research on irrotational motion part a rotating cylinder a thermodynamically admirable motion is contracted in relation between the angular velocity and the circulation around the cylinder, and the lift generated on the cylinder can be obtain by Kutta-Joukowski theorem. The theorem stated that the lift per unit length of the cylinder acts perpendicular to the velocity (v in ft/sec) and given by formula as below.

$$\text{Lift} = C_L q_\infty S \quad (1)$$

$$\text{Drag} = C_D q_\infty S \quad (2)$$

$$\text{Lift}_{\text{rotating vortex}} = p_\infty V_\infty \Gamma \quad (3)$$

Where ; Γ=vortex/rotational strength
C_L=lift coefficient
C_D=drag coefficient
S=lift area
q_∞ =dynamic pressure
p_∞=density of incoming air
V_∞=incoming air velocity

According to NASA publication, the flow field associated with rotating cylinder is two dimensional unlike the three dimension nature of curveball. [2]

Various rotating surfaces' behaviors in fluid are rarely observed by engineers. This is because engineers tend to remove or reduce moving parts in their products because moving parts causes instability. But rotating surfaces, normally of cylindrical shape, had been utilized to power ships before. Due to economic reasons, their applications have not been widespread

3. Methodology

In this section, the setup of the experimental investigations is briefly described. Generally, the design of the cylinder was more concentrated on which profiles can generate better lift. Smooth surface will create less turbulence, hence less drag. On the other hand, according to the Kutta Joukowski theorem, lift will depend on the strength of the vortex created by the lift generator. So naturally rougher surface is needed to create such vortex for higher lift to be generated by the spinning cylinder. We used a small subsonic wind tunnel available in Unikl-MIAT and created variable speed rotating cylinder with four spokes just to get stronger vortex. Prior to that, we test some models on top of a moving van to determine which surface will be tested in the wind tunnel, the smooth or the rough surface. And the rough surface is the clear winner. In the wind tunnel, we also test an airfoil relatively the same size as our rotating model to compare the results between them.

Apparatus used : a. Wind tunnel subsonic (open end)
b. Voltage regulator
c. Electrical wires with clamps
d. Working specimens and its components

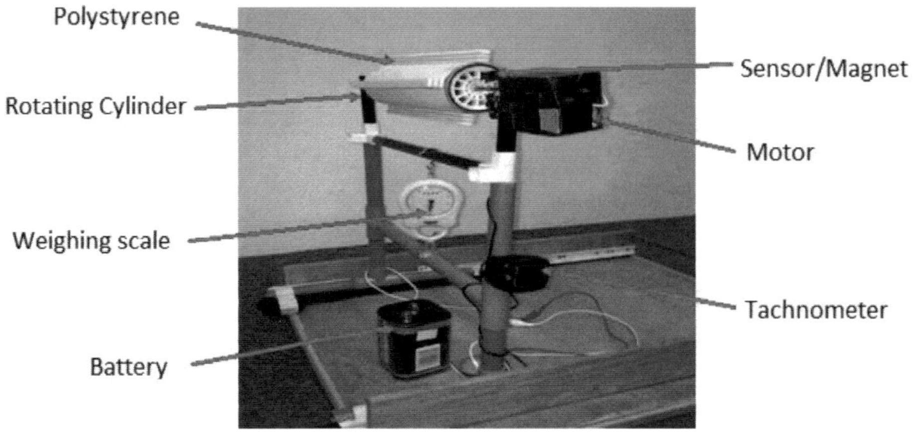

Figure 4. Preliminary setup of the rotating cylinder

For the preliminary setup, the test was conducted on top of a van, moving at specific speed just to verify the feasibility and reliability of this model before it was scaled down for the wind tunnel.

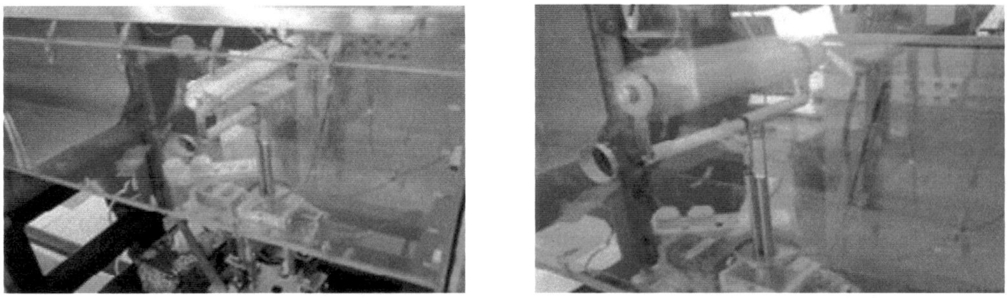

Figure 5. Working model inside the wind tunnel static and rotation.

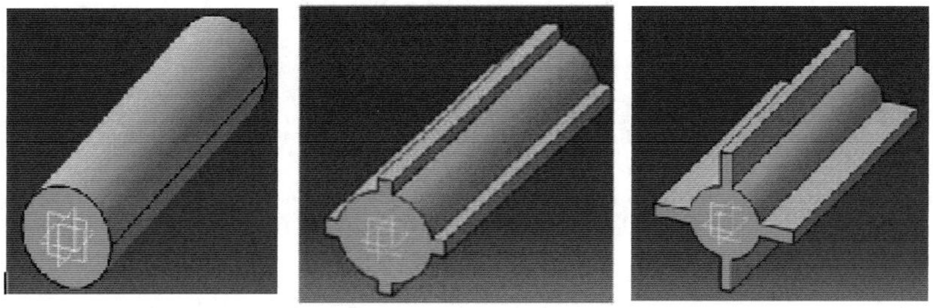

Specimen 1 Specimen 2 Specimen 3

Figure 6. The dimension of the actual tested models in the wind tunnel

Table 1: Dimension of models

Cylindrical specimen	Span (mm)	Diameter (mm)	Spoke height (mm)
1	200	45	0
2	200	30	7.5
3	200	15	15

4. Results and Discussion

Flow past a spinning and translating cylinder has been a subject of numerous computational and experimental studies. Interest in this problem arises not only from the point of view of basic fluid mechanics but also from its applications to flow control. Flow past an isolated rotating cylinder has been studied by various researchers in the past. These include the effect of the aspect ratio and end plates attached to the end of a cylinder that lead to an increase in the overall lift coefficient. Some of the later works on this flow problem include the development of the near-wake behind an impulsively started cylinder via flow visualization. [2][4]

In this project, we used an open loop wind tunnel. In an open-loop wind tunnel, the intake and exhaust ends of the tunnel are not connected and this enough for us to conduct the experiment. The specimen was created based on the dimension stated below. There are three variation of design for our rotating cylinder, and it is tested according to the specific lift generate speed. Using this wind tunnel we gather the experiment for this project. The finding is as below. Below are the results from the wind tunnel:

Table 2. Working model inside the wind tunnel at highest parameters

Cylindrical specimen	Surface condition	Lift generated (N) at 400RPM	Lift generated (N) at 800RPM
1	Almost zero surface roughness	0.03	0.10
2	Adding spokes for roughness	1.00	1.80
3	Adding longer Spokes for better roughness	1.70	2.50

Table 3. Data collected from the cylinders rotating at 448RPM (criteria by surface roughness)

Cylindrical specimen	airspeed (m/s)	Lift generated (N)
1	11	0
	13.67	0
	15.2	0.09
2	10	0.16
	14.19	0.25
	15.37	0.4
3	10.82	0.8
	13.57	1.05
	15.4	1.7

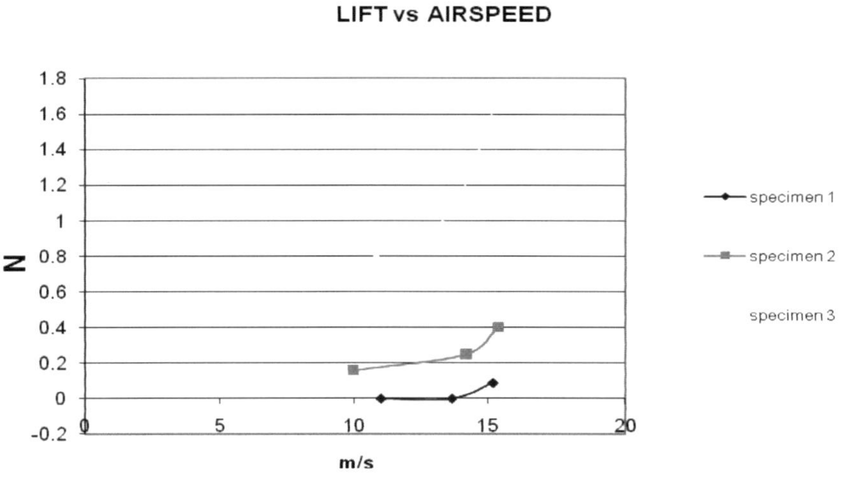

Figure 7. Plotted data from table 1

Table 4: Data collected from the cylinder (criteria by spinning strength) with all the dimensions of the last cylinder in Figure 6 (specimen 3)

RPM	Airspeed (m/s)	Lift generated (N)
0	10.23	-0.29
	11.3	-0.35
	12	-0.43
	14.56	-0.66
	15.54	-0.8
812	10.62	1.07
	11.51	1.33
	12.59	1.4
	13.48	1.7
	14.26	2
	15.15	2.5
448	10.82	0.8
	13.57	1.05
	15.4	1.7

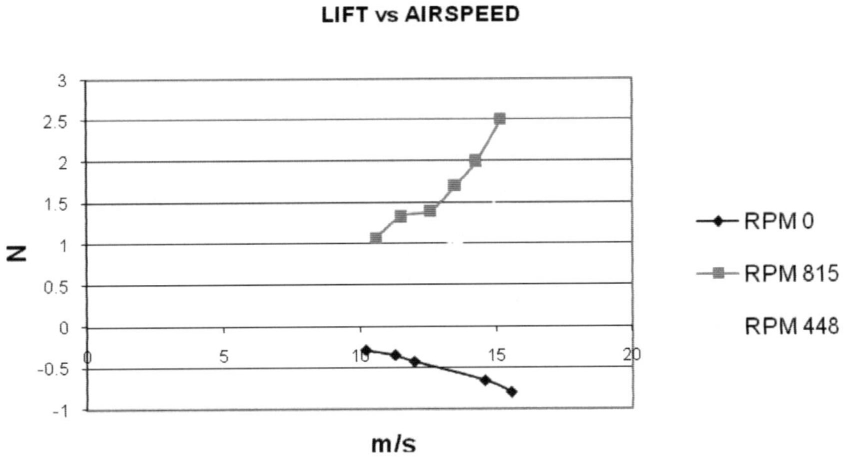

Figure 8. Plotted data from table 2

Based on previous studies done [11], it is proven that the rotating cylinder outperform a conventional airfoil with dimension as follows:

Airfoil:

Length = 280mm

Chord line = 63mm
Total camber = 12mm

Table 5. Results from the conventional airfoil with different angle of attack.

Angle of attack (°)	airspeed (m/s)	Lift generated (N)
5	10.03	0.43
	11.11	0.52
	12.3	0.68
	13.08	0.82
	15.15	1.82
10	10.03	0.5
	11.11	0.74
	12.1	0.91
	13.08	1.11
	15.05	1.74
14	9.93	0.58
	11.02	0.76
	12.1	0.96
	13.08	1.21
	15.15	1.64

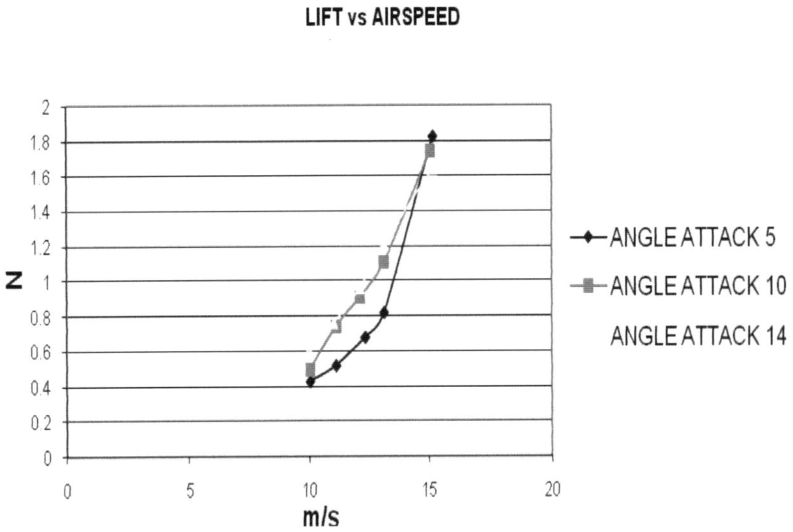

(a)

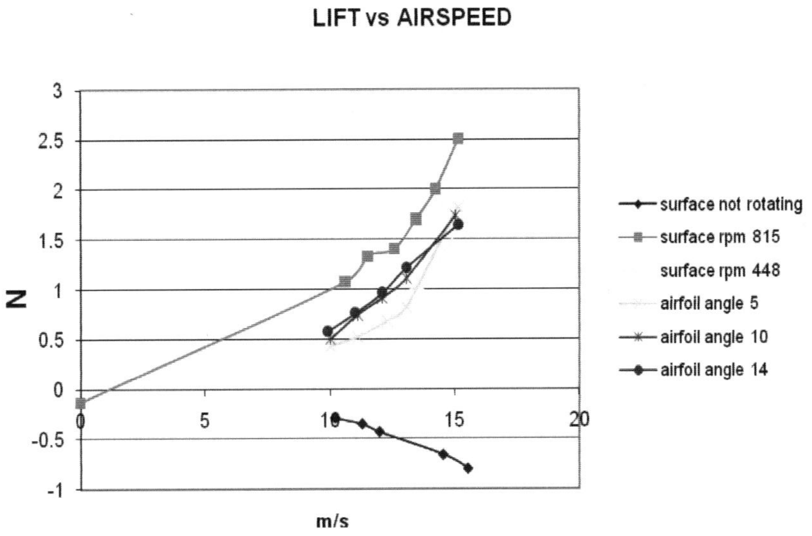

(b)

Figure 9. (a) showing airfoil results in wind tunnel with different angle of attacks and (b) showing combination of results from Table 3 & Table 2

The main results are the following:

1. Table 3 and Figure 7 are shows that rougher surface will have greater influence in producing higher lift when the surface is spinning in moving, laminar airflow. The summary of this result is better represented by Figure 6.
2. Table 4 and Figure 8 on the other hand are the results of the vortex (spinning) strength criteria. The faster the cylinder rotates, the higher the lift force gets. Also, the lift force improves exponentially as the airspeed increases.
3. Table 5 and Figure 9(a) is showing the lift force generated by a conventional airfoil. We can see that as the airfoil increases the angle of attack, the lift improves, but not at higher airspeed as the airfoil experiencing turbulence or stall.
4. From Figure 9(b), we can clearly see that the rotating cylinder has the upper hand to create better lift than the conventional airfoil.

5. Discussion and Conclusion

Generally rotating cylinder generates lift, where velocity is faster over the top of the cylinder than bottom and pressure is higher on the bottom than over the top. It was stated that the lifting force is directed perpendicular to the cylinder velocity.

The data obtained from Table 4 obviously proving that vortex strength is somewhat directly proportional to lift force. And of course, the lift force is exponentially proportional to the airspeed as shown in Figure 7 & 8. But the mass of air being pushed out by the spokes is the main criteria to have good lifting force. Figure 7 shows that cylinder without the spokes have negligible upward force.

When added some spokes, i.e. the one with 7.5mm spoke length from the base cylinder, the lift improved almost immediately. But when doubled the length of the spokes (from base cylinder to the tip), i.e. 15mm, the lift went higher several folds. It means that the deeper the scoop, the more mass to make the air move faster at the topside of the cylinder, hence, higher lift. From Figure 9, it is demonstrated that the airfoil is inferior in performance than the rotating cylinder.

As we know, centrifugal force tends to push everything out from the center, almost perpendicularly. Thus, our models are not forming a true vortex that is ideally what the models were supposed to create. So the actual mass of air conforming to the vortex pattern is not as much as we hoped for. So, it would be recommended that the pattern or the shape of the spoke should take a curvature profile so it can better push more air to move in the intended vortex motion.

References

[1] Lift of a Rotating Cylinder Grc.nasa.gov. Retrieved 2012-07-16
[2] Ou Y R , J A Burns 1992 Optimal *control of lift/drag ratios on a rotating cylinder* Institute for Computer Applications in Science and Engineering, NASA-Langley Research Center, Hampton, VA 23665, U.S.A. Appl. Math. Lett. Vol. 5, No. 3, pp. 57-62
[3] Anderson J D 2001 *Fundamentals of Aerodynamics* (New York: McGraw-Hill)
[4] Smits A J 1994 *A new aerodynamic model of a golf ball in flight* In A J Cochran and M R Farrally, (eds.), Science and Golf II, pp. 340–347. E&FN Spon
[5] Reid E G Tests of rotating cylinders. NACA Technical Note, NACA-TN-209, 1924
[6] Childs P R N 2004 Mechanical design. 2nd Edition.Elsevier
[7] Mittal S and Kumar B 2003 Flow past a rotating cylinder *Journal of Fluid Mechanics* 476 pp 303–334
[8] Theodorsen T and Regier A 1944 *Experiments on drag of revolving discs, cylinders and streamline rods at high speeds* NACA Report 793
[9] Thom A *Effects of discs on the air forces on a rotating cylinder* Aeronautical Research Committee Reports and Memoranda 1623
[10] Ou Y R 1991 *Active control of flow around a rotating cylinde* Presented at the Meeling on Turbulence Stucture and Control, Co-sponsored by AFOSR and Ohio State University, Columbus
[11] Azharrudin A, Mohammad Rizal R and Halim A 2001 *Preliminary study towards the manufacturing of a rotating cylinder surface as a lift generator* ICAM
[12] Tokumaru P T and Dimotakis P E 1993 *The lift of a cylinder executing rotary motions in a uniform flow* J. Fluid Mech. 255, pp. 1-10
[13] N Thouault, C Breitsamter, Jost Seifert, C Badalamenti, S A Prince and N A Adams 2010 *Numerical analysis of a rotating cylinder with spanwise discs* 27TH International congress of the aeronautical sciences (ICAS)
[14] Sridhar Muddada and B S V Patnaik 2010 *An active flow control strategy for the suppression of vortex structures behind a circular cylinder* European Journal of Mechanics - B/Fluids Volume 29, Issue 2, March–April 2010, Pages 93–104

A wavelet decomposition analysis of vibration signal for bearing fault detection

C K E Nizwan[a], S A Ong, M F M Yusof and M Z Baharom
Faculty of Mechanical Engineering, University Malaysia Pahang
26600 Pekan, Pahang, Malaysia

E-mail: eddy@ump.edu.my

Abstract. This paper presents a study of vibrational signal analysis for bearing fault detection using Discrete Wavelet Transform (DWT). In this study, the vibration data was acquired from three different types of bearing defect i.e. corroded, outer race defect and point defect. The experiments were carried out at three different speeds which are 10%, 50% and 90% of the maximum motor speed. The time domain vibration data measured from accelerometer was then transformed into frequency domain using a frequency analyzer in order to study the frequency characteristics of the signal. The DWT was utilized to decomposed signal at different frequency scale. Then, root mean square (RMS) for every decomposition level was calculated to detect the defect features in vibration signals by referring to the trend of vibrational energy retention at every decomposition. Based on the result, the defective bearings show significant deviation in retaining RMS value after a few levels of decomposition. The findings indicate that Wavelet decomposition analysis can be used to develop an effective bearing condition monitoring tool. This signal processing analysis is recommended in on-line monitoring while the machine is on operation.

1. Introduction

Rolling element bearings has vast domestic and industrial applications. Appropriate function of these appliances depends on the smooth operation of the bearings. In industrial applications, bearings are considered as critical mechanical components and a defect in such a bearing causes malfunction and may even lead to catastrophic failure of the machinery.

Presently, vibration monitoring method becomes the most reliable tool as a part of preventive maintenance for rotating machines [1,2]. The vibration data often contain fault signatures where several signal processing techniques; often adapted to a precise defect type, results to online monitoring system.

There are many condition monitoring methods used for detection and diagnosis of rolling element bearing defects such as vibration measurements, temperature measurement, shock pulse method (SPM), and acoustic emission (AE). Various researchers suggested that stator current monitoring can provide the same indications without requiring access to the motor. This technique utilizes results of spectral analysis of the stator current or supply current of any part nearest to the rolling bearing element for diagnosis purpose [3]. Other signal processing technique for condition monitoring method includes averaging technique [4], adaptive noise cancelling [5], and high-frequency resonance technique (HFRT) [6] was developed to improve signal-to-noise ratio for more effective detection of bearing defect. Among all these monitoring methods, the high-frequency resonance technique is more popular for bearing fault detection. However, most of the method requires additional computations and several runs of impact tests to find the bearing resonance frequency. Therefore, extra instruments such as vibration exciters and their controller are needed for HFRT [7].

Wavelet Transform (WT) on the other hand, were proven to be one of the best condition monitoring method/effective tool for detecting single and multiple faults in the ball bearings [7,8,9]. A clear review on using DWT as condition monitoring method and possible early detection was given by Giaccone et al. [10], Tandon and Choudhury [2], Kim et al. [11], Schmitt et al. [12], and Staszewski [13].

Discrete wavelet transform (DWT) provides a time-scale information of a signal, enabling the extraction of features that vary in time. This property makes wavelets an ideal tool for analyzing signal of a transient or non-stationary nature [7]. The continuous wavelet transform (CWT) of *f*(t) is a time-scale method that may be identified as the sum over all time of the signal multiplied by scaled, shifted versions of the wavelet function $\Psi(t)$. Mathematically as proposed by McFadden and Smith [14],

$$WT\,(a,b) = \frac{1}{\sqrt{|a|}}\int_{-\infty}^{\infty} f(t)\Psi^*\left(\frac{t-b}{a}\right)dt \tag{1}$$

Where $\Psi(t)$ denotes the mother wavelet. The parameter a represents the scale index which is a reciprocal of frequency. The parameter b indicates the time shifting (or translation). The DWT is derived from the discretization of CWT (a,b) and the most common discretization is dyadic, given also by McFadden and Smith [15] as,

$$DWT\,(j,k) = \frac{1}{\sqrt{2^j}}\int_{-\infty}^{\infty} f(t)\Psi^*\left(t - \frac{2^j k}{2^j}\right) \tag{2}$$

Where a and b are replaced by 2^j and 2^jk. An effective way to apply this method using filters was devoleped by Mallat in 1989 [16]. *f*(t) passes through two complementary filters and emerges as low and high frequency signals. The decomposition may be iterated with successive approximations begin decomposed in turn, so that signals may be broken down into lower-resolution components.

This paper focus on the analysis of vibration signals at difference bearing condition using time frequency transform approach. The discrete wavelet transform method was used to transform the time domain vibration signal into time frequency domain in order to detect the abnormal features of the faulty bearing.

2. Experimental Setup

The test rig for this experiment was design to investigate failure and vibration characteristic of ball bearings. It was designed specifically to imitate the applications of ball bearings in the industry. The shaft was driven by a variable-speed 0.37kW, 50Hz electric motor with a frequency converter in order to control the speed of the motor. The motor and shaft were connected by using a spring coupling where it could minimize shaft alignment error. The front side of the shaft is fitted with tested bearing and the other will be fitted with a good bearing acting as a dummy.

A set of good bearings and another three bearings with different type of defectives were used for testing. A flywheel is installed at the middle of the spindle in order to apply load to the shaft and at the same time minimizing the speed oscillations of the shaft. The angular speed is set to 287 rpm, 1466 rpm, and 2664 rpm (which are 10%, 50% and 90% of the maximum motor speed respectively), and the vibration signals were acquired by using the Bruel&Kjær (B&K) 4506B accelerometer. Sensors were placed on horizontal directions to collect the data because horizontal bearing types were tested. Accelerations signals are acquired by using a personal computer (PC) based data acquisition system with sampling frequency of 20 kHz (Δt = 0.039 ms). The tools arrangement for this experiment are as shown in Figure 1.

There are four bearings that were tested for this experiment and three of them are defected. Figure 2 shows the bearings used in this research i.e. outer race defect point defect corroded and healthy bearing. The types of defect and the location of each defect specifically were summarized in Table 1.

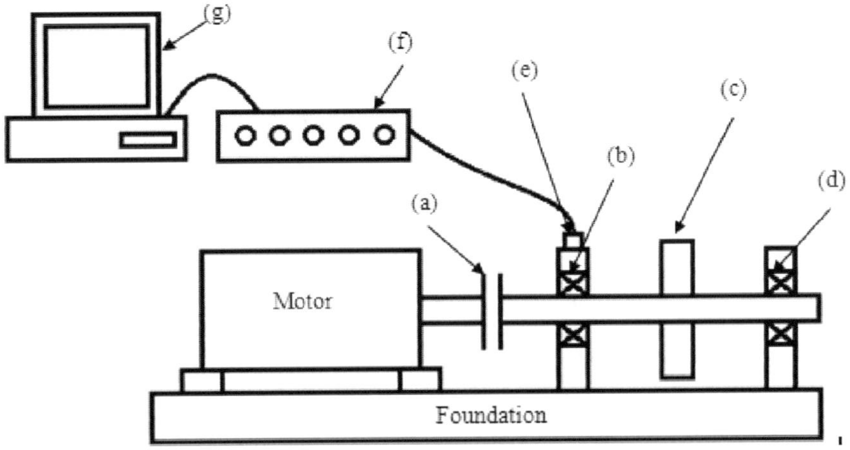

Figure 1. Bearing fault detection test rig consist of: coupling (a), tested bearing (b), flywheel (c), healthy bearing (d), accelerometer (e), data acquisition system (f), and a PC (Analyzer) (g).

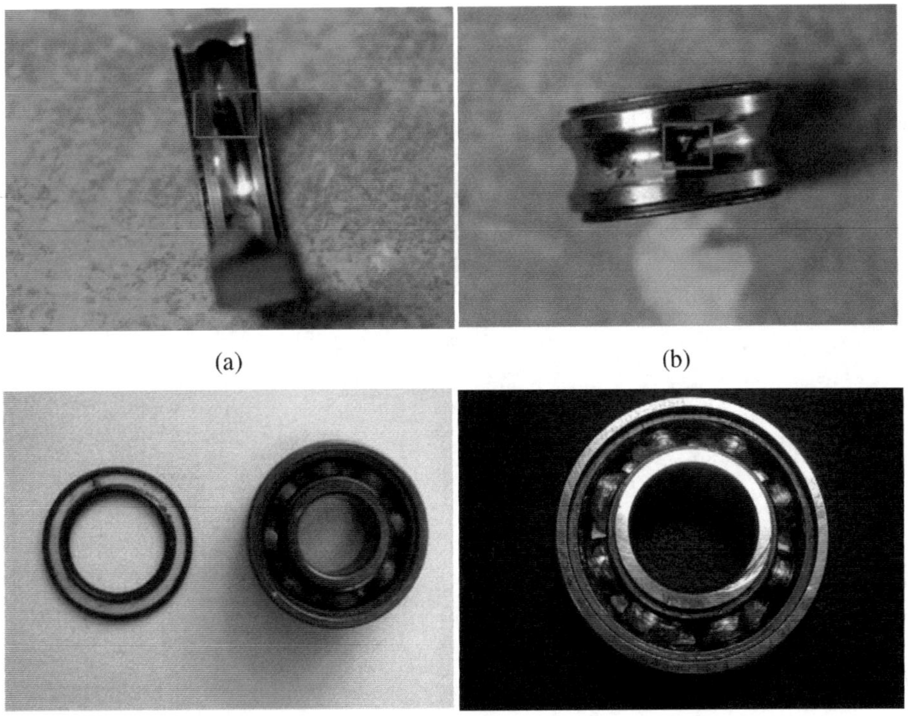

(a) (b)

(c) (d)

Figure 2. (a) Outer race defect (b)Point defect (c) Corroded (d)Healthy bearing

Table 1. Types and location of defect

Types of Defect	Location of Defect
Outer race defect	One scratch mark on the outer race
Point defect	Single point mark on the inner race
Corroded defect	Ball bearing were left to open air and water

3. Results and Discussion

From the experiment, the time series acceleration data at different bearing condition were acquired. Figure 3 shows the time domain acceleration data which recorded at the speed of 1466 rpm, and 2664 rpm. Analysis for this project is done by using MATLAB®'s Discrete Wavelet Transform's Toolbox:Wavelet 1-D. The analysis starts off by reading the desired raw data. Then, the raw data were decomposed into eight level of frequencies scale using 4th order Daubechies wavelet ('db4'). Figure 4 and Figure 5 show the decomposition of vibration signal acquired from healthy and corroded bearing at 1466 rpm. Visually, the defect features can be observed from the decomposition of level 2 and above where the amplitude of decomposed signals shows significant decrement compare to the healthy bearing.

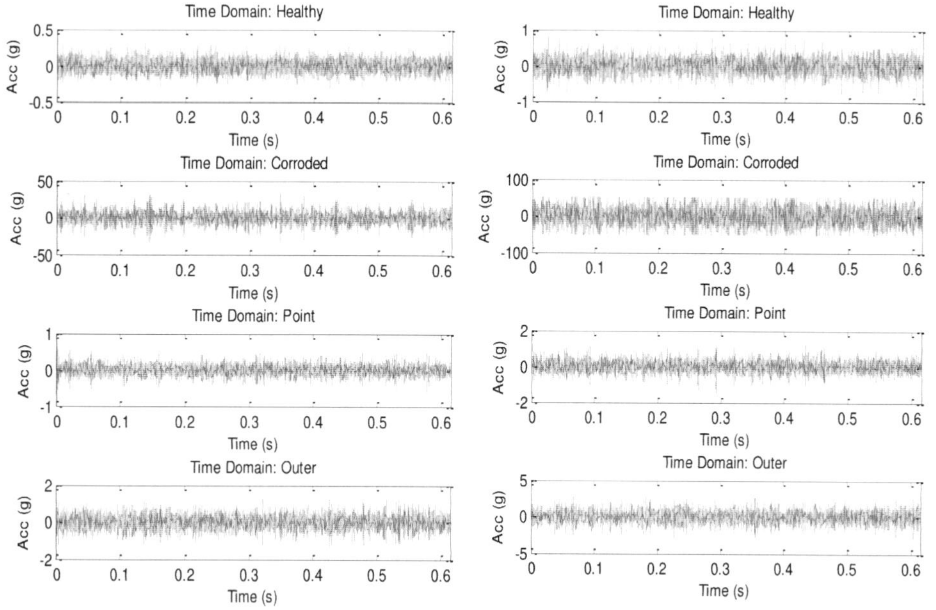

(a) (b)

Figure 3. (a) Time domain data acquired at 1466 rpm. (b) Time domain data acquired at 2664 rpm.

For data of insufficient frequency excitation, calculating the RMS value for each decomposition level and finding the percentage towards the original input seems to yield understandable trend when a graph is plotted. Cartwright [17] proved in his paper the effectiveness of RMS by simple means without calculus. This method uses the definition of RMS of a waveform which is found simply by squaring that waveform, taking the mean of that squared form, and then computing the square root.

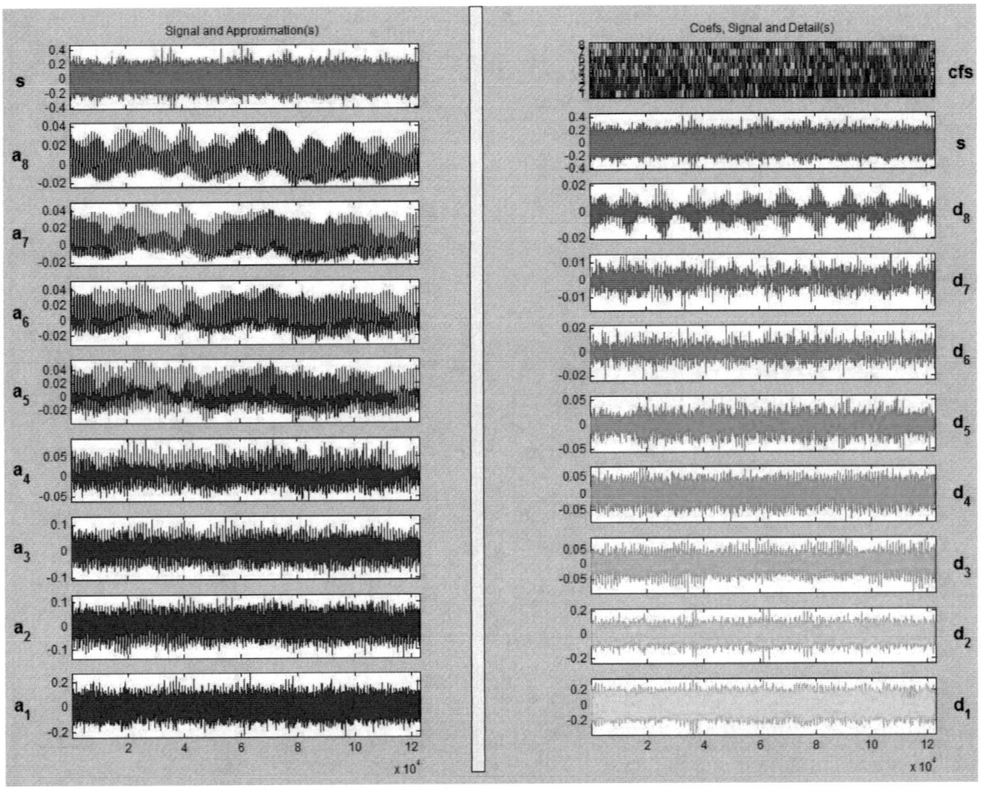

Figure 4. Decomposition result for healthy bearing at 1466 rpm

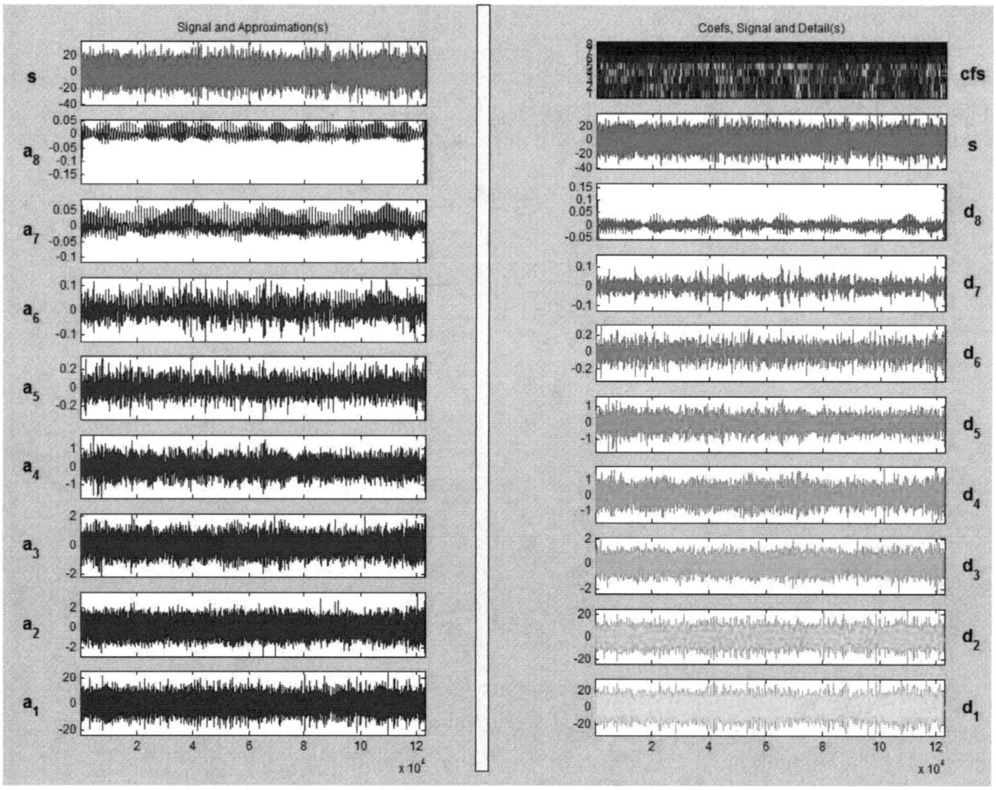

Figure 5. Decomposition result for corroded bearing at 1466 rpm

Any percentage value or graph plotted that is below the healthy line may be considered defected. This is because; decomposition process is where the frequency amount of each level gets smaller as the decomposition level increases. Therefore, if the RMS value of the next level declines at an irregular rate or stays flat on its line, the bearing may be considered as defected.

Figure 6 shows the trend generated by the RMS percentage value of each decomposition level. From the plotted graph, it seems that all of the defected bearing's RMS values for each decomposition level are below the healthy line, red region (except for 287rpm) thus proving that RMS is an excellent statistical tool to support DWT decomposition result during speed-up (2800 rpm) was sufficiently useful as also proven by Kim *et al*, (2007). For the data at 287rpm, it's probably due to low excitation.Research by Tandon&Choudhury [2] yields similar result where it was found out that the direct vibration spectrum from a defective bearing may not indicate the defect at the initial stage.

However, the most defected bearings; corroded, is still below the lines of healthy (red circle) where this is most probably due to giving enough frequency excitation. This is because of how DWT itself decomposes the data. As shown previously in Figure 4, the input was divided into approximate (cA) and detailed (cD) output at each level of decomposition. By referring to raw data transformed into frequency domain, FFT graph of 10,000 Hz in Figure 7; at first level of decomposition, 10,000-length of data were decomposed to 5,000-length data of cA_1 and 5,000-length data of cD_1. As the DWT proceeds with decomposition process, cA_1 was then divided into two parts again leaving 2,500-length data each for cA_2 and cD_2 at level 2. At level 1's decomposition, the energy was divided with most of it towards cD1 rather than cA1. While at level 2's decomposition, most of the energy were left at cA_2 and low amount of energy are seen at cD_2. As the decomposition process continues up to level 8, the energy left at cA_8 would be the lowest thus making it having a low value RMS.

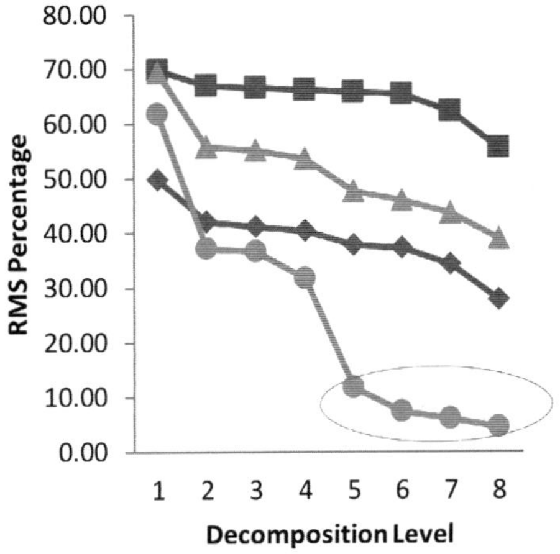

(a)

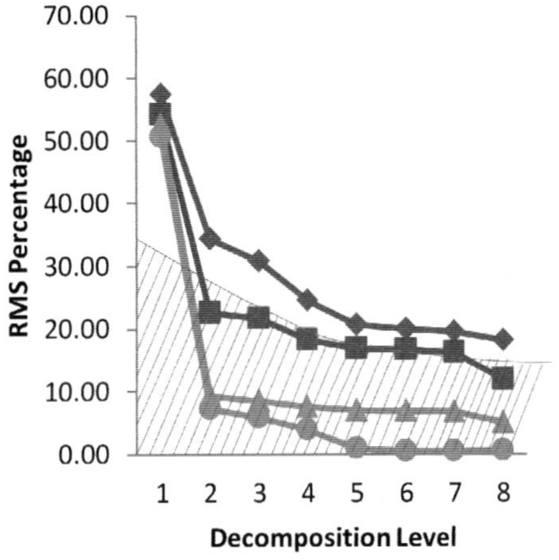

(b)

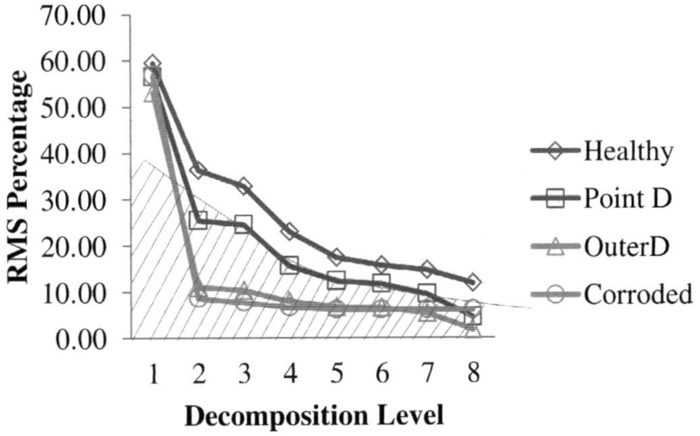

(c)

Figure 6. RMS percentage vs. decomposition level for speed of (a) 287 rpm (b) 1666rpm (c) 2664 rpm

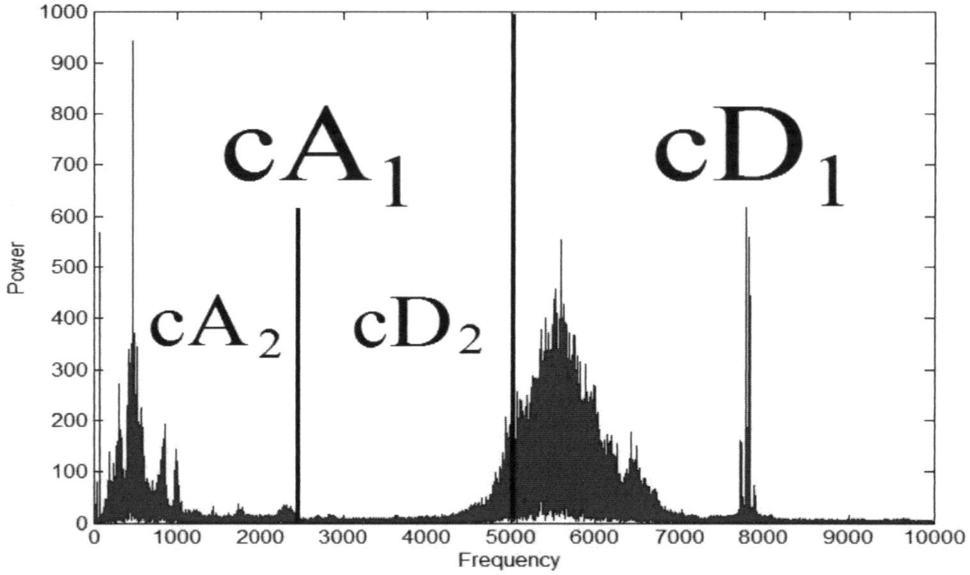

Figure 7. FFT result for corroded bearing at speed of 287 rpm

4. Conclusion

Discrete Wavelet Transform (DWT) was used to detect defect features from defected bearings as they produce vibration signals. Fast Fourier Transform (FFT) and Root Mean Square (RMS) plays an important role in supporting results analyzed by using DWT from MATLAB®Toolbox.Using DWT is indeed appropriate as an effective tool in detecting defect features in bearings. However, on systems of low speed, it is recommended for the DWT results to be enhanced by using RMS value of each decomposition level.All of these are possible for online monitoring without shutting down the machine for maintenance purpose. Therefore, this method is suggested as an alternative technique in bearing fault detection online monitoring, especially for system running on high speed.

5. Acknowledgements

The authors would like to express their gratitude to Universiti Malaysia Pahang especially Faculty of Mechanical Engineering and Automotive Engineering Centerthrough the fund of RDU120322, for supporting this research activity.s

6. References

[1] Yan R. and Gao R X 2009 Base Wavelet Selection for Bearing Vibration Signal Analysis *Int. J. Wavelets Multiresoluton & Inf. Process* **7(4)** 411-26

[2] Tandon N and A Choudhury 1999 A Review of Vibration and Acoustic Measurement Methods for The Detection of Defects in Rolling Element Bearings *Tribolgy International* 469-80.

[3] Schoen RR, Habetler TG, Kamran F and Barthheld R G 1995 Motor Bearing Damage Detection using Stator Current Monitoring *IEEE Transactions on Industry Applications* **31 (6)** 271-82.

[4] Braun S and Datner B 1977 Analysis of Roller/Ball Bearing Vibrations. *ASME paper 77-WA/DE-5*. 1-8.

[5] Chaturvedi GK and Thomas DW 1981 Bearing Fault Detection using Adaptive Noise Cancelling. *ASME paper 91-DET-7*. 1-10.

[6] Prasad H, Ghosh M and Biswas S 1984 Diagnostic Monitoring of Rolling Element Bearings by High Frequency Resonance Technique *ASLE paper 84-LC-3c3* 1-9.
[7] Prabhakar S, Mohanty AR and Sekhar AS 2002 Application of Discrete Wavelent Transform for Detection of Ball Bearing Race Faults *Tribolgy International* 793-800.
[8] Chebil J, Noel G, Mesbah M and Deriche M 2009 Wavelet Decomposition for the Detection and Diagnosis of Faults in Rolling Element Bearings. *Jordan Journal of Mechanical and Industrial Engineering* **3(4)** 260 – 67.
[9] Djebala A, Nouredine O and Nacer H 2007 Detection of Rolling Bearing Defects using Discrete Wavelet Analysis. *Meccanica* 339-48.
[10] Giaccone SJ, Bossio GR and Garcia GO 2011 Wavelet Analysis for Stator Fault Detection in Induction Machines. *International Journal of Wavelets, Multiresolution and Information Processing* **9(3)** 361–74.
[11] Kim BS, Lee SH, Lee MG, Ni J, Song JY and Lee CW 2007 A Comparative Study on Damage Detection in Speed-Up and Coast-Down Process of Grinding Spindle-Typed Rotor-Bearing System *Journal of Materials Processing Technology* 30-36.
[12] Schmitt E Idowu P. and Morales A 2009. Applications of Wavelets in Induction Machine Fault Detection *Ingeniare.Revistachilena de ingeniería* **18 (2)** 158-64.
[13] Staszewski WJ 1998 Structural and Mechanical Damage Detection Using Wavelets *The Shock and Vibration Digest* **30(6)** 457-72.
[14] McFadden PD and Smith JD 1984 Model for the Vibration Produced by A Single Point Defect in A Rolling Element Bearing *Journal of Sound and Vibration* **96(1)** 69-82.
[15] McFadden PD and Smith JD 1985 The Vibration Produced by Multiple Point Defects in A Rolling Element Bearing *Journal of Sound and Vibration* **98(2)** 263-73.
[16] Mallat SG 1989 A theory for Multiresolution Signal Decomposition: The Wavelet Representation *IEEE Transaction on Pattern Analysis and Machine Intelligence* **11(7)** 674-93.
[17] Catwright KV 2007 Determining the Effective Or Rms Voltage Of Various Waveforms Without Calculus *The Technology Interface.Fall.*

Review: Modelling of meniscus of knee joint during soccer kicking

Mohd Azrul Hisham Mohd Adib[a], **Mohd Firdaus Jaafar**
Center for Sport Engineering (CenSE),
Faculty of Mechanical Engineering, Universiti Malaysia Pahang,
26600 Pekan, Pahang, Malaysia.

E-mail: [a]azrul@ump.edu.my

Abstract. Knee is a part of the body that located between thigh and shank is one of the most complicated and largest joints in the human body. The common injuries that occur are ligaments, meniscus or bone fracture. During soccer games, the knee is the most critical part that will easily injure due to the shock from an external impact. Torn meniscus is one of the effects. This study will investigate the effect towards the meniscus within the knee joint during soccer ball kicking. We conduct a literary review of 14 journals that discuss the general view of meniscus and also soccer kicking. The selected topics for this review paper are meniscal function, meniscal movement, meniscal tears and also instep kick. As a finding, statistics show that most meniscal tears (73%) occurred in athletes who were soccer players, basketball players or skiers. The tear is frequently happening at the medial side rather than lateral side with a percentage of 70%.

1. Introduction

Knee is a part of the body that located between thigh and shank is one of the most complicated and largest joint in human body. The knee has to support nearly the whole of human's weight, so that it is easy for the knee to get injured. The common injuries that occur are ligaments, meniscus or bone fracture which are internal part of the knee. For soccer games, knee is the most critical part that will easily injure due to the shock from an external impact. Torn meniscus is one of the effects. This study will learn about the effect towards the meniscus within the knee joint during soccer kicking.

2. Methodology

The method use for this paper is by reviewing 14 journals that discussed about the general view of meniscus and also soccer kicking. The selected topics for this review paper are meniscal function, meniscal movement, meniscal tears and also instep kick. Each topic will be discussed deeply reviewed from the selected journal.

3. Meniscal Function

Out of 15 journals, nine journals discussed about the function of meniscus deeply. The meniscus acts an important role towards the knee in order to protect the joint between the femur and tibia. Meniscus is two crescent shapes that located between the femur and tibia, which its function discussed a lot by researchers. The main function of meniscus is to distribute the load across the joint between femur and tibia. Other than that, meniscus is also act as stabilizer between the joint, joint lubrication and so on.

Many literatures discussed about its main function that is distributing load. By distributing the load across the joint, this will ensure the stability of the joint. This will result the stress experienced by the joint been decreased. Figure 1 below shows the different between the knee with meniscus and without meniscus.

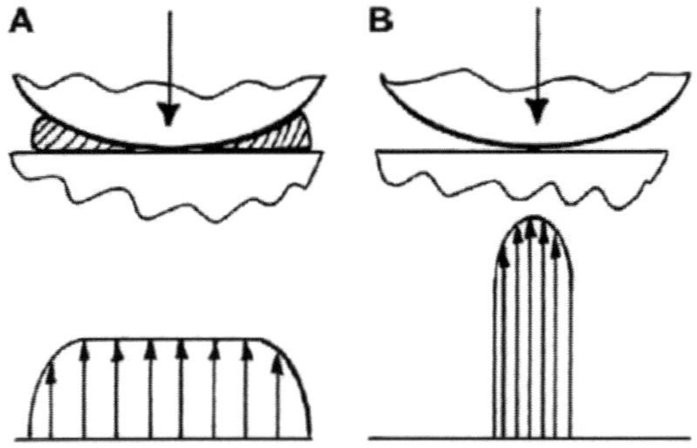

With meniscus Without meniscus

Figure 1. Differences between the knee with meniscus and without meniscus. Taken *from I.D. McDermott et al.*

The common description of the menisci is that they are semi-lunar fibrocartilagenous disks, whose main function is to increase the congruency of the tibiofemoral joint, thereby decreasing the stress in the joint through an increase in the contact area [1]. Since the meniscus is a viscoelastic material, it will tend to compress when the load is applied. As the femur and tibia is moving, the meniscus is also will move, so that the congruency of the joint can be maintained.

When the meniscus move from its original shape, it shows that force is being transferred across the meniscus body. This is called the hoop stresses that act within the meniscus body. The compression force across the knee, as it is transmitted from the femur through the meniscus to the tibia, causes tension along the circumferential collagen fibres within the meniscal tissues [1]. From what been said by I.D. McDermott in his literature about the development of hoop stress within the meniscus body, it is a result from a net resultant force that generated when load is applied. The wedge shape of meniscus after being loaded will make it to extrude from the joint. In order to prevent the meniscus from extrude more, hoop stresses is developed.

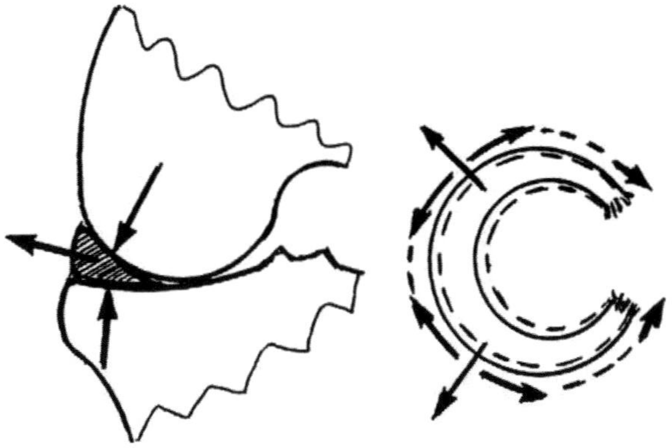

Figure 2. Conversion of axial load into meniscal hoop stresses. Taken from *Ian D. McDermott et al*

IOP Conf. Series: Materials Science and Engineering **50** (2013) 012027 doi:10.1088/1757-899X/50/1/012027

4. Meniscal Movement

Meniscus of the knee is having viscoelastic properties that are described as a combination or mixture of solid and also fluid components. Viscoelastic and elastic has their own properties. For elastic material, Viscoelastic and elastic material is having a different meaning and properties. In an elastic material, the energy used to deform the material is equal to the integral of the force deformation curve [2]. That is mean, when loading and unloading process, there will be no energy loss or any energy absorption. This is opposite with the viscoelastic condition. When loading and unloading process, the energy between those two works will not the same. Fithian et al. said that the energy loss phenomenon is due to the movement of fluid and/or the rearranging of molecule structure of the tissue itself. Since the meniscus is having fluid structure in its cross section, the meniscal water content could be extruded either by compression or by direct application of a pressure differential [1]. This will result with the displacement or deformation of meniscus. P.S. Walker et al. said that the meniscus function is to maintain the congruency of the femur and tibia joint, hence, the meniscus will carry about 40% to 70% of the load across knee. When load is applied, force distribution will go through the femur and the meniscus will experienced the force. The fluid component of meniscus wills slowly to extrude depends on its permeability and the viscosity of fluid. V. Vedi et al have done experimental setup by testing subject's knee joint to determine the meniscal movement through an arc of flexion-extension while weight bearing. The result state that the menisci do move posteriorly as the knee flexes and the lateral meniscus did move more compared to the medial meniscus [3]. From the results, it shows that the movement of the meniscus during the knee- flexion is to ensure the maximum congruency of the joint and also the articulating surfaces in order to avoid it from injury when having load.

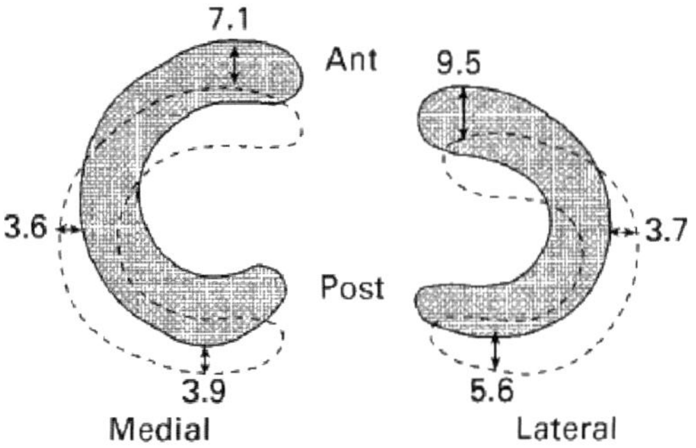

Figure 3 (a). Diagram shows the mean movement in each meniscus with weight bearing. Taken from *V. Vedi et al 1995.*

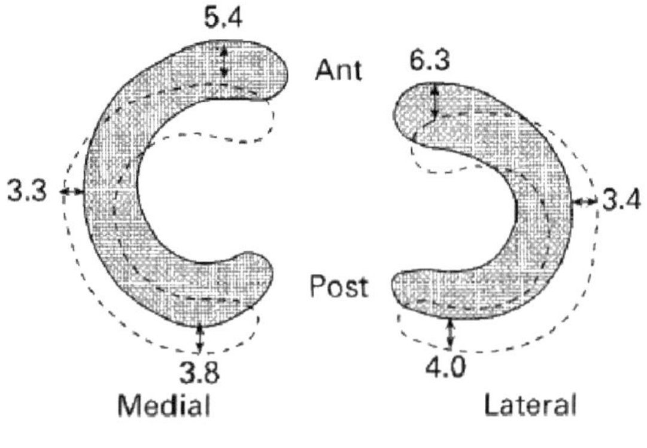

Figure 3(b). Diagram shows the mean movement in each meniscus with unloaded knee. Taken from *V. Vedi et al. 1995*

5. Meniscal Tears

Knee meniscus is a two semicircular pad that can be found between the joint of upper leg bone (femur) and lower leg bone (tibia). It is become the function of meniscus to be an effective load transmission and also shock absorber, so that the joint congruency can be enhanced. All material that is continuously loaded will have defect on its body. Same goes with the meniscus. The common defects that always occur is tear. Meniscal tears can happen to everyone without notice. Whether the age is young or old, everybody have a chance to get it depends on their lifestyle. At the young age, meniscal tears are caused by trauma in which happen while they are doing activity. In case, their knee is pivoted or twisted while they are doing activity, the meniscus can be torn. For old age, meniscus tears are a common thing for them. It is because the fluid structure within the meniscus will degenerate as their age is increased [15].

There are many types of meniscus tears that can be classified. Meniscal tears can be classified according to their location, shape, size, and stability. The classified tears are vertical longitudinal, oblique or called as parrot beaked, displace or called bucket handle, degenerative, transverse, horizontal or complex. Out of those tears, the most common tears that always occur are vertical longitudinal and the oblique tears. Greis et al. said that, about 81% of the meniscus tears are parrot beak or longitudinal tears affecting more often the medial meniscus. Figure 4 below shows a certain pattern of tears that happen at the knee meniscus.

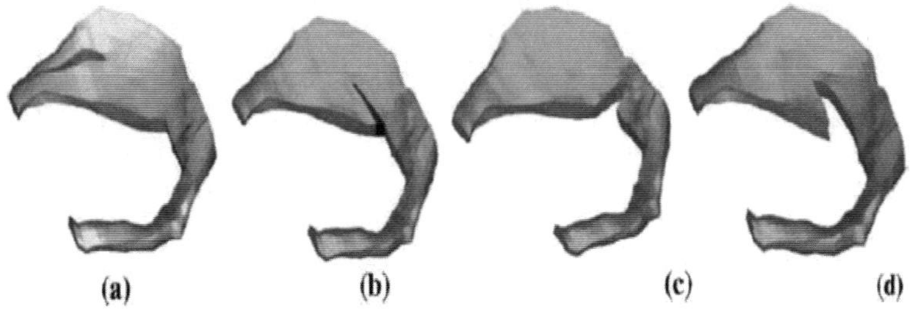

Figure 4. Patterns of meniscal tear: a) longitudinal; b) oblique; c) transverse; d) parrot beak. Taken from *E. Pena et al. 2005.*

Once the meniscus torn, surgical is the best way to prevent any compilation happens towards the knee joint. If not, the continuous knee motion and continuous load applied towards the meniscus can worsen the original tear. The surgical way that is one of the best is meniscectomy. Meniscectomy is a step to alter the shape of meniscus or remove the tear part of the meniscus. There are also different types of meniscectomy which are longitudinal, radial, oblique and total. After meniscectomy is done towards the torn meniscus, it will result with effect towards the joint of femur and tibia. LeRoux and Setton said that meniscectomy dramatically alter the pattern of static load transmission of the knee joint. This is because the fluid structure within the meniscus is being altered and the way of meniscus distributing the load will be different with the healthy meniscus. Hoop stresses are lost when a radial tear occurs or a segmental meniscectomy is performed; the load bearing condition become similar to that after meniscectomy [5]. Figure 5 shows a difference of load distributing between a healthy meniscus and also meniscus after meniscectomy.

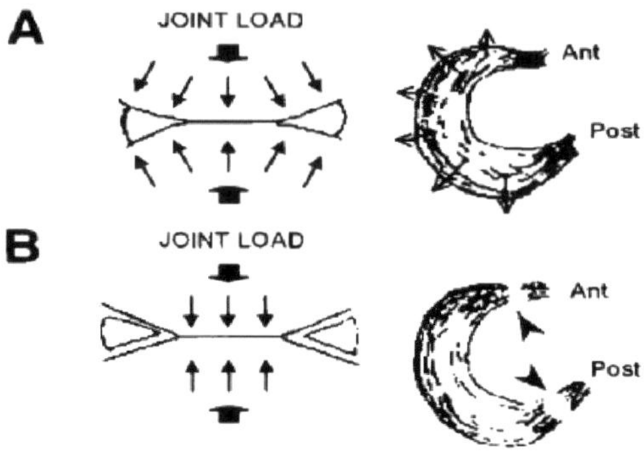

Figure 5. Difference of load distributing between **A**. healthy meniscus and also **B**. meniscus after meniscectomy. Taken from *J.M. Lee 2000.*

Out of four types of meniscectomy, the total meniscectomy is the least type that applied in the surgery. This is because the effect afterwards is worse than others. Total meniscectomy will result with articular wear after a certain years. This is because there is one missing meniscus after the total meniscectomy. Fithian et al. said that, several researchers have reported higher stresses and a decrease in shock absorbing capability after total meniscectomy. That shows the important of meniscus as a load transmission between the joint.

After the meniscectomy, the rehabilitation program is important so that the healing process is smoothly done. Forces that occur during knee motion potentially could distract the meniscal repair and disrupt healing [6]. Basically, the meniscal repair should aim to maintain the structure of the meniscus which is circumferential. This is to maintain the function of meniscus which is load transmission so that the joint surface is not subjected to the excessive stresses.

6. Instep Kick

From the selected journal, the literature is discussed more on instep kick which is the common kick that player use in a soccer game. Instep kick also has become the most widely studied for a soccer kicking [7]. A study about the kicking technique can be important if we want to enhance our understanding regarding the biomechanics of soccer. Other than that, by studying about the biomechanics of kicking, coaching process can be assisted [7]. Figure 6 below shows the kinematic of instep kicking.

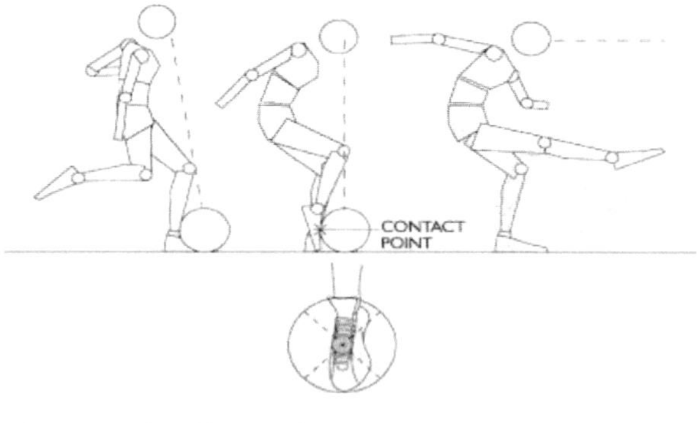

Figure 6. Kinematic of instep kicking. Taken from *http://www.ayso26.org/soccer*

As been discussed by a lot of researchers about the instep kicking, there are a few factors that can ensure the successful of that kicking which are distance of the kick from the ball, type of kick used, the air resistance and also the technique used. All of these can be taken as a biomechanical analysis. Other than the factors above, the foot to ball interaction is also important. This is including the speed of feet before the kicking is done. The higher the speed of the foot before impact, the shorter the foot-ball contact and the highest the ball speed [7]. The best output of instep kicking is depends on the highest velocity of ball after being kicked. In order to get the maximal velocity of ball after being kicked is by applying the highest velocity on the foot before the impact. Kicking with running approach demonstrates higher ball speed values compared with static approach kicks. This is because, the velocity of the foot is increasing since the player applying running approach before kicking. Table below shows the velocity of the foot and coefficient of restitution at impact between the foot and the ball.

Table 1. Summaries of the journal

		Method			Topic			
No	Author of Journal	Experimental	FEA	Theory	Meniscal Function	Meniscal Movement	Meniscal Tears	Soccer
1	Yunfeng Niu et al. (2012)		√		√			
2	T.M. Guess et al. (2010)		√		√	√		
3	J.M. Lee et al. (2000)			√	√			
4	J.R. Meakin et al (2002)	√			√			
5	D.P. Richards et al (2008)	√					√	
6	E. Pena et al. (2005)		√		√		√	
7	I.D. McDermott et al. (2008)			√	√	√		
8	A.A. Allen et al. (1995)			√	√			
9	A.R. Ismail et al. (2010)	√						√
10	S. Andrews et al. (2011)			√	√			
11	A. Lees et al (2010)			√				√
12	E. Pena et al. (2005)	√			√			
13	E. Kellis et al. (2007)			√				√
14	V. Vedi et al (1999)	√				√		

Table 2. The velocity of the foot and coefficient of restitution at impact

	Foot Speed (ms^{-1})		Coefficient of Restitution	
	Min	Max	Min	Max
Preferred Leg	16.1	20.3	0.46	0.68
Non-Preferred Leg	15.5	18.6	0.32	0.66

Based on the pie chart above, it is represented the percentage of injury breakdown by body part for soccer game. There are thirteen body parts that may be injured while playing the soccer game which are thigh, feet, knees, ankles, hips, back, neck, wrists, head, abdomen, elbow, forearms and lastly is shoulders. The factor that may cause the injury may come from external impact towards the body part or from the pivoting or twisting of body parts. From what being discussed in this paper, which is the knee, it is the part that becomes top three commonly injured in a soccer game with 14.6%. It is proven that the knee is a part that can injured easily while player is playing the soccer.

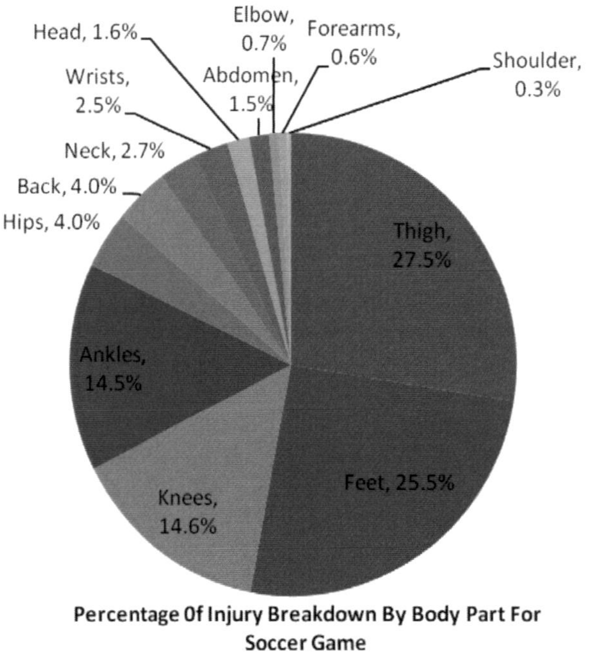

Figure 7. Percentage of injury breakdown by body part for soccer game.

7. Conclusion

Overall, this paper is reviewing the selected 14 journals in study of the knee meniscus and the soccer related. From what can be conclude about the meniscus, it is the small internal part that has big function towards our body. The primary function of the meniscus as the load transmission is to prevent the excessive stress and maintain the congruency to the femur and tibia joint. Based on the pie chart above, it is shows that the knee is in top 3 parts that easily injured in a soccer games and the torn meniscus is one of the injury. Statistical state that the most meniscal tears which is 73% occurred in athletes who were soccer players, basketball players or skiers. The torn is frequently happening at the medial side rather than lateral side with a percentage of 70% [18]. The statistic shows that the injury from playing soccer may injured the meniscus part easily. Related to the study, by choosing either preferred or non-preferred leg will result with different force towards the ball before kicking. This force will directly acting towards the meniscus that located between the joint of the knee.

Acknowledgments
The support of the Universiti Malaysia Pahang (UMP), under grant RDU110331 and under Center for Sport Engineering is gratefully acknowledged. The authors of this paper also would like to express their gratitude to Hospital Pekan, Pahang for supporting these research activities.

References
[1] McDermott I D, Masouros S D and Amis A A 2008 Biomechanics of the menisci of the knee *Basic Science*
[2] Andrews S, Shrive N and Ronsky J 2011 The shocking truth about meniscus *Journal of Biomechanics*
[3] Vedi V, Williams A, Tennant S J, Spouse E, Hunt D M and Gedroyc W M W 1999 Meniscal movement *Journal of Bone & Joint Surgery*
[4] Pena E, Calvo B, Martinez M A, Palanca D and Doblare M 2005 Finite element analysis of the effect of meniscal tears and meniscectomies on human knee biomechanics *Clinical Biomechanics*
[5] Lee J M, and Fu F H 2000 The Meniscus: Basic science and clinical applications
[6] Richards D P, Barber F A and Herbert M A 2008 Meniscal tear biomechanics: Loads across meniscal tears in human cadaveric knee
[7] Ismail A R, Mansor M R A, Ali M F M, Jaafar S and Johar, M S N M 2010 Biomechanics analysis for right leg instep kick *Journal of Applied Science*
[8] Niu Y and Wang F 2012 A finite element analysis of the human knee joint: Menisci prosthesis instead of the menisci and articular cartilage
[9] Guess T M, Thiagarajan G, Kia M and Mishra M 2010 A subject specific multibody model of the knee with menisci *Journal of Medical Engineering & Physics*
[10] Meakin J R, Shrive N G, Frank C B and Hart D A 2002 Finite element analysis of the meniscus: The influence of geometry and material properties on its behaviour
[11] Allen A A, Caldwell G L and Fu F H 1995 Anatomy and biomechanics of the meniscus
[12] Lees A, Asai T, Andersen T B, Nunome H and Sterzing T 2010 The biomechanics of kicking in soccer: A review *Journal of Sports Science*
[13] Pena E, Calvo B, Martinez M A and Doblare M 2005 A three-dimensional finite element analysis of the combined behavior of ligaments and menisci in the healthy human knee joint *Journal of Biomechanics*
[14] Kellis E and Katis A 2007 Biomechanical characteristics and determinants of instep soccer kick *Journal of Sports Science and Medicine*

The influence of football boot construction on ball velocity and deformation

Zahari Taha, Mohd Azri Aris[a], Mohd Hasnun Arif Hassan
Centre for Sports Engineering, Faculty of Mechanical Engineering,
Universiti Malaysia Pahang, 26600 Pekan, Pahang, Malaysia

E-mail: [a]azriaris9@yahoo.com

Abstract. Research on the biomechanical properties of football boots has received little scientific attention. Recently, modern and scientific football boots has been developed that afford excellent ball control, shooting and running ability. Different constructions of football boot deliver different velocity and deformation of the ball. In this study, three commercially available football boots were compared during kicking in terms of, ball velocity and deformation of the ball. A subject performed three repetitive shots of the ball into the net for each football boot model. The footballer's foot and the ball velocity were recorded using a high speed camera. From the recorded video, the kicking velocity, ball velocity and ball deformation were measured. Based on the foot and ball velocity data, the Coefficient of Restitution (COR) was calculated. The results obtained show the COR value varies with different boots (COR_{BootA} = 0.89, COR_{BootB} = 0.59, COR_{BootC} = 0.47) thus delivers different ball velocities at same kicking foot velocity (U_{leg} = 14 ms^{-1}). It is apparent that the Boot A was made with the best upper material (Teijin synthetic leather), design and also construction of the boot.

1. Introduction

Kicking is the most specific technique in football, so it is appropriate to give considerable attention for our understanding of this skill from a biomechanical perspective. Researchers have widened their interest to consider the kick beginning from the way a player approaches the ball to the end of the ball flight. This field of study has encapsulated the overall technique and the influences of the upper body, support leg and pelvis on the kicking action, foot-ball impact and the influences of footwear and soccer balls, ball launch characteristics and corresponding flight of the ball.

When a ball is kicked, the leg is putting energy into the ball and the ball deformed. The part of the ball impacted by the foot was flattened for a few milliseconds. The energy going into the collision is the kinetic energy of the foot plus the stored energy in the deformed ball and the energy coming out is the kinetic energy of the ball. The conservation of energy causes the ball to go faster than the velocity of the foot

As kicking the ball is categorized as a collision mechanics [1], the ball velocity can be determined by using the law of the conservation of momentum and Coefficient of Restitution (COR).

$$M_{leg}U_{leg} + M_{ball}U_{ball} = M_{leg}V_{leg} + M_{ball}V_{ball} \quad (1)$$

$$e = \frac{V_{ball} - V_{leg}}{U_{leg} - U_{ball}} \tag{2}$$

From the above equations, the derivation of the actual formula for the velocity of the ball is [2]:

$$V_{ball} = U_{leg} \frac{M_{leg}}{M_{leg} + M_{ball}} (1 + e) \tag{3}$$

The velocity of the kicking leg (U_{leg}) is controlled by the player before contact. It has been proven that powerful kicking at impact results in differing ball velocities [3]. The term $M_{leg} (M_{leg}+M_{ball})^{-1}$ relates to the muscles involved in the kick, and their strength at impact. The $(1+e)$ term relates to the rigidity of the foot at impact [2]. In this study, the effort to manipulate the mass of leg can be neglected as the player cannot gain the mass of leg while kicking to achieve maximum ball velocity. The focus of this study is on the football boot construction where the COR (e), is the variable that can be measured to increase the ball velocity.

The COR in kicking determine the transfer of velocity from the foot to the ball during the impact phase and is the ratio between relative velocity ($V_{ball} - V_{leg}$) after impact and the relative velocity before impact [4]. The factors determining the COR are the mechanical properties of the foot and ball, i.e. stiffness, contact area and contact point.

There are several studies that had been done by researchers to evaluate the best method for football kicking. In Japan, a study claimed that for a more precise ball, the side-foot kick are the most suitable style while for a faster ball, the instep kick should be used [5]. It was found that the difference in the passive foot angular motion during ball impact was due to the difference of the ball impacting part of the two kicks and that foot motion of both side-foot kicking and instep kicking were different but the quality of the ball impact was similar for both type of kicking. For this current study, the toe kick is selected to be used for the kicking experiment. A study in Denmark [4] had found that the toe kick produce a larger COR compared to the instep kicking.

Several kicking experiments had been done by researchers to evaluate the factors influence the ball velocity in kicking. Previous research had found that football boot also contributed towards the ball velocity. In UK, a study was done to analyze whether football boot mass has an effect on ball velocity [6]. Ten male university students participated in the study to kick a ball with three different masses of boots that was attached with three different lead plates (180.8 g, 260.6 g, and 356.1 g). They had concluded that the foot linear velocity does decrease with heavier boots but the ball velocity was not affected.

A research to quantify the influence of different shoe upper friction properties on ball velocity during kicking was done in Germany [1]. They have used four boots with different friction properties of the upper materials during their experiment. In this study, they concluded that friction materials of football boot uppers did not show any statistically significant influence in achieving different ball velocity during instep kicking.

2. Methods

One male university soccer player with no lower limb injury in the previous six month was chosen as the subject in this study (23 y; 74.3 kg; 1.77 m). Before the experiment, the subject was asked to perform warm up in order to avoid musculo-skeletal injuries during the session. A simple briefing session was carried out to make sure the subject was fully understood the experiment procedures. Three different type of football boot was chosen for the kicking experiment (Nike Mercurial Vapor VIII FG (Boot A); Adidas AdiPower Predator TRX FG (Boot B); Puma King Finale SL I FG (Boot C)). The participant had to perform three maximal toe kicks for each football boot with a mandatory rest period between single kicks to avoid fatigue. The subjects were instructed to take three steps before kicking the ball into the net 2 m from the kick area.

A green Adidas FIFA standard football ball size 5 with diameter of 220 mm and weight 450 g was used for the experiment. The ball was inflated with a pressure of 0.65 bar. Markers were placed on the specific area on the foot and ball to aid in velocity and ball deformation measurement. The kicking action was captured using EPIX SV643C high-speed camera with 2,500 frames per second recording speed. The recorded video was than analyzed using the XCAP high-speed camera build-in software to calculate the foot kicking velocity (U_{leg}), foot after kicking velocity (V_{leg}), and also the ball velocity (V_{ball}). The COR of each football boot was calculated using equation (2). The ball deformation on one selected kicking velocity was measured and grouped based on the type of football boot.

3. Results

The kicking action was captured using a high speed camera and analysed to get foot kicking velocity (U_{leg}), foot after kicking velocity (V_{leg}), and also the ball velocity (V_{ball}). The COR varies with the foot kicking velocity for each boot as shown in figure 1 to figure 3.

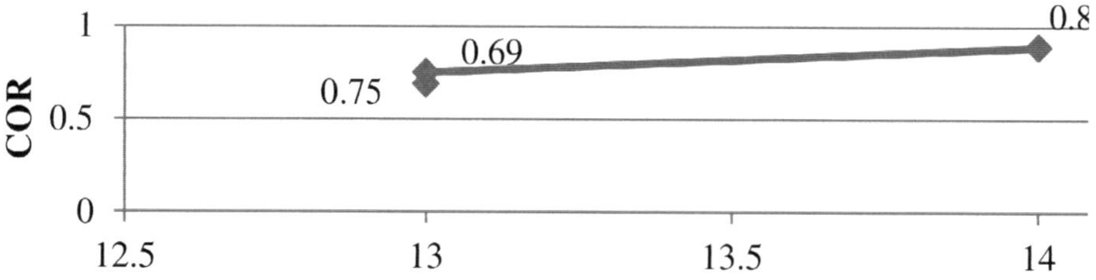

Figure 1. Calculated COR for Boot A with different kicking velocity

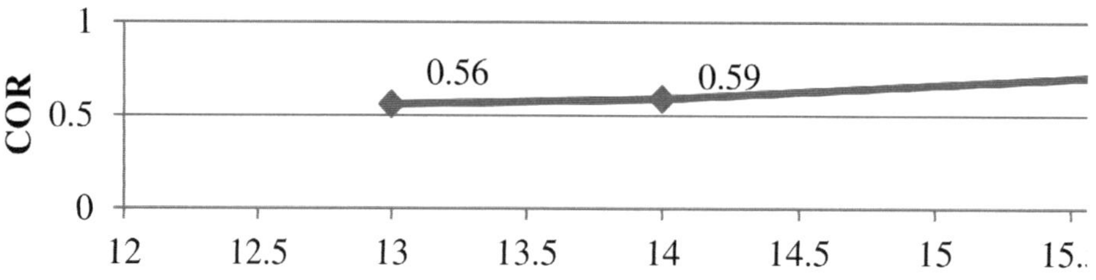

Figure 2. Calculated COR for Boot B with different kicking velocity

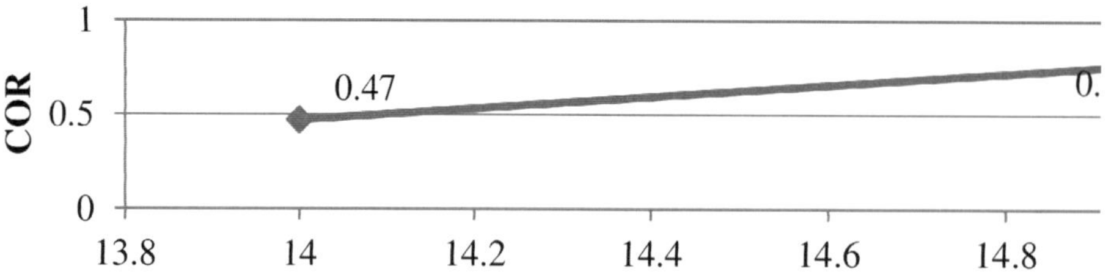

Figure 3. Calculated COR for Boot C with different kicking velocity

From three calculated analysis above, Boot A recorded the highest COR value (0.89) while Boot B and Boot C recorded 0.59 and 0.47 respectively at a specific selected kicking foot velocity (U_{leg} = 14 ms^{-1}) as shown in figure 4.

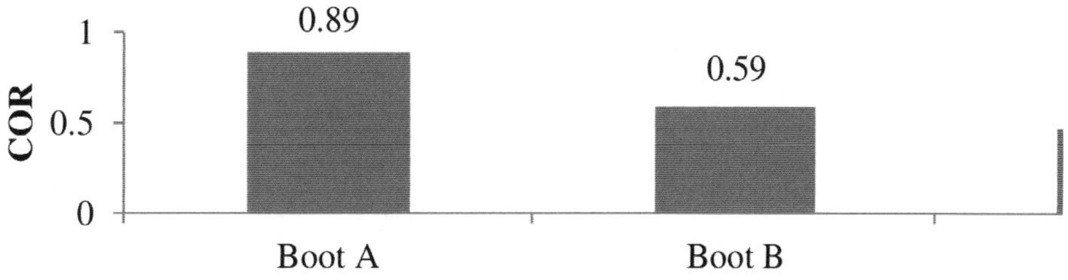

Figure 4. Calculated COR for Boot C with different kicking velocity

The maximum deformation of the ball was measured using the XCAP build-in measuring software. The measured data was grouped into certain kicking velocity based on each football boot model. The largest ball deformation was produced with Boot A (≈36 mm). The maximum ball deformation picture by boot type is shown in figure 5.

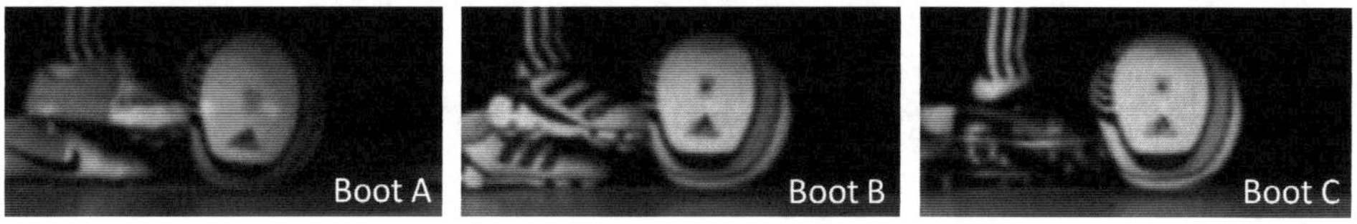

Figure 5. Maximum ball deformation for each boot type

A summary of the foot kicking velocity, maximum ball deformation, ball velocity and also COR are shown in the table 1.

Table 1. Experiment result summary

	Boot A	Boot B	Boot C
Foot kicking velocity (ms^{-1})	14	14	14
Max ball deformation (mm)	36	30	26
Ball velocity (ms^{-1})	20.8	19.2	16.2
COR	0.89	0.59	0.47

4. Conclusion

From the experiment, the COR value varies with different boots (COR_{BootA} = 0.89, COR_{BootB} = 0.59, COR_{BootC} = 0.47) thus delivers different ball velocities at same kicking foot velocity (U_{leg} = 14 ms^{-1}). A high COR results a high ball velocity and also ball deformation. It can be concluded that the Boot A was designed with the best upper material construction properties for toe kicking compared with other two football boots. This type of upper material (Teijin synthetic leather) and the boot construction design with forefront thickness (t) of 1.88 mm would be the best to be use to produce high ball velocity with the lowest kicking velocity compared to the Boot B (full-grain calfskin, t = 2.59 mm) and Boot C (premium kangaroo leather, t = 1.75 mm). An experiment with larger sample sizes could be conduct in future to obtain precise and statistically analysis results.

5. Acknowledgements
The authors would like to thank Universiti Malaysia Pahang and the Ministry of Higher Education, Malaysia for providing the funds for this research under the grant RDU110702.

6. References

[1] Sterzing T and Hennig E M 2007 The influence of friction properties of shoe upper materials on kicking velocity in soccer *Journal of Biomechanics* **40** Supple S195

[2] Lees A, Reilly T, Clarys J P and Stibbe A 1993 Biomechanics of Football *Science and Football* **2**

[3] Kellis E and Katis A 2007 Biomechanical characteristics and determinants of instep soccer kick *Journal of Sports Science and Medicine* **6**

[4] Andersen T B, Kristensen L B, Sørensen H and Cabri I J 2005 Coefficient of restitution (COR) in the toe and instep soccer kicks *Science and Football V: The Proceedings of the Fifth World Congress on Science and Football* **5** 73–7

[5] Shinkai H, Nunome H, Suzuki H, Suito H, Tsujimoto N and Ikegami Y 2008 Comparison of ball impact characteristics between side-foot and instep kicking in soccer *26 International Conference on Biomechanics in Sports*

[6] Moschini A and Smith N 2012 Effect of shoe mass on soccer kicking velocity *30th Annual Conference of Biomechanics in Sports* 150–3

2nd International Conference on Mechanical Engineering Research (ICMER 2013) IOP Publishing
IOP Conf. Series: Materials Science and Engineering **50** (2013) 012029 doi:10.1088/1757-899X/50/1/012029

Active sway control of a gantry crane using hybrid input shaping and PID control schemes

M Z Mohd Tumari[1,a], **L Shabudin**[1], **M A Zawawi**[1] **and L H Ahmad Shah**[2]
[1]Robotics and Unmanned System (RUS) Group
Faculty of Electrical & Electronics Engineering
[2]Faculty of Mechanical Engineering, Universiti Malaysia Pahang,
26600 Pekan, Pahang, Malaysia

E-mail: [a]zaidimt@ump.edu.my

Abstract. This project presents investigations into the development of hybrid input-shaping and PID control schemes for active sway control of a gantry crane system. The application of positive input shaping involves a technique that can reduce the sway by creating a common signal that cancels its own vibration and used as a feed-forward control which is for controlling the sway angle of the pendulum, while the proportional integral derivative (PID) controller is used as a feedback control which is for controlling the crane position. The PID controller was tuned using Ziegler-Nichols method to get the best performance of the system. The hybrid input-shaping and PID control schemes guarantee a fast input tracking capability, precise payload positioning and very minimal sway motion. The modeling of gantry crane is used to simulate the system using MATLAB/SIMULINK software. The results of the response with the controllers are presented in time domains and frequency domains. The performances of control schemes are examined in terms of level of input tracking capability, sway angle reduction and time response specification.

1. Introduction

Cranes are commonly employed in the transport industry for the loading and unloading of freight, in the construction industry for the movement of materials and in the manufacturing industry for the assembly of heavy equipment. In addition, crane is important as a lifting machine, generally equipped with wire ropes or chains and sheaves, which can be used to lift and lower materials vertically and horizontally. It used one or more simple machines to create mechanical advantages and thus move loads beyond the capability of a human.

There are numerous efforts in controlling gantry cranes have been proposed. For instance, fuzzy logic controller has been proposed for controlling the gantry crane system [1-2]. Nevertheless, the fuzzy logic controller design is sophisticated in finding the membership function, satisfactory rules, fuzzification and deffuzification parameter heuristically.

Several researches based on open-loop and closed-loop control schemes have been done for controlling the gantry crane system. For example, open loop time optimal strategies were implemented to the gantry crane system [3]. Unfortunate results were found in these researches because open-loop method could not compensate for the effect of wind disturbance and sensitive to the system parameters. While, closed loop control method has also been adopted for controlling the crane system.

For example, PD controllers has been proposed for both position and anti-swing controls [4]. However, the performance of the controller is not very effective in eliminating the steady state error.

In this study, input shaping is implemented by convolving a sequence of impulses, an input shaper, with a desired system command to produce a shaped input that produces self-canceling command signal [5]. Input shaper is designed by generating a set of constraint equations which limit the residual vibration, maintain actuator limitations, and ensure some level of robustness to modeling errors [6]. In the input shaper, the amplitudes and time locations of the impulses are determined by solving the set of constraints [7]. On the other hand, feedback control which is well known to be less sensitive to disturbances and parameter variations is also adopted for controlling the gantry crane system [8].

2. Modeling of Gantry Crane System

The two-dimensional gantry crane system with its payload considered in this work is shown in figure 1, where x is the horizontal position of the cart, L is the length of the rope, θ is the sway angle of the rope, M and m is the mass of the cart and payload respectively. In this simulation, the cart and the payload can be considered as point masses and are assumed to move in two-dimensional, x-y plane. The tension force that may cause the hoisting rope is also ignored. In this study the length of the cart, $L = 0.5$ m, $M = 2.49$ kg, $m = 0.5$ kg and $g = 9.81$ m/s2 is considered.

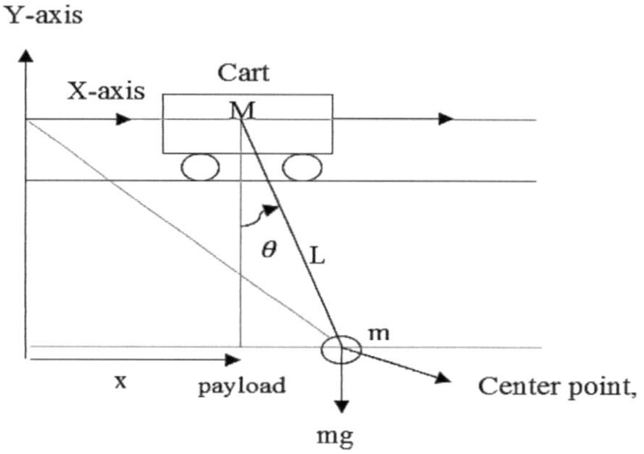

Figure 1. Gantry Crane Model.

The Euler-Lagrange formulation is considered in characterizing the dynamic behavior of the crane system incorporating payload. The kinetic and potential energy of the whole system is given by Eq. (1):

$$T = \frac{1}{2}M\dot{x}^2 + \frac{1}{2}\,m(\dot{x}^2 + i^2 + i^2\theta^2 + 2xisin\theta + 2\dot{x}i\theta cos\theta) \tag{1}$$

The potential energy can be formulated by Eq. (2):

$$U = -mgl\cos\theta \tag{2}$$

Using Lagrange equation in Eq. (3-6):

$$\frac{d}{dt}\left(\frac{dL}{dqj}\right) - \frac{dL}{dqj} = Fj \tag{3}$$

$$u = (m + M)\ddot{x}_1 + ml\big(\theta cos\ddot{\theta} - \dot{\theta}^2 sin\theta\big) \tag{4}$$

$$m\ddot{x}_1 cos\theta + m\,l\theta{=}{-}\mathrm{mg}\sin\theta \tag{5}$$

$$\ddot{x}_1 cos\theta + l\ddot{\theta} = -gsin\theta \tag{6}$$

From Eq. (5) and Eq. (6), the equation can be rewritten in the state space equation form in Eq. (7):

$$\begin{bmatrix} \ddot{x}_1 \\ \ddot{\theta} \end{bmatrix} = \begin{bmatrix} \frac{u+m\sin\theta(l\theta^2+gcos\theta)}{M+msin^2\theta} \\ -\frac{u\, cos\theta+m\sin\theta(g+l\dot{\theta}^2cos\theta)+gM\, sin\theta}{l\,(M+msin^2\theta)} \end{bmatrix} \tag{7}$$

where

M : mass of the trolley (kg)
m: mass of the payload (kg)
L : length of rope (m)
g : gravity acceleration (m/s²)
θ: angle of load swing (rad)
$\ddot{x}$: acceleration of trolley (m/s²)
$\ddot{\theta}$: angular acceleration of the load swing (rad/s²)

The Eq. (7) is in nonlinear function, it can't be used easily for the purpose of analysis, design and other. In order to get a linear model, linearization must be implemented in the above model. The following condition will satisfy the aim:

$$\cos\theta \approx 1; \sin\theta \approx \theta; \sin^2\theta \approx 0; \dot{\theta}^2 \approx 0$$

The linear model of the uncontrolled system can be represented in a state-space form as shown in equation by assuming the change of rope and sway angle are very small.

The state space of gantry crane is represents in Eq. (8-10).

$$\begin{aligned} \dot{x} &= Ax + Bu \\ y &= Cx + Dx \end{aligned} \tag{8}$$

$$\dot{x} = \begin{bmatrix} \dot{x}_1 \\ \dot{x}_2 \\ \dot{x}_3 \\ \dot{x}_4 \end{bmatrix} = \begin{bmatrix} \dot{x}_1 \\ \ddot{x}_2 \\ \dot{\theta} \\ \ddot{\theta} \end{bmatrix} = \begin{bmatrix} \text{cart position} \\ \text{cart velocity} \\ \text{bar's angle} \\ \text{bar's angle rate} \end{bmatrix} \tag{9}$$

$$\dot{x} = \begin{bmatrix} 0 & 0 & 1 & 0 \\ 0 & 0 & 0 & 1 \\ 0 & \frac{m*g}{M} & 0 & 0 \\ 0 & \frac{-((M+m)*g)}{M*L} & 0 & 0 \end{bmatrix} \begin{bmatrix} \dot{x}_1 \\ \ddot{x}_2 \\ \dot{\theta} \\ \ddot{\theta} \end{bmatrix} + \begin{bmatrix} \dot{0} \\ 0 \\ \frac{1}{M} \\ \frac{-1}{M*L} \end{bmatrix}$$

$$C = [1 \quad 0 \quad 0 \quad 0]$$

$$D = [0] \tag{10}$$

3. Input Shaping and PID Control Schemes

3.1 Input Shaping Control Schemes

The design objective of input shaping is to determine the amplitude and time locations of the impulses in order to reduce the detrimental effects of system flexibility. These parameters are obtained from the natural frequencies and damping ratios of the system [9]. Figure 2 shows summarized of positive input shaping techniques. The corresponding design relations for achieving a zero residual single-mode sway of a system and to ensure that the shaped command input produces the same rigid body motion as the unshaped command yields a two-impulse sequence namely positive zero-sway (PZS) with parameter as shown in Eq. (11).

$$t_1 = 0,\ t_2 = \frac{\pi}{\omega_d},\ A_1 = \frac{1}{1+K},\ A_2 = \frac{K}{1+K} \tag{11}$$

where

$$K = e^{\frac{-\xi\pi}{\sqrt{1-\xi^2}}}$$

$$\omega_d = \omega_n \sqrt{1-\xi^2}$$

$$\omega_n = \sqrt{\frac{g}{l}\left(\frac{M+m}{M}\right)}$$

ω_n and ζ representing the natural frequency and damping ratio respectively and *tj* and *Aj* are the time location and amplitude of impulse *j* respectively. This yields a three-impulse and four-impulse sequence namely Positive zero-sway-derivative (PZSD) and positive zero sway- derivative-derivative (PZSDD) with parameter as shown in equations (12) and (13) respectively:

$$t_1 = 0\ ,\quad t_2 = \frac{\pi}{\omega_d}\ ,\quad t_3 = \frac{2\pi}{\omega d}$$
$$A_1 = \frac{1}{1+2K+K^2},\ A_2 = \frac{2K}{1+2K+K^2},\ A_3 = \frac{K^2}{1+2K+K^2} \tag{12}$$

$$t_1 = 0\ ,\quad t_2 = \frac{\pi}{\omega_d}\ ,\ t_3 = \frac{2\pi}{\omega_d}\ ,\quad t_4 = \frac{3\pi}{\omega_d}$$

$$A_1 = \frac{1}{1+3K+3K^2+K^3}\ ,$$

$$A_2 = \frac{3K}{1+3K+3K^2+K^3}\ ,$$

$$A_3 = \frac{3K^2}{1+3K+3K^2+K^3}\ ,$$

$$A_4 = \frac{K^3}{1+3K+3K^2+K^3} \tag{13}$$

Positive Zero Sway (PZS)	Positive Zero Sway Derivative (PZSD)	Positive Zero Sway Derivative-Derivative (PZSDD)
• Contain 2 impulse response include: i. Unity amplitude summation ii. Time optimality constraints	• Contain 3 impulse response include: i. Unity amplitude summation ii. Time optimality constraints iii. First order robustness constraint equation.	• Contain 3 impulse response include: i. Unity amplitude summation ii. Time optimality constraints iii. Second order robustness constraint equation.

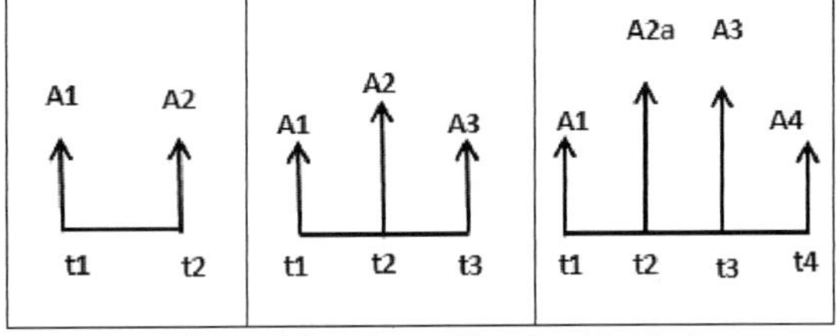

Figure 2. Positive Input Shaping Techniques.

3.2. PID Control Schemes

Proportional Integrated Derivative (PID) control is the most popular feedback controller used within the process industries involved in controlling the crane position. PID also provides a constant system output at a specified set point. The desired closed loop dynamics is obtained by adjusting the three parameters such as Proportional gain (K_P), Integral time (T_i) and Derivative time (T_d), often iteratively by "tuning" and without specific knowledge of a plant model. Stability can often be ensured using only the proportional term. The integral term permits the rejection of a step disturbance. The derivative term is used to provide damping or shaping of the response. The PID control scheme is named after its three correcting terms, whose sum constitutes the manipulated variable (MV) as shown in figure 3.

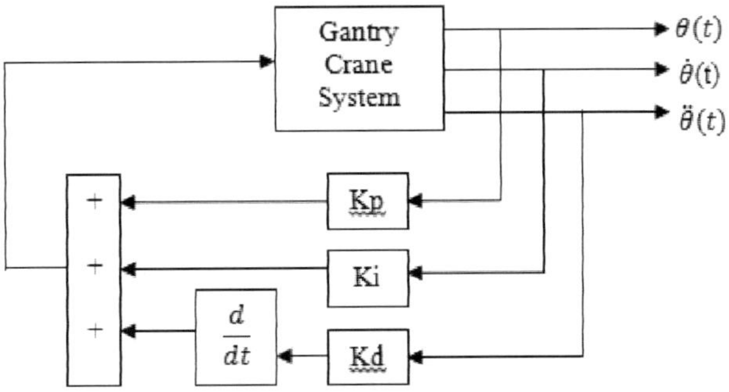

Figure 3. The PID controller structure.

For the Ziegler-Nichols Frequency Response Method, the critical gain, Kcr and the critical period, Pcr have to be determined first by setting the $Ki = \infty$ and $Kd = 0$. Increase the value of Kp from 0 to a critical value, adjust gain to make the oscillations continue with a constant amplitude as shown in figure 3, the value of Kcr is at which the output first exhibits sustained oscillation. Table 1 shows the formula of PID controller tuning parameters for the second method of Ziegler-Nichols. Block diagram of hybrid input shaping and PID control scheme is shown in figure 5.

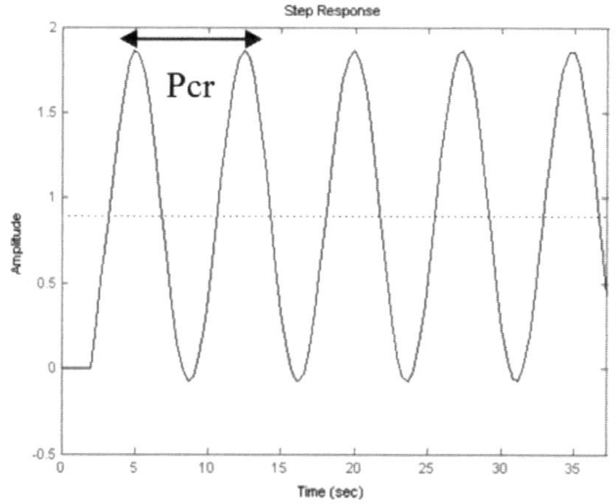

Figure 4. Step Response for Critical Period.

Table 1. Calculation of PID based on Ziegler Nichols (second method)

Value Critical gain and period	**Kp**	**Kd**	**Ki**
$Kcr = 1$ $Pcr = 5$	$0.6 * Kcr$	$1/\left(\frac{Pcr}{2}\right)$	$\frac{Pcr}{8}$
Tuning PID	Kp =0.6	Ti =0.4	Td =0.625

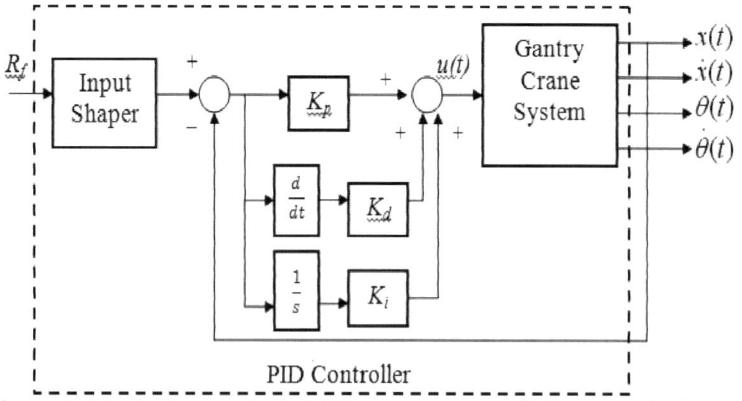

Figure 5. Block diagram for hybrid input shaping and PID control schemes.

4. Results and Discussion

In this investigation, hybrid input shaping and PID control schemes are implemented on gantry crane system and the corresponding results are presented. The bang-bang input voltage of ±1.5V is applied to the gantry crane system. To study the effectiveness of sway suppression, PZS, PZSD and PZSDD shapers with PID controller are designed based on the sway frequencies and damping ratios of the gantry crane system. The first mode, sway of the system is considered, as these dominate the dynamic of the system. The responses of the gantry crane system to the unshaped input were analyzed in time-domain and frequency domain (spectral density). These results were considered as the system response to the unshaped input and will be used to evaluate the performance of the input shaping schemes.

The attenuation level of sway angle in power spectra density (PSD) was analyzed of each types of shaper. Based on a comparison result from the PSD in figure 6, when increasing the value of proportional gain (Kp) at 250, integral time (Ki) at 0.0025 and derivative time at 1000 the swaying frequencies are dominated by the first mode, which are obtained as 0.6868 Hz, 0.6868 Hz and 0.5887 Hz with a magnitude of 2.113 dB, -15.08 dB and -38.18 dB for the hybrid PID control schemes with PZS, PZSD and PZSDD shaper respectively.

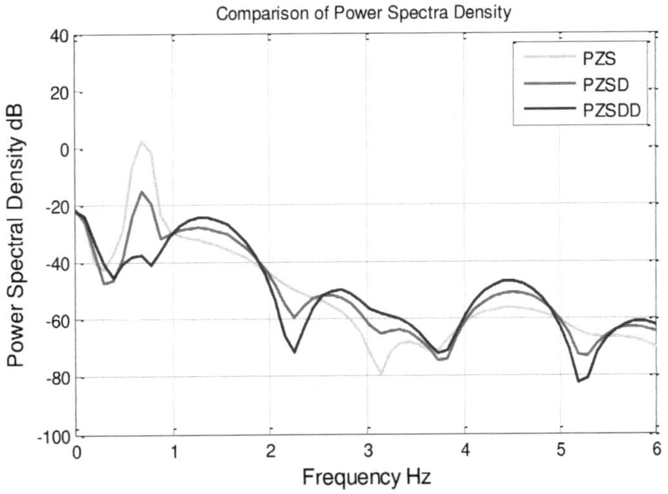

Figure 6. Comparison of Power Spectra Density.

Figure 7 shows the comparison of sway angle of the pendulum between hybrid PID control schemes with PZS, PZSD and PZSDD shaper. It is noted that the level of sway angle was significantly reduced with the increasing of positive input shaping derivative order. Based on the result, highest levels of sway reduction were obtained using PZSDD shaper followed by PZSD and PZS shaper scheme. This is evidenced in the pendulum sway angle responses. The magnitude of sway angle was achieved as ± 0.0358 rad, ± 0.0081 rad and ± 0.0039 rad for the hybrid PID control schemes with PZS, PZSD and PZSDD shaper respectively.

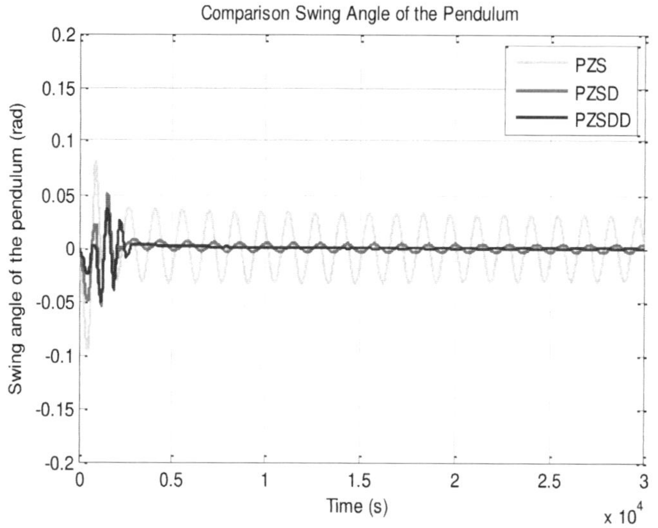

Figure 7. Comparison response of sway angle of pendulum.

Figure 8 shows the comparison of the horizontal position cart between hybrid PID control schemes with PZS, PZSD and PZSDD shaper. Using the hybrid PID control schemes with PZS shapes, the cart motion settles down at 1.50 $x10^4$ ms followed by PZSD and PZSDD shaper with the value of 1.52 $x10^4$ ms and 1.54$x10^4$ ms while for the rise time are at 0.80 $x10^4$ms, 0.92 $x10^4$ ms and 0.94 $x10^4$ ms respectively. It shows that, by incorporating more number of impulses in input shaping schemes resulted in a slower response. However, there is no overshoot in the responses of the cart motion for all hybrid input shaping and PID control cases.

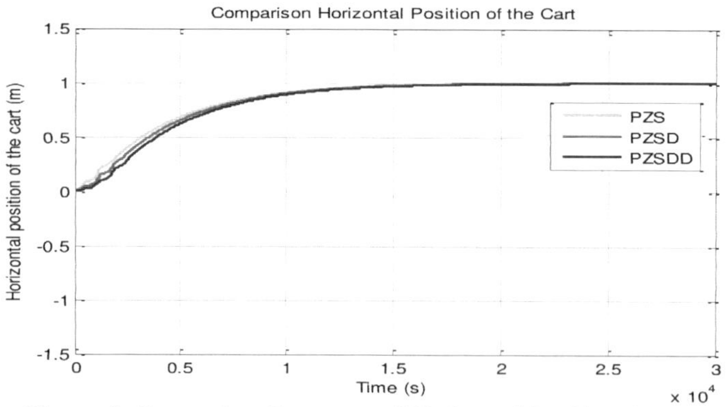

Figure 8. Comparison Response of Horizontal Position Cart.

By comparing the result presented in table 2, it is noted the highest performance in the reduction of the sway of the system is achieved using hybrid input shaping and PID control schemes. The result shows that, highest level of sway reduction is achieved using PZSDD with PID control schemes followed by PZSD and PZS.

Table 2. Level sway reduction of the single pendulum

Feedback Controller	Feed-forward Controller	Attenuation (dB) of sway of the pendulum	Amplitude of sway angle (rad)
Uncontrolled		30.68	± 0.187
	PZS	2.113	± 0.0358
PID	PZSD	-15.08	± 0.0081
	PZSDD	-38.18	± 0.0039

The time response specifications of rise time, settling time and overshoot of the cart position for hybrid input shaping and PID control schemes are depicted in Table 3. For comparative assessment, it shows that the speed of the system response can be improved by using a lower number of impulses compared to the highest number of impulses. It is noted that the combination of feed-forward and feedback controllers are capable of reducing the system sway and at the same time maintaining the steady state position of cart.

Table 3. Time response specification of horizontal position of cart response

Feedback Controller	Feed-forward Controller	Rise Time $x10^4$ (ms)	Settling Time $x10^4$ (ms)	Overshoot (%)
	PZS	0.80	1.50	0
PID	PSDD	0.92	1.52	0
	PZSDD	0.94	1.54	0

5. Conclusion

At the end of this project, active sway control of gantry crane by using hybrid input shaping and PID control schemes has been presented. The main objective in this project which is to reduce the sway of the gantry crane system had been implemented. The effects of the difference derivative order of the positive input shaping in term of a level of sway reduction and time response specification have been studied.

Moreover, the hybrid PID controller and input shaping schemes with higher number of impulses provide a high level of sway reduction. However, in terms of speed of the responses, the input shaping with a low number of impulses, results in a higher speed of tracking response. In overall, the combination feed-forward and feedback controller based on hybrid input shaping and PID control schemes provide better performance in sway reduction as compared to the uncontrolled techniques in the overall.

Acknowledgments

This work was supported by Faculty of Electrical & Electronics Engineering, Universiti Malaysia Pahang, especially Robotics and Unmanned System (RUS) Group under UMP short term Research Grant Schemes RDU100112 and RDU110374.

References

[1] Lee H H and Cho S K 2001 A New fuzzy logic anti-swing control for industrial three-dimensional overhead crane *Proc. of the 2001 IEEE Int. Con. on Robotic and Automation* Seoul pp 2958-2961
[2] Bakhtiari-Nejad F, Nazemizadeh M and Arjmand H 2013 *Inter. J. Automot. Mech. Eng.* **7** 830.

[3] Manson G A 1992 Time-Optimal Control of and Overhead Crane Model *Optimal Control Applications & Methods* **3**(2) 115-120

[4] Omar H M 2003 Control of Gantry and Tower Cranes *M.S. Thesis, Virginia Tech, Blacksburg, VA.*

[5] Singhose W, Crain E and Seering W 1997 Convolved and Simultaneous Two-Mode Input Shapers *IEEE Proc. Control Theory Appl.* vol 144 pp 515-520

[6] Singhose W and Singer N 1996 Effects of Input Shaping on Two-Dimensional Trajectory Following *IEEE Trans. on Robotics and Automation* vol 12 pp 881-887

[7] Gürleyük S S 2007 Optimal Unity-Magnitude Input Shaper Duration Analysis *Archive of Applied Mechanics* **77** 63-71

[8] Wahyudi and Jalani J 2005 Design and implementation of fuzzy logic controller for an intelligent gantry crane system *Proc. of the 2nd Int. Conf. on Mechatronics 2005* pp 345- 351

[9] Ahmad M A, Zulkifely Z and Zawawi M A 2010 Experimental Investigations of Input Shaping Schemes for Sway Control of a Gantry Crane System *2nd Int. Conf. on Computer and Network Technology,* pp 483-386

IOP Conf. Series: Materials Science and Engineering **50** (2013) 012030 doi:10.1088/1757-899X/50/1/012030

Experimental study on the cooling performance of high power LED arrays under natural convection

M S Mohamad[1,a], **M Z Abdullah**[2], **M K Abdullah**[3]
[1]School of Mechatronic Engineering, Universiti Malaysia Perlis, Pauh Putra Campus, 02600, Arau, Perlis, Malaysia
[2]School of Mechanical Engineering, Universiti Sains Malaysia, Engineering Campus, 14300, Nibong Tebal, Penang, Malaysia
[3]Mechanical Section, UniKL MSI, Kulim HiTech Park, 09000, Kulim, Kedah, Malaysia

E-mail: [a]sofwanmohamad@unimap.edu.my

Abstract. This paper presents on the cooling performance of high power light emitting diode (LED) arrays under natural convection condition. A series of experiments with different type of LED array arrangements with a commercial heat sink were performed to evaluate their thermal performance. An analytical thermal resistance model was used to calculate thermal resistance. The results reveal that thermal resistance and junction temperature are affected by the type of array. The triangular array of the high power LED revealed the highest heat transfer coefficient with 3.86% compared to the most common square array. It indicates that array arrangement of the LED significantly affect on the excellent performance.

1. Introduction

High power LED known as the fourth generation of light source and dominated the lighting market nowadays. It offers a numerous advantages over conventional lighting sources due to lower power cost, longer lifetime and high efficiency [1]. LEDs have been widely used in various lighting solutions, including street lighting, traffic light, indoor lighting and LED TV. The luminous efficiency of LEDs has risen to a certain level in response to the applications. However, the output level is still insufficient for some devices. In order to utilize LEDs in universal application, the implementation of multiple LEDs is required. However, it is reported that operating LED beyond the allowable junction temperature could shorten the lifetime, depreciate the light output and decreased the efficacy of the LED[2]. In addition, efficiency and reliability of LED are not only affected from the environment heat, but also the heat generated by the LED itself. In contrast to other conventional lighting, where the majority of the heat generated from the light source is dissipated to the ambient via radiation, approximately 90% of the heat generated by the PN junction in a LED is dissipated away by conduction [3]. Furthermore, due to the small size factor, LED package is unable to effectively dissipate heat trough convection. It is generally known that thermal characterization of LEDs in an array need to account more constraint compared to a single package LED [4]. Apart from the effect of environment temperature, the effect of thermal interaction between LEDs in the array can escalate junction temperature of each LED at the same time.

As the efficiency and reliability of LED are a strong function of temperature, successful thermal management becomes highly important [5]. System with poor thermal design could shorten the lifetime of LED with excellent thermal performance. Despite of many thermal management approaches and solutions, a typical high power LED lighting system must deal with design considerations, such as cost, reliability, energy efficiency and noise, which lead to a limited cooling design options [2, 6-10]. There are a lot of researches reported on thermal management of high power LED at package level focusing on the selection of high performance materials and improvement on package structure in order to enhance heat dissipation [11]. Another direct solution to improved heat dissipation of high power LED is to use high conductivity substrate such as metal and ceramic in printed circuit board (PCB) where LEDs are soldered on [12-14]. In the case of multiple LEDs on a board, several researchers have reported the study on the spacing and placement of LED in order to get uniform luminance and good thermal performance [2,4]. Due to the small form factor, a thermally good design of LED packages and great thermal performance of PCB is insufficient to dissipate enough heat; an additional cooling means is needed [12] such as passive and active cooling. However, passive cooling is the most desired since it offers simple structure and low cost [15].

In consideration of these, present research performs a thermal analysis of high power LEDs soldered on a metal core printed circuit board (MCPCB). A heat sink is attached on LED array system module which acts as a passive cooling solution. The effect of LED placement design in array on thermal removal ability under natural convection condition is then analyzed.

2. Experiments

Commercial high power LEDs (Golden Dragon) manufactured by OSRAM Opto Semiconductors are used in the experiments. The arrays consist of six high power LEDs and soldered on 59 mm x 83 mm MCPCB in four different arrangements as shown in figure 1. The arrays applied the basic grid shapes of square, hexagonal, triangular and circular namely as A, B, C and D. The LEDs are electrically arranged in a typical series/parallel electrical array to account for forward voltage (Vf) variations between the LEDs and ensure a reasonable current distribution. Figure 2 shows the structure of the Golden Dragon LED which is mounted on the MCPCB and heat sink. The size of the heat sink dimension used in the experiments is 59 mm x 83 mm x 33 mm.

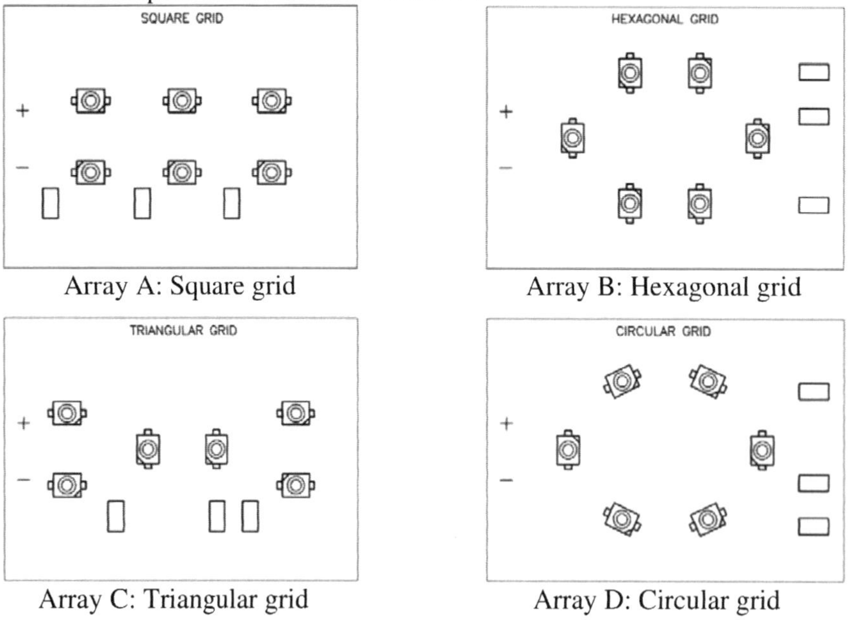

Figure 1. Photograph and schematic of LED arrangement.

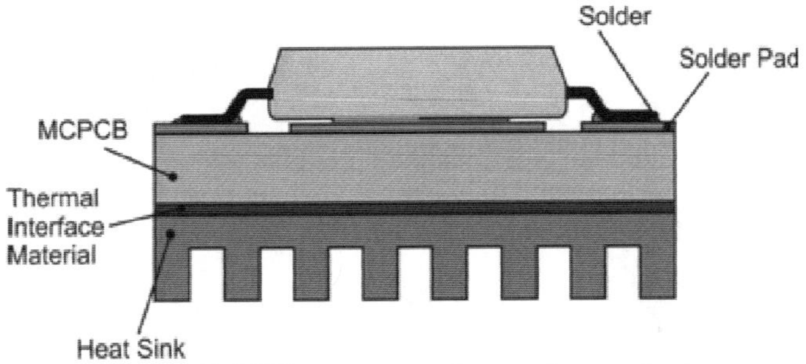

Figure 2. LED structure and mounting method.

Figure 3 shows the experimental setup schematic diagram. The LED arrays are powered using a regulated digital DC power supply, and its power is 15.2 W. One LED on the MCPCB is selected as device under test (DUT). The temperature of solder point, MCPCB and heat sink are measured using K-type thermocouples. The thermocouples are connected to a data acquisition system and then to a computer to record the temperature at 60 seconds interval till the temperature of the system achieved steady state condition.

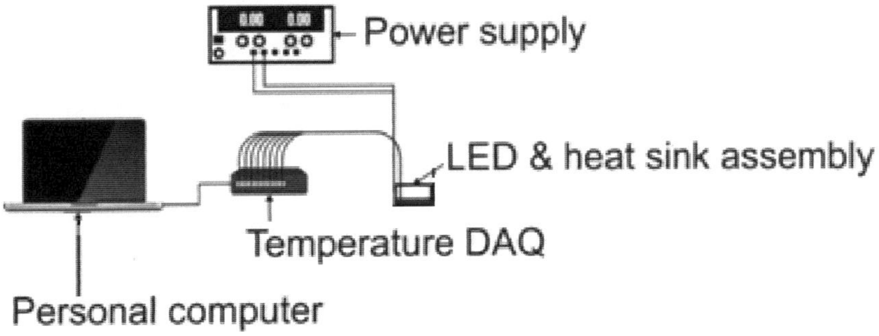

Figure 3. Schematic diagram of experimental set up.

3. Results and discussion

In the experimental set up, junction temperature could not be measured directly. However, it is feasible using solder point temperature to check the thermal behavior of the system since it also increases in same proportion with the junction temperature. Figure 4 shows the temperature variations of solder point with time for each array. It can be seen that the solder point temperatures for each array increase sharply in similar pattern up to 73.9 °C. Figure 5 compares the solder point temperature rise for each array. It is obvious that the solder point temperature rise is higher for array A and lowest for array C. This is due to the effect of better placement shape of LEDs in the array which leads to the reduction in thermal interaction effect between LEDs in the array.

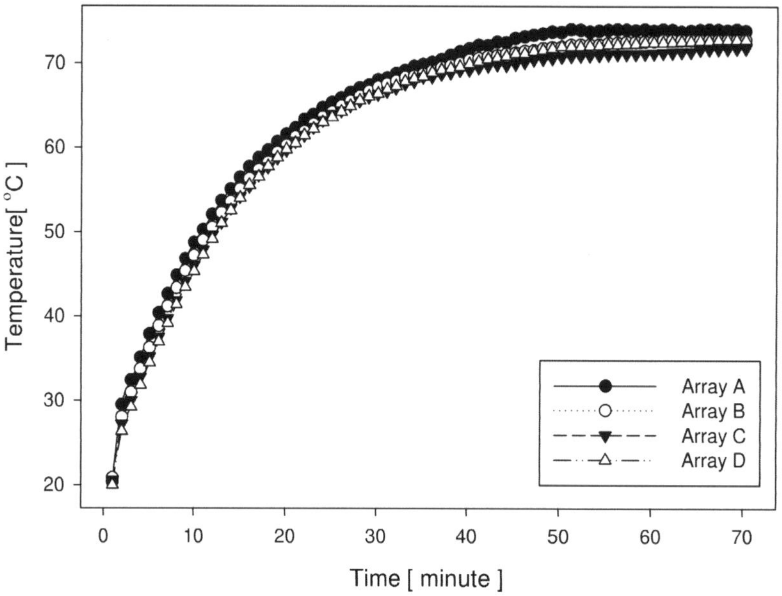

Figure 4. Comparison of solder point temperature variation for each array.

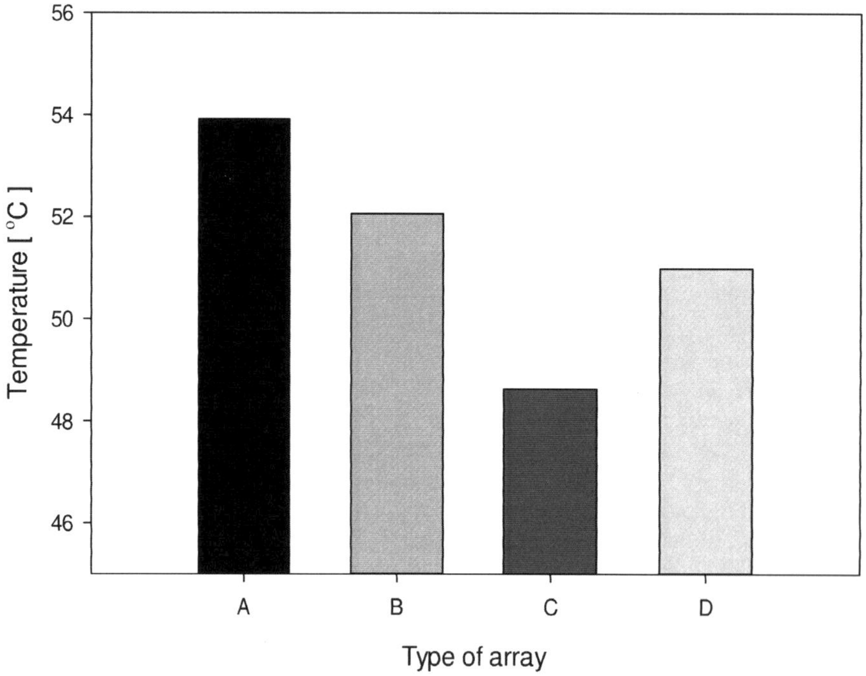

Figure 5. Comparison of solder point temperature rise in the arrays.

The total thermal resistance of multiple LEDs in array is the sum of each individual thermal resistance and can be well explained using parallel thermal resistance model as describes in figure 6. The thermal resistance from junction to board, R_{thJB} for multiple LED array is expressed by using parallel resistance equation as:

$$\frac{1}{R_{thJB,total}} = \frac{1}{R_{thJB,1}} + \frac{1}{R_{thJB,2}} + \cdots + \frac{1}{R_{thJB,n}} \tag{1}$$

Since all of the LEDs used are the same type and exhibit the same characteristics, equation (1) is simplified as:

$$R_{thJB,total} = \frac{R_{thJB}}{n} \tag{2}$$

Thus, the total thermal resistance is expressed as:

$$R_{thJA} = \frac{R_{thJB}}{n} + R_{thBH} + R_{thHA} \tag{3}$$

where R_{thJA}, R_{thJB}, R_{thBH}, R_{thHA} are the thermal resistance from junction to ambient, thermal resistance from junction to board, thermal resistance from board to heat sink and thermal resistance from heat sink to ambient respectively. Equation (3) can be interpreted as the increase in the number of unit LED in a system leads to the decrease in total thermal resistance due to the total power is distributed among the LEDs.

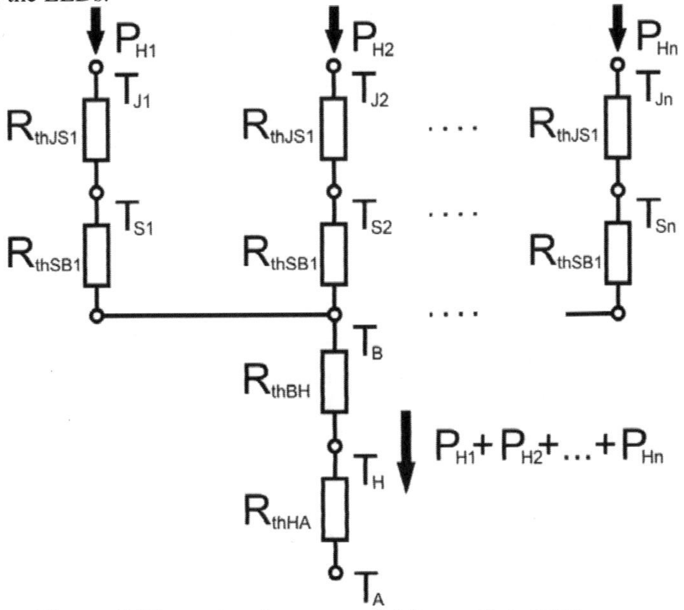

Figure 6. Thermal resistance model for multiple LED array.

Knowing the value of total thermal resistance, junction temperature could be easily calculated by following equation:

$$T_J = T_A + (P_H \times R_{thJA}) \tag{4}$$

where T_J and T_A are the junction temperature and ambient temperature respectively and P_H is the total heat input.

Figure 7 represents the total thermal resistance and junction temperature for each type of array. Array A with the most common square grid shape produced the highest total thermal resistance followed by array B, D and C array with value of 5.49°C/W, 5.33°C/W, 5.25°C/W and 5.32°C/W, respectively. This pattern is similar to the junction temperature whenever it plots against type of array.

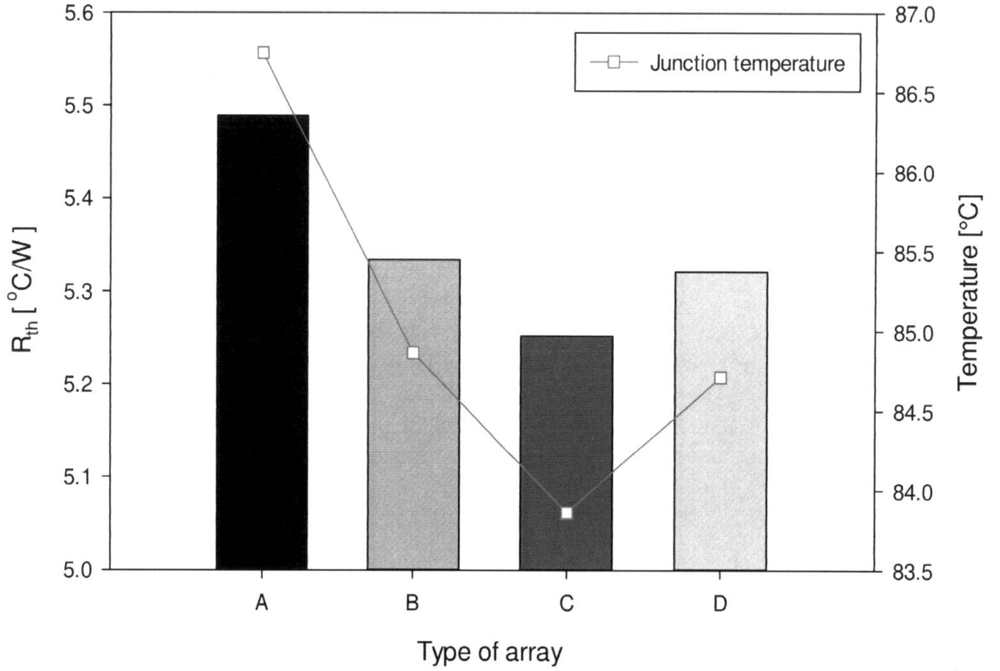

Figure 7. Total thermal resistances for different array.

Heat transfer coefficients for all arrays are then summarized in figure 8. It depicts that the heat transfer coefficient (h) is very limited due to the convection is only relied on the natural movement of hot air, which is very slow. The maximum convection coefficient recorded was as high as 2.65 $W/m^2 \cdot °C$ for the LEDs in array C. This pattern was followed by 2.62 $W/m^2 \cdot °C$, 2.60 $W/m^2 \cdot °C$ and 2.58 $W/m^2 \cdot °C$ for array D, B and A, respectively. It clearly shows that the best thermal convection performance for array C produced the highest increment over array A by about 3.86%.

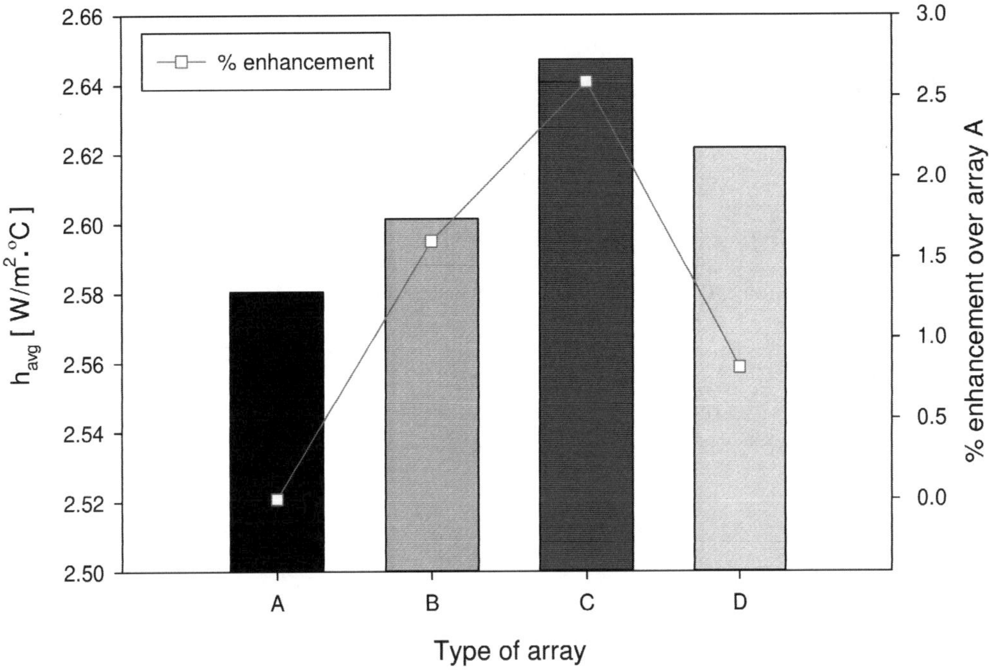

Figure 8. Heat transfer coefficients for different array under natural convection condition.

4. Conclusion

The cooling performance of high power LED arrays attached with heat sink under natural convection has been studied. Four types of array composed with six high power LEDs in different arrangement were considered. Thermal resistance and heat transfer coefficient models were used to calculate thermal resistance and convection from the measured temperature for each array. The triangular array of the high power LED (namely as Type C) shows the highest heat transfer coefficient with 3.86% compared to the most common square array. Therefore, the array arrangement of LED significantly demonstrated has influences the performance of the system.

Acknowledgement

The authors would like to thank the Universiti Malaysia Perlis (UNIMAP), Universiti Sains Malaysia (USM) and UniKL MSI for the financial and facilities support for this research work.

References

[1] R. Wang and Li J 2010 A cooling system with a fan for thermal management of high- power LEDs *J. of Modern Physics* **1** 196

[2] Arik M, Becker C A, Weaver S E and Petroski J 2004 Thermal management of LEDs: package to system *Proc. of the SPIE* (vol 5187) pp 64-75

[3] Petroski J 2004 Spacing of high-brightness LEDs on metal substrate PCB's for proper thermal performance *Proc. of Thermal and Thermomechanical Phenomena in Electronic Systems*

[4] Yung K C, Liem H, Choy H S and Lun W K 2011 Thermal investigation and placement design of high-brightness LED array package on PCB for uniform illuminance *J. of Electronic Packaging* **133** 011006-14

[5] Ken L F Y 2007 *White LED for General Illumination Applications* Master of Engineering, Massachusetts Institute of Technology.

[6] Tahseen T A, Ishak M and Rahman M M 2012 *J Mech Eng Sci* **3** 271

[7] Tahseen T A, Ishak M and Rahman M M 2013 *J Mech Eng Sc* **4** 418
[8] Al- Doori W H A R 2011 *Inter J Automot Mech Eng* **4** 428
[9] Tahseen T A, Ishak M and Rahman M M 2012 *Inter J Automot Mech Eng* **6** 753
[10] Ishak M, Tahseen T A and Rahman M M 2013 *Inter J Automot Mech Eng* **7** 900
[11] Mingsheng M, Zhifu L and Yongxiang L 2011 Thermal performance of high power LED package based on LTCC *Proc of Electronic Packaging Technology and High Density Packaging*
[12] Huaiyu Y, Koh S, Zeijl H V, Gielen A W J and Guoqi Z 2011 A review of passive thermal management of LED module *J. of Semiconductors* **32** 014008
[13] Tsai M Y, Chen C H and Kang C S 2011 Thermal measurements and analyses of low-cost high-power LED packages and their modules *Microelectronics Reliability*
[14] Long X, Liao R and Zhou J 2009 Development of street lighting system-based novel high-brightness LED modules *Optoelectronics* **3** 40
[15] Arik M, Setlur A, Weaver S, Haitko D and Petroski J 2007 Chip to system levels thermal needs and alternative thermal technologies for high brightness LEDs *J. of Electronic Packaging* **129** 328

Burst Pressure Prediction of Multiple Cracks in Pipelines

N A Razak, N A Alang and M A Murad
Faculty of Mechanical Engineering,
Universiti Malaysia Pahang, 26600 Pekan, Pahang, MALAYSIA

E-mail: norhaida@ump.edu.my

Abstract. Available industrial code such as ASME B1G, modified ASME B1G and DNV RP-F101 to assess pipeline defects appear more conservative for multiple crack like- defects than single crack-like defects. Thus, this paper presents burst pressure prediction of pipe with multiple cracks like defects. A finite element model was developed and the burst pressure prediction was compared with the available code. The model was used to investigate the effect of the distance between the cracks and the crack length. The coalescence diagram was also developed to evaluate the burst pressure of the multiple cracks. It was found as the distance between crack increases, the interaction effect comes to fade away and multiple cracks behave like two independent single cracks.

1. Introduction

Engineering structures such as aircraft, pressure vessels and piping may contained cracks like defects. These defects may be inherent in the material or may exist at the beginning of service, being caused by manufacturing and installation process. Alternatively, the defects may develop during service as a result of microstructural damage and reduces its burst strength with increased potential for catastrophic failure. Pipelines are normally subjected to a uniform internal pressure. As part of assessing the integrity, of such structure under design and in service loading, it is necessary to determine their plastic collapses pressure or burst pressure. The burst pressure is the maximum internal pressure that a pipeline can sustain [1].

There are many design codes and standards to determine the burst pressure of pipelines such as ASME B31 [2], Modified ASME B31G [2], DNV RP-F101 [3] and etc. All these codes clearly simplified integrity assessment of in service piping components. According to ASME B31G and DNV-RP-F101 codes, the failure of corroded pipeline is controlled by the defect size as well as the flow stress, S_f of the material. The input parameters are the outer diameter of the pipe, D, wall thickness, t, yield strength of the material, σ_y , ultimate tensile strength, σu, the length of the defect, L and defect depth, d. The equations used to calculate the burst pressure, P_b based on these codes are as follow.

For ASME B31G, the flow stress, S_f and bulging factor M is given by;

$$S_f = 1.1 \times SMYS \quad , \quad z = \frac{L^2}{Dt} ,$$

For $z \leq 20$,

$$M = \sqrt{1 + 0.8z}$$

$$P_b = S_{flow}\frac{2t}{D}\left[\frac{1 - \frac{2}{3}\left(\frac{d}{t}\right)}{1 - \frac{2}{3}\left(\frac{d}{t}\right)\frac{1}{M}}\right] \tag{1}$$

For z > 20

$$P_b = S_F\frac{2t}{D}\left(1 - \frac{d}{t}\right) \tag{2}$$

For DNV-RP S_f and bulging factor M is given by;

$$S_f = SMTS$$

$$M = \sqrt{1 + 0.31\left(\frac{1}{\sqrt{Dt}}\right)^2}$$

$$P_b = \frac{2(S_f)t}{D - t}\left[\frac{1 - \left(\frac{d}{t}\right)}{1 - \left[\left(\frac{d}{t}\right)/M\right]}\right] \tag{3}$$

However, these codes concern about single defect, whereas the multiple crack like defect has a great potential to cause the structure to early fail. The failure of the multiple cracks like defect is usually preceded by the interaction and coalescence of these cracks. In this process, the interacting crack growth rates and direction are significantly affected by their size, shape and proximity [4]. API 579 states that, two cracks can be characterized as a single equivalent cracks following certain rules accounting the flaw shape orientation and interaction. Figure 1 illustrates the characterization rule as in API 579 [5] Table 1 shows the criteria for characterize the multiple cracks into equivalent single cracks.

The knowledge of the multiple cracks is great importance in assessing the structural integrity since the interacting cracks are much more sensitive to fracture. There are several methods for assessing the structural integrity of pipeline steel multiple cracks. These include the body force method, the boundary element method and the finite element method [4]. All these methods employed the stress intensity factor to evaluate the cracks which however were complicated solution.

Table 1. Equivalent single cracks characterization rule

Criterion For Interaction	Effective Dimensions After Interaction
$c_1 + c_2 \geq d$	$2c = 2c_1 + 2c_2 + d$ $a = \max[a_1.a_2]$

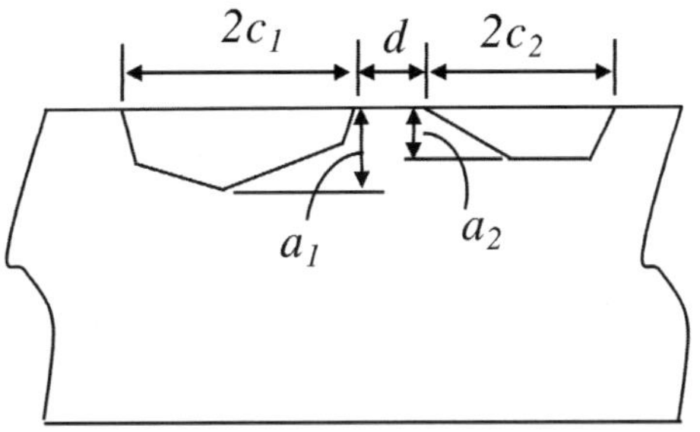

Figure 1. Multiple Crack Like defect configuration

A coalescence diagram evaluation had been proposed for steam generator tube [6].The diagram can be used to determine the burst pressure at the moment the cracks coalescence. It is assumed that the region or ligament between the crack tips cannot sustain the applied load anymore when it is subjected to the fully yielding condition. Later, a total of ten failure pressure prediction model such as flow stress model, necking base model, stress based model, reaction force model, plastic zone contact model had been developed [7]. By comparing the experimental results and with the prediction results, the reaction force model and the plastic zone contact model were selected as the optimum ones to predict the coalescence pressure. The interaction effect of two cracks in pipelines had been investigated in the presence of pressure on the defect [8]. Gonzales employed the condition where the critical lamination pressure defines as the pressure in the lamination that makes the von Mises stress in the interlaminar region to reach or surpass the ultimate strength of the material. This condition is assumed to be the initiation point for the formation of the interconnecting cracks. Thus, while previous research provided promising method, the present paper predicts the burst pressure of collinear cracks and develops the coalescence diagram for pipeline steel. The burst pressure was predicted when the region between cracks fully yielding. The effect of distance between crack and the crack length also the main interest in this research paper.

2. Methodology

2.1. Material

The material used was made of pipeline steel Grade B. Table 2 shows the chemical composition of the material. The mechanical properties were obtained by performing the tensile test according ASTM E8 (2001) as tabulated in Table 3. Figure 2 shows the true stress strain curve for Grade B steel pipe.

Table 2. Chemical Composition of Grade B Steel Pipe

Fe	C	Si	Mn	P	S	Cr	Mo
98.5	0.28	0.38	0.62	0.01	0.01	0.03	0.01

Table 3. Mechanical Properties of Grade B Steel Pipe

	Young Modulus,E (MPa)	Possion Ratio, ν	Yeild Strength,σ_y (MPa)	Tensile Strength,σ_t (MPa)
Experimental	207	0.3	362	464

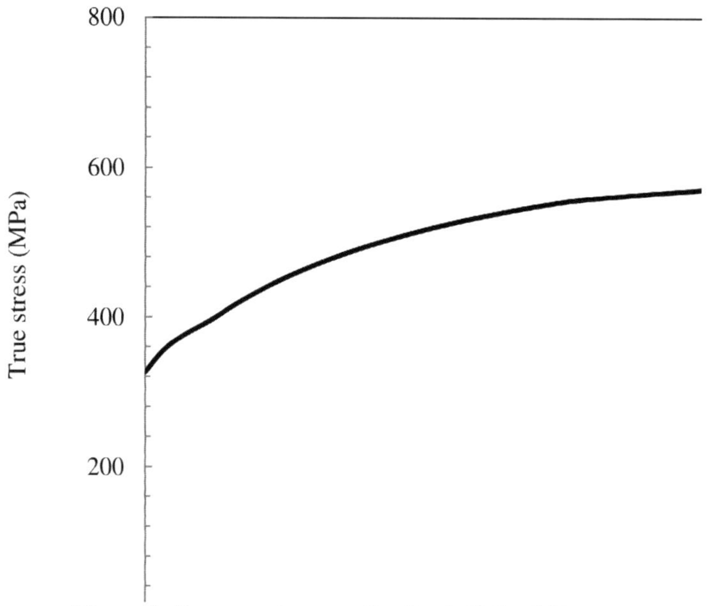

Figure 2. Stress strain curve for Grade B Steel Pipe

2.2. Finite Element ModellingFormatting author names

Figure 3 illustrates the schematic pipe containing two collinear cracks. The outer diameter, D and thickness, t of the pipe are 605 mm and 4 mm respectively. The overall length, L of the pipe was 600 mm. The crack length and the distance between cracks designated by $2c$ and d respectively. 3D nonlinear finite element analysis was performed using MSC Marc 2008r1. Figure 4 shows typical finite element model of pipe containing two collinear cracks. A half of the pipe was modelled by considering symmetrical condition. The finite element mesh was constructed by using 20 node quadratic brick elements with reduced integration point. The boundary condition was applied at the end of the pipe to simulate the closed cap condition and the internal pressure was applied to the inner surface of the pipe. The material is model as an isotropic elasto-plastic material and the true stress strain data were employed. A series of finite element analyses were carried out for the tube containing crack length. to investigate the effect of crack length and the distance between cracks. The crack length of 25mm, 50mm, 75mm and 100 mm and distance between cracks of 0.5mm, 2 mm, 4 mm and 8 mm was used.

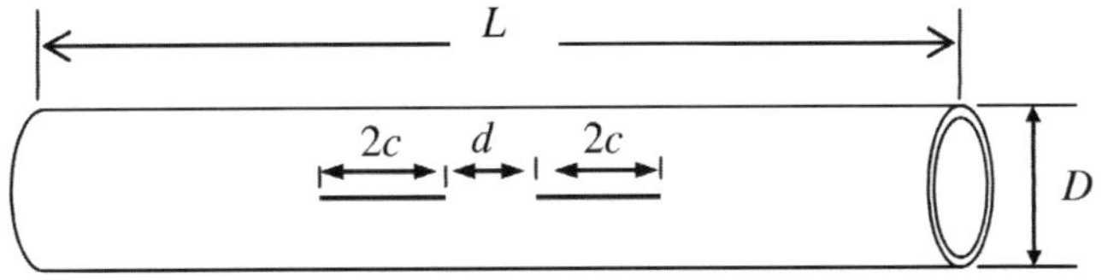

Figure 3. Schematic of pipe containing two collinear cracks.

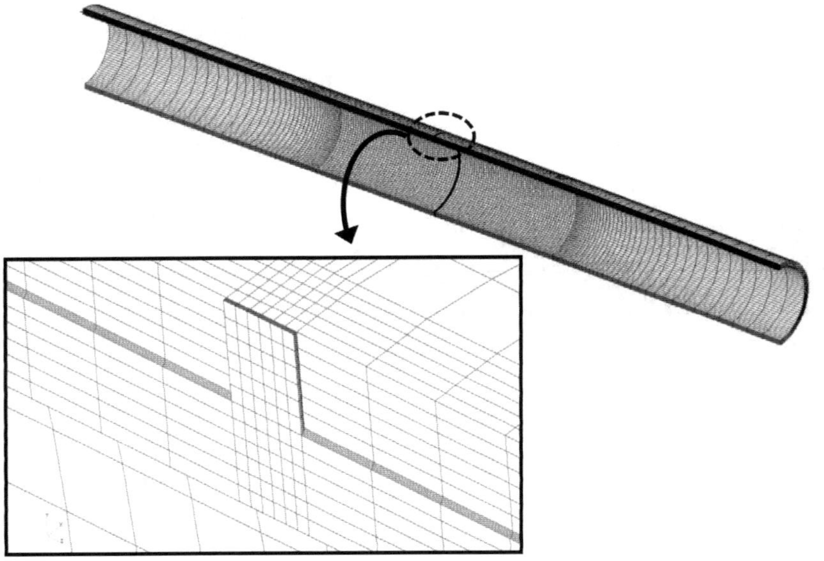

Figure 4. Crack modelling

3. Results and Discussions

In the present paper, the burst pressure was taken when the von Mises stress in at the crack tip reach the ultimate strength of the material. So, it is assumes that the coalesnce occur when the ligament between the cracks is fuly yielding. Figure 5 shows distribution of von Mises stress at the region between the collinear cracks for the case of $2c$ = 100 mm and distance, d = 2 mm with the increase of applied pressure. The maximum stress occurred at the internal tips of the crack indicate there is an interaction of the stress field since the first pressure step.Figure 5(b) shows that at a pressure of 14.4 MPa, there is noticeable yeiding at the internal crack tips, but not across the entire region between the crack. Figure 5(c) shows at a pressure of 32.4 MPa, the von Mises stress distribution are above the yield strength of the material, so it is fully plasticied. Finally, at a pressure in the region of 44.1 MPa, the spread of von Mises stress reached the ultimate strength of the material and therefore the burst pressure was predicted.

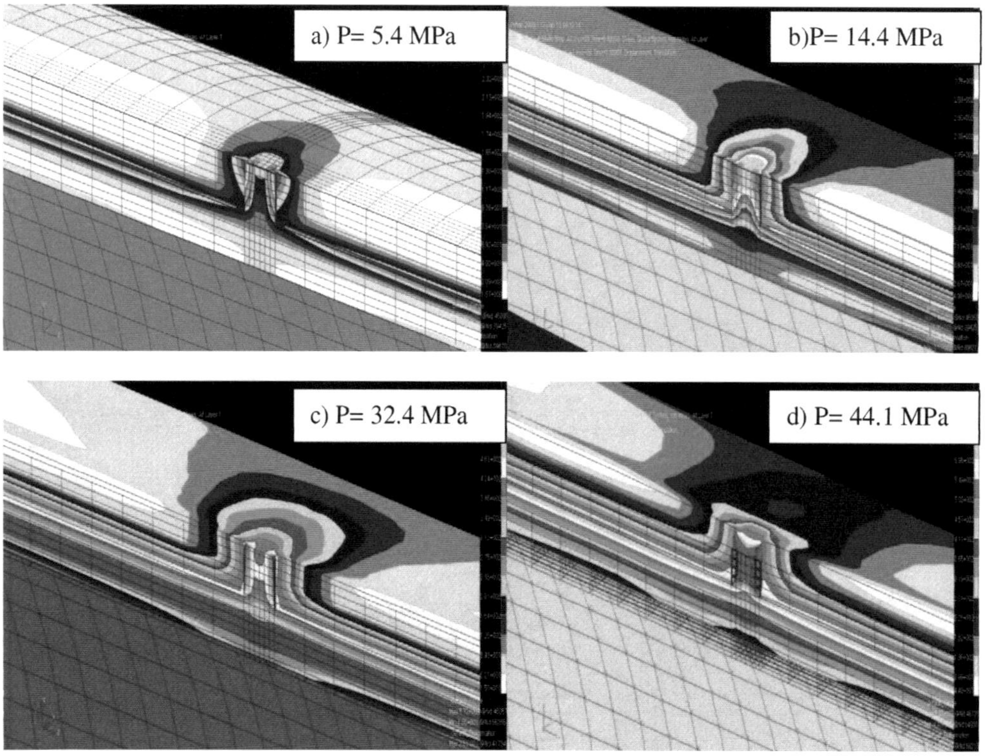

Figure 5. Stress distribution at region between crack tips

Table 4 represent the predicted result obtained by using FEA models and calculated burst pressure according to ASME B31G (ASME B31G, 2009) and DNV (DNV RP-F101, 2010). The burst pressures for single crack were calculated using Eq. (1) and Eq. (3) to acquire a baseline data. For the purposed of calculated single crack, $2c$ = 25, 50, 75 and 100 mm was used.

Figure 6 shows the coalescence load crack length curve obtained from finite element analysis results. This curve can be used to determine the pressure at the moment of crack coalescence or the burst pressure. In this figure, the solid line indicates the burst pressure of the pipe containing single crack calculated using Eq. (3).It is noted that the burst pressure of adjacent collinear cracks with a value d greater than 2mm with crack length greater than 75 mm closely approached to the burst pressure of pipe containing a single cracks. This means that if the distance between a pair of cracks is longer than 2 mm with crack length longer than 75 mm, the interaction effect comes to fade away and collinear cracks behave like two independent single cracks.

Table 4. Burst pressures value

Pipe dimensions		Burst Pressure (MPa)		
2*c*(mm)	*d*(mm)	FEA	ASME	DNV
25	0.5	31.0	23.8	51.47
	2	43.4		
	4	46.4		
	8	52.8		
50	0.5	28.0	23.8	42.70
	2	40.3		
	4	44.2		
	8	49.9		
75	0.5	27.0	23.8	39.1
	2	41.0		
	4	44.0		
	8	49.5		
100	0.5	27.3	23.8	37.3
	2	40.5		
	4	41.4		
	8	49.0		

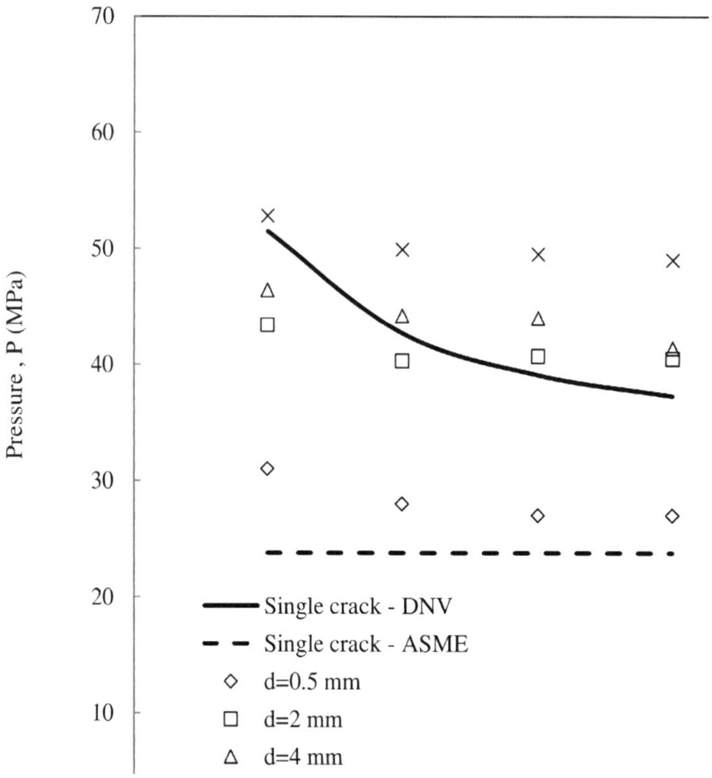

Figure 6. Coalescence diagram

4. Conclusions

In this paper, the burst pressure of pipe containing collinear axial through wall cracks was evaluated based on detail three dimensional elastic plastic finite element analyses and following key finding has been derived:

i) As the crack length increases the burst pressure decreases.
ii) As the distance between cracks increases, the burst pressure also increases..
iii) It is noted that the burst pressure of adjacent collinear cracks with a value *d* greater than 2mm with crack length greater than 75 mm closely approached to the burst pressure of pipe containing a single cracks.
iv) This means that if the distance between a pair of cracks is longer than 2 mm with crack length longer than 75 mm, the interaction effect comes to fade away and collinear cracks behave like two independent single cracks.

5. Acknowledgement

The authors would like to thank the Faculty of Mechanical Engineering in Universiti Malaysia Pahang (UMP) and Universiti Malaysia Pahang for financial support under RDU120311. figure should have a brief caption describing it and, if necessary, a key to interpret the various lines and symbols on the figure.

6. References

[1] Zarrabi.K. 1994. Plastic collapse pressure for defected cylindrical vessels. International Journal Pressure Vessel and Piping 60: 65-69

[2] American Society of Mechanical Engineers (ASME) B31G. 2009. Manual for determining the remaining strength of corroded pipelines. New York USA

[3] DNV Recommended Practice, RP-F101.2010. Corroded pipelines, Det Norske Veritas.

[4] Moussa, W. A., Bell, R., and Tan, C. L., 1999, The Interaction of Two Parallel Semi-Elliptical Surface Cracks Under Tension and Bending. ASME Journal Pressure Vessel Technology, 121: 323–326.

[5] API. Recommended practice for fitness-for-service. API 579.Washington, DC: American Petroleum Institute, 2000.

[6] Lee, J.H., Park, Y.W., Song, M.H., Kim, Y.J. and Moon, S.I., 2001. Determination of equivalent single crack based on coalescence criterion of collinear axial cracks. Nuclear Engineering Design 205

[7] Moon SI, Kim YJ, Lee JH, Park YW, Song MH. 2005 .Optimum local failure model of cracked steam generator tubes with multiple axial through-wall cracks. Nuclear Engineering Design, 235: 2099-2108Qian Z G, Shen W Z, Ogawa H and Guo Q X 2002 J. Appl. Phys. 92 3683

[8] Gonzales, J.L and Morales. A , 2008, Analysis of Lamination in X52 Steel Pipes by Nonlinear by Finite Element, Journal of Pressure Vessel Technology, 130: 021706-1-7.

Comparison of ARX and ARMAX models for thermoelectric refrigerator

M Saifizi[1,a], M Z Ab Muin[2], Sazali Yaacob[1] and M S Mohamad[1]
[1]School of Mechatronics Engineering, University Malaysia Perlis, 02600 Arau, Perlis, Malaysia
[2]School of Manufacturing Engineering, Universiti Malaysia Perlis, 02600 Arau, Perlis, Malaysia

E-mail: [a]saifizi@unimap.edu.my

Abstract. System identification is a procedure to characterize the dynamic behavior of a system, a system component based on measured data. This paper presents a study of the modeling and parameter identification of a refrigerator process by using the mathematical black-box modeling technique- ARX and ARMAX structure. A general ARX and ARMAX structure has been constructed for usual thermoelectric refrigerator systems and the Recursive Least Square (RLS) process is used for model parameters identification. The capabilities of the model constructed compared with that of ARX and ARMAX model. Several compositions of ARX and ARMAX orders and forgetting factor are analyzed for thermoelectric air-conditioning model to examine the optimal adjustments that improve their efficiency of PID controller.

1.0 Introduction

This paper focuses on designing ARX and ARMAX model for thermoelectric refrigerator system. With a purpose to study the capabilities of a system, it is usually important to determine its mathematical models. The selection of system identification method is dependent on knowledge of the method and also knowing the system characteristics. Various studies are available on the identification of engineering systems, as summarized by [1][2]. Approaches for system identification could be generally divided into two aspects: prior knowledge based models (models totally base on physic principles) and data driven models (model completely based on empirical relations).

Methods for the data driven models can also be separated into two main areas, known as grey box models, and black box models. As opposed to direct modelling that is governed by the system's physical laws , it is more suitable to build mathematical model for a system in which there is partial knowledge of the system's and/or and unexpected behaviour of the changing system's characteristics. The grey box method is a composition of the model where the model parameters are traceable to exact physical principles. The black box method is associate mathematically measured inputs to measured outputs when the model parameters are changed without normal physical significance. Supposedly, these methods could defeat the weaknesses of prior knowledge models. Compared with grey box model, the black box technique demands less time and effort to construct.

Current study established a modelling scheme based on a black box mathematical ARX and ARMAX models to deal with the unidentified and random dynamic behaviour of thermoelectric refrigerator systems. This modelling method is capable of taking into account system interference as well as measurement noise. The Recursive Least Square (RLS) method is utilized to predict the parameters of the ARX and ARMAX models. An important factor for the determination of ARX and ARMAX models is the selection of correct orders of the AR, MA and X terms and forgetting factors. Detailed investigation on different orders and forgetting factors will be done as well as the comparison of their properties. Average Error (AE) is employed to evaluate the precision of the model.

An experimental investigation on existing thermoelectric refrigerator system will be performed to test and validate the models and hence to examine how they are used in modelling.

2.0 Experimental Study

The system composed of four main parts: casing characteristic, refrigerator system characteristic, data acquisition for the thermoelectric refrigerator system and measurement instruments.

The thermoelectric module is an electronic component made from semiconductor [3]. It functions as a heat pump. This module has two surfaces that will get hot and cold by applying just a low DC voltage. By reversing the polarity of the power supply, the hot surface will change to cold surface and vice versa. Thus the thermoelectric device can be utilized as a heater as well as a cooler, making the precise temperature control is highly suitable. Thermoelectric module posed numerous advantages such as small size and weight, ability to cool below 0 °C, ability to cool and heat with the same device, precise temperature control, high reliability, electrical quiet operation, convenient power supply, and environmentally friendly.

2.1. Mechanical section

The thermoelectric module is powered by 48 watt power supply. To evaluate the performance of the module, different values of input current are applied, and the temperature is measured. A heat sink is attached to the hot surface of the module in order to carry away the heat pump by the module and the Joule heat from the electrical power supply. The cold surface is also attached with a heat sink which will help in distributing the cold air. This leads to a low temperature different between the hot and cold surfaces of the module which is desired for a better efficiency. A fan is used to forced air flow through the length of the heat sink. The forced air flow creates turbulence which in improve hat transfer. The main body of the Thermoelectric Refrigerator (TR) consists of a 0.17 x 0.34 x 0.38 m metal box as shown in figure 3. The wall of the box is insulated in order to minimize heat loss to the ambient.

2.2. Electrical Section

The thermoelectric module is operated directly from DC power. The suitable power sources can range from batteries to a simple unregulated DC power supply. For the TR, a computer program is utilized to control the current from power supply in order to maintain the desired temperature inside the chamber. The circuit is based on a MOSFET transistor junction, controlled through photodiodes that provide an electric insulation. The TR is equipped with a data acquisition system that records data every second, and also controls and corrects the variations of the temperature by varying the duty cycle of the MOSFET transistor bridge. Integrated circuit temperature transducer LM35 has been used for temperature measurement. It acts as a current source that provides a linear output that is proportional to the absolute temperature (in Degree Celsius).

3.0 Mathematical Models

3.1Generic Model Structures

To realize the ARX and ARMAX modeling structure, the mathematical black box modeling model structures are released. In general, the system is identified by a black box. The input *u(t)* is only to control the system model , and both of the input *u(t)* and output *y(t)* are tracked and recorded. However, the system model is disturbed by disturbance. Generally, the inputs and outputs are completely or partially measureable while the system disturbance is not classified directly. This has influence on the outputs. The modeling error *e(t)* is released as the result of the difference between the system output and model output . [4][5] referred to the mathematical expression of the general structure for black box models in a simple input/output relationship as follows :

$$A(q^{-1})\cdot y(t)=\frac{B(q^{-1})}{F(q^{-1})}\cdot u(t)+\frac{C(q^{-1})}{D(q^{-1})}\cdot e(t) \tag{1}$$

where *y(t)*, *u(t)* and *e(t)* are the output , input , and noise respectively . All A, B, C, D and F are polynomials. The algorithm of the ARX and ARMAX generic structures for black box models are defined below.

3.2 ARX Model

The input-output relationship of an ARX model is given by a linear differential equation as follows:

$$y(t)+a_1y(t-1)+...+a_{n_a}y(t-n_a)=b_0u(t-d)+...+b_{n_b}u(t-d-n_b)+e(t) \tag{2}$$

It is also kwon as an equation error model. Since the noise *e(t)* enters as a direct error in Equation 3. The adjustable parameters of this structure are expressed as below:

$$\theta=\lfloor a_1 \quad a_2...a_{n_a} \quad b_0 \quad b_2...b_{n_b}\rfloor \tag{3}$$

Given that

$$A(q^{-1})=1+a_1q^{-1}+...+a_{na}q^{-na} \tag{4}$$

and

$$B(q^{-1})=b_0+b_1q^{-1}+...+b_{n_b}q^{-n_b} \tag{5}$$

where q^{-1} is the backward shift operator, the following input-output model structure is informed.

$$A(q^{-1})y(t)=B(q^{-t})u(t)+e(t) \tag{6}$$

$$y(t)=\frac{B(q^{-1})}{A(q^{-1})}u(t)+\frac{1}{A(q^{-1})}e(t) \tag{7}$$

3.3 ARMAX Model

The different of the ARX model is the absence of the properties of the error term. Accordingly, with the consideration of the error term, the ARMAX model is developed and given by the following equation.

$$\begin{aligned} &y(t)+a_1 y(t-1)+\ldots+a_{n_a} y(t-n_a)= \\ &b_0 u(t-d)+\ldots+b_{n_b} u(t-d-n_b)+e(t)+c_1 e(t-1)+\ldots+c_{n_c} e(t-n_c) \end{aligned} \tag{8}$$

The ARMAX model is suitable in areas of control, processes and econometrics for both system modelling and control scheme design. The adjustable parameters in this structure are expressed as follows:

$$\theta=\left[a_1 \quad a_2 \ldots a_{n_a} \quad b_0 \quad b_2 \ldots b_{n_b} \quad c_1 \quad c_2 \ldots c_{n_c}\right]^T \tag{9}$$

Given that

$$A(q^{-1})y(t)=B(q^{-1})u(t)+C(q^{-1})e(t) \tag{10}$$

or

$$y(t)=\frac{B(q^{-1})}{A(q^{-1})}u(t)+\frac{C(q^{-1})}{A(q^{-1})}e(t) \tag{11}$$

3.4 Recursive Least Square Estimation

The important step in the improvement of the system identification and adaptive control is the determination of process or model parameters. The Least Square (LS) estimation, RLS and forgetting factor established from previous studies (Wellstead and Zarrop , 1991 ; Wu and Pandit , 1993). With the LS technique , the unknown parameter of a mathematical model are estimated such as that the sum of square of the difference between the parameter values actually identified as well as related values computed are minimized . The LS method is useful to estimate the parameters of differential equation model structures.

Recursive estimation is a technique to arrange the algorithm in such a technique that the results acquired previously can be employed for updating. This way of estimation is acceptable for computing efficiency.

3.5 Forgetting Factor

Limited storing of historical data and fast execution are the characteristic of effective system identification. The technique for forgetting factor is to reduction old measurements to ensure that the model adapts to the changing situations dynamically and efforts to estimate parameters without numeric overflows. In order to make the estimator to follow the change of the process, it is desirable to discount the old measurements in the estimation algorithm.

4.0 Results and Discussions

The system was run for many values of input current. The temperature, time, and input power were measured, and the result are tabulated in the table 1 and plotted in figure 1-5. The TC is equipped with a data acquisition system that records data every second, and also controls and corrects the variations of the temperature by varying the duty cycle of the MOSFET transistor bridge. The duty cycle input is 60%-100% of the MOSFET transistor bridge.

The initial temperature of the air inside the refrigerator was 27.5 degrees Celsius. This environment consisted of an ambient air about 29 Celsius with no solar radiation. Additionally, the ambient temperature was changed so that it was at 29.5 degrees Celsius but the refrigerator decrease linearly 27 degrees Celsius.

Table 1. A cooling test with difference of Mosfet duty cycle input.

Duty Cycle [%]	Voltage [V]	Current [A]	Power [W]
60	6.5	1.8	11.7
70	7.5	2	15
80	8.5	2.3	19.55
90	9.5	2.7	25.65
100	11.11	3	33.33

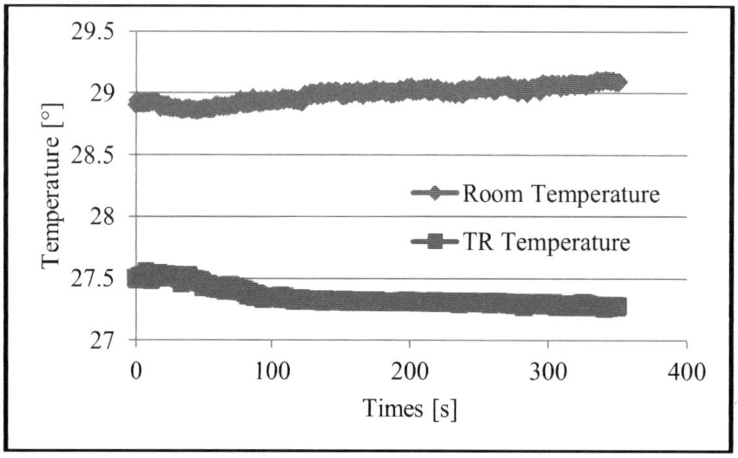

Figure 1. Temperature response of 60% duty cycle MOSFET transistor Bridge.

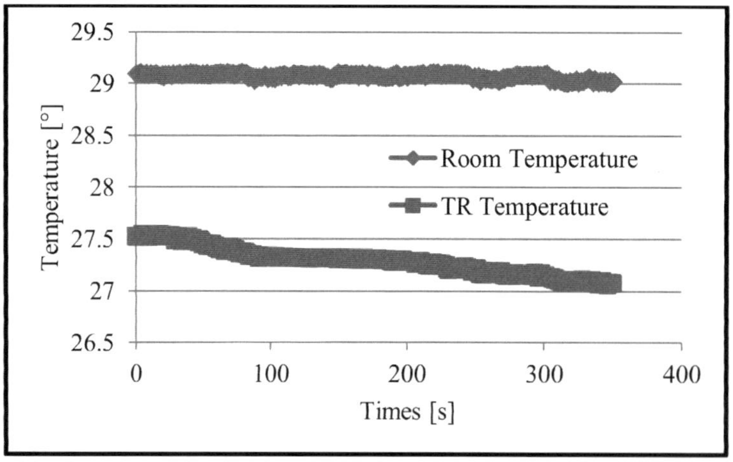

Figure 2. Temperature response of 70% duty cycle MOSFET transistor Bridge.

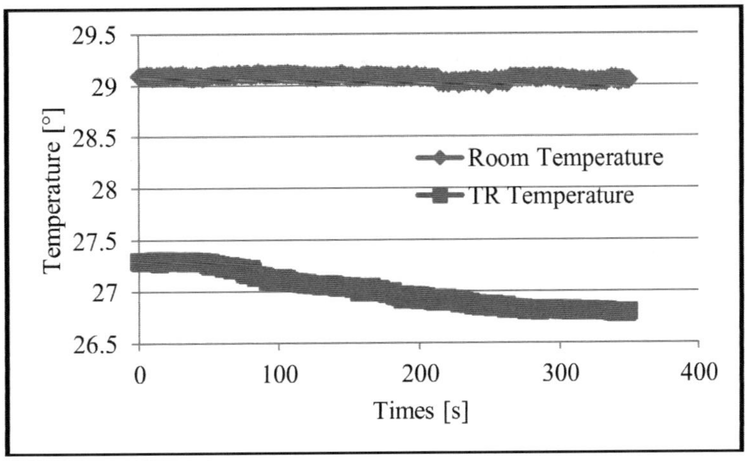

Figure 3. Temperature response of 80% duty cycle MOSFET transistor bridge.

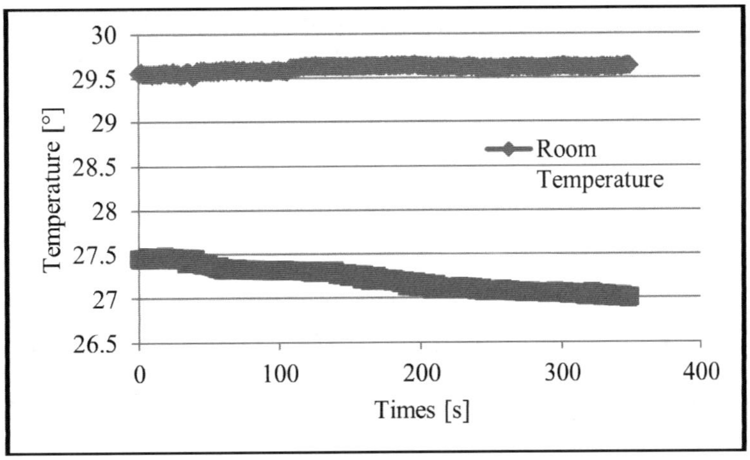

Figure 4. Temperature response of 90% duty cycle MOSFET transistor Bridge.

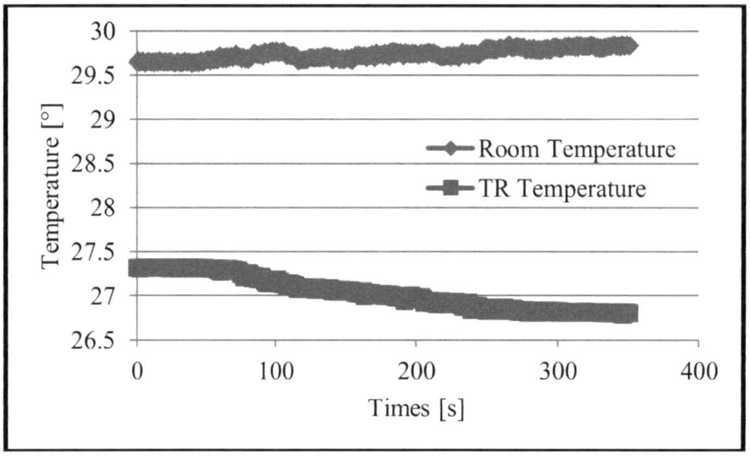

Figure 5. Temperature response of 100% duty cycle MOSFET transistor Bridge.

In a general design setting where the underlying system may have complex dynamics a number of rules of thumb exist which assist in the selection of the controller coefficients k_p, k_d, k_i. In the synthesis situation, however, the requirements on the underlying system are quite strict. In particular, in order to synthesize exactly the PID controller coefficients we must assume that the system to be controlled has the special structure.

$$y(t) = \frac{b_0 z^{-1}}{1 + a_1 z^{-1} + a_2 z^{-2}} \tag{12}$$

To decide proper SISO ARX and ARMAX models for the thermoelectric cooling system are base on special structure of PID controller coefficients in Equation 12. The restriction on the system model form is to ensure that only one set of PID controller coefficients arise from the design. The forgetting factors of 0 .98 and 0 .998 were employed in parameter estimation. Due to the influence of the model order and forgetting factor, all the models were parameterized in order to make evaluations between models. The forgetting factor is to determine the rate of old data discarding, simplifying the computational requirements. A small forgetting factor provides new data to be weighted more heavily than old data, when updating the model parameters. If too small forgetting factor is used, the estimation is affected very easily by spurious variants due to measurement noise.

Table 2. Comparison between Average Errors (AE_{model}) of SISO ARX and SISO ARMAX Models (Forgetting Factor:0.98)

Input Duty Cycles	AE_{model} of ARX (2,1)	AE_{model} of ARMAX (2,1,1)
50%	0.000208254	0.000276909
60%	0.000337975	0.000413136
70%	0.000327246	0.000407147
80%	0.000067462	0.000118472
90%	0.000044062	0.000079502
100%	0.000331795	0.000385762
Total Average Error	0.000263359	0.000336185

Table 3. Comparison between Average Errors (AE_{model}) of SISO ARX for (Forgetting Factor: 0.98) and (Forgetting Factor: 0.998)

Input Duty Cycles	AE_{model} of ARX (2,1), forgetting factor 0.98	AE_{model} of ARX (2,1), forgetting factor 0.998
50%	0.000260553	0.00031384
60%	0.000094028	0.000115523
70%	0.000397896	0.000465734
80%	0.000399621	0.000447136
90%	0.000239474	0.000294001
100%	0.000281872	0.000315692
Total Average Error	0.000334689	0.000390385

Table 2 and 3 show the evaluating results of both models. A comparing was made between the performances of the SISO ARX model and SISO ARMAX with different orders n_a, n_b, n_c for the ARMAX and n_a, n_b for the ARX term and the forgetting factor of 0 .98. The SISO ARX (2, 1) model with different input was discovered to have the lowest total AE_{model} of 0 .000263 compare to SISO ARMAX (2, 1, 1) model. The comparison values of AE_{model} for various input of ARX model with forgetting factor of 0 .98 and 0 .998 are shown in table 3. The lowest total AEmodel for the SISO ARX is 0 .00078 with 0 .98 forgetting factor.

When same forgetting factors were employed, the total AEmodel of the SISO ARX model was discovered to be lower than that the SISO ARMAX model. The results expressed that ARX model has much better performance when a smaller forgetting factor is used.

5.0 Conclusions

This study investigates the way the ARX and ARMAX modeling schemes may be used to Thermoelectric Refrigerator systems to more suitable identify their performance and dynamic characteristics. A varied analysis had been made on several of black box models used Thermoelectric Refrigerator system identification. The overall performance of the ARX model is better to the ARMAX model in terms of model fitness ability in response measured data. The optimal ARX order (2, 1) and forgetting factor of 0 .98 and these parameter are consistent with those used for system identification. The study ensures that ARX models can be applied for thermoelectric refrigerator systems by expecting or correcting the differences between the required and real conditions depending on the system inputs and disturbances. The ability of using ARX models for Thermoelectric Refrigerator system has been investigated. The modeling caused by their accuracy in identifying dynamic behaviors, have capability to be further applied to the associated fields of adaptive control. It is possible to extend the ARX modeling scheme for online parameter estimation involved in adaptive control.

6.0 Acknowledgements

My sincere thanks go to Prof Zaki Ab. Muin for his advice, guidance and support, and for his hard work and dedication to my project. I also thank my friends from the School Mechatronics Engineering for their support. Finally, I would like to dedicate this work to my family and especially to my wife Noor Aneeza.

7.0 References

[1] Schalkoff R J 1992 *Pattern Recognition: Statistical, Structural, and Neural Approaches* (New York: John Wiley & Sons)

[2] Wasserman P D 1989 *Neural Computing: Theory and Practice* ed Van Nostrand Reinhold (New York)
[3] Melcor Aztec 2003 Thermoelectric Handbook, http://www.melcor.com/handbook.htm
[4] Wellstead P E and Zarrop M B 1991 *Self-tuning Systems: Control and Signal Processing* (U.K: Wiley)
[5] Wu S M and Pandit S M 1993 *Time Series and System Analysis with Applications* (Florida: John Wiley & Sons)

IOP Conf. Series: Materials Science and Engineering **50** (2013) 012033 doi:10.1088/1757-899X/50/1/012033

Diversity modelling for electrical power system simulation

R M Sharip[1,a] and M A U A Abu Zarim[2,b]
[1]Faculty of Electrical & Electronic Engineering, Universiti Tun Hussein Onn Malaysia, 86400, Parit Raja, Batu Pahat Johor, Malaysia
[2]Faculty of Science and Defence Technology, Universiti Pertahanan Nasional Malaysia, Kem Sungai Besi, 57000, Kuala Lumpur, Malaysia
Email: [a]rina@uthm.edu.my, [b]amir.a@upnm.edu.my

Abstract. This paper considers diversity of generation and demand profiles against the different future energy scenarios and evaluates these on a technical basis. Compared to previous studies, this research applied a forecasting concept based on possible growth rates from publically electrical distribution scenarios concerning the UK. These scenarios were created by different bodies considering aspects such as environment, policy, regulation, economic and technical. In line with these scenarios, forecasting is on a long term timescale (up to every ten years from 2020 until 2050) in order to create a possible output of generation mix and demand profiles to be used as an appropriate boundary condition for the network simulation. The network considered is a segment of rural LV populated with a mixture of different housing types. The profiles for the 'future' energy and demand have been successfully modelled by applying a forecasting method. The network results under these profiles shows for the cases studied that even though the value of the power produced from each Micro-generation is often in line with the demand requirements of an individual dwelling there will be no problems arising from high penetration of Micro-generation and demand side management for each dwellings considered. The results obtained highlight the technical issues/changes for energy delivery and management to rural customers under the future energy scenarios.

1. Introduction

Recently, the world has been facing severe climate change problems [1- 3].One of the areas identified where savings could be met to reduce the impact on the environment is the mix of electricity generation and the level of use of energy. Thus, the electricity supply industry in the UK, and internationally, is seeing significant changes in order to tackle the climate change problem as well as to ensure a reliable, secure, affordable energy supply and a more efficient use of generation through the adoption of micro renewable and low carbon generation technologies [4-5]. The UK government's Micro-generation Strategy in [6] acknowledged the potential of these technologies in contributing to achieve the above objectives. The combined production of heat and power (e.g. micro CHP) and renewable micro-generators (MG) (e.g. PV panels and micro-wind turbines) are believed to be able to increase the efficiency of local energy production and thusly help reduce carbon emissions [7-8]. Also with the help of widespread demand side management (DSM) such as smart metering and direct load control, storage devices (e.g. electric vehicles (EVs)) and changes in consumer behaviour further impacts can be made on the electricity sector's targets.

This position requires changes to both the operation and management of the electrical system, several options identified will see a gradual shift from large centralised power plants to smaller and more distributed generators [9]. Such a change would have many implications for the existing distribution networks in terms of management and control. It is therefore important to examine how the networks in the UK today would respond to a move towards large-scale deployments of MG and adoption of DSM. Figure 1 shows an example of a future rural network with large penetrations of MG and energy storage. The analysis of the behaviour of such a system requires the modelling of both the electrical distribution system and the coupled MG devices in a realistic context. The remainder of the paper provides a review of relevant literature outlines the modelling approach adopted and applies this to a UK rural Zero Carbon Community network example. Finally simulation results are presented and conclusions and further work are given.

2. Literature review

There are a number of previous studies that have investigated the impacts of a large penetration of MG [8, 10] and DSM [7, 11] on the distribution network system. Those and other studies generally addressed the network concerns in the problem; those of voltage imbalance [7-11], quality of power [10-11], voltage regulation, potential increases in network fault level and the breach of thermal constraints. However, the generation and demand data and how this data has been modelled for all these studies were different. Hence this study provides details about an integrated simulation approach to modelling the network and pays particular attention to how the diversity in the generation and demand profiles of these aspects will affect future system behaviour. In addition, most studies performed to date have been rather simplistic in their approach and considered only the maximum and minimum limit of generation and demand profile as the boundaries conditions for their studies [7-12]. These limits have been widely studied due to their importance in a conventional deterministic approach to network design [7-12].However, such a traditional approach could lead to large capital investment costs across the network that may not be tolerable and the power flow studies may not give an accurate result of the impact of a high level penetration of MG. It is evident from the published material that a range of profiles for both the demand and generation are crucial for understanding the nature of network performance with high penetrations of MG and DSM; hence the paper [12] was used as a foundation for this research but with the key differentiation being improved profiling especially in the stochastic load and generation modelling method (as shown in figure 2).

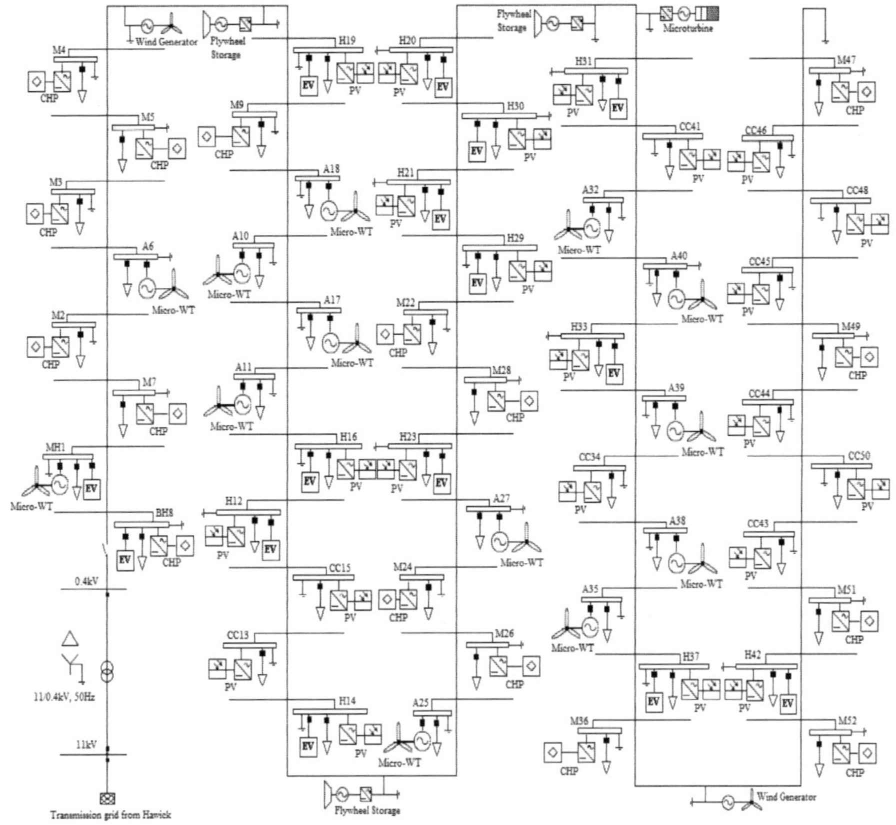

Figure 1. Example of simplified low voltage network with large penetrations of micro-generation and demand side management

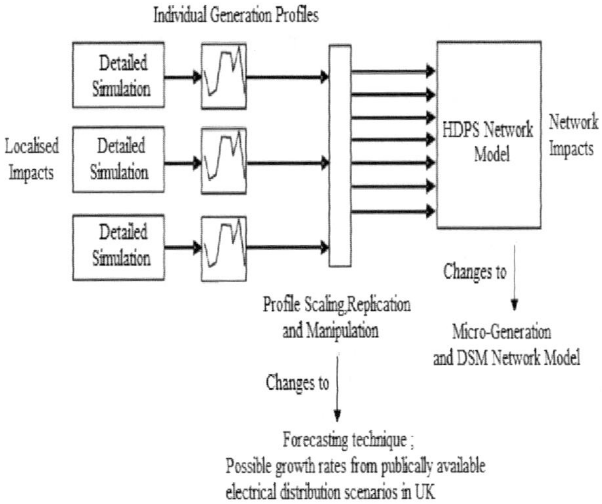

Figure 2. Changes of modelling approach from foundation paper [12]

3. Modelling approach

The method considered in this research utilises a forecasting technique to account for diversity in future generation and load profiles. The new scenarios for 2050 that are applied in this research are gained through considering a range of publicly funded UK future energy scenarios; LENS [13-15], HDPS [12] [16], HiDEF[17-18] and DECC [19-20] scenarios which are all developed by considering several aspects such as the environment, policy, regulation and economic and technical factors.

3.1 Micro-generation and demand profiles; UKGDS

In order to provide a more realistic appraisal of likely system performance, the half-hourly interval data of United Kingdom Generic Distribution Systems (UKGDS) network case study data set [21] was applied as the basis for the profiles used in this research in order to give an accurate result for the load flow simulation studies and form accurate diversity profiles. The demand and generation profiles of UKGDS are then forecasted based on the growth rate(s) which are developed in line with the future distribution scenarios in the UK so as to create a large population of diverse devices and provide appropriate boundary conditions suitable for network studies. The practicality of the approach is demonstrated through its application to the analysis of the Zero Carbon Community network in this paper.

3.2 Publicly available electrical distribution scenarios in the UK

As part of this research work the authors' own future scenarios for 2050 have been considered based on the possible growth rates from publically available electrical distribution scenarios in the UK. Here are the summary of these different future scenarios:

3.2.1 HDPS (Highly Distributed Power Systems) scenarios

These scenarios were created by the consortium of UK Universities and industrial partners funded by the EPSRC which considered the control and design for future power systems [17] with many distributed generation connections. It illustrate the scale of change required to reduce carbon emissions by 60% by 2050 [16] in line with UK targets at the time.

3.2.2 LENS (Long- term Electricity Network Scenarios)

The project was undertaken by OFGEM under its commitment to look at a range of future scenarios for electricity networks that could arise as a consequence of market and policy developments [13]. Together with OFGEM the LENS project team consisted of University of Strathclyde, Policy Studies Institute (PSI), University of Westminster and SPRU (Science and Technology Policy Research), University of Sussex [14].

3.2.3 DECC scenarios (UK Low Carbon Transition Plan)

These scenarios were developed by the Department of Energy and Climate Change (DECC) [19] and contain a comprehensive low carbon transition plan to 2020 which sets out how to deliver emission cuts of 18% on 2008 levels by 2020 and over a one third reduction on 1990 levels.

3.2.4 HiDEF (Highly Distributed Energy Futures) scenarios

This project is made up of UK universities and industrial partners and funded by the EPSRC and is an extension to the SUPERGEN: HDPS Power Systems project. It aims to "develop analytical sustainability and economic evaluation tools, interface technologies and coordination strategies that are required to demonstrate the credibility, test the feasibility and engineer the integrative solutions of a future power system that delivers sustainability and security through the widespread penetration of distributed energy resources (DERs) and thus contributes to global ambition towards a low carbon future" [17-18].

4. Forecasting scenarios for 2050

This research is conducted to improve the stochastic load and generation modelling method used by Galloway *et al* [12], in which sample power generation time series profiles were processed and replicated in order to create a large population of diverse devices and provide appropriate boundary conditions suitable for network simulation as shown in figure 2. Contrastingly in this research, the method considered utilises a forecasting technique to account for diversity in future generation and load that becomes boundary conditions in network simulation study.

Previous studies that have employed forecasting of the load and demand profile in the power system area but most of these studies run on short and medium timescales [22]. The work in this paper will focus on long term forecasting, typically 40 years, which means there is a need to implement a suitable forecasting method for this long term horizon. Based on [23] forecasting methods are broadly classified into two groups i) Quantitative, also known as Univariate and Multivariate and ii) Qualitative, which is known as 'expert judgement 'or subjective.

The qualitative forecasting method, or scenarios writing, was the best choice for this research as this method is suitable for a long term forecasting period of up to 40 years. Moreover this scenarios writing method has been successfully applied in the power system area [24-26]. Based on [24] scenarios writing provide a set of alternative contexts for exploring different ways that a future may unfold. Moreover, it is a unique technique as it can explore a range of possible outcomes resulting from uncertainty; in contrast with the quantitative forecasting method which aims to identify the most likely pathway and estimate uncertainties. Nevertheless, quantitative forecasting models are most efficient when under conditions such that extensive information is available and the understanding of governing is high.

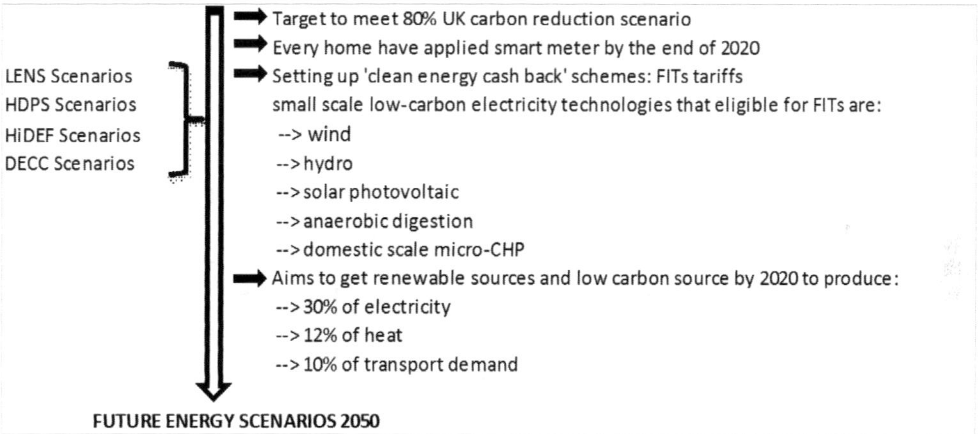

Figure 3. Forecast scenarios for future 'smart networks' 2050 based on publically available electrical distribution scenarios in the UK

However, when the interrelationships between factors are less stable and the predictability of the systems are lower, the quantitative forecasting method has shown to be inadequately equipped to characterise the processes of change [24]. This statement was supported in [27] which asserted that in a recent review of energy forecasting of the US in the last 50 years illustrated that the historical forecasts have regularly failed to represent actual conditions by systematically overestimating consumption and underestimating uncertainty. As stated in modelling approach section, the forecast scenarios were created by considering several aspects such as environment, policy, regulation, economic and technical issues across the published scenarios (LENS, HDPS, DECC and HiDEF). As such, there are four main points for future 2050 scenarios have been successfully created by the author; as shown in figure 3.

IOP Conf. Series: Materials Science and Engineering **50** (2013) 012033 doi:10.1088/1757-899X/50/1/012033

5. Rate forecast for 2020

5.1 Micro-generation profiles

Paper [25] designed a questionnaire to collect opinions on the suitability of different DG technologies to different LV grid condition and from this method the authors have identified the top three ranked technologies in each of the LV grid scenarios for 2010 and 2020. Since this paper more focuses in the future scenarios (2020 until 2050) the result for 2020 only will be considered in this research; based on this paper, photovoltaics emerged as the dominant technologies in 2020 and followed by reciprocating engines and wind turbines. Moreover, as one of the main points for 'future energy scenarios for 2050' in figure 3 above; the 'clean energy cash-back' scheme which also called FITs tariffs will be adopt in the UK and the eligible small-scale low carbon electricity technologies for this FITs tariff are Wind turbine, Solar photovoltaics (PV), Hydro, Anaerobic digestion and Domestic scale micro-CHP. Thus, based on these statements the Photovoltaics, Wind turbine and Micro-CHP have been chosen as the top three ranked technologies that suit the best for this research.

The 'prioritized but arbitrary assignment' method from paper [25] have been applied and altered to be suit with the case of this research in order to produce the growth rate(s) value. From the four mains points for future 2050 scenarios as shown in figure 3, the total percentages of energy value that need to come from the renewable source and low carbon sources by 2020 is 42%. Thus, the total growth rate(s) of each micro-generation that have been considered in this research could be split between them. The split could be based on an arbitrary division, with the top-ranked technology taking, say, half. Then the second ranked taking a third and the third ranked taking a sixth [25]. This arbitrary value of growth rate(s) between technologies would produce realistic and useable scenarios for micro-generation profiles as summaries in more detail in table 1. From these scenarios and the methodologies laid out in [25] and [12], the growth rate(s) of micro-generation profiles for Photovoltaics, Micro-Wind Turbines and Micro-CHP as well as demand profiles for Smart Meters and Electric Vehicles were forecast and created as shown in figure 4.

5.2Demand profiles

Demand profiles that studied in this research includes Smart meters and Electric vehicles were forecast and created based on the four main points for future 2050; as shown in figure 3. As mention above these forecast scenarios were created by considering several aspects such as environment, policy, regulation, economic and technical issues across the published scenarios in the UK. Based on future energy scenarios 2050 created in figure 3, every home have applied Smart meter by the end of 2020; this statement gave an idea that by 2020 the demand profiles for every home in the UK will be reduced to 50% and this reduction rate(s) were predicted to be consistent until 2050. Electric vehicle's percentage for 2020 was also created by considered one of the main points as created in 'future energy scenarios 2050'; as shown in figure 3. By targeting 10% of transport demand need to come from the renewable source and low carbon sources by 2020, this research applied 10% as a growth rate(s) value for demand profiles as represent the adoption of Electric vehicle in the network studies.

Table 1. Defining growth rate(s) for micro-generation penetration in 2020

Micro-generation Technologies	Division	Percentage of growth rates for MG from 42%
Solar Photovoltaic	50%	21%
Micro Wind Turbine	33%	14%
Micro CHP	17%	7%

Rate forecast for energy 2020	**Rate forecast for energy 2030**
➡ Energy produced profile from **Microgeneration**	➡ Energy produces profiles from **Microgeneration**
--> Solar Photovotaic; increase to 21%	--> Solar Photovotaic; increase to 42%
-->Micro Wind turbine; increase to 14%	-->Micro Wind turbine; increase to 28%
-->Micro CHP; increase to 7%	-->Micro CHP; increase to 14%
➡ Energy **demand** profiles	➡ Energy **demand** profiles
-->Electric vehicle; increase to 10%	-->Electric vehicle; increase to 20%
-->Smart meter (100% installed); reduce to 50%	-->Smart meter (100% installed); reduce to 50%
Rate forecast for energy 2040	**Rate forecast for energy 2050**
➡ Energy produced profile from **Microgeneration**	➡ Energy produced profile from **Microgeneration**
--> Solar Photovotaic; increase to 63%	--> Solar Photovotaic; increase to 84%
-->Micro Wind turbine; increase to 42%	-->Micro Wind turbine; increase to 56%
-->Micro CHP; increase to 21%	-->Micro CHP; increase to 28%
➡ Energy **demand** profiles	➡ Energy **demand** profiles
-->Electric vehicle; increase to 30%	-->Electric vehicle; increase to 40%
-->Smart meter (100% installed); reduce to 50%	-->Smart meter (100% installed); reduce to 50%

Figure 4. Micro-generation and demand side management penetration forecast rates for future energy through to 2050.

Since objective of this research is to model and forecast the future profile; 2020 until 2050. Thus, the growth rate(s) in table 1 were used as a base growth rate(s) which it is expected that every 10years the value of growth rate(s) for the profile will be double the rate(s) of the base growth rate(s) as shown in figure 4 aligned with the forecast scenarios for 'future' energy systems in figure 3.

6. Simulations

6.1 Micro-generation and demand profile

Figures 5 and 6 show the original 10 year UKGDS forecast profiles based on growth rate(s) created previously [20] and figures 7 and 8 show the aggregated profiles of the forecasting profiles data for micro-generation and demand in the year of 2050. The methods used to produce figures 10 and 11 were applied in this research in order to produced the final profiles that were used as a boundary conditions in the network simulation studies, which are: i) forecasting the original profile data from UKGDS for every 10 years based on the growth rate of the publically available scenarios of electrical distribution system in the UK and ii) aggregating the forecasting profiles data separately for micro-generation and the demand profile in order to determine the final value of the mixture for both profiles for the year of 2050.

6.2 Overview of the network modelled

Power flow studies were run in order to investigate the effects of high penetrations of MG and DSM on the future 'smart network'. The network was implemented using DIgSILENT Powerfactory V14.0 software and was used as a test-bed study for this research (Figure 9). By modelling individual dwellings network security issues such as under and over voltage, power quality and reverse power flow for each dwelling could be explored in more detail. Here, the example of a Zero Carbon Community network (Figure 1) was designed with the main terminal connected to the three-phase network and each load was connected to single phase network with an underground cable. The load and generation profiles were developed and forecast by the process described above and were installed in each dwelling of the network. The high-level characteristics of this model are its long circuit length, low customer density, radial topology and large overall size.

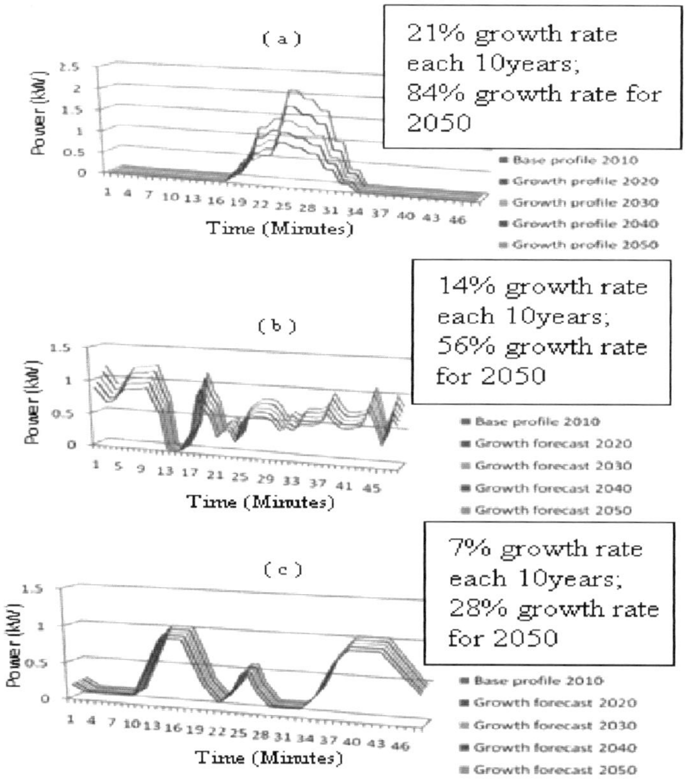

Figure 5. Forecasting growth rates for half-hourly interval of power production data of a day after the adoption of (a) solar photovoltaic (b) Micro-Wind turbine and (c) Combined Heat and Power for 2050

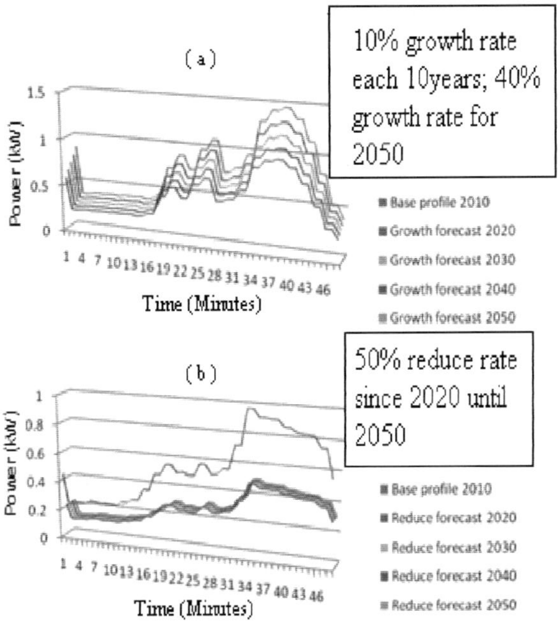

Figure 6. Forecasting rates for half-hourly interval of demand unrestricted data of a day after the adoption of (a) Electric vehicle and (b) Smart meter demand data for 2050

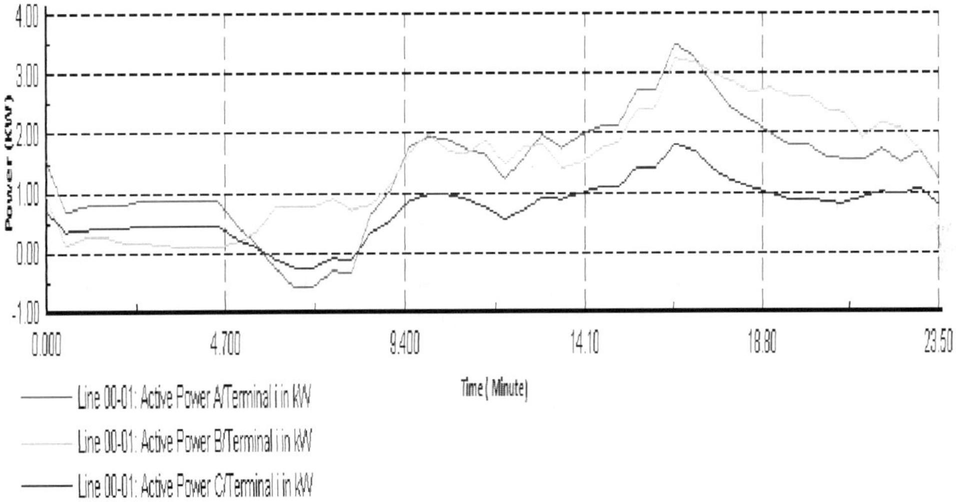

Figure 7. The result of unbalance power; the result was taken at the first substation in the network of Zero Carbon network

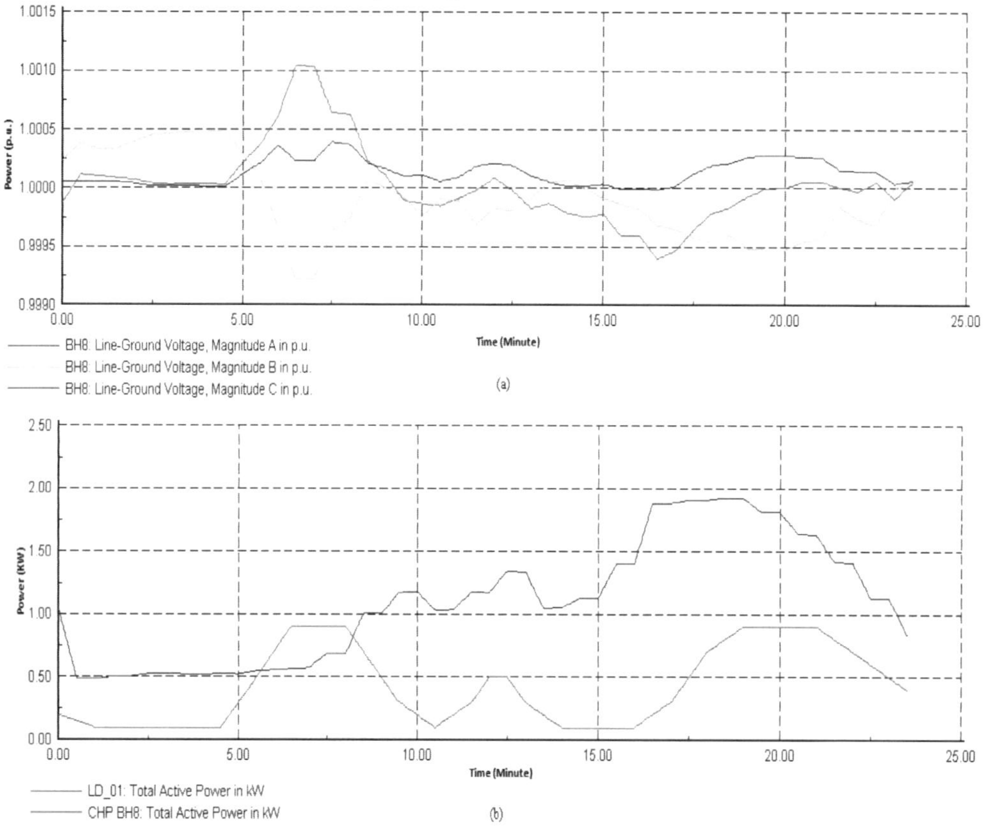

Figure 8. The results of voltages between different phase (A, B, C) and (b) power between load and micro-generators in the bus BH8 in the network

7. Results and discussion

Studies have been conducted for the network as seen in figure 9 and details of the results are shown in figure 10 until 13 and the summaries of the result shown in table 2. The results from table 1 identify that the demand across each of the time periods cannot totally be met by the configuration of onsite generation. Electrically the network is balanced in terms of power quality measures (see figure 10) although loads and generation were installed in a different phase of each bus. This balance is largely achieved because the network still has support from the main grid enabling shortfalls in generation to be covered.

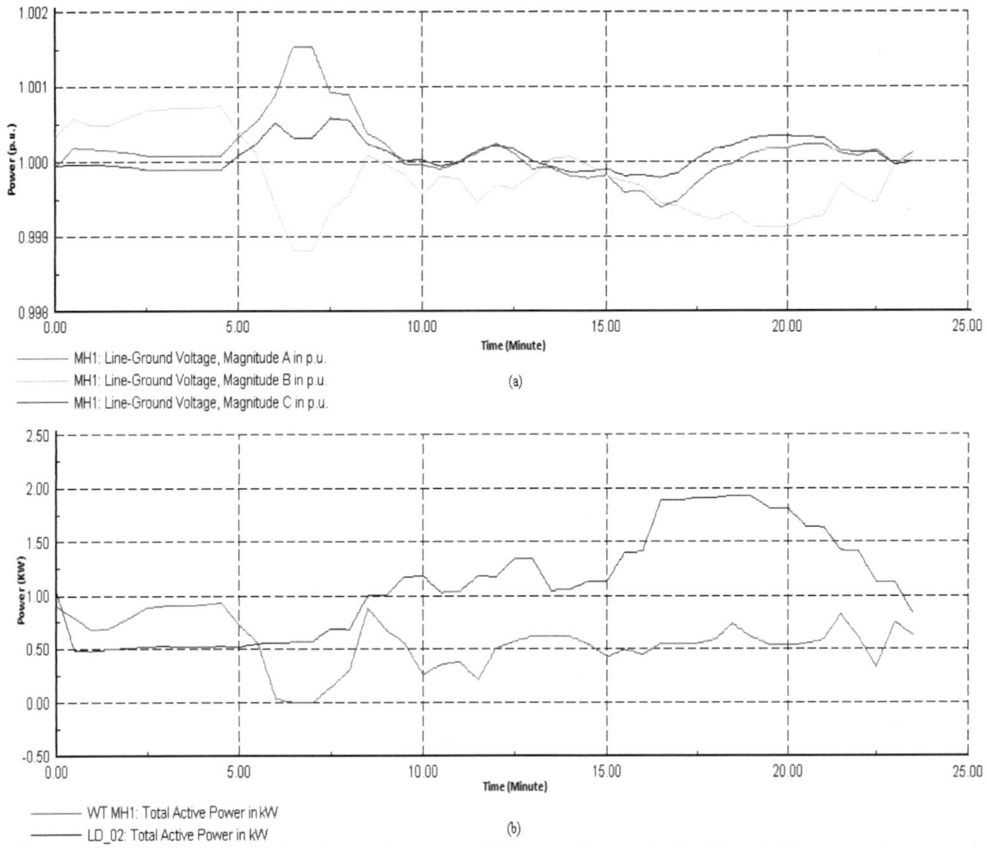

Figure 9. The results of (a) voltages between different phase (A, B, C) and (b) power between load and micro-generators in the bus MH1 in the network

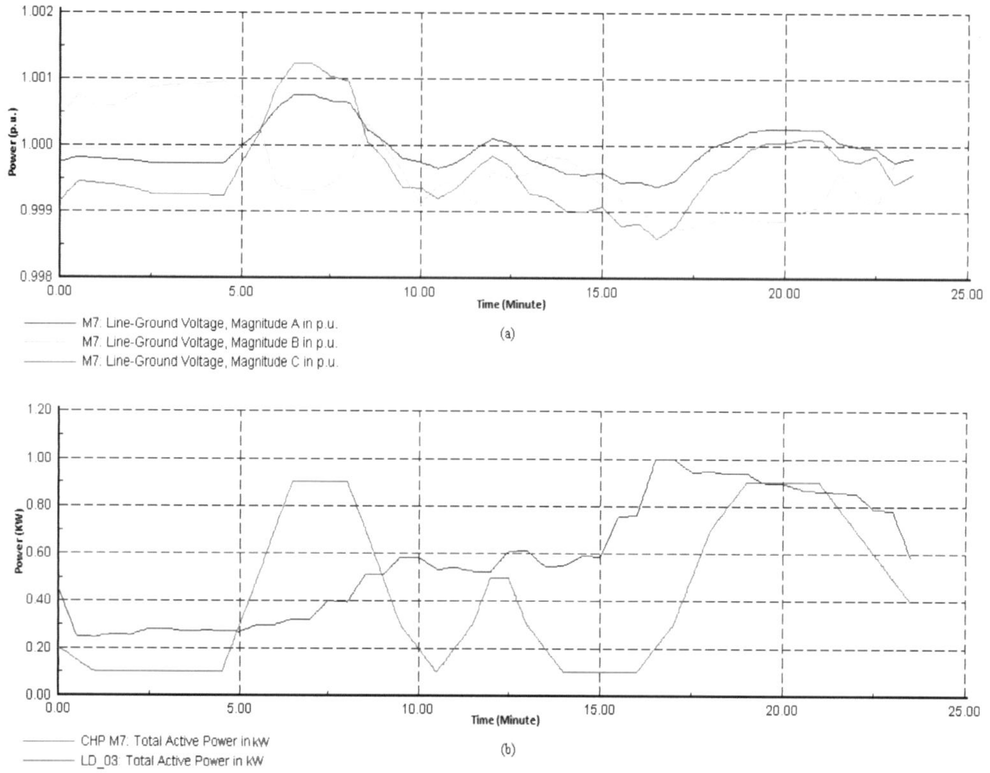

Figure 10. The results of (a) voltages between different phase (A, B, C) and (b) power between load and micro-generators in the bus M7 in the network

Table 2. Result of power flow analysis for selected of dwelling/bus in the network

Bus	On-site generation	00:00 -04:30		05:00-09:30		10:00-14:30		15:00-19:30		20:00-23:30	
BH8	On-site generation could only support 39% of demand needed.	Main Grid	80%	Main Grid	10%	Main Grid	72%	Main Grid	73%	Main Grid	48%
		On-Site Gen.	20%	On-Site Gen.	90%	On-Site Gen.	22%	On-Site Gen.	27%	On-Site Gen.	52%
MH1	On-site generation could only support 52% of demand needed.	Main Grid	0%	Main Grid	37%	Main grid	59%	Main Grid	68%	Main Grid	56%
		On-Site Gen.	Extra 49% of demand needed	On-Site Gen.	53%	On-Site Gen.	41%	On-Site Gen.	32%	On-Site Gen.	44%
M7	On-site generation could only support 74% of demand needed.	Main Grid	59%	Main Grid	0%	Main Grid	54%	Main Grid	48%	Main Grid	12%
		On-Site Gen.	41%	On-Site Gen.	Extra 69% of demand needed	On-Site Gen.	45%	On-Site Gen.	52%	On-Site Gen.	88%

8. Conclusions and further work

In this paper, the energy and demand profiles have been created based on publically available energy scenarios for the UK. These profiles are realistic profiles for 2050 and were used as appropriate boundary conditions for the analysis of the future 'smart networks' Zero Carbon network. It is shown from result in table 2 that even though the value of the power produced from each Micro-generation is often in line with the demand requirements of an individual dwelling there is still a shortfall in supplying the full demand. However, the network results show that for studies considering power and voltage values, each phase of the network remains balanced. As the network still has the support from the main grid enabling shortfalls in generation to be balanced through imports. Thus, this paper shown that there will be no problems arising from high penetration of Micro-generation and Demand side management for each dwelling in the community network as long as the network has the support from the main grid.

The 'smart network' considered here might look advanced compared to the existing infrastructure and management but in line with the future scenarios such systems may appear bringing with them a challenge for the network operator and the future. This research has a special contribution to our country as the energy scenarios in Malaysia have been subject to tremendous change from the 8th

Malaysia Plan (2001-2005) until the current plan, the 10th Malaysia Plan (2011-2015). Oil and gas have traditionally been the main energy sources in Malaysia, however, with its gas reserves estimated to last only another 33 years and oil for only 19 years, the Malaysian government has been strengthening the role of renewable energy (RE), as the fifth main energy source as noted in [28]. Moreover, this research may help the Malaysian government in obtaining a clearer viewpoint regarding the implication of an increased penetration of generation coming from renewable sources to Malaysia's network system.

References

[1] Transition Pathway Consortium Transition pathways to a low carbon economy 2012 Available from: http://www.lowcarbonpathways.org.uk/
[2] Timothy J Foxon Geoffrey P Hammond and Peter J G Pearson 2010 *UK Technological Forecasting and Social Change* **77** 1203-13
[3] Timothy J Foxon Geoffrey P Hammond and Peter J G Pearson *Transition pathways for a low carbon energy system in the UK: assessing the compatibility of large-scale and small-scale options* 7th BIEE Academic Conference 2008
[4] Kamalanath S et. al. 2009 Demand side primary frequency response support through smart meter control *Universities Power Engineering Conference (UPEC), Proceedings of the 44th International*
[5] Willet K et al, 1997 *Electric vehicle as a new power source for electric utilities.* Elsevier Science Ltd. **2** 157-75
[6] Econnect 2005 *Potential for Microgeneration study and analysis* Available from: http://www.Berr.Gov.Uk/Files/File27558.Pdf
[7] Ian R M T David I and Alice D 2009*A modelling framework for the study of highly distributed power systems and demand side management* International Conference on Sustainable Power Generation and Supply 'SUPERGEN'09' 1-6.
[8] Peter R et al 2009 *Impact of high penetrations of micro-generation on low voltage distribution networks.* 20th International Conference On Electricity Distribution Prague
[9] Arturo D A 2009*multi-objective planning framework for analysing the integration of distributed energy resources* University of Strathclyde, Glasgow.
[10] Thomson M and Infield D 2008 *Modelling the impact of micro-combined heat and power generators on electricity distribution networks* Proceeding of the Institution of Mechanical Engineers Part A: Journal of Power and Energy **222** 697-06
[11] Putrus G A S P Johnston D Bentley E C and 2009Narayana M *Impact of electric vehicles on power distribution networks* Vehicle Power and Propulsion Conference 'VPPC '09'827-31
[12] Burt G M et al. 2008 *Journal of Power and Energy.* **222** 657-68
[13] OFGEM 2012 *Long term electricity scenarios (Lens)* Available from: http://www.ofgem.gov.uk/Networks/Trans/Archive/ElecTrans/LENS/Pages/LENS.aspx
[14] OFGEM 2007 *Long term electricity network scenarios (lens), Second Workshop* Available from:http://www.ofgem.gov.uk/Networks/Trans/Archive/ElecTrans/LENS/Documents1/Presentation1.pdf
[15] Thiruchelvam S Tookey J E Samarasinghe D Hock Q W Mustapha K N 2012 *Journal of Mechanical Engineering and Sciences* **2** 148-61
[16] Jardine C N and G W Ault *Journal of Power and Energy* **222** 643-55
[17] Consortium S. 2012 *SUPERGEN: Highly distributed energy future* Available from: http://www.supergen-hidef.org/Pages/Home.aspx
[18] Brenda Boardman 2012 *'Home truths: a low carbon strategy to reduce UK housing emissions by 80% by 2050'* Available from: http://www.foe.co.uk/resource/reports/home_truths.pdf
[19] Department of energy and climate change 2012 *Department of energy and climate change.* Available from: http://www.decc.gov.uk
[20] National strategy for climate and energy 2009 *The UK low carbon transition plan*

IOP Conf. Series: Materials Science and Engineering **50** (2013) 012033 doi:10.1088/1757-899X/50/1/012033

[21] United kingdom generic distribution system (UKGDS) 2012 *United kingdom generic distribution system* Available from: http://Monaco.Eee.Strath.Ac.Uk/Ukgds

[22] Adepoju G A 2007 *Pacific Journal of Science and Technology* **8**.

[23] Chris C *The analysis of time series; an introduction* 6 ed. Chapman & Hall/Crc. P 73-75.

[24] Ghanadan R and J G Koomey 2005 *Energy Policy Using Energy Scenarios to Explore Alternative Energy Pathways in Californi* 1117-42

[25] Foote C E T et al. 2005 *Developing Distributed Generation Penetration Scenarios.* International Conference on Future Power Systems

[26] Elders I M et al. 2005 *Identification of Long-term Scenarios of Electricity Network Development* International Conference on Future Power Systems

[27] Craig, P P 2002 *Annual Review of Energy and the Environment,What Can History Teach Us? A Retrospective Examination of Long-Term Energy Forecasts for the US*

[28] UniMAP 2012 *Renewable Energy and Kyoto Protocol: Adoption in Malaysia* Available from: http://publicweb.unimap.edu.my/~ppkas/home/index.php/news/articles/29-renewable energy-and-kyoto-protocol-adoption-in-malaysia

2nd International Conference on Mechanical Engineering Research (ICMER 2013) IOP Publishing
IOP Conf. Series: Materials Science and Engineering **50** (2013) 012034 doi:10.1088/1757-899X/50/1/012034

CFD analysis of heat transfer and friction factor charaterstics in a circular tube fitted with horizontal baffles twisted tape inserts

Sami D Salman [1, 2,a], **Abdul Amir H Kadhum** [1], **Mohd S Takriff** [1] **and Abu Bakar Mohamad** [1]

[1]Department of Chemical and Process Engineering, Faculty of Engineering and Built Environment, Universiti Kebangsaan Malaysia, Bangi, Selangor 43600, Malaysia.
[2]Biochemical Engineering Department, Al-khwarizmi College of Engineering, University of Baghdad. Baghdad 47024, Iraq

E-Mail:[a]sami.albayati@gmail.com

Abstract. Swirl/vortex flow generator is an important form of passive augmentation techniques. Twisted-tape is one of the most important members of this form which is used extensively in different type heat exchangers. This paper reports the effect of twisted tape inserts on heat transfer and friction factor characteristics in circular tube under constant heat flux and laminar flow conditions using CFD simulation. Plain twisted tape inserts with twist ratios (y = 2.93, 3.91) and baffled twisted tape inserts with twist ratio (y = 2.93) have been used for the simulation using Fluent version 6.3.26. The results obtained by simulation matched with the literature correlations for plain tube with the discrepancy of less than ± 8% for Nusselt number and ± 6.25% for friction factor. The results have also revealed that the heat transfer in term of the Nusselt number enhanced with increases of Reynolds number, decreases of twist ratio and baffle insert. Among the various twist ratios, the twisted tape with twist ratio of y=2.93 and baffle is offered a maximum heat transfer enhancement.

1. Introduction

Heat transfer augmentation techniques can be classified to active and passive technique. The active technique requires an external power facilitate the desired flow modification such as electrostatic field, surface or liquid vibration, etc. The passive technique does not require any external energy, it required fluid additives, special surface geometries or swirl flow devices i.e.; Twisted tape inserts. A great deal of experimental works on heat transfer augmentation studies using twisted tape have been reported in the literature [1-5] Whereas, limited literatures are available in CFD modeling of heat transfer using twist tape inserts, Sivashanmugam et al [6] presented CFD modeling of heat transfer augmentation in a circular tube fitted with helical twist inserts as swirl generator. Mugam et al [7] reported CFD simulations of heat transfer characteristics using different concentrations of Al_2O_3 nanofluid and helical twist tape inserts with different twist ratio in a circular tube. Pathipakka and Sivashanmugam [8] proposed CFD simulation of heat transfer and friction factor characteristics of the circular tube fitted with a right-left helical twist insert with 00 mm spacer based on experimental work. The simulated results of Nusselt number and friction factor were compared with the experimental data with good agreement.

Shabanian et al [9] conducted an experimental and CFD modeling on heat transfer and friction factor characteristics in air cooled heat exchanger using butterfly twist tape insert to investigate the effect of insert configuration on the Nusselt number, friction factor and thermal performance factor. The objective in this paper is to demonstrate the heat transfer and friction factor simulation for the circular tube fitted with Baffled twisted tape (BTT) insert with twist ratio (y=2.93) depend on experimental data from [8]. The longer term objective of this research is to reveal that the insert configuration can play an important role on heat augmentation. This study can be used as guideline for experimental works.

2. Materials and methods

2.1. *Physical models*

The geometry and Grid of the Baffled twisted tape insert (BTT) with a relative twist ratio (y=2. 93) is depicted in figure1 and 2. The material of construction of the tube and twisted tape is Steel and aluminium, respectively. Water is selected as the working fluid and the thermo-physical properties of fluid are selected at 298K and assumed to be temperature independent.

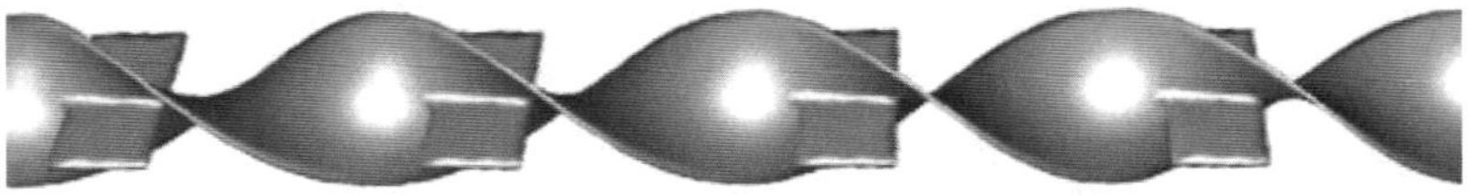

Figure1. Baffled twisted tape insert.

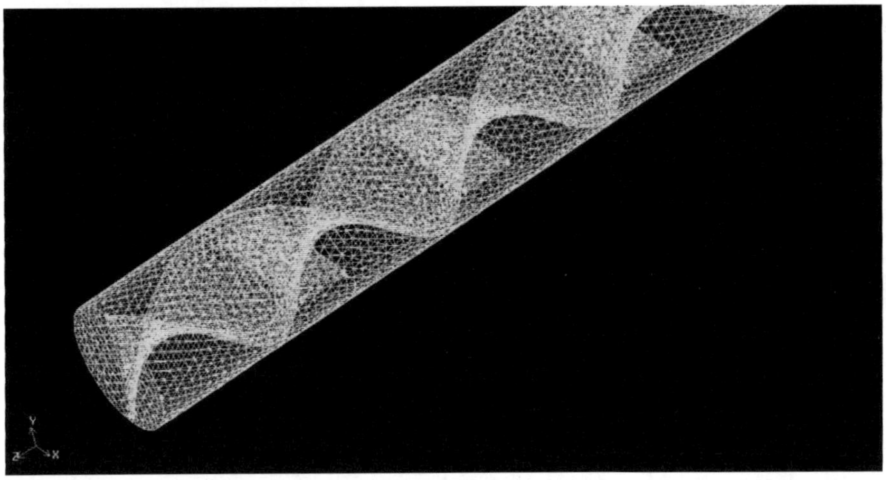

Figure 2. Grid for (BTT) inserts.

2.2. *Modelling Parameters and Numerical method*

Experimental data mentioned in [8] with the commercial software, Fluent 6.3.26 was chosen as CFD tool to solve governing equations. Three dimensional steady state laminar flow through the tube fitted with baffled twisted tape inserts under constant heat flux is investigated by the following model equations.

2.2.1 *Continuity equation for an incompressible fluid*

$$\frac{\partial p}{\partial t} + \nabla.(\rho\vec{\upsilon}) = S_m \quad (1)$$

2.2.2 *Conservation of momentum*

$$\frac{\partial \upsilon}{\partial t} + \rho(\vec{\upsilon}.\nabla)\vec{\upsilon} = -\nabla p + \rho\bar{g} + \nabla.\tau_{ij} + \vec{F} \quad (2)$$

2.2.3 *Conservation of energy*

$$\rho\frac{\partial}{\partial t}(\rho E) + \nabla.\{\vec{\upsilon}(\rho E + \rho)\} = \nabla.\left\{K_{eff}\nabla T - \sum h_i\,(\vec{\tau}_{eff}.\vec{\upsilon})\right\} + S_h \quad (3)$$

3. Results and discussion

3.1. *Validation of plain tube simulation results*

The results obtained by simulation for plain tube are validated using correlations developed by (Tate 1936). The simulated data of the Nusselt number for a tube are compared with these correlations as demonstrated in figure 3 and 4.

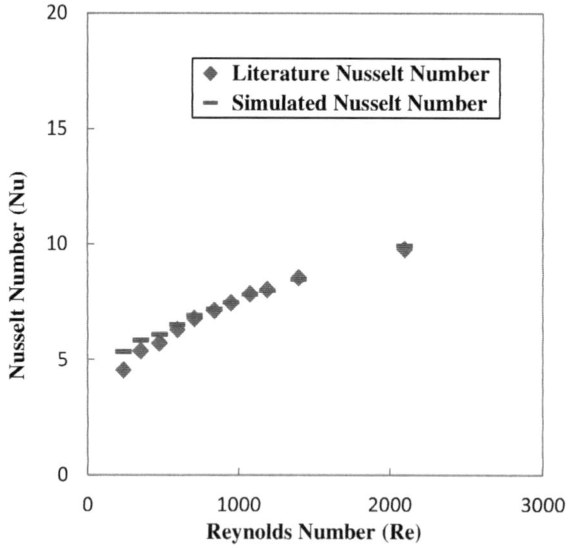

Figure 3. Plain tube simulated Nusselt Number vs Literature data.

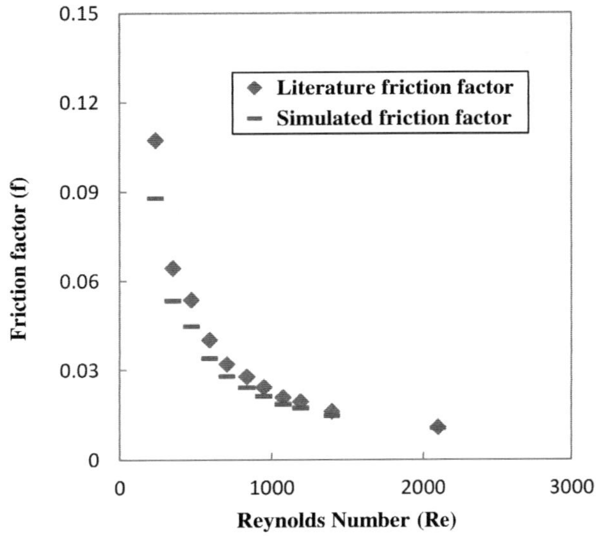

Figure 4. Plain tube simulated friction factor vs Literature data.

3.2. *Effect of twist ratio on heat transfer and friction factor*

Simulated data of the Nusselt numbers, friction factor and their variation with a Reynolds number of plain twisted tape inserts with twist ratio (y=2. 93 and 3.91) are shown in figure 5 and 6.

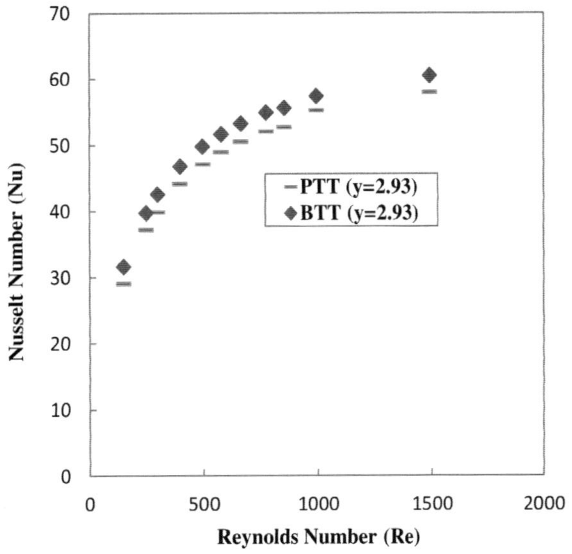

Figure 5. Simulated Nusselt Number for PTT with (y=2.93, 3.91).

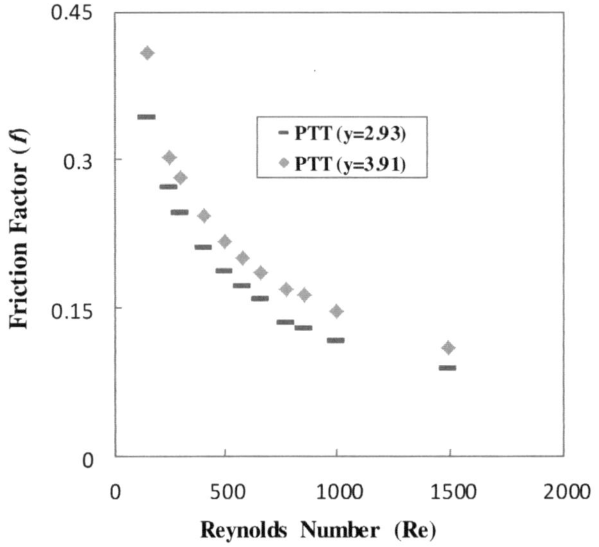

Figure 6. Simulated Froction Factor for PTT with (y=2.93, 3.91).

3.3. *Effect of twist tape configuration on heat transfer and friction factor*

Simulated data for the Nusselt numbers, friction factor and their variation with Reynolds number for Baffled twisted tape insert (BTT) with twist ratio y=2. 93 and is shown in figure 7 and 8.

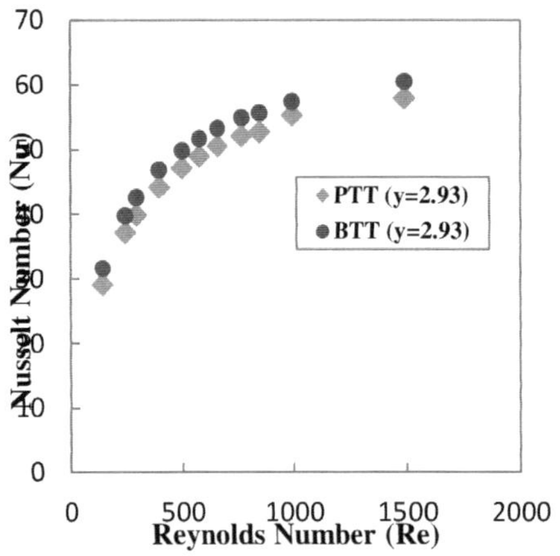

Figure 7. Simulated Nusselt Number for PTT and BTT.

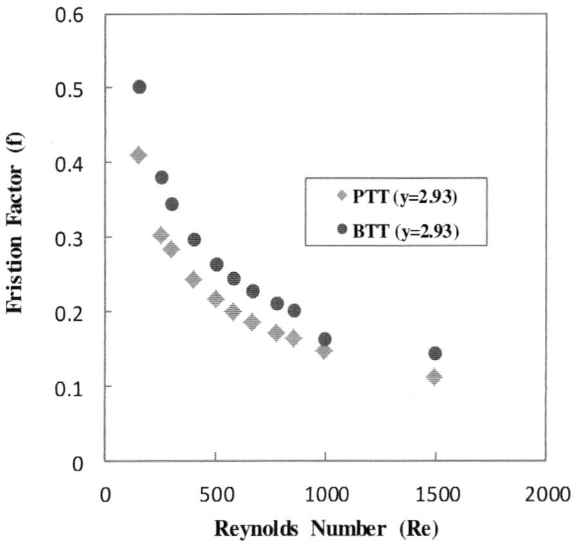

Figure 8. Simulated Friction Factor for PTT and BTT.

4. Conclusion

CFD simulation for the heat transfer augmentation in a circular tube inserted by Baffled twisted tape insert (BTT) with a twisted ratio (y=2. 93& 3.91) and baffle length (2 cm) has been simulated using fluent version 6.2.3.26. The data obtained by simulation are matching with the literature value for plain tube for validation with the discrepancy of less than ± 8% for the Nusselt number and ±6. 25% for friction factor. The data obtained by simulation show that the Baffled twisted tape insert (BTT) offered a higher rate of heat transfer at the same Reynolds number.

Acknowledgement

The authors would like to thank University Kebangsaan Malaysia and MOSTI of Grant research (FRGS/1/2013/TK07/UKM/01/1) for financial support.

Nomenclature

E Energy component in energy equation
F Force component in momentum equation, N
f Fanning friction factor
g Acceleration due to gravity, m/s^2
k_{eff} Thermal conductivity in Energy equation, W/m K
m mass flow rate of fluid, kg/s
Re Reynolds number based on internal diameter of the tube, dimensionless
Nu Nusselt number, dimensionless
p Pressure component in momentum equation, N/m^2
S_m Accumulation of mass, Kg
S_h Accumulation of Energy, J
T Temperature. °C.
v Velocity component in momentum equation, m/s
y Twist ratio (Length of one twist (360°) / diameter of the twist), dimensionless

Greek symbols

ρ Density component in governing equations
$\vec{\tau}_{eff}$ Stress component in momentum equation, N/m^2.

References

[1] Ujhidy A, Nemeth J and Szepvolgyi J 2003 Fluid flow in tubes with helical elements *Chemical Engineering and Processing* **42** 1-7
[2] Ibrahim E Z 2011 Augmentation of laminar flow and heat transfer in flat tubes by means of helical screw-tape inserts *Energy Conversion and Management* **52 1** 250-57
[3] Jaisankar S, Radhakrishnan T K and Sheeba K N 2009 Experimental studies on heat transfer and friction factor characteristics of thermosyphon solar water heater system fitted with spacer at the trailing edge of twisted tapes *Applied Thermal Engineering* **29 5–6** 1224-31
[4] Jian Guo, Aiwu Fan, Xiaoyu Zhang, Wei Liu 2011 Numerical study on heat transfer and friction factor characteristics of laminar flow in a circular tube fitted with center cleared twisted tape *International Journal of Thermal Sciences* 1-8
[5] Wongcharee K and Eiamsa-ard S 2011 Friction and heat transfer characteristics of laminar swirl flow through the round tubes inserted with alternate clockwise and counter-clockwise twisted-tapes *International Communications in Heat and Mass Transfer* **38 3** 348-52
[6] Sivashanmugam P, Nagarajan P K and Suresh S 2008 Experimental studies on heat transfer and friction factor characteristics in turbulent flow through a circular tube fitted with right left helical screw-tape inserts *Chemical Engineering Communication* **195 8** 977-87
[7] Nagrajam P K and Mugam P 2008 CFD Simulation of heat transfer augmentation in circular tube filled with right-left helical inserts with space *International journal of chemical engineering research* **6 1** 1-11
[8] Pathipakka G and Sivashanmugam P 2010 Heat transfer behavior of nanofluids in a uniformly heated circular tube fitted with helical inserts in laminar flow *Superlattices and Microstructures* **47 2** 349-60
[9] Shabanian S R, Rahimi M, Shahhosseini M and Alsairafi A A 2011 CFD and experimental studies on heat transfer enhancement in an air cooler equipped with different tube inserts *International Communications in Heat and Mass Transfer* **38 3** 383-90
[10] Sieder E N and Tate G E 1936 Heat transfer and pressure drop of liquids in tubes. *Industrial and Engineering Chemistry* **28 12** 1429-35.

Simulation study of turbulent convective heat transfer enhancement in heated tube flow using TiO_2-water nanofluid

Adnan M Hussein[1, 2,a], **R A Bakar**[1], **K Kadirgama**[1] **and K V Sharma**[1]
[1]Faculty of Mechanical Engineering, 26600 Pekan, University Malaysia Pahang
[2]Al-Haweeja Institute, Foundation of Technical Educations, Iraq

E-mail: [a]adnanphd2012@gmail.com

Abstract. Simulation by convenient software, the same as FLUENT, was used to predict the friction factor and Nusselt number for forced convection heat transfer of TiO_2-water nanofluid. The range of Reynolds number is from 10000 to 100000 to be turbulent flow in a horizontal straight tube with heat flux 5000 w/m^2 around it. The volume fraction of nanoparticle was (0.25%, 0.5%, 0.75% and 1%) and diameter of particle is 27 nm. The results show that the friction factor and Nusselt number are increasing with increasing of volume fraction. Results compared with the experimental data available in literature and there are good agreements.

1. Introduction

There are number of studies conducted forced convection of TiO_2-water through tube under single phase model [1-10]. Pak and Cho [11] investigated the effect of two different metallic oxide particles, alumina (Al_2O_3) and titanium dioxide (TiO_2), with mean diameters of 13 and 27 nm, respectively, on base fluid (water) experimentally. The Reynolds and Prandtl numbers were varied in the ranges 10000-100000 and 6.5-12.3, respectively. Results showed that Darcy friction factor for the volume concentration (1-3%) was good agreement with Kay's correlation for turbulent flow of a single-phase fluid. The Nusselt number of the dispersed fluids for fully developed turbulent flow is increased with increasing volume concentration as well as the Reynolds number. Duangthongsuk and Wongwises [12] found the heat transfer coefficient and friction factor of the TiO_2-water nanofluids with diameters of 21 nm and volume concentrations of 0.2–2.0 vol.% flowing in a horizontal double tube counter-flow heat exchanger under turbulent flow conditions, experimentally. The results showed that the heat transfer coefficient of nanofluid is higher than that of the base liquid and increased with increasing the Reynolds number and volume concentrations. The pressure drop of nanofluids was slightly higher than the base fluid and increases with increasing the volume concentrations. A theoretical model to predict friction and heat transfer coefficients for a wide range of nanofluids containing Cu, CuO, TiO_2, SiC, ZrO_2, and Al_2O_3 nanoparticles of different sizes, concentration and temperatures dispersed in water [13-14]. The results showed deviation of experimental with theoretical data because of using to different values of properties employed. Forced convection flows of nanofluids consisting of water with TiO_2 and Al_2O_3 nanoparticles in a horizontal tube with constant wall temperature are investigated numerically by [15-17]. The horizontal test section is modeled and solved using a CFD program. Palm et al.'s correlations are used to determine the nanofluid properties. Numerical results showed the heat transfer enhancement due to presence of the nanoparticles in the fluid in accordance with the results of the experimental study used for the validation process of the numerical model. A study of heat transfer enhancement of Al_2O_3 nanofluid using low volume fraction nanofluids in turbulent pipe flow with constant wall temperature was studied by [18-20]. CFD modeling of the nanofluid flow adopting the single phase approach. Nanofluid, up till a volume fraction of 1% is found to be an effective heat transfer enhancement technique. The Nusselt number (*Nu*) and friction factor predicted for the low volume fractions (i.e. 0.02%, 0.1 and 0.5%). Numerical study of turbulent forced convection flow of Al_2O_3-water nanofluid in a circular tube was subjected by [9]. Heat transfer enhancement increases with the particle volume concentration and Reynolds number. Comparisons with correlations present in literature are accomplished and there was a very good agreement. Experimental study of turbulent forced convection heat transfer was conducted by [21] the study was concluded a pipe employing twisted tape inserts with and without Al_2O_3–water nanofluid. The increasing in the heat transfer coefficient was observed both with and without pipe inserts. It then also developed generalized correlations for the estimation of Nusselt number and friction factor for pipes with and without inserts.

In this work, CFD simulation by FLUENT software will be used to predict friction factor and forced convection heat transfer Nusselt number. The study concludes turbulent forced convection for TiO_2-water nanofluid then compared with available experimental work.

2. Mathematical model

CFD model by FLUENT software was used to solve governing equations of turbulent forced convection heat transfer in horizontal tube with constant heat flux. The description of problem under taken graphing by GAMBIT model and mesh the section test with size of (1000 x 20), 1000 with length of pipe and 20 with radius as shown in Figure 1. The governing equations (continuity, Momentum and energy) are written [13] as:

$$\frac{1}{r}\frac{\partial}{\partial r}\left(\rho_{nf}u\right)=0 \tag{1}$$

$$\frac{1}{r}\frac{\partial}{\partial r}\left(\rho_{nf}uu\right)=-\frac{1}{r}\frac{\partial P}{\partial x}+\frac{1}{r^2}\frac{\partial}{\partial x}(\tau) \tag{2}$$

$$\frac{1}{r}\frac{\partial}{\partial r}\left(\rho uT\right)=\frac{1}{r^2}\frac{\partial}{\partial x}\left\{\frac{k_{nf}}{C_p}\frac{\partial T}{\partial x}\right\} \tag{3}$$

Which solved iteratively using finite volume method (FVM) and SIMPLE scheme was adopted for the treatment of pressure. The Reynolds number studied in this work is high, turbulent viscous, i.e. k-ε model has been employed. In this work, converged solutions were considered for residuals lower than 10^{-6} for all the governing equations [22]. The results of simulation for nanofluid were compared with the theoretical data available for the conventional water. The theoretical data of water were simulated in FLUENT software too. Then comparing these data with Dittos-Boelter correlation (5) for heat transfer (*Nu*) as shown [23]

$$f=\frac{0.316}{Re^{0.25}} \tag{4}$$

$$Nu=\frac{h_f}{k_f}D=0.023\,Re^{0.8}\,Pr^{0.4} \tag{5}$$

It is observed that a grid of (1000 x 20) has good agreement with the theoretical results and hence this model is considered as the optimum grid for carrying out all the analysis.

3. Thermal properties of nanofluid

The thermal properties of nanofluid are (density, heat capacity, thermal conductivity and viscosity) have been estimated by equations below [13]

$$\rho_{nf} = \left(\frac{\phi}{100}\right)\rho_p + \left(1 - \frac{\phi}{100}\right)\rho_f \tag{6}$$

$$C_{nf} = \frac{\frac{\phi}{100}(\rho C)_p + \left(1 - \frac{\phi}{100}\right)(\rho C)_f}{\rho_{nf}} \tag{7}$$

$$k_r = \frac{k_{nf}}{k_f} = \left\{0.8938\left(1 + \frac{\phi}{100}\right)^{1.37}\left(1 + \frac{T_{nf}}{70}\right)^{0.2777}\left(1 + \frac{d_p}{150}\right)^{-0.0336}\left(\frac{\alpha_p}{\alpha_f}\right)^{0.01737}\right. \tag{8}$$

$$\mu_r = \frac{\mu_{nf}}{\mu_f} = C1\left(1 + \frac{\phi}{100}\right)^{11.3}\left(1 + \frac{T_{nf}}{70}\right)^{-0.038}\left(1 + \frac{d_p}{170}\right)^{-0.061} \tag{9}$$

Where: k_p (thermal conductivity) of TiO_2 = 8.4 W/m °C, ρ_p (density) of TiO_2 = 4175 kg/m^3, Cp_p (heat capacity) of TiO_2 = 692 J/kg K [2].

4. Boundary conditions

Volume concentration nanofluids (0.25%, 0.5%, 0.75% and 1%) at 25°C were used for TiO_2-water as input fluids. For comparison purposes, water was also employed as working fluid. Simulation study has been carried out with uniform velocity profile at the inlet of the horizontal tube. Turbulent intensity (I) was specified for an initial guess of turbulent quantities (k and ε). The turbulent intensity was estimated for each case based on the formula $I = 0.16Re^{-1/8}$. Outflow boundary condition has been used at the outlet boundary. The wall of the tube was assumed to be perfectly smooth with constant wall heat flux (5000W/m^2) was used at the wall boundary.

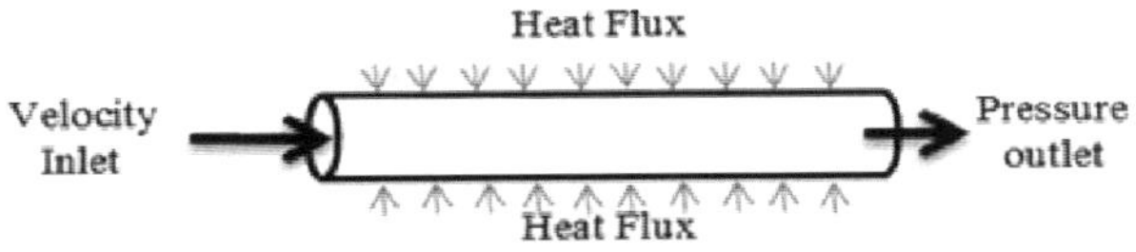

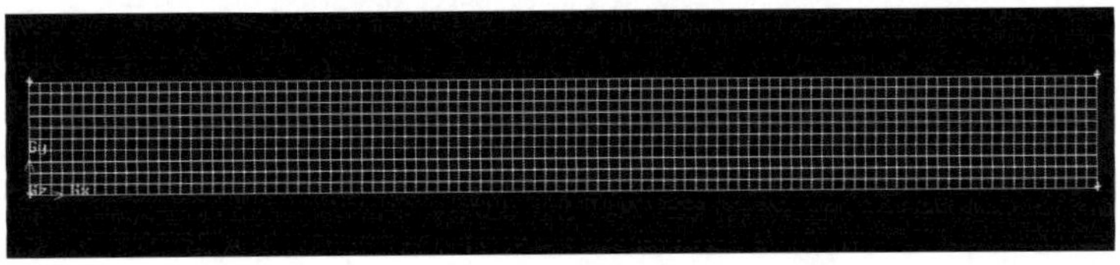

Figure 1. Mesh generated by GAMBIT

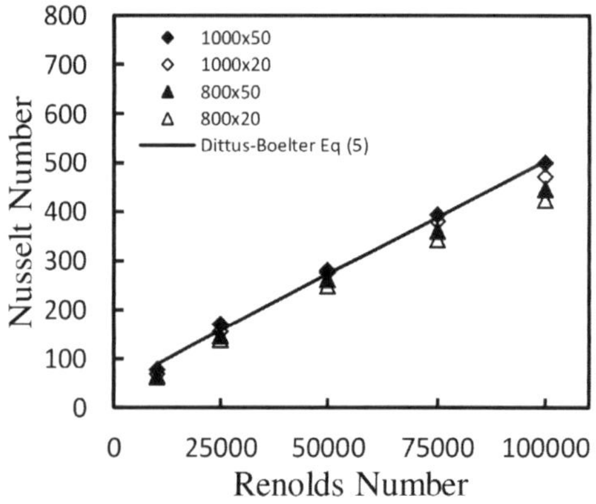

Figure 2. Optimum mesh grid size for Nusselt Number with Reynolds Number for pure water at 25 °C

5. Results and discussion

Simulation results were made reliable by comparing them with available correlation in the literature. Nusselt number for the base fluid (pure water) in the turbulent regime was compared with that of Dittus-Boelter equation (5) as shown in Figure 2. *Nu* from simulation is in very good agreement with that of the correlation values. The effect of nanofluid volume fraction on friction factor and heat transfer enhancement is shown in Figure 3 and Figure 4 There have same behavior of friction factor from Blassius equation and simulation data which decrease when *Re* and volume fraction of nanofluid are increased as shown in Figure 3 *Nu* from Dittos-Boelter equation has same behavior of simulation data which increasing when Re and volume fraction of nanofluid are increased as shown in Figure 4 The reason of increasing of friction factor and Nu with increasing in volume fraction of nanofluid is the high thermal properties of TiO_2 which enhancement of heat transfer and hydrodynamic flow.

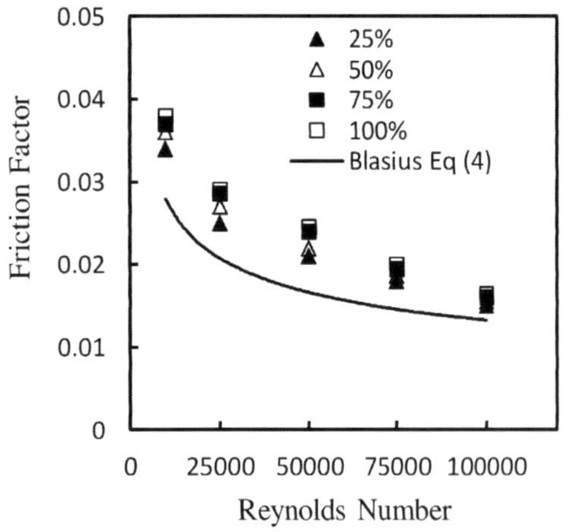

Figure 3. Effect of nanofluid volume fraction on friction factor.

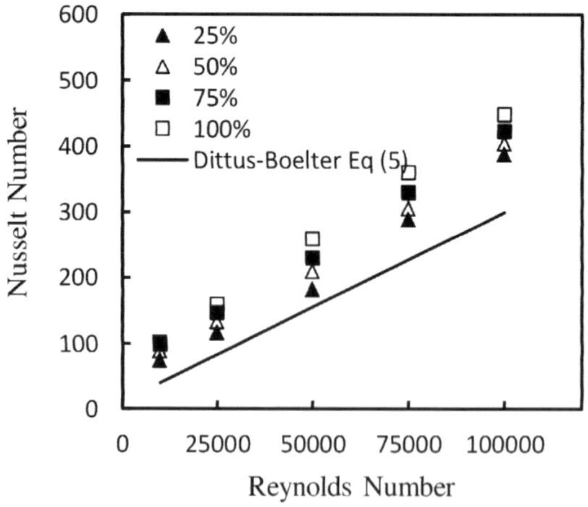

Figure 4. Effect of nanofluid volume fraction on heat transfer enhancement.

Figures 5 and 6 show the results of CFD analysis of friction factor and Nu versus Re compared with [11] experimental data for TiO_2-water nanofluid with particle size is 27 nm and base temperature is 25°C. The results show that good agreement with deviation is 4%.

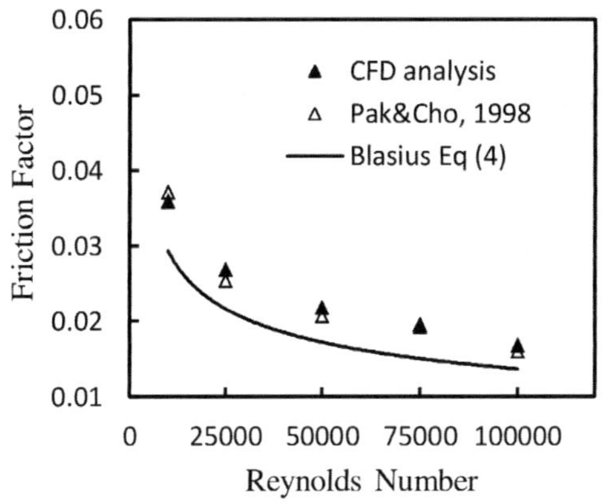

Figure 5. Comparison CFD results of friction factor with Pak and Cho [11] data.

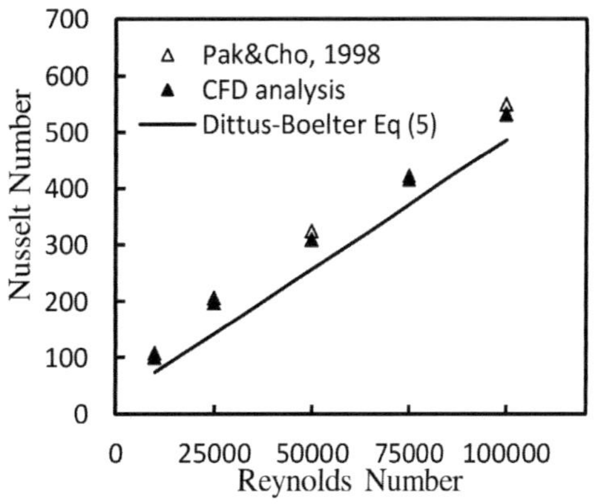

Figure 6. Comparison CFD results of *Nu* with Pak and Cho [11] experimental data.

6. Conclusions

Friction factor and Nusselt number are increased with increasing of volume fraction. When adding solid particle on base fluid it will increase of friction factor and enhancement in heat transfer. The results of friction factor and Nusselt number from simulation showed good agreement with correlations of Blassius and Dittos Boelter equation respectively. The increasing in friction factor and *Nu* is 9.5% and 13.6% respectively. There was a good agreement with [10] experimental data with deviation is 4%.

Nomenclatures

- C - Specific heat capacity [W/ kg.°C]
- D - Diameter [m]
- E - Energy [W]
- f - Friction factor
- htc - Convection heat transfer coefficient [W/m^2.°C]
- k - Thermal conductivity [W/m.°C]
- K_{eff} - Effective thermal conductivity [W/m.°C]
- Nu - Nusselt Number [$htc\ .D/K_{eff}$]
- P - Pressure [N/m^2]
- Pr - Prandtle Number [$C.\mu/K_{eff}$]
- Re - Renolds Number [$\rho.\ u.\ D/K_{eff}$]
- u - Velocity [m/s]
- μ -Viscosity [N.s /m^2]
- ρ - Density [kg/m^3]
- τ - Shear stress [N/m^2]
- ϕ - Volume fraction

Subscripts

- f liquid phases
- p solid particle
- eff effective
- nf nanofluid

Acknowledgment

The financial support to the authors by University Malaysia Pahang is gratefully acknowledged.

References

[1] Heris S Z 2009 *Experimental Heat Transfer* **22** 217–27

[2] Yurong H and Yi J 2007 *Int. J. of Heat and Mass Transfer* **50** 2272–81

[3] Srinivasa Rao G Sharma K V Chary S P Bakar R A Rahman M M Kadirgama K Noor M M 2011 *J. Mech. Eng. Sci.* **1** 1-15

[4] Asirvatham L G 2009 *Energies* **2** 97-19
[5] Gnielinski V 1976 *Int. Chemical Engineering* **16** 359–68
[6] Kim D Kwon Y Cho Y Li C Cheong S Hwang Y Lee J Hong D and Moon S 2009 *Current Applied Physics* **9** 119-23
[7] Xuan Y Li Q 2003 *J. of Heat Transfer* **125** 151–55
[8] Vajjha R S Das D K Kulkarni D P 2010 *Int. J. of Heat and Mass Transfer* **53** 4607 – 18
[9] Bianco V Manca O Nardini S 2011 *Int. J. of Thermal Sciences* **50** 341-49
[10] Mahendran M Lee G C Sharma K V Shahrani A 2012 *J. Mech. Eng. Sci.* **3** 301-10
[11] Pak B C and Cho Y I 1998 *Experimental Heat Transfer* **11** 151-71
[12] Duangthongsuk W and Wongwises S 2010 *Int. J. of Heat and Mass Transfer* **53** 334–44
[13] Sharma K V Sarma P K Azmi W H Mamat R Kadirgama K 2010 *IJMNTFTP* **3** 1-25
[14] Azmi W H Sharma K V Mamat R Anuar S 2013 *J. Mech. Eng. Sci.* **4** 397-408
[15] Syam S L Sharma K V 2011 *J. Mech. Eng. Sci.* **1** 99-12
[16] Demir H 2011 *Int. Communication in Heat and Mass Transfer* **38** 218–28
[17] Syam Sundar L and Sharma K V 2011 *Inter. J. Automot. Mech. Eng.* **3** 265-78
[18] Kumar P 2011 *World Academy of Science Engineering and Technology* **57** 746-50
[19] Hussein A M Bakar R A Kadirgama K Sharma K V 2013 *Inter. J. Automot. Mech. Eng.* **7** 850 63
[20] Hussein A M Sharma K V Bakar R A Kadirgama K 2013 *J. Mech. Eng. Sci.* **4** 452-61
[21] Sundar L S Sharma K V 2010 *Int. J. of Heat and Mass Transfer* **53** 1409 – 416
[22] Das S K Choi S Wenhua Y Pradeep T 2007 *Nanofluids Science and Technology* Wily Inc.
[23] Cimbala J M and Cengel Y A 2005 *Fluid Mechanics Fundamentals and Applications* 1st Edition McGraw-Hill

2nd International Conference on Mechanical Engineering Research (ICMER 2013) IOP Publishing
IOP Conf. Series: Materials Science and Engineering **50** (2013) 012036 doi:10.1088/1757-899X/50/1/012036

Modelling optimization involving different types of elements in finite element analysis

C M Wai[a], Ahmad Rivai and Omar Bapokutty
Faculty of Mechanical Engineering
Universiti Teknikal Malaysia Melaka (UTeM)
Hang Tuah Jaya 76100, Melaka, Malaysia

E-mail: [a]cmwai@live.com.my

Abstract. Finite elements are used to express the mechanical behaviour of a structure in finite element analysis. Therefore, the selection of the elements determines the quality of the analysis. The aim of this paper is to compare and contrast 1D element, 2D element, and 3D element used in finite element analysis. A simple case study was carried out on a standard W460x74 I-beam. The I-beam was modelled and analyzed statically with 1D elements, 2D elements and 3D elements. The results for the three separate finite element models were compared in terms of stresses, deformation and displacement of the I-beam. All three finite element models yield satisfactory results with acceptable errors. The advantages and limitations of these elements are discussed. 1D elements offer simplicity although lacking in their ability to model complicated geometry. 2D elements and 3D elements provide more detail yet sophisticated results which require more time and computer memory in the modelling process. It is also found that the choice of element in finite element analysis is influence by a few factors such as the geometry of the structure, desired analysis results, and the capability of the computer.

1. Introduction

Finite Element Analysis (FEA), which was first developed for the purpose of aerospace structural analysis, is now widely used in other industries, including automotive [1-3], structural engineering [4-7], composite design and manufacturing [8-10], as a convenient and speedy tool for approximation of the solution to a variety of complicated engineering problems. FEA is the practical application of the finite element method (FEM), which is basically an alternative method to solve engineering problems numerically. Instead of using ordinary differential equation and partial differential equation, FEM divides the body into small but finite pieces called elements. Equations are formulated for each element and the results are combined to obtain the final solution to the problem [11-12]. Subdivision of the body into suitable elements is therefore a crucial step in FEA.

There are many different types of elements used in FEA. These elements are developed independently and vary from one finite element (FE) software to another. In general, there are three groups of element which are one-dimensional (1D), two-dimensional (2D), and three-dimensional (3D) elements. Logically, all virtual structures should be modelled as close as possible to the actual structure. However, 3D solid elements are usually not the best choice for FEA. The analysts should learn when other element types can be used and valid in representing the engineering structure.

Since the discussion and also the analytical model to justify the effect of elements in FE modelling is lacking, this paper aims to compare and contrast 1D, 2D, and 3D elements by literature while the advantages and limitations of 1D, 2D and 3D elements are investigated by carrying out a FEA of a simple case study of an I-beam. This paper will provide fundamental information relative to some of the finite elements used in modelling by commercial FE software today. However, it cannot include all the different types of finite elements, which is immense and so specialized that they are beyond the limits of this study.

2. Types of Elements

Elements in FEA are generally grouped into 1D element, 2D element, and 3D element. They are recognised based on their shapes. For example, elements can take on the form of a straight line or curve, triangle or quadrilateral, tetrahedral and many more. The simplest element is a line made of two nodes. All line elements, whether straight or curved, are called 1D element in which they have translational and rotational displacement functions [13]. Examples of 1D element are truss element and beam element [14].

2D elements are typically surface elements with triangle or quadrilateral as their basic shapes [13]. Examples of 2D elements are 3-node triangular element, 6-node triangular element, and many more [14]. These surface elements can have regular or irregular shapes shown in figure 1. 2D elements are plane elements. Therefore, linear approximation of translational displacements considered are u(x,y) and v(x,y) while rotational displacements are θ(x,y) [12,15]. Since they account for plane stress and plain strain, they are often used to solve 2D elasticity problems [13-14].

3D elements are usually used to mesh volumes. They are derived from 2D elements are used when the problem is unable to be simplified [16]. 3D solid elements only accounts for translational displacements. The three translational unknown displacement functions are u(x,y,z), v(x,y,z), and w(x,y,z) [15-16]. Examples of 3D solid elements are 4-node tetrahedral element, 10-node tetrahedral element, 8-node isoparametric element, etc [14].

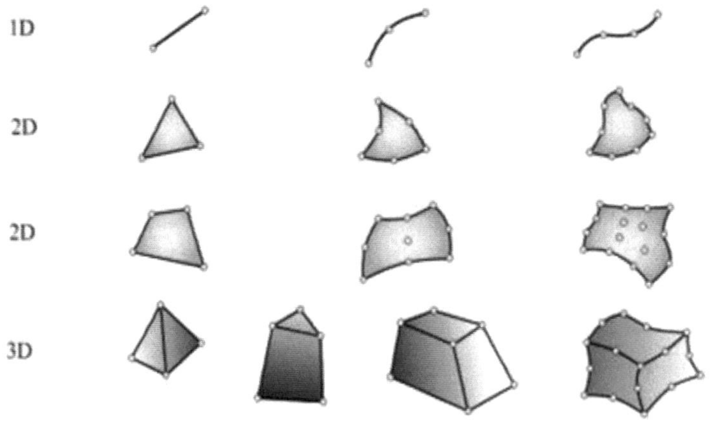

Figure 1. Typical finite element geometries [16].

3. Modelling

A wide flange beam, W460x74, is subjected to an eccentric load of 1000 kN as shown in the figure 2 below. Since the load is applied 200 mm from the neutral axis of the beam, a bending moment is induced at the end of the beam. The other end of the beam is assumed to be fixed with no translational or rotational displacements. The cross section of the beam, which was taken from Beer et. Al. [17], is shown in figure 2. The beam is assumed to be made of steel. Table 1 contains the necessary geometric dimensions and parameters of the beam. The beam is modelled and analyzed with 1D element, 2D element, and 3D element using MSC Patran and MD Nastran. In order to compare the memory and execution time of the solver, the finite elements used are of the same global edge length which is fixed at 100 mm. In this simple case study, the eccentric load applied on the I-beam induced a normal stress in the x-axis and a bending stress which causes the beam to bend upwards. In order to model the beam with 1D element, the load is decomposed into an equivalent force and moment at the centroid of the beam x-axis. The same loading condition is applied to models of 2D and 3D elements to eliminate the effect of loading conditions on the choice of element. Only static analysis which yields the results for stress, displacement, and constraint forces is taken into consideration.

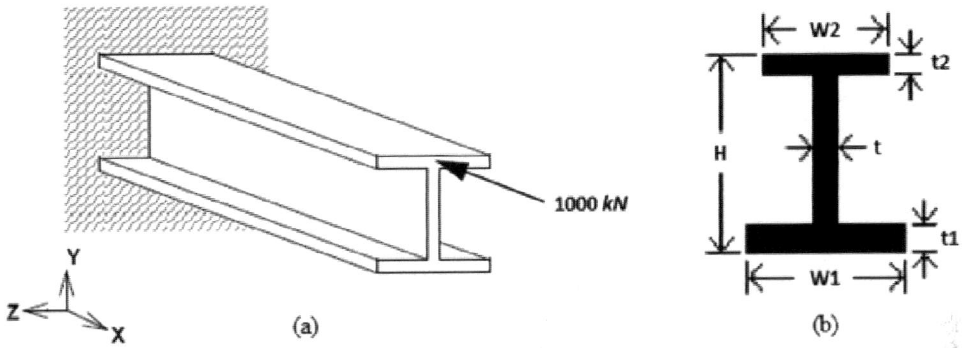

Figure 2. (a) Wide flange beam, W460x74 subjected to an eccentric load; (b) Cross section of wide flange beam, W460x74 [18].

Table 1. Wide flange beam W460x74 dimensions and its properties.

Dimension	
H (mm)	457.00
t (mm)	9.00
t1 (mm)	14.50
t2 (mm)	14.50
W1(mm)	190.00
W2 (mm)	190.00
Properties	
Young's Modulus (N/mm^2)	200000
Poisson ratio	0.3

4. Model Validation

The FE models of the wide flange beam W460x74 are validated by comparing the results of FEA with theoretical calculation. Table 2 shows the comparison between FEA results with theoretical calculation in terms of stresses and displacement. Theoretical values of stresses and displacement are calculated using Eq. (1-6) (Beer et. al., 2006) with dimensions and properties as mentioned in Table 1. The purpose for comparison is also to ensure that the FE model simulated is valid in terms of loading and boundary conditions. It can be seen in Table 2 that the results between the FE models are in good agreement with the theoretical values with acceptable errors. Therefore, all the FE models are considered valid and can be used for this study.

Normal stress:

$$\sigma_n = \frac{F}{A} \tag{1}$$

Bending stress:

$$\sigma_m = \frac{Mc}{I} \tag{2}$$

Maximum combined stress:

$$\sigma_T = \frac{F}{A} + \frac{Mc}{I} \tag{3}$$

Deflection from axial load:

$$\delta_n = \frac{FL}{AE} \tag{4}$$

Deflection from bending moment:

$$\delta_m = \frac{ML^2}{2EI} \tag{5}$$

Maximum deflection:

$$\delta_T = \sqrt{(\delta_n)^2 + (\delta_m)^2} \tag{6}$$

where, F is the axial load,
M is bending moment,
A is the I-beam cross sectional area,
I is the I-beam second moment of area,
σ_n is the normal stress,
σ_m is the bending stress,
σ_T is the maximum stress due to the applied load,
δ_n is the deflection from axial load,
δ_m is the deflection from bending moment,
δ_T is the maximum deflection due to the applied load.

Table 2. Comparison between FEA and theoretical calculation of stresses and displacement for wide flange beam W460x74.

Results	Theory	FE Model		
		1D	2D	3D
Maximum Tensile Stress (MPa)	32.251	32.300	33.950	26.650
Maximum Compressive Stress (MPa)	245.881	246.000	252.500	247.500
Maximum Beam Deflection (mm)	38.132	38.100	38.400	38.700

5. Results and Analysis

Table 3 shows the results available for 1D FEA, 2D FEA, and 3D FEA for wide flange beam W460x74. Stress results for 1D FEA are available in the form of bar stresses and are neatly organized into axial stresses, bending stresses, and combined axial and bending stresses. Axial stresses are stresses due to force acting normal to the surface while bending stresses are stresses due to forces which produce bending moment. Combined axial and bending stresses are further divided into minimum and maximum combined stresses. Stresses in 2D FEA and 3D FEA are displayed in the form of stress tensor. Stresses such as axial stresses, bending stresses and combined axial and bending stresses must be perceived and interpreted by the analyst through the stress tensor.

Table 3. Comparison of FEA results between 1D, 2D, and 3D FE model.

Analysis Results	FE Model		
	1D	2D	3D
Stress tensor	NA	Available	Available
Bar stresses, Axial	Available	NA	NA
Bar stresses, Bending	Available	NA	NA
Bar stresses, Max combined	Available	NA	NA
Bar stresses, Min combined	Available	NA	NA
Displacement	Available	Available	Available
Deformation	Available	Available	Available

Both displacement results and deformation for 1D FEA, 2D FEA, and 3D FEA are available as resultant or as separate components in the x-axis, y-axis, and z-axis directions. Figure 3 shows the deformation of the I-beam modelled using 1D element, 2D element, and 3D elements respectively for the simple case study. 3D element gives a better illustration of deformation of the I-beam than that of 2D element. 1D element gives the simplest but sufficient illustration of the I-beam deformation as the beam elements are modelled as a line at the neutral axis of the actual beam.

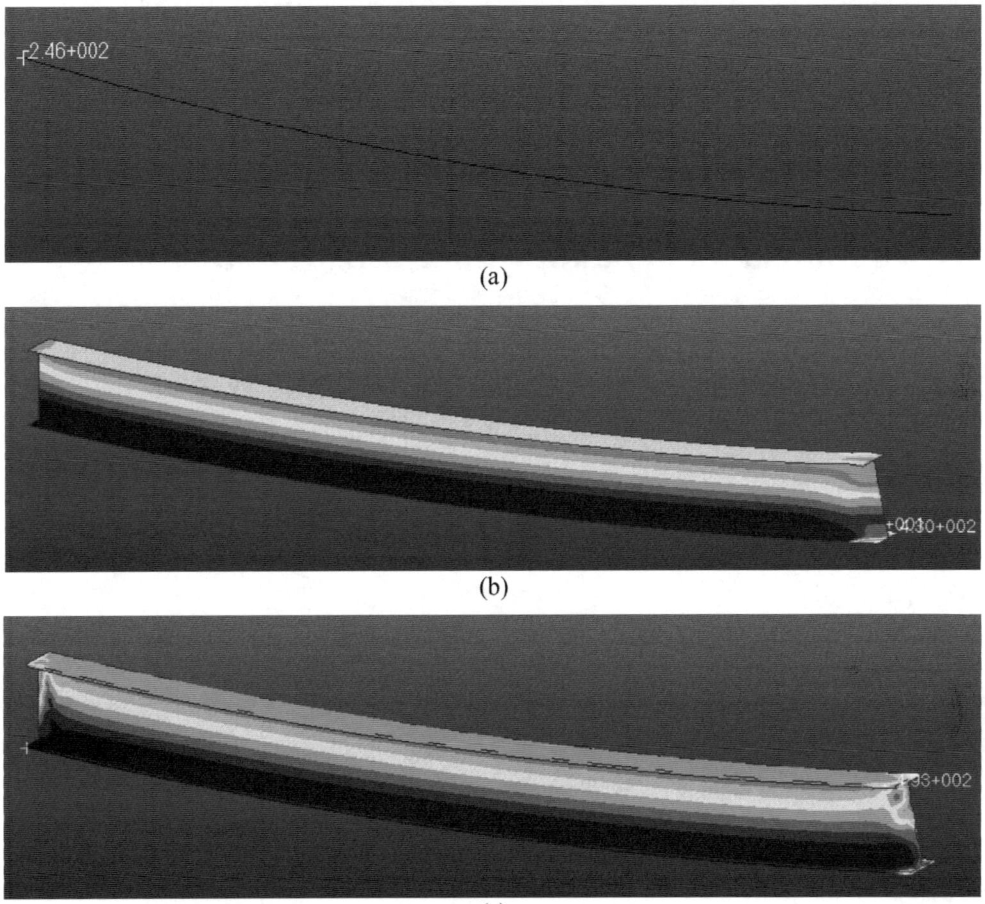

(a)

(b)

(c)

Figure 3. Deformation of wide flange beam W460X74 due to Applied Load: (a) 1D FE model; (b) 2D FE model; 3D FE model.

Table 4 displays the deflection of the beam in the y-axis due to the applied load. The deflection is taken at ten different but constant increment of the length of the beam. Theoretical values are obtained using Eq. (7) [17].

$$\delta_y = \frac{ML}{EI} \tag{7}$$

where, δ_y is the deflection at the end of the beam,
M is bending moment,
L is the length of the beam,
E is the Young's Modulus of the beam,
I is the I-beam second moment of area.

Table 4. Deflection of wide flange beam W460x74 at different beam length.

Length of Beam, x (mm)	Deflection of Beam, y (mm)			
	Theory	1D	2D	3D
500	0.4648	0.4650	0.3520	0.3650
1000	1.6125	1.6100	1.4900	1.5100
1500	3.5159	3.5200	3.3800	3.4100
2000	6.1790	6.1800	6.0400	6.0700
2500	9.6027	9.6000	9.4600	9.4900
3000	13.7869	13.8000	13.6000	13.7000
3500	18.7319	18.7000	18.6000	18.6000
4000	24.4376	24.4000	24.3000	24.3000
4500	30.9040	30.9000	30.7000	30.8000
5000	38.1311	38.1000	38.1000	38.0000

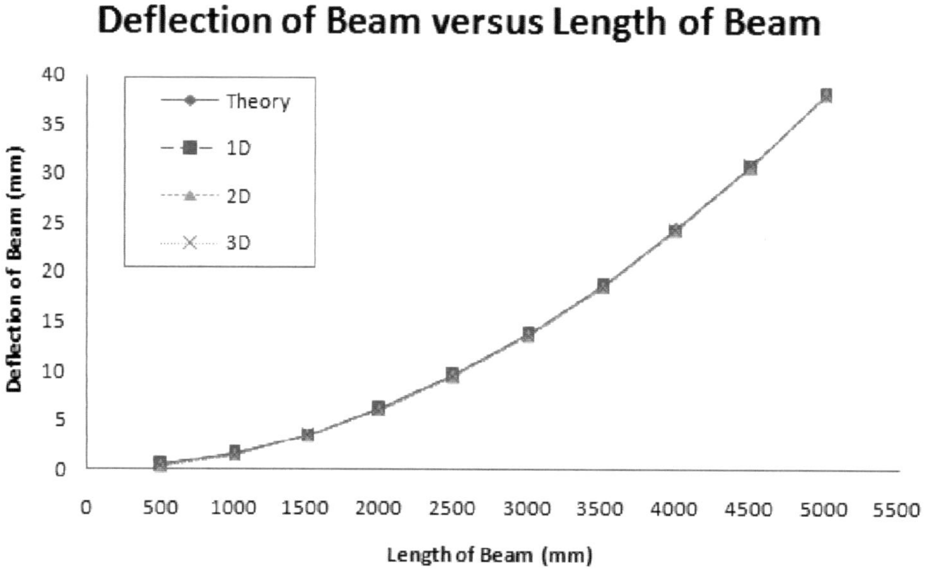

Figure 4. Wide flange beam W460x74 deflection at different length of beam.

6. Discussion

For static analysis in FEA, all types of element will yield the stress and displacement results, which are the two most important parameters in evaluating the structure integrity. However, 1D element provides an easier method of interpretation of FEA results as the results are neatly organized and grouped into individual stresses and displacements as shown in Table 4. 2D and 3D elements, although they provide greater detail, they require some effort in the interpretation of results. The results obtained can be confusing, especially for new users, but if carefully interpreted can be reliable and accurate.

Graphically, 3D element produces the clearest deformation which may be experienced by the beam due to the applied load because the model is simulated closest to the actual I-beam. The importance of the detail results is the existence and location of stress concentration which may cause fatal failure of the structure. This detail stress distribution is only available in 2D FE model and 3D FE model.

In terms of accuracy, 1D element gives the exact stresses and displacements values as those calculated by theory because the assumptions used in 1D element are those of the same as theory. 2D and 3D elements produce stresses and displacement values which deviate slightly from the theoretical values but with acceptable errors. 2D and 3D elements take into account the plane stresses due to shear force between the elements. Due to the consideration of plane stresses in 2D and 3D elements, the results obtained are more detail in terms of stress distribution and displacement.

Since the errors are small and negligible, the choice of elements for FEA depends largely on the geometry of the structure. This is because not all structure can be modelled using 1D element or 2D element. 1D element is used for long and slender structure such as javelins, poles, etc. 2D element is used for plate or shell like structure while 3D element is used for structure with complex geometry which cannot be simplified for analysis. A technique known as dimensional reduction is usually used to reduce the complexity of the structure from 3D or 2D to 2D and 1D respectively. Execution time and computational memory can be saved with this technique.

The type of results desired also plays a role in choosing the right element for FEA. If a detail analysis on the stress distribution to determine the areas of stress concentration is required, then 2D element or 3D element are of better choices. 1D element can still be used for rough estimation and prediction of failure of the structure.

Lastly, the choice of element for FEA also depends on the time and memory capacity of the computer available. If a simple and quick analysis on the structural integrity is required, 1D element is of better choice. If a large computer memory is available, a detail analysis is always the best choice for FEA in which modelling with 2D element or 3D elements is required.

7. Conclusion

The similarity when modelling with 1D element, 2D element, and 3D element is that they are able to generate the same result parameters. For instance, linear static analysis in FEA gives result of stress and displacement though the values might differ from one another. The difference however, does not mean that the results are inaccurate. 1D element will provide a much organize result but less detail which require less effort in the interpretation of results. 2D and 3D FEA results are displayed in the form of tensor, which provide more details for each section of the structure but also require more effort to interpret the results. Besides method of modelling and application of load and boundary conditions, 1D element, 2D element, and 3D element also differ in terms of deformation of the structure. 1D element gives a rough idea of how the structure will deform when load is applied. While 2D element gives a clearer deformation, 3D element gives the best insight to the deformation of structure. With the advantages and limitations in mind, three factors are to be considered when deciding on the types of elements to use in FEA, which are the geometry of the structure, desired results, and time frame as well as the capability of the computer. The most important factor is the geometry of the structure. If the simple, symmetry and uniform such as beam, plate, rod, etc. 1D element is desirable. 3D element should be used only when the structure has complex geometry which cannot be simplified. Next factor to be considered is the results required. 2D element and 3D element are used for detail results such as determination of stress distribution, while 1D element is used for rough estimate. Last but not least is the execution time and memory required for the solver to run the analysis. 1D FEA require the least memory and the fastest to complete while 3D FEA require the most memory and slowest to complete.

8. Recommendation

1D element, 2D element, and 3D element are discussed in general in this study. It would be beneficial to compare and contrast the different types of 1D element, 2D element, and 3D element. The results of the present work are limited to isotropic materials. It is desirable to investigate the effect of material properties on the choice of elements in FEA. Different materials including orthotropic and anisotropic materials such as composites can be used in future study to compare and contrast the result of this study. Three important loadings which may occur in a structure are axial, bending, and torsional loads. The current study only involves axial and bending loads. A case study on the effect of torsional load on the choice of elements would be beneficial to the understanding of the choice of elements used in FEA.

9. References

[1] Wu S R and James C 1997 Advanced development of explicit FEA in automotive applications Computer Methods in Applied Mechanics and Engineering. **149** 189-199

[2] Kamal M, Rahman M M and Rahman A G A 2012 *J Mech Eng Sci* **3** 291-300.

[3] Kamal M, Rahman M M and Rahman A G A 2013 *Inter J Automot Mech Eng* **7** 912-923.

[4] Mahendran M 2007 *Applications of Finite Element Analysis in Structural Engineering* Proc. Int. Conf. on Computer Aided Engineering. 38-46

[5] Zulkifli A, Ariffin A K and Rahman M M 2011 *Inter J Automot Mech Eng* **3** 256-264.

[6] Daud R, Ariffin A K and Abdullah S 2013 *Inter J Automot Mech Eng* **7** 995-1006.

[7] Huzni S, Ilfan M, Sulaiman T, Fonna S, Ridha M and Arifin A K 2013 *Inter J Automot Mech Eng* 7: 1023-1030.

[8] Newberry A L 2000 FEA modeling proves its worth in composite design in RP Asia 2000 Conf. held at Kuala Lumpur Malaysia 20-22 September 2000

[9] Kadirgama K, Rahman M M, Ismail A R and Bakar R A 2011 *J Mech Eng Sci* **1** 37-66.

[10] Najiha M S, Rahman M M, Yusoff A R and Kadirgama K 2012 *Inter J Automot Mech Eng* **6** 766-774.

[11] Department of Aerospace Engineering Sciences University of Colorado 2012 Introduction to finite element methods (ASEN 5007) Ch. 6 FEM Modelling: Introduction. [Accessed 21 December 2012] Available from http://www.colorado.edu/engineering/cas/courses.d/IFEM.d/IFEM.Ch06.d/IFEM.Ch06.pdf

[12] Logan D L 2007 *A First Course in The Finite Element Method 4th ed* (Canada: Thomsom Canada Limited)

[13] Berlioz A and Trompette P 2010 *Solid Mechanics Using The Finite Element Method* (USA: John Wiley & Sons Inc.

[14] Rugarli P 2010 *Structural Analysis with Finite Elements* (UK: Thomas Telford Limited)

[15] Papadopoulos P 2010 ME280A: Introduction to finite element method [Accessed 4 December 2012] Available from www.kochmann.caltech.edu/ae108a/FEnotes.pdf

[16] Boeraeve P 2010 Introduction to the finite element method (FEM) [Accessed 4 December 2012] Available from www.gramme.be/unite9/FEM/Finite%20Element%20Method.pdf.

[17] Beer F P, Johnston Jr E R and DeWolf J T 2006 *Mechanics of Materials 4th ed* (Singapore: McGraw-Hill Education)

[18] MSC Software Corporation MSC Nastran 105 Exercise Workbook: Workshop 13a load analysis of a beam (using a point force and moment).

Acknowledgments

The authors would like to acknowledge CRIM, and Faculty of Mechanical Engineering, Universiti Teknikal Malaysia Melaka (UTeM) for supporting the present work through PJP/2012/FKM(18C)S1112 Research Fund.

2nd International Conference on Mechanical Engineering Research (ICMER 2013) IOP Publishing
IOP Conf. Series: Materials Science and Engineering **50** (2013) 012037 doi:10.1088/1757-899X/50/1/012037

The use of MRT-lattice Boltzmann method for the prediction of fluid solid flow

Nor Azwadi Che Sidik[a], Aman Ali Khan, Emad Kermani, Akmal Hamizi Manshor and Leila Jahanshaloo
Faculty of Mechanical Engineering, Universiti Teknologi Malaysia
81310 UTM Skudai, Johor Malaysia

E-mail: [a]azwadi@fkm.utm.my

Abstract. Computation of fluid solid flow requires an accurate scheme to capture the moving boundary problems for solid objects moving in a fluid. In this paper, a multi-relaxation time lattice Boltzmann method is coupled with the Newton's second law to predict the flow characteristics around obstacle placed in a channel. The treatment of the curved boundary is based on the conventional bounce-back boundary scheme and interpolations. Good agreement with the previous studies by other methods indicates good applicability of the present scheme.

1. Introduction

Lattice Boltzmann method (LBM) is a powerful numerical technique for simulating fluid flows and modelling the physics in fluids [1-3]. This method is as an extension of the lattice gas automata (LGA) or as a special discretisation from the Boltzmann equation of the kinetic theory [4-8]. In lattice Boltzmann simulations, collision and propagation routines are performed to model the flow density and momentum which can be determined via distribution functions in phase space. The resulting information then enables the fundamentals quantities such as the drag or lift forces to be computed. Overall, the LBM with single-relaxation Bhatnagar-Gross-Krook (BGK) model provides a convenient and straightforward approach for solving continuum flow problems describes by the Navier-Stokes equation at low Reynolds numbers. Recently, the LBM has been improved to its stability and has been successfully extended for the application of flows at high Reynolds [2,8].

A major advantage of lattice Boltzmann method is its ease and accurate treatment at complicated boundary geometries. The conventional bounce-back (BB) scheme provides a particularly straightforward approach for modeling no-slips conditions on solid surfaces in macroscopic flows. In this scheme, the outgoing directions of the distribution function at the boundary sites are simply specified as the reverse of the incoming directions. In practice, there are two basic types of BB scheme, namely the "on-site" scheme [9] and the "mid-plane" scheme [10]. It is well-known that the on-site BB scheme is simpler, but has only first-order accuracy, whereas the mid-plane BB scheme provides second-order accuracy in both space and time [11].

Curved boundary treatments conventionally used in LB simulation when the uniform lattice is mapped in space and has been suggested as a means of improving the accuracy of the stair-shaped approximation. Several strategies have been proposed for dealing with complex geometry, curved boundaries in LBM. The first approach is to use a body-fitted (arbitrary) mesh and to execute the distribution functions throughout the entire computational domain [12-13]. The second strategy also applies an interpolation-based approach, but under a uniform Cartesian mesh, to track the position of boundary. Then, the on-site or off-site BB scheme is executed at the boundary surface depending on the location of the boundary relative to the lattice nodes [14-15]. The latest one is based upon the utilization of the immersed boundary treatments [15-17]. Due to their characteristics of a superior numerical accuracy, an intuitive approach and an inherent reliability, the interpolation type models are the most commonly employed technique for resolving curved boundary problems in LB simulations. Therefore, in the present study, MRT-LBM will be coupled with the interpolation type of boundary model to predict the flow around obstacles.

2. Mathematical Modelling of MRT-LBM

In MRT-LBM model, the physical space is discretized into uniform lattice nodes. Every node in the network is connected with its neighbors through a number of lattice velocities that are to be determined through the selected model.

Recently, Lallemand and Luo [17] suggested that the use of a MRT model could improve the numerical stability. The collision step in velocity space is difficult to perform; it is more convenient to perform the collision process in the momentum space [2].

The multi-relaxation-time lattice Boltzmann equation reads

$$f_i(\mathbf{x}+\mathbf{c}_i\Delta t, t+\Delta t) - f_i(\mathbf{x},t) = -M^{-1}S\left[m(\mathbf{x},t) - m^{eq}(\mathbf{x},t)\right] \tag{1}$$

where is the distribution function for particles with velocities c at time t. Velocity is equal to rate of change of position dt per unit time dt. $m(x,t)$ and $m^{eq}(x,t)$ and are vectors of moments, $m = (m_0, m_1, m_2, ..., m_n)^T$. The relaxation matrix is a diagonal matrix.

The mapping between velocity and moment spaces can be performed by linear transformation as follow

$$m = Mf \tag{2}$$

$$f = M^{-1}m \tag{3}$$

The matrix M for two-dimension nine-velocity model (D2Q9) is:

$$M = \begin{bmatrix} 1 & 1 & 1 & 1 & 1 & 1 & 1 & 1 & 1 \\ -4 & -1 & -1 & -1 & -1 & 2 & 2 & 2 & 2 \\ 4 & -2 & -2 & -2 & -2 & 1 & 1 & 1 & 1 \\ 0 & 1 & 0 & -1 & 0 & 1 & -1 & -1 & 1 \\ 0 & -2 & 0 & 2 & 0 & 1 & -1 & -1 & 1 \\ 0 & 0 & 1 & 0 & -1 & 1 & 1 & -1 & -1 \\ 0 & 0 & -2 & 0 & 2 & 1 & 1 & -1 & -1 \\ 0 & 1 & -1 & 1 & -1 & 0 & 0 & 0 & 0 \\ 0 & 0 & 0 & 0 & 0 & 1 & -1 & 1 & -1 \end{bmatrix}$$

The moment vector m is:

$$m = \left(\rho, e, \in, j_x, q_x, j_y, q_y, P_{xx}, P_{xy}\right)^T \tag{4}$$

where, ρ is the fluid density, ∈ is related to the square of the energy e, j_x and j_y are the mass flux in two directions, and P_{xx} and P_{xy} correspond to the diagonal and off-diagonal component of the viscous stress tensor.

The equilibrium of the moment m^{eq} is:

$$m_0^{eq} = \rho \tag{5a}$$

$$m_1^{eq} = -2\rho + 3\left(j_x^2 + j_y^2\right) \tag{5b}$$

$$m_2^{eq} = \rho - 3\left(j_x^2 + j_y^2\right) \tag{5c}$$

$$m_3^{eq} = j_x \tag{5d}$$

$$m_4^{eq} = - j_x \tag{5e}$$

$$m_5^{eq} = j_y \tag{5f}$$

$$m_6^{eq} = - j_y \tag{5g}$$

$$m_7^{eq} = \left(j_x^2 - j_y^2\right) \tag{5h}$$

$$m_8^{eq} = j_x j_y \tag{5i}$$

where

$$j_x = \rho u_x \text{ and } j_y = \rho u_y \tag{6}$$

The diagonal matrix S is

$$S = \begin{bmatrix} s_0 & 0 & 0 & 0 & 0 & 0 & 0 & 0 & 0 \\ 0 & s_1 & 0 & 0 & 0 & 0 & 0 & 0 & 0 \\ 0 & 0 & s_2 & 0 & 0 & 0 & 0 & 0 & 0 \\ 0 & 0 & 0 & s_3 & 0 & 0 & 0 & 0 & 0 \\ 0 & 0 & 0 & 0 & s_4 & 0 & 0 & 0 & 0 \\ 0 & 0 & 0 & 0 & 0 & s_5 & 0 & 0 & 0 \\ 0 & 0 & 0 & 0 & 0 & 0 & s_6 & 0 & 0 \\ 0 & 0 & 0 & 0 & 0 & 0 & 0 & s_7 & 0 \\ 0 & 0 & 0 & 0 & 0 & 0 & 0 & 0 & s_8 \end{bmatrix}$$

In compact notation S can be written as;

$$S = diag(1.0, 1.4, 1.4, s_3, 1.2, s_5, 1.2, s_7, s_8) \tag{7}$$

where $s_7 = s_8 = 2/(1+6\upsilon)$, s_3 and s_5 are arbitrary, can be set to 1.0.

What should we mention here, it is possible to recover the SRT-LBM solution from MRT-LBM by setting

$$s_1 = s_2 = s_4 = s_6 = s_7 = s_8 = 1/\tau \tag{8}$$

Furthermore, we use D2Q9 model (figure 1) and the nine discrete velocities are given by

$$\mathbf{c} = \begin{bmatrix} 0 & 1 & 0 & -1 & 0 & 1 & -1 & -1 & 1 \\ 0 & 0 & 1 & 0 & -1 & 1 & 1 & -1 & -1 \end{bmatrix} \tag{9}$$

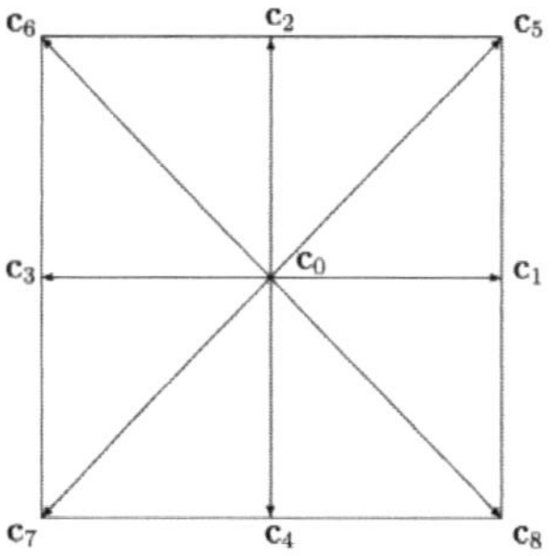

Figure 1. 9-velocity LBE model on the 2D square lattice.

The macroscopic density and velocities can be computed simply by moment integration as

$$\rho = \sum_i f_i \tag{10}$$

$$\mathbf{u} = \frac{1}{\rho}\sum_i \mathbf{c}_i f_i \tag{11}$$

3. Treatment on Fluid Solid Boundary

In this present work, we applied the interpolated bounce-back boundary-condition to model no slip boundary-condition with curved boundaries [17]. The method was initially proposed by Ladd [8] which is the modified version of classical BB method by adding an adjustment term to the bounce back. However, this method has some limitation which suffers low resolution for general surface geometry. In order to overcome limitation, Lallemand and Luo [17] proposed a curve boundary treatment as follow;

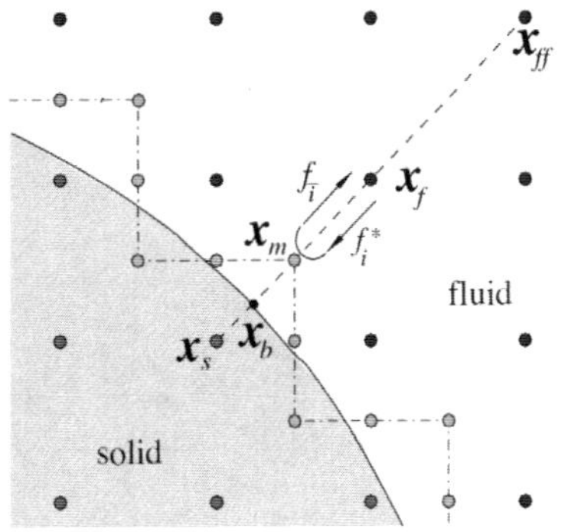

Figure 2. Illustration of Curve Boundary.

Lallemand's model applied the combination of simple bounce-back and quadratic interpolation. Generally, this technique of quadratic interpolation model will give more accurate and more stable results. This model is an extension of Bouzidi et al. [18]. Figure 2 illustrates boundary lattice link, with two end node X_S in a solid domain and X_f in a fluid domain. X_b is the intersection node between solid surface and fluid surface, with midpoint of lattice link is denoted by X_m.

The location of the solid boundary can be determined from the value of the fraction of boundary lattice link in the solid domain, Δ based on Lallemand's model where to

$$\Delta = \left| \frac{X_S - X_b}{X_S - X_f} \right| \quad (13)$$

For $\Delta \le 1/2$, where the midpoint X_m locates between X_b and X_f, the midpoint velocity V_m can be obtained as bellow

$$V_m = \frac{0.5V_b + (0.5 - \Delta)V_f}{1 - \Delta} \quad (14)$$

while for $\Delta \ge 1/2$,

$$V_m = \frac{3/2V_b + (\Delta - 0.5)V_{ff}}{2 - \Delta} \quad (15)$$

The hydrodynamic force exerted on the solid particle at any boundary nodes can be obtained by using equation (16)

$$\vec{F}(x,t) = e_{\alpha'}\left[f_{\alpha'}(x_s,t) + f_{\alpha}(x_f,t+\Delta t)\right] \quad (16)$$

4. Results Validation and Discussion

The physical domain of the problem is shown in figure 3. The proposed model applied the MRT-LBM model to simulate the fluid flow past a circular static solid in a channel. Figure 3 shows the setup in the simulation with length, height and inlet velocity.

The value of Reynolds number can be obtained by using the following equation:

$$Re = \frac{U \max H}{\upsilon} \tag{17}$$

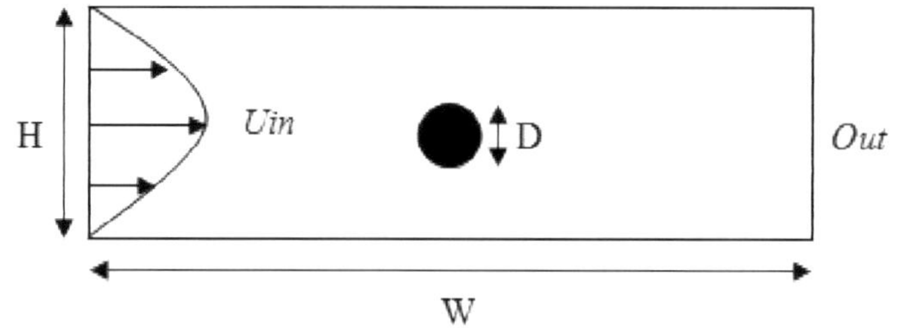

Figure 3. The problem geometry.

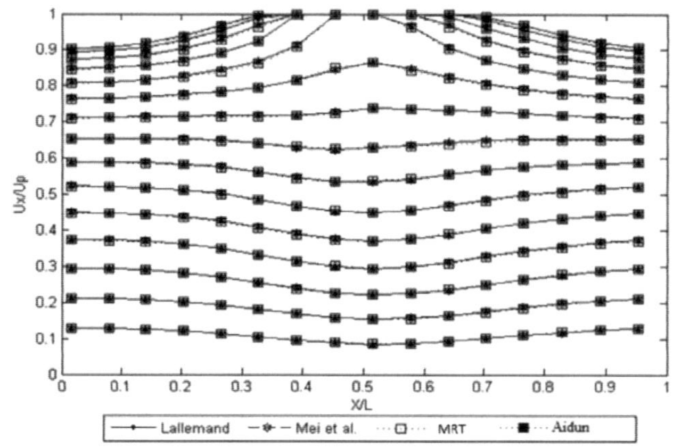

Figure 4. x-component velocity profile along horizontal channel.

For the purpose of code validation, the computed results in terms of velocity distribution along the channel length were compared with Aidun et al. [19-20]. Excellent agreement was obtained showing the applicability of the numerical scheme.

In our next computation, the flow characteristics around a static cylinder were studied at Reynolds number of 90 by using MRT-LBM model couple with curve-boundary treatment. Figure 5a - 5d shows the plot of velocity contour at different computational times.

Figure 5 clearly demonstrates the flow development in terms of velocity contour until steady state condition. The flow velocity becomes higher near both upper and lower side of the blockage to satisfy the requirement of conservation of mass. Negative value of velocity downstream of the cylinder indicating the presence of vortices resulted from inverse pressure on the flow field.

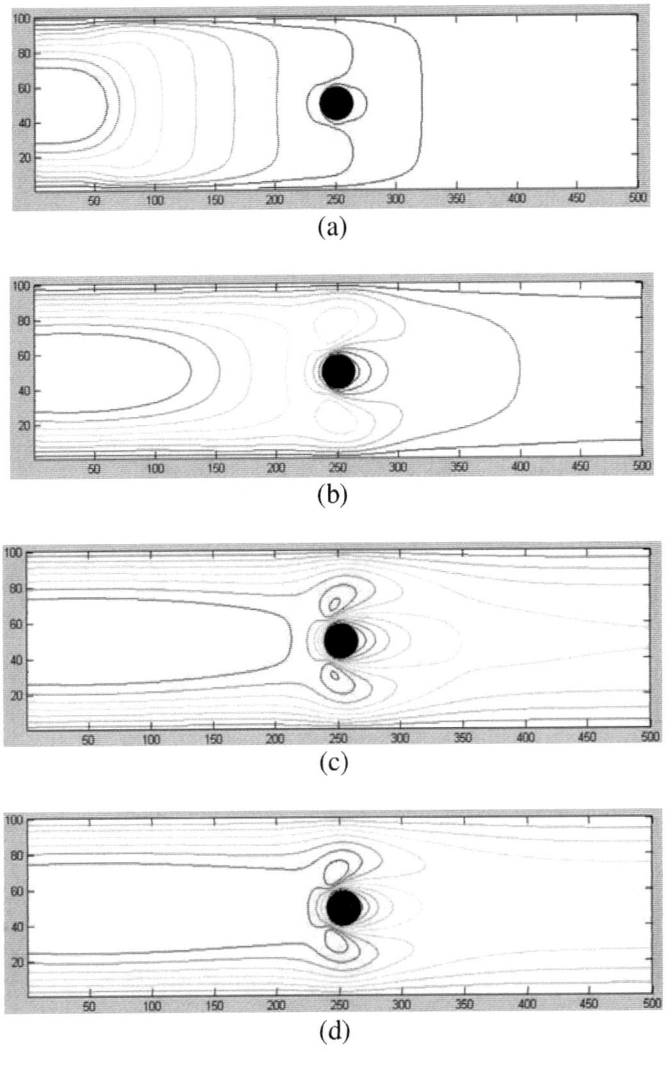

Figure 5. Snapshots of velocity contour for flow around obstacle.

5. Conclusion

Numerical computations of flow around cylindrical solid obstacle were out using MRT-LBM method coupled with curve-boundary model. It is observed that the proposed method was capable in predicting all the flow features for the presence case. Future efforts need to extend the current formulation for investigation at various types of solid fluid flow related to real engineering problems.

Acknowledgements

The authors would like to thanks the Universiti Teknologi Malaysia and Ministry of Higher Education, Malaysia for sponsoring this research. This research is supported by research grant Vot. 4F114 and 4L074.

References

[1] Chen S and Doolen G 1998 *Annu. Rev. Fluid Mech.* **30** 329
[2] Nor Azwadi C S, Aman A K, Manshor A H, Safdari A and Jahanshaloo L 2012 *Int. Review of Mechanical Engineering* **6** 1661
[3] Rosdzimin A R M, Zuhairi S M and Azwadi C S N 2010 *Int. J. Automot. Mech. Eng.* **2** 130
[4] Najiha M S, Rahman M M, Kamal M, Yusoff A R and Kadirgama K 2012 *J Mech. Eng. Sci* **3** 340
[5] Tahseen T A, Ishak M and Rahman M M 2012 *Int. J. Automot. Mech. Eng.* **6** 753
[6] He X and Luo L S 1997 *Phys. Rev. E* **56** 6811
[7] Frisch U, Hasslacher B and Pomeau Y 1986 *Phys. Rep.* **56** 1505
[8] Yu H, Girimaji S S and Luo L S 2005 *J. Comput. Phys.* **209** 599
[9] Behrend O 1995 *Phys. Rev. E* **52** 1164
[10] Ladd A J C 1994 *J. Fluid. Mech.* **271** 285
[11] Lallemand P and Luo L S 2003 *J. Comput. Phys.* **184** 406
[12] Filippova O and Ha¨nel D 1998 *J. Comput. Phys.* **147** 219
[13] Mei R, Luo L S and Shyy W 1999 *J. Comput. Phys.* **155** 307
[14] Mei R, Shyy W and Luo L S 2000 *J. Comput. Phys.* **161** 680
[15] Feng Z G and Michaelides E E 2004 *J. Comput. Phys.* **195** 602
[16] Peng Y, Shu C, Chew Y T, Niu X D and Lu X Y 2006 *J. Comput. Phys.* **218** 460
[17] Lallemand P and Luo L S 2000 *Phys. Rev. E* **61** 6546
[18] Bouzidi M, Firdaouss and Lallemand P 2001 *Phys. Fluids* **13** 3452
[19] Aidun C K, Lu Y and Ding E J 1998 *J. Fluid Mech.* **373** 287
[20] Zin M R M and Sidik N A C 2010 *Int. Review of Mechanical Engineering* **4** 719

Modelling and validation of magnetorheological brake responses using parametric approach

Zainordin A Z[1,a], **Abdullah M A**[2] **and Hudha K**[3]
[1]LectPoliteknik Sultan Haji Ahmad Shah, Kuantan
[2]Faculty of Mechanical Engineering, Universiti Teknikal Malaysia Melaka, Malaysia
[3]Faculty of Mechanical Engineering, Universiti Pertahanan Nasional Malaysia, Malaysia

E-mail: [a]zaifazlin@polisas.edu.my

Abstract. Magnetorheological brake (MR Brake) is one x-by-wire systems which performs better than conventional brake systems. MR brake consists of a rotating disc that is immersed with Magnetorheological Fluid (MR Fluid) in an enclosure of an electromagnetic coil. The applied magnetic field will increase the yield strength of the MR fluid where this fluid was used to decrease the speed of the rotating shaft. The purpose of this paper is to develop a mathematical model to represent MR brake with a test rig. The MR brake model is developed based on actual torque characteristic which is coupled with motion of a test rig. Next, the experimental are performed using MR brake test rig and obtained three output responses known as angular velocity response, torque response and load displacement response. Furthermore, the MR brake was subjected to various current. Finally, the simulation results of MR brake model are then verified with experimental results.

1. Introduction

X-By-Wire is one of the system that had potential to improve performance by minimized the number of part used in the system. Usually, by-wire systems have been employed in several segments such as steering system, suspension system, braking system and medical equipments [1-8]. MR brake has introduced as an actuator of brake-by-wire system in automotive industries. MR brake employs with MR fluid where this fluid solidifies once applied with magnetic field. This fluid is also known as smart fluid where it solidifies by increasing the strength of magnetic. Once the fluid is free from magnetic field, it represents as a Newtonian fluid behaviour. The MR brake consists of a rotating disc immersed with Magnetorheological Fluid (MR fluid) in an enclosure of an electromagnetic coil. MR fluid is developed using micron sized suspensions measured between 20-50 microns [9-10]. The rheological behaviour of MR fluid is similar with the carrier fluid when there is no external field is occurred in the fluid [11-12].

The application of a MR fluid in braking system is related with current research work. Li & Du (2003) presented the design and experimental evaluation of a Magnetorheological brake and introduced an amplifying factor to evaluate brake performance [13]. Meanwhile, Park *et al.* (2008) presented a design optimization procedure using simulated annealing combined with finite element simulations involving magnetostatic, fluid flow and heat transfer analysis [14]. MR fluid selection for MR brake application, such as magnetic circuit design and torque requirements for automotive application was also studied. Karakoc *et al.* (2008) focussed on the investigation of practical MR brake design criteria such as material selection, sealing, working surface area, viscous torque generation and MR fluid selection for basic automotive braking system [15]. Additionally, Tan *et al.* (2007) studies braking response of inertia/load by using an electro-rheological (ER) brake for ER-robotic application in term of ER braking velocity response in order to halt the robot arm rapidly [16]. In 2009, Nam and Ahn (2009) is proposed the new structure of MR brake with the waveform boundary of rotary disk that generated more resistance torque compare to the conventional MR brake [17]. Furthermore, the MR brake system had been implemented to other application such as joystick and prosthetic knee [18, 4].

The outline of this paper is as follows. In section I, brief introduction on the current research of MR brake. Section II represents the experimental facilities and continued with mathematical model of MR brake with test rig in Section III. Section IV describe the model validation between simulation and experimental at constant loads. Finally, the conclusion of this work is presented in Section VII.

2. Experimental Apparatus

Figure 1(a) shows the testing equipment used in the experiment with the load of 50 N and 100 N that shown in Figure 1(b). The inertia load will be attached to the load shaft and coupled with brake shaft to generate a constant or falling load that resultant the net torque produce by MR brake. The function of AC electric motor is to drive the MR brake shaft to desired velocity where it is coupled to the input shaft/rotor of MR brake via pulley an A-type V-belt.

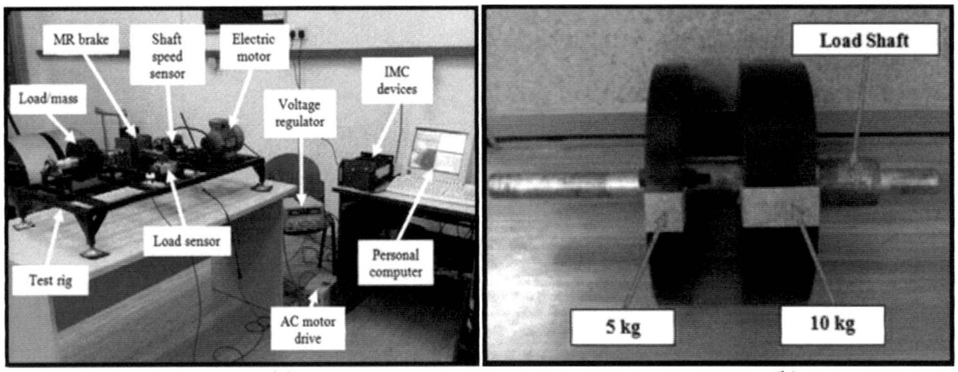

(a) (b)

Figure 1. a) Mechanical assembly of the MR brake test rig, and b) Load

The speed from the motor is transmitted to the MR brake shaft using belt tensional and well fitted beside the electric motor. The pulley shaft is connected to the MR brake shaft using jaw coupling and same concept also was applied at the load shaft. The pulley shaft and load shaft used a pillow block bearing to support the rotating shaft where the inner bearing will allow the shaft to rotate in free direction. The MR brake housing is coupled to a load cell via an arm of length of 238 mm. In this equipment, the load cell is employed to measure the braking torque. The load cell was calibrated at 1V: 53.4 Nm and the maximum torque measured using this sensor is 534 Nm. Next, the load cell is connected with bridge amplifier as the signal conditioning. Meanwhile, the rotational speed of the MR brake shaft was measured by using an ABS speed sensor.

The MR brake test rig is equipped with an I/O device for data processing. Next, the Integrated Measurement and Control (IMC) device provides signal processing of the sensory system. The IMC device that is only capable to received analogue voltage signal. Then, the signals are digitally processed and stored in a personal computer using FAMOS control software. IMC device is connected to a personal computer using NetBEUI protocol. A DC power supply manufactured by GW-INSTEK is used to supply electric currents to the MR brake electromagnetic coil. All the measured data are displayed in Personal Computer (PC) for the further analysis.

Mathematical Model

The characteristics of MR fluid can be described by using a simple Bingham plastic model (Philips 1969). The constitutive equation for a Bingham plastic fluid where the total shear stress (τ) is written as below [19]:

$$\tau = \tau_H + \mu_p \dot{\gamma} \tag{1}$$

where, τ_H is the yield stress due to the applied magnetic field, H, μ_p is the constant plastic viscosity which is considered equal to the non-field viscosity of the fluid, and $\dot{\gamma}$ is the shear strain rate. Based on the equation (1), the braking torque generated by the friction of the interface between static and moving parts in the MR fluid inside the MR brake can be written as equations [12][15]:

$$T_b = 2\pi N \int_{r_i}^{r_o} (kH^{\beta} + \mu_p \frac{rw_s}{h}) r^2 dr \tag{2}$$

Where r is the radius of the disk, w_s is the angular velocity of the rotating disk, h is the thickness of the MR fluid gap between rotor and enclosure, H is the magnetic field intensity corresponding with k and β. The values of k and β are constant by considering the relationship between the magnetic field intensity and the yield stress of the MR fluid.

An integration of equation (2) will determine the two types of components of braking torque which are torque generated due to applied magnetic fields (T_H) and torque due to friction and viscosity of the fluids (T_μ). Both torque elements are expressed as follows [12][15].

$$T_\mu = \frac{\pi}{2h} N \mu_p (r_o^4 - r_i^4) w_s \tag{3}$$

$$T_H = \frac{2\pi}{3} Nk\alpha (r_o^3 - r_i^3) i \tag{4}$$

Therefore, the total braking torque produced by MR brake can be written as follow,

$$T_b = T_\mu + T_H \tag{5}$$

An effective MR brake torque generated when applied current to the magnetic coils. This will decelerate the dynamics of all inertia (J_{all}) that coupled rigidly to the MR brake shaft. The (T_{input}) term is the torque motor (T_m) is combined with loading torque (T_L) that is coupled rigidly on the MR brake shaft. The loading torque is generated based on the weight of the load (N) and the effective radius (r_L) of the load. Then, the mathematical expression in terms of (T_{input}) can be written in equation (6) as:

$$T_{input} = T_m + T_L = (P_m / w_m) + mgr_L \tag{6}$$

Where P_m = power of electric motor and w_m = angular speed of electric motor pulley. Then, the equation motion of MR brake with a test rig is derived and stated in equation (7) as follow:

$$T_{input} - T_b - T_c = J_{all}\alpha_s \quad (7)$$

Where T_c = viscous damping of bearing and α_s = angular acceleration of MR brake shaft. Then, the total inertia of MR brake (J_{all}) consists of rotor, shaft, four bearing inner parts, a sprocket, a pulley and the load can be written in equation (8).

$$J_{all} = J_{disk} + J_{sprocket} + J_{pulley} + 4J_{bearing} + J_{Load} + J_{shaft} \quad (8)$$

The total inertia is approximately about 0.25 Nm^2 for 50 N, 0.45 Nm^2 for 100 N and 0.65 Nm^2 for 150 N respectively. By substituting equation (6) into equation (7), the equation (9) can be written as follow:

$$((P_m / w_m) + mgr_L) - (T_\mu + T_H) - 4cw = J_{all}\alpha_s \quad (9)$$

3. Model Validation

The MR braking response was tested by using a constant load of 150 N in order to energized MR brake when applied current to the brake coils. The response of the MR brake was investigated at four constant applied currents that are 1 A, 2 A, 3 A, and 4 A. These three responses will be obtained from the experiments which are shaft angular velocity response, torque response and load displacement response. The measured of three responses in time domain are shown in Figure 2. Based on the Figure 2(a), shaft angular velocity decreased very fast when applied current is increase. This is due to the maximum braking torque generated by MR brake that is shown in Figure 2(b). The MR brake torque increase proportionally when the current is increased. Furthermore, the load displacement takes longer displacement when lower current is applied in this system which is shown in Figure 2(c). Basically, the load displacement responses are determined by integrating the MR brake shaft angular velocity.

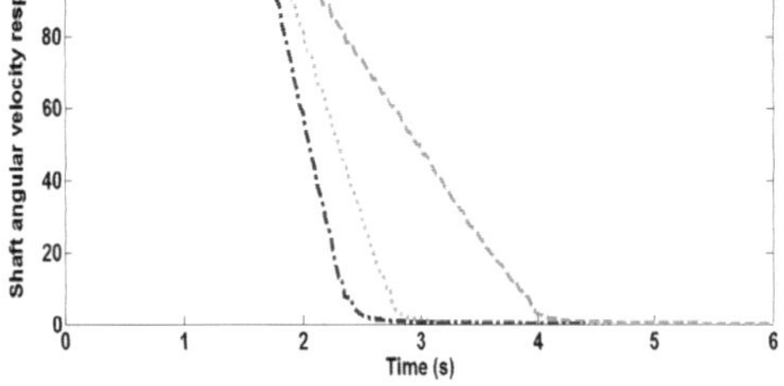

(a)

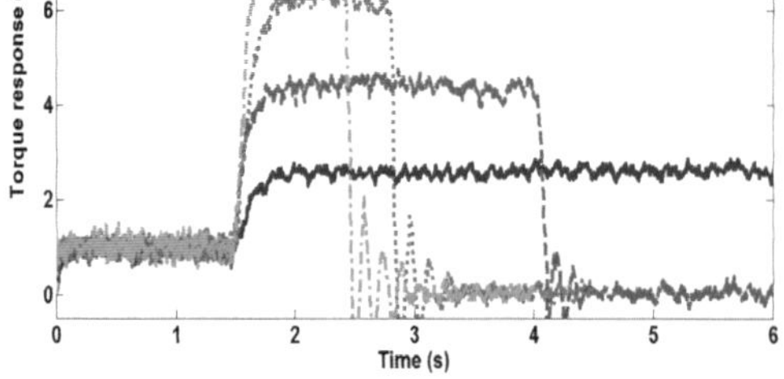

(b)

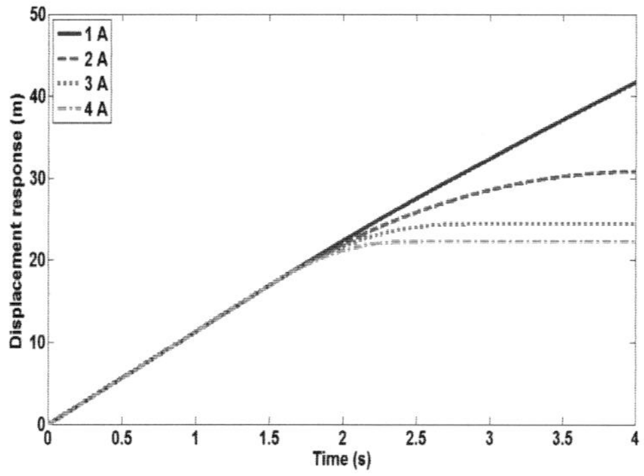

(c)

Figure 2. Measured responses at four constant currents; a) Shaft angular velocity response, b) Torque response and c) Load displacement response

In this model validation, a constant load of 150 N is considered at various current. The current are varies from 1 A to 4 A with increment of 1 A. The triggering time for the electric current in modelling is 1.5 s which is same as experimental. Figure 3, Figure 4 and Figure 5 show the model validation result at various current. The simulation model was developed using MATLAB based on the equations of MR brake with test rig in the previous section and the parameters given in table 1.

Table 1. MR Brake Parameter

Parameter	Value
Outer radius of disc, r_z	0.04 m
Inner radius of disc, r_i	0.01 m
Number of surfaces, N	2
MR fluid gap, h	0.0025 m

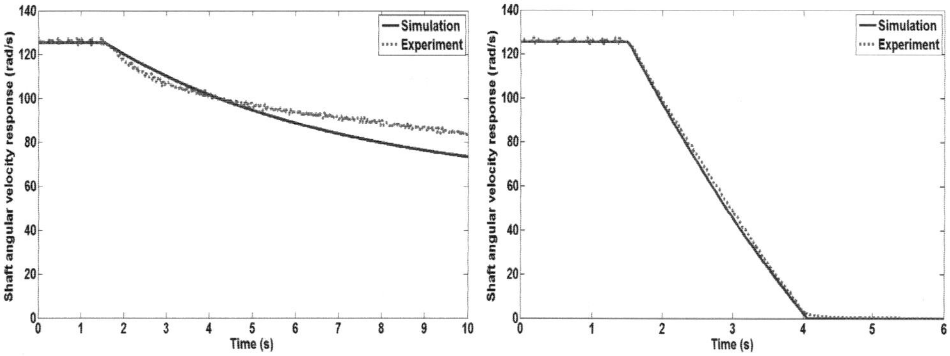

(a) (b)

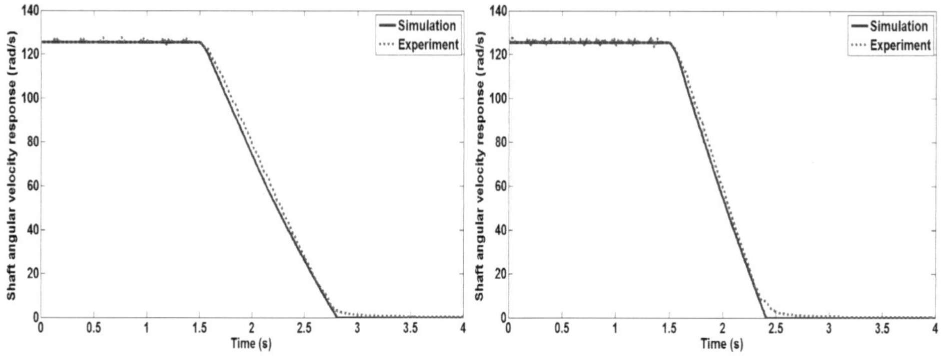

(c) (d)

Figure 3. Comparison between model and experimental at several applied current for shaft angular velocity response; a) 1 A, b) 2 A, c) 3 A and d) 4 A

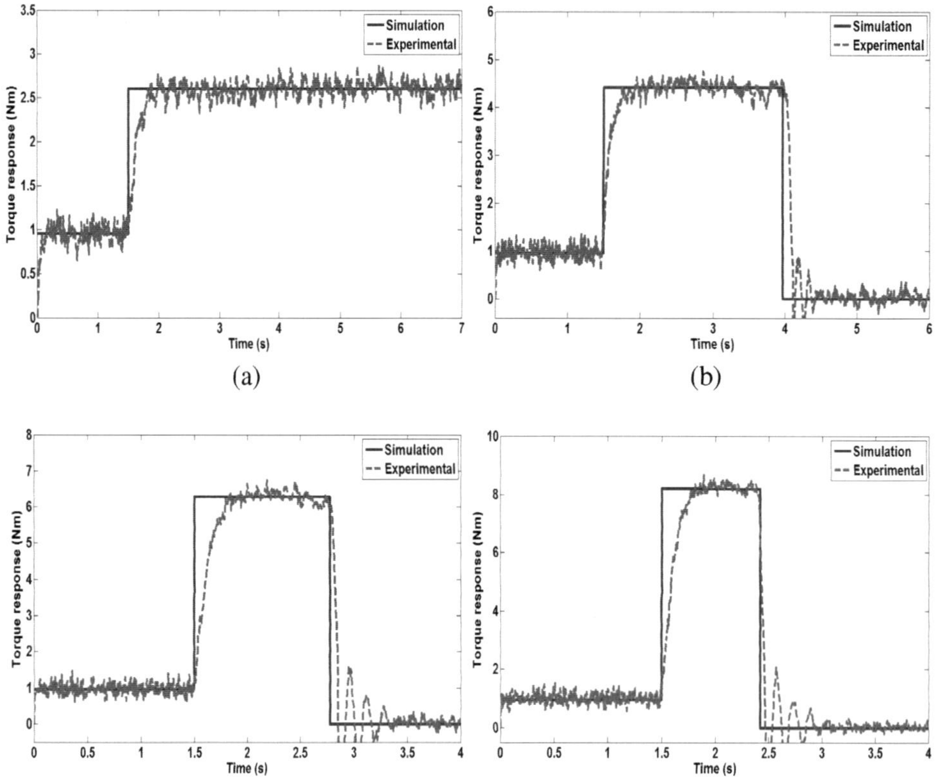

(a) (b)

(c) (d)

Figure 4. Characteristic of torque response between model and experimental; a) 1 A, b) 2 A, c) 3 A and d) 4 A

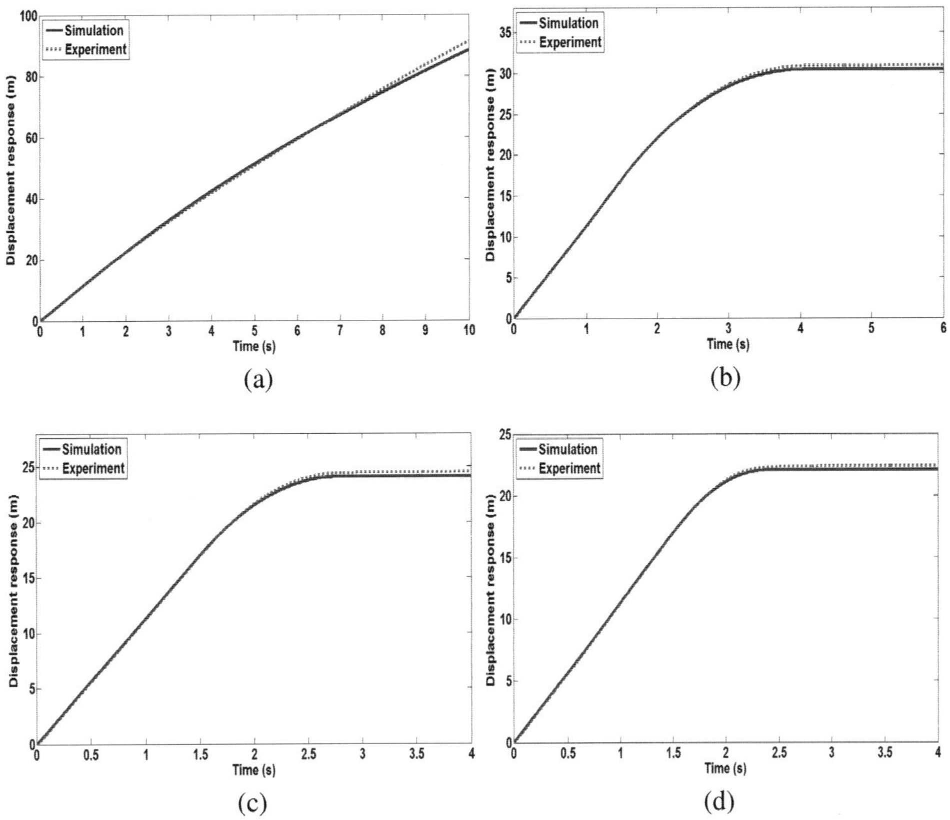

Figure 5. Comparison between model and experimental of load displacement response at several applied current; a) 1 A, b) 2 A, c) 3 A and d) 4 A

The model and experimental results are compared and it shows a close relationship between both results. This response indicates that the model of MR brake actuator with a test rig is valid. However, the response time of MR brake torque to reach constant steady-state has a delay time which is shown in Figure 4. This is due to the delay response of material and sensory system. However, the trend of experimental follow the model is similar.

4. Conclusion

The model validation of MR braking response in term of shaft angular velocity responses, torque responses and load displacement responses was obtained and discussed in this paper. The MR brake with a test rig was developed which is contains some instrumentation sensors such as ABS speed sensor and load sensor. The MR brake model that been simulated by using MATLAB software to obtained the model result are validated with the experimental result. The results validation shows good agreement between both graphs in this study. When increased the load, it will caused the MR brake takes longer settling time and the load displacement response becomes longer to constant steady-state displacement. Furthermore, the heavier load will reduce the effectiveness of MR brake to halt all inertia to stationary condition.

Acknowledgments
The authors gratefully acknowledged the financial support from Universiti Teknikal Malaysia Melaka and The Ministry of Higher Education, Malaysia (MoHE) under Exploratory Research Grant Scheme (ERGS), grant no.: ERGS/1/2012/TK08/UTEM/02/1/E00007.

References

[1] Poynor J C and Reinholtz C 2001 Innovative Designs for Magneto-Rheological Dampers *Expedition*
[2] Wang X and Gordaninejad F 2003 LYAPUNOV-BASED CONTROL OF A BRIDGE USING MAGNETO-RHEOLOGICAL FLUID DAMPERS *Intelligent Material System and Structure*, 4331 1-14.
[3] Diep T, Thanh C and Ahn K K 2006 Intelligent phase plane switching control of pneumatic artificial muscle manipulators with magneto-rheological brake *October* 16(May 2005) 85-95
[4] Gudmundsson K H, Jonsdottir F and Thorsteinsson F 2010 A geometrical optimization of a magneto-rheological rotary brake in a prosthetic knee *Smart Materials and Structures IOP* 11-19
[5] Karakoc K, Park E J and Suleman A 2008 Design considerations for an automotive magnetorheological brake *Mechatronics* **18** 434-447
[6] Kamal M, Rahman M M and Rahman A G A 2013 Inter J Automot Mech Eng **7** 912.
[7] Nagarkar M P, Vikhe G J, Borole K R and Nandedkar V M 2011 Inter J Automot Mech Eng **3** 364
[8] Kamal M, Rahman M M and Rahman A G A 2012 J Mech Eng Sci **3** 291
[9] Carlson J D 2001 What Makes A Good MR Fluid? In *Electrorheological (ER) Fluids and Magneto-rheological (MR) Suspensions* USA 1-7
[10] Bossis G 2002 Magnetorheological fluids *Magnetism and Magnetic Materials* **252** 224-228
[11] Jolly M R, Bender J W and Carlson J 1998 Properties and applications of commercial magnetorheological fluids In *Proc. of SPIE, 5th Annual Intern. Symposium on Smart Materials and Structure* Cary NC 262-275
[12] Park E J 2006 A performance evaluation of an automotive magnetorheological brake design with a sliding mode controller *Mechatronics* 16 405-416
[13] Li W Hl 2007 A 2-DOF MR actuator joystick for virtual reality applications *Sensors And Actuators* 137 308-320
[14] Park E J, Falcao L and Suleman A, 2008 Multidisciplinary design optimization of an automotive magnetorheological brake design *Computers and Structures* **86** 207-216
[15] Karakoc K, 2008 Design of a Magnetorheological Brake System Based on Magnetic Circuit Optimization *Engineering*
[16] Tan K P, Stanway R, and Bullough W A 2007 Braking responses of inertia / load by using an electro-rheological (ER) brake *Mechatronics* **17** 277-289
[17] Nam, T H and Ahn K K 2009 A new structure of a magnetorheological brake with the waveform boundary of a rotary disk *Smart Materials and Structures IOP* **18** 1-4
[18] Liu B et al 2006 Development of an MR-brake-based haptic device *Smart Materials and Structures* **15** 1960-1966
[19] Philips, 1969. Engineering applications of fluids with a variable yield stress PhD thesis (University of California: Berkeley, CA)

IOP Conf. Series: Materials Science and Engineering **50** (2013) 012039 doi:10.1088/1757-899X/50/1/012039

Computational Analysis of an effect of aerodynamic pressure on the side view mirror geometry

P Murukesavan, M A N Mu'tasim and I M Sahat
Faculty of Mechanical Engineering,
University Malaysia Pahang, 26600, Pekan, Pahang, Malaysia

E-mail: ammar@ump.edu.my

Abstract. This paper describes the evaluation of aerodynamic flow effects on side mirror geometry for a passenger car using ANSYS Fluent CFD simulation software. Results from analysis of pressure coefficient on side view mirror designs is evaluated to analyse the unsteady forces that cause fluctuations to mirror surface and image blurring. The fluctuation also causes drag forces that increase the overall drag coefficient, with an assumption resulting in higher fuel consumption and emission. Three features of side view mirror design were investigated with two input velocity parameters of 17 m/s and 33 m/s. Results indicate that the half-sphere design shows the most effective design with less pressure coefficient fluctuation and drag coefficient.

1. Introduction

Reducing fuel consumption, and therefore reducing the carbon dioxide emissions, is one of the most important goals in today's car industry. One way this can be achieved is by reducing the engine size, using an electric motor with a combustion engine, reducing the weight of the car and reducing the aerodynamic drag of the car. The latter is of great importance when it comes to velocity over 60km/h. Above this velocity, the aerodynamic resistance is higher than the rolling resistance [1]. Streamlined body design in a passenger car helps reducing the aerodynamic drag and eventually improves the engine mileage. On the contrary, accessories attached to the body skin of a car cause the unfavourable aerodynamic examples. In order to obtain the rear sight, unfortunately the mirror does not pay only the aerodynamic penalty which increases drag from the body, but also causes the acoustic noise thus causes mirror fluctuations to the cabin crews. While the aerodynamic body styling of the passenger car has been upgraded with a lot of efforts, the defects caused by important accessory such as the side view mirror have been ignored. The main stream meets a side flow which has the flow direction tangent to the windshield surface near the A-pillar. And a conical vortex sheet is generated along the pillar and merges into the mainstream. Therefore, very complicate flow pattern appears by combining these flow patterns near the driver side window. Moreover, since the side mirror is mounted on the driver door near hinge, the wake flow behind this obstacle become much complicated. [2].

Numbers of research has been done to study the effect of flow and noise towards the front side view mirrors. Some studies have been done experimentally using pressure measurement devices on automobile side mirror to evaluate the effect of pressure fluctuations on mirror vibrations [3]. Same approach was also done by using hot wire anemometry and the pressure scanning system in a wind tunnel at high Reynolds Number [4]. This resulted in a vortex sheet with a conical shape developed between the centres of the side mirrors. Studies conducted using computational fluid dynamics on the noise generation on side mirrors have been done using Detached Eddy Simulation, DES, and different turbulent methods [5-8]. Another investigation was also done using the same method with the different approach by increasing the diameter of the side mirror in a high constant Reynolds Number [9]. Three turbulent models were studied on flow around two mirrors to measure the surface pressure fluctuations and resulted in underestimated sound pressure levels along the side mirrors [10].

2. Mathematical Model

The K-epsilon model is one of the most common turbulence models, although it just doesn't perform well in cases of large adverse pressure gradients. It is a two equation model that means, it includes two extra transport equations to represent the turbulent properties of the flow. This allows a two equation model to account for history effects like convection and diffusion of turbulent energy. K-epsilon model has been shown to be useful for free-shear layer flows with relatively small pressure gradients. Similarly, for wall-bounded and internal flows, the model gives good results only in cases where mean pressure gradients are small; accuracy has been shown experimentally to be reduced for flows containing large adverse pressure gradients.

$$\frac{\partial}{\partial t}(\rho k)+\frac{\partial}{\partial t_j}(\rho k u_j)=\frac{\partial}{\partial x_j}\left[(\mu+\frac{\mu_t}{\mu_k})\frac{\partial k}{\partial x_j}\right]+P_k+P_b-\rho\in-Y_M+S_k \tag{1}$$

$$\frac{\partial}{\partial t}(\rho\in)+\frac{\partial}{\partial x_j}(\rho\in u_j)=\frac{\partial}{\partial x_j}\left[(\mu+\frac{\mu_t}{\mu_\in})\frac{\partial\in}{\partial x_j}\right]+\rho C_1 S_\in-\rho C_2 S_\in-\rho C_2\frac{\in^2}{k+\sqrt{v\in}}$$
$$+C_{1\in}\frac{\in}{k}C_{3\in}P_b+S_\in \tag{2}$$

Where

$$C_1=\max[0.43,\frac{n}{n+5}], n=S\frac{k}{\in}, S=\sqrt{S_i S_i} \tag{3}$$

In these equations, P_k represents the generation of turbulence kinetic energy due to the mean velocity gradients, calculated in same manner as standard k-epsilon model. P_b is the generation of turbulence kinetic energy due to buoyancy, calculated in same way as standard k-epsilon model.

3. Design and Modelling

3.1. Wind Tunnel

The wind tunnel is considered is made up of a straight floor and roof with an inlet up stream of the car 8.145m distance ahead of the front of the car and the outlet 20m downstream of the car. So that, as the tunnel outlet is far from the car, it can be assumed that it will have a little effect on the air flow close to the car. The tunnel height is taken as 5m. Then the boundaries if the simulation model is the car surface, side mirror, velocity inlet, pressure outlet, symmetry, symmetry upper, symmetry side and road.

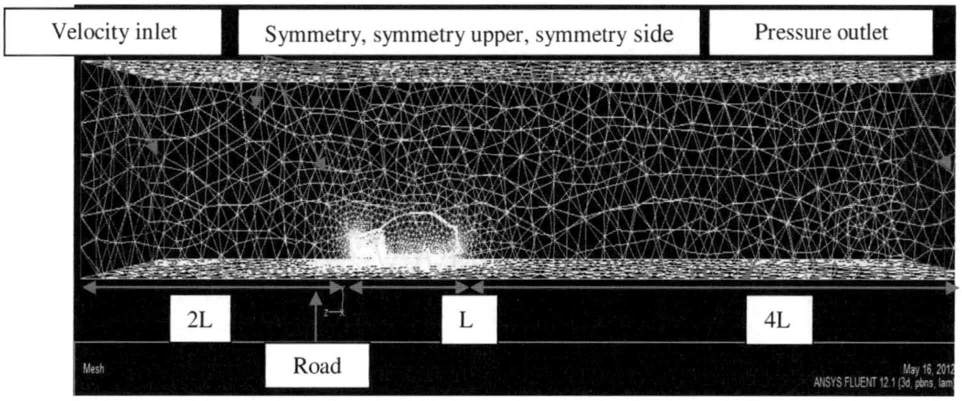

Figure 1. Wind Tunnel Simulation Setup.

3.2. Generic Side Mirror Designs

The generic model of the side mirrors is shown in the figure below with common dimensions. The model is using the actual common dimension of side mirror which is the length of the model is 1.310 m, the width of the model is 0.540 m, and the height of the model is 0.930 m. Solid Works was used to produce the models with the required dimensions. Design 1 was done based on a semi-sphere shape while Design 2 was done based on a sharp end with triangular shape and the Design 3 was done by combining rectangular shape with triangular edges.

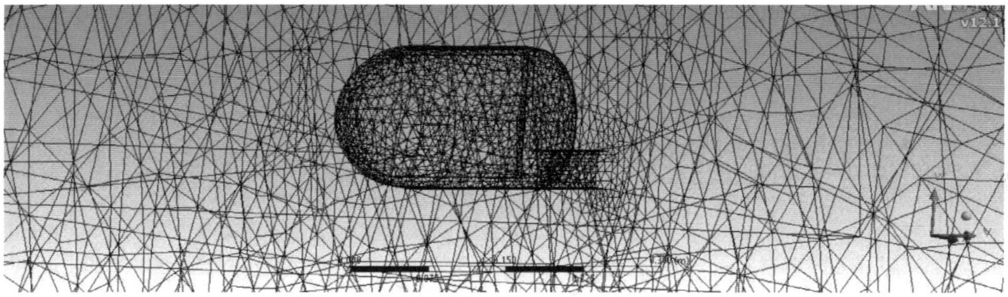

Figure 2. Meshing Generic Side Mirror Setup.

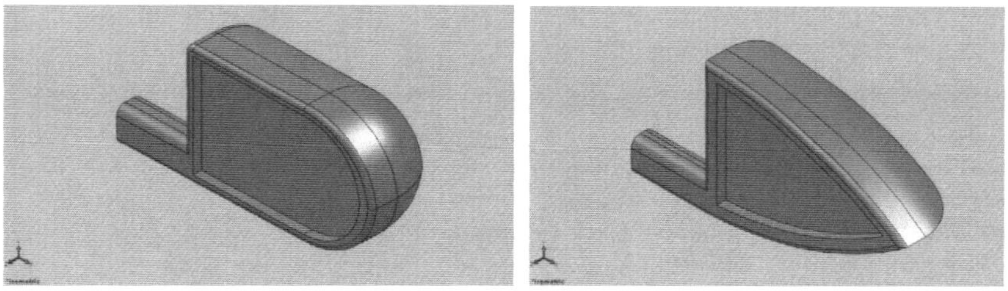

(a) (b)

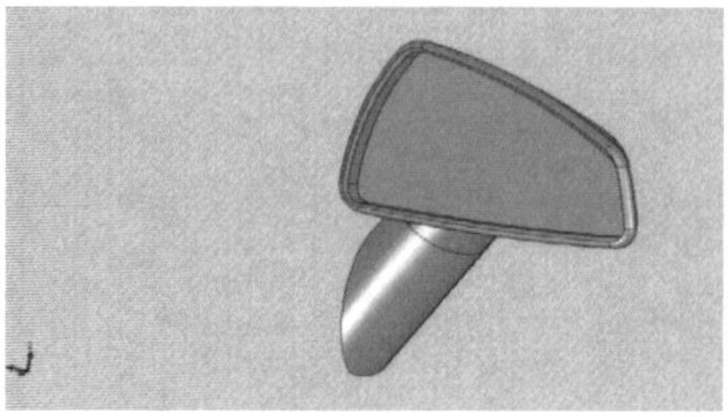

(c)

Figure 3. Side Mirror Designs (a) Design 1 (b) Design 2 and (c) Design 3.

Figure below shows the selection of 100 critical points across the side mirror for all the designs used. This is done to extract the results which are pressure coefficient, total pressure, drag coefficient and lift coefficient at the required area. The critical points are chosen across the frontal area of the side mirror which faces the simulated flow.

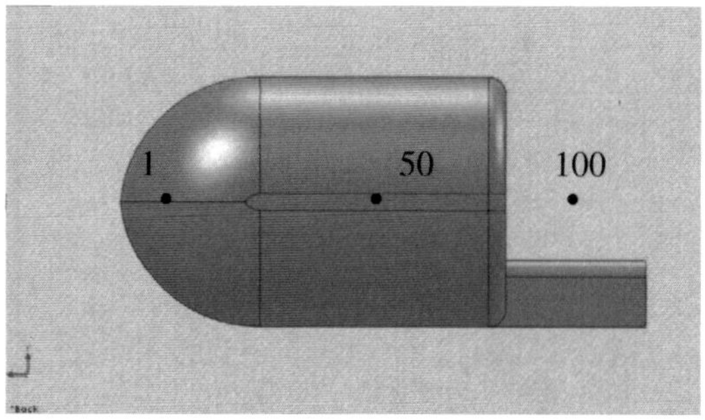

Figure 4. 100 critical points across the side mirror frontal area.

4. Validation

In order to validate with the same reference, an experiment analysis was conducted by [4] with the same design 1 with a velocity of 25 m/s. Figure 5 shows comparison between the present works with the experimental results. It can be seen that a good agreement is achieved between the results obtain from computational and the experimental results for the mirror housing.

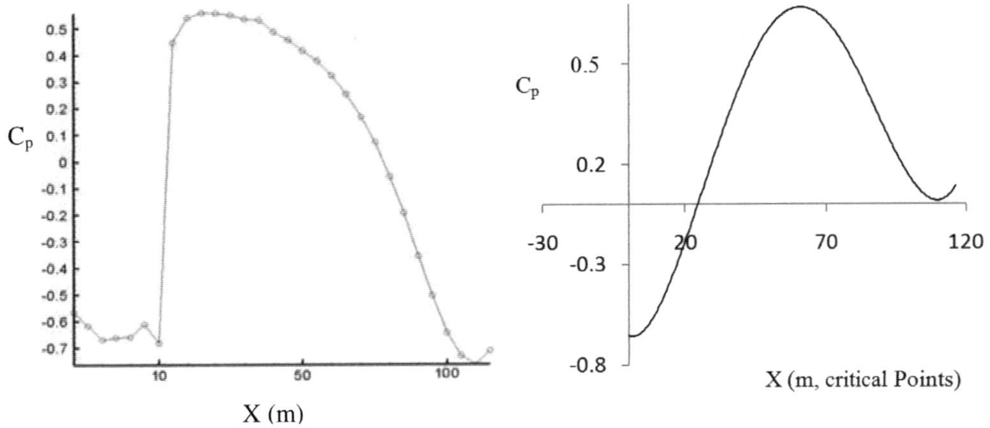

Figure 5. Comparison between the experimental results with Present Study for velocity of 25 m/s.

5. Result and Discussion

The pressure coefficient at a point near a body is independent with the body size. To determine the model, it is tested in a wind tunnel or water tunnel, pressure coefficients is determined at critical locations around the model and these pressure coefficients is used with confidence to predict the fluid pressure at those critical locations around a passenger car.

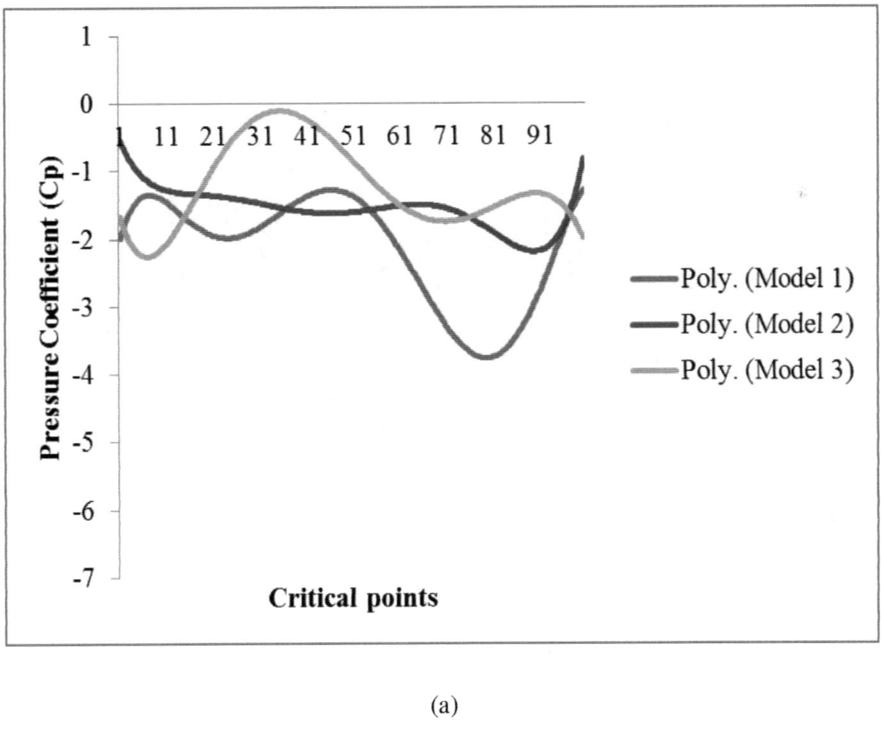

(a)

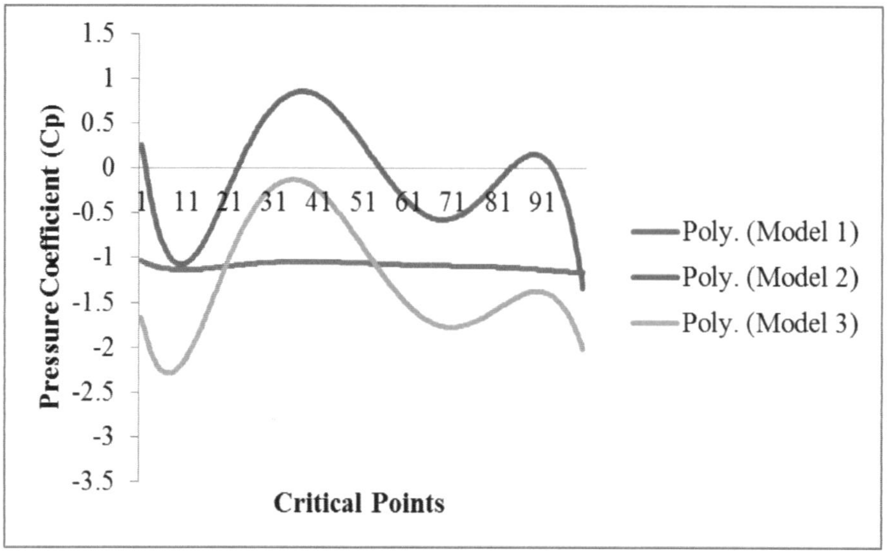

(b)

Figure 6. Side Mirror Designs velocity at (a) 16.7 m/s (b) 33.3 m/s.

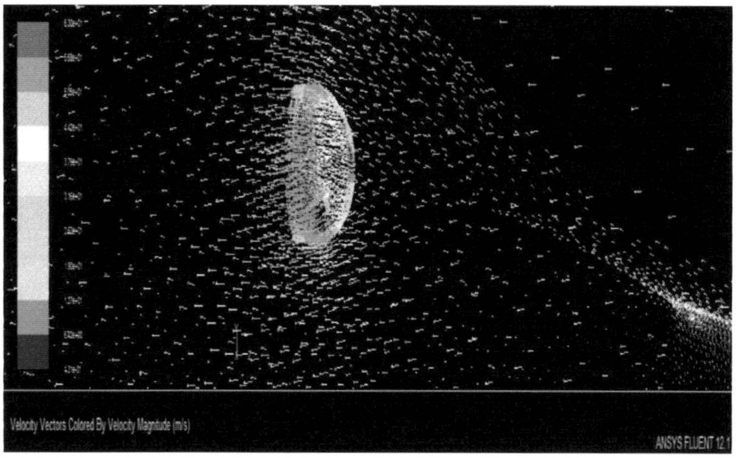

(a)

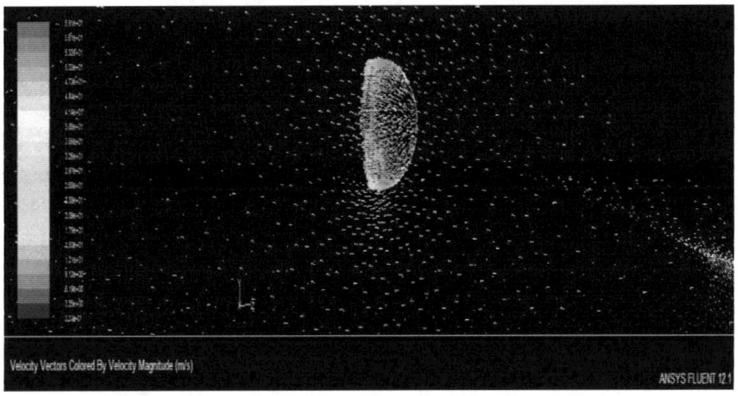

(b)

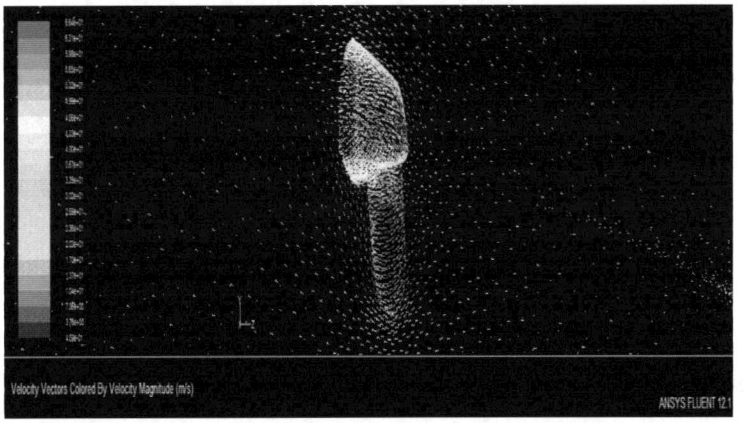

(c)

Figure 7. Side Mirror flow contour for (a) Design 1 (b) Design 2 and (c) Design 3

Before the simulation was done, boundary conditions such as environment pressure and outlet pressure of 1atm are applied. The speed used for the simulation is 16.67 m/s (60 km/h), 25 m/s (90 km/h) and 33.33 m/s (120 km/h). The speed used is same as the experiment method done by Alam et al., 2007. Non-slip condition was applied on the entire solid surface. The discussion done shows the pressure coefficient, total pressure distribution, drag coefficient and lift coefficient value around the car side mirror. Therefore the effect of different side mirrors designs based on the pressure coefficient, drag coefficient and lift coefficient can be estimated. For all the cases, the pressure coefficient around the passenger car was negative but that associated to in front of the side mirror which is always positive. The CFD simulation of flow over passenger car side mirror was in good agreement with that of the experiment method done by Alam et al., in 2007. The comparison of the drag coefficient respectively for the nine cases was studied. All the cases show positive results in a range value of 0.4 to 0.6. The pressure coefficients fluctuate based on the design and speed. Whereas, lift coefficient show the lift coefficient of the nine cases studied. All the cases show negative results in a range value of -0.05 to -0.3. Side Mirror 3 shows highest drag coefficient with the highest lift coefficient and Side

Mirror 1 show the lowest drag coefficient with the smallest lift coefficient. Therefore, Mirror 1 is suitable to be used in passenger cars due to less mirror fluctuations which could lead to fuel saving.

6. Conclusion

This conclusion was made from the work presented here. Fluctuating aerodynamic pressures are not uniformly distributed over an automobile mirror surface. The highest magnitude of fluctuating pressure coefficient was found at the central across the back of the side mirror while the lowest magnitude of fluctuating pressure coefficient was found at the sharp edges around the side mirror frontal area. As the yaw angle is known to affect mirror noise and vibration, this should be considered in future work.

7. Acknowledgements

The authors would like to thank Universiti Malaysia Pahang (UMP) for financial support under RDU110379.

8. References

[1] Versteeg H K and W Malalasekera 2007 *An Introduction to Computational Fluid Dynamics, The Finite Volume Method* (Pearson Prentice Hall, second edition)

[2] Thomas D G 2000 *Fundamentals of Vehicle Dynamics* pp 74-96

[3] Jaitlee R, Alam F and Watkins S 2001 Pressure Measurements on an Automobile Side Rear View Mirror, School of Aerospace, Mechanical & Manufacturing Engineering RMIT University, Melbourne, VIC 3083, AUSTRALIA, 15th Australasian Fluid Mechanics Conference

[4] Jeong-H K, Yong O H, Min H L, In H H and Ui H J Surface Flow And Wake Structure Of A Rear View Mirror Of The Passenger Car, School Of Mechanical *Int. Colloquium On: Bluff Bodies Aerodynamics & Applications* Engineering Yeungnam University, Gyeongsan, Gyeongbuk 712-749, Korea, Bbaa Vi ()

[5] Heisler H 2002, *Advanced Vehicle Technology Elsevier Butterworth Heinemann*, (Jordan Hill, Oxford) pp 121-123

[6] Vijaya Lakshmi B, Subrahmanyam T, Dharma Rao V and Sharma K V 2011 *Int. J. Automot. Mech. Eng.* **4** 455

[7] Naga Sarada S, Sita Rama Raju A V, Kalyani Radha K and Syam Sunder L 2012 *Inter. J. Automot. Mech. Eng.* **6** 795

[8] Tahseen T A, Ishak M and Rahman M M 2012 *J. Mech. Eng. Sci.* **3** 271

[9] Jonas Ask and Lars Davidson 2006 The Sub-critical Flow past a Generic Side Mirror and its Impact on Sound Generation and Propagation *12th AIAA/CEAS Aero acoustics Conference, (8-10 May, Cambridge, Massachusetts)*

[10] Snegirev A Y, Lupulyak S V and Yu K S *Use* Of High Performance Computing For Simulations Of Aerodynamically Generated Noise In Turbulent Separated Flows, Laborator For Applied Mathematics And Mechanics Saint-Petersburg State Polytechnic Universit Polytechnicheskaya, *West-East High Speed Flow Field Conference.*

[11] Yunus A C and John M C 2006 Fundamentals of Fluid Mechanics, 1st Edition in SI Units pp 562-600

[12] Michele Ferlauto and Roberto Marsilio 2006 A Viscous Inverse Method for Aerodynamic Design *Computer & Fluids* pp 304-325

[13] Yiping W, Zhengqi G, Weiping L and Xiaohui L 2010 Evaluation of Aerodynami Noise Generation by a Generic Side Mirror *World Academy of Science, Engineering and Technology* **61** 364-571

Finite Element Modeling of the effect of tool rake angle on tool temperature and cutting force during high speed machining of AISI 4340 steel

S Sulaiman, A Roshan[a], M K A Ariffin
Department of Mechanical and Manufacturing Engineering, Faculty of Engineering, Universiti Putra Malaysia, 43400 UPM, Serdang, Selangor, Malaysia, Phone: +603-98466336; Fax: +603-86567122

E-mail: amirroshan25@yahoo.com

Abstract. In this paper, a Finite Element Method (FEM) based on the ABAQUS explicit software which involves Johnson-Cook material model was used to simulate cutting force and tool temperature during high speed machining (HSM) of AISI 4340 steel. In this simulation work, a tool rake angle ranging from 0° to 20° and a range of cutting speeds between 300 to 550 m/min was investigated. The purpose of this simulation analysis was to find optimum tool rake angle where cutting force is smallest as well as tool temperature is lowest during high speed machining. It was found that cutting forces to have a decreasing trend as rake angle increased to positive direction. The optimum rake angle observed between 10° and 18° due to decrease of cutting force as 20% for all simulated cutting speeds. In addition, increasing cutting tool rake angle over its optimum value had negative influence on tool's performance and led to an increase in cutting temperature. The results give a better understanding and recognition of the cutting tool design for high speed machining processes.

1. Introduction

As HSM has been one of the modern technologies which in comparison with traditional machining can produce parts with the best accuracy, efficiency and quality at lower manufacturing and production costs, as well as can lead to a significant reduction in the machining time. In addition, HSM has many advantages such as leading to excellent dimensional accuracy, improve the quality of surface finishing and remove the difficult to cut materials with high hardness. Recently, HSM technology has been widely used in many industrial fields such as automotive, aeronautics, die/mold making and light industry, etc [1-4].

The high temperatures generated in metal cutting processes are due to many reasons: heat generation is concentrated over a small area, more heat is generated; and, the adiabatic nature of the cutting process where heat generated in cutting zone cannot be easily conducted away from deformation zone due to the very short contact time. In addition, tool-chip interface temperature is closely connected to tool geometry, feed rate, cutting speed and depth of cut [5-7]. Cutting force is one of the important other factors that may have either direct or indirect influence on other variables such as machined surface characteristics, tool wear rate and machining cost. Many parameters can effect on increase or decrease of cutting force such as tool material and wear factor [8], tool rake angle, feed rate and cutting speed [9-10]. Although, there have been many studies concerning different factors effect on tool temperature and cutting force, it seems that tool rake angle can influence on both of them [11].

In this paper, a FEM based on Johnson-cook material model was used to model the effect of tool rake angle on tool temperature and cutting force which are difficult to measure by experiment. The simulation of metal cutting process is performed using ABAQUS/Explicit software which can be handled for variety of non-linear problems. The purpose of this simulation analysis is optimum tool rake angle where cutting force is smallest as well as tool temperature is lowest during HSM of AISI 4340 steel.

2. Simulation details

The model proposed in this study uses ABAQUS software to simulate the cutting process. The workpiece material and cutting tool use four-node bilinear displacement and temperature (CPE4RT) for the coupled thermo-mechanical calculation. It is assumed that the value of cutting depth is bigger than that of feed rate.

2.1. Hypothesis and simulation software

In this simulation work, an adaptive meshing is used since metal cutting process is an extremely dynamic event with substantial changes in geometry. The 2D finite element model of orthogonal cutting before and after deformation is shown in figure 1. The FE of the tool is modeled using 1287 nodes and 1365 elements and the mesh division of the workpiece performed into 3536 nodes and 3380 elements. The sliding and sticking friction model between tool rake face and chip is based on Coulomb's friction law [12]. The machining parameters used in this case were: 0.2 mm/rev feed rate, 2.5 mm depth of cut, 3.0 width of cut, and a range of cutting speeds between 300-550 m/min. Carbide cutting tool has the following geometrical angles: 5° clearance (or flank) angle and tool rake angle ranging from 0° to 20°.

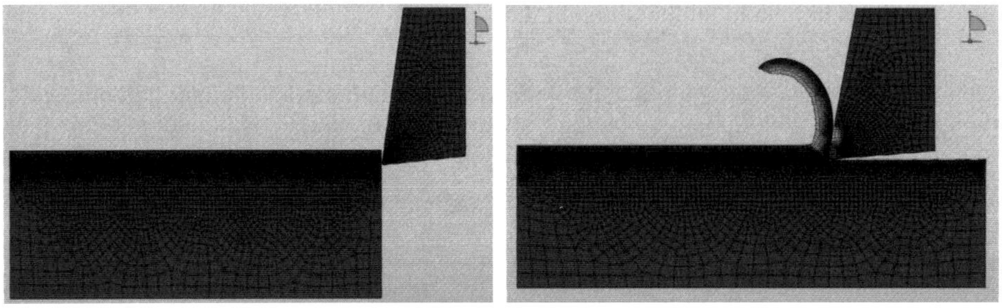

(a) (b)

Figure 1. 2D FEM simulation of metal cutting (a) before and (b) during deformation.

2.2. Material behavior and chip separation criterion

The constitutive model proposed by Johnson and Cook [13] describe the relationship of stress, strain, strain rate and temperature fields that are individually determined as shown in Eq. (1). This particular plasticity model is suitable in FEM simulation of materials at high strain rates and the high speed cutting process [14].

$$\sigma = (A + B\varepsilon^{n}) [1 + C \ln\frac{\dot{\varepsilon}}{\dot{\varepsilon}_0}] [1 - (\frac{T - T_r}{T_m - T_r})^{m}] \quad (1)$$

In above equation, the parameters *A, B, C, n* and *m* are material constants that are found by material tests. In definition of temperature softening term, *T* is temperature of material, T_r is room temperature or reference and also T_m is melting temperature of a given material. A ε represents the true strain, $\dot{\varepsilon}$ is the equivalent plastic strain rate and normalized with reference strain rate $\dot{\varepsilon}_0$. Table 1 shows the J-C material model constants use for deformation behavior of AISI 4340 steel [15].

Table 1. The Johnson-Cook Parameters for AISI 4340 Steel.

Material	A [MPa]	B [MPa]	n	c	m	T_m (°C)	T_r (°C)
AISI 4340	792	510	0.26	0.014	1.02	1520	20

In this simulation work, the Carbide cutting tool was modeled as an elastic material where workpiece material was defined as elasto-plastic material. Table 2 shows the properties of cutting tool and workpiece [15].

Table 2. Thermal property of workpiec material and cutting tool.

Thermal Parameters	AISI 4340	Carbide tool
Density, (g/cm^3)	7.85	15
Young's modules, (GPa)	207	800
Poisson ratio	0.2	0.2
Specific heat, (J/Kg/° C)	477	200
Thermal conductivity, (W/ m° C)	44	46
Expansion, (μm/m° C)	1.3	4.7

The chip separation from workpiece was simulated using Johnson-cook failure model which is most suitable for deformation of material at high speed and high strain-rate. The Johnson-Cook dynamic failure is allowed to occur when the fracture damage criteria D exceed 1 and the concerned element are removed from computation. The damage parameter used in this simulation is defined in Eq. (2):

$$D = \Sigma \left(\frac{\Delta \bar{\varepsilon}^{pl}}{\bar{\varepsilon}_f^{pl}}\right) \tag{2}$$

where $\Delta\bar{\varepsilon}^{pl}$ is the increment of equivalent plastic strain and $\bar{\varepsilon}^{pl}$ is strain at fracture, under current conditions. According to Johnson-Cook damage law, the cumulative Strain $\Delta\bar{\varepsilon}^{pl}$ is updated at every analysis increment and the general expression for the strain at fracture criteria $\bar{\varepsilon}_f^{pl}$ is assumed to be separable and given as Eq. (3):

$$\bar{\varepsilon}_f^{pl} = [d_1 + d_2 \exp(d_3 \frac{p}{q})]\, [1 + d_4 \ln(\frac{\dot{\varepsilon}^{pl}}{\dot{\varepsilon}_0})]\, [1 + d_5 \frac{T - T_r}{T_{melt} - T_r}] \tag{3}$$

where d_1 - d_5 are failure parameters were experimentally determined and the ratio of p/q is defined where p is the pressure stress and q is the Von Mises stress. Physical shear failure occurred in the model when the damage parameter reached unity. The damage parameters of AISI 4340 [16] used in this simulation are presented in table 3.

Table 3. The Johnson-Cook damage law parameters for AISI 4340 steel.

Damage law parameters	AISI 4340
Initial failure strain, d_1	0.05
Exponential factor, d_2	3.44
Triaxiality factor, d_3	- 2.12
Strain rate factor, d_4	0.002
Temperature factor, d_5	0.61

3. Results and discussion

3.1. Effect of tool rake angle on tool temperature

Temperature distributions for machining of AISI 4340 steel at three different cutting speeds are obtained as shown in figure 2. This temperature distribution is the consequence of shear-localized deformation regions which cause temperature rise in the separation zone. However, the heat generation at tool-chip interface and at secondary deformation zone is conducted to the cutting tool.

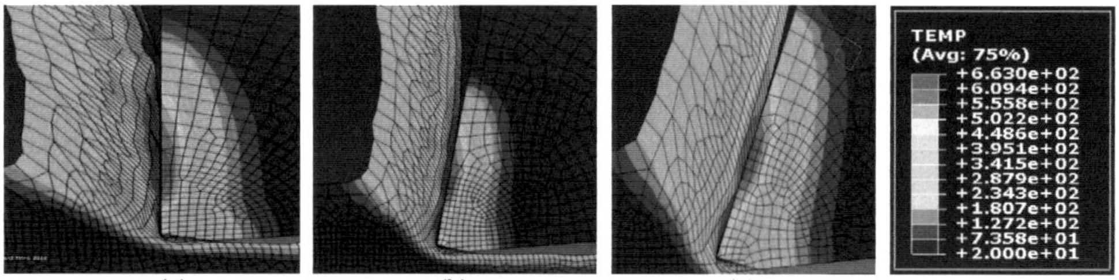

(a) (b) (c)

Figure 2. Temperature distribution between tool and chip using different rake angles a) 0°, b) 10°, c) 20°, at the cutting speed of 433 m/min.

Figure 3 shows the variation of temperature along the tool-chip interface with different rake angle. The increase of temperature along the tool/chip interface due to frictional sliding and plastic deformation has an appreciable dependence on the rake angle. The contact is high at low rake angle. For cutting speeds of 300 and 433 m/min the lowest temperature occurred at rake angle between 10° and 18°. The excessive rake angle weakens the tool, therefore increase tool temperature. The result agrees well with the equivalent and analogous work [8]. From the simulation results, it is observed that with increase cutting speed, temperature generated on tool increase. Temperature generated on tool face was in the range of 620-655°C at cutting speed 550 m/min. As a result, there is optimum rake angle for cutting speeds where tool temperature is the lowest.

3.2. Effect of tool rake angle on cutting force

The most remarkable results concluded from figure 4 are that the tool rake angle has a significant effect on the cutting force where cutting force decrease with increasing tool rake angle for all cutting speed investigated. The cutting tool can easily dive into the workpiece material since more positive rake angle produces higher shear angle, therefore it helps to reduce cutting force. In this way, chip flows easier and also lower plastic deformation work required when using cutting tool with positive rake angle. The results agree well with the previous works. The obtained results had a good agree with the experiment results obtained from literature [9] where that change of rake angle in positive direction had positive effect on cutting force. In this simulation work, the extent of cutting force reduction is most evident as the rake angle increase from 0° to 14° at cutting speeds 300 and 433 m/min. On the other hand, when rake angle increased 14° to 20°, the cutting force had only a small reduction. The results agree well with [8]. In addition, simulation results show that force changes due to changing of cutting speed. The major decrease of cutting force was for cutting speed of 550 m/min where rake angle changing from 0° to 18°.

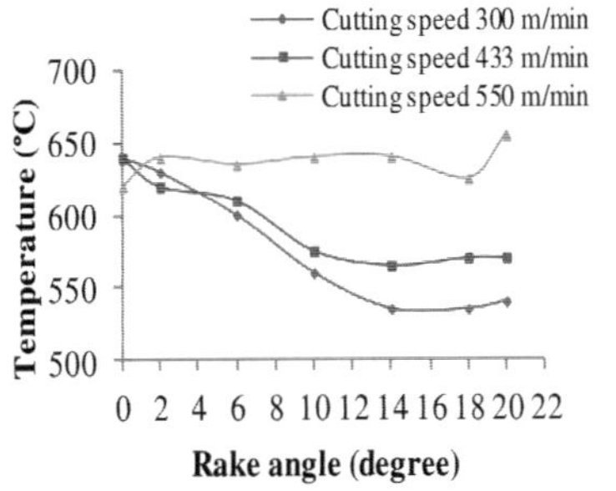

Figure 3. The effect of tool rake angle and cutting speed on predicted tool temperature.

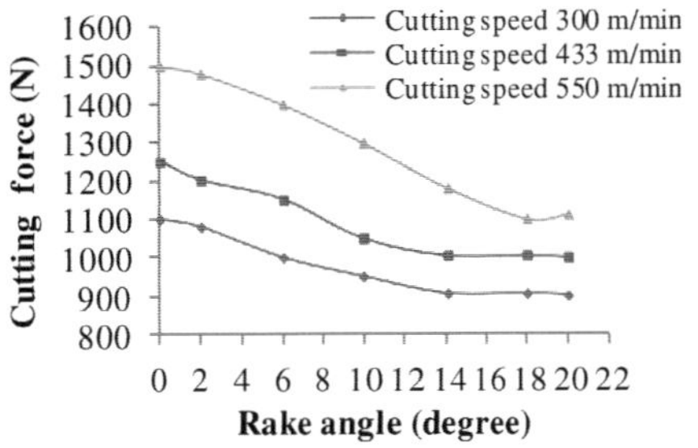

Figure 4. The effect of tool rake angle and Cutting speed on cutting force.

4. Conclusion

A finite element model based on ABAQUS software was used to study the effect of tool rake angle on tool temperature and cutting force. From the FEM simulation results when used with the Johnson-Cook material model and the Coulomb's friction law for HSM of AISI 4340 steel was observed that increasing cutting tool rake angle over its optimum value had a negative effect on tool's accelerate and performance which led to an increase in cutting temperature. It was observed that cutting forces were decrease as rake angle was changed from 0° to 20°. The reduction was recognized up to 20% for all simulated cutting speed. Additionally, with increase cutting speed, more cutting force required, therefore it led to an increase in cutting temperature. As a result of this simulation work, by evaluating temperature, cutting force and cutting speed together, optimum rake angle can be considered between 10° and 18° where the optimum cutting speed was obtain at 433 m/min.

Acknowledgments

The authors would like to express their deep gratitude and sincere thanks to staff and laboratory technicians of the Department of Mechanical and Manufacturing Engineering, Universiti Putra Malaysia (UPM) for their help and encouragements.

Reference

[1] Duan C Z, Dou T, Cai Y J and Li Y Y 2009 Finite element simulation and experiment of chip formation process during high speed machining of AISI 1045 hardened steel *International Journal of Recent Trends in Engineering* **1** 46–50

[2] Davoodi B and Hosseinzadeh H 2012 A new method for heat measurement during high speed machining *Measurement* **45** 2135–2140

[3] Najiha M S, Rahman M M, Kamal M, Yusoff A R and Kadirgama K 2012 *J. Mech. Eng. Sci.* **3** 340

[4] Najiha M S, Rahman M M, Yusoff A R and Kadirgama K 2012 Inter J Automot Mech Eng **6** 766

[5] Abhang L B and Hameedullah M 2010 Chip-tool interface temperature prediction model for turning process *International Journal of Engineering Science and Technology* **2** 382–393

[6] Kadirgama K, Rahman M M, Ismail A R and Bakar R A 2011 *J. Mech. Eng. Sci.* 1 37

[7] Yusoff A R 2011 *J. Mech. Eng. Sci.* **1** 47

[8] Dogra M, Sharma V S and Dureja J 2011 Effect of tool geometry variation on finish turning – A review *Journal of Engineering Science and Technology Review* **4** 1-13

[9] Satyanarayana K, Venugopal A and Venkateswaraao G 2011 Effect of rake angle and feed rate on cutting forces in an orthogonal turning process *Int. Conf. on Trends in Mechanical and Industrial Engineering* Bangkok, p 150-154

[10] Nicoleta L and Marian B 2012 Effect of cutting speed and feed rate on tool geometry, temperature and cutting forces in machining AISI 1045 carbon steel using FEM simulation *Proc. in Manufacturing Systems* **7** 245–252

[11] Haci S, Faruk U and Suleyman Y 2006 Investigation of the effect of rake angle and approaching angle on main cutting force and tool tip temperature *International Journal of Machine Tools and Manufacture* **46** 132–141

[12] Haglund A J, Kishawy H A and Rogers R J 2008 An exploration of friction models for the chip–tool interface using an Arbitrary Lagrangian–Eulerian finite element model *Wear* **265** 452–460

[13] Johnson G R and Cook W H 1985 Fracture characteristics of three metals subjected to various strain, strain rates, temperature and pressure *Eng. Frac. Mech.* **2** 31-48

[14] Iqbal S A, Mativenga P T and Sheikh M A 2007 Characterization of machining of AISI 1045 steel over a wide range of cutting speeds. Part 2: evaluation of flow stress models and interface friction distribution schemes *Proc. of the Institution of Mechanical Engineers - Part B: Journal of Engineering Manufacture* **221** 917–926

[15] Ozel T and Zeren E 2007 Finite element modeling the influence of edge roundness on the stress and temperature fields induced by high speed machining *International Journal of Advanced Manufacturing Technology* **35** 255–267

[16] Mabrouki T and Rigal J F 2006 A contribution to a qualitative understanding of thermo-mechanical effects during chip formation in hard turning *Journal of Materials Processing Technology* **176** 214–221

Automated a complex computer aided design concept generated using macros programming

Mohammad Rizal Ramly[a], **Azharrudin Asrokin**[b], **Safura Abd Rahman**[d] **and Nurul Ain Md Zulkifly**
Universiti Kuala Lumpur,Malaysia Institute Of Aviation Technology, Lot 2891, Jalan Jenderam Hulu, Jenderam Hulu,43800, Dengkil, Selangor Darul Ehsan, Malaysia

E-mail: [a]mdrizal@miat.unikl.edu.my, [b]azharrudin@miat.unikl.edu.my, [c]safura@miat.unikl.edu.my

Abstract. Changing a complex Computer Aided design profile such as car and aircraft surfaces has always been difficult and challenging. The capability of CAD software such as AutoCAD and CATIA show that a simple configuration of a CAD design can be easily modified without hassle, but it is not the case with complex design configuration. Design changes help users to test and explore various configurations of the design concept before the production of a model. The purpose of this study is to look into macros programming as parametric method of the commercial aircraft design. Macros programming is a method where the configurations of the design are done by recording a script of commands, editing the data value and adding a certain new command line to create an element of parametric design. The steps and the procedure to create a macro programming are discussed, besides looking into some difficulties during the process of creation and advantage of its usage. Generally, the advantages of macros programming as a method of parametric design are; allowing flexibility for design exploration, increasing the usability of the design solution, allowing proper contained by the model while restricting others and real time feedback changes.

1. Introduction

In the past, CAD was referred as Computer Aided Drafting and nowadays, it is commonly referred as Computer Aided Design, where it refers to software that helps designer in a wide variety of industries to design and manufacture various types of products such as cameras, cars and aircraft.

The history of CAD started in the early 1950s with a developer of the first graphical system. It was designed for US Air Force's SAGE (Semi Automatic Ground Environment) air defence system that was developed at Massachusetts Institute of Technology. Generally, the system was used to coordinate radar station and directed airplane to intercept incoming planes [14]. In 1957, the first commercial numerical control program system was developed which called PRONTO and it became the pioneer contribution to computer aided design and manufacture [13]. From 1970s to 2000s, the evaluation of CAD software was vast in development and it has been used in many fields of areas [9]. Nowadays, CAD software is not only used for drawing only but various other areas have utilized the modules from CAD software for the purpose of analysis, manufacturing, and visual-prototyping. Some of the CAD software is considered low end and easy to use, other offer as a higher order modelling and designing that are costly but more efficient and it's also capable as collaborative design where multiple users can work with the same design at the same time.

Generally parametric modelling is a capability to edit and update graphics components. This is done by controlling certain modelling attributes and behaviour of the object design. The model maintains consistent relationships between elements as it is manipulated. The objectives of this study are; firstly, to understand the concept of parametric design methods, hierarchical constraints and rules. Secondly, to develop a model by understanding the concept of the design and configuration of the programming line as a fundamental of macros programming. Lastly, to generate a complex model of commercial aircraft by implementing parametric design using macros programming.

2. Literature Review

The study of parametric design has been discussed in previous study done by [1] [2] [3] .[4] stated that design procedures are important questions bringing the validity of design with respect to design languages. His studies look into simple analogy, where the variations of parametric model created instances which normally grouped in a category named as a family of design bringing up a greater number of design instances. Another important aspect to consider is the evaluation of the design instances. [12] stated that there are three types of evaluation of the design, which are; performances based, aesthetic and compliance. Performances based, look into a design instances with respect to an ideal result so that model is modified with respect to the ideal one. Aesthetic on the other hand will determine if an instance satisfies a set of value determined by the designer. Lastly, compliances assert if a design instance fulfils a predetermined set of requirements. The paper by [8] presents a surface generation tool designed for the construction of aircraft geometry. The generated of complex geometries is based on Partial Differential Equations where it can generate a number of different configurations of the aircraft shapes interactive in the real time by setting a small set of design variables, instead of hundreds of control points. The shape of the surface is defined through boundary conditions and using a small set of design parameters. From above studies, it shows how important and complex parametric design towards CAD part creation.

In design level, it is important to decide between which levels of detail of the features needed to be presented in a parametric form. Based on the frequently design requirement, the designer will propose varies concept of design for future modification. The study done by [6] explains that the terminologies of parametric modelling are based on; design unit, functions features and form feature. In contrary, dimensions are considered as a constraint that affects a particular set of points in the model and therefore mathematical equations can also be set up within dimensions and features that can have a hierarchical relationship. The other consideration is the design interest should be error free from under-constrained or over-constrained.

3. Methodology

The most important aspect of creating an automated parametric design is to understand the design concept. For example, if creating a cylinder with a centre hole, how do you control the requirement of object dimension changes? Will the requirement of flexibility needed be just outer diameter, inner diameter or both? The standard rule is to create an understanding of design concepts that stated what the requirements needed to be parametric, and what will stay unchanged? This therefore not a problem of an issue for simple design as most of the CAD software is able to reconstruct a design method by simply click a mouse and enter the appropriate new value. The issue of design will arise if there are more requirements needed to be controlled such as the design of the piston head, with parameter requirement, for instance, the size of the object such as piston height, pin diameter, piston thickness, size of compression ring groove and other configuration needed. So, before start creating a design of the part, the understanding of design terminology of the model is considered important because the end parametric model creates will only reflect on the current setting that the user created. As basic rules, the ground of automated design is the generation of geometry from the definition of a family of initial parameters and the design of the formal relations they keep with each other's. It is about the use of variables and algorithms to generate a hierarchy of mathematical and geometric relations that allow us to generate a certain design, but to explore the whole range of possible solutions that the variability of the initial parameters may allow.

Generally a CATIA V5, the CATScripts / CATvba structure consists of:

a- The part body
b- Sketch plane
c- Set the axis system of the sketch
d- Sketch geometry and feature

The structure of the programming line is created by a CAD software are based on the software itself. CATIA V5 program structure start with the part body, then sketch plane, define the axis of the sketch and lastly apply add on feature to convert a two dimension sketch into a three dimensional object.

The programmer, need to familiar with the variety of design interest before recording the macros. The general questions that the designer needed are: what do the parametric design information required, understand the flow of the macros programming and how to overcome the problem with error from the program line or command.

3.1. Case study of macros programming (Design of a commercial aircraft)

The purpose of having an automated design of commercial aircraft is to look into the variety of designs of the aircraft under various specifications. In this study the model created is used to run various problem of design. Example: What are the effects of different configurations of the wing design towards the air flow, using CFD software? With this model the analysis can be done with ease because the data or specification that is required is by key-in the required value and the new model will be created.

Figure 1, show the parameter required for a commercial aircraft design that consist of the fuselage, wing and tail section.

Fuselage length
Fuselage radius
Wing dihedral angle
Wing sweepback angle
Wing span length
Wing root chord length
Horizontal stabilizer dihedral angle
Horizontal stabilizer sweepback angle
Horizontal stabilizer span length
Horizontal stabilizer root chord length
Vertical stabilizer sweepback angle
Vertical stabilizer height
Vertical stabilizer root chard length

Figure 1.Requirement of commercial aircraft design

After the requirements have been set, the flow chart was created to cater the design of the surface. The design of the fuselage, wing and tail section (vertical and horizontal section) will follow the same flow chart that is shown on figure 2.

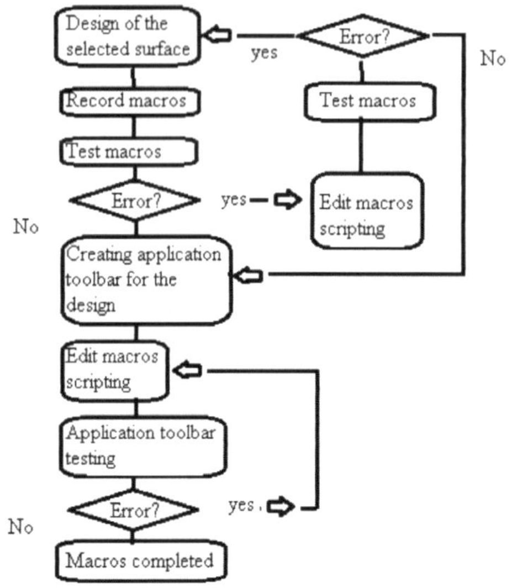

Figure 2.Flow chart of decision making for macros programming.

Generally the design of the selected surface will start by sketching the requirement of the design according to certain known requirement. The data is recorder and later testing is done. If there is an error, the process will have to start again. If successful, the macros are edit and application toolbar is created. The toolbar is test for the last time, to check for some errors, if no errors, the macros of the design are completed.

It is important to ensure that all feature created will follow the requirement that is set by looking at the structure and its contents. The only best solution if error still occurs on the scripting part is to start and recorder the sketch step as per pervious procedure. Observations of the sketches are the most important consideration. Any value that is created, the sketches have to be written down so that the previous value is notified, later for the purpose of editing the macros.

The macros are edit by adding a message box functions. There are 2 basic internal function added on the macros line such as MsgBox() functions: the format such as written below:

yourMsg=MsgBox(Prompt, Style Value, Title) and

myMessage=InputBox(Prompt, Title, default_text, x-position, y-position)

This message box is use to create a pop-up message box and this allow the user to enter the command. Another statement of the macros, such as element value, other function such as formula, and if, else statement. Below is the sample of the line program that is added or edit and the message that will pop up when the program is running.

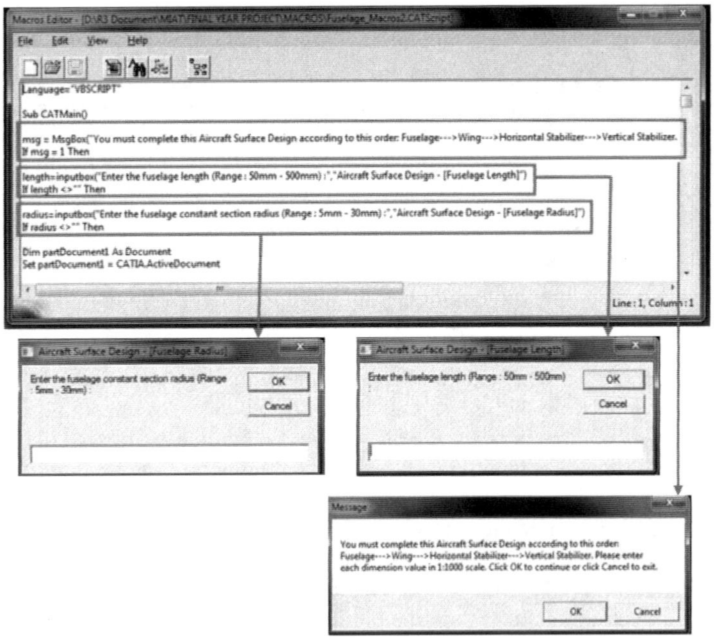

Figure 3. Shows the program line a Massage box pop-up.

In order to look whether the macros works, the macros were tested for it functions. A toolbar icon is created to ease the user using the macros programming. By clicking a toolbar icon, a massage window will pop up. The user has to follow all of the requirement of the design before the surface is generate; any parameter that is not enter will result in error to generate the surface design.

Below is the two variation of the design created using macros programming.

DESCRIPTION	VALUE (m)
Fuselage Length	36.450
Fuselage Radius	2.005
Wing Dihedral Angle	6°
Wing Sweepback Angle	25°
Wing Span Length	28.910
Wing Root Chord Length	7.320
Horizontal Stabilizer Dihedral Angle	7°
Horizontal Stabilizer Sweepback Angle	30°
Horizontal Stabilizer Span Length	12.700
Horizontal Stabilizer Root Chord Length	3.840
Vertical Stabilizer Sweepback Angle	35.000
Vertical Stabilizer Height	6.140
Vertical Stabilizer Root Chord Length	7.500

DESCRIPTION	VALUE (m)
Fuselage Length	63.700
Fuselage Radius	3.100
Wing Dihedral Angle	6°
Wing Sweepback Angle	31.640°
Wing Span Length	60.900
Wing Basic Chord Length	15.300
Horizontal Stabilizer Dihedral Angle	7°
Horizontal Stabilizer Sweepback Angle	35°
Horizontal Stabilizer Span Length	21.500
Horizontal Stabilizer Root Chord Length	7.028
Vertical Stabilizer Sweepback Angle	40°
Vertical Stabilizer Height	9.800
Vertical Stabilizer Root Chord Length	9.400

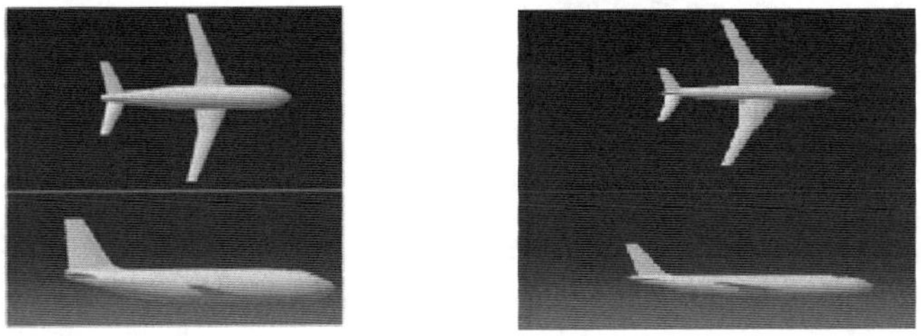

Figure 4. Automated aircraft surface design for Boeing 737-400 and Boeing 777

4. Results and Discussion

The studies compare between two different configurations between Boeing 737-400 and Boeing 777, where both design required 13 parameters with different value. Both have been create successful using the macros. The scale of the design has been set to a ratio of 1:1000, so value are entered in the unit of mm because the measure scale of the software are in a unit of mm instate of meter. These macros can also be used with different configuration of the design that fall under same requirement. The important aspect that should take into account is the parameter that is required for the design. Basically there are some changes of the design requirements from one model to another, where a new parameter need to be added to the macros structure to create a new set of the aircraft configuration. The automated application is using macros programming helps the designer by reducing the time to redesign or alter the complex problem of the design. The reason of macros programming is to create a different group of configuration from the previous design, therefore create a group of similar design with different variation of configuration. This will help what if questions from the design expect of view when tailor the specification of the aircraft. Basically having an automated design helps a user from different area of field to generate the model that is required without the full knowledge of the CAD software because by simple click of the toolbar, entering the requirement of the design, a new design concept is created.

Parametric models have the general purpose of providing a framework for manipulation of geometrical components that perform transformation during the design process. Such advantage discuss by [4] are, be able to perform changes of a geometrical components without erasing a redrawing, increase reusability of the design, allow some transformations, while restricted others and it provide real time feedback when changes in the parametric model to another of the design take place. In reality, the approach has proven to be very useful to capture different variation of the design in the early stage of conceptual design beside reduce time of recreate a new variation of design.

The problem with macros programming is the knowledge of the user towards the creation of the macros. Basic knowledge of programming using visual basic and the structure line of CATIA V5 help a lot toward understands the syntax from the command. Observation and learning will be important so that the structure of the program purely understood; the concept of the design must be gathered so before the sketch of the object is take into account. There is no short cut in programming, because the more complex the requirement of the design, the more difficult the macros need to be edit and created, this is where the requirement to master the CAD software is important.

The most difficult task that the designer has to solve is how to overcome the initial setup such as deciding what variable need to be parameter which can be a time consuming but worthwhile. Perhaps a carefully and accurate analysis of the pre-conditioning of the setup would provide some solution in the early stage.

5. Conclusion

The design of two different configurations of the commercial aircraft has been done successfully. There is no problem to create a different model from the set of requirement as long the design fall under the same group of designers. In the early stage of the macros programming of a model, if can be found that most of the syntax of the program fall under missing or wrong command from the program line and writing out the set of rules where it is impossible.

The surface generation using macros programming has brought many benefits to designer by reducing a time and improve the conceptual design stage. Improper assumptions made at the conceptual design stage may result in difficulties in meeting the requirement of the design and therefore will require the designer to start the process all over again.

References

[1] E Mermoz, J M Linares and A Bernard 2011 Benefits and limitations of parametric design implementation in helicopter gearbox design phase CIRP Annals – Manufacturing Technology 60 199-02

[2] C Ledermann, C Hanske, J Wenzel, P Ermanni and R Kelm 2005 Associative parametric CAE methods in the aircraft pre-design *Aerospace science and technology* **9** 641-51
[3] Daniel M Fudge and David W Zingg 2004 A CAD-Free and a CAD-Based geometry control system for aerodynamic shape optimization AIAA 1-15
[4] Carlos Roberto Barrios Hernandez 2006 Thinking parametric design: Introducing parametric Gaudi *Design Studies* **27** 309-24
[5] A Veroust, F Schonek and D Roller 1992 Rule oriented method for parameterized computer-aided designs *Computer Aided Design* 531-40
[6] Guk-Heon Choi, Duhwan Mun and Soonhung Han 2002 Exchange of CAD part models based on the macro-parametric approach *International Journal of CAD/CAM* **2**
[7] Michael Athanasopoulas, Hassan Ugail and Gabriela Gonzales Castor 2009 Parametric design of aircraft geometry using partial differential equations *Advances in Engineering Software* **40** 479-86
[8] Narayan K Lalit 2008 *Computer Aided Design and Manufacturing* New Delhi: Prentice Hall of India
[9] Andrews P T J, Shahin T M M and Sivaloganathan S 1999 Design reuse in CAD environment *Computer & Industrial Engineering* **37** 105-09
[10] A Veroust, F Schonek and D Roller 1992 Rule oriented methods for parametrized computer-aided designs *Computer Aided Design* **24** (10) 531-40
[11] Stiny G and Gips J 1978 Algorithmic aesthetics: computer models for criticsm and design in the arts 220 p
[12] Ibrahim Zeid 2005 *Mastering CAD/CAM* Mc Graw Hill
[13] C Bill 1998 Development and application of a computer-based system for conceptual aircraft design Delft University Press

IOP Conf. Series: Materials Science and Engineering **50** (2013) 012042 doi:10.1088/1757-899X/50/1/012042

Computational analysis on plug-in hybrid electric motorcycle chassis

S J Teoh[1,a], R A Bakar[1,2], and L M Gan[1,2]
[1] Faculty of Mechanical Engineering, Universiti Malaysia Pahang, 26600 Pekan, Pahang, Malaysia
[2] Automotive Engineering Centre, Universiti Malaysia Pahang, 26600 Pekan, Pahang, Malaysia

E-mail: [a]jacky.mucklow@iop.org

Abstract. Plug-in hybrid electric motorcycle (PHEM) is an alternative to promote sustainability lower emissions. However, the PHEM overall system packaging is constrained by limited space in a motorcycle chassis. In this paper, a chassis applying the concept of a Chopper is analysed to apply in PHEM. The chassis 3dimensional (3D) modelling is built with CAD software. The PHEM power-train components and drive-train mechanisms are intergraded into the 3D modelling to ensure the chassis provides sufficient space. Besides that, a human dummy model is built into the 3D modelling to ensure the rider's ergonomics and comfort. The chassis 3D model then undergoes stress-strain simulation. The simulation predicts the stress distribution, displacement and factor of safety (FOS). The data are used to identify the critical point, thus suggesting the chassis design is applicable or need to redesign/ modify to meet the require strength. Critical points mean highest stress which might cause the chassis to fail. This point occurs at the joints at triple tree and bracket rear absorber for a motorcycle chassis. As a conclusion, computational analysis predicts the stress distribution and guideline to develop a safe prototype chassis.

1. Introduction

A plug-in hybrid electric motorcycle (PHEM) is equips with two or more energy source, typically a gasoline internal combustion engine is coupled with an electric motor to propel the PHEM. The electric system in PHEM operates at high efficiency, allow diversification of energy resources, zero local emission and work silently [1,2]. However, PHEM require more components and complex drive-train mechanism as compare to conventional internal combustion engine motorcycle [3,4]. These cause the design of PHEM overall system packaging is challenging as constraint by limited space in a motorcycle chassis [5].

For instant, there are several motorcycle chassis design available with varies features and applications. These include Cruiser, Sport Bike, Touring, Standard, Dual-Sport, Scooters and Off-Road type [6]. The chassis is the central frame of the motorcycle which upholds the components and loads. These include the weight of each component and the forces manifest during acceleration, deceleration and cornering [7]. The chassis design must be able to withstand above loads besides consideration for the rider safety, fuel efficiency and ergonomics [8-10].

Chopper is type of cruising motorcycle, which originally means custom made motorcycle ("chopped") or built from scratch to have unique hand-crafted appearance. The main features include extended frame, raked fork, hardtail-style, chrome finishes, ape hanger handlebars and feet forward riding posture. Figure 1 illustrates the features in Chopper type motorcycle.

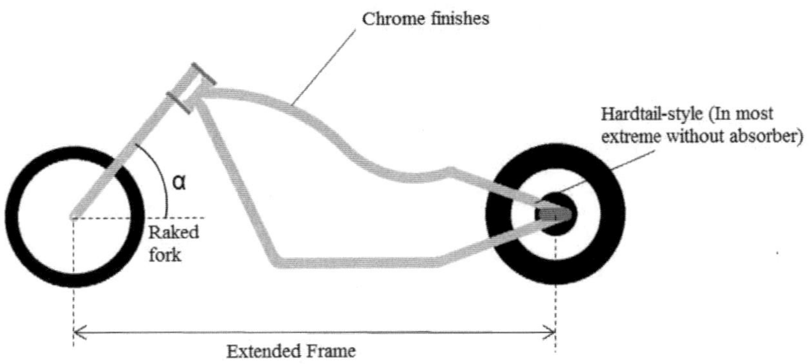

Figure 1. Features in Chopper type motorcycle

2. Chassis design

A Chopper type motorcycle chassis is chosen due to increasing demand [11] but market is lack of hybrid electric Chopper type motorcycle besides providing sufficient space for the PHEM components and drive-train mechanism. Material selected in this study is round tube hollow stainless steel 304 diameter 1.5 inches thickness 1.2mm and diameter 1.0 inch thickness 1.0mm. Material stainless steel 304 is chosen due to its manufacturability, strength and reflective appearance which suit to build a Chopper type chassis. Figure 2 shows the chassis 3D modelling with CAD software SolidWorks 2012.

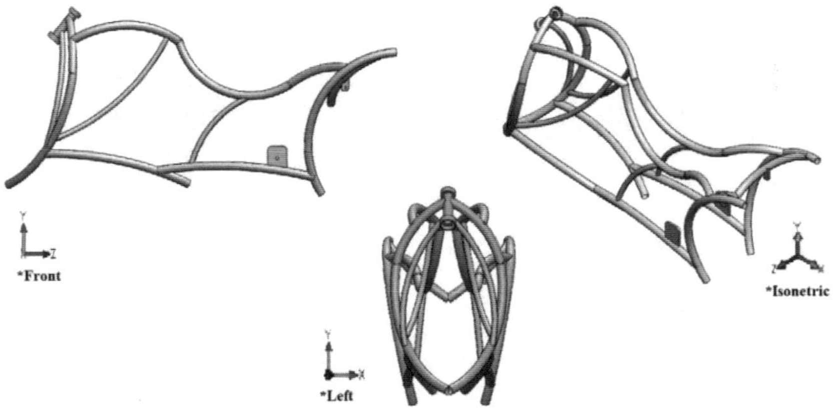

Figure 2. Chassis 3D modelling with CAD software SolidWorks 2012

The PHEM components and drive-train mechanisms are intergraded into the 3D model to ensure the chassis provides adequate space. Figure 3 shows chassis 3D modelling intergraded with PHEM components and drive-train mechanism.

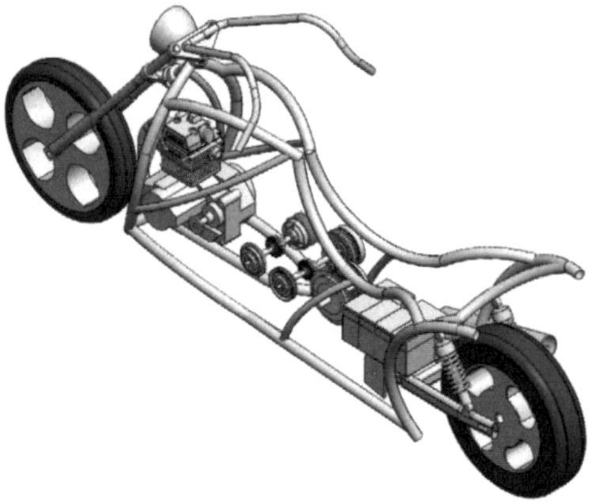

Figure 3. Chassis 3D modelling intergraded with PHEM components and drive-train mechanism.

Then, an adult human dummy model is built into the 3D modeling to verify the rider's ergonomics and comfort. Figure 4 shows the 3D modeling with an adult human dummy.

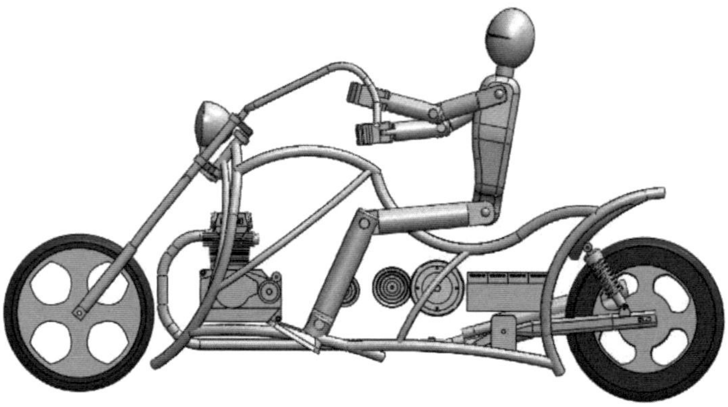

Figure 4. 3D modelling with an adult human dummy.

3. Finite element analysis

Stress-strain simulation feature is available in SolidWorks 2012. In static load, triple tree and bracket absorber rear are set as fixed geometry, forces are applied on the chassis as the load from the components and the weight of the rider. Force F1 is the weight from the petrol tank, force F2 force is the weight of the rider. F3 includes the weight from engine, control box and drive-train mechanism, force F4 include weight from drive-train mechanism, batteries and electric motor. Finite Element Analysis (FEA) provides reliable numerical technique for analysing chassis design. The FEA software mesh the model with a number of small pieces of standard shape or element, connected at common points or nodes. Figure 5 illustrates chassis 3D model with fixtures and loads.

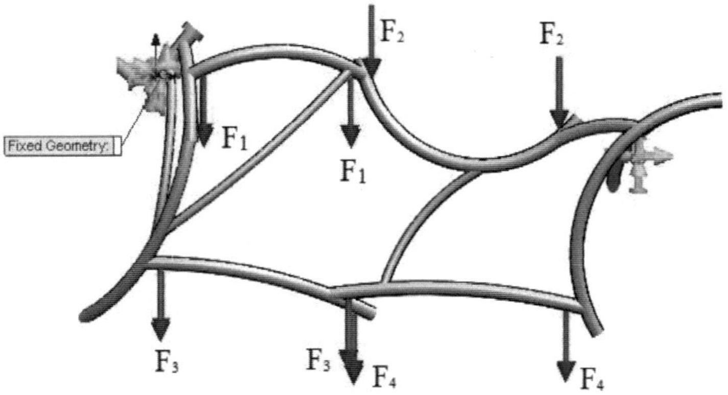

Figure 5. Chassis 3D model with fixtures and loads.

4. Governing equation

The study of mechanics of materials is to analyse and design of load-structures. Both the analysis and design of a given structure involve the determination of stresses and deformations. When loads are applied to the chassis while being held fix, the chassis deforms both locally and throughout the body. These induce internal forces as well as reaction forces to render the body in state of equilibrium [12].

In general, the stress analysis software package offers four failure criteria to access the safety of design and list in table 1.

Table 1. Summary of failure criteria

Criterion	Material Type
Maximum von Mises stress	Ductile
Maximum shear stress	Ductile
Mohr-Coulomb stress	Brittle material with different tensile and compressive strengths
Maximum normal stress	Ductile

The maximum von Mises stress criterion is based on the von Mises-Hencky theory, also known as Shear-energy theory or the maximum distortion energy theory. In terms of principal stresses σ1, σ2 and σ3, the von Mises stress is expressed as:

$$von\ Mises = \left(\frac{(\sigma_1-\sigma_2)^2+(\sigma_2-\sigma_3)^2+(\sigma_1-\sigma_3)^2}{2}\right)^{1/2} \quad (1)$$

The theory states that a ductile material starts to yield or at a location when the von Mises stress becomes equal to the stress limit:

$$von\ Mises > limit \quad (2)$$

In most cases, the yield strength is to be used as the stress limit. The factor of safety (FOS) at a location is calculated with below equation. The chassis is safe when the FOS value is more than 1.

$$FOS = \frac{\sigma_{limit}}{\sigma_{von\ Mises}} \quad (3)$$

5. Results and discussion

Figure 7 shows the simulation result for von Mises Stress.

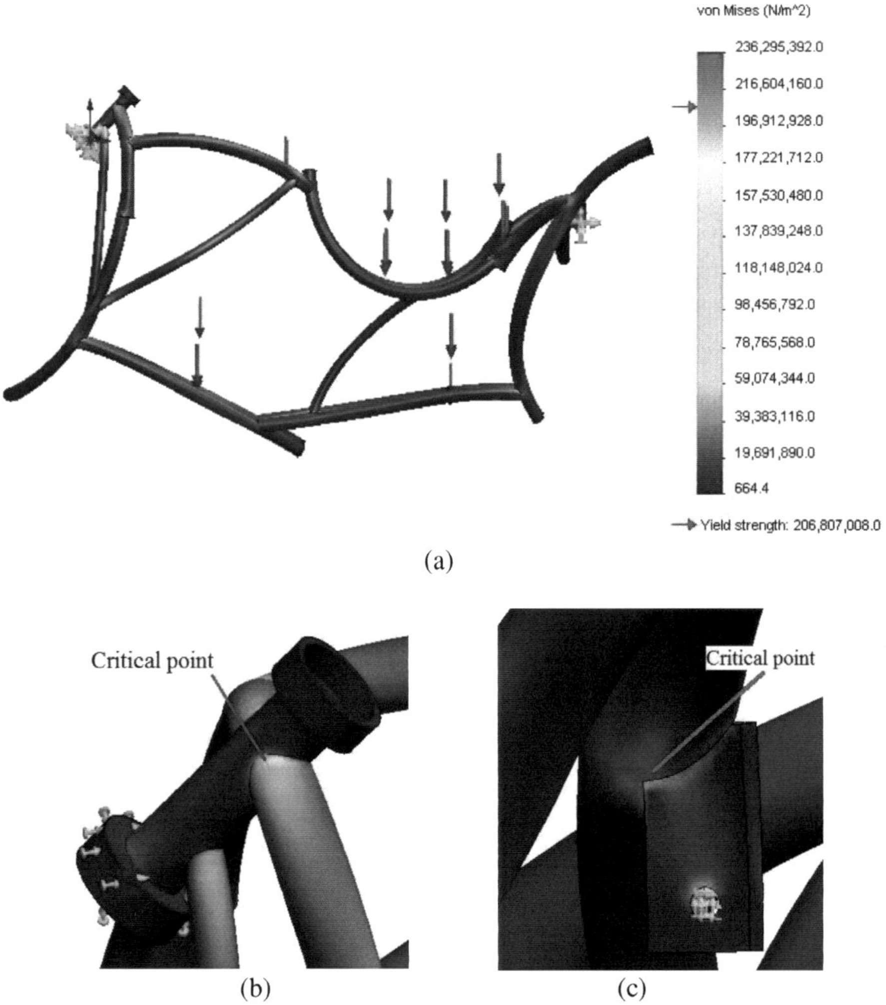

Figure 7. Simulation result (von Mises Stress).

The simulation predicted the stress distribution. The area with greenish colour shown in figure 7(b) and 7(c) suggested the highest stress occurs at the triple tree joint and bracket for rear absorber. In other word, the critical point where the chassis will likely to fail at excessive load.

Apart from predicting stress distribution, the FEA software also calculates for displacements, FOS and strain. Figure 8 shows the simulation results for displacements, FOS and strain.

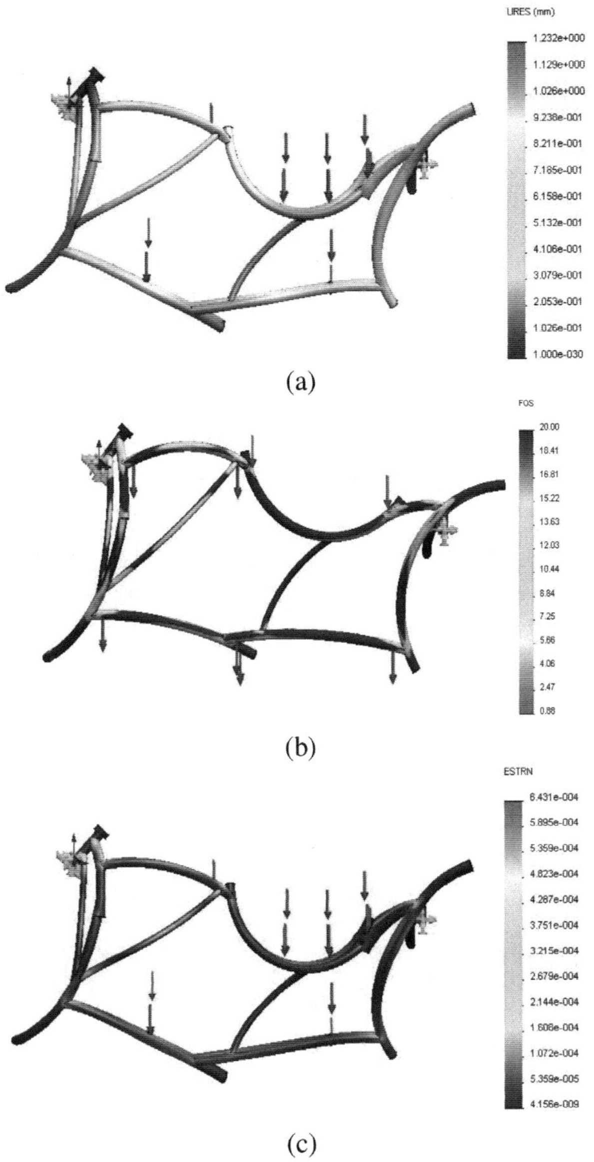

Figure 8. Simulation results (a) Displacement, (b) FOS and (c) Strain

The simulation result for displacement show highest displacement is at the frame where the rider sits and the frame supporting the weight of engine, electric motor, controller, batteries and drive-train mechanism. The maximum displacement is 1.232mm, thus providing insignificant undesirable effect neither chassis nor rider. The simulation result for FOS shows foremost of the chassis is having FOS value > 20 and part of the chassis FOS value range 5 – 20. There is little area showing FOS value < 1 suggesting strengthening is require.

After strengthening and achieving required FOS, the PHEM chassis will be undergoing construction stage. Computational analysis predicted the stress distribution, displacements, factor of safety (FOS) and strain for the PHEM chassis.

6. Conclusion

Computational analysis is widely used in engineering design especially in chassis or framework design. This is because data from computational analysis is able to provide guideline to develop a safe structure in shorter time. In this study, the computational analysis predicted the stress distribution, displacements, factor of safety (FOS) and strain for the PHEM chassis, the data are analysed and used to strengthen the critical area, as a result, a safe PHEM chassis design is developed.

Acknowledgments

The authors would like to thank the Faculty of Mechanical Engineering in Universiti Malaysia Pahang (UMP) and Universiti Malaysia Pahang for financial support under RDU120352.

References

[1] Nakamura S, Senoh M, Nagahama S, Iwase N, Yamada T, Matsushita T, Kiyoku H and Sugimoto Y 1996 *Japan. J. Appl. Phys.* **35** L74

[1] Chau K T and Wong Y S 2002 Overview of power management in hybrid electric vehicles *Energy Conversion and Management* **43** 1953–68

[2] Amjab S, Neelakrishnan S and Rudramoorthy R 2010 Review of design considerations and technological challenges for successful development and deployment of plug-in hybrid electric vehicles *Renewable and Sustainable Energy Reviews* **14** 1104-10

[3] Sheu K B and Hsu T H 2006. Design and implementation of a novel hybrid-electric-motorcycle transmission *Applied Energy* **83** 959-74

[4] Wu X, Cao B, Li X, Xu J and Ren X 2011 Component sizing optimization of plug-in hybrid electric vehicles *Applied Energy* **88** 799-04

[5] Bayindir K C, Gozukucuk M A and Teke A 2011 A comprehensive overview of hybrid electric vehicle: Powertrain configuration, powertrain control techniques and electronic control units *Energy Conversion and Management* **52** 1305-13

[6] Teoh E R and Campbell M 2010 Role of motorcycle type in fatal motorcycle crashed *Journal of Safety Research* **41** 507-12

[7] Mat M H and Ghani A R 2012 Design and analysis of 'Eco' car chassis *Procedia Engineering* **41** 1756-60

[8] Kamal M, Rahman M M and Rahman A G A 2012 *J Mech Eng Sci* **3** 291

[9] Robertson S A and Minter A 1996 A study of some anthropometric characteristics of motorcycle riders *Applied Ergonomics* **27** 223-29

[10] Kamal M, Rahman M M and Rahman A G A 2013 *Inter J Automot Mech Eng* 7: 912

[11] Nagarkar M P, Vikhe G J, Borole K R and Nandedkar V M 2011 *Inter J Automot Mech Eng* **3** 364

[12] Micro D C 2009 *Hogs & Highways* (England: Abbeydale Press)

[13] Beer F P, Johnston E R, Dewolf, J T and Mazurek D F 2009 *Mechanics of Materials* (New York: McGraw-Hill)

IOP Conf. Series: Materials Science and Engineering **50** (2013) 012043 doi:10.1088/1757-899X/50/1/012043

Heat transfer simulation of motorcycle fins under varying velocity using CFD method

K.Shahril[a], Nurhayati Binti Mohd Kasim, and M.Sabri[b]
Mechanical Section, Universiti Kuala Lumpur Malaysian Spanish Institute, Kulim Hi-Tech Park, 09000, Kulim, Kedah, Malaysia

Email: [a]khairuls@msi.unikl.edu.my, [b]msabri@msi.unikl.edu.my

Abstract. Motorcycle engine releases heat to the atmosphere through the mode of force convection. To solve this, fins are provided on the outer of the cylinder. The heat transfer rate is defined depending on the velocity of vehicle, fin geometry and the ambient temperature. Increasing the temperature difference between the object and the environment, increasing the convection heat transfer coefficient, or increasing the surface area of the object increases the heat transfer. Many experimental methods are available in literature to analyze the effect of these factors on the heat transfer rate. However, CFD analysis will be use to simulate the heat transfer of the engine block. ANSYS software is selected to run the simulation.

1. Introduction

This project focuses on studying the simulation of heat transfer of motorcycle fin under varying climatic condition. Fin act as the important part that help to reduce the distribution of overheat by the motorcycle engine block. The study includes modelling the motorcycle fin using solid model as the references used for CATIA software. Furthermore, the scope of the study had two model engine block of motorcycle will to simulated, where is Kriss 110 and Lagenda 110.

Heat transfer simulation will be conducted using ANSYS software. This simulation is used to determine the relationship between velocity, temperature and heat transfer coefficient of motorcycle fin. The purpose of the selection two model engine block is to make the comparison which is more effective to help the engine block maintain the heat is needed.

The rate of heat transfer that will distribute depends upon the engine block model, number of fin, fin pitch and wind velocity. There is an optimal cooling rate of an engine for its efficient operation. If the cooling rate decreases, it results in overheating leading to seizure of the engine. At the same time, an increase in cooling rate affects the starting of the engine and reduces efficiency.

The objective of this project is to study relationship between velocity and heat transfer coefficient (h). Others objective are to compare motorcycle fin modeling with different type of motorcycle engine block that have different design and to analyze relationship between velocity and varying temperature. The existing of motorcycle engine block was used in this study.

The problem of this part is it contribution of heat transfer coefficient (h), Temperature (T), Velocity, Fin Geometry (thickness, gap), material and surface roughness, as shown in Figure 1. Heat transfer rate that contribute while running the engine depends upon the velocity of vehicle, fin geometry and the ambient temperature. If the cooling rate decreases, it results in overheating leading to seizure of the engine. At the same time, an increase in cooling rate affects the starting of the engine and reduces efficiency. It is means that the efficiency of heat transfer rate will be improved with the result of this study.

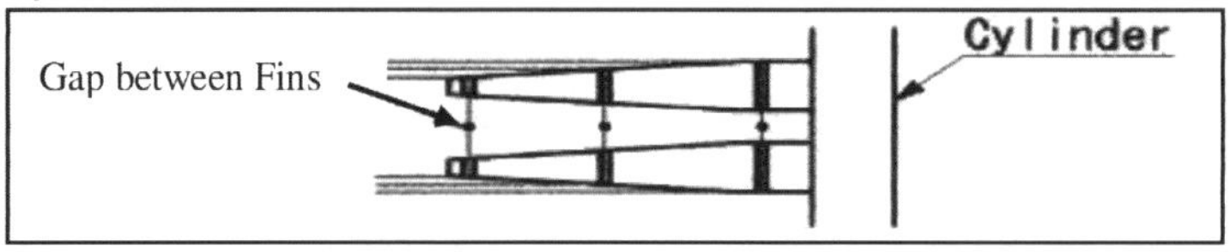

Figure 1. Gap between Pitches.

This study consist two type of motorcycle engine block, which is Modenas Kriss 110 and Yamaha Lagenda 110z, as shown in Figure 2. This selection is necessitated caused to get the differences of fin performance based on different brand. Engine block was designed with different geometry based on brand of motorcycle. Besides that, shape and material of engine block are considered.

Figure 2. Modenas Kriss 110 and Yamaha Lagenda 110z engine blocks.

In study of heat transfer, a fin is a surface that it is extends from an object to increase the rate of heat transfer to or from the environment by increasing convection. The amount of conduction, convection or radiation of an object determines the amount of heat transfer. Increasing the temperature difference between the object and the environment, increasing the convection heat transfer coefficient, or increasing the surface area of the object increases the heat transfer. Sometimes it is not economical or it is not feasible to change the first two options. Adding a fin to an object, however, increases the surface area and can sometimes be an economical solution to heat transfer problems.

Motorcycle fin is designed with several type of it. It is constant area straight fin, variable area straight fin, pin fin and annular fin. The reason of designing is commonly to analysis the effect of the heat transfer wherein the type of design is one of the factors that gave effect. The purpose of designing fin is to obtain the effectiveness of the fin. Although the fins significantly increase heat transfer from the cylinder, considerable improvement could still be obtained by increasing the number of fins.

Besides that, the thickness of fin, gap between fins are important in order the designing. Based on the type of fin (Figure 3), motorcycle engine block is normally design with the constant area straight fin. But, it also has certain motorcycle fin have been designed with variable area straight fin.

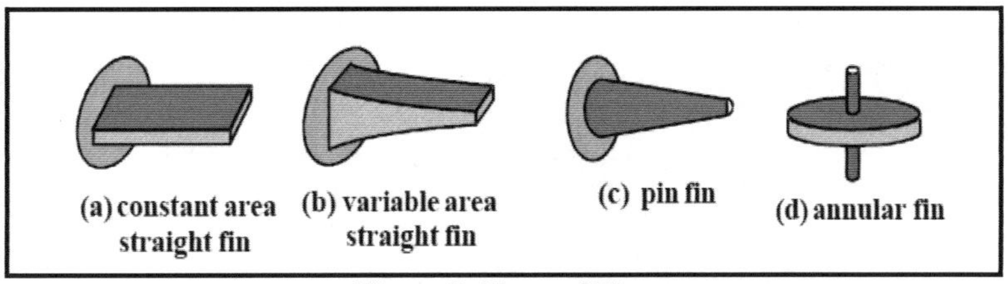

Figure 3. Type of Fins.

A fine mesh has been created near the fins to resolve the thermal boundary layer which is surrounded by a coarse external mesh for better results and fast solution. A face mesh has been done by quad element and pave scheme with size function as shown in Fig.5. The volume was then meshed by hex element and cooper scheme in GAMBIT to obtain the 3D mesh. The metallic fins have also been meshed using the same principles for obtaining accurate temperature distribution.

2. Design and Modeling

The engine block was designed based on existing engine block which is engine block Modenas Kriss 110 and Yamaha Lagenda 110z. The design was created in CATIA software as shown in Figure 4 and Figure 5.

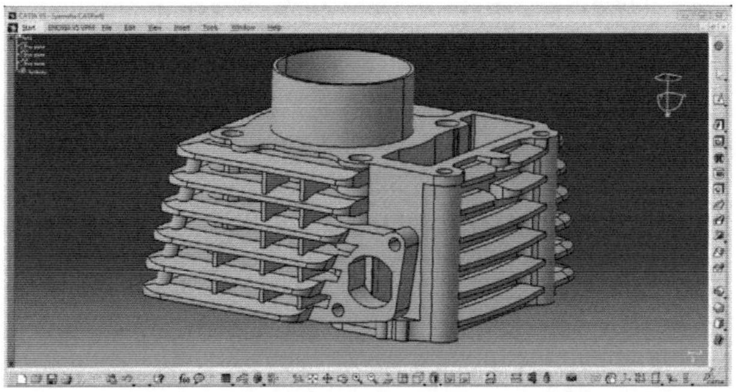

Figure 4. 3D design for Yamaha Lagenda 110z.

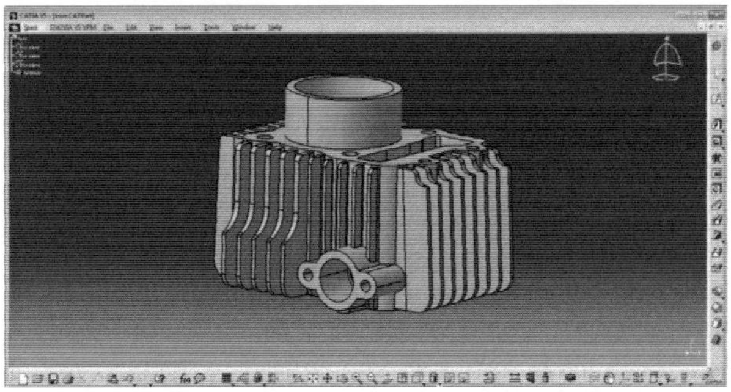

Figure 5. 3D design for Modenas Kriss 110.

The engine block was modeled as aluminum and was created in ANSYS software. Meshing has been done before run the simulation, as shown in Figure 6.

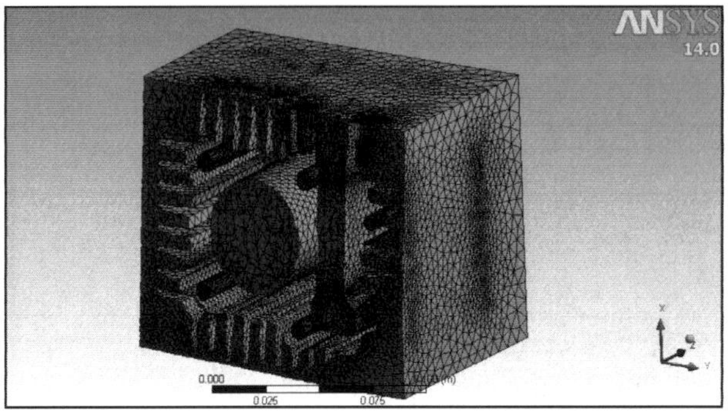

Figure 6. Meshing at engine block (Modenas Kriss).

3. Result and Discussion

The engine block of motorcycle for Modenas Kriss 110 and Yamaha Lagenda 110z were simulated by using ANSYS software. In order to study relationship between velocity and heat coefficient (h), a various value of air velocity of wind were determined. The air velocity of wind is selected from 60km/h, 80km/h and 100km/h. The results of air velocity of wind against heat transfer coefficient (h) are shown in table 1.

$$h = 5.108\ u^{0.643} \tag{1}$$

Where:-
h: Average fin surface heat transfer coefficient.
u: wind velocity in km/hr.

With the heat transfer coefficient being independent of the ambient temperature, Eq. (1) relates the heat transfer coefficient with the wind velocity.

Table 1. Result of wind velocity against heat transfer coefficient for Modenas Kriss 110.

Wind velocity (m/s)	Total heat transfer coefficient (h)
16.67 (60km/h)	15.802
22.22 (80km/h)	16.532
27.28 (100km/h)	17.520

The result shows that the total heat transfer coefficient (h) increased when air velocity of wind increased. Value of heat transfer coefficient increased upon on a several reason, one of the reason is, high velocity. Table 2 shows the result of wind velocity against heat transfer coefficient for Yamaha Lagenda 110z. The values of heat transfer coefficient for Yamaha Lagenda was same with Modenas Kriss 110, which is the values of heat transfer coefficient increased when the air velocity of wind increased.

Table 2. Result of wind velocity against heat transfer coefficient for Yamaha Lagenda 110z.

Wind velocity (m/s)	Total heat transfer coefficient (h)
16.67 (60km/h)	16.013
22.22 (80km/h)	17.368
27.28 (100km/h)	18.002

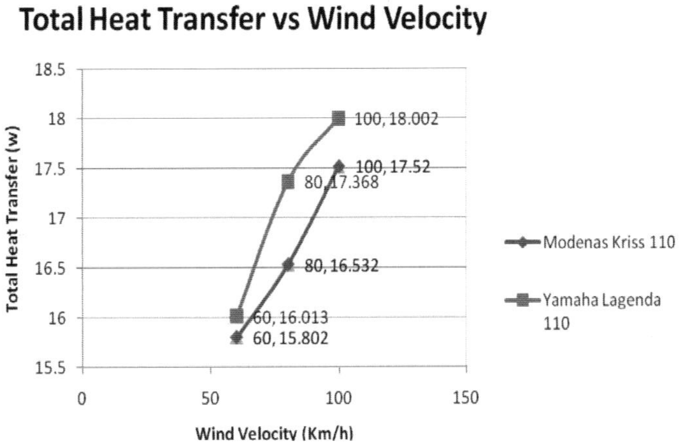

Figure 8. Wind Velocity against Heat Transfer Coefficient (Modenas Kriss 110 and Yamaha Lagenda 110z).

Comparison between Kriss and Lagenda shown in Figure 8. Yamaha Lagenda performance better than Modenas Kriss. Even though, the difference of total heat transfer between two models are not large. The values of wind velocity were determined to gain the value of heat transfer coefficient. The values of heat transfer coefficient increased when the wind velocity increased.

4. Conclusion

As the conclusion, the simulation of two type of engine block which is Modenas Kriss 110 and Yamaha Lagenda 110z has been done accordingly to the requirement. The values of wind velocity have been determined to gain the heat transfer coefficients. This simulation is proving that the wind velocity is one important part that can affected the total of heat transfer and the value of heat transfer coefficient. Besides that, the design of motorcycle fins need to considered in order to measure total heat transfer, because fins are work to trap the air to maintain the heat of engine block, if the fins design are not too appropriate with the requirement, it can cause to overheating.

Acknowledgements
The author would like to acknowledge Universiti Kuala Lumpur, Malaysian Spanish Institute (UniKL-MSI) for providing equipments and facilities to carry the project.

References

[1] Pulkit A Mayur S and P Srinivasan 2011 Heat Transfer Simulation By CFD From Fins Of An Air Cooled Motorcycle Engine Under Varying Climatic Conditions

[2] Travis J Schneider Using Finite Element Analysis to Investigate Lattice Fins for Thermal Behavior Department of Mechanical Engineering Milwaukee School of Engineering 1025 N. Broadway Milwaukee, WI 53202
[3] Al- Doori W H A R 2011. *Inter J Automot. Mech. Eng.* **4** 428
[4] Fin Types for Finned Tube Heat Exchangers. [Online]. Available http://www.deltathx.com/uploadsDocs/Fintypes.pdf
[5] Motorcycle engine. [Online]. Available: http://en.wikipedia.org/wiki/Motorcycle_engine
[6] Extended surfaces – fins. [Online]. Available http://ottp.fme.vutbr.cz/vyuka/prenostepla/Zebra.pdf
[7] Extended surfaces – fins. [Online]. available http://ottp.fme.vutbr.cz/vyuka/prenostepla/Zebra.pdf

IOP Conf. Series: Materials Science and Engineering **50** (2013) 012044 doi:10.1088/1757-899X/50/1/012044

Flow and heat distribution analysis of different transformer sub-stations

H Hasini[1,3,a], N H Shuaib[2], S B Yogendran[2] and K B Toh[2]
[1]Centre for Advanced Computational Engineering, College of Engineering, Universiti Tenaga Nasional, Jalan IKRAM-UNITEN, 43000 Kajang, Selangor, Malaysia
[2]TNB Research Sdn. Bhd., Jalan Ayer Hitam, 43000 Kajang, Selangor, Malaysia

E-mail: [a]hasril@uniten.edu.my

Abstract. This paper describes CFD investigation on the flow and heat transfer in transformers at different sub-station buildings. The analysis aimed to determine the cooling capability of the existing transformer building employing natural ventilation system to dissipate heat sufficiently when new dry-type transformer operating under full load condition is used. The transformer and building models were developed based on the actual transformer configuration in operation at three different locations in Malaysia. The calculation was carried out on three different types of sub-stations namely stand-alone, attach-to-building and underground. The effect of natural ventilation speed and building volume on the transformer surfaces temperature are also investigated. It was predicted that the existing sub-station configuration is able to dissipate heat produced from the dry type transformer by using its natural ventilation system regardless of the sub-station types. However, the smallest building case shows relatively high surrounding temperature.

1. Introduction

Currently the majority of all distribution transformers in Malaysia were of the oil-cooled type at which cooling of the transformers are provided by liquid oil and in some cases gas. The use of oil-cooled transformer has proven to be effective in helping to ensure that the transformer surface temperature does not exceed the safe temperature limit of any particular transformer. Nevertheless, the oil-cooled transformer poses several disadvantages such that it needs to be maintained frequently and the need for cooling oil-handling facility that requires higher maintenance costs. With the advent of oil-free transformer, the maintenance of transformer could be made minimal and most importantly, the maintenance cost could be significantly reduced e.g. cast resin transformer. However, it is known that the dry type transformer dissipates large amount of heat as compared to the oil-cooled one. The cooling of the dry-type transformer thus depends solely on the natural ventilation system or forced cooling using different medium such as air. Using this method, air has to be blown, usually from the bottom of the transformer in order to increase the rate of cooling of the transformer. One of the most important factor that need to be taken into account when replacing transformer of different characteristics is the ability of the existing sub-station (SS) to dissipate heat effectively when new transformer is installed.

Studies on transformer development are abundant in the open literature. The majority of these works emphasized on the detail cooling of the transformer windings [1-5]. Very limited works were given to the investigation of the cooling system for efficient heat dissipation in room containing the transformer [6]. Thus this paper aims to qualitatively investigate the ability of the existing SS to dissipate heat resulted from the replacement of traditional oil-cooled to dry-type transformer. The work focused on three different type of SS commonly found in Malaysia and the detail of the SS model is explained in the following section.

2. Sub-station model descriptions

The SS models were developed based on the full scale stations located in Malaysia. Three different types of sub-stations were tested and the description of each case is shown in table 1. Different sub-stations (SS) have different building sizes and experience different air circulation. Figure 1 shows the stand-alone SS model which has an approximate dimension of 8m×4m×3m. This is the largest SS under investigation and there are 2 transformers with rating of 1000 kV. Figure 2 shows the model of an underground-type SS which dimension is slightly smaller than the stand-alone type SS. Its dimension is approximately 6m×3m×2.5m and this SS is located at underground car park close to the vehicle entrance area. The last SS model was based on compact/attached-to-building type which dimension is the smallest among all the three SS. The dimension is approximately 4m×3.5m×2.5m and the transformer placed in this building is a single core with multiple cylindrical surface areas. This is shown in figure 3. For Case 3, detail model was developed where air and window louvers were also taken into account. The amount of air entering the SS will be calculated based on the inlet boundary conditions imposed at one end of the SS. This is a more realistic scenario whereby the volume of air entering the SS is calculated based on the pressure difference between the inside and outside of building.

Table 1. Case descriptions

Case	Descriptions
Top	Stand-alone SS
Bottom	Underground SS
Left	Compact/Attach-to-building SS

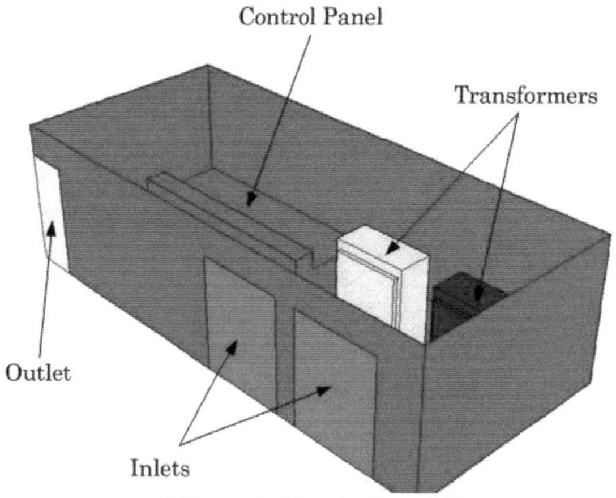

Figure 1. Stand-alone SS

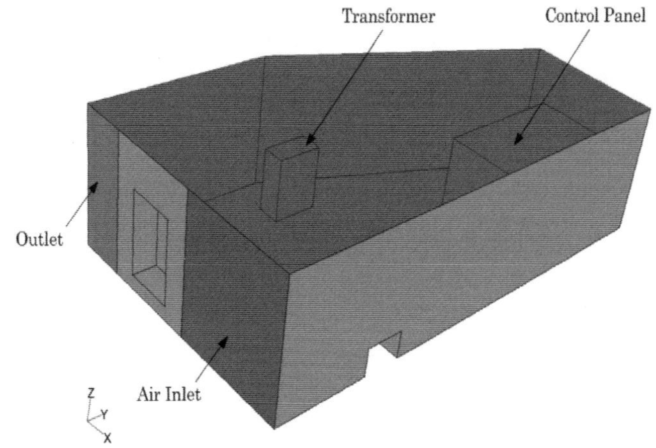

Figure 2. Underground SS

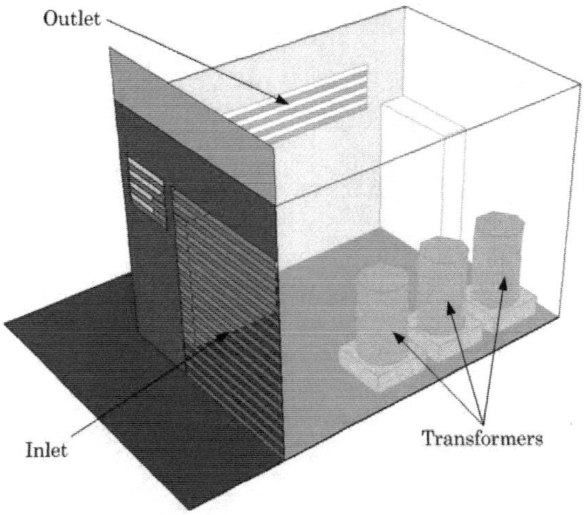

Figure 3. Attached-to-building SS

3. Governing equations

The flow inside the sub-station is fully turbulence despite the relatively weak flow as air passes through door and window louvres from the surrounding environment. The mathematical equations describing steady, flow can be casted into general form as follows:

$$\frac{\partial(\rho u_j \phi)}{\partial x_j} = \frac{\partial}{\partial x_j}\left(\Gamma_\phi \frac{\partial \phi}{\partial x_j}\right) + S_\phi \quad (1)$$

where ϕ is the three momentum components, Γ_ϕ is the diffusion coefficient of the transported variable ϕ and S_ϕ is the source term. For mass conservation equation, the variable ϕ is set to unity and thus the right hand side of the equation is zero. The introduction of turbulence into the Navier-Stokes equation resulted in extra terms that require additional equations for closure. In order to close the equations, the standard k-ε turbulence model is used to model the turbulence fluctuations (Launder and Spalding, 1972; Launder and Spalding, 1974). The two-equation models consists of one differential transport equation for the turbulence kinetic energy, k and one for its dissipation rate, ε and is given by the following equations:

$$\frac{\partial(\rho u_j k)}{\partial x_j} = \frac{\partial}{\partial x_j}\left[\left(\mu + \frac{\mu_t}{\sigma_k}\right)\frac{\partial k}{\partial x_j}\right] + \rho(P_k - \varepsilon) \quad (2)$$

$$\frac{\partial(\rho u_j \varepsilon)}{\partial x_j} = \frac{\partial}{\partial x_j}\left[\left(\mu + \frac{\mu_t}{\sigma_\varepsilon}\right)\frac{\partial \varepsilon}{\partial x_j}\right] + S_\varepsilon \quad (3)$$

Where μ_t is the eddy viscosity and is computed from:

$$\mu_t = \rho C_\mu \frac{k^2}{\varepsilon} \quad (4)$$

where C_μ is the turbulence constant. In this work, SIMPLE calculation algorithm is used for the coupling of continuity and momentum equations. The numerical equations are solved using finite volume solver, FLUENT which is able to handle different types of structured, unstructured and hybrid grids.

4. Mesh generation and boundary conditions

Generating a good mesh is extremely important in CFD in order to accurately predict the flow and heat transfer characteristics in complex geometry. In this study, a combination of structured and unstructured grid was applied to all models. Figure 4(a) to 4(c) illustrates the mesh scheme applied to all SS geometries. Higher mesh concentration was applied to critical regions and in this work, near the transformer surface and volume for resolving the conjugate heat transfer due to conduction and convection. The number of cells generated for Case 1, Case 2 and Case 3 are 1 445 298, 762 823 and 444 771 respectively. This is sufficient to resolve the flow field and grid dependency study has revealed that the current mesh is sufficient to accurately predict the flow and heat transfer characteristics inside the flow domain.

To accurately emulate the existing natural ventilation flow inside the SS, air flow measurement was taken on-site and the value obtained was imposed as the boundary condition for the computational simulation. In order to compare the heat dissipation ability for different SS, the transformer load is assumed to be identical for all cases. In this case, different loads are tested and in the range of 2.7 kW to 10.8 kW, representing 25% to 100% actual load. In reality, the transformers are operated at lower load depending on the demand requirement surrounding the area. The geometrical model of the transformers were simplified where array of fins for heat transfer enhancement were omitted and combined into a single block for each case. The effect of heat transfer enhancement was recalculated by increasing the heat transfer constant on the transformer surface.

5. Results and discussions

In general, very weak flow was predicted in all cases since the only driving force for the flow was due to natural air flow into the sub-station building. Despite the weak flow condition, the prediction of transformer surface temperature show acceptable temperature magnitude. This is extremely important to ensure that the surface temperature to be below the acceptable limit. Despite various security measures taken at each SS such as fire alarm and automatic sprinkler system, prior knowledge in estimating the maximum temperature of transformer surface temperature is vital. Figure 5 shows the variation of average transformer surface temperature against transformer load. Comparison is made between two transformers place at different locations within the same SS. In this case, transformer #1 is placed near to the corner while transformer #2 is placed next to transformer #1 towards the SS centre. The average surface temperature was higher for transformer #1 and this could be attributed to the lack of air ventilation penetration since transformer #1 is located at the SS corner. Thus, the heat dissipation for this transformer is less as compared to transformer #2. This trend is applied regardless of the transformer load. It could also be observed that at higher load, the temperature difference between the two transformers is higher. It is predicted that at full load condition, the maximum average transformer surface temperature is approximately 71°C and 74°C for transformer#2 and transformer#3 respectively.

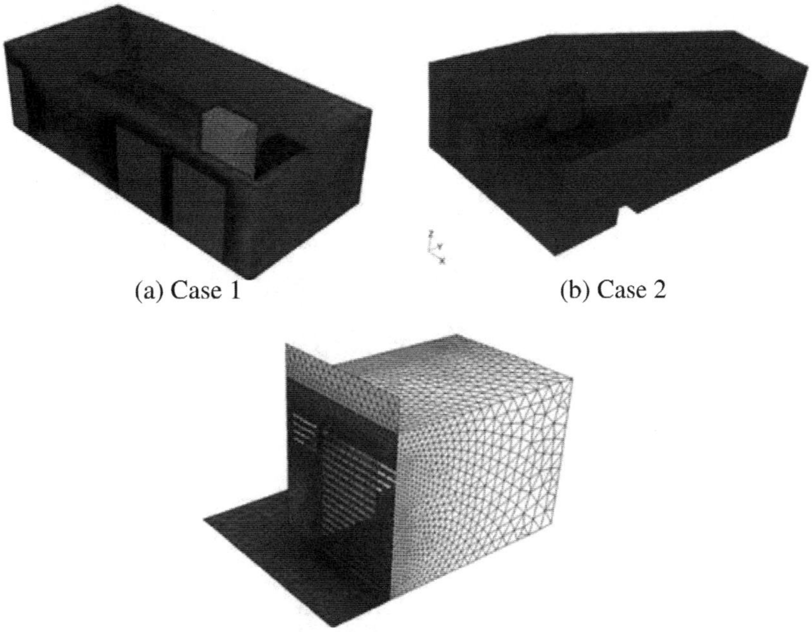

(a) Case 1 (b) Case 2

(c) Case 3

Figure 4. Mesh scheme for all SS models

The average transformer surface temperature for Case 2 is illustrated in figure 6. For Case 2, the average air speed entering the SS is slightly higher as compared to Case 1 and thus heat dissipation is more pronounced. The predicted transformer surface temperature is approximately 69°C. Another factor that might contribute to the lower transformer surface temperature in Case 2 is due to the fact that the SS only contains single transformer thus, the total heat needs to be dissipated by ventilating air is less compared to Case 1. Figure 7 shows the temperature profiles of transformer against the load for Case 3. The SS was the smallest among all the stations under investigation and thus, heat dissipation is not as good as in Case #1 and #2. The lack of ventilation air entering the critical regions within the SS causes inefficient heat dissipation on transformer surface. In addition, due to the smaller SS size, the rate of heat accumulated is greater than the heat dissipation for this SS. This results in higher surface temperature at full load condition with a magnitude of 80°C. Nevertheless the temperature is within an acceptable range. The measured local air temperature just above the top transformer surface at 25% load condition show good agreement with the predicted one.

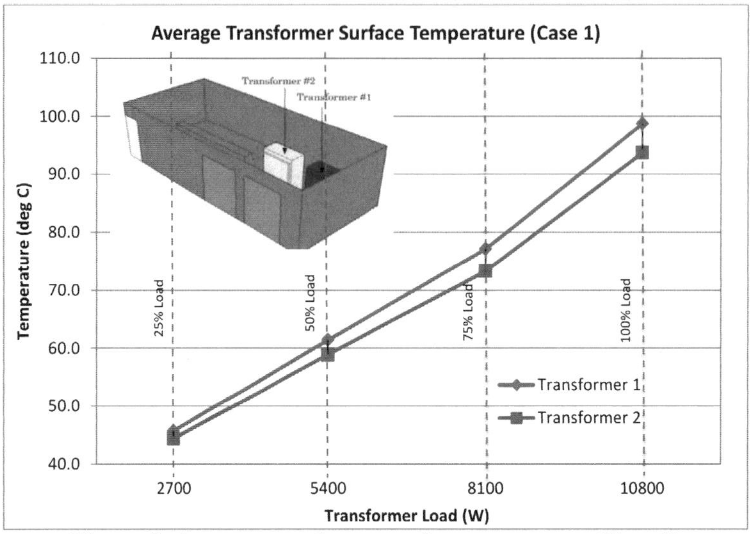

Figure 5. Average transformer surface temperature (Case 1)

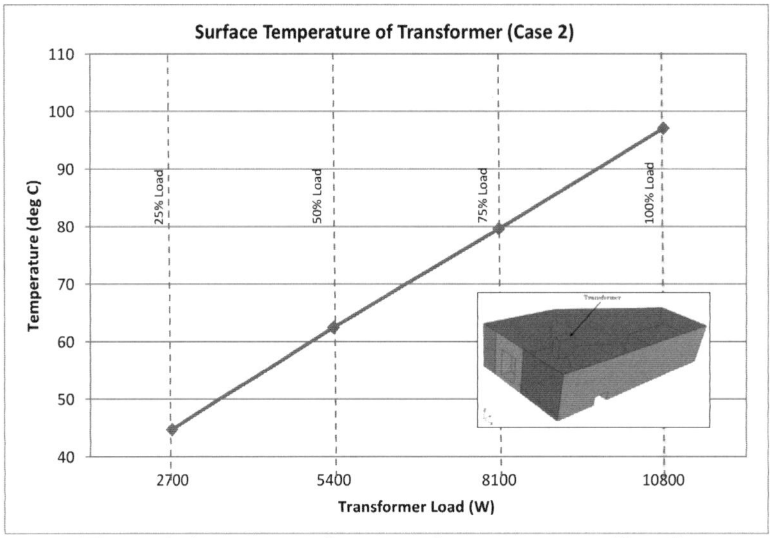

Figure 6. Average transformer surface temperature (Case 2)

6. Conclusions

The CFD calculation of flow and heat transfer in different SS models has been carried out. Emphasis is given to the prediction of transformer surface temperature when only natural ventilation air was present inside the building. Based on the prediction, the following conclusions can be withdrawn:

- The flow inside the different stations was very weak due to the slow naturally circulating air. Despite the low velocity air, the heat could be dissipated effectively due to reasonably large volume air change in each station.
- The transformer location inside a sub-station plays an important role in determining the surface temperature. Transformer located away from the ventilating air path has higher temperature as compared to the one located at the air flow path. The average transformer surface temperature in all stations increases almost linearly with the transformer load.

- The prediction of transformer surface temperature in all stations show acceptable temperature magnitude at full load conditions. Thus, the existing natural ventilation system in the stations under investigation was able to dissipate heat when the oil-cooled transformer was replaced by the dry-type.

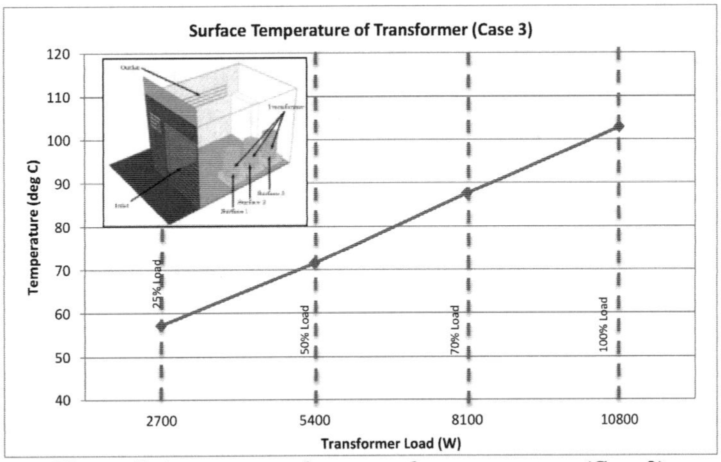

Figure 7. Average transformer surface temperature (Case 3)

Acknowledgements
The authors wish to acknowledge Tenaga Nasional Berhad (TNB) for providing the research grant to carry out this project.

References

[1] Smolka J Nowak A J and Wrobel L C 2004 Numerical modeling of thermal processess in an electrical transformer dipped into polymerized resin by using commercial CFD package fluent *Computers & Fluids* **33** 859

[2] Smolka J 2013 CFD-based 3-D optimization of the mutual coil configuration for the effective cooling in an electrical transformer *Applied Thermal Energy* **50** 124

[3] Skillen A Revell A Iacoviden H and Wu W 2012 Numerical prediction of local hot spot phenomena in transformer windings *Applied Thermal Energy* **36** 96

[4] Eslamiah M Vahid B and Eslamian A 2011 Thermal analysis of cast resin dry-type transformer *Energy Conversion and Management* **52** 2479

[5] Gastelurrutia J Ramos J C Larraona G S Rivas A and Izagirre J 2011 Numerical modeling of natural convection of oil inside distribution transformers *Appled Thermal Engineering* **31** 493

[6] Ramos J C Beiza M Gastelurrutia J Rivas A Anton R Larraona G S de Miguel I 2013 Numerical modelling of natural ventilation of underground transformer substations *Applied Thermal Engineering* **51** 852

[7] Launder B E and Spalding D B 1972 *Lectures in mathematical models of turbulence London England* Academic Press

[8] Launder B E and Spalding D B 1974 The numerical computation of turbulent flows *Comput. Methods Appl. Mech. Eng.* **3** 269

Numerical simulation of cross-flow in a bank of tubes with three rows in the subcritical region of Reynolds

Suzairin[1,2,a], **Mohd Faizal**[1,2,b], **V R Raghavan**[3,c] **and Zainal Ambri**[1,d]
[1] Department of Mechanical Engineering, Universiti Teknology Petronas, Bandar Seri Iskandar, 31750, Tronoh, Perak, Malaysia
[2] Faculty of Mechanical and Manufacturing Engineering, Universiti Tun Hussein Onn Malaysia, 86400, Parit Raja, Johor, Malaysia
[3] OYL R&D Center Sdn. Bhd., 47000, Sungai Buloh, Malaysia

E-mail: [a]suzairin@uthm.edu.my, [b]mfaizal@uthm.edu.my, [c]vijay@oyl.com.my, [d]ambri@petronas.com.my

Abstract. The present work focused on 2-dimensional unsteady numerical simulation in predicting hydrodynamics and thermal characteristics of air flow across circular tube banks with integral wake splitters. The tube banks studied consist of three rows of tubes in staggered arrangement. The lengths of the splitter are 0, 0.5, 1.0, 1.5 and 2.0 times the tube diameter. The range of Reynolds number investigated is in the range of 1000 to 10000, which is in the sub-critical region of Reynolds number. The flow condition within this range is incompressible since the maximum Mach number is less than 0.3. The numerical approach was validated against the experimental works of Zukauskas (1985) and Anderson (1997). Local pressure coefficient for flow around a single tube with integral wake splitter is also presented for comparison. It was found that the present of the wake splitters was able to improve the overall heat transfer of the system.

1. Introduction

Cross-flow heat exchanger is a common type of heat exchanger with wide range of applications. Cross-flow means that one fluid flow perpendicular to the second fluid flow direction. The most common tube used is circular in shape. The fluid flow across this tube is very complex and was subjected to many research works [1-3]. The hydrodynamics of the flow give effect to thermal characteristics of the system. Rigorous studies on this have been done by many in order to understand the underlying phenomena. Many have attempted to improve its overall performance [3-5]. Flow across a circular cylinder is a classical problem, and has been studied experimentally, visually and numerically covering various aspects of flow behavior. The development of the vortex street, many studies on the cylinder flow problem have been undertaken. Such flow in a heat exchanger has been studied for the purpose of getting high heat transfer efficiency i.e. high heat transfer rate at low required pumping power. For flow around a single tube, at very low Reynolds number in the order of one, the laminar boundary layer that starts from the forward stagnation point may not separate from the tube surface up until the rear stagnation point. As the Reynolds number increases, the boundary layer starts to detach and form symmetrical recirculation regions or vortices. These vortices which initially small begin to increase in size and shed alternately at Reynolds number in the order of two. This particular region is known as the wake region. The vortices periodically shed in a deterministic manner and the point of separation maintained at an angle of 80° from the forward stagnation point at large range of Reynolds number until critical value of 2 x 10^5.

Many researchers have done a lot of studies to improve the performance of cross-flow heat exchanger. Some added additional feature to heat exchanger in order to improve the heat transfer performance. Zhang et al. [6] added vortex generators to promote higher turbulence flow in order to enhance heat transfer rate. Tiwari et al. [7] integrated longitudinal plate to circular tube in order to split the wake region for reducing form drag. Jacimovic et al. [8] investigated the effect of plate fin to the pressure drop. Kongkaitpaiboon et al. [9] used circular-ring turbulator to improve the efficiency of heat exchanger. Some researchers modified the shape of the traditional circular tube to enhance the overall heat exchanger performance. Yajima and Sano [10], and Valera et al. [11] studied the effect of circular tube perforation towards pressure drag and heat transfer rate. Khan et al. [12] and Matos et al. [13] investigated elliptical tube array for cross-flow cooling. Shirani and Nasibi [14] studied the hydrodynamics of flow across in-line tube banks at low Reynolds number ranging from 25 to 2500 numerically. They suggested that there is an optimum value for longitudinal pitch at which drag is at minimum. Khoddamrezzaee et al. [15] investigated numerically the heat transfer characteristics of nanofluid flowing across tubes in a rectangular arrangement. It was predicted that the separation points were postponed and the heat transfer coefficients were higher than that of base fluid without nano particle. High performance cross-flow heat exchanger means that the heat transfer capability is high. However heat transfer rate almost always proportional to high pressure drop. All previous works in attempt to achieve heat exchanger design with high efficiency concentrated in improving heat transfer rate without significant increase in pressure loss, thus the focus of the present work.

2. Approach and Methods

2.1. The Cases Studied

The tube was arranged in staggered formation with the pitch-to-diameter ratio of 2. The splitter plate length-to-tube diameter ratio was varied from 0 to 2 in the increment of 0.5. The splitter plate thickness was 1/25 the tube diameter.

2.2. The Computational Domain and Boundary Conditions

The computational domain is shown in figure *2*. Uniform free-stream velocity of air enters the domain at a temperature of 300 K. Meanwhile the temperature of the tubes and splitter plates were set to be 400 K. Periodic boundary conditions were set at both side of the domain in order to reduce the computational cost. Each periodic pairs experiences the same flow and thermal conditions. The outlet boundary was distanced 20 tube diameters from the center of the 3rd row tube to minimize the effects of the outlet boundary condition on the flow in the vicinity of the tubes.

An example of a discretized domain is shown in figure *3*. Distance to the nearest tube wall, Y_{wall}, is 0.004 times the tube diameter. Structured grid was used to resolve the near-wall region. Unstructured grid with triangular cells was implemented elsewhere to ease the meshing process. This hybrid meshing technique can reduce the processing cost by reducing the total number of cells. Finer cells were created near the wall and the wake region to give better resolution on the large property gradient region.

2.3. The Governing Equations

The present study simulations were performed for unsteady, 2-dimensional flow past unconfined bank of tubes with and without wake splitters using laminar model. The followings are the conservation equation of mass, momentum and energy involved in tensor form.

$$\frac{\partial u_i}{\partial x_i} = 0 \tag{1}$$

$$\rho\left(\frac{\partial u_i}{\partial t}+u_j\frac{\partial u_i}{\partial x_j}\right)=-\frac{\partial p}{\partial x_i}+\frac{\partial \tau_{ij}}{\partial x_j} \tag{2}$$

where

$$\tau_{ij}=\left[\mu\left(\frac{\partial u_i}{\partial x_j}+\frac{\partial u_j}{\partial x_i}\right)\right]-\frac{2}{3}\mu\frac{\partial u_j}{\partial x_i}.\delta_{ij}$$

is the stress tensor

$$\rho c_p\left(\frac{\partial T}{\partial t}+u_j\frac{\partial T}{\partial x_j}\right)=\frac{\partial}{\partial x_j}\left(k\frac{\partial T}{\partial x_j}\right) \tag{3}$$

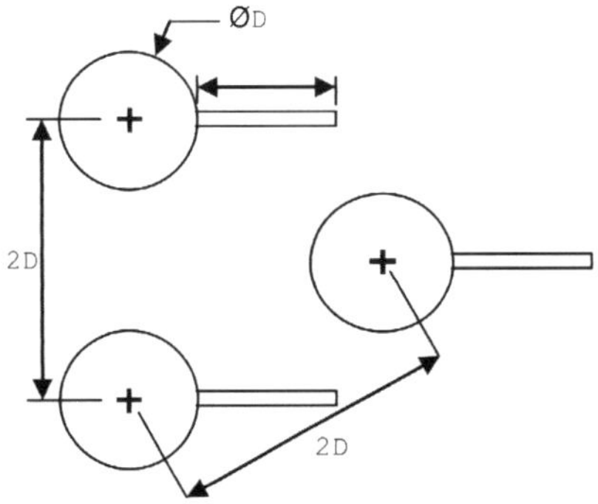

Figure 1. The investigated configuration.

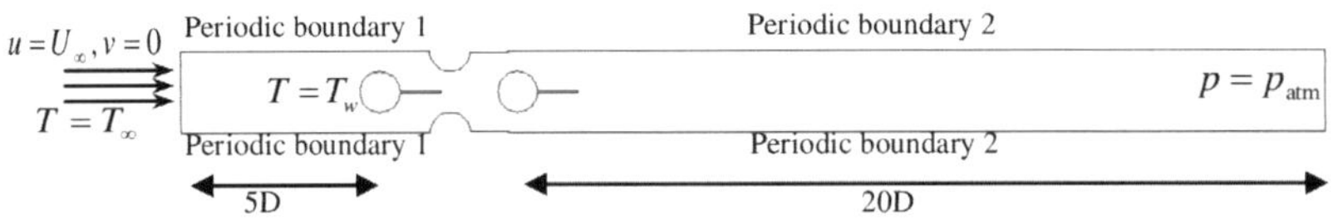

Figure 2. The computational domain.

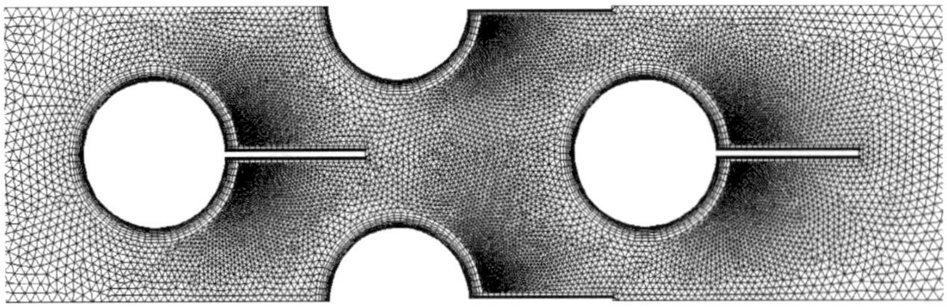

Figure 3. The discretized computational domain.

3. Numerical Results and Discussions

3.1. The Validations

The present numerical approach was tested by validating it against the experimental works of Anderson and Szewczyk [16], and Zukauskas and Ziugzda [17]. Good agreements were obtained between the numerical and the experimental data. They were evaluated quantitatively by calculating the root mean square of error, RMSE. The RMSE are 0.23 and 0.31 respectively.

$$\mathrm{RMSE} = \sqrt{\frac{\sum_{i=1}^{n}(x_{1,i} - x_{2,i})^2}{n}} \tag{4}$$

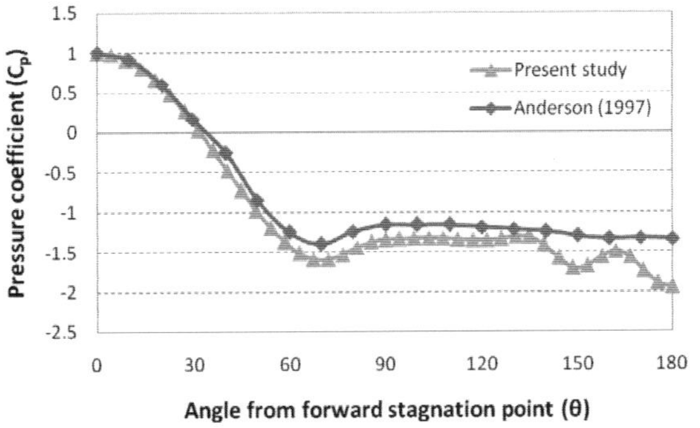

a) C_p distribution around the cylinder for Re = 4.2 x 10^4 and BR = 1/4

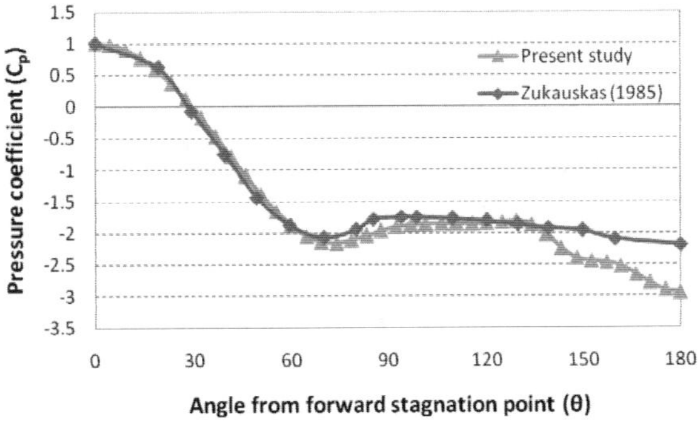

b) C_p distribution around the cylinder for Re = 4.6 x 10^4 and BR = 1/17

Figure 4. Validation against a) Anderson et. al. 1997 and b) Zukauskas et. al. 1985.

3.2. The Grid Sensitivity Study

The sensitivity of the numerical results can be observed from figure 5. Apparently, the results are not sensitive to change of grid or cell density. According to table 1, the maximum RMSE of the predicted local pressure coefficient is below 0.1.

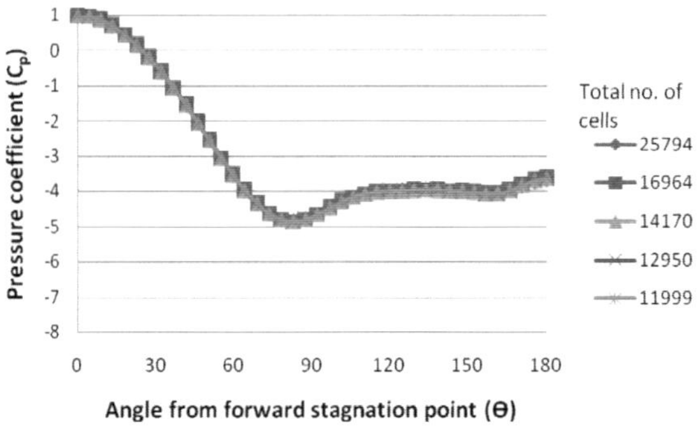

a) 1st row

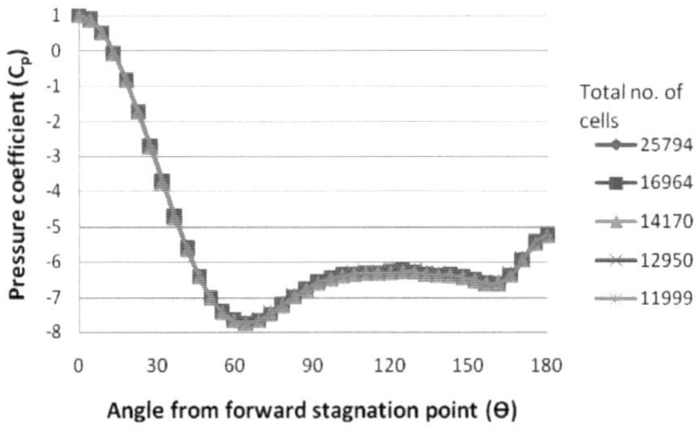

b) 2nd row

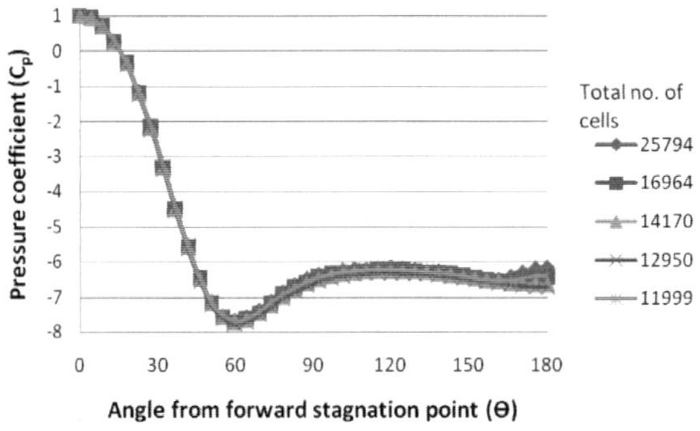

c) 3rd row

Figure 5. C_p distribution around the tubes at 1st, 2nd and 3rd row for Re = 1000 and L/D = 0.0 for various no of total elements

Table 1. RMSE of C_p distribution around the tubes at 1st, 2nd and 3rd row for Re = 1000 and L/D = 0.0 compared to its mean result for various total no. of cells

Total no. of cells	RMSE (compared to mean value)		
	Row no.		
	1	2	3
25794	0.007574	0.01112	0.096945
16964	0.001859	0.005152	0.009259
14170	0.002595	0.006167	0.04767
12950	0.005521	0.007286	0.089357

3.3. Local Pressure Coefficient

For flow around single tube, shown in figure 6, the base pressure (pressure near the rear stagnation point) of tubes with splitter plate are higher than that of plain tube. This indicates that the splitter plates have effectively reduced the pressure loss, hence reducing drag. However, for flow across bank of tubes, illustrated in figure 7, the splitter plates, in most cases increase the pressure loss. Only at the 3rd row, splitter plate of L/D = 1.5 and L/D = 2.0 at Re = 1000, and splitter plate of L/D = 0.5, L/D = 1.0, L/D = 1.5 and L/D = 2.0 at Re = 10000, give positive effects.

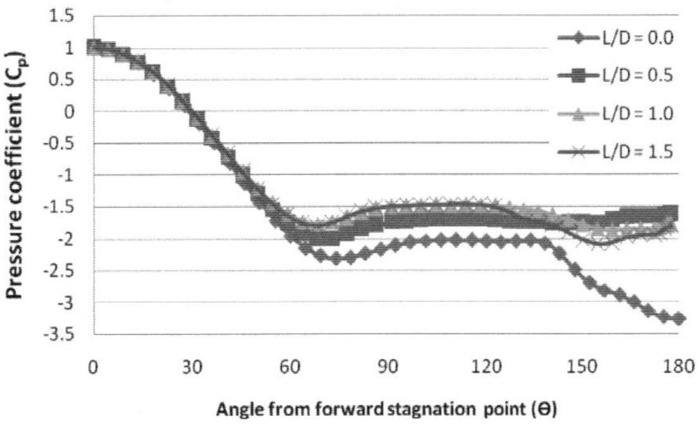

Figure 6. C_p distribution around a single tube with splitter plate of various L/D for Re = 4.6 x 10^4 and BR = 1.

3.4. Total Pressure Drop

Energy loss is evaluated by calculating the difference in total pressure at the inlet and the outlet. From figure 8, it can be seen that the total pressure loss increases as the Reynolds number increases. The same trend can be seen as the splitter plate length increases, with the exceptional of L/D =0.5 at Re = 8000.

3.5. Total Heat Transfer

From figure 9, it is observed that the attachment of wake splitter does not improve the heat transfer process at the surface of the tubes. Since mixing intensity is constrained by the splitter plate, the heat transfer rate is thereby attenuated. However, according to figure 10, the overall heat transfer improves since the splitter plates also act as fins, which increases the surface contact area for better overall heat transfer process.

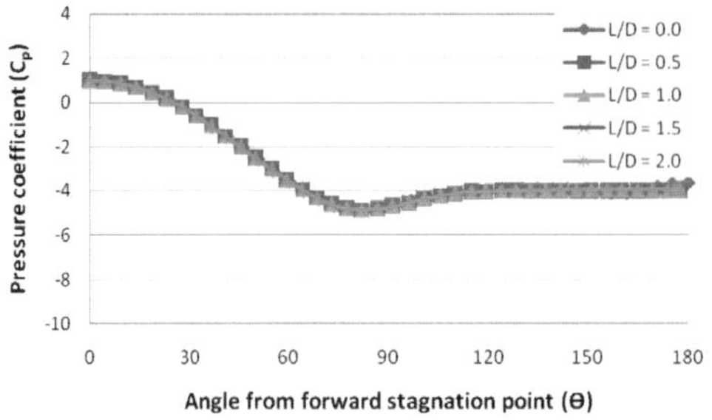

a) Re = 1000; 1st row

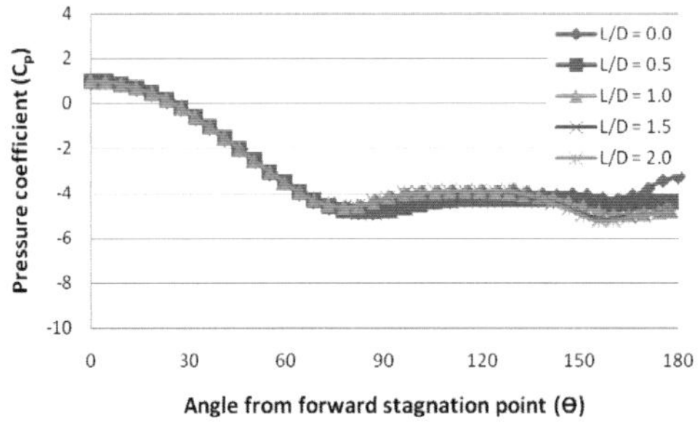

b) Re = 10000; 1st row

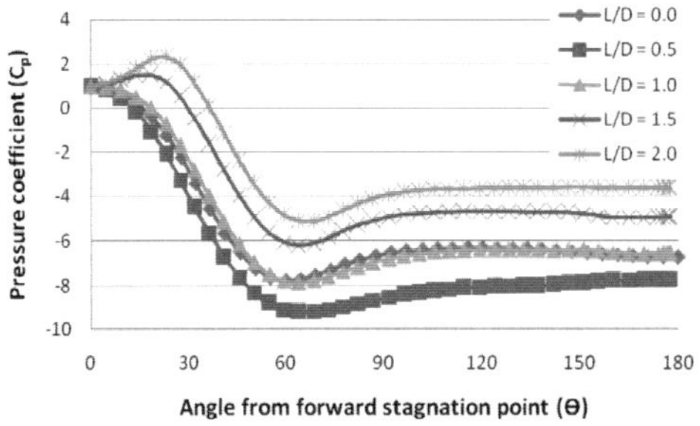

c) Re = 1000; 3rd row

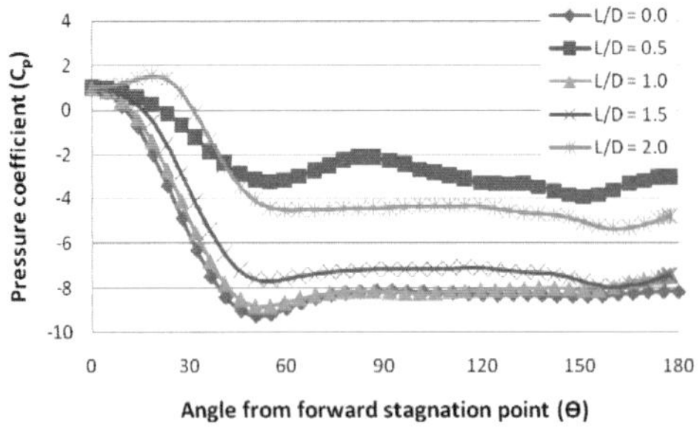

d) Re = 10000; 3rd row

Figure 7. C_p distribution around the tubes at 1st and 3rd row for Re = 1000 and Re = 10000.

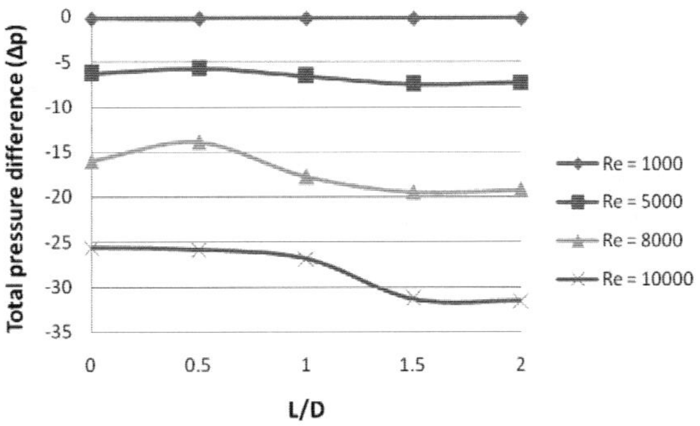

Figure 8. Total pressure drop, Δp for various L/D and Reynolds number.

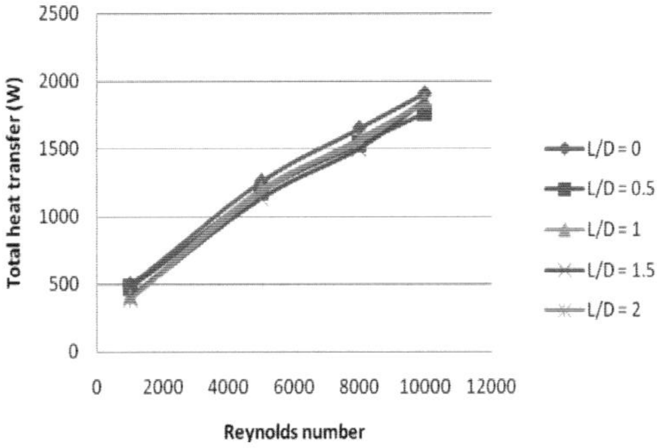

Figure 9. Total Surface Heat Transfer from The Tubes (excluding wake splitters).

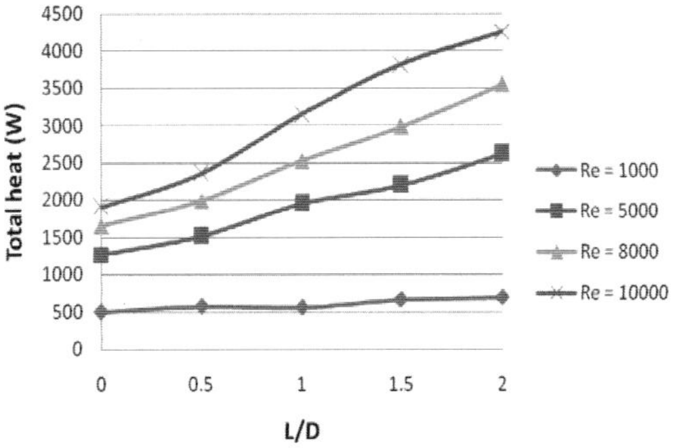

Figure 10. Total Surface Heat Transfer from The Tubes (including wake splitters).

4. Conclusions

The present study has arrived at the following conclusions:

- Good agreements between the present numerical work and the existing experimental data of Zukauskas and Ziugzda [17], and Anderson and Szewczyk [16] for flow around a single tube have been obtained.
- For flow around a single tube, wake splitters of all studied length were predicted to be able to reduce pressure loss significantly
- For cross-flow in a bank of tubes, the addition of splitter plates increases the pressure loss, especially for tubes at the 1st and 2nd row.
- The numerical simulations predicted that the overall heat transfer process can be intensified by attaching wake splitter plates to the tubes.

Acknowledgements
The authors would like to express gratitude towards the Malaysia Ministry of Higher Education and Universiti Tun Hussein Onn Malaysia as this work is a part of the research projects supported under the top-down Fundamental Research Grant Scheme.

References

[1] Tahseen T A, Ishak M and Rahman M M 2012 *J. Mech. Eng. Sci.* **3** 271
[2] Tahseen T A, Ishak M and Rahman M M 2013 *J. Mech. Eng. Sc.* **4** 418
[3] Al- Doori, W.H.A.R. 2011 *Inter. J. Automot. Mech. Eng.* **4** 428
[4] Tahseen T A, Ishak M and Rahman M M 2012 *Inter. J. Automot. Mech. Eng.* **6** 753
[5] Ishak, M., Tahseen, T.A. and Rahman, M.M. 2013 *Inter. J. Automot. Mech. Eng.* **7** 900
[6] Zhang Y H, Wu X, Wng L B, Song K W, Dong Y X and Liu S 2008 Experimental Thermal and Fluid Science **33** 58.
[7] Tiwari S, Chakraborty D, Biswas G and Panigrahi P K 2005 *International Journal of Heat and Mass Transfer* **48** 439.
[8] Jacimovic B M, Genic S B and Latinovic B R 2006 *International Journal of Refrigeration* **29** 1138.
[9] Kongkaitpaiboon V, Nanan K and Eiamsa-ard S 2010 *International Communications in Heat and Mass Transfer* **37** 568.
[10] Yajima Y and Sano O 1996 *Fluid Dyn. Res.* **18** 237.
[11] Valera G G A, Castellano N E and Avila E E, 2007 *International Journal of Heat and Mass Transfer* **50** 4098.
[12] Khan M G, Fartaj A and Ting D S K 2004 *International Journal of Heat and Fluid Flow* **25** 636.
[13] Matos R S, Vargas J V C, Laursen T A and Bejan A 2004 *International journal of Heat and Mass Transfer* **47** 1347.
[14] E Shirani and H Nasibi 2001 *Journal of Applied Sciences* **1** 275.
[15] F Khoddamrezaee, R Motallebzadeh and D Jajali Vahid 2010 *Journal of Applied Sciences* **10** 500.
[16] Anderson E A and A A Szewczyk 1997 *Experiments in Fluids* **23** 161-174.
[17] Zukauskas A and J Ziugzda 1985 Hemisphere Publishing Corporation, Washington D.C.

Numerical analysis of three dimensional flow around marine propellers in restricted water

M Nakisa, A Maimun[a], Y M Ahmed and F Behrouzi
Mechanical Eng. Faculty, Universiti Teknologi Malaysia (UTM), 31810, Johor Bahru, Malaysia

[a]E-mail: adi@fkm.utm.my

Abstract. Marine propeller blades have complicated geometries and as a consequence, the flow pattern around these propellers is very complex. The efficiency of the marine propeller is strongly dependent on propeller performance, thrust force and torque of propeller. In channels, canals, harbors and other types of restricted waters, flow inlet to the propellers is asymmetric and non-uniform, therefore hydrodynamic characteristics of the propeller is affected greatly by the presence of lateral restrictions of the navigation area, such as banks, quay walls and bottom. This research has approached the propeller hydrodynamic performance related to study on wake pattern behind of propeller affected by bank via numerical modeling using a finite volume code. Finally, the results of simulation the propeller's wake pattern and 3D flow around the propeller, with and without bank are compared. The influence of these parameter changings in the working propeller performance are carefully considered and analyzed.

1. Introduction

Nowadays, with increasing the main particular size of certain ships such as LNG vessels, oil tankers and container ships, ship navigators are considered to navigate in confined waters such as channels and ports, because the access channel dimensions are not increased at the same rate. Moreover, the restricted vertically and laterally in waterways are affected on ship maneuvering and behavior [1-3]. It is well known that when a ship travels along a bank, the asymmetric flow around the ship induced by the bank causes a pressure difference between the port and starboard sides of the hull. As a result, a lateral force directed towards the bank and a yaw moment pushing away the bow will act on the ship. The fact is the so-called bank effects. Bank effects have a significant impact on the navigation safety when a ship sails along a bank. If the distance between ship and bank is very small, ship-bank collision may happen. When the water becomes shallow, the hydrodynamic forces induced by bank effects will change greatly. The sway force acting on a ship moving along a bank in an extremely shallow water way may be towards or away from the bank [3-8]. So, the study of bank effects for a ship sailing in shallow channel is very important to ensure the navigation safety. [9] found that the hydrodynamic coefficients were severely affected by the presence of lateral banks. He investigated the bank effects on ship manoeuvring behavior by varying the hull form, water depth, ship speed, ship-bank distance, bank slope and propeller rotation rate. They found that the direction and magnitude of the bank-induced sway force and yaw moment were dependent on the above-mentioned parameters when water depth to draught ratio is less than 1.5.

With the fast development of the computer technology in recent decades, great progresses have been made in application of computational fluid dynamics (CFD)-based numerical methods to predict ship manoeuvring hydrodynamic forces. Some numerical methods were developed and applied to study the bank effects. Yasukawa developed a method by combining the slender body theory with panel method to calculate the lateral force and yaw moment acting on a ship in the proximity of arbitrary shaped bank [10-11]. [12] calculated sway force and yaw moment acting on a ship traveling off the centerline of a rectangular channel by using a numerical approach based on Dawson's method. [13] performed a series of numerical simulations to analyse the bank effects for a container ship in restricted waters by using commercial CFD software.

In this paper, in the numerical study, a LNG carrier' propeller sailing along the bank in a shallow channel is taken as example. A vertical bank and shallow water are considered. The influence of restricted water on thrust force, torque and efficiency of the propeller are calculated and analysed. The numerical results are compared with the available open water simulation that verified by experimental results.

2. Propeller Model

In this study one propeller in model scale, which is extensively used for the validation of CFD code in open water, is considered then, in according to this modeling the bank effect on propeller performance is considered. The propeller has a diameter D of 7.7 m and hub diameter D_{hub} of 0.17D, and is right handed when mounted on a pulling thrust. Design pitch ratio (P/D) is 0.94 and the blade ratio subjected to expanded area ratio (EAR) is 0.88. It's rake degree is 15degrees as well.

A propeller geometric characteristics and drawing are depicted in table 1 and figure 1, respectively.

Table 1. Propeller geometric characteristics

Parameter	Dimension
Z	5
D	7.7 m
D_{hub}	1.28 m
Br	0.17
P/D	0.94
A_e/A_0	0.88
R	15 Deg.

Figure 1 shows the propeller's front view which is can see in behind of ships and calculate the thrust and torque results of the kinetic energy of flow motion as well. That means propellers generate thrust through the production of lift by their rotating blades. This considered propeller is rotating right hand for making the axial force by push the water.

Figure 1. Front view of propeller

3. Boundary Condition

The numerical predictions presented in this work were carried out with the ANSYS-FLUENT 13, commercial CFD solver. It employs the node centred finite volume method. Figure 2(a), 2(b) and table 2 show the schemes and dimensions of computational domain for modelling and simulation of the propeller with and without bank, respectively. Table 2 shows the rotational and stationary computational domain dimensions respect to propeller diameter. The stationary domain dimension should be far from rotational domain to decrease the effect of wall side effects on the results.

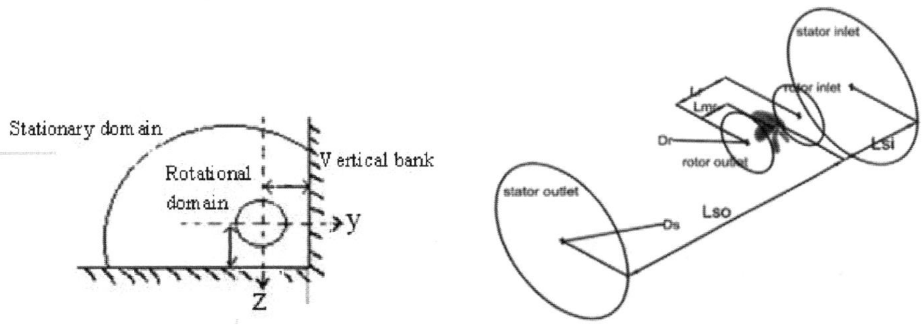

(a) Open water condition **(b)** Restricted water conditions

Figure 2. Schemes of computational domain

Table 2. Dimensions of computational domain (Based on Propeller diameter)

Value	Rotational	Stationary
D_r	1.44 D	
L_{mr}	1.5 D	
L_r	3 D	
D_s		10 D
L_{si}		3 D
L_{so}		5 D
d		1.7 D

In to a rotating part, called Rotating domain and in to a stationary part, called Stationary domain. The dimensions of the domains of propeller are given in table 2 that the variable L_{mr}, visible in figure 2 is the axial length of outlet the Rotating part, D is the propeller diameter and d is the distance between bank and centre of shaft. The inlet, outlet and outer boundaries of the Stationary part were placed far enough from the propeller in order to not affect the results. The distances were set through a domain independence study, carried out considering more shapes of the Stationary part, defined varying systematically L_r, L_{si}, D_s that shows in figure 3(a) and 3(b). Table 3 is shown the mesh information of domains with and without bank that are known as open water and restricted water.

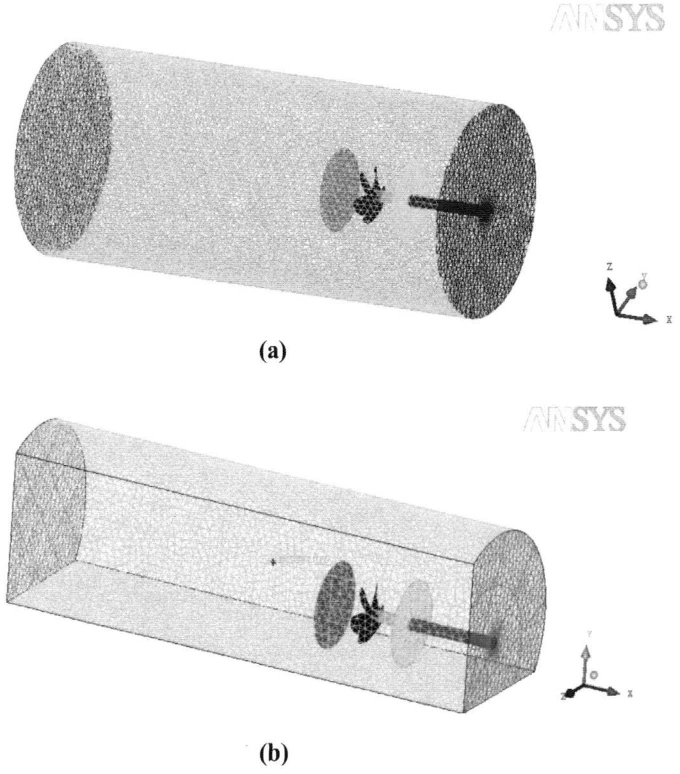

Figure 3. Rotational and Stationary domains for propeller working in (a) Open water and (b) Restricted water, generated using ICEM-CFD

Table 3. Mesh information of domains in open and confined water

Computational Domain	Total Elements	Total Nodes
Open water	393893	80634
Restricted water	171629	51389

4. Numerical Simulation

In Cartesian tensor form the general RANS equation for continuity can be written as:

RANS equation:
$$\frac{\partial \rho}{\partial t} + \frac{\partial(\rho u_i)}{\partial x_i} = 0 \tag{1}$$

The equations for momentum become:

$$\frac{\partial(\rho u_i)}{\partial t} + \frac{\partial(\rho u_i u_j)}{\partial x_j} = \tag{2}$$

$$-\frac{\partial p}{\partial x_i} + \frac{\partial}{\partial x_j}\left[\mu\left(\frac{\partial u_i}{\partial x_j} + \frac{\partial u_j}{\partial x_i} - \frac{2}{3}\delta_{ij}\frac{\partial u_i}{\partial x_i}\right)\right] + \frac{\partial}{\partial x_j}(-\rho\overline{u_i' u_j'})$$

Where δ_{ij} is the Kronecker delta, $(-\rho\overline{u_i' u_j'})$ are the unknown Reynolds stresses that have to be modeled to close the momentum equation. With Boussinesq's eddy-viscosity assumption and two transport equations for solving a turbulence velocity and turbulence time scale (turbulence modeling), RANS equations are closed.

Only steady incompressible equations are solved in the present work. The shear-stress transport SST k-e model with the "transitional flow option" is employed here due to its good performance for wall-bounded boundary layer flows. The FLUENT 13 code is used to solve the three dimensional viscous incompressible flow. The parallel version of FLUENT simultaneously computes the flow equations using multiple processors. The software can automatically-partitions the grid into sub-domains, to distribute the computational job between available numbers of processors.

Simulations in this paper are completed using FLUENT software, SST (k-w) model and standard wall function. Some major parameters are set as shown in table 4 while the other options remain default. Force and momentum data are recorded every step.

Table 4. Computational setting

Parameter	Setting
Computing	64-bit Desktop pc 6GB of RAM
Simulation type	Steady state
Mesh type	Unstructured hybrid(tetrahedral/prism)
Turbulence model	k-w (Shear stress transport)
Wall modeling	Automatic wall function based on a law of the wall formulation
Advection scheme	CFX high resolution

The shear-stress transport SST (k-w) model is used for turbulence closure in this research. The SST (k-w) model is an empirical model transport equations solving for the turbulence kinetic energy (k) and the specific dissipation rate (w) and remained as one of the most accurate and reliable turbulence models for external hydrodynamics.

κ function:
$$\frac{\partial}{\partial t}(\rho k) + \frac{\partial}{\partial x_i}(\rho k u_i) = \frac{\partial}{\partial x_j}\left(\Gamma_k \frac{\partial k}{\partial x_j}\right) + G_k - Y_k + S_k \tag{3}$$

ω function:
$$\frac{\partial}{\partial t}(\rho\omega) + \frac{\partial}{\partial x_i}(\rho\omega u_i) = \frac{\partial}{\partial x_j}\left(\Gamma_\omega \frac{\partial \omega}{\partial x_j}\right) + G_\omega - Y_\omega + D_\omega + S_\omega \tag{4}$$

Where G_κ and G_ω represents the generation of turbulence kinetic energy due to mean velocity gradients and the generation of ω. Γ_κ and Γ_ω represent the effective diffusivity of κ and ω. Y_κ and Y_ω represent the dissipation of κ and ω due to turbulence. D_ω represents the cross-diffusion term, S_κ and S_ω are user-defined source terms [13].

5. Results and Discussion

Pressure distribution on the surface of blade is shown in figure 4 that the face and back are experiencing high pressure and low pressure respectively. This explains the development of thrust by propeller at high pressures whereas the propeller is contributing to resistance and friction coefficient that shows in figure 5(a) and 5(b). It is evident that there is a concentration of high-pressure region near the leading edge of the propeller.

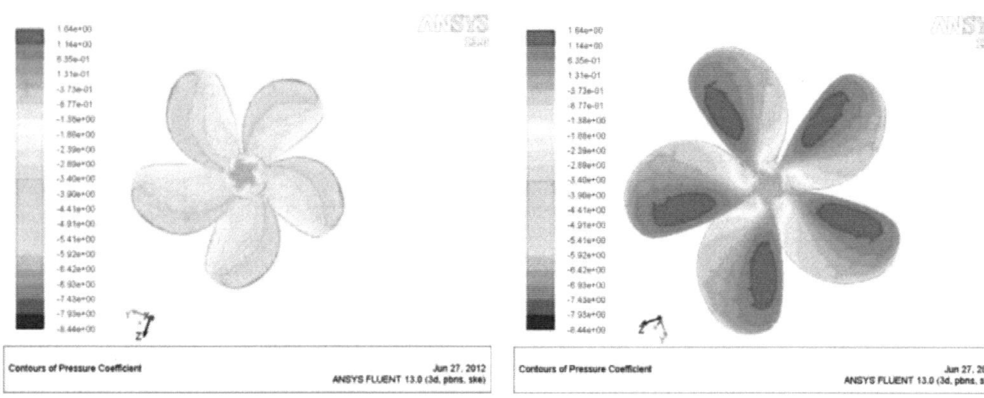

(a) Back side of the propeller (b) Face side of the propeller

Figure 4. Propeller's blades pressure distributions

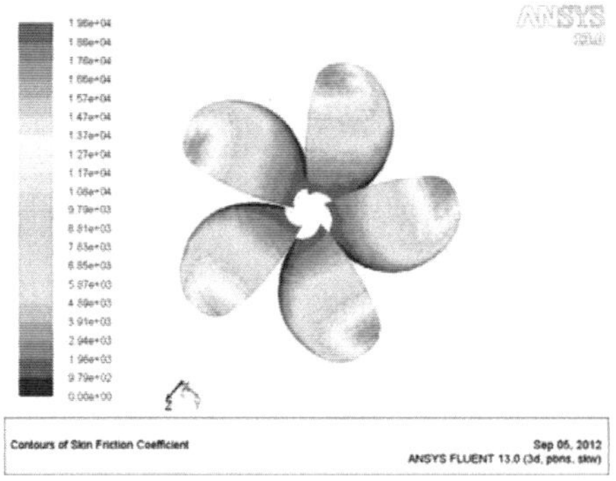

Figure 5. Contours of skin friction coefficient

Figure 6(a) and 6(b) show the distribution of pressure coefficient on face and back surface of one blade at three radial section r/R = 0.80. It is clearly that the low pressure region is happen on back surface and the high pressure region is occurred on face surface of blade. Figure 7(a) and 7(b) show the contours and velocity distribution on face of blades that the high velocity is occurred near the tips of blades.

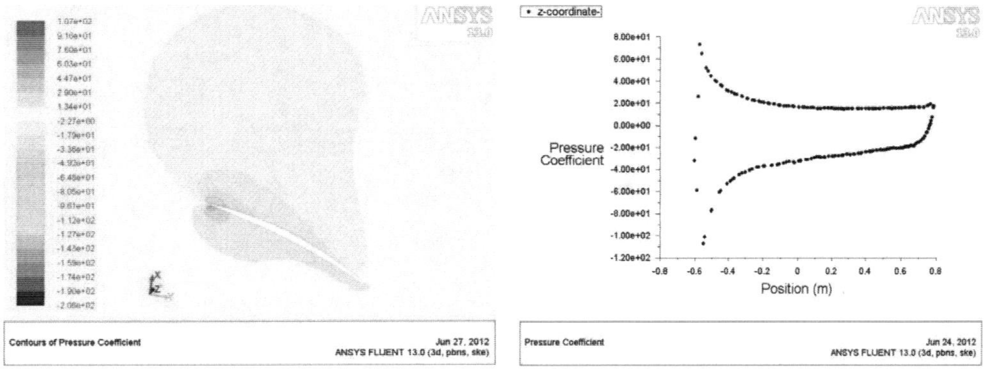

(a) Pressure distribution contours (b) Pressure coefficient

Figure 6. Pressure coefficient on blade section

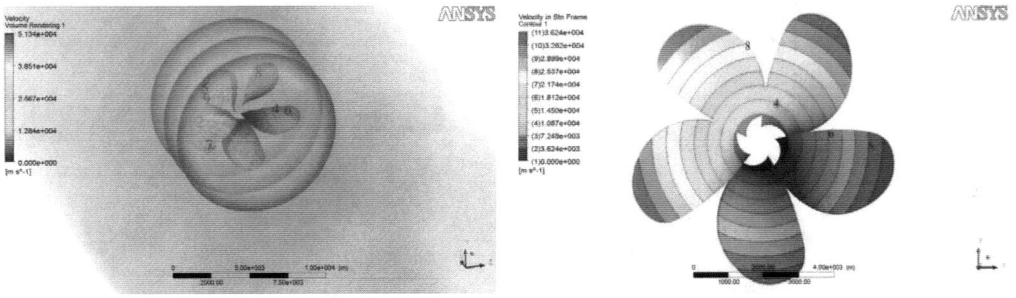

(a) (b)

Figure 7. Velocity magnitudes around the propeller blades in (a) and (b) views

When a ship sails on a canal or other restricted waterway the propeller of ship will experience a higher velocity inlet than sailing in open water. The effects of bank and shallow water on the longitudinal force and momentum are taken into account by the thrust and torque coefficient. As a consequence the influence of the bank and shallow water are not calculated independently.

A ship sailing within the horizontal reach of a bank will be attracted to the closest bank when the under keel clearance exceeds a minimum value. In according to velocity vectors and contours in restricted water in comparison with open water condition that are shown in figure 8(a), 8(b), 9(a) and 9(b), While she sails in shallow water the propeller will be pushed the stern of ship near the bottom because the outlet vectors go up than in open water.

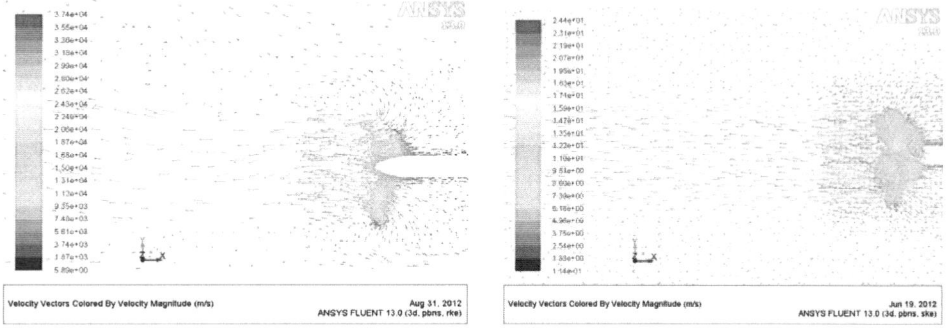

(a) Shallow water condition (b) open water condition

Figure 8. Velocity vectors behind the propeller

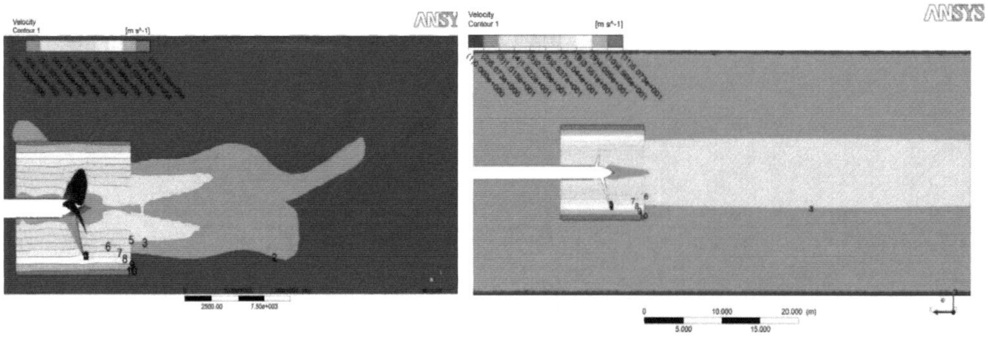

(a) Shallow water (b) Open water

Figure 9. Velocity contours behind the propeller with and without bank

The forces and moments produced by the propeller are expressed in their most fundamental form in terms of a series of non-dimensional characteristics: these are completely general for a specific geometric configuration. The non-dimensional terms used to express the general performance characteristics are as follows:

Thrust coefficient: $$K_t = \frac{T}{\rho n^2 D^4} \quad (5)$$

Torque coefficient: $$K_Q = \frac{Q}{\rho n^2 D^5} \quad (6)$$

Advance coefficient: $$J = \frac{V_a}{nD} \quad (7)$$

The results from the Numerical simulation of propeller in open and restricted water based on RANS equation at full scale compared to the model test results can be seen in figure 9 Various J-values are obtained by keeping a same revolutions (n=90 rpm) but varying the flow speed. The trends of results with varying advance ratio are well predicted. It should be noted that K_Q and K_t are slightly over predicted all the way. The maximum hydrodynamic propeller efficiency will be occurred in j=0.84 and j=0.87 for open water and restricted water, respectively.

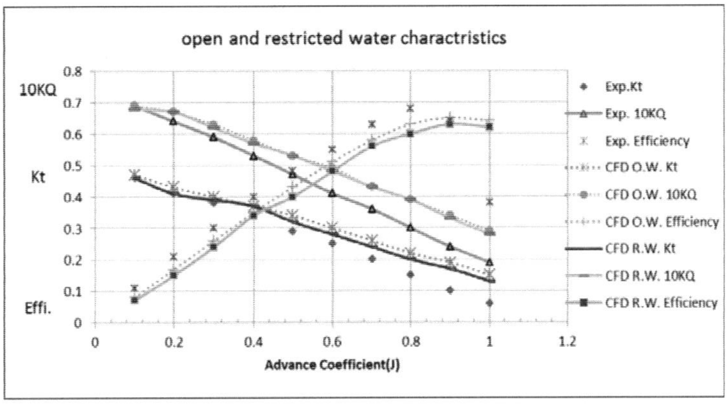

Figure 10. Open and restricted water diagram of propeller performance at full scale

6. Conclusion

In according to the presented computational results based on RANS equations, the following conclusions are taken:

- The hydrodynamic propeller performance is influenced by the proximity of bank and shallow water.
- The forces and moments acting on the propeller when operating in a uniform fluid stream, hence the propeller generate the thrust and make pressure and velocity in the fluid particles on face blades surface and behind the propeller.
- The fluid velocity in the tips of blades is higher than other areas and the pressure distribution in the face blades near leading edges is higher than other areas.
- The maximum hydrodynamic propeller efficiency will be occurred in j=0.84 and j=0.87 for open water and restricted water, respectively

Acknowledgements

The authors would like to thank the Faculty of Mechanical Engineering in Universiti Teknologi Malaysia (UTM) for financial support under MOHE-UTM-Vote No. 00G44.

References

[1] Ch'ng P W, Doctors L J, Renilson M R 1993 A method of calculating the ship-bank interaction forces and moments in restricted water *International Shipbuilding Progress* **40**(421) 7-23.
[2] Li Da-qing, Ottosson P, Tr¨agardh P, 2003 Prediction of bank effects by model tests and mathematical models [C]/ *Proceedings MARSIM'03, International Conference on Marine Simulation and Ship Maneuverability*. Kanazawa, Japan: [s.n.], RC**30**.1-12.
[3] Tahseen T A, Ishak M and Rahman M M 2012 *J. Mech. Eng. Sci.* **3** 271.
[4] Tahseen T A, Ishak M and Rahman M M 2013 *J. Mech. Eng. Sc.* **4** 418.
[5] Al- Doori, W H A R 2011 *Inter. J. Automot. Mech. Eng.* **4** 428.
[6] Tahseen T A, Ishak M and Rahman M M 2012 *Inter. J. Automot. Mech. Eng.* **6** 753.
[7] Ishak, M., Tahseen, T A and Rahman, M M 2013. *Inter. J. Automot. Mech. Eng.* **7** 900.

[8] Vantorre M, Delefortrie G, Eloot K, et al. 2003 Experimental investigation of ship-bank interaction forces *Proceedings MARSIM'03, International Conference on Marine Simulation and Ship Maneuverability*. Kanazawa, Japan, RC**31**.1-9.

[9] Norrbin N H 1974 Bank effect on a ship moving through a short dredged channel [C]// *Proceedings 10th Symposium on Naval Hydrodynamics*. Cambridge, MA, USA 71-87.

[10] Yasukawa H 1991 Bank effect on ship maneuverability in a channel with varying width [J]. *Transactions of the West-Japan Society of Naval Architects*, **81**: 85-100.

[11] Miao Quang-ming, Xia Jin-zhu, Chwang A T, et al. 2003 Numerical study of bank effects on a ship traveling in a channel [C]/*Proceedings 8th International Conference on Numerical Ship Hydrodynamics*. Busan, Korea 266-273.

[12] LoDC, Su Dong-taur, Chen Jan-ming 2009 Application of computational fluid dynamics simulations to the analysis of bank effects in restricted waters *The Journal of Navigation*, **62** 477-491.

[13] Maimun, A, Yasser M Ahmed., Priyanto A, Y S Ang., M Nakisa 2012 Assessment of Ship-Bank Interactions on LNG Tanker in Shallow Water, University Technology Malaysia, Johor Bahru, Malaysia, *Proceeding of 6th Asia-Pacific Workshop on Marine Hydrodymics-APHydro12 Malaysia*, September 3-4.

Nomenclature

D: Propeller diameter, (m)
Z: Number of blade
R: Rake of Blades
C_p: Pressure coefficient
p_0: Reference pressure at infinity
J: Advance ratio
K_Q: Torque coefficient
A_E/A_0= Expanded Area Ratio (EAR)
η = Open water efficiency
L_r= Length of rotational domain
D_{hub}: Hub diameter, (m)
P/D: Pitch Ratio
n: Rate of revolutions of propeller, (rpm)
p: Static pressure at point of interest
V_a: Advance velocity, (m/s)
K_T: Thrust coefficient
B_r: Boss ratio
ρ = Density of water
D_r= Diameter of Rotational domain

2nd International Conference on Mechanical Engineering Research (ICMER 2013) IOP Publishing
IOP Conf. Series: Materials Science and Engineering **50** (2013) 012047 doi:10.1088/1757-899X/50/1/012047

Control strategies to optimise power output in heave buoy energy convertors

M A U A Abu Zarim[1,a] **and R M Sharip**[2,b]
[1]Faculty of Science and Defence Technology, Universiti Pertahanan Nasional Malaysia, Kem Sungai Besi, 57000 Kuala Lumpur, Malaysia
[2]Faculty of Electrical & Electronic Engineering, Universiti Tun Hussein Onn Malaysia, 86400 Parit Raja, Batu Pahat Johor, Malaysia

E-mail: [a]amir.a@upnm.edu.my, [b]rina@uthm.edu.my

Abstract. Wave energy converter (WEC) designs are always discussed in order to obtain an optimum design to generate the power from the wave. Output power from wave energy converter can be improved by controlling the oscillation in order to acquire the interaction between the WEC and the incident wave.The purpose of this research is to study the heave buoys in the interest to generate an optimum power output by optimising the phase control and amplitude in order to maximise the active power. In line with the real aims of this study which investigate the theory and function and hence optimise the power generation of heave buoys as renewable energy sources, the condition that influence the heave buoy must be understand in which to propose the control strategies that can be use to control parameters to obtain optimum power output. However, this research is in an early stage, and further analysis and technical development is require.

1. Introduction

Renewable energy comes from a variety of resources, all of which are based on self-renewing energy sources such as wind, flowing water, sunlight, biomass such as energy crops, agricultural, industrial and municipal waste, and also from the earth's internal heat. These resources have the potential to be used as a fuel for transportation, electricity for all sectors of the economy, and heat for the purpose of industrial processes and buildings purposes [1]. Consequently, in order to meet the rapidly growing demand for energy while also maintaining clean global environment, the world requires a clean energy to power the economy [2].

Furthermore, rising on oil- crisis and environmental safety concerns will eventually induce the transition from oil-based fuels to renewable fuels. Demands of distributed energy resources, international market, ongoing technical progress, and public opinion would increase the use of renewable electric technologies to replace power plants based on coal, nuclear, and possibly natural gas. Significant emergence of renewable energy use will result in energy production and use will no longer be harm to the environment [2].

One of the renewable power generators that actively developing in research and application is the heave buoy. The idea not new at all, but development has been quite slow due to constraints and limitations in term of cost, raw material, knowledge, and technology [3].

Continue what have done by Budal and Falness, and Falcoa [3,4] on conversion of wave energy, they proposed to apply control equipment such as 'a combined motor and generator' to maximise power take off in regular wave. Not finish like that, a few years later he proposed the latching control by means of a mechanical latching device or hydraulic machinery. However in 1993, Salter S.H and Rampen suggested hydraulic machinery, a combination of pump and motor used to optimum control of the device [5].

The appearance of heave buoy using renewable technologies will impact upon the operation, protection and control in such systems. Precisely, maintaining the required voltage and frequency stability within the defined limits is a more onerous task in such scenarios and requires thorough analysis.

2. Related work

In the last decade, much research has emerged regarding small generators and renewable technology from countries all over the world such as Norway and the United Kingdom. The topic of optimisation the heave buoy and how to control the power generated is fairly new, but the idea behind it is not new at all.

Falnes and Lillebekken [6] in their paper, study the three different versions of latching controlled intended to operate in the heave mode, and to force-react against anchor and the sea bed. One form of control uses hydraulic machinery for control and power take off, while the two others use a latching mechanism for control and pneumatic power take-off. In theory, if the ocean wave had the same amplitude and the same wave period, it would be easy to develop a feasible wave energy converter (WEC). But in reality, real ocean waves perform in form of stochastic that much more difficult to predict, in which there is a large variation in wave periods and wave amplitude. In this paper, the authors mention that optimum control of the amplitude may be an even more challenging problem than the phase control problem. Budal, and Falnes, (1980) in Falnes and Lillebekken [6], in their research, advocated that WEC units should be relatively small and point absorber diameters should preferably be in the range of five to ten percent of prevailing wavelength.

Falnes and Falcoa [3,5] stated that the ability of the wave energy converter (WEC) may increase by controlling oscillation to reach the optimal interaction between the WEC and the incident wave. The author accentuation most of WEC discussed utilised only one mode of oscillation, say as heave. An upper boundary refer to the ratio changed from wave power conferred on the geometrical volume of the converter. One of the strategies on maximizing the power converter is based on measuring the incident wave, and the other involving the measured quantities to predict some second in future because of the non-causal control function. The author concludes that applying optimum control over WECs designed to operate on a large scale with full capacity, may significantly improve the economic prospects for wave power. These methods may stand alone or in combination of both. However further research is required to compare these two control methods in order to determine which is the best system to be implement. Furthermore the author also suggests that further study is necessary including on more modes of wave characteristic and constraints to get an accuracy or else, as close as possible result on the analysis [5].

Falnes [7] describes the principles of wave absorption. Basically oscillating body will generate waves. A large body may generate equally large wave, meanwhile a small body will oscillates with a larger amplitude. This information can be used for wave energy conversion, especially by response of heaving motion of small floating body to the incident wave. The device may absorb an energy to maximum by displace a water in an oscillatory manner with correct phase to the incident wave. Theoretically, only 50 % absorption is possible if symmetrical radiated wave, when a wave is generated by a symmetrical body oscillating in only one mode of motion (heave). However, more than 50 % absorption is possible if an anti-symmetrical radiated wave is applied. But if a sufficiently non-symmetrical body oscillates in only one mode of motion, , its ability to absorb nearly all the incident wave energy is high [7,8].

K. Budal et al [9] were discussed a spherical buoy, anchored by strut joined on the sea bed. The buoy is supplied with latching control system for optimum phase control and with an air turbine mechanism for power take-off. This research studies the function of the buoy including controlled motion, power transmission, and the estimated energy price by labour and investment in materials. This research concludes that wave energy was not competitive with hydroelectric power in Norway. The cost may be reduced by optimising of the size of buoy, mass producing units in specialised factories, developing cheaper mechanical components with lower maintenance and repair costs, using alternative construction materials, and improving the installation procedure.

3. Condition influence of heave buoy

Optimum motion of the body is fairly easy to obtain in regular wave. Then it can be explained using the simple damped mass-spring system. The spring system should be tuneable so that the natural period of the system is related to that of the wave period. In reality, ocean waves are in irregular form. So it may be difficult to predict the perfectly of the converter to be optimum [10]. However, the parameters of the wave may understand and be tackle to create a good design, so optimum conditions can be achieved or the output power can be as near as possible to the optimum value. Basically, the important parameters that influence heave buoys as follows;

3.1 Unconstrained amplitude

Previous studies by Evans and Newman [11,12] clearly show that in the open sea for resonant point or any axisymmetrical body, the maximum power that can be absorbed equal to the incident wave associated with the width of the wavelength divided by 2π. However for a two-dimensional case, only half of the incident wave energy can be absorb. If the two-dimensional were not quite symmetrical, i.e Salter Duck, possibility to absorb all the incident wave are high.

The following two conditions must be fulfilled for the case when only one oscillation mode is used to obtain maximum absorption from a regular wave [5]:

- The velocity of oscillation must be in phase with excitation force (described as wave force acting on fixed oscillating body). If the body is in resonance with the wave, it is automatically satisfied the condition.
- The optimum value is obtained by adjusting the amplitude of the oscillations, achievable when maximum absorbed power equals the power re-radiated into the sea by the oscillating system.

These two conditions may be referred to as conditions for optimum phase and optimum amplitude respectively. However, in practice, these two conditions occur and only certain fragments of the harvested power can be a useful power, where the remaining power lost due to friction, viscous effects, and other dissipative effects.

3.2 Constrained amplitude

It is necessary for convertor device limit the power capacity for the case of wave heights exceeding the upper boundary of the oscillation amplitude. The reason is to ensure that the converter does not surpass the highest capacity of power converters. Higher waves must be optimised under constraints, where the optimum amplitude are set to be a design amplitude of the devices, while optimum phase condition remains without any changes as before. This also can be avoid by applying the suitable controllable power take-off device. This is in contrast to when the wave height is below a certain value, and then the device needs to optimise the output.

For a small oscillating body, the constraints which come into role are lower than a large one. In circumstances of unconstrained amplitude , problems faced when the wave height relatively larger than the design amplitude resulting the re-radiated power is much lesser than the absorbed power. In this condition, only a small fraction of wave energy is absorbed even the incident wave transport a large wave energy. However the power absorbed is large when the wave amplitude is in range of the design amplitude (Falnes, 2001).

Table 1. Five conditions satisfied the upper boundary of a wave [5]

No	Five condition satisfied upper bound	Effect/ Constraint
1	The oscillating body is relatively small (velocity, V→ 0).	In circumstances of the wave height is moderate or large, only a small portion power can converted by wave power plant from incident wave
2	Most of the incident (free) wave power remains in the ocean.	
3	The oscillation has an optimum phase.	Power take-off device require certain requirements
4	The oscillation amplitude equals the design amplitude.	
5	The oscillation mode has a source-i.e. radiation of wave.	Only two things can satisfy this condition which is a) Oscillating Water Column (OWC) *b)* heave mode being utilised by floating body

4. Control strategies

J Falnes [5] outlines a few methods of controlling the parameters that influence heave buoys to obtain optimum power output. The three methods are:

- Unconstrained continuous method
- Causal control function
- Discrete control

4.1 Unconstrained continuous control

Most research on waves normally uses regular waves in calculation to fully understand the attitude of the wave (monochromatic, harmonical or sinusoidal) or in other word to make it simple as possible rather than irregular wave. Hence, in the sea irregular waves require modified formulation, to achieved optimum condition for maximum power conversion.

In the Fourier series, physical quantities consist of wave and oscillation are transformed into harmonious components. For unconstrained cases, linearity may assumed. Hence, the optimum conditions are articulated in terms of convolution in the time domain by applying inverse Fourier transform. However the limitation for this transformation in non-causal, which means at least one of the physical quantities must be predict. The better this quantity can be predicted, the closer the converted power may approach the maximum theoretical power [5].

The conditions of phase and amplitude can still be considered valid if the wave spectrum is relatively narrow. The limitation for this method is that applying these conditions requires information about the excitation force during a time gap, within a few seconds in the past and the future. Hence instead of measuring the excitation force, the incident wave may assume and pertain transfer function. The measuring wave is must be correct with the re-radiated wave, so substantial fragment of incident wave power can fully harvested by the convertors. The beauty of this method is excitation force is known as input, where as the controller may provide as output the optimum oscillating velocity.

For irregular wave conditions, the optimum load force is given by the product in the frequency domain. To achieve optimum control, this condition must be satisfied. The measured and predicted velocity of the oscillating body will act as an input signal to the controller, rather than the excitation force due to the incident wave. Hence, the controller will generates optimum load force based on computational of convolution integrals as output. However, only one mode acting on WEC must be assumed i.e heave motion, without amplitude constraint to formulate this optimum control strategy [5].

4.2 Causal control function

In reality, maximum power cannot be achieved exactly as calculated (this only approach) because of prediction of physical variables. Some controls strategies have been propose with causal control functions which means the prediction of physical quantities is not required. In this method, the input of the controller is the excitation parameters. Above, we have discussed how to maximize the absorbed power when wave predictions involved. However, the transfer function of the controller is the inverse of twice of the body's radiation resistance.

For the causal control, the past values from measurement are utilised and pertain transfer function will chosen to optimise the absorption power under the constraint. This transfer function however, must be rational, and may involve very complex mathematical calculation. This limitation may resulting the harvested power less than theoretical calculation [5].

4.3 Discrete control

Both control methods above are known as continuous control, which means that the controller can operate at any time when necessary. However in discrete control, there are limited number for controller to act for each wave cycle. A good example of discrete control is a latching technology. For instance, if the condition of buoy is outside resonance, the latching system will act to obtain an optimum phase by re-match the phase velocity to the best possible of the predicted excitation force.

In practice, wave energy devices are constantly retuned to resonate with the waves that incident at any time. Hence the large variety of real wave condition will put extra constraint to the control system. Therefore, the system may require a strong, heavy-duty and flexible power take-off to fulfil this job.

The advantage of using latching control is it eased the requirement of high conversion efficiency of the power take-off, rather than continuous power control. However it is restricted only to a certain approximation of the phase condition.

5. Discussion

Table 2 below clearly shows the advantages and disadvantages of these three methods. Hence, it suggested the most suitable method to apply in a device. The types of control method must be carefully chosen depends on limitation and constraint on the WEC such as heave buoy design, knowledge, technology, cost and efficiency.

Table 2. Comparison between control methods [5]

Type of control	Advantage	Disadvantage
Discrete control	1) Only a limited number of times the controller can act on the system for each wave cycles (not continuous) 2) Can be apply even in the condition outside resonance by latching method 3) Any negative energy flow to the power take-off is eliminated 4) Easy and more efficient to control and use	1) Optimum amplitude can be obtained simply by adjusting the damping of oscillations correctly. 2) Only efficient at certain approximations of the phase condition
Causal control function	1) No need to predict of physical quantities 2) The input of the controller	1) The absorbed power may be lower than the theoretical maximum absorbed power

	is the excitation parameters, where the excitation volume fluxes for an Oscillation Water Column or excitation force in the case of an oscillating body	2) Transfer function must be rational function, and may involved complex mathematical theory
Unconstrained continuous control	1) Can predict optimum conditions in terms of convolutions in the time domain 2) WEC utilising body assume as one mode without amplitude constraint	1) The function of wave and oscillation need to transform to harmonical form 2) The transformation is non causal, future quantity is required 3) Information of excitation force is needed during a time gap, within a few seconds in the past and the future. 4) Much more complicated if it involves irregular wave computational

6. Conclusion

In reality, maximum power captured from the wave cannot be achieved exactly as calculated because of prediction of physical variables. For instance, the sea irregular waves require modification in term of formulation, say linearity to fully understand the attitude of the wave. Hence, control strategy is one of the approaches to propose the best functional device to capture as close as possible wave energy from the incident wave. Every single control strategies have advantages and disadvantages.

Unconstrained continuous control method expressed optimum condition in terms of convolution in time domain. This convolution is a linear transformation between two physical quantities which introduce the simplicity. However, this transformation is non-causal which means that some future information is required, at least one of the physical quantity. Hence, the optimum power can be converted is totally depends on how close this quantity can be predicted.

Contradict with unconstrained continuous control, causal control functions do not require the prediction of physical quantities. This method utilised past values information and transfer function is chosen in order to increase the absorption power into optimum under the constraint of the transfer function being causal. However, transfer functions require rational function and may be involve a complex mathematical formula.

Both control strategies above are known as continuous methods where the controller can act at any time. Discrete control however allows controller to act on the system only a limited number during each wave cycle. The advantage of using discrete control is it eased the requirement of high conversion efficiency of the power take-off, compare to continuous power control methods. Bare in mind, optimum amplitude can be obtained simply by adjusting the damping of oscillations correctly. For real sea condition, large variety of wave may obligate a very flexible power take-off . Hence, this method may restrict only to a certain approximation of the phase condition.

Acknowledgement

The authors would like to thank the Universiti Pertahanan Nasional Malaysia (UPNM) for financial support.

References

[1] Paolo F 2012 Renewable Energy coming of age iea energy *Journal of the international energy agency*

[2] Stanley R Bull 2001 *Renewable Energy Today and Tomorrow* Proceedings of the IEEE **89** 8
[3] Falcao A F D O 2010 Wave energy utilization: A review of the technologies *Renewable and sustainable Energy Reviews* **14**.
[4] Budal K and Falnes J 1978 *A system for the conversion of sea wave energy* British patent application
[5] Falnes J 2001 *Optimum control of oscillation of wave-energy converters* Proceedings of tile Eleventh (2001) International Offshore and Polar Engineering Conference
[6] Falnes J and Lillebekken P M 2003 *Budal's latching-controlled-buoy type wave-power plant.* Institute for Fysikk, Noregsteknisk-naturvitskaplegeuniversitet (NTNU), Trondheim Norway.
[7] Falnes J n d *Principle for capture of energy from ocean waves, phase control and optimum oscillation.* Department of Physics, NTNU, N-7034 Trondheim, Norway.
[8] Farley F J M 2012 Far-field theory of wave power capture by oscillating systems *Philosophical Transactions of The Royal Sociaty A* 278-87.
[9] Budal K Falnes J Iversen L C Lillebekken P M Oltedal G Hals T and Onshus T 1982 *The Norwegian wave power buoy project* 2nd International symposium on wave energy utilization, Trondheim.
[10] Simon L and Leijon M 2011 Offshore wave power measurements A review *Renewable and sustainable Energy Reviews* **15** 4274– 85.
[11] Evans D V 1976 Wave power absorption by oscillating bodies **J. Fluid Mec. 77**
[12] Newman J N 1976 *The Interaction of Stationary Vessels with Regular Waves* Proc. 11th Symp Naval Hydrodynamics London UK 491-01.

IOP Conf. Series: Materials Science and Engineering **50** (2013) 012048 doi:10.1088/1757-899X/50/1/012048

Development of strand burner for solid propellant burning rate studies

A Aziz[1,a], R Mamat[1] and W K Wan Ali[2]
[1]Faculty of Mechanical Engineering, University Malaysia Pahang
26600, Pekan, Pahang, Malaysia
[2]Department of Aeronautical Engineering, Universiti Teknologi Malaysia
81310, UTM Skudai, Johor.Malaysia

E-mail: [a]amiraziz@ump.edu.my

Abstract. It is well-known that a strand burner is an apparatus that provides burning rate measurements of a solid propellant at an elevated pressure in order to obtain the burning characteristics of a propellant. This paper describes the facilities developed by author that was used in his studies. The burning rate characteristics of solid propellant have be evaluated over five different chamber pressures ranging from 1 atm to 31 atm using a strand burner. The strand burner has a mounting stand that allows the propellant strand to be mounted vertically. The strand was ignited electrically using hot wire, and the burning time was recorded by electronic timer. Wire technique was used to measure the burning rate. Preliminary results from these techniques are presented. This study shows that the strand burner can be used on propellant strands to obtain accurate low pressure burning rate data

1. Introduction

A strand burner is an apparatus that provides burning rate measurements of a solid rocket propellant at elevated pressure. It was used to evaluate new propellant formulations and ensuring quality control for a large propellant production, due to its low cost, simplicity and ability to produce good results in a short time period compared to sub-scale and full scale motor [1]. For burning rate measurement, the length of the strand is the importance parameter to be measured, but the size and shape of the strand is less significant [2-3]. Actually, there were no specific lengths, size or shape for a strand in measuring burning rate. Donald et al. [4] reported that, the lengths of propellant strand did not have a significant effect on burning rates. While Matthew et al. [5] reported that sample size did not have a large affect on burning rates except for the smallest size tested, 3.2 mm square. Studies made by Nelson et al. [6] used the propellant strand which have circular cross-sectional area while several researcher [7-8] applied square shape. Strand surface usually burning inhibited with an external coating such as cured HTPB [2, 9], vinyl resin solution [6], bituminous compound [4] or water based acrylic paint [7] to protect from the heat of combustion and ensure the burning only occur in one dimensional or ‘cigarette type’.

Commonly, a strand burner has a mounting stand to mount the strand either horizontally [10-11] or vertically [12-13]. It also equipped with ignition and timing circuit. Work by Nelson et al. [6, 14-17] used nickel-chrome wire as an ignition wire. For the timing wire, Kirk et al. [16] used 30 S.W.G tin /lead (63/37). In order to simulate the condition of high pressure in a chamber of a rocket motor, inert gas such as helium, carbon dioxide, argon or nitrogen is used to pressurize the strand burner. Rodolphe et al. [2] reported that, there are no significant different in the burning rate resulted when applying these gases. Study made by several researchers [2, 6] used the high chamber pressure ranging from 40 atm to 360 atm for burning rate test. However, there were also some studies used low chamber pressure [18]. The information from low-pressure combustion commonly used to optimize a base bleed-propellant design for high-altitude projectile [19-20]. The previous studies showed that, varying chamber pressures will give different burning rate [21-23]. The objective of this work is to determine the effect of chamber pressure to the burning rate of aluminized AP/HTPB propellant and along the way establishing the strand burner facility. The scope included measuring burning rate using the strand burner for five different chamber pressures ranging from 1 atm to 31 atm.

2. Methodology

2.1. Facility hardware

The strand burner in figure is designed for combustion of a propellant strand in continues gas flow up to 31 atm. The body, flange and both end cap are made of low carbon steel. The 23 cm long cylinder has an inner diameter of 10 cm and an outer diameter of 13 cm, offering a wall thickness of 1.5 cm thickness. Each end cap is 1.5 cm thick, making the overall length of the burner 26 cm. Both end caps are square with side length of 21 cm. A 1.2 mm black gasket was inserted between the end caps for a gas tight seal.

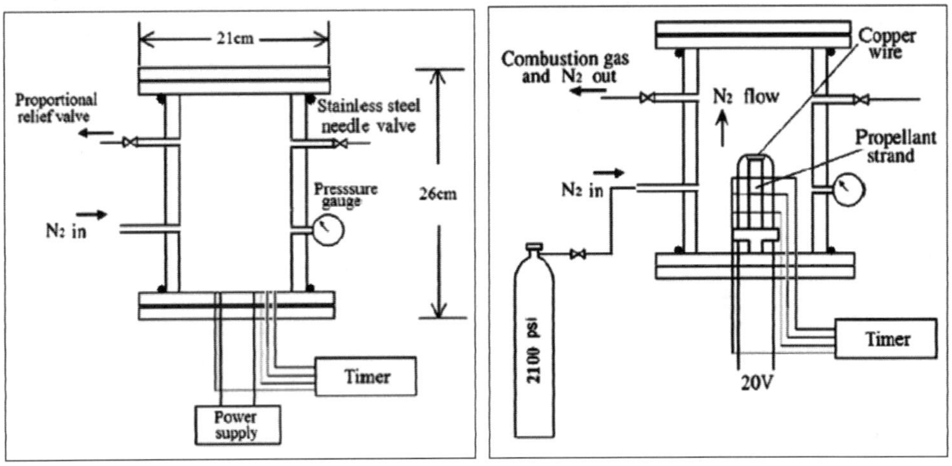

Figure 1. Schematic diagram of the strand burner facility

The strand burner was pressurized using nitrogen gas, which is same as the working fluid used by Shigeyuki et al. [24]. Nitrogen was chosen due to its low cost, availability and low density relative to air. Nitrogen gas was supplied from 22100 psi nitrogen tank at a suitable rate of flow. This will pressurize the burner as well as allowing steady flow of nitrogen at the outlet. Measurement of pressure within the strand burner was made by using a standard bourdon-type pressure gauge. At one of the end cap, there is a mounting stand used to mount the propellant strand. It was made from 5mm low carbon steel nut as shown in Figure 2.

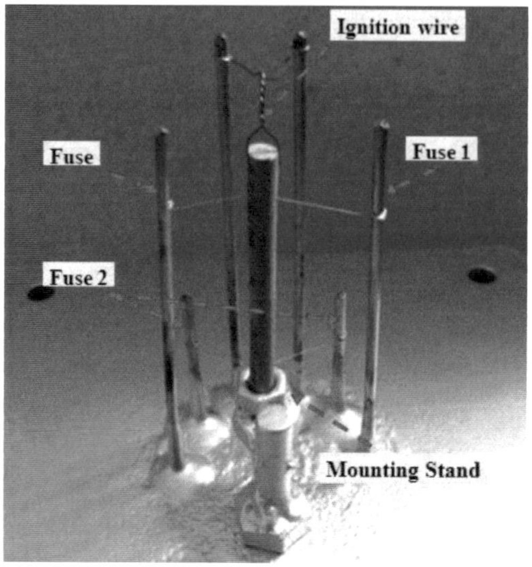

Figure 2. Example of propellant strand mounting

Around the mounting stand, there are six stainless steel rods with a diameter of 2mm which were used to hold wires/fuses for ignition system and instrumentation timer. The attachment between the rod and the end cap was air sealed using epoxy glue. To ensure the propellant strand could stand firmly and withstand the inlet and outlet flowing of the gas, the end of it was wrapped with white tape. During the experiment, the chamber pressure increased slightly due to combustion effects, and heat released. To stabilize it, two relief valves were installed at the upper end of the strand burner.

2.2. Propellant Selection

In this study, four specific propellant formulations have been selected for study. Table 1 present these four formulations where each propellant comprises a multimodal ammonium perchlorate (AP) as an oxidizer fraction, aluminum powder and HTPB binder. The first formulation, p60 subsequently referred as the baseline propellant, consists of 60 percent (by mass) AP and 25 percent aluminum powder. The O/F ratio (the ratio of AP fraction to the binder and aluminum fraction) is 1.5. The next three formulations were formulated by varying the O/F ratio to study their effect on the burning characteristics of aluminized propellant.

All propellants were manufactured manually at the Universiti Teknologi Malaysia (UTM) Propulsion Laboratory. To ensure safe practice, all propellants were prepared in 100gram batches. After finish the mixing procedure, the next step is to pressed the mixture into the soda straws, similar works as reported by Howard G. Cutforth [25]. The soda straws have 6mm diameter and they were cut to a length of 80mm. The strands were then transferred to the oven and cured at 64°C for five days. The strands were visually inspected and were rejected if cracks, pores, or shape irregularities were seen.

Table 1. Formulations of the propellants.

Propellant	AP (%)	AL (%)	HTPB (%)	O/F
p80	80	5	15	4.00
p74	74	11	15	2.84
p66	66	19	15	1.94
p60	60	25	15	1.50

2.3. Procedure of testing

Wire technique was used to measure the burning rate. Three small holes were accurately placed along the strand length using a needle. An igniter and two fuses wires were passed through these holes and connected to a power supply and electronic timer respectively. Similar method was applied by Jolley [26] which used two fuses wire threaded through the propellant strand, while Clark [27] used three fuses wire. All wires used were of the same type of 38 S.W.G. tinned copper wire with 0.152mm thickness, this is the same type of igniter wire used by Rodolphe et al. [2]. The strand is mounted vertically and is ignited at the top end using electrical current. Depends on the resistance of igniter, it took about 1 to 2s to ignite. No inhibitor is used on the side faces of the propellant similar with the work done by several other researchers [15, 28]. The burning rate was measured by using electronic timer. It was determined from the period it took for both fuses separated at a distance of 50 mm apart to cut-off as shown in figure 3.

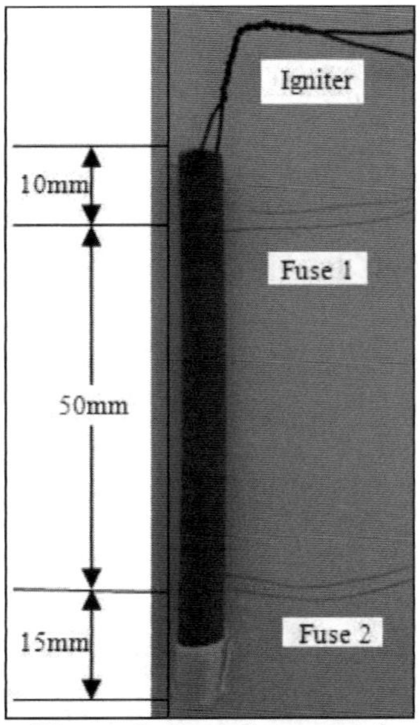

Figure 3. Propellant strand

The electronic timer was integrated with Data Acquisition Device (DAQ) in acquiring and recording the data [29]. The LabVIEW® software was then being used to analyze the data acquired as shown in figure 4.

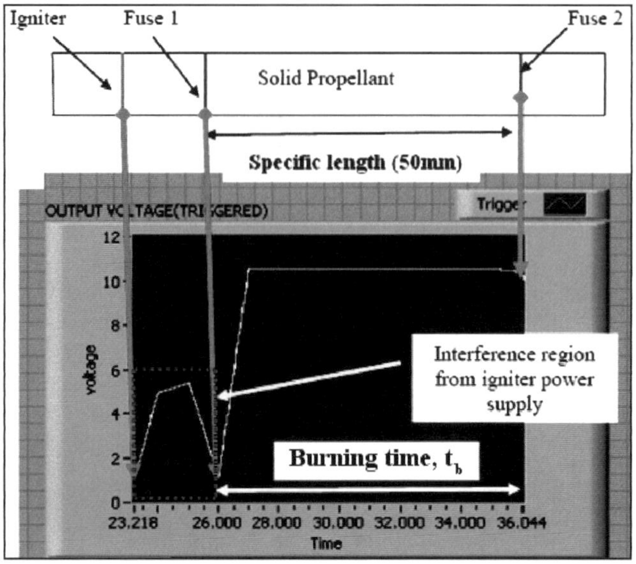

Figure 4. Actual burning time of the solid propellant at specified length.

$$r = \frac{L_p}{t_b} \tag{1}$$

Where,
L_p = Specific propellant length, (mm)
t_b = Burning time from recorded data, (sec)

3. Result and Discussion

The strands were burned at six different combustion pressures ranging from 1atm to 31atm. Generally, three strands established the burning rate at a single pressure level and repeatability of the burning rates was observed within 5% and is acceptable according to Jayaraman et al. [30]. From the observations, the increase in pressure inside the strand burner was on average about 20% of the initial pressure and the burn rates were assigned to pressures equaling half the pressure rise added to the initial level as mention by Matthew et al. [9]. Throughout this study, nitrogen gas was allowed to flow slowly past into strand burner in order to prevent possible preheating of the strand by reverse flow of the hot combustion products and also to prevent the flame from flashing down the side of the strand as mention by Summerfield et al. [3]. Table 2 shows the average burn rate for the selected compositions.

Table 2. Average burn rate for four selected composition.

Propellant	1 atm	11 atm	21 atm	31 atm
p80	1.776	6.684	10.045	11.991
p74	1.630	6.113	9.520	10.749
p66	1.565	5.343	8.249	9.944
p60	1.527	4.610	7.327	8.425

3.1. Uncertainty Analysis for Burning Rate Test

Consider the calculation of burning rate from p80 at $\left(r = 1.776\,\text{mm}\,\text{sec}^{-1}\right)$ 1atm,

$$r = \frac{L_p}{t_b}$$

Where, L_p and t_b are measured as;

$$L_p = 50\,\text{mm} \pm 1\,\text{mm}$$

$$t_b = 28.153\,\text{sec} \pm 0.010\,\text{sec}$$

The sensitivity of electronic timer is 1/1000. However, the burning time is taken to 1/100, considering a worst case condition. Thus, by taking the worse possible variations in propellant length and burning time, the error analysis could calculate as follows;

$$r_{\max} = \frac{(50+1)\,\text{mm}}{(28.153+0.010)\,\text{sec}} = 1.811\,\text{mm}\,\text{sec}^{-1}$$

$$r_{\min} = \frac{(50-1)\,\text{mm}}{(28.153-0.010)\,\text{sec}} = 1.741\,\text{mm}\,\text{sec}^{-1}$$

Thus, according to the method suggested by Holman [31], the uncertainty in the burning rate is $\pm 1.97\%$.

4. Conclusion

This paper has described the equipment being used in an intensive study of burning rate characteristics of ammonium perchlorate based solid propellant in order to get a better understanding on the effect of varying combustion chamber pressure to the burning rate. A strand burner has been developed which is capable of holding the surface of a burning propellant strand in pressurized condition. With this constrains satisfied, the burning rate of a propellant can be successfully investigated.

Acknowledgments

The authors gratefully acknowledge the support for this work through Universiti Malaysia Pahang and Ministry of Education (MOE) for funding this project under Research Acculturation Collaborative Effort (RACE) grant scheme (RDU121307).

References

[1] Fry R S 2001 *Solid Propellant Test Motor Scaling* CPTR 73 Chemical Propulsion Information Agency Columbia The John Hopkins University

[2] Rodolphe C M S Jason A A P and Eric P *High Pressure Testing of Composite Solid Propellant Mixtures: Burner Facility Characterization* 41st AIAA/ASME/SAE/ASEE Joint Propulsion Conf. & Exhibit 2005 Tucson Arizona

[3] Steinz J A P L S and Surmerfield M 1969 *The Burning Mechanism of Ammonium Perchlorate Based Composite Solid Propellants* Princeton University Department of Aerospace and Mechanical Sciences

[4] Donald C A G D D 1984 *Closed Vessel Combustion of Stick Propellant*, in Technical Report U.S. Army Armament Research and Development Center Large Caliber Weapon Systems Laboratory Dover New Jersey

[5] Matthew S E P David R Rodolphe C and Sudipta S *Nano Additives and Plateau Burning Rates in Composite Solid Propellants* 44th AIAA/ASME/SAE/ASEE Joint Propulsion Conf. & Exhibit 2008 Hartford CT

[6] Nelson B G C W 1977*Burning Rates of Standard Army Propellants in Strand Burner and Closed Chamber Tests* U.S.A Army Ballistic Research Laboratory Aberdeen Proving

Ground Maryland

[7] Pietrobon A H H R A 2003*Improved Strand Burn Rate Reproducibility Using a New Preparation Methodology for Paint-Based Inhibitors* Australian Goverment Department of Defence

[8] Kimura N K *Oscillatory Burning of High Pressure Exponent Double-Base Propellants* AIAA /SAE 12th Propulsion Conference 1976 Palo Alto California

[9] Matthew S R C Steven W Thomas S and Eric P *Performance of AP-Based Composite Propellant Containing Nanoscale Aluminum* 41st AIAA/ASME/SAE/ASEE Joint Propulsion Conf. & Exhibit 2005 Tucson Arizona

[10] Rizalman M Nazri M J and Wan K W A *Pembakaran Propellan Roket Kimia Berasaskan Kalium Nitrat* 2001 Malaysia science and technology congress

[11] Matthew S T S Rodolphe C Alexander L and Eric P *Comparison of Hand and Mechanically Mixed AP/HTPB Solid Composite Propellants* 43rd AIAA/ASME/SAE/ASEE Joint Propulsion Conf. & Exhibit 2007 Cincinnati OH

[12] Kubota N 2001*Propellants and Explosives* (WILEY-VCH)

[13] Gilbert P D 2008 *Multifunctional Additives for Composite Solid Propellant* in Mechanical Engineering Purdue University

[14] Yoshida M K Burning rate characteristics of AP-based composite propellant using bimodal AP 42nd AIAA/ASME/SAE/ASEE Joint Propulsion Conf. & Exhibit 2006 Sacramento California

[15] Fenn R E C and J B 1968 *Strand Size and Low Pressure Deflagration Limit In a Composite Propellant* Project Squid School of Mechanical Engineering Purdue University

[16] Kirk W H J and D R 1977 *Description of the W.R.E. Strand Burning Facilities and Instructions for Use* Weapon Research Establishment W R E Adelaide Australia

[17] Meda G M L Galfetti F Severini L D L 2007 *Journal Materials Science and Engineering* **27** 1393 - 96

[18] Holmes M S M and H E 1989 *Subatmospheric Burning Rates and Critical Diameters for AP/HTPB Propellant* U.S. Army Ballistic Research Laboratory Aberdeen Proving Ground

[19] Tanaka M C N and Hayakawa S *Combustion of Solid Propellents at Low Pressure* 36th AIAA/ASME/SAE/ASEE Joint Propulsion Conf. & Exhibit 2000 Huntsville AL

[20] Tae H L I C K and Young K P *A Study on the Low Pressure Burning Characteristics of the Composite Propellants* 31st AIAA/ASME/SAE/ASEE Joint Propulsion Conf. and Exhibit 1995 Diego CA

[21] Sutton G P 1992 *Rocket Propulsion Element* 6th Edition (John Willy & Sons MIT)

[22] Mamat R 2002 *Ciri-Ciri Propelan Berasaskan Kalium* Faculty of Mechanical Engineering Universiti Teknologi Malaysia

[23] Bozic M V M *Small Motor Measurement method for Determining Burning Rate of Solid Rocket Propellants* 34th AIAA/ASME/SAE/ASEE Joint Propulsion Conf. & Exhibit 1998 Cleveland OH

[24] Hayakawa S M T *An Effect of Oxidizer Particle Size on Combustion Stability in Composite Propellants* 36th AIAA/ASME/SAE/ASEE Joint Propulsion Conf. & Exhibit 2000 Huntsville AL

[25] Cutforth H G 1971 *Solid propellant selection and characterization* NASA.

[26] Jolley W H 1977 *Timing Instrumentation for A Crawford Type Strand Burner* Defence Science and Technology Organisation Weapon Research Establishment

[27] Hermance C E 1961 *Comparison of Solid Propellant Burning Rates In Strands And Rocket Motors* Department of Aeronautical Engineering Princeton University Princeton New Jersey

[28] Jolley D K and W H A 1991*Windowed Bomb for Propellant Combustion Studies* Technical Memorandom Defence Science and Technology Organisation Australia

[29] Khalid S I A 2009 *Development of Crawford Bomb Time Measuring Instrumentation* Faculty of Mechanical Engineering Universiti Teknologi Malaysia

[30] Jayaraman K KVA Chakravarthy S R and Sarathi R *Combustion and Flame* 2009 **156** 1662 73

[31] Holman J P 2001 *Experimental Methods for Engineers* 7th Edition (McGraw Hill)

2nd International Conference on Mechanical Engineering Research (ICMER 2013) IOP Publishing
IOP Conf. Series: Materials Science and Engineering **50** (2013) 012049 doi:10.1088/1757-899X/50/1/012049

Investigation of potential waste material insulating properties at different temperature for thermal storage application

T Z S Ali[1], A B Rosli, L M Gan, A S Billy and Z Farid
Faculty of Mechanical Engineering, Universiti Malaysia Pahang, 26600 Pekan, Pahang. Malaysia

E-mail: [a] tanty.ali@gmail.com

Abstract. Thermal energy storage system (TES) is developed to extend the operation of power generation. TES system is a key component in a solar energy power generation plant, but the main issue in designing the TES system is its thermal capacity of storage materials, e.g. insulator. This study is focusing on the potential waste material acts as an insulator for thermal energy storage applications. As the insulator is used to absorb heat, it is needed to find suitable material for energy conversion and at the same time reduce the waste generation. Thus, a small-scale experimental testing of natural cooling process of an insulated tank within a confined room is conducted. The experiment is repeated by changing the insulator from the potential waste material and also by changing the heat transfer fluid (HTF). The analysis presented the relationship between heat loss and the reserved period by the insulator. The results show the percentage of period of the insulated tank withstands compared to tank insulated by foam, e.g. newspaper reserved the period of 84.6% as much as foam insulated tank to withstand the heat transfer of cooking oil to the surrounding. The paper finally justifies the most potential waste material as an insulator for different temperature range of heat transfer fluid.

1. Introduction

In a solar energy power plant, four elements are required to have which are concentrator, receiver, transport/storage media system and power conversion device. Of all the components, thermal storage is a key component. However, it is also the less developed. Only a few plants in the world have tested high temperature thermal energy storage systems [1].

Energy storage systems in commercial use today can broadly categorised as showed in Figure 1. Energy storage became a dominant factor in economic development to make electrical generation power plant reliable. Thermal energy storage can be divided into three main categories: sensible heat storage, latent heat storage and chemical storage [2].

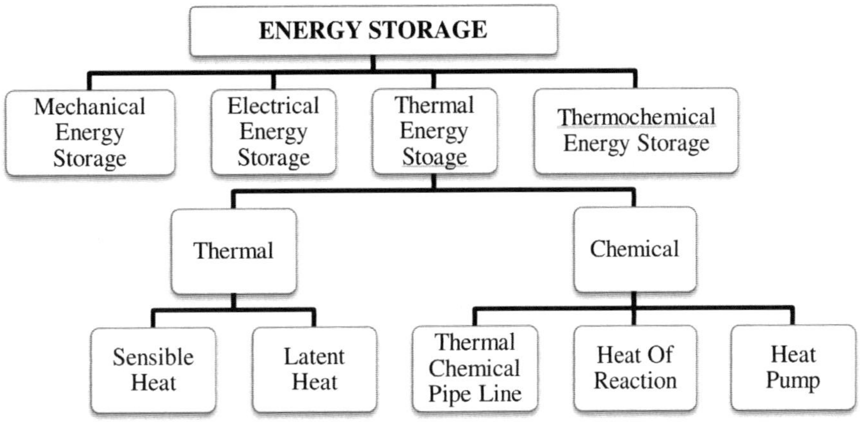

Figure 1. Classifications of energy storage

Today, the development of the energy storage system is growing as to fulfil the demand of users. An energy storage design needs to consider a lot of criteria (applicable laws, standards, technology and cost) in order to establish the system. Researches have been conducted in proposing not only a safe and efficient energy storage system but also an economic and green environment configuration.

2. MATERIALS

2.1 Waste Materials

Waste refineries focusing on various outputs of material resources, energy carriers, and nutrients may potentially provide more sustainable utilization of waste resources than traditional waste technologies. On the basis of research, it is recommended that a narrow focus on global warming aspects should be avoided as most waste technologies may provide comparable performance.

Presently the environmental permits will include necessary regulations on wastes and reduction of their quantity and harmfulness. Table 1 explained the problems associated upon waste and alternatives to land disposal for different waste type. This information is taken from Trash Management Guide 1992.

Table 1. Waste type problems and alternatives.

Waste Type	Problems	Alternatives
Paper	Litter. Fuel source for uncontrolled fires	Ask stores to reuse packaging for other purposes.
Leather & rubber	Minimal concerns	Set up salvage area
Metals	Corrosion slowly releases metals and causes landfill area	Recycle
Yard wastes/wood	Shelters rodents and other animals	Compost yard waste

In addition, efficiency in the use of materials must be taken into account as required. Thus, in TES development, there are a form of materials can be used as an insulator in order to absorb heat. The insulator ability is measured by its R-value, which was the resistance to heat flow. The higher the number, the more blocking capacity or insulating it provides. Table 2 summarized the insulating capabilities, densities and thermal conductivities of materials used in this study.

Table 2. Thermal properties of natural materials.

Material	Insulating Capability (R-value) $K.m^2/W$	Density, ρ kg/m^3	Thermal conductivity, k W/m.K
Foam	1.23-1.41	30-150	0.025-0.035
Sand	1.11	1515	0.2-1.0
Rubber	1.23	1100	0.13
Newspaper	0.52-0.7	930	0.18
Wire mesh	3.8-4.1	7850-7950	0.22-0.65

Gil suggested to used a material that is low cost, good thermal conductivities, comfort during handling and have good structural strength [1]. Hence using the waste materials as insulators, it is not only reducing the price but also reduce the pollution.

2.2. Heat Transfer Fluid (HTF)

Fluid which flows neither through nor around a device to prevent overheats, transfer the heat produced to other devices that use or disperse it is known as coolant. The term coolant commonly used in automotive applications, in industrial processing, heat transfer fluid (HTF) is one technical term more commonly used, in high temperature as well as low temperature manufacturing applications. HTF can either maintain its form or undergo a phase transition with the latent heat add to the cooling efficiency.

The most common HTF for low temperature is water. It contained high heat capacity and low cost makes it a proper heat-transfer medium. Water appears to be the best liquid available because it is inexpensive and has a high specific heat. However for temperature above 100°C, oils, molten salts and liquid metals, etc. are used.

Oils are used for applications where water is incompatible. Oil can be raised to a higher temperature above 100°C without introducing high pressures within the container. Oil has higher boiling points than water. Other than that mineral oils serve as both coolants and lubricants in various mechanical gears. Castor oil is also used. Due to their high boiling points, mineral oils are used in portable electric radiator-style space heaters in residential applications and in closed-loop systems for industrial process heating and cooling.

For example, Mawire [3] has been using oil in the glass tube TES system and its thermal performance is evaluated. In his study, he presented the results of energy and exergy charging rates during the experiments at temperatures of 200^0C, 250^0C, and 300^0C. The results indicate an optimal charging temperature when exceeded the usage and the thermal performance due to increased heat losses.

3. Thermal study of materials

Insulating capability is measured with thermal conductivity, k. Low thermal conductivity is equivalent to high insulating capability, R-value. Thermal conductivity, k represents how well a material conducts heat.

A useful; material is the one that reasonable, have good thermal capacity and the rate for heat to be released and extracted (thermal diffusivity). Thermal diffusivity measures on how fast the heat diffuses through a material. Thermal energy can be stored in solid or liquid materials. The ratio of the thermal conductivity and heat capacity defined the thermal diffusivity.

$$\alpha = \frac{k}{\rho c} \tag{1}$$

High values of thermal conductivity indicate that the material is a potential heat conductor, and a low value indicates that the material is an insulator.By considering steady heat conduction,

$$\dot{Q}= -kA\frac{dT}{dx} \tag{2}$$

Here, dT/dx is the temperature gradient, which is the slope of the temperature curve on T-x diagram (the rate of change of T with x, e.g. time), at location x [4]. Therefore, the amount of thermal energy stored in a mass of material can be calculated as:

$$Q = \rho \,.\, \bar{c_p} \,.\, V \,.\, \Delta T \tag{3}$$

where, Q : amount heat stored (J)
ρ : density of the storage material (kg/L)
$\bar{c_p}$: specific heat over the temperature range of operation (J/(kg.K))
ΔT : temperature range of operation (°C)

4. Experimental testing setup

A small-scale storage tank undergoes natural cooling process in a confined room without any force cooling system, e.g. fan. In the end, the experiments will justify the most potential waste material as an insulator in different HTF based on the comparison.

Firstly, a cylindrical container, Tank A, is filled with the insulator material with thickness of 40mm compactly. Thermocouples (T1 and T2) are placed at the position illustrated in Figure 2. Then a litre of HTF; water or cooking oil, is prepared in a cylindrical container, Tank B. The HTF is heated and maintained at 100^0C for water and 300^0C for cooking oil. Then Tank B is placed in the Tank A.

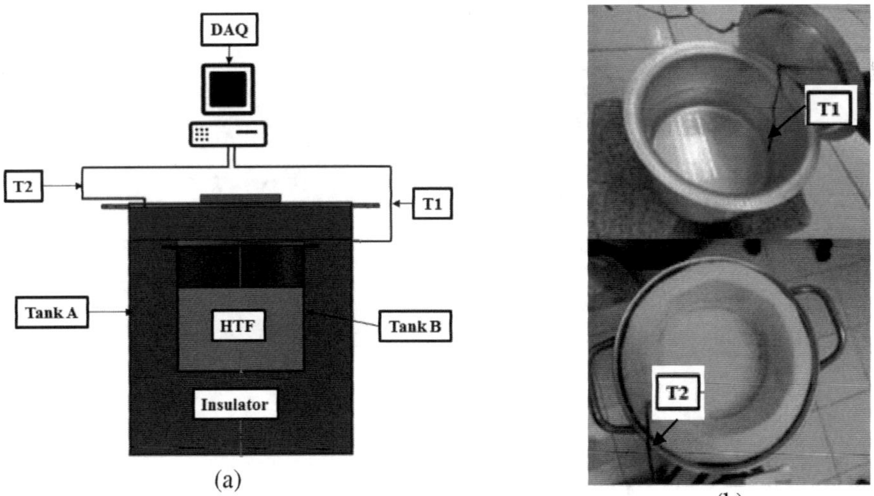

(a) (b)

Figure 2. Experimental setup (a)section view and (b)actual conditions.

Data acquisition system was setup to measure the temperatures and collect data. T1, measurement point was set in the middle of the container, Tank B, and T2, was set at a point between Tank A and Tank B. The process of testing the storage materials was based on monitoring the temperature and time of the experimental period. The testing is measured until T1 reached 40^0C and reported in plotted graphs.

This system is developed for the study on dispersal of materials act as an insulator in thermal storage system. This process was setup also based on the combination testing done by other researchers [5][6][7]. Visibly, the experiment was designed to validate the theoretical approach on heat losses and reserved period.

5. Results & discussion

Figure 3 shows exponential decay pattern graphs which are governed by the equation of:

$$f(x)=ae^{-kx} \tag{4}$$

where,

x:	time
k:	thermal conductivity(k = 0.0016 and 0.0035)
a:	mean temperature (a = 90.59 and 309.1408)
(-) sign:	cooling process

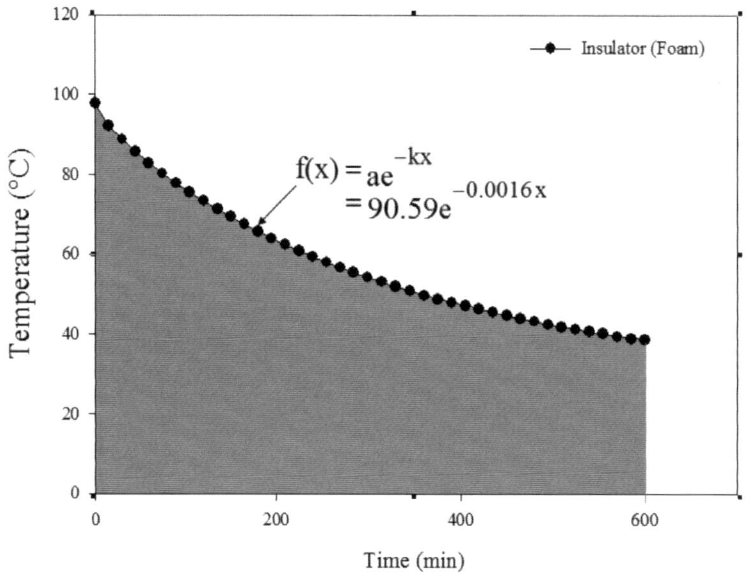

(a)

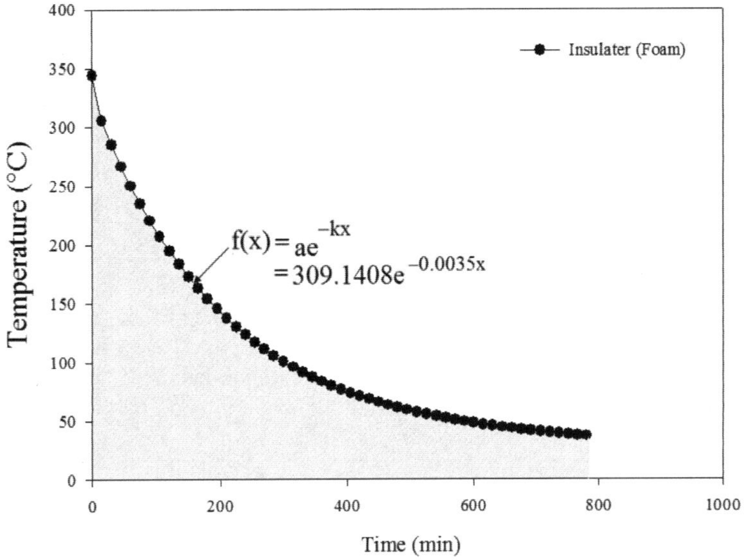

(b)

Figure 3. Effect of temperature at different time with (a)water HTF and (b)oil HTF.

As shown in Figure 3(a) and 3(b) are the distribution of temperature at the variation of time for foam insulated tank at different HTF. The figures show the heat of insulators decay decreased and cooling towards the temperature of 40^{0}C. The areas under the graphs are the reserved period of the insulated tank using a litre of water (Figure 3(a)) and oil (Figure 3(b)) as HTF respectively.

Based on the results from experiments, it is proved the experimental endured the process of cooling. The experiment is repeated using different waste materials as insulators, which were results in different k-values. Then, the k-values are pointed in Figure 4.

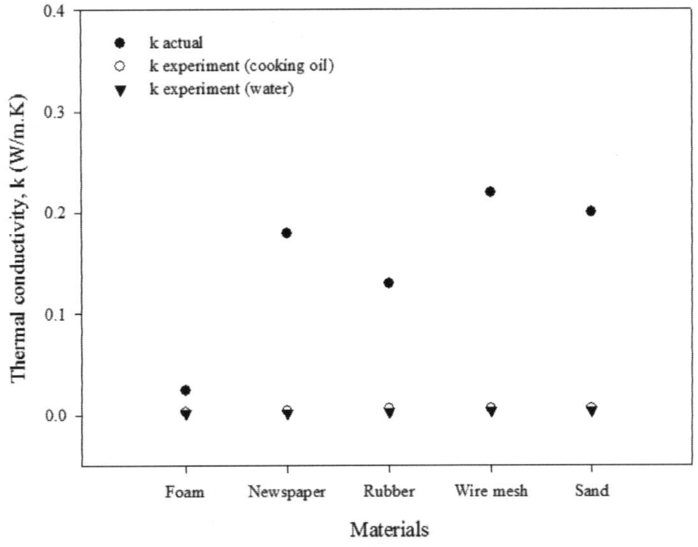

Figure 4. Effect of thermal conductivity, k at different materials.

From the Figure 4, the values for water and oil testing result almost the same, but there is a difference between the experimental and reference values are due to the experimental setup of the containers. It is occurred that containers did not reach the level of vacuum.

Subsequently, based on the results of time, the overall percentage of the reserved period of the insulator is calculated by:

$$\text{Percentage of Reserved Period} = \frac{t_{insulator}}{t_{insulator(foam)}} \text{ x } 100\% \tag{5}$$

The reserved period for the insulators is shown in Figure 5. The results show the percentage of the reserved period for potential waste insulating materials in water and oil testing. This figure also showed the value differences between an insulated tank and non-insulated tanks.

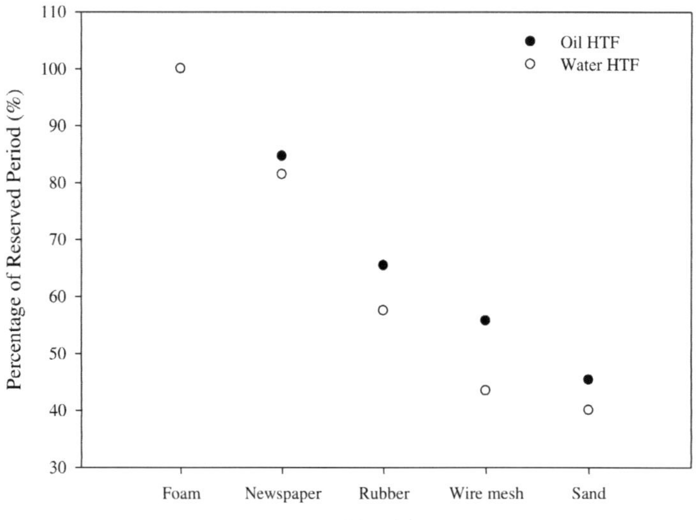

Figure 5. Effect of percentage of the reserved period at different potential waste materials.

The highest percentage of the reserved period for the insulated tank to withstand the heat as good as foam for water and oil as HTF is newspaper. It resulted as well as foam with just about 84.6% in water and 81.4% in oil testing.

This is followed by rubber, wire mesh and sand in both HTF tested. The percentages of the reserved period in water and oil for rubber are 65.4% and 57.5%, wire mesh are 55.7% and 43.4%, and sand are 45.3% and 40% respectively.

The experimental testing proven that the waste materials can be used as insulator which resulted as well as the common insulator used is foam. Based on other researchers [5][7], a suitable material is the one with a low thermal conductivity with low heat loss and high insulating capability.

6. Conclusion

Developing a storage system is an implementation of heat transfer knowledge. As the basis of designing TES is the insulator, selecting the insulator materials are also complicated and costly if based on the application. In this study, it shown that the waste materials could act as well as foam in preserve period of the HTF. Based on the point of reference of the insulated tank by foam, the percentage of reserved period can be calculated for each waste materials calculated.

Thus for future improvement, the waste materials could be mix and organize surrounded the HTF tank for the insulation test. By conducting the testing could also prove that arrangement of variation insulators could increase the insulation system reserved period.

7. References

[1] Gil A, Medrano M, Martorell I, Lazaro A, Dolado P, Zalba B and Cabeza L F 2009 State of the art on high temperature thermal energy storage for power generation. Part 1 – Concepts, materials and modellization. *Renewable and Sustainable Energy Reviews.* **14** 31-35

[2] Barlev D, Vidu R And Stroeve P 2010 Innovation in concentrated solar power. *Solar Energy Materials and Solar Cells.* **95** 2703-2725

[3] Mawire A, McPherson M and Heetkamp R R J 2009 Thermal performance of a small oil-in-glass tube thermal energy storage system during charging. *Energy* **34** 838-849

[4] Cengel Y A 2006 Heat and Mass Transfer. New York: McGraw-Hill.

[5] Canbazoglu S, Sahinaslan A, Ekmekyapar A, Aksoy Y G and Akarsu F 2005 Enhancement of solar thermal energy storage performance using sodium thiosulfate pentahydrate of a conventional solar water-heating system. *Energy and Buildings.* **37** 235-424

[6] Adinberg R, Zvegilsky D and Epstein M 2010 Heat transfer efficient thermal energy storage for steam generation. *Energy Conversion and Management.* **51** 9-15

[7] Royon L, Guiffant G and Flaud P 1997 Investigation of heat transfer in a polymeric phase change material for low level heat storage. *Energy Conversion and Management.* **38** 517-524

[8] Arteconi A, Hewitt N J and Polonara F 2012 State of the art of thermal storage for demand-side management. *Applied Energy.* **93** 371-389

[9] Bazmi A A, Zahedi G and Hashim H 2011 Progress and challenges in utilization of palm oil biomass as fuel for decentralized electricity generation. *Renewable and Sustainable Energy Reviews.* **15** 574-583

[10] Greiner H and Bussick F 2000 Mechanical Sealing for Heat Transfer Fluids. *Lubrication Engineering.* **56** 17-24

[11] Ng W P Q, Lam H L, Ng F Y, Kamal M and Lim J H E 2012 Waste-to-wealth: green potential from palm biomass in Malaysia. *Journal of Cleaner Production.* **34** 57-65

[12] Solangi K H, Islam M R, Saidur R, Rahim N A and Fayaz H 2010 A review on global solar energy policy. *Renewable and Sustainable Energy Reviews.* **15** 2149-2163

Acknowledgement

The authors would like to thank the Faculty of Mechanical Engineering in Universiti Malaysia Pahang (UMP) and Universiti Malaysia Pahang for the support.

2nd International Conference on Mechanical Engineering Research (ICMER 2013) IOP Publishing
IOP Conf. Series: Materials Science and Engineering **50** (2013) 012050 doi:10.1088/1757-899X/50/1/012050

Circle grid fractal plate as a turbulent generator for premixed flame: an overview

M F Othman , B Manshoor[1] and A Khalid[1]
[1]Centre for Energy and Industrial Environment Studies (CEIES),
Faculty of Mechanical and Manufacturing Engineering,
Universiti Tun Hussein Onn Malaysia 86400 Pt. Raja, Batu Pahat, Johor ,Malaysia.

E-mail: m.fahmi1989@gmail.com

Abstract. This review paper focuses to ascertain a new approach in turbulence generation on the structure of premixed flames and external combustion using a fractal grid pattern. This review paper discusses the relationship between fractal pattern and turbulence flow. Many researchers have explored the fractal pattern as a new concept of turbulence generators, but researchers rarely study fractal turbulence generators on the structure premixed flame. The turbulent flow field characteristics have been studied tand investigated in a premixed combustion application. In terms of turbulence intensity, most researchers used fractal grid that can be tailored so that they can design the characteristic needed in premixed flame. This approach makes it extremely difficult to determine the exact turbulent burning velocity on the velocity fluctuation of the flow. The decision to carry out additional research on the effect circle grid fractal plate as a turbulent generator for premixed flame should depends on the blockage ratio and fractal pattern of the grid.

1.Introduction

Turbulent flow is defined as formation of chaotic and eddies motions which do not contribute to the volume flow rate. It's also described as moving in irregular patterns, while the overall flow is in one direction . Turbulent flow is common in nonviscous fluids moving at high velocities. Turbulent flow considered as most exited topic to explore, there are many researchers had been studied about it [8-13], over past decade or so. They purpose fractal grid/multiscale space-scale structure as a new type of turbulent flow generator. These new approach depend on correlation between fractal behaviour and turbulent flow. The turbulent flow field characteristics have been studied to be used to investigate in a premixed combustion application. Turbulent flows were distinguished from laminar flow by Reynolds number [14] characterized qualitatively as irregularity in space and time (chaotic behaviour with coherent structures) ,continuous spectrum of time and length scales, higher Reynolds number, domination of vortical motion and enhance both, mass and heat transfer [15].

The idea of the fractal/multiscale as a turbulent generator establish from a fractal object obstructing a steady laminar stream that turbulence force by a 3-D fractal grid [16]. Fractal can create space-filling fractal devices for high efficient fluid mixing. The invention was changed in the fractal dimension of the grid alters the scaling of the structure function of the structure functions of turbulence generated by fractal grid. The fractals potential also identified by distinguishes between the scaling properties of turbulence stirred by fractal object [16, 17].

In the structure of premixed flame, fractal grid produced different characteristic compared to the regular grid. At the same burner, same mean and turbulent condition, the turbulent burning velocity fractal grid produced is higher than standard turbulent grid. Although with same turbulent intensity background, the turbulent burning velocity is different depends on the different decay law of turbulent. [18].

2. Relation between fractal pattern and turbulence

2.1 Blockage Ratio

Blockage ratio is defined as ratio of the area blocked by the grid to the whole area of the channel/tunnel square or circle section before it become turbulent or laminar flow. Experiment investigation on a turbulence generation system with high blockage ratio that compared two families of turbulent generator (multi circular jet plate and non circular jet plate) show the multi circular jet have high turbulent level at same blockage ratio, flow rate and preserving radial uniformity.[8].

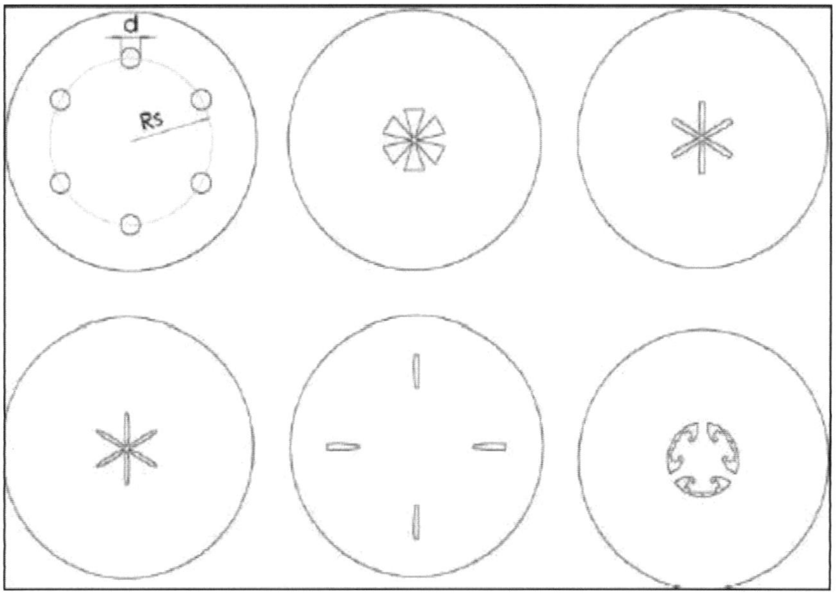

Figure 1. Families multi-circular jet (MCJ) plate (top left) and non-circular jet (NCJ) plates (single-jet (top center); central-jet (top right); central wake (bottom left), multi-jet (bottom center), wake (bottom right))[8].

Non-circular geometry considered as a good candidate for turbulence generation in a system compared to the circular geometry. It is because, in the same percentage of blockage ratio, the maximum turbulence intensity of non-circular geometry is much higher than circular geometry. There have been several studies looking at the effect of blockage ratio [19]. Depending on blockage ratio Castro demonstrated experimentally that two well distinguishable flow regimes exist. As the blockage ratio decreases, more bleed air is allowed to pass resulting in the re-circulation region moving further downstream until at a certain blockage, it disappears all together. As the blockage ratio increased from 70% to 80% suggesting that at some point in this range, the bleed air is not allowing the flow to come around to create the re-circulating region. To increase the blockage ratio, some geometry altered such as decreasing the bar thickness ratio of the fractal grid, but when lower blockage close to the grid centre and therefore produce large velocities in the central region of the grid [20].

Even though same blockage ratio=0.507, the average normalised pressure drop between upstream and far downstream of the fractal grid is about half that of the regular grid. [12, 21, 22]. Figure 2 satisfied that pressure drop fractal grid higher compare to the regular grid. But in turbulent intensity graph, the fractal grid lower than regular grid [21].

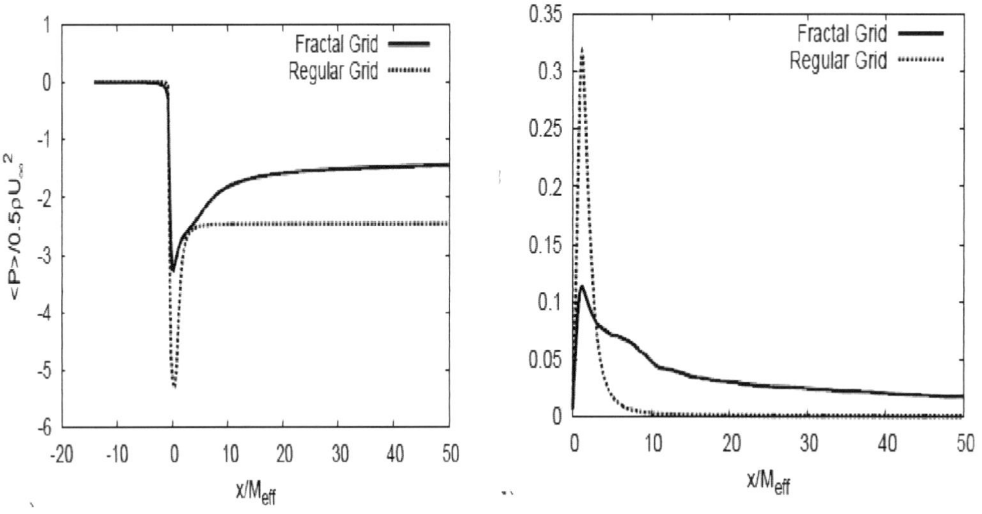

Figure 2. Comparison between fractal grid and regular grid.

From figure 2 it's shown that , there are differences between grid turbulence generation is distributed in the streamwise direction causing the turbulence to be less and the pressure drop smaller very near the grid by compare to the regular grid. By using fractal grid of larger blockage ratio, particularly because the relation between blockage ratio and pressure drop is very nonlinear and a small increase in blockage ratio can thereby lead to a significant increase in turbulence intensity, which is proportional to the square root of the pressure drop [12, 13, 22.]. Multiple perforated plates used to achieve high turbulence close to the grids. Their grids are shifted in space such that the blockage ratio of the plates increases the further downstream the grids are located. By measuring the blockage ratio, the maximum turbulence intensity that plays important role in premixed flames combustion could be determined.

2.2 Fractal pattern

Fractal pattern is variable that important to measure the turbulent burning velocity, pressure drop and turbulent intensity. Reference such as F. C.G.A. Nicolleau [3] used different pattern in their experiment, they found that pressure drop from a lower value to higher value by increasing the distance. The blockage ratio used in their experiment is 33.6%. The experiment that had done to compare pressure drop and velocity on the centerline normalized by inlet velocity by using different patterns such as circle, triangle and pentagon with same blockage ratio.

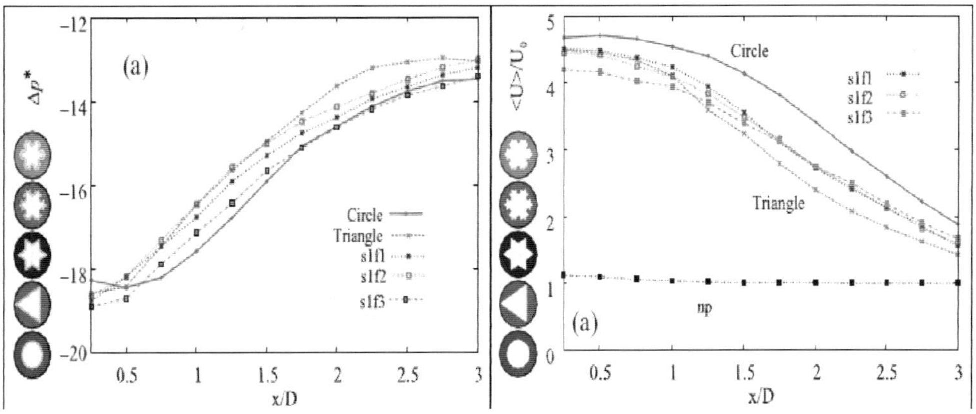

Figure 3. Comparison between different fractal patterns (pressure drop and turbulence intensity).

From figure 3, circle pattern shows the lower pressure drop compare to the others pattern. Meanwhile, velocity on the centerline normalized by inlet velocity for circle pattern is higher compare to the other pattern such as triangle and others fractal orifice plate. It's shown in figure 4. In our research, we try to introduce combination of circle grid and perforated plate in a plate. The suitable blockage ratio is about 50% to 70% because of the some limitation during fabrication the plate.

2.3 Turbulence intensity in premixed flame.

Turbulent intensity also often referred to as turbulence level. For fractal generated turbulent case, the level classified as medium turbulent case (no complex device) usually in between 1% and 5% turbulent intensity. The location maximum turbulence intensity determined by the blockage ratio (the ratio between the area occupied by the grid and the duct area). By using similar blockage ratio, the fractal grids have higher turbulence intensity compare to the regular grids [12, 20, 22, and 23].

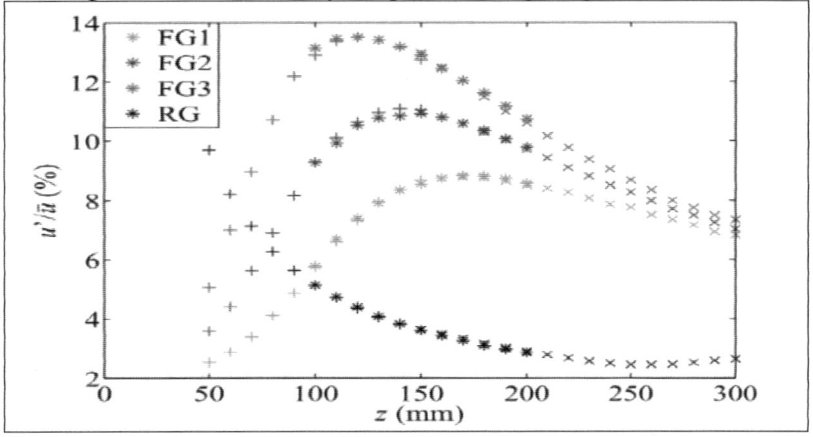

Figure 4. Comparison of turbulent intensity varied by different fractal grid.

Figure 4 significantly shown the turbulence intensity for regular grid is lower compare to the fractal grid. By compare fractal grid plate and without fractal grid, R. E. Seoud decided that using the fractal grids as higher as three times turbulence intensities produced than classical non fractal grids [22]. They also concluded that their space filling fractal square grids generated a kind of homogeneous isotropic turbulence which decays and locked into a single length scale. But N. Mazellier [13] stated that the grids are shifted in space such that the blockage ratio of the plates increases the further downstream the grids are located by using multi-scale injector (tested on a premixed v-shape flame) then they assume injectors double the turbulence intensity of normal grids. According to the need of combustion application, some researchers [12, 20] use the grid that can „tailor" so that they could design the position of the location maximum turbulence intensity. In term of position of fractal grid, the turbulence intensity depends on the position of the fractal grid located.

Griebel [24] studied the differences of turbulence intensity by used different position of the fractal grid. The result shown, the turbulence intensity for fractal grid located near to the chamber have more turbulence intensity compare to fractal grid located far away from the chamber.

3. Conclusion

In this present review paper, the aim is to introduce the circle fractal grid perforated as one of the option to generate the turbulence in order to stabilize on the structure of premixed flames. There are many researchers done their studied (experiment and simulation) about fractal grid as turbulence generator, but, a circle fractal grid perforated was introduce to stabilize on the structure of premixed flames.

References

[1] Mohd Adib M A H, Mohd Hasni N N T N H and Osman K 2013 *J Mech Eng Sci* **4** 496.
[2] Vijaya Lakshmi B, Subrahmanyam T, Dharma Rao V and Sharma K V 2011 *Inter J Automot Mech Eng* **4** 455.
[3] Idris M S, Irwan M A M and Ammar N M M 2012 *J Mech Eng Sci* **2** 206.
[4] Naga Sarada S, Sita Rama Raju A V, Kalyani Radha K and Syam Sunder L 2012 *Inter J Automot Mech Eng* **6** 795.
[5] Tahseen T A, Ishak M and Rahman M M 2013 *J Mech Eng Sci* **4** 418.
[6] Tahseen T A, Ishak M and Rahman M M 2012 *Inter J Automot Mech Eng* **6** 753.
[7] Syam Sundar L and Sharma K V 2011 *Inter J Automot Mech Eng* **3** 265.
[8] Coppola G and Gomez A 2009 *Exp. Thermal and Fluid Science* **33** 1037–1048
[9] Vallikivi M, Hultmark M, Bailey S C C and Smits A J 2012 *Exp. Fluids* **53** 1713-1722
[10] Nicolleau F C G A , Salim S M M and Nowakowski A F 2011 *J. Turbulence* **12** 44
[11] Salim S M M, Nicolleau F C G A and Borkowski M 2009 *Int. Conf. Mech Eng.*
[12] Hurst D and Vassilicos J C 2007 *Phys. Fluids* **19**, 035103
[13] Mazellier N and Vassilicos J C 2010 Phys. Fluids **22** 075101
[14] O Reynolds 1883 *Phil.Trans. R. Soc. London* **174** 935–982.
[15] Tennekes, Lumley H and J L 1972 *The MIT Press Cambridge* 223-229
[16] Mandelbrot B 1982 *The fractal geometry of nature (W.H. Freeman & Co, San Francisco)* 98-114.
[17] Queiros-Conde D and Vassilicos J C 2001 *New York, Cambridge University Press* 136-167
[18] Adrian S, Biagio M, Vassilicos J C and Willem 2003 *Phys. Review* E **67** 066306
[19] Sponfeldner T, Henkel S, Soulopoulos N, Beyrau F, Hardalupas Y, Taylor A M K P and Vassilicos J C 2011*1stUK-Japan Bilateral Workshop on Turbulent flow generated in fractal ways*
[20] Castro I P 1971 *J. Fluid Mechanics* **46** 599–609
[21] Laizet S and Vassilicos J C 2012 *Phys. Review* E **86(4)**
[22] Seoud R E Vassilicos J C 2007 *Phys. Fluids 19* **10** 105108
[23] Nedic J , Ganapathisubramani B , Vassilicos JC, Boree J, Brizzi L E and Spohn A, 2012 *AIAA Journal 50* **12** 2697-2710
[24] Griebel P, Siewert P and Jansohn P 2007 *Proceed. Combustion Institute* **31** 3083–3090.

Performance evaluation of an actual building air-conditioning system

M F Othman[1], H Abdullah[2,a], N A Sulaiman[1,3] and M Y Hassan[2]
[1]Centre of Artificial Intelligent and Robotic (CAIRO), Universiti Teknologi Malaysia, Kuala Lumpur, Malaysia
[2]Centre of Electrical Energy System (CEES), Universiti Teknologi Malaysia, 81310, Skudai, Johor, Malaysia
[3]Faculty of Electronic and Computer Engineering, Universiti Teknikal Malaysia Melaka, Hang Tuah Jaya, 76100, Durian Tunggal, Melaka, Malaysia

Email: [a]hayati@fkm.utm.my

Abstract. Air-conditioning contributes significantly to high energy consumption in commercial buildings. It is essential to monitor and evaluate the performance of air-conditioning systems to avoid unnecessary energy wastage. This paper presents the results of a study carried out to evaluate the performance of an actual building air-conditioning system with reference to the Malaysian Standard MS1525. The results of the air-conditioning system's performance and thermal comfort analysis are presented and a comparison is made to the results of a survey conducted among the occupants of the building. Although the results indicate a satisfactory coefficient of performance for the building despite the age of the system, recommendations on system improvements are proposed to further save energy.

1. Introduction
Building consume about one-third of the world's energy and it is expected that the trend will grow further until the year 2025. Among the buildings in Malaysia, office buildings consume the most energy for air-conditioning system compared to shopping complexes, hotels and residential buildings[1]. Thus, office building owners should take action to save more energy to avoid unnecessary energy wastage. In general, building air-conditioning system can be classified into the following categories:

i) Split/Window units
ii) Package units
iii) Centralised air-conditioning systems

The split unit consist of the indoor unit and outdoor unit. The indoor unit is installed inside the area to be cooled. It can be ceiling or wall suspended and comprises of the evaporator and blower. The outdoor unit is installed outside the building and comprises of the compressor, condenser and blower. Package units are for higher capacity air-conditioning loads in the range of 3 to 15 tonnes. They comprise of the filtering, cooling and dehumidifying as well as air handling components with water-cooled or air-cooled condensers. The centralised system can be of the direct expansion system or the indirect chilled water system. This study is focused on an actual building installed with a package unit air-conditioning system and some individual split units.

In order to implement energy saving initiatives, the performance of the air-conditioning system should be a major consideration. The performance of an air-conditioning system is measured by the Coefficient of Performance or COP. It is the ratio of the desired output over the required input of the system which is a ratio of the heat removed from the cooled space over the input work or electrical energy consumed. The COP can be calculated as in Eq. (1).

$$COP = \frac{total\_heat\_load}{total\_electrical\_load} \tag{1}$$

The heat removed from the cooled space can be determined by considering the chilled water heat rejection or condenser water heat gain or the heat rejection in the building Air Handling Units (AHU). For the package unit in this study, the heat rejection in the building AHU will be considered.

Thermal comfort is also an important consideration in evaluating the performance of an air-conditioning system. The main factors that affect thermal comfort are dry bulb temperature, relative humidity and air velocity. Air cleanliness, odour, noise and radiation effect also affect the comfort level of occupants in a conditioned space [2].

2. Case study

The study was carried out at the P02 Building which is one of the Faculty of Electrical Engineering's (FKE) buildings in Universiti Teknologi Malaysia (UTM). The objectives of this study are to determine the performance of the air-conditioning system of the building and to identify energy saving potentials for future improvement.

The P02 Building is a four storey building and consists of administrative offices, lecture rooms, lecturers' rooms and laboratories. The building is mainly installed with Water-Cooled Package Unit (WCPU) air-conditioning system. The condensers of the WCPU are cooled by cooling water from a nearby cooling tower. The cooling tower also serves two other buildings in the faculty. Apart from the WCPU air-conditioning system, there are also some rooms in the building which are installed with individual split units.

3. Methodology

The performance analysis of the air-conditioning system was carried out using data collected through an energy audit and a thermal comfort survey. Each level of the building was divided into several zones depending on the activities of the occupants.

3.1. Energy Audit

Data were collected from the rooms in the P02 building for temperature, relative humidity, air velocity and flow rate as well as carbon dioxide concentration. The audit was carried out to include both types of air-conditioning systems installed in the building. The air-conditioning system heat rejection load for the WCPU was obtained by assuming that the heat load rejected through the AHU is equal to the amount of cooling distributed by the AHU and can be determined using the following equations:

$$\text{Total Heat Load} = \text{Total Sensible Heat} + \text{Total Latent Heat} \tag{2}$$

$$\text{Sensible heat [W]} = 1.21 \times Q\ \text{[l/s]} \times \Delta T\ [^{o}\text{C}] \tag{3}$$

where,

Q = supply air flow rate [in litres/sec]

ΔT = mixed air temperature – supply air temperature

$$\text{Latent heat [W]} = 3 \times Q\ \text{[l/s]} \times \Delta M\ \text{[g/kg]} \tag{4}$$

where,

Q = supply air flow rate [in litres/sec]

ΔM = mixed air moisture content – supply air moisture content

Data collected from the AHU includes temperature, flow rate and relative humidity for the supply air, return air and mixed air. The total electrical load of the WCPU was obtained by considering the electric consumption of the compressor and blower of the refrigeration plant as well as the electric consumption of the cooling water pumps and cooling tower fan. The electrical power input for the compressor and blower of the refrigeration plant was obtained from the distribution board in the AHU rooms. The electrical power input for the cooling water pumps and cooling tower fans was obtained from the cooling tower building. A similar calculation and analysis were also used for the individual split units. The COP of the building's air-conditioning system was then determined using Eq. (1) and compared to the standards published in MS 1525:2007.

A heat load analysis was also carried out for the building by taking hourly measurements from the AHU rooms. Equations (3), (4) and (2) were used to calculate the sensible heat, latent heat and total heat rejected by the AHU respectively. The heat load was plotted against time to study the hourly trend of the heat load.

3.2. *Thermal Comfort Survey*

A thermal comfort survey was conducted among the occupants in P02 Building, FKE. The survey was distributed among a total of 55 occupants. The questions were focused on the comfort level and the effect of temperature and air quality on their office activities.

4. Results and discussion

This section presents the findings of the energy audit and thermal comfort survey carried out at level 3 of the P02 Building. The rooms located on level 3 include the dean and deputy dean's offices, meeting room and lecturer's rooms. The area of the audited zone is 503 m^2 and the floor layout is as shown in figure 1.

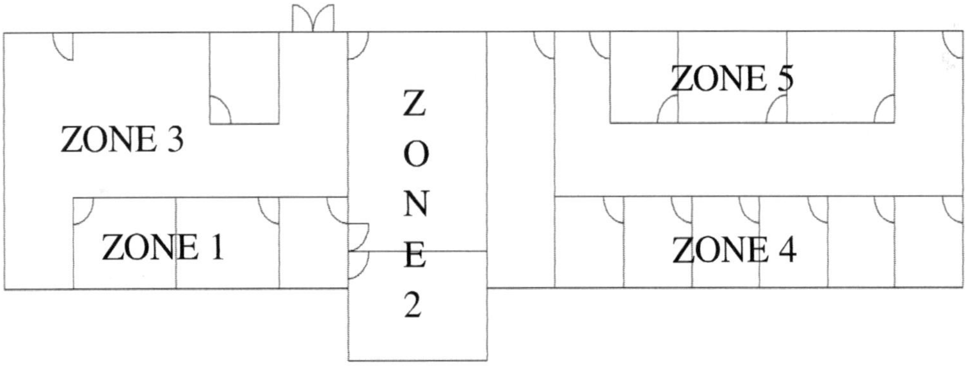

Figure 1. Layout of Level 3

Level 3 is divided into five zones according to the location of the area, air-conditioning usage pattern and source of supply air. Zone 1 consists of two deputy dean offices and personal assistant's office. Meanwhile, Zone 2 comprises of dean's office and meeting room, and Zone 3 is an administrative office area. Lastly, Zones 4 and 5 consist of lecturers' rooms.

4.1. *Questionnaire Survey Analysis*

A thermal comfort survey was conducted among the occupants on level 3 to get their feedback on the comfort level and the effect of air-conditioning on carrying out their office activities with regards to temperature and air quality. 13 occupants on level 3 participated in the survey and the analysis is presented according to the assigned zones as shown in figure 2 and 3.

4.2. Feedback on room comfort.

Figure 2 shows the feedback on room comfort from the occupants in the respective zones. The feedback obtained varied according to the zones. 15% of the occupants surveyed felt that the air-conditioning in their office is comfortable while the majority or 46% felt neutral on the issue of room comfort. Those that felt comfortable with the air-conditioning system were mainly occupants in Zone 1 and Zone 3. There were no occupants surveyed in Zone 2. The occupants that were uncomfortable especially in Zones 4 and 5 attribute their feedback due to the low temperature in their zones.

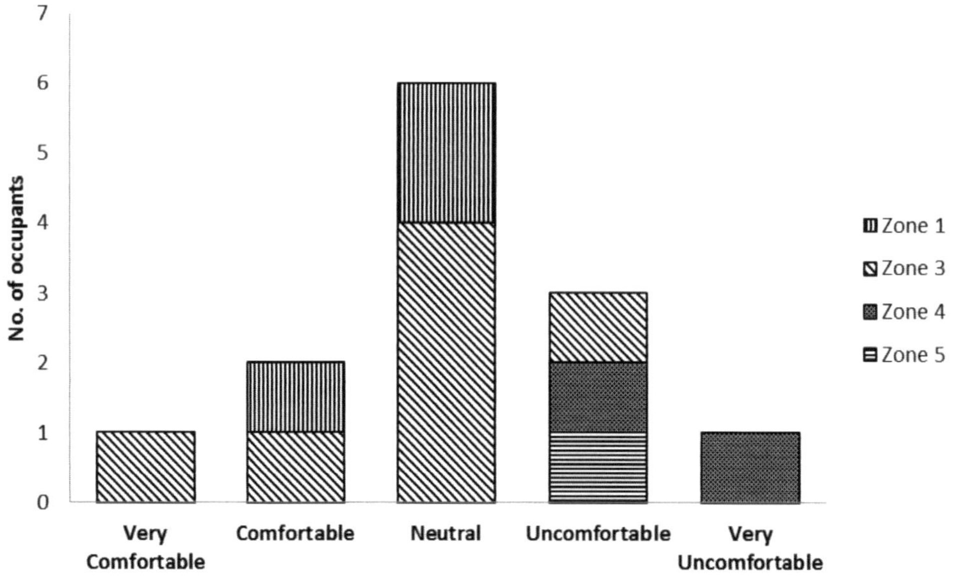

Figure 2. Feedback on room comfort at Level 3.

4.3. Feedback on the effect of air-conditioning on office activities.

Figure 3 shows the effect of the air-conditioning system on the office activities of the occupants. The majority of the occupants surveyed did not feel that the air-conditioning system was helpful to them in carrying out their office activities. 38% of the occupants felt slightly disturbed due to the low air temperature making them uncomfortable and affecting their productivity. They had to put on a jacket almost all the time to keep warm. Those that were slightly disturbed by the air-conditioning system were from zones 4 and 5 and these feedbacks seem to be in agreement with the findings on room comfort as observed in figure 2.

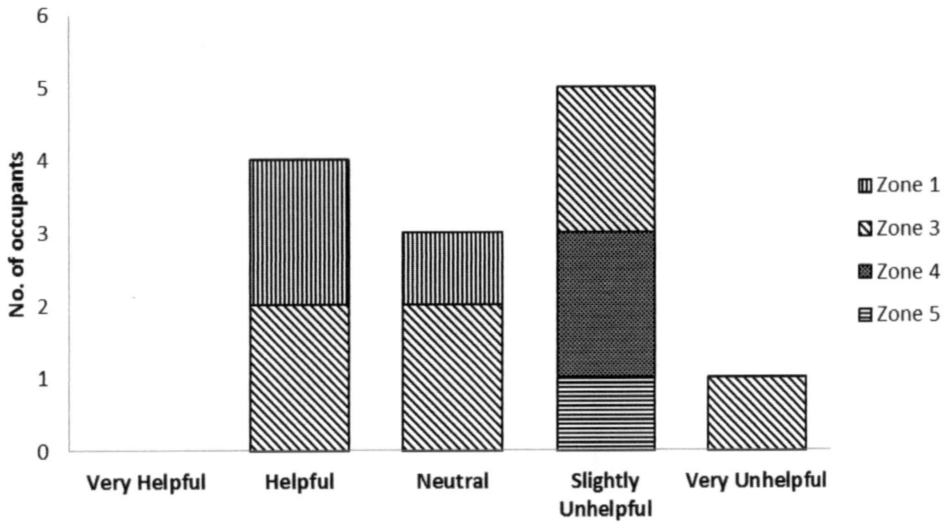

Figure 3. The effect of air-conditioning on office activities at Level 3

4.4. Comfort Level Analysis

An analysis on comfort level for the building was carried out based on the data measured and tabulated in table 1.

Table 1. Measured comfort level data for Level 3

Level	Zone	No. Of Occupants	Average Temp [°C]	Average RH [%]	Average room air velocity [m/s]
3	1	4	23.7	66.9	0.19
3	2	1	23.3	67.5	0.16
3	3	7	24.7	63.8	0.15
3	4	6	20.4	74.1	0.14
3	5	3	20.6	73.3	0.16

Table 1 shows the average measured values taken during the audit. The level was divided into five zones and the zone was divided according to the location of the area, air-conditioning usage pattern and source of supply air. The data in table 1 were then plotted on a psychrometric chart to identify if the zones were within the comfort range according to the Malaysian Standard (MS 1525:2007) as shown in figure 4 below.

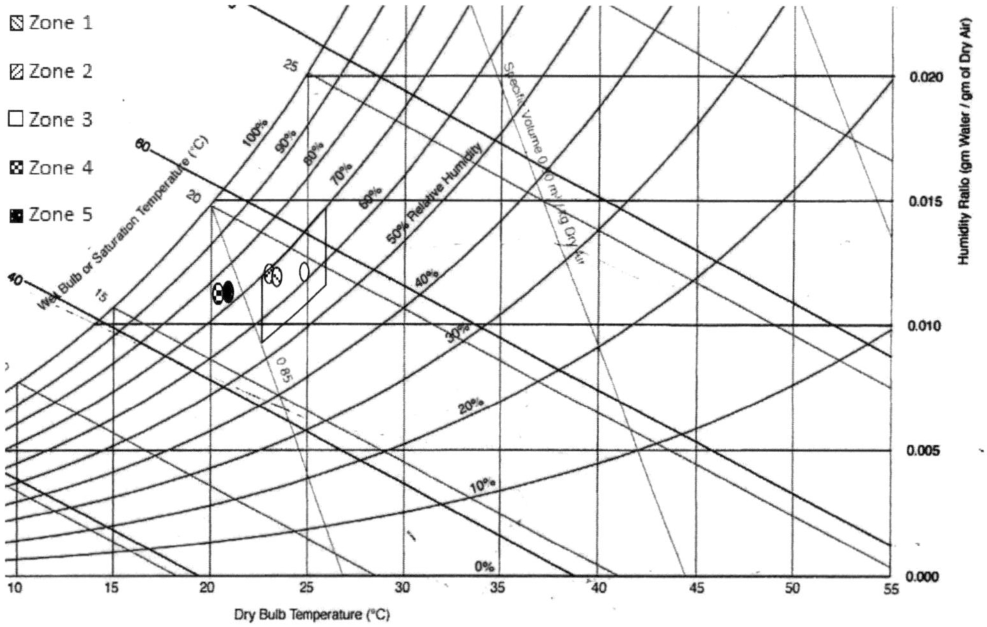

Figure 4. Comfort zones for Level 3 [3]

According to figure 4, Zone 4 and Zone 5 were a bit colder where the temperatures in these zones were much lower than the recommended range of 23°C – 26°C. The percentage of relative humidity was also beyond the recommended design relative humidity of 55% -70%. This is because Zone 4 and Zone 5 consist of rooms that are always vacant.

Heat Load Analysis

The heat load during office hours for level 3 is shown in figure 5. It can be observed that the heat load was high at the start of the day but remained almost constant until noon. The heat load decreased gradually from 12 noon to 2 pm due to the lunch break which mean fewer activities and less people were inside the office. The heat load can be seen to increase again after 2 pm until 4 pm. The centralized air-conditioning was switched off at 4.30 pm.

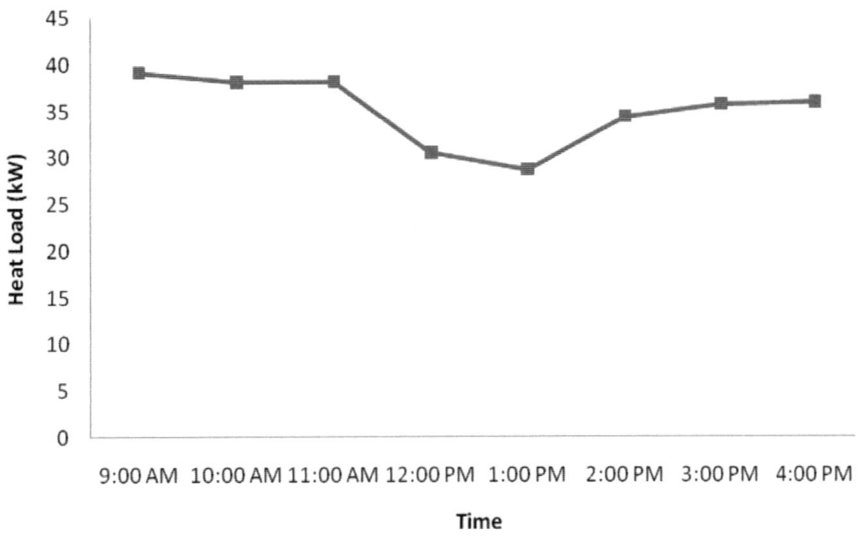

Figure 5. The heat load profile of air-conditioning system at Level 3.

Coefficient of Performance, COP determination

The COP of the air-conditioning system on level 3 was obtained by determining the heat removed from the conditioned area and the electrical power input to the system. Table 2 gives the measured data obtained at the AHU room. The data required for the determination of the heat load includes temperature, relative humidity and flow rate of the air at various locations in the AHU. The electrical power was measured from the distribution board to identify the power consumed by the AHU. Electrical power input for the cooling tower water pumps and fans were also taken into consideration to determine the electrical load of the AHU.

Table 2. Measured parameters at AHU Level 3

Field Measurement	Return Air	Mixed Air	Supply Air	Distribution Board
Temperature (°C)	23.7	24.2	20.0	-
Humidity (%)	70.6	74.1	84.4	-
Flow rate (m^3/s)	5.59	5.45	5.87	-
Voltage (V)	-	-	-	408
Current (A)	-	-	-	17.7
Power (kW)	-	-	-	9.38
Power Factor	-	-	-	0.75

By using Equation (1), the COP was determined to be 2.82 which is slightly lower than the minimum requirement COP of 3.0 as in the Malaysian Standard [4].

Conclusion

This paper has presented the results of a study carried out on an air-conditioning system of an actual building. The COP obtained is slightly lower than the minimum recommended COP of 3.0 but the result is as expected since the system has been installed for more than 20 years. Zones 1 and 3 were found to be within the recommended comfort zone according to the Malaysian Standard MS

1525:2007. Zones 4 and 5 were found to be outside the recommended comfort zones. But the results were in agreement with the survey carried out on thermal comfort based on the feedback from the occupants. The results of the study also showed that there is room for improvement in terms of thermal comfort and energy efficiency.

Recommendations

There is a need to study further on how to use energy more efficiently especially as can be seen from the results for Zones 4 and 5. Both these zones were not occupied most of the time but since both zones were installed with centralized air-conditioning system, this would result in energy wastage. A control system that takes into consideration variation in heat load could be installed to save energy for the air-conditioning system. Installing a Fault Detection and Diagnosis (FDD) system to detect possible failures and localize faults [5] would also benefit the air-conditioning system. According to [6], a successful FDD may save up to 40% of heating, ventilating and air-conditioning (HVAC) energy consumption and reduces unnecessary energy wastage.

Acknowledgements

The authors would like to thank the Ministry of Higher Education, Malaysia and Universiti Teknologi Malaysia for financially supporting this research project under the GUP Flagship Grant (Vot 00G18; Vot 00G20). The authors would also like to thank Universiti Teknikal Malaysia Melaka for the support.

References

[1] Aziz M B A et al 2012 Air-conditioning energy consumption of an education building and it's building energy index: a case study in engineering complex, UiTM Shah Alam, Selangor, in *IEEE Control and System Graduate Research Colloquium (ICSGRC)* 175-80

[2] Cengel Y A and Boles M A 2007 Thermodynamics: An Engineering Approach Sixth ed. McGraw-Hill Higher Education

[3] Ogawa A 2009 Psychrometric chart for sea-level pressure using SI units. [cited 2012 Nov 14]; Available from: http://en.wikipedia.org/wiki/File:PsychrometricChart.SeaLevel.SI.svg.

[4] MS1525:2007 *Code of practice on energy efficiency and use of renewable energy for non-residential buildings*, D.o.S. Malaysia, Editor

[5] Zimmermann G, Lu Y and Lo G 2012 Automatic HVAC fault detection and diagnosis system generation based on heat flow models *HVAC&R Research* **18**(1-2): 112-25

[6] Wu S and Sun J Q 2011 A top-down strategy with temporal and spatial partition for fault detection and diagnosis of building HVAC systems *Energy and Buildings* **43**(9): 2134-39

Palm olein ozonation as a renewable resourse: spectroscopic analysis for monitoring the degree of saturation

Bilal AWasmi[a], Abdul Amir H Kadhum, Abu Bakar Mohamed
Chemical and Process Engineering Department, Faculty of Engineering and Built Environment, National University of Malaysia, Malaysia, Bangi.

E-mail: [a]bilal.chem.82@gmail.com

Abstract: The manufacturing of organic compounds by environmentally-friendly methods has been intensively reexamined in recent years. Several excellent methods have been devised to produce organic compounds from renewable resources. The ozonation is one of the high active oxidation methods which lead to producion of organic compounds by the breaking of double bonds. Palm olein as a renewable source is subjected to the ozonation process to break the double bond which leads to the formation of two carbonyl groups as well described by Criegee mechanism. The monitoring of the degree of saturation was obtained by the Fourier transform infrared spectroscopy (FTIR) by observing the change in function groups as a result of ozone consumption and heat of reaction. The reaction time was 2 hrs at different temperatures and without any solvent or participating catalyst. The complete cleavage of a double bond occurred at 150 °C temperature of reaction.

1. Introdcution

Palm Olein (Palm Oil fractionation product) considers one of the most important used oils for industrial application in Malaysia as it has huge amounts of fatty acids i.e. Saturated and unsaturated fatty acids[1]. The Unsaturated fatty acids are subjected to a variety of reactions with several compounds to form a wide range of products[2] . These reactions may involve either the carboxyl group or the carbon–carbon double bond in an unsaturated fatty acid [3, 4] .The oxidation of unsaturated fatty acids results in partial or complete cleavage of the carbon–carbon double bond[5, 6] . One of the most active oxidation methods is the Ozonolysis; the process by which ozone (O3) reacts with alkenes to break the double bond and form two carbonyl groups which could be monitored simply by Spectroscopic analysis.

A number of works was done by some researchers on the monitoring of the degree of saturation by using a Spectroscopic analysis. Nestor , at.el , was investigate the degree of saturation for the sunflower oil and sunflower methyl esters by using FT-IR and H1NMR [7, 8], the study had monitored the consumption of ozone by the unsaturated fatty acids and the changes in the functional groups. Other work was investigated by N. Vlachos and his colleagues [9], their work was used FT-IR spectroscopy to monitor the oxidation process of corn oil samples undergone during heating or/and exposure to ultraviolet radiation. In our work, the ozonation of palm olein was performed under different temperature. All the experiment was conducted without using a catalyst and participating solvent. The degree of saturation for palm olein was monitored throughout the change in function groups by the Fourier transform infrared spectroscopy (FTIR).

2. Materials and Methods

2.1 Materials

Palm Olein (Fractionated palm oil) as starting material was purchased from LAMSOON Company (Malaysia). IR spectra were obtained on a Thermo Scientific, Nicolet 6700 FT-IR spectrometer. The gas bubbling reactor was fabricated by Dura Glass (Malaysia). Ozone was generated by use of an AC series ozone generator from IN (USA).

2.2 Methods

Oxygen gas at a flow rate of 1 L/min and pressure of 0.03 MPa was fed into the electric ozone generator to generate ozone gas which was directed as fine bubbles into the reactor containing Palm Olein. The reaction temperature was maintained at 50, 80, 120, 150 C° for time of 2hr with vigorous stirring. Unreacted ozone was decomposed before emission of gas to atmosphere.

3. Results and discussion

The Ozonolysis reaction proceeds by the three-step Criegee mechanism [4] which includes, first, the formation of a primary ozonide (PO), second, the decomposition of the PO to aldehydes (or ketones) and carbonyl oxides (or Criegee intermediates, CI) [4], then, finally, recombination of the carbonyl oxide and the aldehyde (or ketone) to form secondary ozonide (SO). The suggested mechanism of this study which described in scheme .1 explores the mechanism of the reaction regarding to the high percentage content of unsaturated fatty acid (Oleic Acid 45% approximately) [5] which may influence by the attack of ozone according to the ozonation process. This mechanism observes the breakage of the double bond, Cycloaddition and Retro Cycloaddition to form the aldehyde and the carboxylic group; otherwise the hydrogen peroxide was subjected to the reaction for further oxidation and obtaining the final products i.e. Azelaic acid and Pelargonic acid after the extraction and drying [10] .

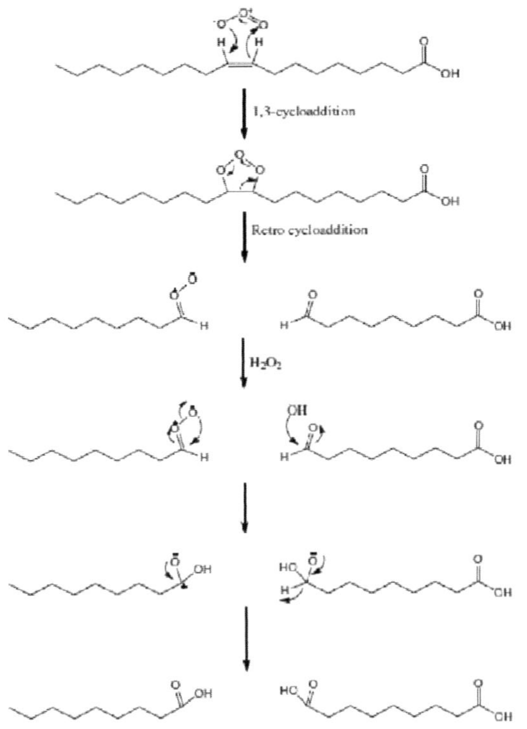

Figure 1. The suggested mechanism of the reaction.

The Major changes in the functional groups found in Palm olein and ozonized samples were analyzed by FTIR. Figure 1 shows the IR spectra of palm olein, while the figure 2, 3, 4 and 5 shows the ozonized samples at 50, 80, 120, 150 C° respectively with increasing amounts of consumed ozone and increasing the heat of reaction. One characteristic band of Palm Olein at 3006.0 cm−1 due to the presence of C=C bond, this bond reduced with increasing amounts of ozone consumed and increasing the temperature of reaction. At the 50 and 80 C° reaction temperatures, there is no change had observed for the double bond as shown in figure 2, 3 respectively, while its starts slightly decreasing with increasing of the temperature i.e. 120 C° as shown in figure 4.

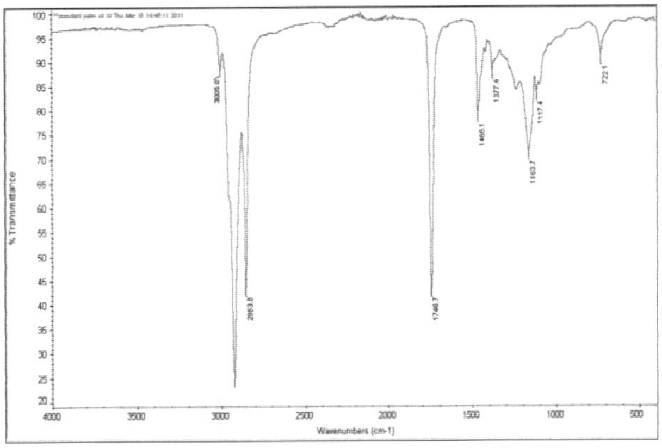

Fig.1 Palm Olein before Ozonation

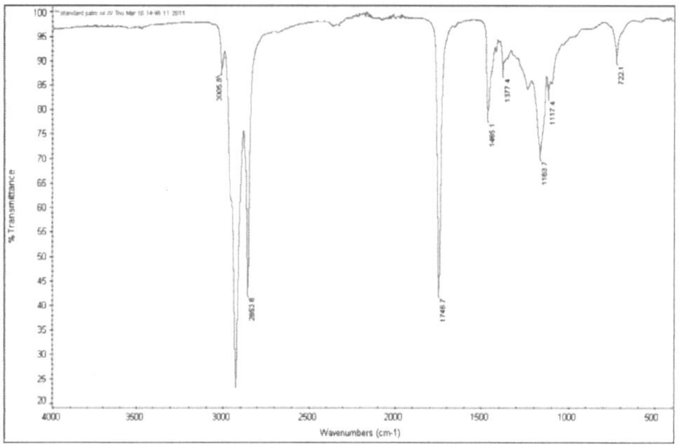

Fig.2 Palm Olein at 50 C° of Ozonation

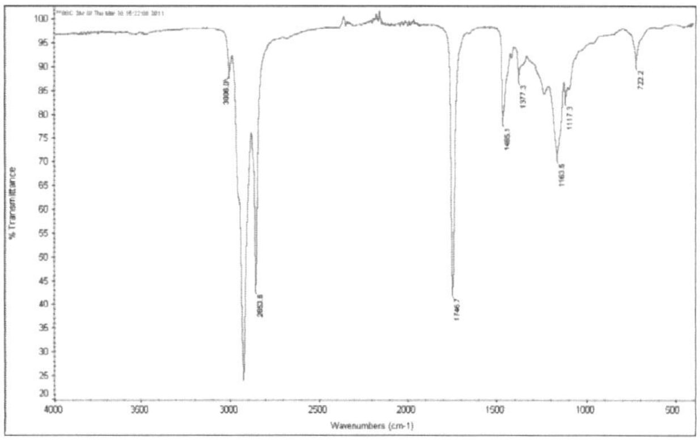

Fig.3 Palm Olein at 80 C° of Ozonation

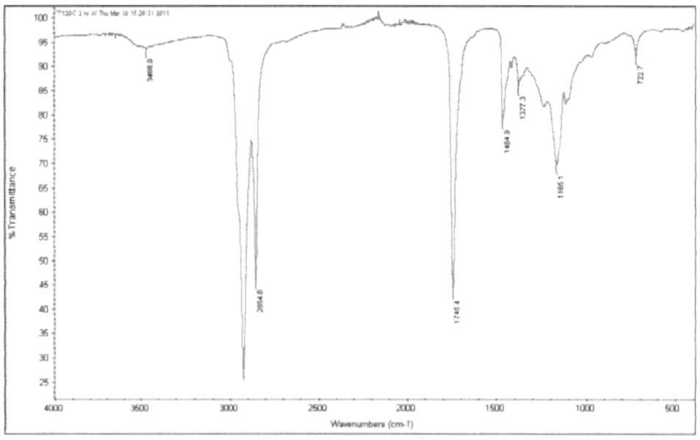

Fig.4 Palm Olein at 120 C° of Ozonation

After the increasing of reaction temperature to 150 C°, the double bond was disappeared as a result of completely ozone Consumption and by reacting with the double bond at high temperature as illustrated at figure 5.

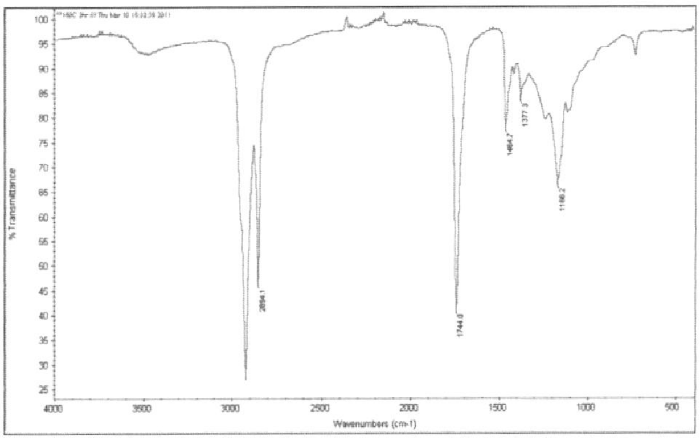

Fig.5 Palm Olein 2hr, 150 C° Ozonation

4. Conclusion

Ozonolysis of Palm olein was performed at different temperatures without a solvent and participating catalyst. The reaction was performed with very fine gas bubble size. The contact time between gas and liquid was increased by vigorous stirring; at the same time the temperature of reaction was increased as well to enhance reaction of ozone with the double bond of unsaturated fatty acids in Palm olein. The Fourier transforms infrared spectroscopy (FTIR) results illustrate clearly the change in function group and the completely cleavage of the double bond at 150 C° temperature and 2hr time of the reaction.

ACKNOWLEDGEMENTS

The authors would like to extend their appreciation to the University Kebangsaan Malaysia, and to the CRIM for research grant UKM-KK-02-FRGS0196-2010.

REFERENCES

[1] 2005 Edible Oil And Fat Products: Processing Technologies *Bailey's Industrialoil And Fat Products* Ed F Shahidi (Memorial University of Newfoundland: John Wiley & Sons)
[2] Metzger J O and U B 2006 *Applied Microbiology and Biotechnology* **71** 13
[3] Kadhum A H W, Bilal A, Abubakar M, Ahmed A and Mohd Sobri T 2012 *Research on Chemical Intermediates* **38** 659
[4] Bailey P S 1982 *Ozonation in Organic Chemistry* **1**
[5] Scrimgeour C 2005 Edible Oil and Fat Products: Chemistry, Properties, and Health Effects *Bailey's Industrial Oil and Fat Products* Ed F Shahidi (Hoboken: Memorial University of Newfoundland: Wiley)
[6] Hammond E G and P J W 2011 *J. of the American Oil Chemists' Society* **88** 891
[7] Nestor U S, Migo J V P and Matsumura M 2003 *J. of the American Oil Chemists' Society* **80**
[8] Nestor U S and Migo J V P 2003 *Chemistry and Physics of Lipids* **126** 133

[9] Vlachos N, Psaroudaki Y S, Konstantinidou M V, Chatzilazarou A and Tegou E 2006. *Analytica Chimica Acta* **573** 459
[10] Ckritz A K and A M 2011 *European Journal of Lipid Science and Technology* **113** 83

IOP Conf. Series: Materials Science and Engineering **50** (2013) 012053 doi:10.1088/1757-899X/50/1/012053

Voltaic turbine – A paradigm shift proposed in generating electricity

Meera V[a]
Department of Aeronautical Engineering,
Sri Ramakrishna Engineering College, Vattamalaipalayam-641022, Tamil Nadu, India

E-mail: [a]meerakumar72@gmail.com

Abstract. Sustainable demand for electric power results in a warehouse that naturally favors development of wind turbines that are significantly quieter and more efficient for generating electricity than today's fleet. Achieving this will require a revolutionary new concept, in particular "Magneto Voltaic Power plant" that generate electricity with the help of repulsive force which cause the turbine blades to rotate. The magneto voltaic power plant triggered the original development of wind turbines raising its energy and output power. This paper outlines the main issues involved in replacing the wind source to that of the repulsive force. It's shown that implementation of repulsive force generates electric power at a feasible cost and this would clearly be the enabling technology for generating electricity of the future.

1. Introduction

Generation of electricity using wind power has received considerable attention worldwide in recent years. First wind turbines were built by the end of 19th century by Prof James Blyth in Scotland (1887), at which 72 wind driven electric generator produce the electric power of range 5-25 kW. Later improvisation in this technology leads to the gaining of electricity at a rate of 7.58 Mw by Enercon E-126. Advance development in the field of windmill looks out for alternating source for generating power. At present year the solar based wind turbines has filled the position for generating electric power, but the electric power that has been generated are of less power output and the cost efficiency behind the solar power is high. To meet such demand of gaining larger power generation and also to meet the cost efficient here establishes a new concept called "Magneto voltaic power plant". The main concept behind this criterion is that uses of permanent magnet. The classic property of the magnet is that the like poles attract each other and the opposite poles repel each other. The maglev trains have been the most powerful trains that use repulsive force between the tracks and the wheels which causes the trains to levitate from the tracks and travel at a speed of 300 miles per hour, implementing this principle in the turbine blades and that makes the blades to rotate at the constant rpm, which helps in generating larger power. Due to the significant drawback of generating continuous power from the past decades starting out from the ancient turbines to that of the present solar turbines the results of generating the power at the feasible cost paves to be of high risk. To overcome those recession techniques numerical and experimental study of the optimal geometric arrangement of magnets and the repulsive forces generated between them shows that there is an existence of forces between them to turn the turbines blades.

In the first part of the study, it is established experimentally that there exists an optimal forces between the two adjacent pair of blades, depicting the architecture and the working of the turbines. The second part of the research is carried out of collecting the most powerful magnets which is capable of withstanding very higher temperature and the gauss power produced by the magnet and also the lifetime property of the magnets. From the mathematical it is found that the generation of electricity will be around 0.12 Kw. The setup has the greater advantages over the resulting wind turbines is that for generating of power at home appliance the magnetic turbine setup is placed at the certain height as it generates electric power due to the repulsive power between the blades, when ordinary wind turbine that depends on wind when placed below the magnetic setup, the velocity of air the generates from the magnetic turbine can used for generating the ordinary wind turbines so that the dual amount of power could be generated. A repulsive force between the magnets is calculated by the Coulombs inverse square law and also has analyzed the performance of the magnets by the magnetic theory. Detailed description of the turbines arrangement, selection of magnets, mathematical model, power generation are depicted.

2. Architecture

The magneto turbines consist of two blades in which two blades are separated at a suitable distances facing each other, where one of the blades remains stationary and other has the ability to rotate. The arrangement of magnets is an important aspect for rotating the blades at the required rpm. The system consists of stationary turbine blades that are placed in a slot of 15 degrees. Across the diameter, the magnets are placed at an equal interval of angle 15 degrees, thus covering the space with 24 magnets, ranging up to 30 cm and the rotary turbine has three blades separated at an equal angle. The turbine blades consist of the magnets placed at the faces of the blades using suitable fixing mechanism. The power plant is designed with three turbines blades of which each blade of 1m weighs 16.285 kg made up the of fiber glass materials. Since for the generation of electricity, magnets are needed to rotate at an rpm of 1200-1800. So the setup is designed for rotating the blades at a required rpm, which could be implemented in the industry and dwelling for daily uses of power.

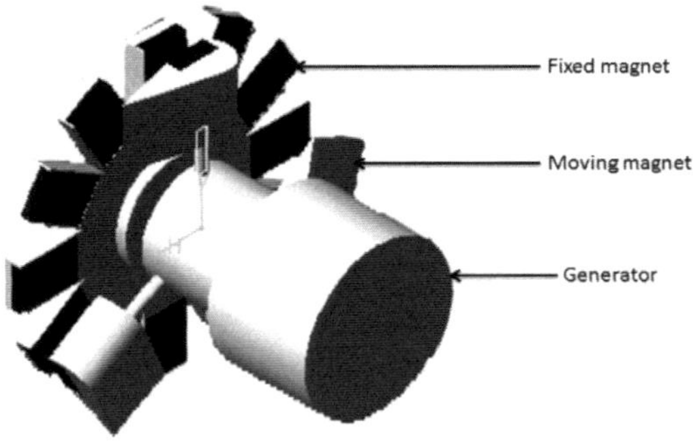

Figure 1. Arrangement of magnets in a wind turbine setup.

The starting of the rotary turbine blades requires a small amount of starting current. When the turbine begins to rotate, the repulsive force is created between the magnets placed in the rotary turbine blades in proximity to that of the stationary turbine blades. The 24 magnets that are placed at the stationary blades are shown in figure 1, which produces repulsive force causing the blades to rotary blades to rotate such that the torque is induced at the set up creating rotation motion. When the blades start to rotate it transfers some energy to the rotor. The wind-turbine shaft is connected to the center of the rotor. When the rotor spins, the shaft spins as well. In this way, the rotor transfers its mechanical, rotational energy to the shaft, which enters an electrical generator on the other end.

The generator is a simple device. It uses the properties of electromagnetic induction to produce electrical voltage - a difference in electrical charge. Voltage is essentially electrical pressure - it is the force that moves electricity, or electrical current, from one point to another. So generating voltage is in the effect of generating current. A simple generator consists of magnets and a conductor. The conductor is typically a coiled wire. Inside the generator, the shaft connects to an assembly of permanent magnets that surrounds the coil of wire. In electromagnetic induction, if you have a conductor surrounded by magnets, and one of those parts is rotating relative to the other, it induces voltage in the conductor. When the rotor spins the shaft, the shaft spins the assembly of magnets, generating voltage in the coil of wire. That voltage drives electrical current (typically alternating current, or AC power) out through power lines for distribution. The arrangement of the magnets must be in such a way that the magnets placed between the turbine blades must be either North Poles or South Poles so that same poles always repel each other.

Magnets in the rotary blades are inclined at an angle of 15 degree, so that there will be easy movement of blades. The repulsive forces produced will be too high which results in movement of the turbine blades. The stationary blades have the capability to glide with the hub. When the wind turbines are started it rotates at the constant rpm generating constant power output. To makes the wind turbines to rest the distance between the magnets in stationary blades is rotated in anticlockwise along with the hub so that the distances between the turbine blades get increased, so that the speed of the turbine blades decreases dramatically, on further increasing the distance results the turbine blades to stop. As we all know that speed of the blade is inversely proportional the distance between them.

3. Preferred Magnets

The selection of magnets plays the vital role because the speed of the turbine blades depends on the gauss of the magnet. For a larger gauss power the repulsive force produced is higher so that the blades rotate at a larger rpm. The magnet chosen for this operation is the neodymium iron boron magnet, which is the world powerful magnets whose gauss power varies from (11- 14) k Gauss. This has the larger lifespan of 10,000 years and has the operating temperature of 280 K.

For the operation of turbine blades the neodymium magnet with gauss 1450 is assumed.

Length	= 2 in
Width	= 5 in
Thickness	= 4 in
Weight of one magnet	= 4.990 kg

The stationary blades are placed with 24 magnets therefore the weight of 24 magnets would be 119.76 kg. The turbine is made up of fiber glass material and hence

The weight of single blade of 30cm = 16.285 kg.

The weight of three blades would be = 48.855 kg.

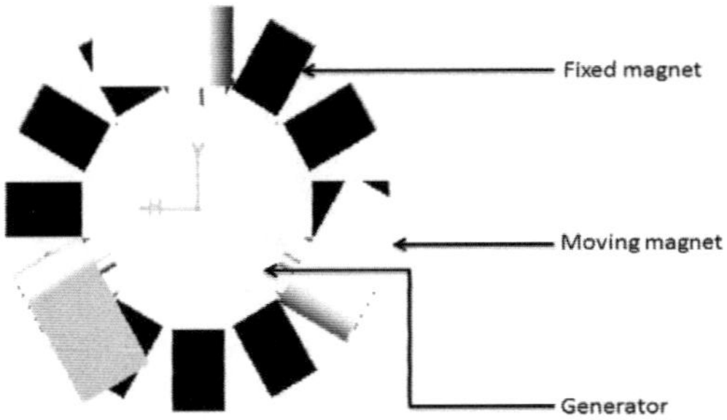

Figure 2. Front view of the arrangement of magnets in the Magneto turbine.

When the distance between the magnets gets increased the repulsive force is decreased, which shows that the distance between the magnets and the repulsive force are inversely proportional. Each magnet between the blades is inclined to an angle 30 degrees, so the repulsive forces generated would not affect the blades and also the mean time it prevent the opposite rotation of the magnets. Due to the rotation the blades the centrifugal forces gets acted towards the centre of the rotary blades since the force generated from the blades is transformed to the centre, the force that drives the turbines is the repulsive force towards the centre the forces is calculated by the Coulomb's inverse square law.

$$F = \left[\frac{B0^2 A^2 (L^2 + R^2)}{\pi \mu 0 L^2}\right]\left[\frac{1}{x^2} + \frac{1}{(x+2L)^2} - \frac{1}{(x+L)^2}\right] \quad (1)$$

Where

$B_0 = \frac{\mu_0}{2} M$

A = Area of each pole magnet in m^2
L = Length of the magnet in m
M = Magnetization magnitude (100,000 A/m)
From equation (1), for the different value of x, the corresponding repulsive forces are calculated.

Table 1. Repulsive force for various magnet positions.

Distance 'x' (cm)	Repulsion Force Between both magnets	
	Lbs.	N
1	23.8	105.867
2	12.1	53.823
3	6.2	27.578
4	3.4	15.123
5	1.9	8.451
6	1.1	4.893
7	0.7	3.113
8	0.4	1.779
9	0.3	1.344
10	0.2	0.889
11	0.1	0.444
12	0.1	0.444
13	0.1	0.444
14	0	0

From the tabulated values shown in table 1, between the repulsive forces and the distance between, the graph is plotted, where the distance between the magnets (x) are taken along the x axis and the repulsive forces generated are plotted along y axis as shown in figure 3.

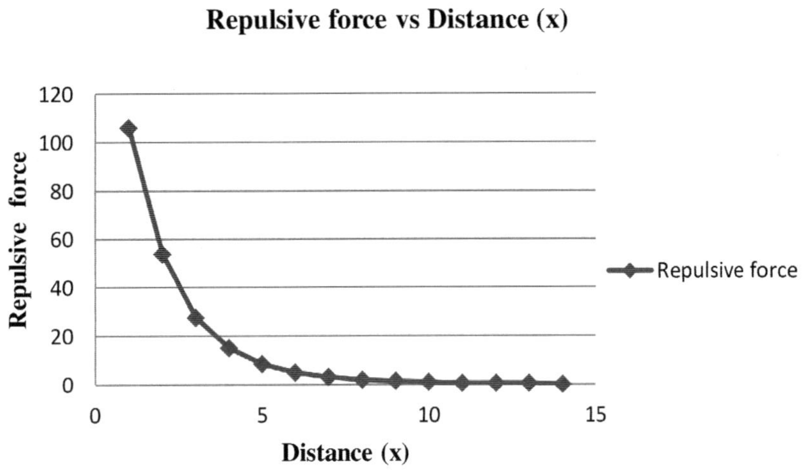

Figure 3. Graph between the repulsive force and the distance between the magnets.

The repulsive forces plays the main criteria in rotating the turbines, for the object to move from the particular places certain forces is required, same as that for moving a turbine we requires certain forces, Here the force that are used to rotate the turbine is the repulsive force. Since the forces are acting at the tip of the blades drives the forces at the centre, hence the forces acts from the tip of the blade to the centre we call this force to be the centrifugal forces, which acts towards the centre.

4. Mathematical Model

Equations are written for the amount of torque that has been produced when the rotating motion is induced and also the amount of power that is obtained by the rotation of the turbines are derived and are calculated.

Assumption made for the calculations are:

Consider

Length of the turbine blade = 0.3 m

Weight of the turbine blade = 16.285 kg for one blade (fiber glass material)

Radius of the turbine blade = 0.35 m.

For the turbines to rotate it is necessary that some forces are acting between the blades, Centrifugal forces are the forces that are acted towards the centre, hence for this magneto turbine the repulsive force at the tip of the blades drives towards the centre.

$$\text{Centrifugal force:} \quad F = m\frac{v^2}{r} \tag{2}$$

F is the repulsive forces calculated from the table 1. Since the distance between the magnets are considered to de x=1 cm the corresponding repulsive force from the figure 3 is 105.867 N, v is the velocity of the blades that are created.

$$v = 0.225 \text{ ms}^{-1}$$

Since when the rotation motion is produced in the blades, the turbine moves along the circumference of a circle of radius r and the angular velocity of the blades is calculated (ω) from the "Equation (3)", where v is the linear velocity.

$$\omega = \frac{v}{r} \tag{3}$$

$$\omega = 0.75 \text{ rad sec}^{-1}$$

Wind turbines work by converting the kinetic energy generated from the blades into rotational kinetic energy in the turbine and then electrical energy that can be supplied. Kinetic energy is the energy posssed by the body by virtue of its motion. Kinetic energy is calculated from equation (4)

$$K_{rotation} = 0.5\, I\omega^2 \tag{4}$$

Where I is the moment of inertia and ω is the angular velocity calculated from the "Eq. (3)", is substituted in "Eq. (4)", to calculate the kinetic energy induced due to the rotation of the blades. The inertial role is played when the blades continues in its state of rest or of uniform motion unless it is compelled by force. The inability of the blades to change its state of rest or motion by itself is called inertia. The inertia at rotational motion is called moment of inertia I, hence I is calculated from the "Eq. (5)", where m is the mass of the blades since the magnets are attached to the blades the mass includes the sum of mass of the blades and the mass of the magnet, hence weight of the one blade = 16.285 kg, since the turbine contains three blades its weight = 48.855 kg. Since the magnets are attached on the blades the weight of one magnet = 4.990 kg and hence for three magnet it is 14.97 kg. Total mass = 63.82 kg

$$I = mr^2 \tag{5}$$

$$I = 63.82 \times (0.3)^2$$

$$I = 5.74 \text{ kg m}^2$$

The value of I obtained from "Eq. (5)", is substituted in "Eq. (4)",

Hence $$K_{rotation} = 0.5\, I\omega^2$$

$$K_{rotation} = 1.6 \text{ rad sec}^{-1}$$

The power in the wind is given by the rate of change of kinetic energy,

$$P = \frac{dE}{dT} = 0.5v^2\frac{dm}{dt} \tag{6}$$

As mass flow rate is given by:

$$\frac{dm}{dt} = \rho A\frac{dx}{dt} \tag{7}$$

And the rate of change of distance is given by:

$$\frac{dx}{dt} = v$$

We get

$$\frac{dm}{dt} = \rho A v$$

Hence from "Eq. (6)", the power can be defined as the

$$P = 0.5\rho A v \tag{8}$$

The swept area of the turbine can be calculated from the length of the turbine blades using the equation for the area of a circle:

$$A = \pi r^2$$

Where r is the radius of the turbine blades (0.3 m)

$$A = 3.14 \times (0.3)^2$$

$$A = 0.282 \text{ m}$$

Sub the value of A and the density of the neodymium magnet (ρ) is the 7.5gcm^{-3} which is 7500 kg m^{-3} in "Eq. (6)" we get the power produced from the turbine blade is

$$P = 0.5\rho A v^3$$

$$P = 12.04 \text{ MW}$$

For turbine blades to generate electricity it is necessary for the blades to rotate at (1200-1800) rpm, hence the rpm is assumed to be 1200. From the blade momentum theory weight of the turbine blades changes cubic of their length and the power increases by square of their length. Which is evident that

the largest amount of power is produced from the repulsive force between the magnet and by this power plant it does not depends on the climatic factor for generating electricity and due to continuous repulsion constant amount of power is developed.
To avoid attraction force from the magnet to the other parts of the total section, Demagnetizers are used. Except the face from which the Repulsive force is produced must be cover with demagnetizers (like demagnetizing paint or with advanced wooden covers).This helps in obtaining better performance.

5. Results
The result shows that the repulsive forces between the magnets are based on the distance between the magnets and also at the gauss power of the magnet. When large amount of blades is used the repulsive forces produces is enormous so that the power produced is also enormous in this case of blades selection. As we know from figure 4.

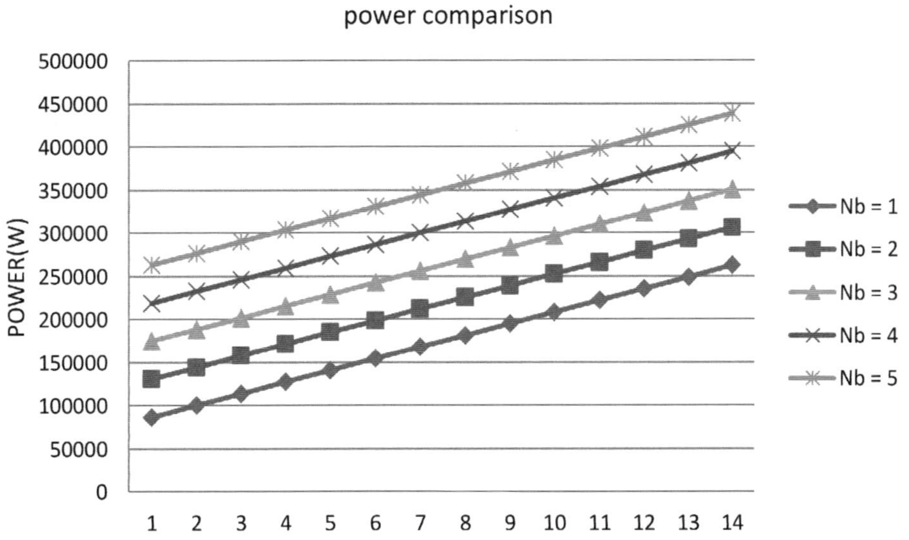

Figure 4. Relationship between power and the distance between the magnets.

The increase in weight of the blades produces larger amount of power thus producing tremendous amount of power thus only with repulsive forces between the magnets larger amount of electricity is produced, This could be implemented in wide places for generating electricity at a feasible cost and less transportation. If the rotary turbine is arranged with larger blades the magnets used are more to generate larger power. Thus the larger amount of magnet contributes to the larger electric power.

6. Conclusion
The magneto turbine which could generate large amount of power than that of the present wind turbine generating power based on the source of the wind. This paper outlines the alternative way of generating power using magnets in replacing the wind source, so that the power obtained is continuous and efficient. When there is continuous rotation of wind turbine at a constant rpm. The power produced would be continuous and hence with this technology due to its continuous rotation the speed produced would be higher so that velocity of the air is at high so ordinary wind turbine can be implemented beside so that power produced would be high.

Acknowledgement.
I take this opportunity to express my heartfelt gratitude to my beloved father for his exemplary guidance, monitoring and constant encouragement throughout the project without him which is not possible, and also I take this opportunity to thank the faculty for Aeronautical Engineering in Sri Ramakrishna Engineering College, Coimbatore.

References

[1] Alan Wyatt 1986 *Electric Power: Challenges and Choices* (Toronto, Book Press Ltd).
[2] Charles F. Brush 2008 *A Wind Energy Pioneer* (Danish, Danish Wind Industry Association).
[3] Donald Routledge Hill 1991 *Engineering in the Medieval Near scientific American* (America, Donald Routledge Hill).
[4] Furlani, Edward P 2001 *Permanent Magnet and Electromechanical Devices: Materials, Analysis and Applications.* (London: Academic Press).
[5] Gharabegian, Areq 2008 *"Maglev—A super-fast train"*(America: The Journal of the Acoustical Society of America).
[6] Morthorst, Poul Erik; Redlinger, Robert Y.; Andersen, per 2008 *" Wind energy in the 21st century: economics, policy, technology and the changing electricity industry "*(America: The Journal of the Acoustical Society of America).
[7] Gharabegian, Areq 2008 *"Floating Wind Turbines to Be Tested"*(IEEE Spectrum).

Analysis of carbon dioxide emission of gas fuelled cogeneration plant

Adzuieen Nordin[a] and M Amin A Majid
Mechanical Engineering Department, University Teknologi Petronas, Bandar Seri Iskandar, 31750 Tronoh, Perak, Malaysia.

E-mail: [a]adzuieen_G01249@utp.edu.my

Abstract. Gas turbines are widely used for power generation. In cogeneration system, the gas turbine generates electricity and the exhaust heat from the gas turbine is used to generate steam or chilled water. Besides enhancing the efficiency of the system, the process assists in reducing the emission of CO_2 to the environment. This study analyzes the amount of CO_2 emission by Universiti Teknologi Petronas gas fuelled cogeneration system using energy balance equations. The results indicate that the cogeneration system reduces the CO_2 emission to the environment by 60%. This finding could encourage the power plant owners to install heat recovery systems to their respective plants.

1. Introduction

Gas turbines (GT) are widely installed at the gas fuelled power plant to generate electricity. It is integrated into the power generation systems either as an open cycle system, a combined cycle system or a cogeneration system [1-6]. Besides generating electricity, the GT generates exhaust heat. The exhaust heat released to the environment consists of carbon dioxide (CO_2) and other air pollutant emission. Studies on CO_2 emission have been undertaken by a number of authors. [7] reported the amount of CO_2 intensity released using power and heat method by gas-fuelled power generating system is 404 g/kWh. While [8] noted that the major contributor of GHG is CO_2 emissions. Changing the type of fuel percentage from fossil fuel to renewable fuel for electricity generation reduced the CO_2 emission [9]. [10] have also studied the CO_2 emission with regards to fuel consumptions in Iran covering until 2025. The results show that the CO_2 emission will increase to 2.1 times by 2025. Other studies on CO_2 emission have also been undertaken by various authors. The studies are; emission assessment [11], analysis of CO_2 emission factor [12] and the capture and storage of CO_2 [13].

For the district cooling plant at Universiti Teknologi Petronas (UTP), the exhaust heat from GT is used to generate chilled water by absorption process. A study on the conversion process has been undertaken by [14] . Findings from the study indicate only maximum of 66.6% of the exhaust heat can be used to generate steam while the remaining 33.4% was emitted to the environment. In another study by [15], it is reported that the total amount of exhaust heat produced power generation is very large and to store all of the exhaust heat is not an option. This study is to focus on the evaluation of CO_2 emissions by GT for the cogeneration system.

2. Methodology

A typical GT for a cogeneration system is shown in figure 1. The GT consists of a compressor, combustor and turbine section. Ambient air is first compressed in the compressor. Fuel is added and combusted in the combustor. Most GT applications rely on natural gas or fuel oil for fuel. The combustion products exit the combustor and expand in the turbine section. The expansion drives an electric generator to generate electricity. The exhaust heat from the turbine is normally used to produce process steam or hot water. The steam in turn could either be used to power steam turbine or to produce chilled water for absorption process.

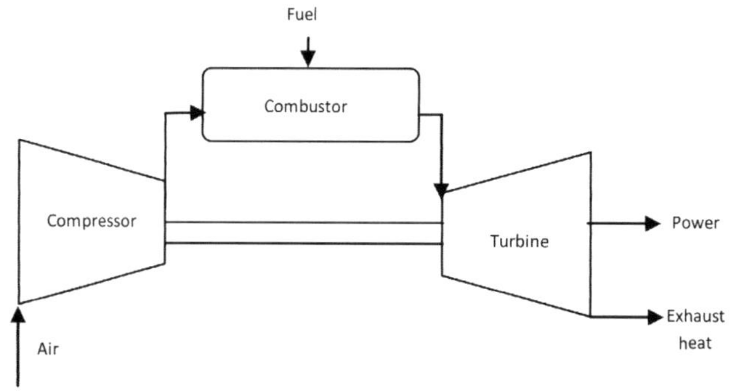

Figure 1. Typical gas turbine system.

The energy balance equation was used to evaluate the energy input, energy output and the difference between them. Eq. (1) [16] is used for analysis:

$$\dot{m}_i\left(h_i + \frac{V_i^2}{2} + Z_i\right) + \dot{Q} = \dot{m}_o\left(h_o + \frac{V_o^2}{2} + Z_o\right) + \dot{W} \tag{1}$$

where $\dot{m}_i$ is mass flow rate (kg/s), h is enthalpy (kg/kJ), V is velocity (m/s), Z is elevation (m), $\dot{Q}$ is the rate of heat and $\dot{W}$ is power generated by the system. Mass and energy balance for any control volume with negligible potential and kinetic energy is expressed by Eq. (2);

$$\dot{m}_i(h_i) + \dot{Q} = \dot{m}_o(h_o) + \dot{W} \tag{2}$$

For the cycle outline in figure 1, the thermodynamic models and related balance equations for the compressor, combustor and the turbine section are covered in the following sections:

2.1 Compressor

The model for the compressor is shown in figure 2. By applying the energy equations for the compressor, the power output from the compressor according to [17] is given as;

$$\dot{W}_c = \dot{m}_{a1}(Cp_{a2'}T_{2'} - Cp_{a1}T_1) \tag{3}$$

where $\dot{W}_c$ is power of the compressor (kW), $\dot{m}_{a1}$ is mass flow rate of air entering the compressor (kg/s), T_1 is air temperature entering the compressor (K), $T_{2'}$ is actual temperature air leaving the compressor (K) and Cp_a is the specific heat of air (kJ/kg.K).

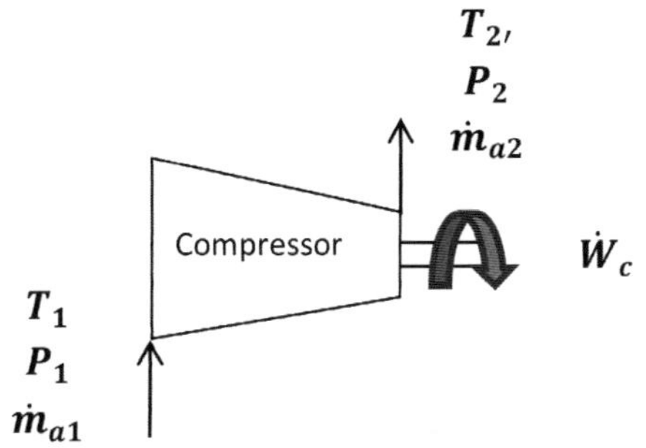

Figure 2. Thermodynamic model for compressor.

Figure 2 shows air enters the compressor at T_1, P_1 in atmospheric condition. Using the ideal gasses relation for air, the ideal temperature of T_2 is obtained by Eq. (4) [18].

$$\frac{T_2}{T_1} = \left(\frac{P_2}{P_1}\right)^{\frac{k-1}{k}} \tag{4}$$

where P_1 and P_2 are pressure entering and leaving the compressor respectively and k is the specific heat ratio of air. With an isentropic efficiency, the actual air temperature leaving the compressor is calculated using Eq. (5) [19];

$$T_{2'} = \frac{(T_2 - T_1)}{\eta_c} + T_1 \tag{5}$$

where the η_c is the compressor isentropic efficiency.

2.2 Combustor

The compressed air from the compressor enters the combustion section. In the combustor, the heat is supplied by fuel as shown in figure 3.

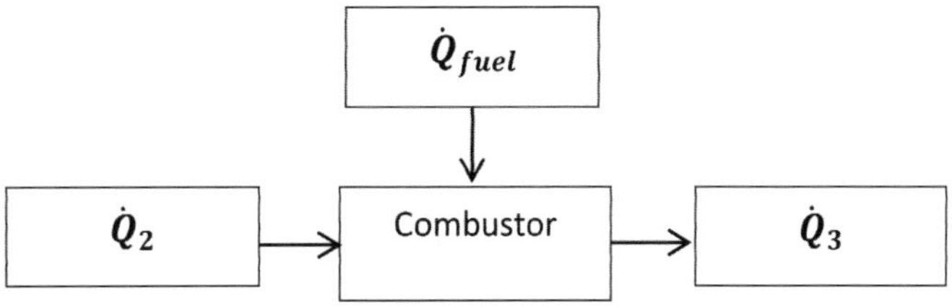

Figure 3. Thermodynamic model for combustor.

The energy input, fuel energy input and energy output from the combustor were calculated using Eq. (6), (7) and (8) respectively [20];

$$\dot{Q}_2 = \dot{m}_{a2} Cp_{a2\prime} T_{2\prime} \tag{6}$$

$$\dot{Q}_{fuel} = \dot{m}_f \times LHV \tag{7}$$

$$\dot{Q}_3 = \dot{m}_{g3} Cp_{g3} T_3 \tag{8}$$

where $\dot{m}_{a2}$ is mass flow rate of air leaving the compressor (kg/s), $\dot{m}_f$ is mass flow rate of fuel (kg/s), LHV is lower heating value, $Cp_{a2\prime}$ is specific heat of air at $T_{2\prime}$, $\dot{m}_{g3}$ is mass flow rate of flue gas leaving the combustor, T_3 is temperature outlet of the combustor section and Cp_{g3} is specific heat of flue gas at constant pressure (kJ/kg.K). The energy balance equations model with reference to figure 3;

$$\dot{Q}_3 = \dot{Q}_2 + \dot{Q}_{fuel} \tag{9}$$

where the mass flow rates of flue gas is denoted by Eq. (10);

$$\dot{m}_{g3} = \frac{\dot{Q}_3}{Cp_{g3} T_3} \tag{10}$$

2.3 Turbine

Figure 4 shows the thermodynamic model of the turbine. The energy output from combustor entered the turbine to generate power. During the process, exhaust heat is emitted to the environment.

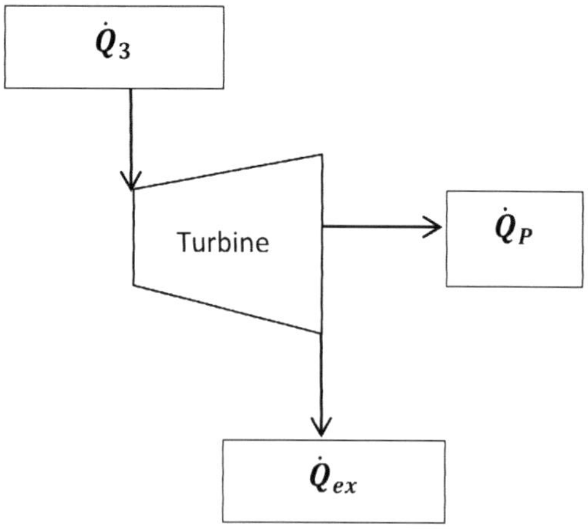

Figure 4. Thermodynamic model for turbine.

The power of the gas turbine was calculated by Eq. (11) [19];

$$\dot{Q}_P = \dot{m}_{g3} (Cp_{g3} T_3 - Cp_{gP} T_P) \tag{11}$$

The exhaust heat released by the turbine was calculated as [19];

$$\dot{Q}_{ex} = \dot{m}_{g3} Cp_{gex} T_{ex} \tag{12}$$

where T_{ex} is exhaust heat temperature (K), T_P is temperature for power generating (K), Cp_{gex} is specific heat of gas for the exhaust temperature (kJ/kg.K) and Cp_{gP} is specific heat of gas for power generating temperature (kJ/kg.K). The temperature of exhaust heat can be calculated as [19];

$$T_{ex} = T_3 + \eta_t T_3 \left(\left(\frac{1}{r_c} \right)^{\frac{\gamma-1}{\gamma}} - 1 \right) \tag{13}$$

where η_t is the isentropic efficiency of the turbine, γ is the specific heat ratio of gas and r_c is the pressure ratio of the compressor.

Gas District Cooling (GDC) plant at UTP was selected for the case study. The gas turbine is examined for operating conditions shown in table 1 with the fluid properties shown in table 2. The gas turbine performance was specified for continuous duty rating in standard ISO condition of 15°C and 60% relative humidity and at sea level ambient pressure.

Table 1. Operating conditions of the UTP GDC gas turbine.

Components	Operating condition	Value
Compressor	Mass flow rate of air inlet and outlet	19.0 kg/s
	Inlet temperature	300K
	Pressure Ratio	11.7
	Isentropic efficiency	0.89
	Specific heat ratio of air	1.4
Combustor	LHV	41 000 kJ/kg.K
Turbine	Specific heat ratio of gas	1.33
	Isentropic efficiency	0.91

Table 2. Physical properties of fluids [20].

Substance	Property		
Air	Specific heat	$Cp_a =$	$1.04841 - (3.8371 \times 10^{-4} \times T) + (9.4537 \times 10^{-7} \times T^2) - (5.49031 \times 10^{-10} \times T^3 + (7.9298 \times 10^{-14} \times T^4)$
Flue Gas	Specific heat	$Cp_g =$	$0.991615 + (6.99703 \times 10^{-5} \times T) + (2.7129 \times 10^{-7} \times T^2) - (1.22442 \times 10^{-10} \times T^3)$

For the analysis, the following assumptions were used [19]:

- The process at compressor section was an adiabatic process.
- The total energy output from the compressor was to be taken as energy input for combustor section and remained constant.
- The total energy output from the combustor was to be taken as energy input for the turbine.

- The outlet temperature of the compressor was the temperature inlet of the turbine.
- A turbine engine was at steady state.
- The pressure ratio was constant both in compressor and turbine
- Heat lost due to lubrication was eliminated.

For this study, historical data from 12 September 2011 to 16 September 2011 was used. In order to evaluate the amount of CO_2 emission released to the environment, the calculation was based on the earlier researches as well as on published literature.

3. Results and discussions

Using historical data for 12/9/2011 to 16/9/2011 and Eq. (12), the amount of exhaust heat generated is as shown in figure 5.

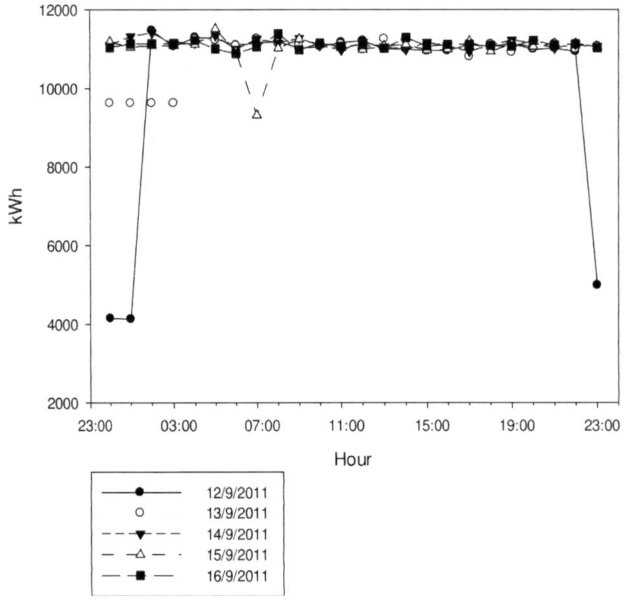

Figure 5. Estimated Exhaust heat generated by GT for 12/9/2011 to 16/9/2011.

Figure 5 shows the total hourly exhaust heat from GT from 12 a.m. to 11 p.m. for five working days in September 2011. The average amount of estimated exhaust heat released to the environment during this period is equivalent to 10 875 kWh daily. The results are summarized in table 3.

Table 3. Summarized results of estimated exhaust heat in terms of equivalent kWh from energy analysis of GT.

	12/9/2011	13/9/2011	14/9/2011	15/9/2011	16/9/2011
Min (kWh)	4127	9615	10 940	9307	10 887
Max (kWh)	11 467	11 262	11 418	11 513	11 390
Mean (kWh)	10 267	10 843	11 138	11 020	11 105

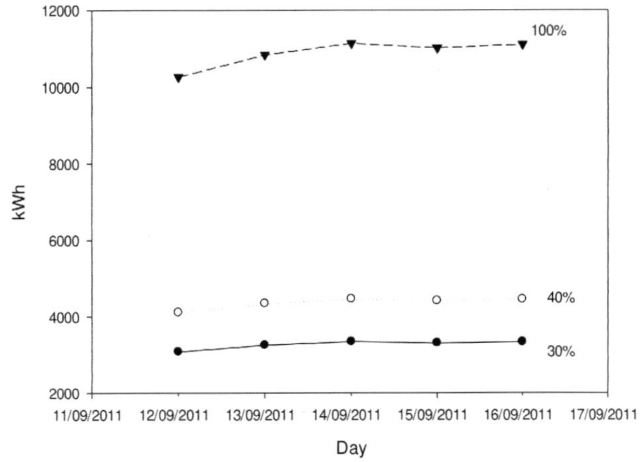

Figure 6. Amount of estimated exhaust heat based on variation percentage of exhaust heat released to the environment for 12/9/2011 to 16/9/2011.

Figure 6 and table 4 shows the calculated amount of estimated exhaust heat released to the environment. The case of 30 percent to 40 percent refers to the maximum amount of exhaust heat released taking into account with the amount captured which is in between 60 percent to 70 percent for heat recovery process. While the case for 100 percent implies all of the exhaust heat is emitted to the environment which means there is no heat recovery process.

Table 4. Summarized results of estimated exhaust heat from GT released to the environment (kWh) based on 30%, 40% and 100%.

	12/9/2011	13/9/2011	14/9/2011	15/9/2011	16/9/2011
30%	3080	3253	3341	3306	3332
40%	4107	4337	4455	4408	4442
100%	10267	10843	11138	11020	11 105

Using conversion factor from [16], that is 474g of CO_2/kWh will be released to the environment. figure 7 shows the plotted results for the amount of CO_2 emitted to the environment by the GT during the study period.

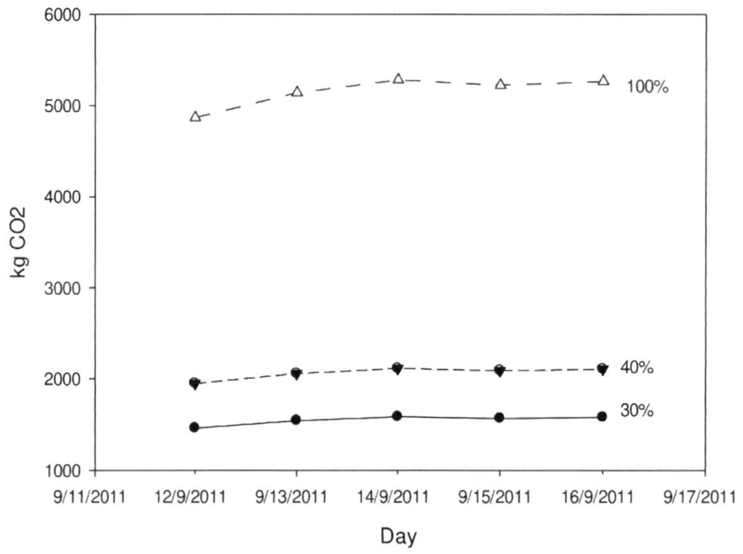

Figure 7. Estimated amount of CO_2 equivalent to exhaust heat based on 30%, 40% and 100% released to the environment for 12/9/2011 to 16/9/2011.

For the case of 100 percent of exhaust heat released to the environment, the total estimated amount of CO_2 emitted is 5155 kg. While if only 30 percent of the exhaust heat is released to the environment, the estimated amount of CO_2 emitted is 1546 kg. Thus, the steam conversion process helps to save about 60 percent of CO_2 that is 3600 kg from being released to the environment.

4. Conclusions

The energy balance analysis shows that the estimated exhaust heat in terms of equivalent per kWh from the GT is 10 875 kWh daily for district cooling plant at UTP. It contributed 5155 kg of CO_2 emission released to the environment for the GT with 8.4 MW. Increasing the amount of exhaust heat will increase the amount of CO_2 released to the environment. Thus, the heat recovery process avoided about 60 percent of CO_2 from being released to the environment. These findings could be useful for power plant owners as well as the researchers.

Acknowledgements

The authors would like to express their appreciation to the support by Fundamental Research Grant Scheme (FRGS) and Universiti Teknologi Petronas (UTP) for the project.

References

[1] Ibrahim T K and Rahman M M 2013 *J. Mech. Eng. Sci.* **4** 383
[2] Ibrahim T K and Rahman M M 2012 *Adv. Sci. Lett.* **13** 263
[3] Rahman M M, Ibrahim T K, Kadirgama K, Mamat R and Bakar R A 2011 *Adv. Mater. Res.* **189-193** 3007
[4] Ibrahim T K and Rahman M M 2012 *J. Appl. Res. Technol.* **10**(4) 567
[5] Ibrahim T K and Rahman M M 2012 *Inter. J. Energy Eng.* 2(1) 9-14.
[6] Ibrahim T K, Rahman M M and Ahmed N. Abdalla 2011 *Inter. J. Phys. Sci.* **6**(4) 620.
[6] W Graus and E Worrell 2011 *Energy Policy* **39** 613-627

[7] H Ong, T Mahlia, and H Masjuki 2011 A review on energy scenario and sustainable energy in Malaysia *Renewable and Sustainable Energy Reviews* **15** 639-47
[8] H Masjuki, T Mahlia, I Choudhury, and R Saidur 2002 *Energy Conversion and Management,* **43** 763-70
[9] A Mazandarani, T Mahlia, W Chong, and M Moghavvemi 2011 *Renewable and Sustainable Energy Reviews* **15** 1575-92
[10] P Mancarella and G Chicco 2009 *Applied Energy* **86** 2096-06
[11] S Nazari, O Shahhoseini, A Sohrabi-Kashani, S Davari, R Paydar, and Z Delavar-Moghadam 2010 *Energy* **35** 2992-98
[12] C Hendriks, E de Visser, D Jansen, M Carbo, G J Ruijg, and J Davison 2009 *Energy Procedia* **1** 1497-1504
[13] S A S Mohd Amin Abd Majid, Idris Ibrahim and Zuhairi Baharddin 2012 *Journal of Applied Sciences* **12** 2636-40
[14] M Othman, R Zakaria, and W Fernando 2009 *Energy Policy* **37** 1718-35
[15] G F M de Souza 2012 *Thermal Power Plant Performance Analysis*: Springer
[16] Y A Cengel and M A Boles 2006 *Thermodynamics: an engineering approach*: McGraw-Hill Higher Education
[17] W K a A K Elsaied, Z A 2008 *International Conference on Plant Equipment and Reliability (ICPER)*
[18] Z A A Karim and P W Yongo 2011 *Journal of Applied Sciences* **11** 1519-27
[19] P Ahmadi and I Dincer 2011 *Applied Thermal Engineering* **31** 2529-40
[20] R Kannan, K Leong, R Osman, H Ho, and C Tso 2005 *Energy conversion and management,* **46** 2145-57

An overview of palm, jatropha and algae as a potential biodiesel feedstock in Malaysia

S Yunus[1,a], N R Abdullah[1], R Mamat[2], A A Rashid[1]
[1]Faculty of Mechanical Engineering, Universiti Teknologi Mara, 40450 Shah Alam, Selangor, Malaysia
[2]Faculty of Mechanical Engineering, University Malaysia Pahang, 26600, Pekan, Pahang, Malaysia

E-mail: [a]sya_mechys@yahoo.co.uk

Abstract. The high demand to replace petroleum fuel makes renewable and sustainable sources such as Palm oil, Jatropha oil and Algae a main focus feedstock for biodiesel production in Malaysia. There are many studies conducted on Palm oil and Jatropha oil, however, the use of Algae as an alternative fuel is still in its infancy. Malaysia already implemented B5 based Palm oil as a feedstock and this biodiesel has been proven safe and can be used without any engine modification. The use of biodiesel produced from these feedstock will also developed domestic economic and provide job opportunities especially in the rural area. In addition, biodiesel has many advantages especially when dealing with the emissions produce as compared to petroleum fuel such as; it can reduce unwanted gases and particulate matter harmful to the atmosphere and mankind. Thus, this paper gathered and examines the most prominent engine emission produced from Palm oil and Jatropha feedstock and also to observe the potential of Algae to be one of the sources of alternative fuel in Malaysia.

1. Introduction

Alternative fuels for diesel engine, boilers and other combustion equipment becoming increasingly important due to the depletion of petroleum reserves in the near future. The dependence and the consumption of petroleum products is increasing day-by-day and it is resulting in higher prices of petroleum fuel. The global primary petroleum fuel consumption has grown up from 6630 million tons in 1980 up to 12,002.4 million tons (it is almost double) in 2010 [1, 2]. Malaysia is also one of the countries that suffered from this crisis, as according to British petroleum statics, oil production in Malaysia in 2001 has decrease from 32.9 million tons to 32.1 million tons in 2010. However, the corresponding consumption of oil has increased from 22 million tons in 2001 to 25.3 million tons in 2010 [1].

The emission emitted from the petroleum fuel may cause various negative impacts on global climate of greenhouse gases and also lead to acid rain and air pollution. It is predicted that if the average global temperature increases by more than 2°C, more than million species could extinct and hundreds of millions people could lose their lives [3, 4]. Living people may also suffer from various diseases such as lung cancer, breathing difficulties, poisoning and skin cancer etc. This concern led to find the alternative fuel to overcome these problems and Malaysia is no exception.

Biofuels from various biomasses seems to be a realistic alternative renewable fuel to resolve twin crisis of petroleum depletion and environmental degradation. The implementation of biomass energy will contribute to sustainable development in various fields of environmental, social and economic [5-6]. Biomass feedstocks used for biodiesel production include Algae, animal fats and vegetable oils like Palm, Jatropha, rapeseed, sunflower etc. Brazil and US for example are promoting ethanol as a potential biofuel derived from sugar cane and corn, and Asian countries like Malaysia, Indonesia and India have been promoting Palm oil and Jatropha as biodiesel throughout widely [7-11].

There are so many advantages and disadvantages of using biodiesel as compared to diesel fuel. Figure 1 show some of the literature gathered by past researcher [4].

Advantages
Biodiesel has 10-11% of oxygen; this makes biodiesel a fuel with high combustion characteristics.
Biodiesel reduces net carbon-dioxide emissions by 78% on a lifecycle basis when compared to conventional diesel fuel and reduces smoke due to free soot.
Biodiesel has superior better lubricity properties. This improves lubrication in fuel pumps and injector units, which decreases engine wear, tear and increases engine efficiency.
Biodiesel has higher cetane number (about 60-65 depending on the vegetable oil) than petroleum diesel (53) which reduces ignition delay.
Biodiesel may not require engine modification up to B20. However, higher blends may need some minor modification.
Production can be raised easily and is less time consuming.
Disadvantages
Biodiesel has 12% lower energy content than diesel; leads to increase fuel consumption of about 2-10%.
Biodiesel causes excessive carbon deposition and gum formation (polymerization) in engine and oils get contaminate and suffer from flow problem.
Biodiesel has higher cloud point and pour point, higher nitrogen oxide emissions, lower volatilities that cause formation of deposits in engine due to the incomplete combustion characteristics.
Transesterification process is expensive (cost of fuel increases), these oil require expensive fatty acid separation or use of less effective (or expensive acid catalyst).
Use of biodiesel in internal combustion engine may lead to engine durability problems including injector cocking, filter plugging and piston ring sticking, etc.
Lower engine speed and power, high price, high engine wear, engine compatibility.

Figure 1. Advantages and disadvantages of biodiesel.

Malaysia had launched the National Biofuel Policy to encourage more sources from renewable energy [12]. The policy was eventually launched in March 2006 and is underpinned by five strategic thrusts, with short-term, medium-term and long-term implementation periods. Figure 2 show the summary of the National Biofuel Policy [13].

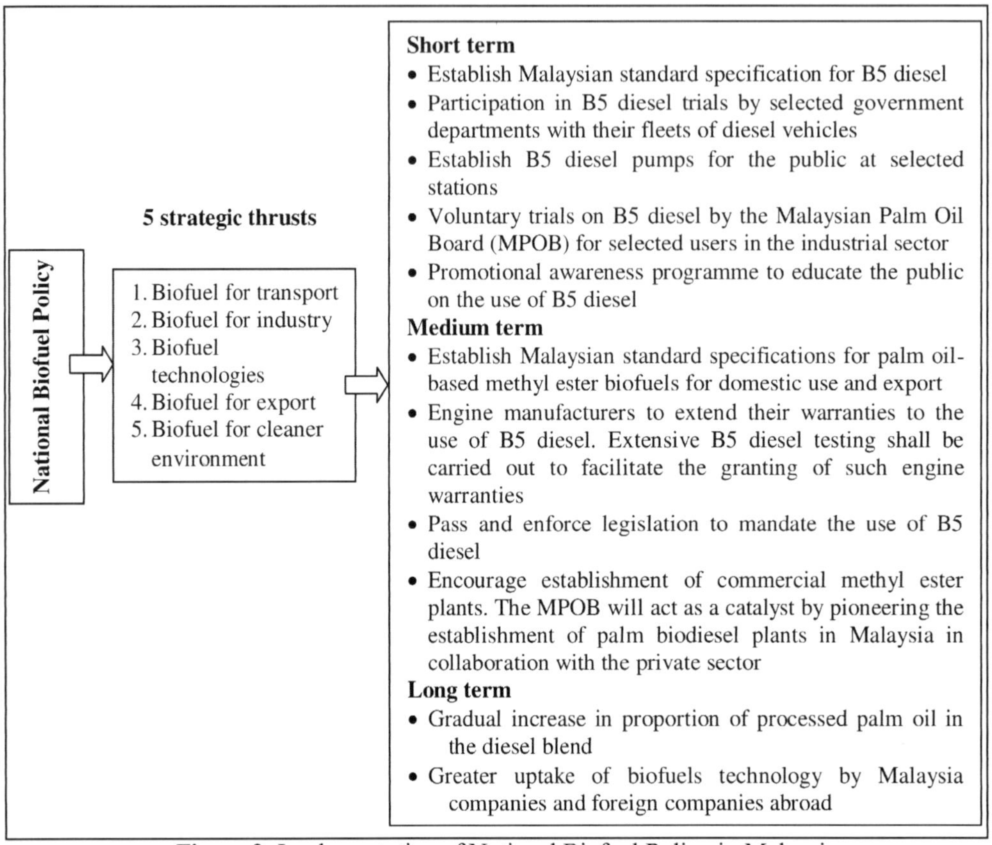

Figure 2. Implementation of National Biofuel Policy in Malaysia.

Malaysia has introduced a 5 percent biodiesel blended composition which is 5 percent consist of Palm oil methyl ester (POME) and known as B5 since mid-year 2011. Currently, extensive research also has been conducted in Malaysia to promote the usage of Jatropha and Algae biodiesel as they are easily obtained [3, 14-18]. Compatible with climate change in Malaysia make Jatropha a promising alternative fuel. Apart from that, Algae also has attracted many researchers as it is easy to be found especially in Sabah and Sarawak. However, the use of Algae as an alternative fuel is still in its infancy [3].

This paper gathered and discusses the work carried out by past researcher on vegetable oil - Palm, and non-edible oil - Jatropha curcas, and also potential of Algae to be one of the sources of alternative fuel. The engine performances from Palm and Jatropha feedstock are reported comparable as compared to petroleum fuel [19-21]. However, the emission results indicate variation in emission with the use of these biodiesel [22-25]. Thus, effort has been made in this study to review the most prominent engine emissions produced from Palm and Jatropha feedstock.

2. Potential of Palm Oil as a Feedstock for Biodiesel source

Malaysia is currently implementing biodiesel known as B5 (5% Palm oil methyl ester and 95% diesel fuel) based Palm oil as a feedstock. The usage of this B5 in diesel engine is safe and has been used without any engine modification. The performance of biodiesel is also comparable to diesel engine reported by other researchers [10, 24, 26]. Some types of emissions also reduced when used this Palm oil biodiesel [2, 4, 27].

Malaysia is listed as the largest world exporter of Palm oil and the second largest of Palm oil producer after Indonesia. Malaysia as 2010, the area of oil Palm plantation was 4.85 million hectares covering 14% of the total land area in Malaysia [13]. Currently, Malaysia encourages private companies to further intensify in this project, providing adequate subsidies to the parties involved and increase programs to encourage farmers to nurture the growth of the biodiesel industry. This could indirectly provide job opportunities, especially in rural areas.

3. Potential of Jatropha Oil as a Feedstock for Biodiesel source

In the future, Jatropha can be one of the sources biodiesel productions in Malaysia. Jatropha oil has been found promising for biodiesel production due to the attracting characteristics and it is claimed not to compete with food, not to compete with agriculture land and nature. The biodiesel produced from this feedstock also reduce the danger of emissions compared with the petroleum fuel. This biodiesel also has a comparable cetane number and calorific value with petroleum diesel and it can be used in the diesel engine without requiring any major modifications in the mechanical system of the engines [10, 19, 20, 24].

The advantages of climate suitability and availability of land in Malaysia has made Malaysia took the opportunity to explore and expand Jatropha production. It is reported that a total of 1712 hectares of land has been identified as a start of primary production of Jatropha and expected to increase to 57,601 hectares by 2015 [28]. A few international leading oil company are also interested to investing and develop the Jatropha projects in Malaysia [1, 28].

4. Potential of Algae oil as a Feedstock for Biodiesel source

Algae as an alternative fuel are still in its infancy in Malaysia. Currently, extensive studies have been carried out to identify the most suitable Algae species to be a source of alternative fuel. Much effort also has been conducted in order to produce biodiesel from this feedstock mainly in Sabah and Sarawak area.

Algae is a promising source of alternative fuel as it will not compromise with food, fodder and other products derived from crops [3]. Algae also provide a reduction in greenhouse gases because it absorbs or utilize carbon dioxide as it grows [29]. It can grow quickly at a large scale and potential to generate up to 50 times more than the amount of other crops [30]. Due to these advantages, Algae has attracted the attention of many researchers to make one of the alternative fuels to replace the petroleum fuels.

5. Emissions of Biodiesel

The diesel and also biodiesel fuelled engine emits the unwanted gases in the atmosphere that can be hazardous to mankind. There are two types of emissions that have been produced by this diesel engine. One type is known as regulated emissions as it is regulated by the government of many countries and another one is known as unregulated emissions as it is not-regulated by the government. The regulated emissions examples are carbon monoxide (CO), carbon dioxide (CO2), nitrogen oxide (NOx), unburned hydrocarbon (HC) and smoke opacity, whereas, the unregulated emissions examples are formaldehyde, acetaldehyde, acetone and toluene [23]. However, this study is only limited to observe the most prominent regulated emissions pollutant form Palm and Jatropha feedstock.

5.1. Palm Oil's Emissions

T.H. Lim et al. [31] conducted the study of CO and NO emissions of a diesel engine operated on $60^{o}C$ heated crude Palm oil (CPO). It was observed that higher CO and NO emissions produced by 2.8% and 25.7% respectively as compared to diesel fuel. This is due to the chemical reaction at the core of fuel spray of CPO which produces heavy compounds that are less volatile. These heavy compounds resulted in local-rich mixtures that cause more CO to produce. Higher pressure in the engine cylinder during CPO combustion lead to the higher temperature that caused more NO formation resulted in increasing of NO emission produced.

On the other hand, J. Narkpakdee et al. [32] evaluated the emission characteristics of diesel-crude Palm oil-water emulsion blend in a small diesel engine. The result shows that the emission parameters such as CO, NOx and black smoke were significantly lower than diesel fuel. This is due to the emulsified water content (that is up to 10%) in the blend biodiesel which reduces the viscosity and heating value.

A.M. Liaquat et al. [27] carried out experimental study on exhaust emissions of a diesel engine using Palm diesel blend with diesel fuel. The experiment was conducted at constant speed of 2000 rpm at varied load conditions between 5 Nm to 17.5 Nm with intervals of 2.5 Nm. The result shows that the CO emission of the Palm diesel blends produced was less than the diesel fuel at all loads. However, CO_2 and NOx emissions produced were higher as compared to diesel fuel at the higher load applied. This is indication of complete combustion of more amount of CO_2 produced by this Palm diesel fuel in exhaust emissions and the increasing of NOx is caused by its higher heating value and also the increasing of combustion chamber temperatures.

The influence of Palm oil biodiesel on the chemical composition of the particulate matter emissions produced in an automotive diesel engine operated with Palm biodiesel and its blends with diesel fuel were carried out by Maurin Salamanca et al. [33]. The study shows that the particulate matter emissions and smoke opacity amount decreases in a non-linear way as the proportion of Palm biodiesel increases as compared to diesel fuel. This reduction is caused by the chemical nature of the fuels as biodiesel has high level of oxygen and the absence of aromatics molecules it its composition that lead to reduction of particle nucleation and increase the oxidation rate.

5.2. Jatropha Oil's Emissions

Bhupendra Singa Chauhan et al. [34] studied the effects of fuel inlet temperature (FIT) on engine emissions using a dual fuel engine test rig with tube heat exchanger (with exhaust bypass arrangement) using Jatropha oil as a fuel. The results show lower NOx emission produced by Jatropha oil as compared to petroleum fuel; nevertheless, increasing FIT caused the increasing of NOx emission. Other gases produced by Jatropha oil resulted in higher CO, HC and CO_2 than petroleum fuel. However, preheated Jatropha oil, the emissions of CO, HC and smoke opacity were reduced but it is slightly increased for CO_2 emission. The higher amount of CO_2 produce from the diesel engine is an indication of the complete combustion of the fuel [27]. S. Jindal et al. [35] investigated the emissions of Jatropha methyl ester on DI-CI engine. The results show that the HC and NOx emission produced were less than petroleum diesel. The reduction of HC emission is caused by the higher oxygen content present in the biodiesel that leading to more complete combustion. This is also due to higher cetane number that reducing combustion delay and advancing the injection and combustion timing [2, 36].

Iman K. Reksowardo et al. [37] found that the emissions of CO, CO_2 and HC of Jatropha oil were lower as compared to petroleum diesel and slightly increase of NOx emission. The experiment was conducted on a Direct Injection (DI) diesel engine. The result from this study was with the agreement of other studies, and the slightly increasing of NOx is caused by the oxygenated in biodiesel fuel [20, 23]. The increasing of NOx occurred especially when the engine load increases, as more fuel is injected and burned in the engine cylinder which causes the gas to a higher temperature and resulted in a more NOx formation in the engine cylinders and produce higher NOx emissions from the engine [38].

On the other hand, T. Elango and T. Senthilkumar [39] evaluated the emissions characteristics of Jatropha oil blends in a CI engine. The results show that all the Jatropha oil blends emitted less CO, CO_2 and HC than the diesel fuel. However, the smoke opacity was increased with increasing proportion of Jatropha oil in the blends as compared to diesel fuel. The increasing of smoke opacity amount is caused by the poor atomization and combustion due to the higher viscosity of the blends [39]. Table 1 shows related research findings of emissions from Palm oil and Jatropha oil.

Table 1. Summary emissions produce from Palm oil and Jatropha oil.

Emissions	Palm oil	Jatropha oil
CO increase	Preheated: [31]	
CO decrease	Water-emulsion: [32]	Blended: [39], [37] Preheated: [34]
CO_2 increase	Blended: [27]	Preheated: [34]
CO_2 decrease		Blended: [39], [37]
NO increase	Preheated: [31]	
NOx increase	Preheated: [31]	Blended: [37]
NOx decrease	Water-emulsion: [32]	
HC decrease		Blended: [39], [37] Preheated: [34]
Smoke opacity increase		Blended: [39]
Smoke opacity decrease	Water-emulsion: [32] Blended: [33]	Preheated: [34]
Particulate matter decrease	Blended: [33]	

6. Conclusions

This paper has presented a potential of biodiesel feedstock from Jatropha, Palm and Algae. The implementation of B5 based Palm oil as a feedstock in Malaysia had proven that this biodiesel blend can be safely used without any major modification of diesel engine. Currently, due to the climate suitability and land availability in Malaysia make Jatropha as a promising alternative to replace petroleum fuel. The advantages of Algae also attracted attention from many researchers and it is another promising source of alternative fuel. However, the current stage of Algae biodiesel is still in its infancy. Therefore, further research and studies on Algae biodiesel are needed before it can prove to be used in a diesel engine.

This paper also reviews some emissions from Palm and Jatropha feedstock as a fuel in diesel engine. There are many ways to be used in diesel engine such as emulsified with water, added catalyst, blended with diesel fuel at certain proportion or directly used in the engine. Therefore, many studies need to be done in order to understand the characteristic of these biodiesel before it can be fully implemented in diesel engine.

Acknowledgement
The author would like to thank the Ministry Of Higher Education (MOHE), Research Management Institute (RMI) of Universiti Teknologi Mara (UiTM), Malaysia through FRGS grant awards (600-RMI/ST/FGRS 5/3Fst (32/2011), (600-RMI/ST/FRGS 5/3/FST (103/2010)) and also to MyBrain PHD.

References

[1] M Mofijur, H H Masjuki, M A Kalam, M A Hazrat, A M Liaquat, M Shahabuddin, and M Varman 2012 *Prospects of biodiesel from Jatropha in Malaysia* Renewable and Sustainable Energy Reviews, **16** 5007-20

[2] H C Ong, T M I Mahlia, H H Masjuki, and R S Norhasyima 2011 *Comparison of palm oil, Jatropha curcas and Calophyllum inophyllum for biodiesel: A review,* Renewable and Sustainable Energy Reviews **15** 3501-15

[3] A L Ahmad, N H M Yasin, C J C Derek, and J K Lim 2011 Microalgae as a sustainable energy source for biodiesel production: A review *Renewable and Sustainable Energy Reviews* **15** 584-93

[4] A E Atabani, A S Silitonga, I A Badruddin, T M I Mahlia, H H Masjuki, and S Mekhilef 2012 A comprehensive review on biodiesel as an alternative energy resource and its characteristics *Renewable and Sustainable Energy Reviews* **16** 2070-93

[5] Rahim R, Mamat R, Taib M Y and Abdullah A 2012 *J. Mech. Eng. Sci.* **2** 226

[6] Nematizade P, Ghobadian B and Najafi G 2012 *Inter. J. Automot. Mech. Eng.* **5** 639

[7] A Z Abdullah, B Salamatinia, H Mootabadi, and S Bhatia 2009 Current status and policies on biodiesel industry in Malaysia as the world's leading producer of palm oil *Energy Policy* **37** 5440-48

[8] S Mekhilef, S Siga, and R Saidur 2011 A review on palm oil biodiesel as a source of renewable fuel *Renewable and Sustainable Energy Reviews* **15** 1937-49

[9] H C Ong, T M I Mahlia, H H Masjuki, and D Honnery 2012 Life cycle cost and sensitivity analysis of palm biodiesel production *Fuel* **98** 131-39

[10] P K Sahoo and L M Das 2009 Combustion analysis of Jatropha, Karanja and Polanga based biodiesel as fuel in a diesel engine *Fuel* **88** 994-99

[11] N N A N Yusuf, S K Kamarudin, and Z Yaakub 2011 Overview on the current trends in biodiesel production *Energy Conversion and Management* **52** 2741-51

[12] Ministry of plantation industries and commodities, Malaysia, *M Ministry of plantation industries and commodities, Ed,* *ed* http://wwwamericanpalmoilcom/pdf/biodiesel/Malaysia%20Biofuel%20Policypdf

[13] M Chin 2011, *Biofuels in Malaysia: An analysis of the legal and institutional framework Working Paper 64*

[14] Azhari, M Faiz, R Yunus, T I M Ghazi, and T C S Yaw 2008 Reduction of free fatty acids in crude jatropha curcas oil via an esterification process *International Journal of Engineering and Technology* 92-98

[15] J C Juan, D A Kartika, T Y Wu, and T Y Hin 2011 Biodiesel production from Jatropha oil by catalytic and non-catalytic approaches: an overview *Bioresour Technol* **102** 452-60

[16] M Y Koh and T I Mohd Ghazi 2011 A review of biodiesel production from Jatropha curcas L oil *Renewable and Sustainable Energy Reviews* **15** 2240-51

[17] E Akbar, Z Yaakob, S K Kamarudin, M Ismail, and J Salimon 2009 Characteristic and composition of jatropha curcas oil seed from Malaysia and its potential as biodiesel feedstock *European Journal of Scientific Research* **29** 396-403

[18] A R Choudry, R Karmakar, KKundu, and V R Dahake 2011 Algal biodiesel : Future prospects and problems *Water & Energy International (Renewable Energy Section)*

[19] T Baitiang, K Suwannakit, T Duanmukpanao, C Sukjamsri, S Topaiboul, and N Chollacoop 2008 Effects of biodiesel and jatropha oil on performance, black smoke and durability of single-cylinder diesel engine *Journal of Metals, Materials and Minerals* **18** 181-85

[20] W Kamei, I P Singh, M L Sharma, K Singh, and S K Singal 2010 Performance and emissions of biodiesel blends in an automotive diesel engine *Journal of Biofuels* **1** 92-96

[21] R Mamat, R Rahim, A A Abdullah, A Aziz, and N R Abdullah 2012 Characteristic of biodiesel fuel derived from palm oil *Journal of Biobased Materials and Bioenergy* **6** 1-4

[22] S K Hoekman and C Robbins 2012 Review of the effects of biodiesel on NOx emissions *Fuel Processing Technology* **96** 237-49

[23] P-q Tan, Z-y Hu, D-m Lou, and Z-j Li 2012 Exhaust emissions from a light-duty diesel engine with Jatropha biodiesel fuel *Energy* **39** 356-62

[24] P K Sahoo, L M Das, M K G Babu, P Arora, V P Singh, N R Kumar, and T S Varyani 2009 Comparative evaluation of performance and emission characteristics of jatropha, karanja and polanga based biodiesel as fuel in a tractor engine *Fuel* **88** 1698-1707

[25] A F Yusop, R Mamat, A Sudrajad, and M H M Yasin 2012 Characteristics of Particulate Matter (PM) of Diesel engine using Palm Oil Methyl Ester (PME) fuel *Journal of Biobased Materials and Bioenergy* **6** 1-4

[26] D K Bora, D C Baruah, L M Das, and M K G Babu 2012 Performance of diesel engine using biodiesel obtained from mixed feedstocks *Renewable and Sustainable Energy Reviews* **16** 5479-84

[27] A M Liaquat, M A Kalam, H H Masjuki, and A Rezza, *An experimental study on exhaust emissions of a diesel engine using jatropha oil and palmdiesel with diesel fuel*, SciInt(Lahore), pp 245-249, 2010

[28] M A Kalam, J U Ahamed, and H H Masjuki 2012 Land availability of Jatropha production in Malaysia *Renewable and Sustainable Energy Reviews* **16** 3999-4007

[29] *Algae-based Biofuel: Pros And Cons* Available: http://wwwtriplepunditcom/2012/04/algae-based-biofuel-pros-cons/

[30] *Economic Benefits of Algae* Available: http://algaeucsdedu/potential/economic-benefitshtml

[31] T H Lim, S Bari, and C W Yu, *Using crude palm oil (CPO) as diesel engine fuel*, ASEAN Journal on S&T for Development (AJSTD), vol 19, pp 1-13, 2002

[32] J Narkpakdee, A Permsuwan, T Deethayat, and T Kiatsiriroat 2012 Performance and Emission of Small Diesel Engine Using Diesel-Crude Palm Oil-Water Emulsion as Fuel E*nergy Science and Technology* **3** 38-45

[33] M Salamanca, F Mondragón, J R Agudelo, and A Santamaría 2012 Influence of palm oil biodiesel on the chemical and morphological characteristics of particulate matter emitted by a diesel engine *Atmospheric Environment* **62** 220-27

[34] B S Chauhan, N Kumar, Y Du Jun, and K B Lee 2010 Performance and emission study of preheated Jatropha oil on medium capacity diesel engine *Energy* **35** 2484-92

[35] S Jindal, B P Nandwana, and N S Rathore 2010 A comparative study of performance, combustion and emissions of Diesel, Jatropha Methyl Ester and Karanj Methyl Ester in a DI-CI Engine *Journal of Biofuels* **1** 20-29

[36] M Lapuerta, O Armas, and J Rodriguezfernandez 2008 Effect of biodiesel fuels on diesel engine emissions *Progress in Energy and Combustion Science* **34** 198-223

[37] I K Reksowardoj, I H Lubis, W M SA, T P Brodjonegoro, Tatang H Soerawidjaja, W Arismunandar, N N Dung, and H Ogawa, *Performance and Exhaust Gas Emissions of Using Biodiesel Fuel from Physic Nut (Jatropha Curcas L) Oil on a Direct Injection Diesel Engine (DI)*, Copyright © 2007 Society of Automotive Engineers of Japan, Inc and Copyright © 2007 SAE International, 2007

[38] S Pandey, A Sharma, and P K Sahoo 2012 Experimental investigation on the performance and emission characteristics of a diesel engine fuelled with ethanol, diesel and jatropha based biodiesel blends *International Journal of Advances in Engineering & Technology*

[39] T Elango and T Senthilkumar 2011 Performance and emission characteristics of CI engine fuelled with non-edible vegetable oil and diesel blends *Journal of Engineering Science and Technology* **6** 240 - 50

Review on utilization of the pervaporation membrane for passive vapor feed direct methanol fuel cell

N F I Fauzi[1,a], U A Hasran[1] and S K Kamarudin[1,2]
[1] Fuel Cell Institute, Universiti Kebangsaan Malaysia
43600 UKM Bangi, Selangor, Malaysia
[2] Department of Chemical and Process Engineering, Universiti Kebangsaan Malaysia
43600 UKM Bangi, Selangor, Malaysia

E-mail: [a]nurfarahida.engukm@gmail.com

Abstract. The Direct Methanol Fuel Cell (DMFC) is a promising portable power source for mobile electronic devices because of its advantages including easy fuel storage, high energy density, low temperature operation and compact structure. In DMFC, methanol is used as a fuel source where it can be fed in liquid or vapor phase. However, the vapor feed DMFC has an advantage over the liquid feed system as it has the potential to have a higher operating temperature to increase the reaction rates and power outputs, to enhance the mass transfers, to reduce methanol crossover, reliable for high methanol concentration and it can increase the fuel cell performance. Methanol vapor can be delivered to the anode by using a pervaporation membrane, heating the liquid methanol or another method that compatible. Therefore, this paper is a review on vapor feed DMFC as a better energy source than liquid feed DMFC, the pervaporation membrane used to vaporize methanol feed from the reservoir and its applications in vapor feed DMFC.

1. Introduction

A Direct Methanol Fuel Cell (DMFC) is an electrochemical energy conversion device that converts the chemical energy of methanol fuel directly into electricity and a little of heat for as long as fuel and oxidant are supplied, without combustion as an intermediate step. The DMFC is an attractive candidate for mobile power sources, such as Notebooks PCs, handheld PCs, cellular phones, PDAs, etc. The advantages of DMFC are easy fuel storage, low operating temperature and simple design [1]. In spite of these advantages, the commercialization of DMFC is still hindered by several technological obstacles, namely low active catalyst for methanol oxidation, methanol crossover (MCO) through the polymer electrolyte membrane (PEM) in the membrane electrode assembly (MEA), mass transport and water management [2]. In this review, the focus was on the vapor feed DMFC as a better energy source than liquid feed DMFC. Pervaporation membrane is looked into as a medium to vaporize methanol fuel and its applications in vapor feed DMFC is reviewed.

2. Active and passive DMFC

In terms of fuel and oxidant delivery schemes, DMFC can be classified either as active-feed or as passive-feed DMFC [3]. Table 1 shows the differences between active and passive DMFC system [4]. Active DMFC operates with the help of external devices. In active-feed DMFCs, methanol solution is usually delivered by a liquid pump, while the oxidant, oxygen from surrounding air, is supplied by a gas blower or fan [3]. The active system has more reliability, gives a better performance and is more convenient for controlling operating conditions such as temperature, methanol concentration and methanol flow rate [5]. However, parasitic energy losses are inevitable due to the power consumption from pump operation [6].

In contrast, the concept of a passive DMFC is that a DMFC can perform stand-alone operations, which eliminates the need for external devices to pump methanol and blow air into the cell. Therefore, a passive system has some advantages over an active system as they are simpler and more compact. Furthermore, parasitic energy losses can be reduced to enhance the overall system efficiency by using a concentrated methanol solution, and this potentially results in high reliability, low cost, high fuel utilization and high energy density which is favorable in portable device application [4]. Every movement of methanol and oxygen in passive systems occurs naturally. Oxygen diffuses into the cathode from the ambient air due to the air-breathing action of the cell and methanol diffuses into the anode driven by a concentration gradient between the anode and the fuel reservoir.

Table 1. Differences between active and passive DMFC.

	Active	Passive
Definition	Operate with auxiliary devices to deliver methanol at anode and to supply air at cathode	Operate without any auxiliary devices to deliver methanol at anode and to supply air at cathode
Cost	High	Low
Volume	Requires large volumes	Compact cell design
Application	Incompatible for portable devices	Compatible for portable devices
Design	Complex	Simple

3. Vapor feed DMFC

One of the fundamental factors that prevent faster development of DMFC technology is MCO from the anode to the cathode. The MCO rate is roughly proportional to the methanol concentration at the anode [7]. A higher methanol concentration leads to a higher methanol permeation rate [8]. Therefore, in order to reduce MCO, it is necessary to regulate the methanol feed concentration [7]. The DMFC usually shows the highest performance at low concentrations of methanol from in the range of 2-3 M under active conditions and about 5 M under passive conditions due to the MCO [9, 10]. However, a low methanol concentration used in DMFC can lead to a low energy density of the fuel cell system and thus a short runtime, which cannot meet the requirement of the commercialization [11]. To ensure that DMFC are compact enough for portable devices, produce adequate power and are lightweight, fuel should be supplied as concentrated methanol solutions, or even neat methanol [12]. Vapor feed fuel delivery system is an alternative to operate the DMFC with high methanol concentration or even pure methanol without suffering high MCO because MCO in vapor feed DMFC is lower compared to liquid feed DMFC [13].

Since methanol is a very volatile substance, it can easily be evaporated and delivered to the fuel cell through the vapor phase. In a vapor feed passive DMFC, methanol from reservoir tank will be transported to the vaporizer to generate methanol in vapor phase due to the high volatility of liquid methanol [14]. The vaporizer can be a pervaporation membrane or a heated porous pad [2]. Next, methanol vapor will go through the vapor transport layer (VTL), which is a hydrophobic porous layer used to transport methanol vapor to the anode diffusion layer (DL), and simultaneously to prevent liquid from being transported through it. Then methanol is transported through the ADL to the reactive sites in the catalyst layer (CL), on which the electrochemical oxidation of methanol takes place [2]. In methanol oxidation reaction, methanol will react with water to produce proton, electron and carbon

IOP Conf. Series: Materials Science and Engineering **50** (2013) 012056 doi:10.1088/1757-899X/50/1/012056

dioxide. Hydrogen ions formed from this oxidation pass through the PEM to reach the cathode where more electrons are used to reduce oxygen [15]. In oxygen reduction reaction, oxygen then reacts with the electrons and protons to produce water.

3.1. Differences between vapor feed and liquid feed DMFC

The electrochemical oxidation reaction of methanol for both vapor feed and liquid feed DMFC takes place mainly in liquid phase [2]. However, there are some differences between the vapor feed DMFC and liquid feed DMFC [2], as shown in table 2.

Table 2. Differences between active and passive DMFC.

Vapor feed DMFC	Liquid feed DMFC
. Methanol fuel fed in vapor phase	. Methanol fuel fed in liquid phase
. Uses concentrated methanol solution or neat methanol as fuel	. Uses diluted methanol solution as fuel
. Methanol fuel supplied to the fuel reservoir is converted to vapor phase before being fed at anode	. Methanol fuel fed directly into anode from methanol tank in liquid form
. Methanol vapor condenses to form diluted methanol solution in the anode DL and CL	. Liquid methanol vaporizes from the diluted solution into the gas phase with the removal of gas CO2
. Water needed in the anode for methanol oxidation needs to be recovered from the cathode	. Water needed for the anodic reaction may be directly supplied from the diluted methanol solution feed

3.2. Advantages of vapor feed DMFC

A vapor feed DMFC has potential over a liquid feed system in several ways. One of its advantages is its potential to have a higher operating temperature [2, 14] as it exists in vapor state, thus increasing the reaction rates and in turn results in relatively low MCO.

Rice and Faghri [14] found that the vapor feed DMFC also has the potential for shorter start-up time because the mass diffusivity is several orders of magnitude greater than in liquid phase. Thus, the vapor feed system has an advantage in the mass transfer process [16], which results in higher power outputs and leading to the better cell performance. Kamarudin et al. [5] supported this by stating that the cell voltage and power density achieved for liquid feed are significantly lower than those obtained with a vapor feed system, while Fukunaga et al [16] reported that a vapor feed has higher activity than liquid feed due to the higher methanol concentration used as fuel.

Vapor feed DMFCs on working at higher temperature also has the following advantages [17]: (1) Methanol ionization step carried out to produce protons proceeds more rapidly in these cells; (2) Higher temperature results in higher catalytic electrode activity and the faster reaction leads to reduction in methanol fuel crossover; (3) Higher operation temperatures could drastically reduce or eliminate carbon monoxide poisoning of platinum or possibly even allow platinum to be replaced by much less expensive catalysts.

4. Critical issue of vapor feed DMFC

Vapor feed DMFC need to overcome some major challenges in order to operate passively in high methanol concentration, namely MCO through the PEM and water management in the fuel cell [4].

4.1. Methanol crossover (MCO)

MCO is the process by which unreacted methanol is transported from the anode through the PEM to the cathode, where it reacts directly with oxygen, so that no current is produced from the cell [7]. It is caused by the driving forces of concentration gradient, pressure gradient and electro-osmosis [18]. Fuel crossover phenomenon is one of the issues that can reduce the voltage and the fuel utilization efficiency of a DMFC [19], as well as contributing to fuel wastage. Hence, the rate of MCO needs to be reduced to alleviate its impacts on the DMFC performance.

In order to reduce the MCO, hydrogels was used as the fuel-diffusion-rate-controlling agent in passively operated flat pack type DMFCs [20]. The agent retards the fuel diffusion rate from the fuel reservoir remarkably, thus suppressing the MCO even at high fuel concentrations. Most of the liquid methanol that soaked into the hydrogel was possibly in vapor-liquid equilibrium [4]. Nakagawa et al. used the porous carbon late (PCP) as a support to prevent their passive DMFC from undergoing a significant loss of methanol due to the crossover and also being out of temperature control [21]. Nevertheless, degree of reduction in MCO was shown to depend on the properties of the PCP [21]. Flux of methanol through the PEM can be controlled by the diffusion resistance of the porous plate.

Abdelkareem et al. [22] used different methanol concentrations at different cell voltages and found that the effect of the pore structure and thickness of the porous plate, as well as the gas barrier thickness, influence the methanol transport and cell performance of their passive DMFC. As a result of mass transfer restrictions by employing the PCP, high methanol concentration over 20 M could be efficiently used to produce the relatively high power density of 30 mW cm^{-2} for more than 10 h. MCO could be decreased by changing the open ratio of the anode vaporizer because the mass transfer resistance of methanol from the fuel reservoir to the anode CL increases [12].

4.2. Water management

Recently, water management system is more critical and challenging in a passive vapor feed DMFC than in a conventional liquid feed DMFC. The objectives of water management are to provide water for the anode reaction, remove water generated in the cathode layer and keep a high water content in the PEM to lower the proton conductive resistance [23]. Depletion of water inside the anode catalyst layer, especially at higher current densities, decreases the cell performance substantially [24]. Ideally, water needed for the anodic reaction should be completely recovered from water generated from the electrochemical reaction at the cathode, especially when fed with neat methanol [25].

According to Tsujiguchi et al [10], water required for the reaction at the anode side of a vapor feed DMFC, which operates with neat methanol as fuel, can be supplied from the cathode side through the PEM by back diffusion, resulting in a simple system with an increased energy density. Water management becomes a key issue at a higher operating temperature because the evaporative losses are much greater [14]. However, humidified MEA with a composite membrane of lower MCO and diffusion layers of hydrophilic nanomaterials can force the water with back diffusion from the cathode to the anode through the composite membrane [1].

Besides that, Guo and Faghri [7] used a micro porous sub-layer bonded to the cathode catalyst that effectively repels liquid from the cathode and into the PEM to manage water in miniature vapor feed DMFCs without the need for a complex micro-fluidics subsystem. These pores are too small and have a hydrophobic surface. Hence, liquid water that accumulates is pushed back in the other direction across the sub-layer to the anode, as is required for the anode reaction that uses even strictly neat methanol as the fuel.

Xu et al. [25] studied the effects of different water management approaches, that is the addition of a water management layer (WML) and a hydrophobic air filter layer (AFL), and the use of thinner PEM in a passive vapor feed DMFC to improve its water management and cell performance when neat methanol was directly used. They found that adding a WML and an AFL or thinning the membrane thickness greatly improved the water management capability as the water recovery from the cathode to the anode could be enhanced, leading to lower internal resistance, better membrane hydration and better cell performance. The MCO was also curbed and the fuel efficiency was increased.

Li and Faghri [12] studied a change in the cell structures of passive DMFC fed with neat methanol to improve water management system by adopting perforated covers with different open ratios at the cathode without increasing the cell volume or weight. The water management of the fuel cell could be improved by decreasing the open ratio at the cathode side as the water loss from the cathode side was reduced and the water back flow from the cathode to the anode was increased.

5. Pervaporation

Vapor feed direct methanol fuel cell was reported since early 1990s by utilizing an external device, such as an electric vaporizer, heater and so on. However, the external devices consume more additional energy and eventually reduce the total system efficiency. Hence, other methods to produce methanol vapor has been explored to overcome this problem. The most common way to deliver methanol vapor without any additional device is by using a pervaporation membrane which affects the phase change from the liquid methanol, which is contained within the fuel reservoir, to vapor phase [25].

Pervaporation is a membrane technical method that allows only one component of a particular mixture to pass through a non-porous or porous membrane. The phase separation membrane is employed between fuel reservoir and anode diffusion layer to separate mixture of liquids by partial vaporization. There are various types of pervaporation membrane used to vaporize methanol in the fuel reservoir such as silicone membrane [26], Nafion membrane [1, 12, 25, 27], and polydimetylsiloxane (PDMS) [24]. The pervaporation membrane is typically located between the fuel reservoir and anode DL in the DMFC system.

5.1. Mechanism of pervaporation membrane

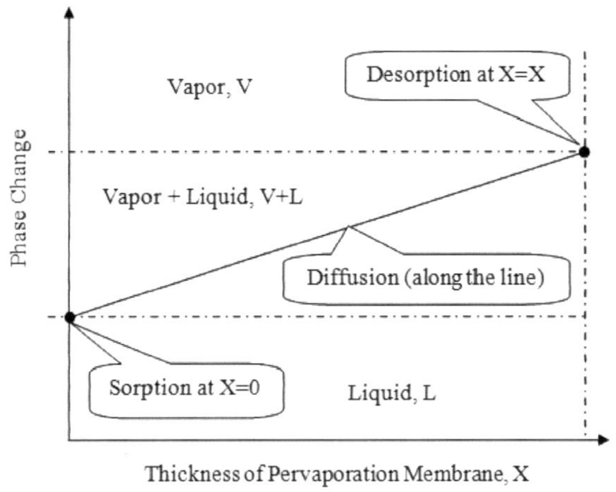

Figure 1. Physical process in mechanism of pervaporation membrane.

There are three physical processes that are involved in the permeation of methanol through a pervaporation membrane, as shown in figure 1: (1) sorption of liquid methanol molecules from the reservoir tank at the feed side of the membrane; (2) diffusion of the dissolved methanol through the membrane; and (3) desorption of methanol vapor from the permeate side [28].

6. Application of pervaporation membrane in vapor feed DMFC

Ren et al. [26] developed a passive vapor feed DMFC in which both the passive fuel delivery system and water management system were presented. Silicone membrane was used as the pervaporation membrane and its surface properties were modified to make it more hydrophobic. The cathode structure was optimized to enable the supply of water from the cathode to the anode by back diffusion and back convection.

Kim et al. [1, 27] developed a semi passive vapor feed DMFC with a humidified MEA and proposed to achieve high energy density and high fuel cell efficiency simultaneously. Liquid methanol was supplied to the system by a syringe pump at a rate of 0.3 $cm^3\ h^{-1}$ for the continuous supply of methanol from the fuel reservoir. Methanol was vaporized by Nafion 112 membrane and then diffused through a hydrophobic water barrier layer and buffer layer to get to the anode electrode. The pervaporation membrane, hydrophobic water barrier and buffer layer verify that the delivery of methanol is as a vapor phase and control the diffusion rate. The vapor feed passive DMFC with humidified MEA maintained the power density at 20-25$mWcm^{-2}$ for 360 h and performed with a 70% higher fuel efficiency. The energy density achieved was 1.5 times higher than a liquid feed passive DMFC.

Li and Faghri [12] used Nafion 117 as their pervaporation membrane for a passive DMFC with neat methanol. The passive DMFC also consisted of methanol barrier which was made from a hydrophobic PTFE membrane placed between the fuel reservoir and pervaporation membrane. PTFE films were placed in both sides of anode and cathode as perforated covers. Furthermore, a gas mixing layer was added between the pervaporation membrane and anode current collector to increase the mass transfer resistance. The cell performance was found to be improved, the MCO was decreased and the fuel efficiency was increased, especially during high current density operation because more carbon dioxide gas was generated at a higher current density [12].

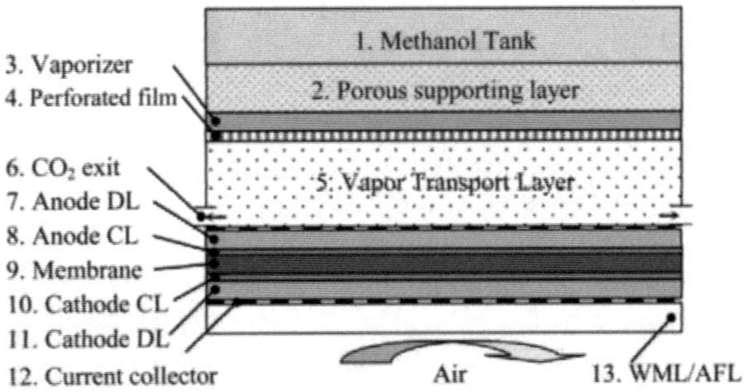

Figure 2. Schematic of the passive vapor feed DMFC [25]

Xu et al. [25] also used Nafion 117 membrane for their passive vapor feed DMFCs with neat methanol. As shown in figure 2, a supporting layer made of porous polyethylene was inserted into the methanol tank and attached tightly to the pervaporation membrane to avoid serious swelling deformation to the Nafion 117. A perforated PTFE film was inserted between the pervaporation membrane and the VTL to restrict the generation rate of methanol vapor. VTL have a hydrophobic surface so that it could function to adjust the transport rate of methanol vapor and simultaneously prevent the transport of liquid.

Meanwhile, Eccarius et al. [24] investigated the impacts of structure parameters and operating conditions for a passive vapor-feed DMFC with neat methanol. Their pervaporation membrane is the PDMS membrane, which is a well known hydrophobic membrane for removing organics from aqueous mixtures [29]. The evaporation rate of methanol into the vapor chamber was controlled by different open ratios of the solid plate attached to the PDMS membrane. In addition, several membranes with differing thickness, ion exchange capacity and specific conductivity were investigated and compared to study their role to reduce MCO in a vapor feed DMFC by Fokken and Lutter. However, PDMS was found to be too soft, flexible, easy to tear and viscoelastic, which means that the methanol liquid may permeate through this membrane in a long-term cell operation.

A novel porous carbon plate (PCP) with better durability than the pervaporation membranes was introduced to vaporize the methanol [9, 22, 30]. Abdelkareem et al. [9] investigated the performance of the passive vapor feed DMFC with and without the PCP under closed circuit conditions with different methanol concentrations ranging from 1 M to neat methanol. The MCO flux and water flux through the MEA with and without PCP were also analyzed. The maximum power density obtained at room temperature was 24 mW cm^2 for methanol concentration of 2 M without the PCP, and this was reproduced at 16 M with PCP. In addition, the mass transfer for both methanol and water from the reservoir to the anode were significantly restricted by the PCP. Thus, high methanol concentrations could be efficiently used in order to get high power density. The back diffusion of water from the cathode to the anode through the PEM also occurred and this resulted in no flooding at the cathode, which was contrary to the case without the PCP.

7. Conclusion

Vapor feed DMFC have a better performance compared to liquid feed DMFC. MCO and water management are currently the main challenges faced by vapor feed DMFC because they have the most effect on the cell performance. Several common approaches to reduce MCO are modification of the PEM, addition of fuel-diffusion-rate-controlling agents such as hydrogel and PCP, usage of neat methanol as fuel and addition of water management layer on the cathode side to push more water backward from cathode to anode for membrane hydration in the methanol oxidation reaction. The most common way to deliver methanol vapor without any auxiliary device is by using a pervaporation membrane, which affects the phase change from methanol liquid contained within the fuel reservoir to methanol vapor.

Acknowledgments

The authors gratefully acknowledge the financial support given for this work by the Ministry of Science and Technology Innovation under the Science Fund Grant Number of ERGS/1/2012/TK07/UKM/03/2.

References

[1] Kim H 2006 Passive direct methanol fuel cells fed with methanol vapor *Journal of Power Sources* **162**(2) 1232-35

[2] Xu C and Faghri A 2010 Mass transport analysis of a passive vapor-feed direct methanol fuel cell *Journal of Power Sources* **195**(20) 7011-24

[3] Chen R and Zhao T 2005 Mathematical modeling of a passive-feed DMFC with heat transfer effect *Journal of Power Sources* **152** 122-30

[4] Halim F A, Hasran U A, Masdar M S, Kamarudin S K and Daud W R W 2012 Overview on Vapor Feed Direct Methanol Fuel Cell *APCBEE Procedia* **3**(0) 40-5

[5] Kamarudin S K, Achmad F, and Daud W R W 2009 Overview on the application of direct methanol fuel cell (DMFC) for portable electronic devices *International Journal of Hydrogen Energy* **34**(16) 6902-16

[6] Chang I, Ha S, Kim S, Kang S, Kim J, and Choi K 2009 Operational condition analysis for vapor-fed direct methanol fuel cells *Journal of Power Sources* **188**(1) 205-12

[7] Guo Z and Faghri A 2007 Vapor feed direct methanol fuel cells with passive thermal-fluids management system *Journal of Power Sources* **167**(2) 378-90

[8] Liu J G, Zhao T S, Chen R, Wong C W 2005 The effect of methanol concentration on the performance of a passive DMFC *Electrochemistry Communications* **7**(3) 288-94

[9] Ali Abdelkareem M, Nakagawa N 2006 DMFC employing a porous plate for an efficient operation at high methanol concentrations *Journal of Power Sources* **162**(1) 114-23

[10] Tsujiguchi T, Abdelkareem M A, Kudo T, Nakagawa N, Shimizu T and Matsuda M 2010 Development of a passive direct methanol fuel cell stack for high methanol concentration *Journal of Power Sources* **195**(18) 5975-79
[11] Chan Y H, Zhao T S, Chen R and Xu C 2008 A self-regulated passive fuel-feed system for passive direct methanol fuel cells *Journal of Power Sources* **176**(1) 183-90
[12] Li X and Faghri A 2011 Effect of the cathode open ratios on the water management of a passive vapor-feed direct methanol fuel cell fed with neat methanol *Journal of Power Sources* **196**(15) 6318-24
[13] Xu C, Faghri A and Li X 2010 Effect of the Structure Design on the Performance of a Passive Vapor-Feed DMFC Fed With Concentrated Methanol *ASME Conference Proceedings* 51-7
[14] Rice J and Faghri A 2008 Analysis of a passive vapor feed direct methanol fuel cell *International Journal of Heat and Mass Transfer* **51**(3–4) 948-59
[15] Ismail A, Kamarudin S K, Daud W R W, Masdar S and Yosfiah M R 2011 Mass and heat transport in direct methanol fuel cells *Journal of Power Sources* **196**(23) 9847-55
[16] Fukunaga H, Ishida T, Teranishi N, Arai C and Yamada K 2004 Impedance of vapor feed direct methanol fuel cells—polarization dependence of elementary processes at the anode *Electrochimica Acta* **49**(13) 2123-29
[17] Lulu S, Jiusheng G, Aruna Z and Bor Z J 2007 Controlled-release vapor fuel cell *US Patent* Pub No US 2007/0190379 A1
[18] Han J and Liu H 2007 Real time measurements of methanol crossover in a DMFC *Journal of Power Sources* **164**(1) 166-73
[19] Chiu Y-J 2010 An algebraic semi-empirical model for evaluating fuel crossover fluxes of a DMFC under various operating conditions *International Journal of Hydrogen Energy* **35**(12) 6418-30
[20] Kim W-J, Choi H-G, Lee Y-K, Nam J-D, Cho SM and Chung C-H 2006 Suppression of the methanol crossover by hydrogels in passively operated flat-pack type DMFCs and its application for the power source of cellular phone *Journal of Power Sources* **163**(1) 98-102
[21] Nakagawa N, Abdelkareem M A and Sekimoto K 2006 Control of methanol transport and separation in a DMFC with a porous support *Journal of Power Sources* **160**(1) 105-15
[22] Abdelkareem M A, Morohashi N, Nakagawa N 2007 Factors affecting methanol transport in a passive DMFC employing a porous carbon plate *Journal of Power Sources* **172**(2) 659-65
[23] Li X, Faghri A and Xu C 2010 Water management of the DMFC passively fed with a high-concentration methanol solution *International Journal of Hydrogen Energy* **35**(16) 8690-98
[24] Eccarius S, Krause F, Beard K and Agert C 2008 Passively operated vapor-fed direct methanol fuel cells for portable applications *Journal of Power Sources* **182**(2) 565-79
[25] Xu C, Faghri A and Li X 2011 Improving the water management and cell performance for the passive vapor-feed DMFC fed with neat methanol *International Journal of Hydrogen Energy* **36**(14) 8468-77
[26] Ren X, Becerra J J, Hirsch R S, Gottesfeld S, Kovacs F W and Shufon K J 2005 Controlled direct liquid injection vapor feed for a DMFC *US Patent* Pub No 2005/0170224 A1
[27] Kim H, Oh J, Lee J and Chang H 2011 Fuel cell system comprising vapor phase fuel supplying system *US Patent* Pub No 2011/008043765B2
[28] Faghri A and Guo Z 2009 Vapor feed fuel cells with a passive thermal-fluids management system *US Patent* No Pub US 7625649 B1
[29] Mohammadi T, Aroujalian A and Bakhshi A 2005 Pervaporation of dilute alcoholic mixtures using PDMS membrane *Chemical Engineering Science* **60**(7) 1875-80
[30] Abdelkareem M A and Nakagawa N 2007 Effect of oxygen and methanol supply modes on the performance of a DMFC employing a porous plate *Journal of Power Sources* **165**(2) 685-91

Analysis of failure in manufacturing machinery

S.Sulaiman[a] and N.Abidin Ismail
Department of Mechanical and Manufacturing Engineering, Faculty of Engineering, Universiti Putra Malaysia, 43400 UPM, Serdang, Selangor, Malaysia, Phone: +603-98466336; Fax: +603-86567122

E-mail: [a]suddin@eng.upm.edu.my

Abstract. This paper presents information about how to interpret the machine failure in suitable ways. The concept in this paper focuses on the methodology and creates active thinking with positive attitude to solve the machine failure. It is also described few topic on elaborate the technique towards investigate and develop a better understanding of using this concept in manufacturing industry. Failure analysis (FA) in manufacturing has its own value for each characteristic and to communicate, it needs specific data. Two methods were discussed to perform FA i.e. Why-why analysis and PM analysis (is a philosophy which aims to clarify the mechanism behind the chronically defective phenomena by analyzing the phenomena in terms of physical principle). For Why-why analysis, it is mainly effective in preventing the recurrence of failures that occur at the initial period. Whereas PM analysis is mainly effective for resolving on a fundamental problem that occur in the accidental period.

1. Introduction

The purpose of FA is entirely positive; to prevent further failures. Failures occur when some system or part of a system fails to perform up to the expectations for which it was created such as a transmission fails, a pipeline leaks or a cell phone explodes. The concept of failure is easy to understand intuitively. But underneath that intuitive understandings are important conceptual principles which are commonly either misunderstood or not considered at all [1-3].

Failure itself is a human impression. Materials do not fall through in and of themselves. They follow the laws of nature faultlessly. If a part is loaded beyond its tensile strength, it wills breakdowns. While waiting for that pressure level is reached, it does not break. When a part fails in service, it was under-designed or poorly mass-produced for the situations in which it was used [4].

1.1 Ideas behind Equipment Failure Reductions

Nowadays in manufacturing, the failure is repeating and mostly it is happens due to many factor such as modification of equipment (planning and implementation), feed back to similar equipment and equipment of the next generation, daily maintenance (add or review the contents of daily maintenance, lubrication and cleaning such as parts and frequency) or periodical inspection/service (adds or reviews the contents such as parts and frequency).

In cases same breakdown occurs, the routine or method that being carried out to reduce breakdown is usually done by disassembly and service manual (disassembling procedures, check points etc. The thing needs to keep towards this activity is spare parts development (keep parts as spare parts).

Therefore, FA has being created as a method to pursue the true cause abnormality and prevent recurrence as an improvement after the current practise which is only restore the parts after detect an abnormal part at an early stage. The purpose of FA being implemented is described as in table 1.

Table 1. Purpose of FA depending on division

Division	Purpose of FA
Company	Product quality, delivery and price that meet customer's request.
Plant	Improve productivity and zero defect delivery
Maintenance	Stable operation of equipment (produce no defects) and increase efficiency.

In general, failures of machine come from two main types which is functionality stops failure and decline in functionality failure. Functionality stops failure occur when the operation of equipment is totally stop due to breakdown while decline in functionality failure occur when the equipment function shows in decline of speed, accuracy or short stoppage.

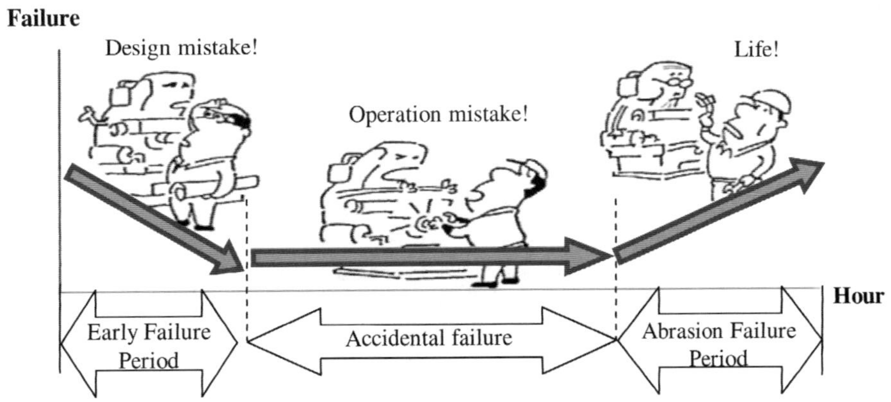

Figure 1. Summary of machine life cycle

The causes include mistakes in the planning stage or manufacturing stage, wrong way to use; wrong way to conduct daily maintenance, negligence of maintenance group and repair mistakes as illustrated in Figure 1.

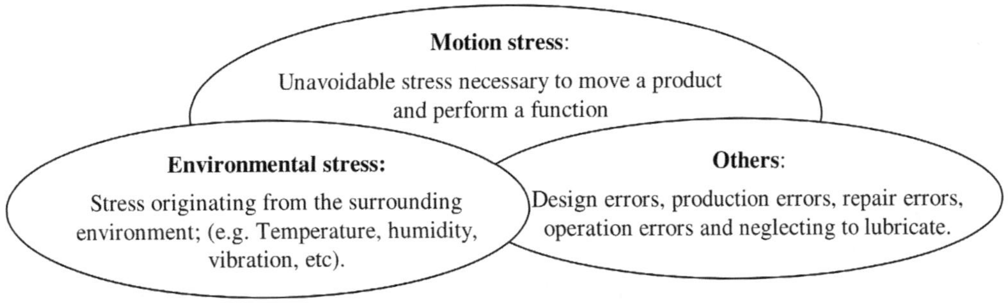

Figure 2. Three main reason of breakdown

Failure is happen due to many reasons. Generally it can be divided to three main reasons as illustrated in Figure 2.

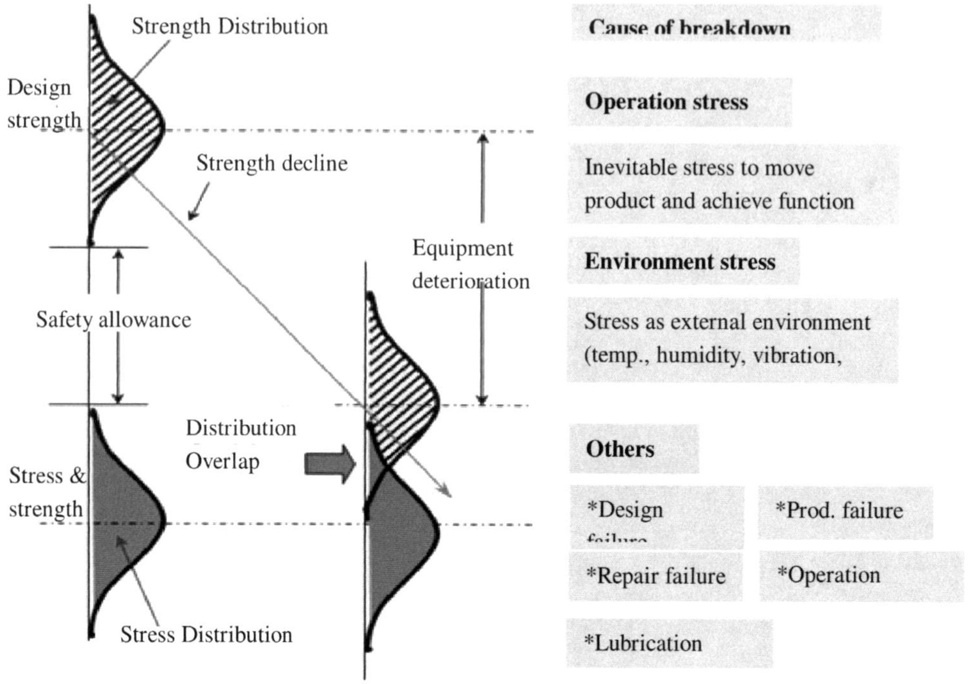

Figure 3. Design strength versus Stress

Like human beings, equipment is constantly under stress whether it is operated or left unattended and its life comes to an end in due course as shown in Figure 3. Therefore, health maintenance and physical strength enhancement (Kaizen) are important for the equipment.

2. Failure Analysis

Failure of a component indicates it has become completely or partially ineffectual or has deteriorated to the point that it is unreliable or unsafe for normal sustained service. FA is an engineering methodology to determining how and why equipment or a component has failed. The goal of a FA is to understand the root cause of the failure so as to avoid similar failures in the future. In addition to confirming the failure mode it is important to determine the factors that explain the how and why of the failure event. By recognising the roots cause of the failure occurrence, how and why of failure can be clarifying [5-7].

It is not always necessary to prevent the first, or root cause, from happening. It is merely necessary to break the chain of events at any point and the final failure will not occur. Frequently the root cause analysis identifies an initial design problem. Then a redesign is commonly enacted. Where the root cause analysis leads back to a failure of procedures it is necessary to either address the procedural weakness or to develop an approach to prevent the damage caused by the procedural failure.

2.1 WHY-WHY Analysis

The answers to the "Why" questions form a chain of causes leading to the root cause. The answer to the first Why is the direct cause. The logical end of each chain is a root cause (each chain will have its own root) and the causes in between the direct cause and the root cause are contributing causes. There may be no contributing causes, but there is always a root cause – the best and logical place to stop as identified by the team. This place is where continuing to ask Why adds no value to prevention of repetition, inconsistency reduction, or cost savings [8].

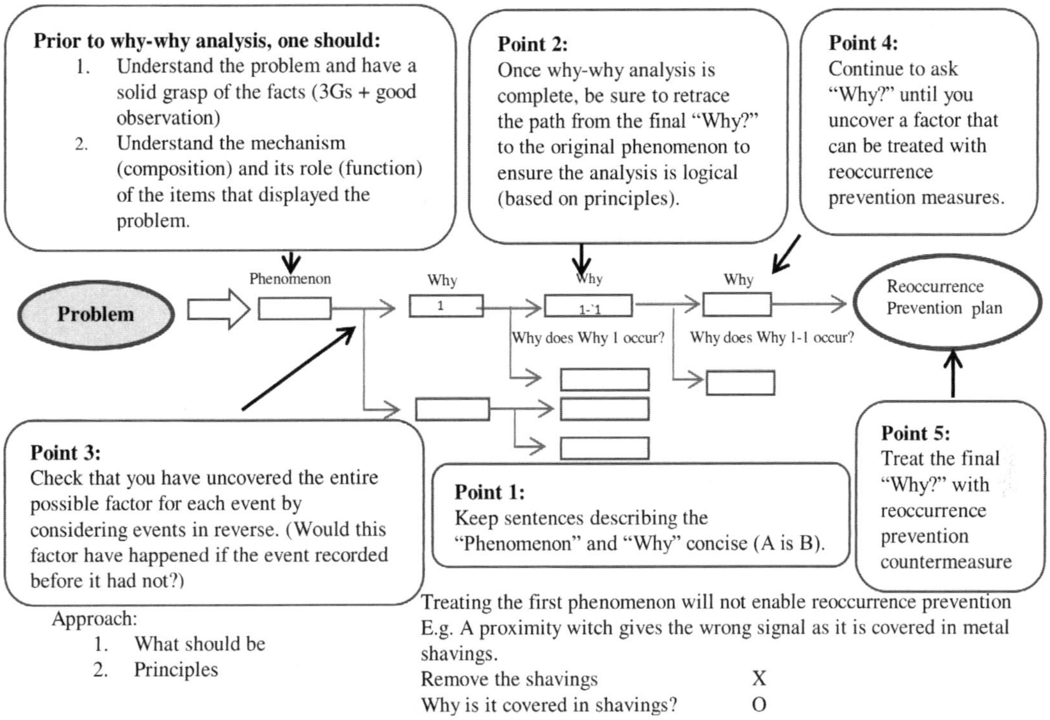

Figure 4. Points for WHY-WHY Analysis

When the problem is identified, and preliminary data has been gathered and verified, the analysis can begin. This process uses the Why-Why analysis method to build a cause chain because it is a natural, logical progression for thinking through a problem as described in Figure 4.

Using the 5-Why approach provides a structured approach to corrective action and can form the basis for a broad-based continual improvement and preventive action plan. This formal process should capture and prioritize causes and address them as the basis for continual improvement efforts. The issues acknowledged through this process are difficulties within the organization that are costing time, money, and frustration [9]. This cause identification process, coupled with a structured prioritization process will also satisfy the requirements for:

- Corrective action to a degree appropriate to the magnitude of the problems and commensurate with the risks encountered
- Effective handling of customer complaints and reports of product nonconformities
- Investigation of the cause of nonconformities relating to product, process and quality system, and recording the results of the investigation.

2.2 PM Analysis

PM Analysis is a '**Philosophy**' which aims to clarify the mechanisms behind the chronically defective phenomena by analyzing the phenomena in terms of physical principles. This can be done by using 5 M's, which are the first physical principles is to analyze and understand the mechanism of a machine.

Then analyze how the machines, man, materials and method are interrelated with the phenomena. This is the main ideas behind analyzing the factors of failure.

PM Analysis also can be described as chronic problems are physically analysed based on principles and laws to clarify the mechanism for the phenomenon. PM Analysis is suitable for problems with clear causes and non-effective counter-measures. Through PM analysis process, confidence is build up and "Kaizen" abilities and techniques improve as described in table 2. Table 3 shows the method on how to perform PM analysis.

Table 2. Differential between Conventional Kaizen and Kaizen for Zero Defects (PM Analysis)

	Conventional Kaizen	Kaizen for zero defects
Goal	½ Substantial reduction	As close to) as possible
Ways of thinking	1. Focus on priority thinking 2. Resolve issues that have a substantial effect (from the biggest point) 3. Carry out counter measures on a limited number of factors	1. Cannot focus on priority thinking 2. Think rationally and list up all possible factors which may have had an effect 3. Thoroughly investigate each factor 4. Fix all defects 5. When possible, fixes together 6. Implement intensively
Approach	Fishbone diagram (factors and effects)	PM analysis
Other	When problem rates are high	When problems rates are low; When problems are chronic

Table 3. Step to perform PM Analysis

Step 1	Clarify the phenomena
Step 2	Understand the structure/composition
Step 3	Physical analysis of the phenomenon
Step 4	Review the conditions that lead to the phenomenon
Step 5	Review the 5 Ms and correlations
Step 6	Review the investigation methods
Step 7	Determine the problem
Step 8	Restoration and Kaizen
Step 9	Maintenance Control

Figure 5 shows of example by using CAD; the mechanism of the structural drawing can easily being understood. This is described as in step 2 in Table 3.

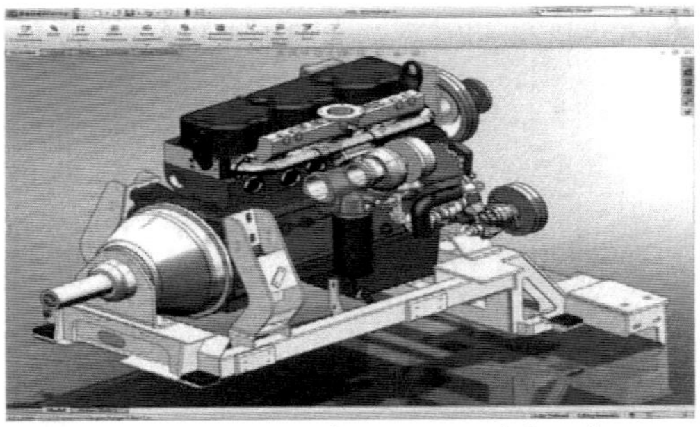

Figure 5. Using CAD software to create the structural drawing to understand the mechanism.

3. Summary of Why-why analysis and PM analysis.

Why-why analysis is used to respond to sudden failures and to reviews how to prevent their recurrence. It's mainly effective in preventing the recurrence of failures that occur during the initial failure period. The analysis of true cause of the problem is by repeating the question **"Why".** It's also needed to understand the fact of what is happening, and repeatedly analyse why this fact is occurring.

PM analysis is used to reduce chronic losses, especially when reducing defects and short-period stops to "zero". It's mainly effective for resolving on a fundamental level the problems that occur in the accidental failure period. The analysis is by grasp the problem as a "physical phenomenon" and analyzes the factors that lead to the phenomenon. It can be done by list the factors that may affect the occurrence of the problem from the point of view of man, machine, materials and methods.

4. Conclusion

When a new case of machine failure keeps happening, truly reducing the failure in true sense means gigging down to its root cause and answering the question," What really must be stopped?" Before digging into the search for a root cause, organizational Kaizens can often be devised for problem arising from how daily maintenance & periodic machine adjustments are performed, or problems from design (how safety factors are calculated, how devices are used.

Failures will return and countermeasures will need to be repeated without doing failure analysis to this extent. For these reasons, while digging down in the search for root causes is important, the series of motions presented here cannot be performed by the Maintenance Department alone, and will require the cooperation of all related departments.

Acknowledgments

This work was supported by Department of Mechanical and manufacturing Engineering, Faculty of Engineering, Universiti Putra Malaysia. The authors would like to thanks to staff of the Department for their helps and encouragement for this project.

References

[1] Andrew Muno M and Ibrahim R N 2013 *J. Mech. Eng. Sci.* **4** 488-95
[2] Mansor M A Ohsato A and Sulaiman S 2012 *Inter. J. Automot. Mech. Eng.* **5** 612-21
[3] Zamanzadeh M E Larkin and D Gibbon 2004 A Re-Examination of Failure Analysis and Root Cause Determination 1-33
[4] William T Becker and Roch J Shipley 2002 *Failure Analysis and Prevention* Volume 11 Ohio ASM International
[5] Jones D R H 2001 *Failure Analysis Case Studies II* Pergamon Elsevier Science
[6] Kamal M Rahman M M and Rahman A G A 2012 *J. Mech. Eng. Sci.* **3** 291-00
[7] Kamal M, Rahman M M and Rahman A G A 2013. *Inter J Automot Mech Eng* **7** 912-923.

[8] Performance Review Institute 2006 *Root Cause Corrective Action Supplemental Help Nadcap Style* – Report, www.pri-network.org.
[9] Mike S 2004 LRS Consultant *Understanding How to Use the 5-Whys for Root Cause Analysis* Report, www.lifetime-reliability.com.

Development of plant maintenance management system (pmms): a case study

N A Che Azhar[1] and M A Mansor[2]
[1] Faculty of Mechanical Engineering, University Malaysia Pahang, 26600, Pekan, Pahang, Malaysia
[2] Faculty of Technology, University Malaysia Pahang, 26300, Gambang, Pahang, Malaysia

Abstract. In large plant industry, it is not easy to maintain machine performance without using any method such as checklist system. Manual checklist is a common maintenance checklist used in industry. All machine, equipment and parts that need to be checked will be written down for the employee to do maintenance checks. Converting the manual checklist to the Plant Maintenance Management System (PMMS) can improve the way of employees work and make plant management easier. Therefore, a new system was designed to maintain the equipment so that the activities are more efficient and cost effective. The system consists of three frames that connect to each other. The frames divide to section, equipment and checklist. This system also builds to prevent data from arbitrarily changes. Only certain officers or staffs are permitted to make modifications to data. Using this system, a company can make the office environment a paperless environment.

1. Introduction

In today's environment, Maintenance Management System (MMS) is wide expands for process plants to increase profitability by improving reliability and achieving maintenance excellence. Maintenance is accepted as a means of gaining additional control of operational budgets and significantly increasing a company's bottom line. When we consider that maintenance costs can make up 40 to 50% of operational budgets in capital-intensive industries, the effect of a reduction in maintenance costs is both obvious and impressive (Daryl Mather, 2002). In addition, maintenance is generally the largest controllable operating cost of a capital investment industry.

Maintenance can be clarify as activities required or undertaken to conserve as nearly, and as long, as possible the original condition of an asset or resource while compensating for normal wear and tear. In the engineering world, the word maintenance can be simplifies as actions necessary for retaining or restoring a piece of equipment, machine, or system to the specified operable condition to achieve its maximum useful life. It includes corrective maintenance and preventive maintenance. The objective of maintenance is to try to maximise the performance of equipment by ensuring that, items of equipment function regularly and efficiently, by attempting to prevent breakdowns or failures, and by minimising the losses incurred by breakdowns or failures (C.D. O'Donoghue et al., 2004). In fact, to maintain or increase the reliability of the operating system taken as a whole is the objective of maintenance. Properly implemented of an integrated maintenance management can lessen emergencies by 75%, cut the purchasing by 25%, increase warehouse accuracy by 95% and improve preventative maintenance by 200%. Then, in the short and long term the potential saving is high if the maintenance cost rising from 9% to 11% per Annum and if maintenance can minimise the costs by as much as 35% it can count as good management (R. Sivaligam, 1997).

Profitability and competitiveness are the overall goal that a company need. The overall goal can be defined in different level of control using the term of efficiency, effectiveness and cost-effectiveness.) The word of efficiency is defined as the ratio of output to inputs of a system and a measure of how well the system's output contributes to the business goals and objectives is defined for term effectiveness (R.N. Anthony, 1998. Collecting all the cost information incurred during equipment life can be determined as Life Cycle Costing (LCC). The measure of a system in terms of mission fulfilment and total LCC is defined as cost-effectiveness. The operational maintenance goals and is a combination of technical and organisational aspects can be relate to high resource efficiency (B. Al-Najjar, 1997).

New technologies bring us improvement toward job and live. Electronic checklist nowadays can provide an external memory of pending, completed, and skipped steps. There is no debating it anymore with a solid foundation in place; electronic checklist processing is growing because it virtually ensures substantial benefits for any organization that implements it. Computerized maintenance management systems (CMMS) are one of the ways to improve manual checklist. The goal of a maintenance manager is to employ a management system that optimizes the use of scarce resources such as manpower, equipment, material, and funds to maintain the facilities and equipment that are the responsibility of the maintenance organization. The system should provide for integrated processes giving the manager control over the maintenance of all facilities and maintainable equipment from acquisition to disposal.

The facilities maintenance manager with work reception, planning, control, performance, evaluation and reporting was requirement and assists in modern CMMS nowadays. Besides that, this system will also maintain historical information for management use. The manager should evaluate management data requirements and establish electronic data needs prior to acquiring a system or additions to, or replacement of, an existing system. The evaluation should include a return on investment (ROI) analysis before investing in additional or new CMMS capabilities. To accomplish the maintenance organization's goals, manager should only acquire what is necessary for company.

2. Convert manual checklist to electronic checklist

The development of a system, the main thing that should be implemented is phases of creating new Plant Maintenance Management System (PMMS). The system design will be develop trough out many stage before can be running up. The stages of this phase have been planned in developing this system such as system planning stage, the stage of systems analysis, systems design stage, implementation stage and the stage of the operation and support system. Things that need to be addressed in this stage are to understand the problem, determine the existing constraints and finally identify the benefits derived from the system that will be developed. The problems that have been detected and gathered will be analyzed and then a plan to solve the problems be created.

The main stage is system design stage. In this stage, the display screen, files and procedures manual is designed. Then, the design for input data need to be choose for each of available data whether use button or need to be write. To ensure accurate system, controlled, reliable and easy maintained this system need to be planned according to the needed of company. When start creates the system, LAMP (Linux, Apache, MySQL and PHP) will be best software use to complete the PMMS. Figure 1 at the following page show the step of build up the system.

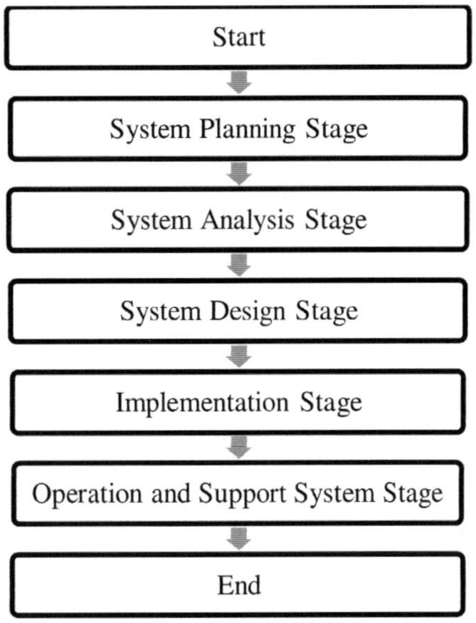

Figure 1. System development flow.

PMMS was created for one of the largest cement industry in Malaysia. This plant has 7 main sections consist of crusher, raw mill, core line, coal mill, cement mill, cement silo and packing plant. This plant also consists of the other section such as blower, palletizer, air and water service and fire fighting. All of these sections contain 17 to 37 equipments and machines. Each of the equipments and machines consists 2 to11 parts that need to check every week. Using PMMS can help maintenance management maintain this plant so that they can run this plant 24 hour per day without having any problem.

3. Result

Figure 2 at the following page illustrates the main page of the system. Users that want to enter this system must key in the user data first. The user must enter the staff id, password and section from the screen. Section consists of mechanical, electrical or production where user must select one of the sections. User needs to click the 'login' button to go to the next screen. User can change their new password by click 'change password' button. The 'admin' button design for administration to modified system such as to update or remove data by a specific person.

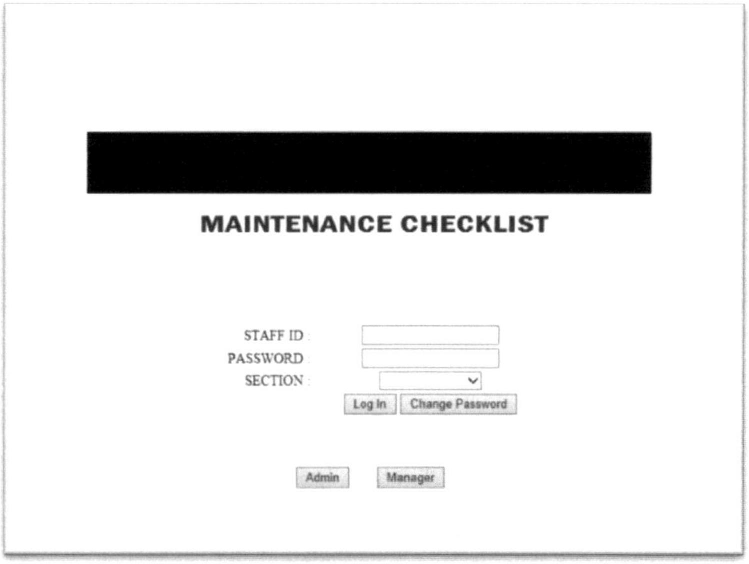

Figure 2. Main page.

Figure 3 below shows the front view of the system has been developed. At the top of the screen display there are several main sections of crusher, raw mill, core line, coal mill, cement mill, cement silo, packing plant, blower, air & water service, fire fighting and log out. User must click one of the sections, and then the equipment listed will be reflected at the left side of the web page. By clicking one of the equipment lists, the checklist list will be reflected to the right side of the web page. Users just need to click or type the requirement information. All the data collected will store at the data based then it can be review by the manager as illustrated in figure 4 at the following page.

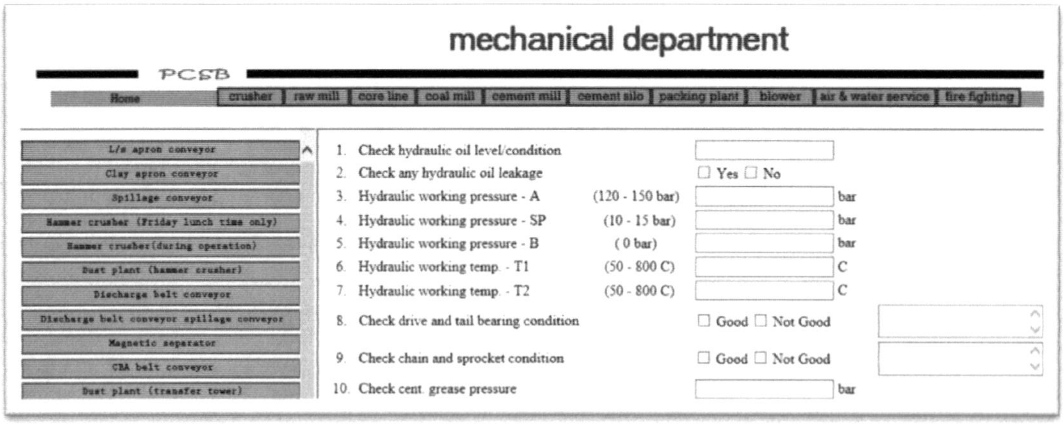

Figure 3. Maintenance checklist menu

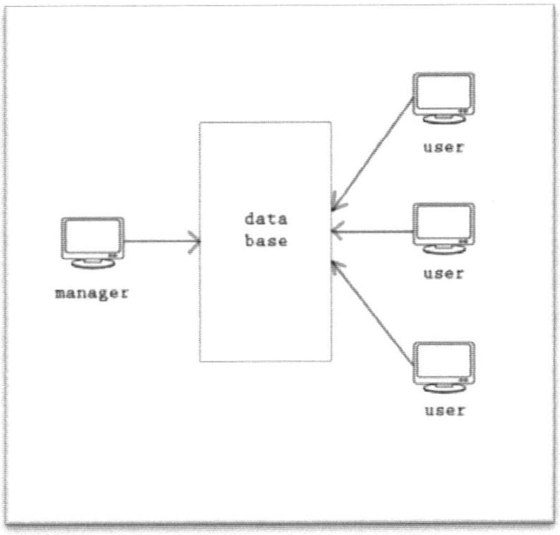

Figure 4. Context diagram

4. Discussion and conclusion

PMMS is convenient, reliable and advantageous to the organization as a whole. Significant time saved in every level of an organisation which can be utilized for other project, which can bring greater benefit and productivity. With the implementation of PMMS, need for sorting and filling is fully eliminated thus saving time and space, which indirectly creates a more conclusive working environment. Abnormal operating conditions of plant equipment is quickly detected since the data obtained is immediately and easily uploaded into the database that does not rely on manually data entry which is time consuming and usually not immediate. Although the initial cost of implementation might be somewhat higher compared to traditional methods, return of investment (ROI) can be seen as early as within 1 year to 2 years. The bulk of returns will come from time saved and spent on more productive projects, significant reduction in the use of paper which in turn result in less cost and significant reduction in environment impact.

5. Acknowledgements

The authors acknowledge the research grant RDU 120371 received from University Malaysia Pahang.

References

[1] Mather D 2002 *CMMS: Journal of a timesaving implementation process*
[2] O'Donoghue C D and Prendergast J G 2004 *Journal of implementation and benefits of introducing a computerised maintenance management system into a textile manufacturing company*
[3] Sivaligam R 1997 A12 step programme for moving down the road to recovery, Plant Engineering *Journal of applying best practices to maintenance* **51(6)** 120
[4] Anthony R N 1998 *Management Control Systems*, ninth edition, Irwin/McGrawHill, USA
[5] Al-Najjar B 1997 Condition-based maintenance selection and improvement of a cost-effective vibration-based policy for rolling element bearings, PhD thesis, Lund University, Department of Industrial Engineering.

Investment casting using multi-jet modelling patterns: the thermogravimetric analysis of visijet® SR200 UV curable acrylate plastic

M N Hafsa[1,a], **M Ibrahim**[2,b] **and S Sharif**[3,c]
[1]Faculty of Mechanical and Manufacturing Engineering, Universiti Tun Hussein Onn Malaysia, Batu Pahat, Johor, Malaysia
[2]Faculty of Mechanical and Manufacturing Engineering, Universiti Tun Hussein Onn Malaysia, Batu Pahat, Johor, Malaysia
[3]Faculty of Mechanical Engineering, Universiti Teknologi Malaysia, Skudai, Johor, Malaysia

E-mail: [a]hafsa.mpputhm@gmail.com, [b]mustaffa@uthm.edu.my, [c]safian@fkm.utm.my

Abstract. Rapid Prototyping (RP) technology is actively studied to be implemented in Investment Casting (IC) process. Nowadays RP techniques are studied for their feasibility as IC master patterns, in terms of pattern collapsibility and drainage during burnout. The purpose of the study is to determine the characteristic of Visijet® SR200 acrylate material during burnout process. Traditional IC patterns made from wax have properties that limit their application in precision casting, especially for parts with thin geometries that readily break or deform when handled or dipped in the refractory slurry. Furthermore, it is not economical when producing a small number of parts. Non wax patterns fabricated for IC process, revealed ceramic shell cracking due to excessive thermal expansions, incomplete collapsibility of pattern during burnout, residual ash and poor surface finish. Thermogravimetric analysis (TGA) was used to measured the weight loss of acrylate material as the temperature was increased. TGA measured the change of material's mass as it is heated. It represents the decomposition temperature after being subjected to varying temperatures, as well as the amount of residual ash. In this experiment, the temperature range was from 20°C to 700°C with 5°C increment. Experiment results show the values of material's optimum reaction temperature and decomposing temperature of Visijet® SR200 acrylate. The percentages of remaining materials were also monitored throughout the process to obtain the amount of residual ash. All of the temperature values obtained is a resemblance for the actual burnout process and can be used as references.

1. Introduction

Traditional IC process would require the master pattern which is made of wax to create the ceramic mould. RP technology has evolved the possibilities of implementing non-wax material as the master pattern of IC. Along with the ability to produce such complex geometry in shorter time, RP also has wide selection of material and approach to choose from. By using direct method of RP technique, a significant reduction of cycle times and production lead time in IC process [1-6]. Wax was traditionally used as it is rather easy to be shaped in various forms, either solid or liquid, for mass production, but it has quite low strength and could easily affected if exposed to high temperature surrounding thus limiting the applications of wax pattern when it comes to high precision process [7-8].

One of the common RP process for IC implementations is Selective Laser Sintering (SLS) technique [9-16]. Usually wax is used to make IC pattern among the other SLS material because it's thermal properties do compliments IC burn out process requirement. Pattern material should burn out completely without leaving any residue during the dewaxing process. Unfortunately, wax product may distort if it's to be made using SLS process, affecting the final IC product. Plus, one of the serious problems of SLS is that since the solidified part is cooled rapidly, the model tends to be deformed and cracked due to the thermal stress [17].

The RP polymer pattern materials have a higher melting temperature and higher-melting viscosity thus it is very difficult to dewax [16,18]. To replace wax, it is necessary to use other material that is also easy to shape but have greater strength. Furthermore, the RP patterns need to have an excellent surface finish and must precisely recreate fine details with minimal burnout residue [14]. Besides SLS, other RP systems, such as Stereolithography (SLA) and Laminated Object Manufacturing (LOM) were tested for IC applications. However, since the epoxy material used in the SLA is a thermoset material, it does not melt like wax. It would expand and creates thermal stresses to the ceramic shell causing it to crack [9,19,20]. For the LOM application in IC, the paper which is the main material of the system would create very high amount of residual ashes during the burn out process [19].

Recent studies have shown the capabilities of pattern built from ABS using FDM technique. ABS pattern are capable to produce a clean burn out, it is robust and could be handled without damage, it also easy to be prepared and has good dimensional stability. However, the surface conditions are quite rough due to its surface layer and built condition, thus requiring surface finishing of the pattern before it could proceed with IC as the quality of final IC product are highly dependent to the quality of the pattern [21,22].

Another technique to produce the IC pattern is by using MJM. MJM technique could produce better surface finish compared to FDM [22]. MJM also be implemented in building part with internal structure as sacrificial IC pattern [23]. During burnout process, the main purpose is to completely remove the master pattern, leaving the vacant space for metal to be poured in. The temperature for burn out process depends on the type of pattern material. This paper will be focused on MJM part based on acrylate material. Thermal decomposition and weight loss are measure by using Thermogravimetric instrument. TGA is a technique in which, upon heating a material measure whether the weight increases or decreases.

2. Methodology

The implementation of RP in IC was on the part of the pattern making. The pattern is fabricated by the RP machine along with normal post processing of RP product. The RP pattern then will be dipped into a slurry then covered ceramic grain. The RP pattern will stucco a couple of the ceramic coating before the ceramic harden and turn into a shell. During burnout process, the acrylic pattern must be completely burnt out leaving the ceramic shell a perfect vacant for metal pouring. In this process, the acrylic material being heated and will start to decompose leaving just residual ashes. This part of the process is important as it is the last step before metal is being poured into a mould.

Since the materials reacts differently as it is subjected to heating, using Thermogravimetric (TG) will help to predict and understand how a material weight changes as temperature changes. It is a technique which the mass of the sample is monitored against time or temperature while the temperature of the sample, in a specified atmosphere is programmed.

Thermal analysis was carried out using Linesis Thermobalance TGA equipment. The samples were prepared in form of tiny flakes. For the testing, 70.0 mg of MJM Visijet® SR200 acrylate flakes sample was used. The sample was put into the crucible before it was inserted into the measuring system. In a thermobalance equipment system, a schematic diagram of the specific balance and furnace assembly is shown in figure 1. The sample was burnt under air atmosphere with the heating rate of 5°C/min. The scale has to be balance with the least possible percentage of tolerance using the potentiometer. It is very desirable to get the 0% of balance tolerance. The sample was heated to slightly 700ºC with 5ºC increment. The machine was left to run for 2 hours and 20 minutes.

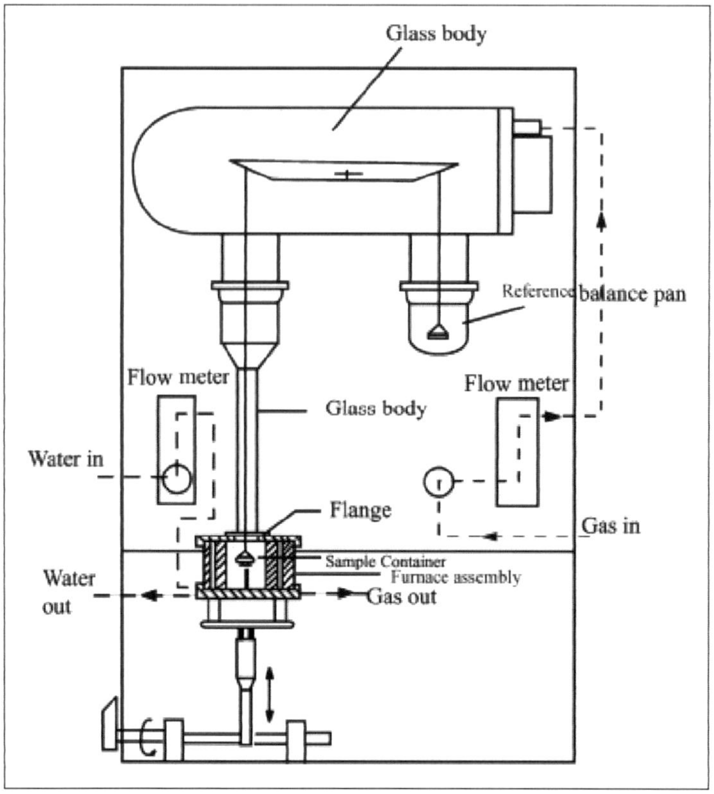

Figure 1. Schematic diagram of a typical balance and furnace assembly.

The data obtained from the experiment is the raw data and will be manipulated using Simultaneous Thermal Analysis (STA) Software. The raw data include very high error thus it is necessary for the value to be corrected. Reference sample that was used to calibrate the system is Alumina Oxide (Al_2O_3) burnout under air atmosphere. The X-axis value will be corrected according to the weight absolute $1x10^{-4}$. For the Y-axis value, the value will be smoothed according to the weight absolute $1x10^{-4}$. The smoothing process should be done accordingly as the data that was too smooth will not resemble the actual character of the whole burning process. The data could then be derived into thermogravimetric data to obtain the decomposition value.

3. Result and Discussion

3.1. Thermal decomposition

A derivative thermogravimetric (DTG) curve presents the rate of mass change (dm/dT) as a function of temperature against temperature (T) on the x-axis. Figure 2 shows typical DTG analysis curves for a sample of Visijet® SR200 acrylate heated in air atmosphere at a rate of 5ºC/min. In this figure, the derivative of the curve is shown by brown line. A TG curve which is blue line represents the variation in the mass (m) of the sample with the temperature.

Within the graph, the line descends to resemble the lost of mass along the process. The material decomposition occurs in three stages. It could be seen from the graph by identifying the three curves from the DTG that was derived from the initial TG curve. The first stage of decomposition (point 2-3) is the initial degradation in primary of the decomposition of the hard segment. During this stage, the rate of the temperature change is as much as 208.5°C, with the total weight loss of -33.85 mg. The second stage (point 4-5) is the polycondensation and poly of degradation mechanism and effected by the soft segment content [24]. For this stage, the rate of the temperature change is by 25.9°C with total mass lost of -14.15 mg. The third stage (point 6-7) is only attributing to the ash formation [21,25].

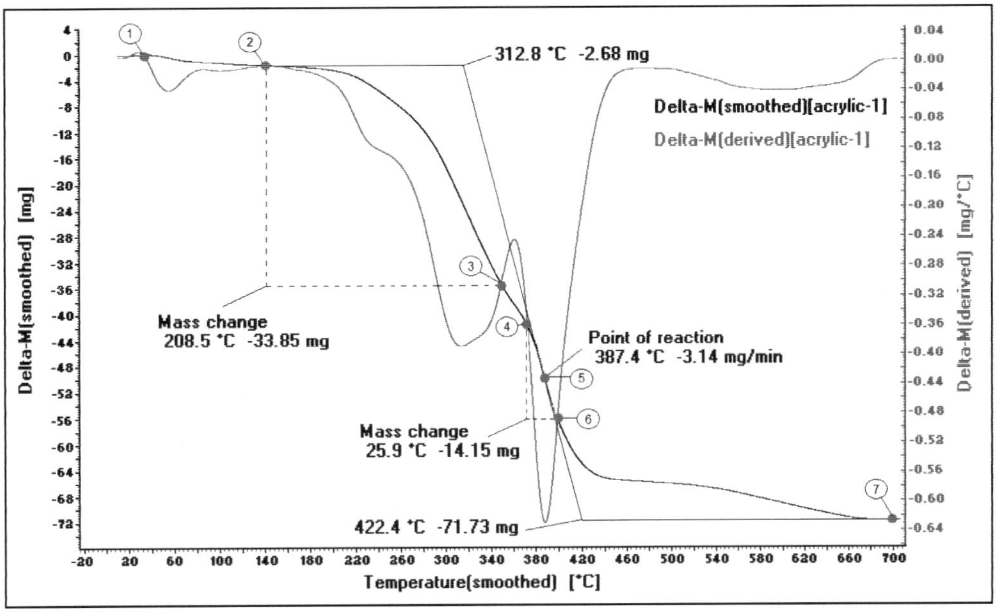

Figure 2. Derivative thermogravimetric (DTG) curves for Visijet® SR200 acrylate material.

In the early stage (point 1-2) of the decomposition, the material was not in stable conditions due to the presence of some volatile components [24,26]. The graph shows that there are two step of reaction which occurs at 312.8°C and 422.4°C. At 312.8°C, the rate of mass loss is 2.68mg/min while at 422.4°C the rate is -71.73mg/min. The crossing point between the both reactions created a point (point 5) which called the point of reaction which is at 387.4°C with the rate of mass loss of -3.14 mg/min. This particular point is where the maximum point of Visijet® SR200 acrylate material significantly decomposes. It caused by the break of polymer chain and polymer thermal decomposition [26,27].

3.2. Residual Ash

TG curves represent the variation in the mass (m) of the sample with the temperature as shown in figure 3. From the graph, the early stage of the experiment shows little changes by only -3.4% from the 70 mg sample. The line pose very small gradient up to around 200°C. From there towards 500°C, nearly -95% of the 70 mg sample has already decomposed which were resembled by a huge gradient in the graph within 200°C to 500°C and during this stage, only 3.9 mg of the sample remains. Further temperature increase completely burn the sample without leaving any residue. Though it took up to 622.1°C for the material to be completely burnt out, it is still well below the expected temperature of 700°C.

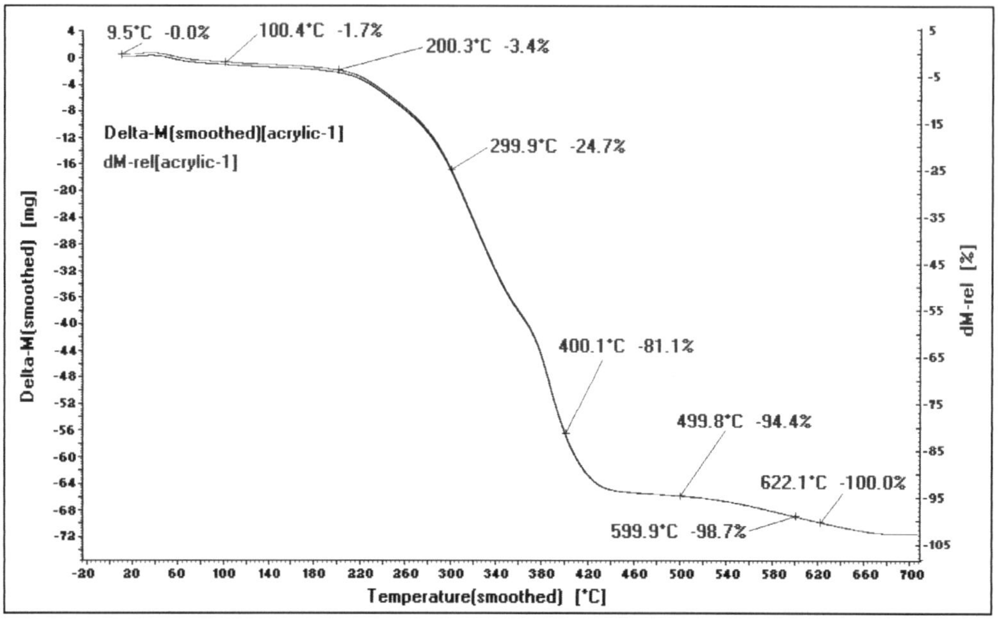

Figure 3. Thermogravimetric (TG) curves for Visijet® SR200 acrylate material.

There are many factors which influence a TGA curve. These factors may be due to instrumentation or nature of sample which could affects the shape, precision and accuracy of the experimental results, for example like the sample holder and the effect of the particle size [25]. Different TGA instrument may carry ranges of sample holder such as flat plates and deep crucibles of various capacities. The shape of the TGA curve may vary due to dissimilar heating conditions. Normally the thermocouple will be placed as near as possible without touching the sample, thus it leads to source of error due to the thermal lag and partly due to the finite time taken to cause detectable mass change. If the composition of the sample contains is such that it reacts either with the sample, or product formed or the evolved gases then this will cause error in recording the mass change of the sample.

Other than that, the particle size of the sample could also vary the result of TGA [21]. The smaller the particle size, the greater the extent of decomposition at any given temperature could occur, and the use of large crystal may result in apparent vary rapid mass loss during heating due to mechanical loss of part of the sample.

4. Conclusion

By using TGA, the burn out temperature to be used during IC for pattern made from Visijet® SR200 acrylate material could be determined. It also gives an idea how the material characteristic during the actual burn out process. It was found out from the TGA curves that the material would be able to decompose within the temperature of 700°C without leaving any residual ash which is exactly at 622.1°C. The material significantly decomposes at 387.4°C when the polymer chain breaks and the polymer thermal decompositions occur. As TGA could resemble the actual burn out process in IC, it shows the suitability of this material to be used as the pattern for IC. The material are able to be completely burnt out to make sure the quality of the casting will not be affected.

Acknowledgements

The authors would like to thank the Faculty of Mechanical & Manufacturing Engineering in Universiti Tun Hussein Onn Malaysia (UTHM) and the Faculty of Mechanical Engineering in Universiti Teknologi Malaysia (UTM) for their financial support and facilitating in this research programme.

References

[1] Dickens P M 1993 *Foundry International* **16** 6
[2] Cheah C M, Chua C K, Lee C W, Feng C and Totong K 2005 *International Journal Advance Manufacturing Technology* **25** 308-20
[3] Singh R 2011 *Inter J. Automot Mech Eng* **3** 293
[4] Singh R 2011 *Inter J. Automot Mech Eng* **4** 397
[5] Singh R and Singh R 2012 *Inter J Automot Mech Eng* **6** 680
[6] Singh R 2012 *Inter J Automot Mech Eng* **6** 775
[7] Marwah O M F, Sharif S and Ibrahim M 2012 *International Journal of Automation Technology.* **6** 570-5
[8] Chhabra M and Singh R 2011 *Rapid Prototyping Journal* **17** 328-50
[9] Hague R and Dickens P M 1995 *Proceedings of the Solid Freeform Fabrication Symposium (University of Texas at Austin, Austin)* ed H Marcus pp 242-52
[10] Wirtz H and Freyer C 2000 *Foundryman* **93** 63-5
[11] Wuis M, Schadlich S J, Wagner I and Sahm P R 2000 *Journal of Manufacturing.* **52** 45-7
[12] Zhao J F 2000 *Spec. Cast. Nonferr. Alloys* **5** 9-10
[13] Dimov S S, Pham D T, Lacan F, Dotchev K D 2001 *Assembly Automotive.* **21** 296-302
[14] Ferreira J C and Mateus A 2003 *Journal of Materials Processing Technology.* **134** 135-44
[15] Norouzi Y, Rahmati S and Hojjat Y 2009 *Rapid Prototyping Journal.* **15** 255-63
[16] Yang J, Shi Y, Shen Q and Yan C 2009 *Journal of Material processing Technology.* **209** 1901-8
[17] Matsumoto M, Shiomi M, Osakada K and Abe F 2002 *International Journal of Machine Tools and Manufacture* **42** 61-7
[18] Dickens P M, Stangroom R, Greul M, Holmer B, Hon K K B, Hovtun R, Neumann R, Noeken S and Wimpenny D 1995 *Rapid Prototyping Journal* **1** 4-11
[19] Rosochowski A and Matuszak A 2000 *Journal of Materials Processing Technology* **106** 191-8
[20] Yao W L and Leu M C 2000 *Material and Design* **21** 101-9
[21] Gouldsen C and Blake P 1998 Investment casting using FDM/ABS rapid prototype patterns http://www.stratasys.com, accessed on 3 February 2012
[22] Omar M F M, Sharif S, Ibrahim M, Hehsan H, Busari M N M and Hafsa M N 2012 *Advanced Materials Research* **463-464** 226-33
[23] Hafsa M N, Ibrahim M, Sharif S, Omar M F M and Zainol M A 2013 *Applied Mechanics and Materials.* **315** 587-91
[24] Rau S R, Vengadaesvaran B, Naziron N N and Arof A K 2013 *Pigment and Resin Technology.* **42** 111-6
[25] Sepe M P 1997 *RAPRA technology Ltd.* **8** 13-8
[26] Salmoria G V, Gonzalez V J, Ahrens C H, Soldi V and Pires A T N 2005 *Journal of Materials Processing Technology* **168** 164-71
[27] Wang Y, Liu F and Xue X 2013 *Progress in Organic Coating* **76** 863-9

IOP Conf. Series: Materials Science and Engineering **50** (2013) 012060 doi:10.1088/1757-899X/50/1/012060

Sustainable manufacturing: Effect of material selection and design on the environmental impact in the manufacturing process

Mohd Hazwan Syafiq Harun[1,a], **Zahari Taha**[2] **and Hadi Abdul Salaam**[1]
[1]Faculty of Mechanical Engineering,
Universiti Malaysia Pahang, 26600, Pekan, Pahang
[2]Faculty of Manufacturing Engineering,
Universiti Malaysia Pahang, 26600, Pekan, Pahang

E-mail: [a]hazwansyafiq@gmail.com

Abstract. The environmental impact of a manufacturing process is also dependent on the selection of the material and design of a product. This is because the manufacturing of a product is directly connected to the amount of carbon emitted in consuming the electrical energy for that manufacturing process. The difference in the general properties of materials such as strength, hardness and impact will have significant effect on the power consumption of the machine used to complete the product. In addition the environmental impact can also be reduced if the proposed designs use less material. In this study, an LCA tool called Eco-It is used. Evaluate the environmental impact caused by manufacturing simple jig. A simple jig with 4 parts was used as a case study. Two experiments were carried out. The first experiment was to study the environmental effects of different material, and the second experiment was to study the environmental impact of different design. The materials used for the jig are Aluminium and mild steel. The results showed a decrease in the rate of carbon emissions by 60% when Aluminium is use instead from mild steel, and a decrease of 26% when the-design is modified.

1. Introduction

The increasing world population has created a greater demand for production goods. The resulting increase in productivity has resulted in more pollution (Franci, 2009). From a manufacturing point of view, an increase in production is good because it reduces manufacturing cost, but in terms of environmental impact, the increase in energy consumption leads to increase in carbon dioxide emission. Energy generation as driven by consumption demand is a key contributor to carbon dioxide and climate change. Nicholas, 2006 said in the year 2000, energy related carbon dioxide equivalent (CO_e2) emissions represented about 65% of the global greenhouse gas emissions. Due to this problem, an increase in awareness by the community to reduce pollution on the environment and a more sustainable production has been proposed. Sustainable production is an approach to improving environmental performance in manufacturing production. Sustainable production means that products are designed, produced, distributed, used and disposed with minimal (or none) environmental and occupational health damages, and with minimal use of resources (materials and energy), (Leo, 1993).
This approach is primarily driven by rapidly increasing problems concerning environmental damages, huge amounts of waste, occupational health damages and increasing use of non-renewable resources.

These problems coupled with expected exponential growth in world consumption during the next 50 years where the world population will grow to about 12 billion people, requires prompt solutions. There are many methods to improve sustainability in manufacturing. Some of them as suggested by Pusavec in 2010 include:

- Reduce machining processes energy consumption,
- Minimize waste (generate less waste and increase waste re-usage or recycling),
- Use resources efficiently,
- Use recyclable materials or reuse machine-tool components,
- Improve the management of metalworking fluids, swarf, lubricating oils, and hydraulic oils (improved environmental, health, and safety performance,) and
- Adopt life cycle assessment methods.

In this paper the reduction of machining processes energy consumption is studied. A key element of sustainability is the prudent use of natural resources, such as energy. Energy generation as driven by consumption demand is a key contributor to carbon dioxide emissions and climate change. In the year 2000, energy related carbon dioxide equivalent (co_2e) emissions represented about 65% of the global greenhouse gas emissions, from this percentage, about 24% and 14% of co_2e emissions were attributed to power generation and industrial activity,(Nicholas, 2006). Hence reducing energy consumed by machining processes is one of the strategies to improve the performance sustainability by reducing the energy footprint, (Rajemi, 2010).

2. Life Cycle Assessment (LCA)

According to G.Martinopoulos in 2006, the simplest LCA tool that is available in market today is Eco-It. Eco-It or Eco-Indicator tool is a simple tool for Eco-design. It works with Eco-Indicators, these are single scores that express the seriousness of the environmental load of a process or material. The higher the score, the more serious is the impact. Eco-It allows a designer to model a complex product and most of its life cycle in a few minutes. It calculates the environmental load, and shows which parts of the product's life cycle contribute most. With this information user can target own creativity to improve the environmental performance of product. The program structure is simple. There is a Main window with four pages (figure 2):

- Life cycle page, which is allows the user to describe the product life cycle under investigation.
- Production page, in which the user needs to enter the hierarchical structure of the product, and specify the materials and production processes per part.
- Use page, where user needs to enter the energy and transport components that use during jigs manufacture process.
- Disposal page, where the user needs to specify the waste scenario for the product or for different parts and materials.

The working philosophy of Eco-It is based on Eco-Indicator 95 and 99, which is are earlier Eco design tools before Eco-It. The first step in any further interpretation consists of comparing the scores with another value. In LCA terminology this is called the normalisation step, and the values are normalised to the average European. A weighting step is therefore necessary to achieve an overall result. After detailed analysis of the options the so-called Distance-to-Target principle is chosen for determining the weight factors. The evaluation method used for calculating the Eco-Indicator99 is in "three steps". These three steps are

- Inventory of all relevant emissions, resource extractions and land-use in all processes that form the life cycle of a product. This is a standard procedure in Life Cycle Assessment (LCA)
- Calculation of the damages these flows cause to Human Health, Ecosystem Quality and Resources
- Weighting of these three damage categories

The above steps are illustrated in figure 1.

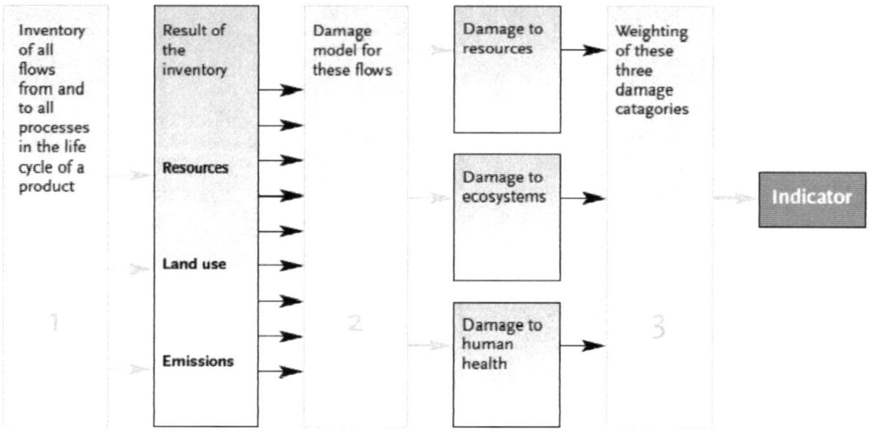

Figure 1. General procedure for the calculation of Eco-indicators. The light coloured boxes refer to procedures, the dark coloured boxes refer to intermediate results.

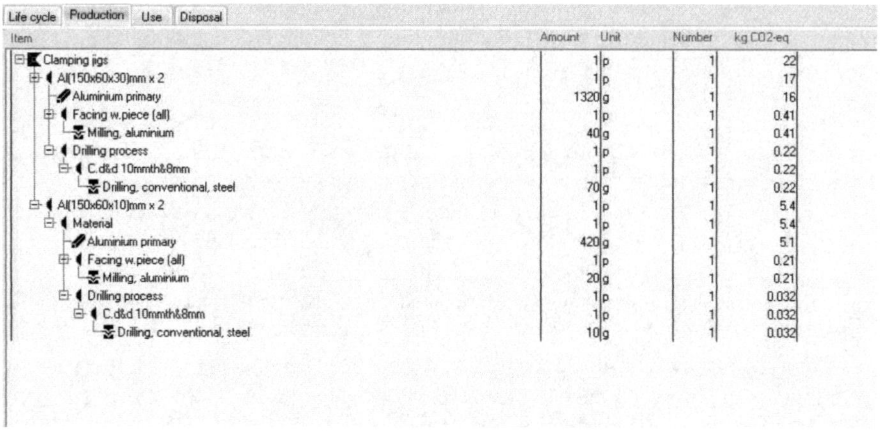

Item	Amount	Unit	Number	kg CO2-eq
Clamping jigs	1	p	1	22
Al(150x60x30)mm x 2	1	p	1	17
Aluminium primary	1320	g	1	16
Facing w.piece (all)	1	p	1	0.41
Milling, aluminium	40	g	1	0.41
Drilling process	1	p	1	0.22
C.d&d 10mmth&8mm	1	p	1	0.22
Drilling, conventional, steel	70	g	1	0.22
Al(150x60x10)mm x 2	1	p	1	5.4
Material	1	p	1	5.4
Aluminium primary	420	g	1	5.1
Facing w.piece (all)	1	p	1	0.21
Milling, aluminium	20	g	1	0.21
Drilling process	1	p	1	0.032
C.d&d 10mmth&8mm	1	p	1	0.032
Drilling, conventional, steel	10	g	1	0.032

Figure 2. Eco-It interface.

In this study, the jig data were taken from the experiment such as data of power consumption of the machine and is analysed by Eco-It in order to get the rate of carbon emissions.

3. Case study of a jig

In order to study the environmental impact in production tooling, the manufacturing of a simple jig is used as a case study. Two experiments were conducted which are different material and different design of the jig. The two different materials used are mild steel and aluminium. The main objective is the calculation of the environmental impact due to different materials and designs. The jig is made up of four parts, two parts form the body holder and other two parts are for clamping. In the first design, five holes are drilled in the sides of the body holder surface. Two holes on the left and right of the body holder are through holes to tie the parts together. The other three holes are for locator pins. The dimension of the holder body is 150 mm x 60 mm, with a thickness of 30 mm. While for the clamping parts is 150mm x 60 mm with a thickness of 10mm. While for the second design for the body holder is 150 mm x 60 mm, with a thickness of 20 mm. While for the clamping parts is 150mm x 60 mm with a thickness of 10mm. This design uses two pin locating holes and two through holes.

The jig manufacturing processes starts with facing the surface of the material by using a milling machine. The inserts use for both materials is Carbide coated with a cutting speed use of 800 m/min depth of cut of 0.20mm and feed rate of 0.20mm/tooth. After surface facing, the process continues with drilling. The process starts with setting the coordinate edge finder and the speed of the edge finder at 200rpm. Next, a center drill is used to drill five holes for design 1 and four holes for design 2 at a speed 1000rpm. The drill diameter is 10mm for the through holes and 8mm with 10mm depth for the 3 holes of design 1 and 2 holes for design 2. The speed use for drilling is 800rpm. The energy used by the milling machine was recorded by a three phase power analyzer. Data obtained from the experiments is included in Eco-It for the calculation of the rate of carbon emissions. Figure 3 below shows design 2 of the simple jig.

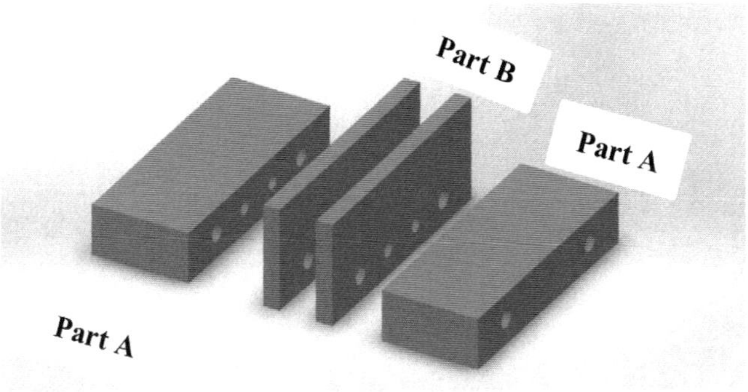

Figure 3. Design of simple jig.

4. Result and Discussion

This section presents the results of the experiments. Table 2 and table 3 below shows the total power consumption for each process and the total weight of material and weight of material that have been removed for both experiments. The data are entered into Eco-It to generate the amount of carbon emissions produced for each material.

Table 2. Total power consumption for each process used for both experiment.

Power consumption for each process.	Design 1		Design 2	
Processes	Mild steel	Aluminium	Mild Steel	Aluminium
Tool Setup and facing for part A plate	0.267 kWh	0.379 kWh	0.263 kWh	0.375 kWh
Total of drilling operation part A plate	0.530 kWh	0.424 kWh	0.495 kWh	0.399 kWh
Facing part B plate	0.228 kWh	0.296 kWh	0.228 kWh	0.296 kWh
Total of drilling operation part B plate	0.302 kWh	0.104 kWh	0.277 kWh	0.089 kWh

Table 3. Total weight of material and material removed after specific process

	Design 1		Design 2	
Weight	Mild steel	Aluminium	Mild Steel	Aluminium
Part A plate	3850 gram	1320 gram	2360 gram	840 gram
Part B plate	1180 gram	420 gram	1180 gram	420 gram

Facing part A plate	125 gram	40 gram	121 gram	40 gram
Facing part B plate	80 gram	20 gram	80 gram	20 gram
Drilling operation for part A plate	190 gram	70 gram	160 gram	60 gram
Drilling operation for part B plate	40 gram	10 gram	30 gram	7 gram

For design 1, the results indicate that mild steel has a high rate of carbon emissions by 60% compared to aluminium. (figure 4 (a) and (b)).

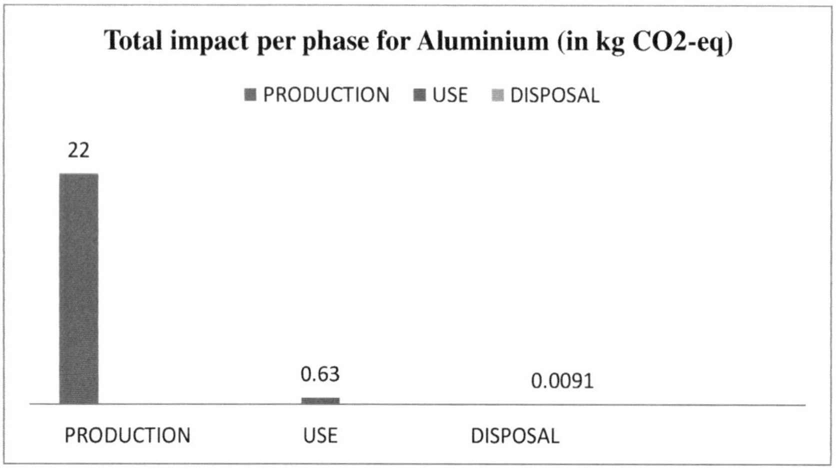

Figure 4(a). Total impact per phase for Aluminium of design 1

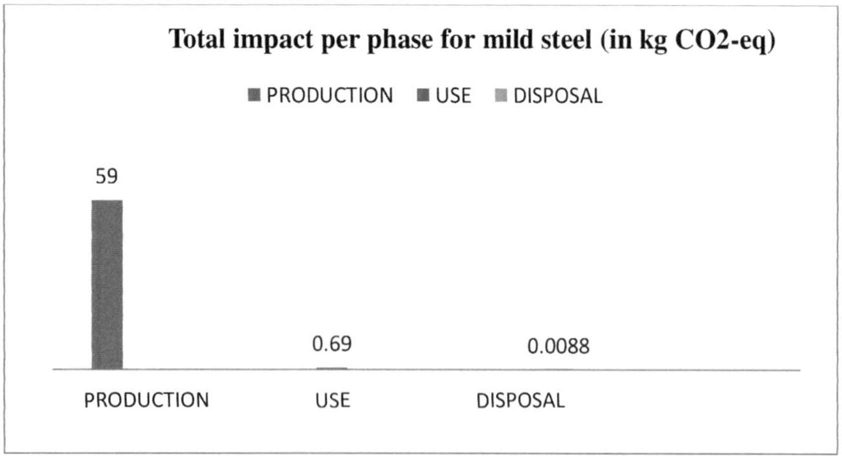

Figure 4(b). Total impact per phase for Mild Steel of design 2.

With design 2 there is a decrease of 26% of the carbon emission rate. For both materials as shown in Table 4

Table 4. Summary of Comparison

	DESIGN 1		DESIGN 2	
Material/ Phase	Mild Steel (in kg CO2-eq)	Aluminium (in kg CO2-eq)	Mild steel (in kg C02-eq)	Aluminium (in kg CO2-eq)
Life Cycle	60	23	43	17
Production	59	22	42	16
Use	0.69	0.63	0.66	0.61
Disposal	0.0088	0.0091	0.0062	0.0066

The life cycle and production carbon emission is higher for mild steel than Aluminium because of the higher carbon content in mild steel. Mild steel contains 0.16% to0.29% carbon. There is not much difference in term of electrical consumption Mild steel and Aluminum jigs the material hardness. The small changes in the design, in this case reducing 5 holes to four and reducing the thickness of the material of the holder body from 30 mm to 20 mm, have resulted in a reduction of the carbon emission by 28.33% for mild steel and 26.09% for aluminium. This clearly shows that design has a significant impact on the carbon emissions.

5. Conclusion

It can be concluded that the use of the appropriate material and design for a product has a profound impact on carbon emissions. Future works includes the use of metal heuristics approaches such as genetic algorithm to optimize the material and design of a product.

Acknowledgement

The authors would like to thank Universiti Malaysia Pahang and the Ministry of Higher Education, Malaysia for providing the funds for this research under the grant RDU121223.

References

[1] Cristiano A da Silva, Paulo M C Ferrao, Manuel Freitas 2008 Using Life Cycle Assessments on environmental management projects: a case study of a Brazilian vehicle development.
[2] ECO-it 1.3 manual, Pre Consultants, the Netherlands, www.pre.nl
[3] Franci Pusavec, Peter Krajnik and Janez kopac 2009 Transitioning to sustainable production-Part 1: application on machining technologies *Cleaner Production*
[4] Franci Pusavec,Peter Krajnik and Janez kopac 2009 Transitioning to sustainable production-Part 11: evaluation of sustainable machining technologies *Cleaner Production*
[5] Flore Vallet, Benoit Eynard and Stephanie Glatard 2012 Using eco-design tools: An overview of experts' practices *Journal of Manufacturing Technology*
[6] G Martinopoulos, G Tsilingiridis and N Kyriakis 2007 Three Eco-Tool comparisons with the example of the environmental performance of domestic Solar Flat Plat hot water systems.
[7] J Jeswiet, and S Kara 2008 Carbon emissions and CES™ in manufacturing *CIRP annals-Manufacturing Technology*
[8] Kalpakjian S, Schimid S 2010 Manufacturing Engineering Technology 6th edition *Pearson Prentice Hall* Upper Saddle River, NJ.
[9] Leo Alting and Jorgen Jorgensen 1993 The life cycle as a basis for sustainable industrial production
[10] M F Rajemi, P T Mativenga and A Aramcharoen 2010 Sustainable machining: selection of optimum turning conditions based on minimum energy considerations *Journal of*

cleaner Production

[12] M F Rajemi, P T Mativenga and A Aramcharoen 2010 Sustainable machining: selection of optimum turning conditions based on minimum energy considerations *Journal of cleaner Production*

[13] Nicholas S 2006 Stern Review: The Economic of Climate Change. Office of Climate Change UK

[14] P T Mativenga and M.F Rajemi 2011 Calculation of optimum parameters based on minimum energy footprint *Journal of Manufacturing Technology*

[14] VanLoon G, Duffy SJ 200 Environmental Chemistry: A Global Perspective Gary vanLoon and Steven J. Duffy *Oxford University Press*, oxford, UK

[15] Wenzel H, Hauschild M, Alting L Environmental Assesment of Products, Methodology Tolls and Case Studies in Product Development, Institute for Product Development *Chapman & Hall* vol 1: pp. 244-47

Finite element analysis of filament-wound composite pressure vessel under internal pressure

S Sulaiman[a], S Borazjani and S H Tang

Department of Mechanical and Manufacturing Engineering
Faculty of Engineering, Universiti Putra Malaysia
43400 UPM , Serdang , Selangor, Malaysia
Phone: +603-89466336, Fax: +603-86567122

Email: [a]suddin@eng.upm.edu.my

Abstract. In this study, finite element analysis (FEA) of composite overwrapped pressure vessel (COPV), using commercial software ABAQUS 6.12 was performed. The study deals with the simulation of aluminum pressure vessel overwrapping by Carbon/Epoxy fiber reinforced polymer (CFRP). Finite element method (FEM) was utilized to investigate the effects of winding angle on filament-wound pressure vessel. Burst pressure, maximum shell displacement and the optimum winding angle of the composite vessel under pure internal pressure were determined. The Laminae were oriented asymmetrically for $[0^0,0^0]_s$,$[15^0,-15^0]_s$,$[30^0,-30^0]_s$,$[45^0,-45^0]_s$,$[55^0,-55^0]_s$, $[60^0,-60^0]_s$,$[75^0,-75^0]_s$,$[90^0,-90^0]_s$ orientations. An exact elastic solution along with the Tsai-Wu, Tsai-Hill and maximum stress failure criteria were employed for analyzing data. Investigations exposed that the optimum winding angle happens at 55^0 winding angle. Results were compared with the experimental ones and there was a good agreement between them.

1. Introduction

The wide range of pressure vessel applications turned it into one of the most important equipment of industry. Today composite pressure vessels known as new generation vessels have been widely used in many industrial zones. The superior characteristics of a composite vessel such as light weight, high stiffness, corrosion resistance and long lifetime make it a perfect replacement for metallic vessels [1-2]. Although composite vessels were first used for military cases and aerospace functions, nowadays they have many civilian applications including oxygen and hydrogen gas storage cylinders, scuba and fire-fighter tanks [3-7].

Filament winding is the usual considered process for fabricating composite structures. It is the process in which continuous filaments of fiber are wound on a supporting mandrel. The mandrel rotates with the spinning wheel on a horizontal axis then carriage begins to move linearly so fibers are laid down in the predetermined path. The most conventional uses of filament winding process are in high-pressure storage tanks, rocket motor cases, launch tubes, and for commercial applications, such as golf club shafts [8].

A thin-walled (the ratio of outside to the inside diameter of the vessel less than 1.1) composite vessel has been studied in this present work. The composite vessel is made of inside aluminum liner overwrapping with Carbon fiber/Epoxy composite. This type of vessel is placed in the third type of composite vessel usually used has higher corrosion resistance, more safety and longer lifetime [9].

Barboza Neto et al. [10] investigated the behaviour of LLDPE and HDPE composite pressure vessel under burst testing. The experimental and finite element analyses were applied to determine the proper composite thickness based on the value of burst pressure. Studies estimated the required vessel thickness and proper stacking sequences according to the burst pressure. Liu et al. [11] focused on predicting the burst pressure and lifetime of composite pressure vessels. Damage modeling and progressive failure analysis were utilized to determine the strength reliability, stiffness degradation and failure properties of composite vessel. Sayman et al. [12] did the experimental study at GRP composite cylinders to survey the effects of temperature and winding angle on strength of composite vessels. Studies demonstrated the optimum stacking sequence and also proved that the strength of composite vessels decreases with increasing temperature. Gohari et al. [13] compared finite element analysis with theoretical studies to ascertain that static internal pressure and fiber angle orientation have the direct effect on stress distribution of composite vessel. Tabakov and Summers [14] introduced a 3-D elasticity solution to study the behaviour of multi-layered cylinder subjected to asymmetric internal and external pressure along with the axial force. The analyses demonstrated that for thin-wall and thick-wall single-layered cylinders the optimum winding angle is in the range of 54^0-57^0. Xu et al. [15] utilized finite element analysis to predict burst pressure of the composite storage vessel. Tsai-Wu failure theory was applied to predict the damage evolution and failure strength in composite vessel subjected under internal pressure. Bhavya et al. [16] investigated the effects of diameter-to-thickness ratio and surveyed the variation of failure pressure with respect to fiber angle using finite element software. Marzbanrad et al. [17] employed "unit load method" to predict the burst pressure and fatigue lifetime of composite vessel. The simulation results obtained from ABAQUS were in good agreement with the experimental ones. In this present research attempts have been made to:

- Determine the burst pressure and the maximum shell displacement for asymmetric fiber orientations of composite vessel.
- Determine optimum winding angle according to diverse failure criteria.
- Compare simulation results with the experimental ones.

2. Theoretical Studies

2.1. Winding Angle

Winding angle has significant effects on the structure of filament-wound vessels; so determining the appropriate angle for each part of the vessel is an important issue. The winding angle is defined for the two types of geodesic and non-geodesic winding based on the need for friction between fibers and the shell as in Eq. (1). The second part of the following relation relates to the non-geodesic winding method [11, 18-21]:

$$\alpha(R)=sin^{-1}\left(\frac{R_0}{R}\right) \mp \delta\left(\frac{R-R_0}{R_{tl}-R_0}\right)^n \tag{1}$$

Here, R is the radial distance from the center line to a point in the layer, R_0 is the radius of the polar axis, and R_{tl} is the radius at the dome-cylinder tangent line. A geodesic winding pattern, as in this research is obtained by choosing $\delta = 0$.

2.2. Optimum Winding Angle

Netting analysis is one of the most popular analytical techniques used for investigating the behavior of multi-layered composite materials, especially for filament-wound pressure vessels. The main assumption in netting analysis is that all loads are carried by the fibers neglecting the stiffness of the matrix and internal pressure subjected to the vessel produces a hoop-to-axial-stress ratio of 2:1 [22].

Figure 1 indicates the thin-wall cylindrical vessel with the thickness of "t" and length of "L" which is subjected into the internal pressure of "P". In order to calculate the axial (longitudinal) and hoop stresses at the cylinder with the radius of "R", Eq. (2) can be utilized [23-25]:

$$N_{\emptyset} = \frac{PR}{2} \quad , \quad N_{\theta} = PR \tag{2}$$

Axial and hoop stresses can be calculated as in Eq. (3):

$$\sigma_{axial} = \frac{N_{\emptyset}}{t} = \frac{PR}{2t} \quad , \quad \sigma_{hoop} = \frac{N_{\theta}}{t} = \frac{PR}{t} \tag{3}$$

Axial and hoop stresses can be calculated according to equilibrium across the cut section too as in Eqs. (4) and (5):

$$PL\,(2R) = 2\sigma_{hoop}Lt \;\Longrightarrow\; \sigma_{hoop} = \frac{PR}{t} \tag{4}$$

$$P\pi r^2 = \sigma_{axial}(2\pi Rt) \;\Longrightarrow\; \sigma_{axial} = \frac{PR}{2t} \tag{5}$$

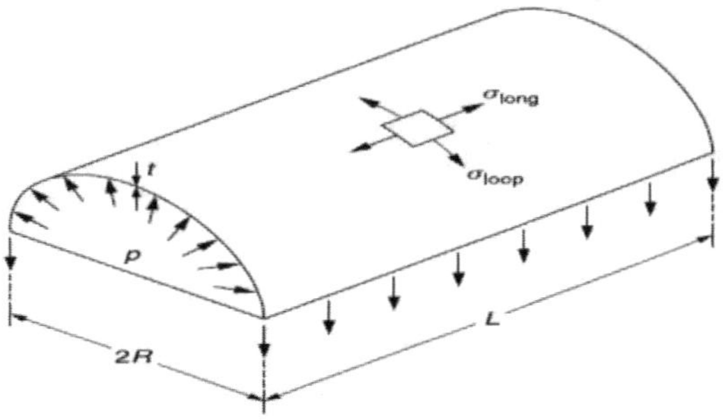

Figure 1. Thin cylindrical shell [23].

Optimum winding angle can be estimated by using ultimate tensile strength along with applying netting analysis respect to 2:1 hoop-to-axial stress ratio as it is shown at Eqs. (6) and (7)., as it is shown in figure 2. The body diagram of the cylinder consists of axial and hoop forces wrapped with the fibers at the angle of "α" is indicated at figure 2.

$$N_{\theta} = \sigma_u t \sin^2\alpha \quad , \quad N_{\emptyset} = \sigma_u t \cos^2\alpha \tag{6}$$

$$\frac{N_{\theta}}{N_{\emptyset}} = \tan^2\alpha = 2 \;\Longrightarrow\; \alpha = \arctan(\sqrt{2}) = 54.7^0 \tag{7}$$

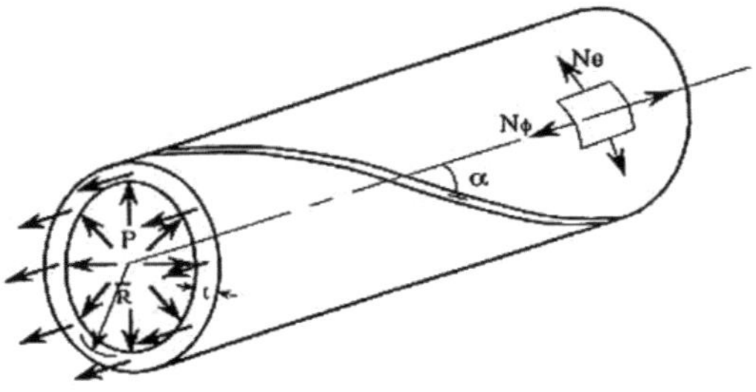

Figure 2. Body diagram of axial and hoop forces and internal pressure [26].

2.3. Failure Theories

Precise analyzing the strength of composite layers requires employing failure criteria. Different types of failure analyses can be used for evaluating the strength of composites laminae.

2.3.1. Tsai-wu failure criterion

Satisfying the Eq. (8) is essential for Tsai-Wu theory to predict failure in an orthotropic lamina under plane stress condition [27]:

$$F_{11}\sigma_1^2 + F_{22}\sigma_2^2 + F_{66}\tau_{12}^2 + F_1\sigma_1 + F_2\sigma_2 + 2F_{12}\sigma_1\sigma_2 \geq 1 \quad (8)$$

Elastic characteristics consist of four independent elastic constants (E_{11},E_{22},G_{12},υ_{12}). Strength properties are divided into five independent strength properties:

X_t = longitudinal tensile strength, Y_t = transverse tensile strength
X_c = longitudinal compressive strength, Y_c = transverse compressive strength
S = in-plane shear strength

F_1, F_2 are the strength coefficients and are given by, it should be considered that F_1, F_2, F_{11}, F_{22}, F_{12}, F_{66} can be calculated by using the tensile, compressive and shear strength properties in the principal material directions at the Eq. (9):

$$F_1 = \frac{1}{X_t} - \frac{1}{X_c}, \quad F_2 = \frac{1}{Y_t} - \frac{1}{Y_c}, \quad F_{11} = \frac{1}{X_t X_c},$$

$$F_{22} = \frac{1}{Y_t Y_c}, \quad F_{66} = \frac{1}{S^2}, \quad F_{12} = -\frac{1}{2}\sqrt{F_{11}F_{22}} \quad (9)$$

2.3.2. Tsai-hill failure criterion

Tsai-Hill failure criterion can be explained at Eq. (10) [28]:

$$F_{11}\sigma_1^2 + F_{22}\sigma_2^2 + F_{66}\tau_{12}^2 + 2F_{12}\sigma_1\sigma_2 \geq 1 \quad (10)$$

The strength parameters F_{11}, F_{22}, F_{66} and F_{12} are given by at Eq. (11):

$$F_{11} = \frac{1}{X^2}, \quad F_{22} = \frac{1}{Y^2}, \quad F_{66} = \frac{1}{S^2}, \quad F_{12} = -\frac{1}{2}\left(\frac{1}{X^2} + \frac{1}{Y^2}\right) \quad (11)$$

2.3.3. Maximum stress failure criterion

This theory, Eq. (12), expresses that the failure occurrences when at least one of the following criteria is satisfied [15]:

$$\left|\frac{\sigma_1}{X}\right| \geq 1, \left|\frac{\sigma_2}{Y}\right| \geq 1, \left|\frac{\tau_{12}}{S}\right| \geq 1 \tag{12}$$

3. Simulation Details

The considered pressure vessel is the vertical one designed for low capacity applications. The vessel is sketched with 1200 mm length, 300 mm diameter at the center point and the wall thickness of (0.3-4) mm based on the geometry .The material of this cylindrical vessel consist of the aluminum alloy inside layer reinforced with six layers of Carbon/Epoxy T300/LY5052 fiber. The wall thickness of vessel in center section of the vessel is 0.3 mm and each lamina thickness is 0.762. The inner radius of the vessel is 150 mm, considering six composite layers (4.572 mm); the outer radius of the vessel becomes 154.872 mm. Tables 1 and 2 indicate material properties of aluminum alloy [29] and CFRP [3] respectively.

Table 1. Mechanical properties of Al 6061 and CFRP.

	Density(kg/m^3)	E_1 (GPa)	E_2 (GPa)	υ_{12}	G_{12} (GPa)	τ (MPa)
AL 6061	2750*	70	70	0.3	27	600
CFRP	1570*	135	8	0.27	3.8	-

*.Yingjun et al., 2010

Table 2. Strength parameters for CFRP composite.

X_t (MPa)	X_c (MPa)	Y_t (MPa)	Y_c (MPa)	S (MPa)
1860	1470	76	85	98

In this research finite element analysis software has been used to collect the proper data. In order to analyze the effects of variables on filament wound structures "ABAQUS" software is utilized for modeling the vessel. In order to reduce the time of calculation, the finite element model has been made for half the composite pressure vessel as it is indicated at the figure 3.The mesh is constant for the whole part and because the model has been made due to the revolving process, the "Quad-dominated" meshing type has been selected. For obtaining precise results meshing with global size of 0.01 has been used. This mesh consists of 4290 nodes and 7872 elements.

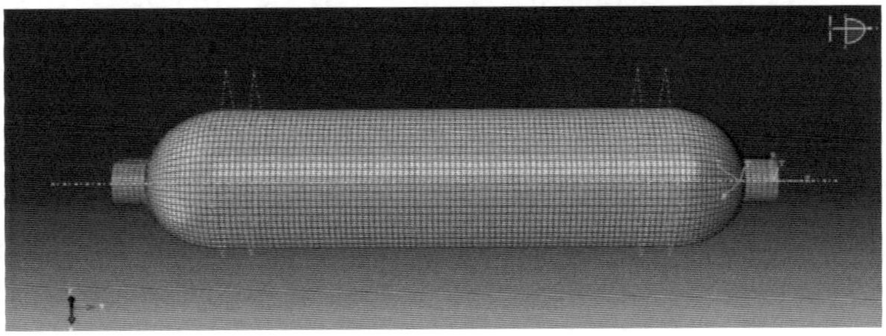

Figure 3. The finite element model of composite vessel.

Internal pressure load has been applied for subjecting to model to estimate the burst pressure and maximum shell displacement in two cases; first non-constant pressure to find the burst pressure and second constant pressure to find the shell displacement. In order to consider the effects of winding angle, the data should be evaluated based on failure theories. According to what was mentioned before Tsai-Wu, Tsai-Hill and maximum stress failure theories have been utilized for analysing data. In composite pressure vessels the burst pressure is determined by investigating the first-ply failure strength. Although the ultimate burst strength is usually higher than the first-ply failure strength, determining the first-ply failure of composite layer is safe enough to design the vessel [30].

According to the mentioned failure theories the method for obtaining the outputs are different. In order to analyse the model based on Tsai-Wu and Tsai-Hill failure criteria, after defining the Winding Angle it is essential to select the appropriate amount of internal pressure. This means that the quantity of internal pressure should be set to reach the proper failure coefficient (slightly higher than one) and then the outputs can be calculated. Subsequently the internal pressure can be submitted as burst pressure which is one of the main outputs of research. Maximum stress failure theory follows the philosophy that the maximum stress of composite first-ply should be always less than the ultimate stress ($\sigma_{max} \leq \sigma_u$) [3,31].

In this present work, asymmetric fiber orientation,[θ /- θ / θ /- θ / θ /- θ], has been presented for $[0^0,0^0]_s$,$[15^0,-15^0]_s$,$[30^0,-30^0]_s$,$[45^0,-45^0]_s$,$[55^0,-55^0]_s$, $[60^0,-60^0]_s$,$[75^0,-75^0]_s$,$[90^0,-90^0]_s$. Figure 4 shows the stacking sequences for 55^0 winding angle:

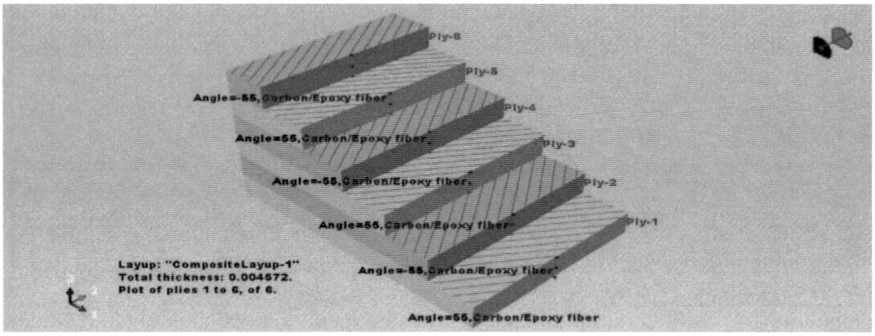

Figure 4. The stacking sequences for 55^0 winding angle.

4. Results And Discussion

The following results are extracted from analysing aluminum alloy which is strengthened by six layers of Carbon/Epoxy fiber with constant thickness of 0.762 mm for each layer. Figure 5 shows that in asymmetric fiber orientation the graph is not following a steady manner. According to failure criteria, the burst pressure displays increasing trend for 0^0 to 55^0 angle degrees, after 55^0 angle degrees the graph follows the decreasing trend up to reach 90^0 angle degrees. As it is clear at figure 5, the maximum burst pressure which vessel can withstand happens in 55^0 degrees (The optimum winding angle) based on failure theories. The maximum burst pressures have been obtained 15.5 MPa, 15.6 MPa and 16 MPa for Tsai-Wu, Tsai-Hill and maximum stress theories respectively. This angle has a good agreement with the winding angle calculated with netting analysis for composite laminates (54.7^0 degrees) [22].

Internal pressure is a combined load which produces axial and hoop stresses in materials. In order to find the optimum angle which withstands the maximum burst pressure, it is essential to search for the fiber angle which can tolerate hoop and axial stresses under the internal pressure. The burst pressure increases with increasing winding angle up to 55^0 degrees because laminates show more resistance to hoop stress than the axial stress. As the curve reaches to its maximum in 55^0 degrees, the amount of burst pressure decreases due to fiber higher resistance to axial stress and its lower resistance to hoop stress [32].

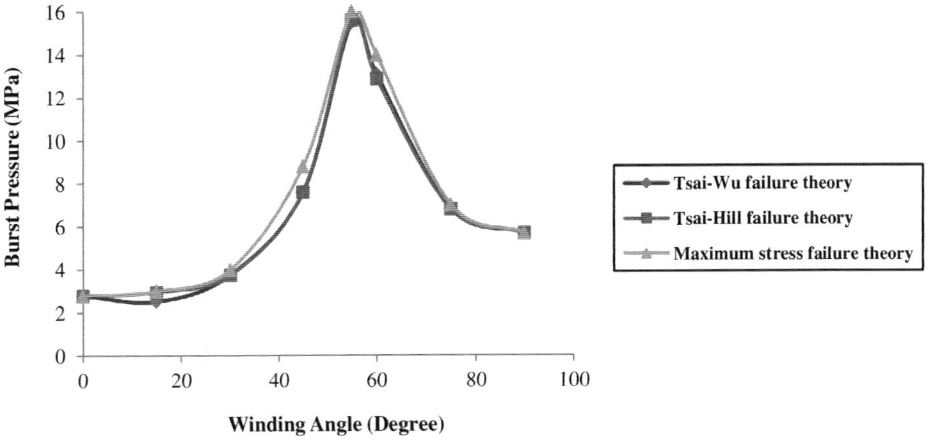

Figure 5. Variation of burst pressure with increasing winding angle.

As it is obvious at figure 5 Tsai-Wu and Tsai-Hill failure theories present more conservative results than maximum stress failure criteria but the difference can be negligible. Figure 6 exposes the maximum normal stress and the maximum stress failure failure coefficient for the 55^0 fiber angle orientation. The results based on failure theories are in good correlation with the experimental ones [4,12, 32].

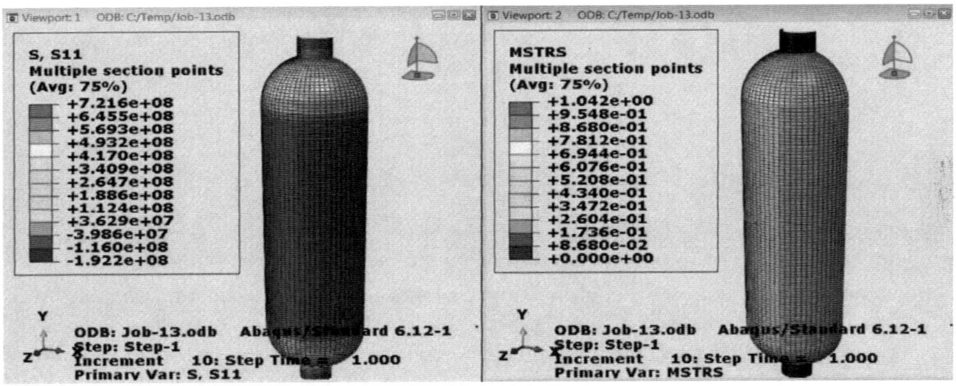

Figure 6. The maximum stress and failure coefficient at 55^0 winding angle.

At the second case with applying the constant pressure in model, the 15 MPa internal pressure has been subjected in vessel. As it is shown at figure 7 the minimum shell displacement belongs to 55^0 winding angle which proves the claim that 55^0 is the optimum angle and this is in good correlation with the experimental results [34-35].The FE model for the vessel shell displacement at 55^0 is presented at figure 8.

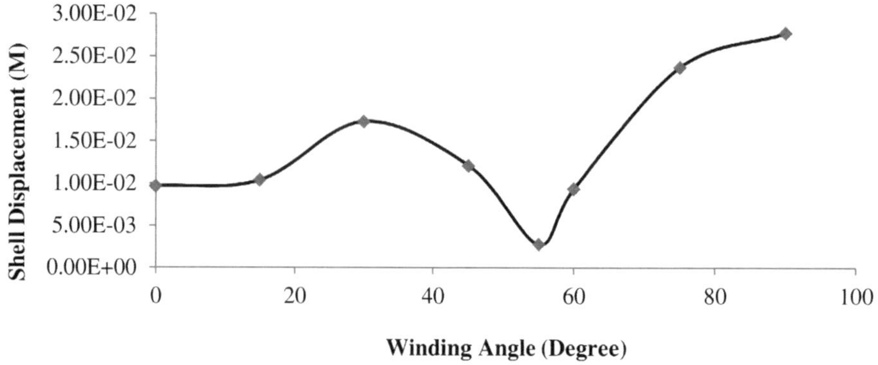

Figure 7. Variation of shell displacement with increasing winding angle

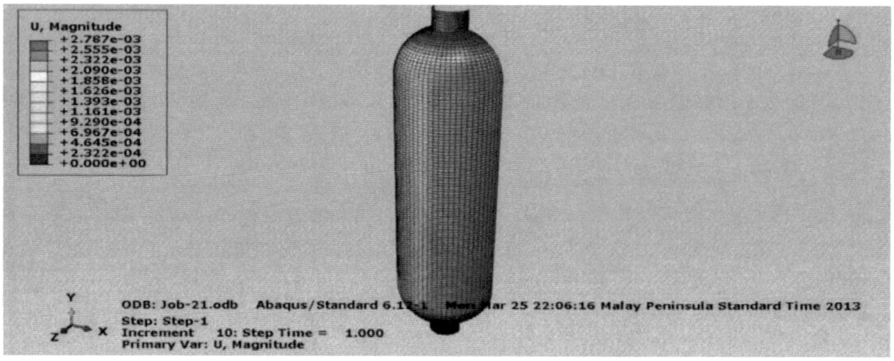

Figure 8. The shell displacement at 55^0 angle.

5. Conclusion

In this work, Finite element analysis was employed for investigating the structural behavior of pressure vessels. Aluminum alloy was utilized as the inside layer covered with Carbon/Epoxy fiber which was roving at different winding angles. FEA employed failure criteria such as Tsai-Wu, Tsai-Hill and maximum stress to predict the burst pressure, maximum shell displacement and determine the optimum winding angle.

Results and discussions were resulted in the following findings:

- Based on Tsai-Wu and Tsai-Hill failure criteria, 55^0 winding angle was approved as the optimum winding angle due to its maximum burst pressure and minimum shell displacement. This optimum angle was in good correlation with the experimental results trend and the netting analysis (54.7^0).
- Determining burst pressure using maximum stress failure theory was lead to less conservative results and higher burst pressure due to estimating the burst pressure based on ultimate strength of Carbon/Epoxy fiber.

Acknowledgements
The authors would like to thank the Faculty of Mechanical and Manufacturing Engineering in Universiti Putra Malaysia (UPM) for financial support and encouraging researchers to publish this paper.

References

[1] Tsai S W 1987 *Composite Design, Think Composites* USA
[2] Babu M S, Srikanth G & Biswas S 2007 Composite Fabrication by Filament Winding - An Insight. http://www.tifac.org.in/news/pub.htm
[3] Sinha M & Pandit S 2012 Maximum Stress and Burst Pressure Analysis of CFRP Composite Pressure Vessel *International Journal of Emerging trends in Engineering and Development* **4(2)** 714-21.
[4] Onder A, Sayman O, Dogan T & Tarakcioglu N 2009 *Composite structures* **89(1)** 159.
[5] Adebisi A A, Maleque M A and Rahman M M 2011 *Inter. J. Automot. Mech. Eng.* **4** 471.
[6] Jeffrey K J T, Tarlochan F and Rahman M M 2011 *Inter. J. Automot. Mech. Eng.* **4** 504.
[7] Bachtiar D, Sapuan S M and Hamdan M M 2010 *Inter. J. Automot. Mech. Eng.* **1** 79.
[8] Shen F C 1995 A filament-wound structure technology overview *Materials Chemistry and Physics* **42(2)** 96-100.
[9] Legault M 2012 Next-generation Pressure Vessels *HIGH-PERFORMANCE Composites* **20(4)** 40-45.
[10] Barboza Neto E, Chludzinski M, Roese P, Fonseca J, Amico S, & Ferreira C 2011 Experimental and numerical analysis of a LLDPE/HDPE liner for a composite pressure vessel *Polymer Testing* **30(6)** 693-00.
[11] Liu P, Chu J, Hou S, Xu P, & Zheng J 2012 Numerical simulation and optimal design for composite high-pressure hydrogen storage vessel: A review *Renewable and Sustainable Energy Reviews* **16(4)** 1817-27.
[12] Sayman O, Deniz M E, Dogan T, & Yaylagan E 2011 Failure pressures of composite cylinders with a plastic liner *Journal of Reinforced Plastics and Composites* **30(10)** 882-88.
[13] Gohari S, Golshan A, & Ayob A 2011 Theoretical analysis and finite element simulation of behavior of laminated hemispherical GRP dome subjected to internal pressure *International Conference on Computer and Software Modeling* pp. 111-117.
[14] Tabakov P Y & Summers E 2006 Lay-up optimization of multilayered anisotropic cylinders based on a 3-D elasticity solution *Computers & Structures* **84(5)** 374-84.
[15] Xu P, Zheng J & Liu P 2009 Finite element analysis of burst pressure of composite hydrogen storage vessels *Materials & Design* **30(7)** 2295-01.
[16] Bhavya S, Kumar P R & Kalam S A 2012 Failure Analysis of a Composite Cylinder *Journal of Mechanical and Civil Engineering* **3(3)** 01-07.
[17] Marzbanrad J, Paykani A, Afkar A. & Ghajar M 2013 Finite element analysis of composite high-pressure hydrogen storage vessels *J. Mater. Environ. Sci.* **4(1)** 63-74.
[18] ABAQUS/Standard User's Manual, Version 6.12 ABAQUS, Inc. Providence, RI.
[19] Zulkifli A, Ariffin A K and Rahman M M 2011 *Inter. J. Automot. Mech. Eng.* **3** 256.
[20] Kamal M, Rahman M M and Rahman A G A 2013 *Inter. J. Automot. Mech. Eng.* **7** 912.
[21] Kamal M, Rahman M M and Rahman A G A 2012 *J. Mech. Eng. Sci.* **3** 291.
[22] Evans J T, Gibson A G, Evans J & Gibson A 2002 Composite angle ply laminates and netting analysis *Proceedings of the Royal Society of London.Series A: Mathematical, Physical and Engineering Sciences* **458(2028)** 3079-88.
[23] Onder A 2007 First failure pressure of composite pressure vessels Master of Science Thesis, Graduate School of Natural and Applied Sciences of Dokuz Eylül University.
[24] Umar A H, Zainudin, E S and Sapuan S M 2012 *J. Mech. Eng. Sci.* **2** 198.
[25] Bhaskar H B and Sharief A 2012 *J. Mech. Eng. Sci.* **3** 281.

[26] Kabir M Z 2000 Finite element analysis of composite pressure vessels with a load sharing metallic liner *Composite Structures* **49(3)** 247-55.

[27] Tsai S W & Wu E M 1971 A general theory of strength for anisotropic materials *Journal of Composite Materials* **5(1)** 58-80.

[28] Azzi V & Tsai S 1965 Anisotropic strength of composites. Experimental Mechanics **5(9)** 283-88.

[29] Liu P, & Zheng J 2008 Progressive failure analysis of carbon fiber/epoxy composite laminates using continuum damage mechanics *Materials Science and Engineering: A* **485(1)** 711-17.

[30] Kam T, Liu Y & Lee F 1997 First-ply failure strength of laminated composite pressure vessels *Composite structures* **38(1)** 65-70.

[31] Yingjun W, Zixiong Z, Minqing S & Sirong Z 2010 Finite element modeling of carbon fiber reinforced polymer pressure vessel *International Conference on Educational and Network Technology (ICENT)* pp. 259-262.

[32] Balya B 2004 Design and Analysis of Filament Wound Composite Tubes Master of Sience Thesis, Graduate School of Natural and Applied Sciences of Middle East Technical University.

[33] Martins L A L, Bastian L F & Netto T A 2012 Structural and functional failure pressure of filament wound composite tubes *Materials and Design* **36** 779-87.

[34] Chang R 2000 Experimental and theoretical analyses of first-ply failure of laminated composite pressure vessels *Composite structures* **49(2)** 237-43.

[35] Ibrahim M S, Sapuan S M and Faieza A A 2012 *J. Mech. Eng. Sci.* **2** 133.

Electrical discharge machining (EDM) of Inconel 718 by using copper electrode at higher peak current and pulse duration

S Ahmad[a] and M A Lajis[b]
Faculty of Mechanical and Manufacturing Engineering, Universiti Tun Hussein Onn Malaysia, 86400, Parit Raja, Batu Pahat, Johor, Malaysia

Email: [a]said@uthm.edu.my, [b]amri@uthm.edu.my

Abstract. This experimental work is an attempt to investigate the performance of Copper electrode when EDM of Nickel Based Super Alloy, Inconel 718 is at higher peak current and pulse duration. Peak current, I_p and pulse duration (pulse on-time), t_{on} are selected as the most important electrical pulse parameters. In addition, their influence on material removal rate (MRR), electrode wear rate (EWR), and surface roughness (Ra) are experimentally investigated. The ranges of 10 mm diameter of Copper electrode are used to EDM of Inconel 718. After the experiments, MRR, EWR, and Ra of the machined surfaces need to be measured in order to evaluate the performance of the EDM process. In order to obtain high MRR, higher peak current in range of 20A to 40A and pulse duration in range of 200μs to 400μs were used. Experimental results have shown that machining at a highest peak current used of 40A and the lowest pulse duration of 200μs used for the experiment yields the highest material removal rate (MRR) with value 34.94 mm^3/min, whereas machining at a peak current of 20A and pulse duration of 400μs yields the lowest electrode wear rate (EWR) with value -0.0101 mm^3/min. The lowest surface roughness (Ra) is 8.53 μm achieved at a lowest peak current used of 20A and pulse duration of 200μs.

1. Introduction

Nickel based super alloy, Inconel 718 is one of the most difficult-to-machine material which attributed to its ability to maintain hardness at elevated temperature and consequently it's very useful for hot working environment. Formation of complex shapes by this material along with reasonable speed and surface finish is not possible in traditional machining. This alloy is characteristically difficult to machine due to its poor thermal properties, high toughness, high hardness, and high work hardening rate. Usually, a nonconventional machining method like electrical discharge machining (EDM) is chosen for machining Inconel 718 in order to overcome such limitations. However, due to the great physical properties of Inconel 718, the cutting process for this material is become an issue in order to improve the speed of machining process. This alloy has attracted many researchers because of its increasing applicability and the machinability of aerospace alloys will continually decline as service demands increase in order to satisfy the demand for higher temperature capability for structural engine alloys [1,2].

For several decades, EDM has been an important for manufacturing process. It has proved for the machining of super-tough, electrically conductive materials such as the aerospace materials that are difficult to machine by conventional methods [3-6]. The main influence to the machining in EDM will be determined by electrical parameters such a current, pulse duration and voltage, and material properties of work piece and electrode like the material's melting temperature, as well as its electrical and thermal conductivity [7].

Copper became the metallic electrode material of choice for EDM due to its high in electrical and thermal conductivity properties. Copper can produce very fine surface finishes, even without special polishing circuits [8]. Bharti et al. [9] explained the machining characteristics of die-sinking EDM on Inconel 718 with copper as tool electrode. He found that the discharge current is the most influential input parameter on each performance measure. Discharge current and pulse-on-time are identified as common influencing parameters for MRR, Ra and EWR. According to Kuppan et al. [10], the MRR and Ra increases with the increase in peak current for EDM of Inconel 718. The effect of pulse duration is insignificant on MRR because long pulse duration can decrease the MRR. However, result from an experiment done by Kumar et al. [11] when EDM of Inconel 718 shows that an Increase in pulse duration up to 750μs has improves MRR. The primary objective in EDM of materials is always having higher material removal rate (MRR) in order to improve the productivity. Thus, by using higher peak current and pulse duration in this experimental work, hopefully capable to increase the performance and productivity of EDM Inconel 718.

2. Experimental details

In this study, Nickel base super alloys, Inconel 718 were selected as the material for the work piece (specimens 40mm x 30mm x 10mm) and Copper as a tool electrode with diameter of 10mm. The experiments were carried out on a standard CNC EDM machine, *Sodick AQ55L* with positive electrode polarity. The EDM experimental conditions and parameters are summarized in table 1.

Table 1. Experimental conditions and parameters.

Parameters	Levels
Work piece material	Inconel 718
Tool electrode	Copper
Peak Current, I_p (A)	20, 30, 40
Pulse duration, t_{on} (μs)	200, 300, 400
Pulse interval, t_{off} (μs)	Based on 80% duty factor
Voltage, V	120
Electrode polarity	Positive
Dielectric fluid	Kerosene
Depth of cut	3mm

Before experimentation, the work piece top surface was flattened using a surface grinding machine. The initial weight of the work piece and electrode was weighed using a 0.1mg accuracy digital weight balance. The work piece was held on the machine table using a fixture shown in figure 1. The time taken for machining and the weighed of work piece and electrode after machining are noted. All the gathered information from machining time, mass loss after machining process for both tool electrode and work piece were used to determine the values of MRR and EWR.

Figure 1. Experimental setup before machining.

The MRR and EWR were calculated using the equation (1) and (2). The surface finish was measured in terms of arithmetic mean roughness (Ra in µm). Ra was measured in the top of machining surface using Mitutoyo's surf test instrument SJ-400. In this research, a constant duty factor was used for the purpose of maintaining machining efficiency. The formula for duty factor is stated in equation. (3).

$$\text{MRR (mm}^3\text{/min)} = \frac{\text{Mass loss of workpiece (g)}}{\text{Density of work piece (g/mm}^3\text{) x machining time (min)}} \quad (1)$$

$$\text{EWR (mm}^3\text{/min)} = \frac{\text{Mass loss of electrode (g)}}{\text{Density of electrode (g/mm}^3\text{) x machining time (min)}} \quad (2)$$

$$\text{Duty factor (\%)} = \frac{\text{Pulse duration } (\mu s)}{\text{Pulse duration } (\mu s) + \text{pulse interval } (\mu s)} \quad (3)$$

3. Result and discussion

Table 2 shows the experimental results for EDM of Inconel 718 by using copper electrode.

Table 2. MRR, EWR, and Ra of EDM Inconel 718.

Peak Current, A	Pulse duration, µs	MRR, mm^3/min	EWR, mm^3/min	Ra, µm
20	200	18.6097	-0.0034	8.53
20	300	16.6653	-0.0058	9.72
20	400	14.7250	-0.0101	10.44
30	200	32.7401	0.0391	13.39
30	300	31.3544	0.0098	13.46
30	400	30.3742	-0.0050	14.04
40	200	34.9412	0.0438	14.27
40	300	31.4583	0.0079	14.07

40	400	30.5636	0.0030	14.61

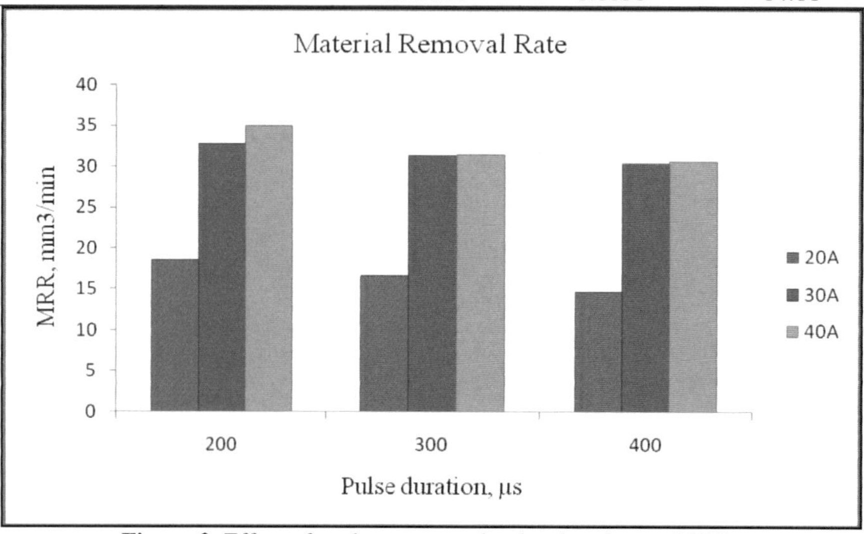

Figure 2. Effect of peak current and pulse duration on MRR.

3.1. Peak current and pulse duration against MRR

The effect of peak current and pulse durations on the MRR is shown in figure 2. It is shown that the peak current affects the MRR significantly. An increased in peak current MRR increase for all setting of pulse duration. MRR increases with the increase in peak current due to the increases of the energy per pulse causes temperature raises sharply that leads to rapid melting of work piece material at sparking area. However, higher pulse duration decreased MRR for all peak current used. With a pulse duration longer than 200μs, the MRR start decreases because of the exceeding value of pulse interval. High ignition delay due to high pulse interval in each cycle reduces the machining rate at a constant machining efficiency [12]. The highest MRR is achieved at 40A of peak current and 200μs of pulse on with value approximately 34.94mm^3/min.

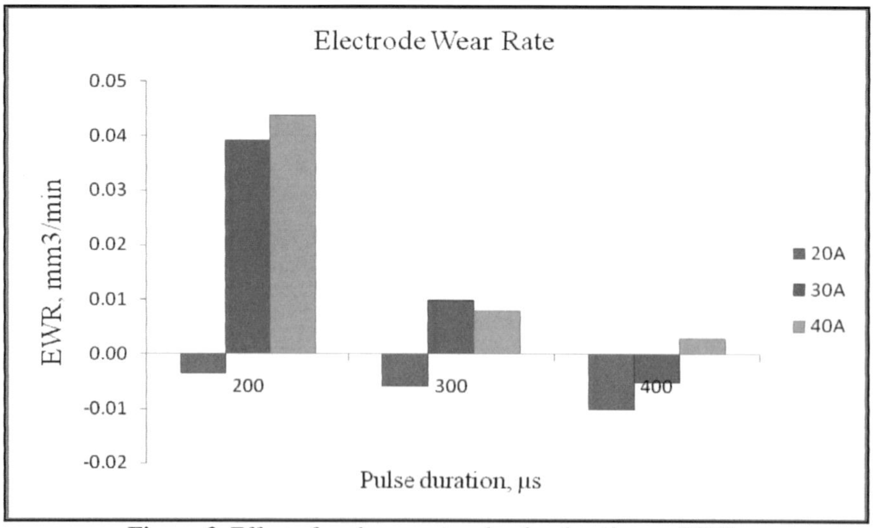

Figure 3. Effect of peak current and pulse duration on EWR.

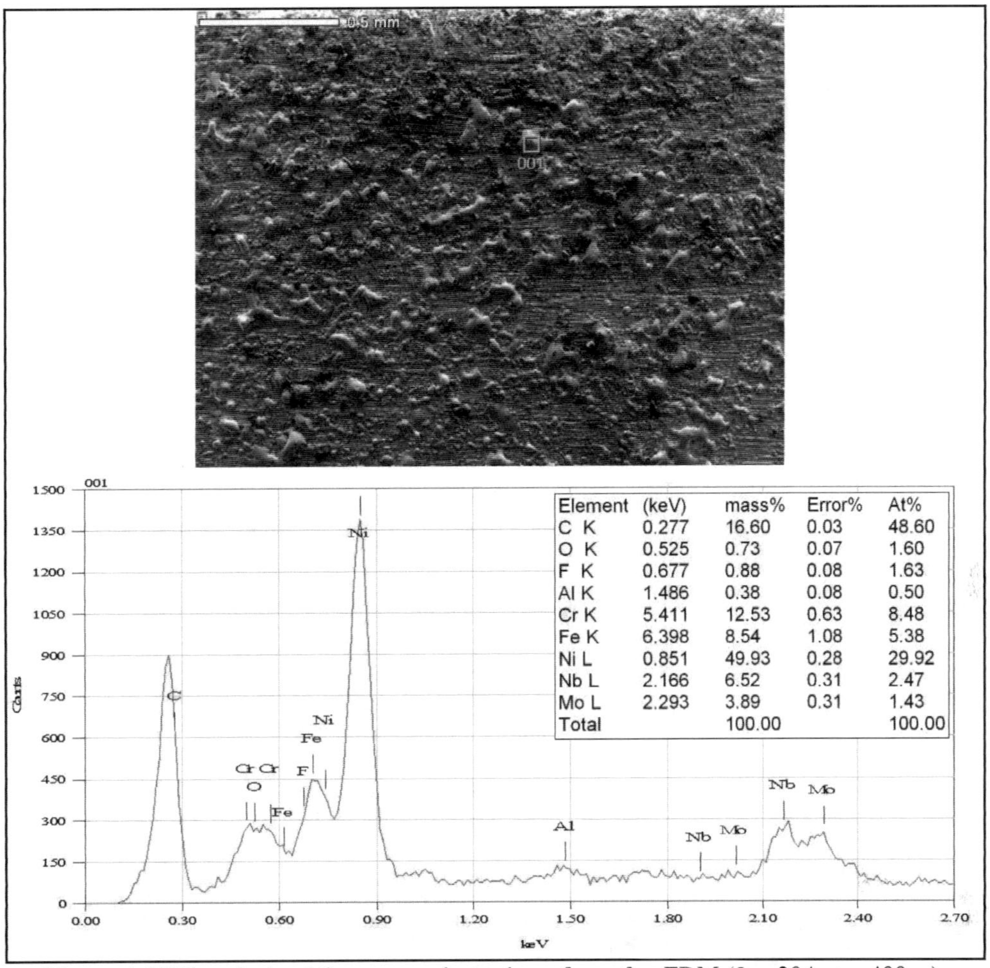

Element	(keV)	mass%	Error%	At%
C K	0.277	16.60	0.03	48.60
O K	0.525	0.73	0.07	1.60
F K	0.677	0.88	0.08	1.63
Al K	1.486	0.38	0.08	0.50
Cr K	5.411	12.53	0.63	8.48
Fe K	6.398	8.54	1.08	5.38
Ni L	0.851	49.93	0.28	29.92
Nb L	2.166	6.52	0.31	2.47
Mo L	2.293	3.89	0.31	1.43
Total		100.00		100.00

Figure 4. EDS analysis of the copper electrode surface after EDM (I_p = 20A, t_{on}=400μs).

3.2. Peak current and pulse duration against EWR

The effect of peak current and pulse durations on the EWR is shown in figure 3. The higher peak current resulting an increasing in EWR for constant pulse duration. High discharge current leads to high spark energy causes more material removal from work piece and tool electrode which in effect increases the EWR. EWR decrease when increasing of pulse duration for each of peak current used respectively. This is because of deposition of carbon on tool electrode at a high temperature for a longer pulse on time. The negative value for the lowest EWR is indicating that the electrode is deposited by the carbon is more compare to the wear of electrode. Longer pulse duration tends to increase the possibility of carbon deposition on the electrode surface, which function as wear resistant layer for copper electrode and helps to decrease the electrode tool wear. The lowest EWR is approximately -0.01mm^3/min at 20A of peak current and 400μs of pulse duration. A dissolved metal from the work piece also revealed deposited on the copper electrode. This is proved by the EDS test as shown in figure 4. The increment in the electrode weight after machining can be explained with this deposition effect. The similar result also found by Kang and Kim [13].

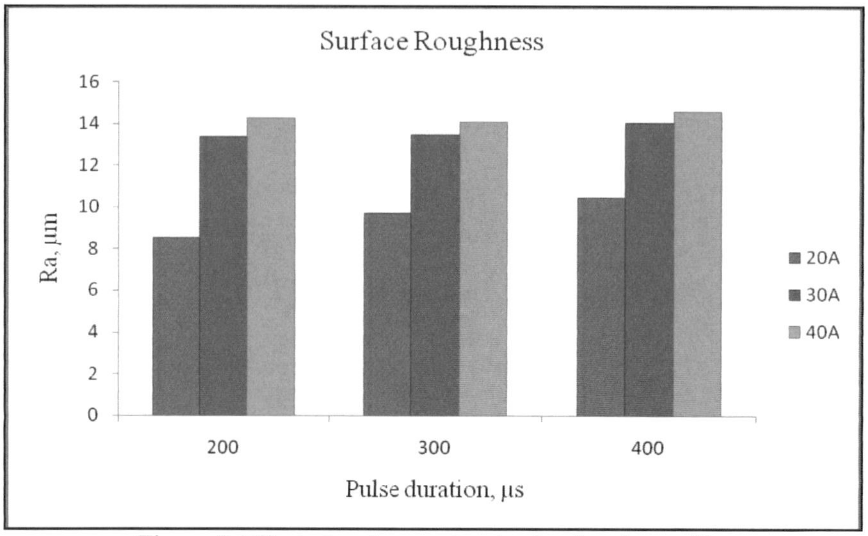

Figure 5. Effect of peak current and pulse duration on Ra.

3.3. Peak current and pulse duration against Ra

Figure 5 shows the effect of peak current and pulse durations on surface roughness of Inconel 718. Ra increases with the increase in discharge current and pulse duration. Figure 6 shows the differences in crater size for the lowest and highest Ra. When the discharge current is high, then the spark intensity are more, as a result, a larger crater depth on the surface of the work piece are produced, hence Ra is high. Pulse duration also strongly influences the Ra. An increase in pulse duration results in proportional increase in spark energy and consequently melting boundary becomes deeper and wider, and hence increases the roughness value [14]. The lowest Ra with value of 8.53μm is achieved at a peak current of 20A and pulse duration of 200μs.

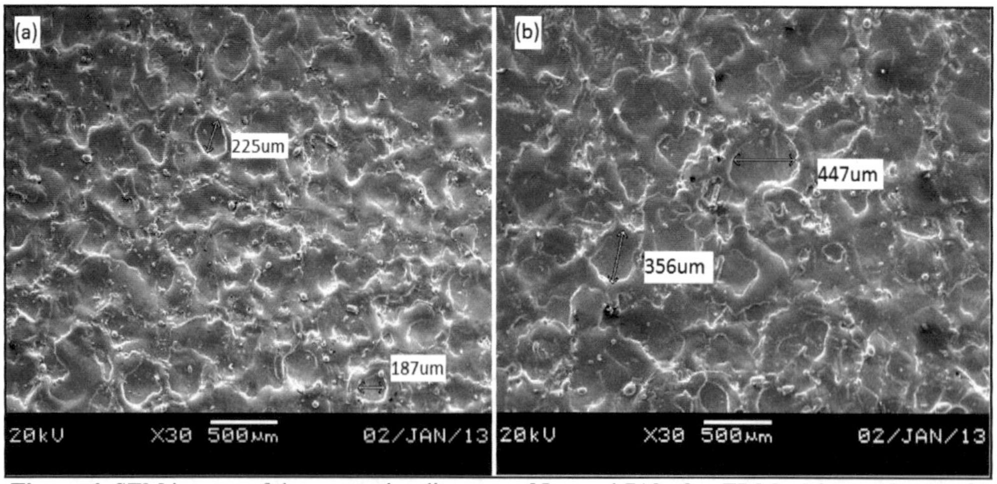

Figure 6. SEM images of the crater size diameter of Inconel 718 after EDM; a) lowest Ra (I_p=20A, t_{on}=200μs), b) Highest Ra (I_p=40A, t_{on}=400μs).

4. Conclusion

Material removal rate is the most important performance measure in this study. In order to improve productivity in EDM machining of Inconel 718, the higher peak current and pulse duration up to 40A and 400μs respectively are used. The conclusion can be made the peak current is the most influence parameter for achieving high MRR while for pulse duration it shows insignifant for improving MRR when EDM of Inconel using Copper electrode. For electrode wear rate, the longer pulse duration used may improved the EWR but affect adversely when higher peak current used. For surface roughness, lowest peak current and the lowest pulse duration is suggested in order to achieve good surface finish.

Acknowledgements
The authors would like to thank the Universiti Tun Hussein Onn Malaysia (UTHM) and Ministry of Higher Education (MOHE) for financial support under the Exploratory Research Grant scheme (ERGS) Vot 0886.

References
[1] Ezugwu E O 2005 Key improvements in the machining of difficult-to-cut aerospace superalloys *International Journal of Machine Tools & Manufacture* **45** 1353-67
[2] Rajesha S, Sharma A and Kumar P 2011 On Electro Discharge Machining of Inconel 718 with Hollow Tool *Journal of Materials Engineering and Performance* 1-10
[3] Tzeng Y-f 2008 Development of a flexible high-speed EDM technology with geometrical transform optimization *Journal of Materials Processing Technology* **203(1–3)** 355-64
[4] Khan M A R, Rahman M M, Kadirgama K, Maleque M A and Ishak M 2011 *J Mech Eng Sci* **1** 16
[5] Kumar S and Singh R 2010 *Inter J Automot Mech Eng* **2** 181
[6] Rupinder Singh and Bhupinder Singh 2011 *Inter J Automot Mech Eng* **3** 239
[7] Lee S H and Li X P 2001 Study of the effect of machining parameters on the machining characteristics in electrical discharge machining of tungsten carbide *Journal of Materials Processing Technology* **115(3)** 344-58
[8] Kern R 2008 Techtips Sinker Electrode Material Selection *EDM Today* (July/August 2008 Issue)
[9] Bharti P S Maheshwari S and Sharma C 2010 Experimental investigation of Inconel 718 during die-sinking electric discharge machining *International Journal of Engineering Science and Technology* **2(11)** 6464-73
[10] Kuppan P, Rajadurai A and Narayanan S 2008 Influence of EDM process parameters in deep hole drilling of Inconel 718 *International Journal Advance Manufacturing Technology* **38** 74-84
[11] Kumar A, Maheshwari S, Sharma C and Beri N 2011 Analysis of machining characteristics in additive mixed electric discharge machining of nickel-based super alloy Inconel 718 *Materials and Manufacturing Processes* **26(8)** 1011-18
[12] Hamid F E A and Lajis, M A 2012 High performance in EDM machining of AISI D2 hardened steel *Advanced Materials Research* **500** 259-65
[13] Kang S and Kim D 2003 Investigation of EDM characteristics of nickel-based heat resistant alloy *Journal of Mechanical Science and Technology* **17(10)** 1475-84
[14] Patel K M, Pandey P M and Venkateswara R P 2009 Surface integrity and material removal mechanisms associated with the EDM of Al2O3 ceramic composite *International Journal of Refractory Metals and Hard Materials* **27(5)** 892-899

IOP Conf. Series: Materials Science and Engineering **50** (2013) 012063 doi:10.1088/1757-899X/50/1/012063

Effect of adhesive thickness on adhesively bonded T-joint

A R Abdullah[1,a], Mohd Afendi[2] and M S Abdul Majid[2]
[1]Petronas Chemicals Fertiliser Kedah Sdn. Bhd., 08300 Gurun, Malaysia
[2]Universiti Malaysia Perlis, 02600, Arau, Malaysia

E-mail: [a]abdullah_rm@petronas.com.my

Abstract. The aim of this work is to analyze the effect of adhesive thickness on tensile strength of adhesively bonded stainless steel T-joint. Specimens were made from SUS 304 Stainless Steel plate and SUS 304 Stainless Steel perforated plate. Four T-joint specimens with different adhesive thicknesses (0.5, 1.0, 1.5 and 2.0 mm) were made. Experiment result shows T-joint specimen with adhesive thickness of 1.0 mm yield highest maximum load. Identical T-joint specimen jointed by spot welding was also tested. Tensile test shows welded T-Joint had eight times higher tensile load than adhesively bonded T-joint. However, in low pressure application such as urea granulator chamber, high tensile strength is not mandatory. This work is useful for designer in fertilizer industry and others who are searching for alternative to spot welding.

1. Introduction

In fertilizer industry, urea granules are formed inside a granulator while floating on air cushion. Thin stainless steel perforated plate is an essential component to this process. Spot welding is mostly used to join perforated plate to its supporting frame. Welders often find difficulty in producing good quality welds on thin stainless steel plate. Due to its thinness, it sensitive to high heat input which can cause significant reduction in material strength, e.g. burn through, cracking, porosity and plate distortion.

Many researchers studied adhesive bonding that covers broad spectrum of parameters such as joint configuration, adherend material, surface preparation, temperature, corrosion and humidity. Others studied mechanical behavior such as shear, tensile and peel strength. Kant and Narayanan [1] experimented epoxy adhesive lap joint with adhesive thicknesses of 0.5, 1.3 and 2.4 mm. Costa Mattos et al. [2] studied the effect of adhesive thickness on butt joint. Both researchers concluded that thinner adhesive yield higher tensile strength. Although the conclusion is indisputable, study on adhesive thickness is limited to lap joint and butt joint.

Some researchers studied T-joint in various combinations of stiffeners and adhesive parameters. Khalili and Ghaznavi [3] examined T-joint stiffened with triangular block on each side. Various specimens with different triangular angle were subjected to tensile load. Guo and Morishima [4] constructed T-joint specimen stiffened with adhesive cleat. Silva and Adams [5] experimented T-joint with thin and thick perpendicular adherend together with various configurations of stiffeners. Hu et al.

[6] studied the effect on bond-line length of a T-joint. Tensile test were performed with varied bond-line length. Their T-joint specimen was jointed with two 90° angle aluminum sheet, whereas this study was focused on construction of T-joint with two plates in perpendicular to each other. Although these researchers studied T-Joint, all specimens were stiffened. Various adhesive thicknesses were also not examined.

In this study, adhesively bonded T-joint specimens were constructed with thicknesses of 0.5, 1.0, 1.5 and 2.0 mm. Adherend were SUS 304 Stainless steel plates and SUS 304 perforated plates, similar to material used in most urea granulator. Tensile tests were performed to investigate maximum tensile load and type of adhesive failure. Welded specimen was subjected to similar test for comparison.

2. The experiment

To simulate actual construction of granulator, 10 mm thick SUS 304 Stainless Steel plate and 1.25 mm thick SUS 304 perforated plate were used as adherend. In this study, two-part epoxy adhesive, Araldite 2011 were used to join both adherend to form a T-joint as shown in Fig.1. The dimension of bonded area is 10 mm width and 100 mm length. Special jig and shim plates were used to control adhesive thickness during fabrication as shown in Fig. 2.

Four specimens were fabricated with different thickness of 0.5, 1.0, 1.5 and 2.0 mm, respectively. For each thickness, testing were performed 6 times to ensure consistency, where average value were taken. Grease and dirt were cleaned by acetone before jointing. For each specimen, time taken for curing was 16 hours. T-joint specimen was bolted to a special platform before it can be mounted to Autograph universal testing machine as shown in Fig.3. To ensure consistency in result, 6 specimens were fabricated for each adhesive thickness. Tensile testing was conducted at room temperature with cross head speed of 1 mm/min. Welded specimen was also tested in similar method to analyze maximum tensile load in comparison with adhesively bonded specimens.

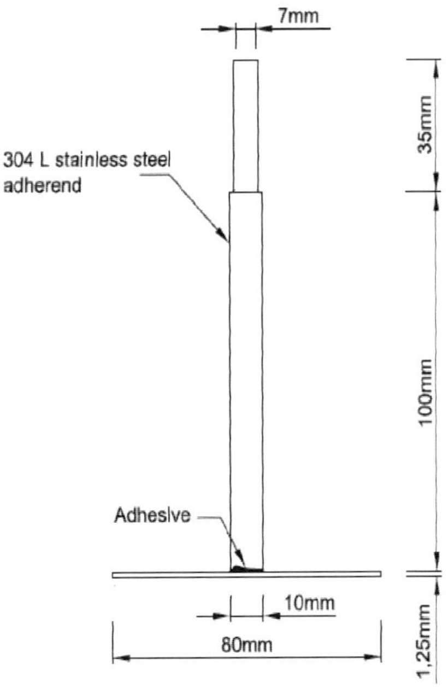

Figure 1. T-Joint specimen.

Figure 2. Special jig to control adhesive thickness by using shim plate.

Figure 3. Specimen mounted on UTM machine.

3. Results and discussion

Figure 4 shows failure load of all specimens tested. Specimen thickness of 1.0 mm yields highest tensile load among adhesively bonded specimen. In comparison, welded specimen failure load was 8 times higher than adhesive specimen. The results of this experiment show that there is an optimum thickness that yield highest tensile load. Figure 5 shows peak tensile load was observed at 1.0 mm adhesive thickness before it drops.

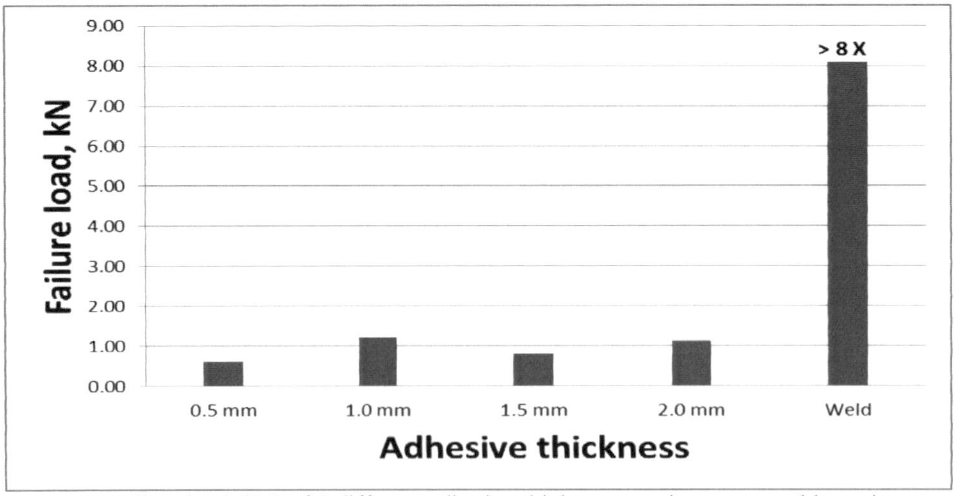

Figure 4. Failure load (kN) of 4 different adhesive thickness specimens vs. weld specimen.

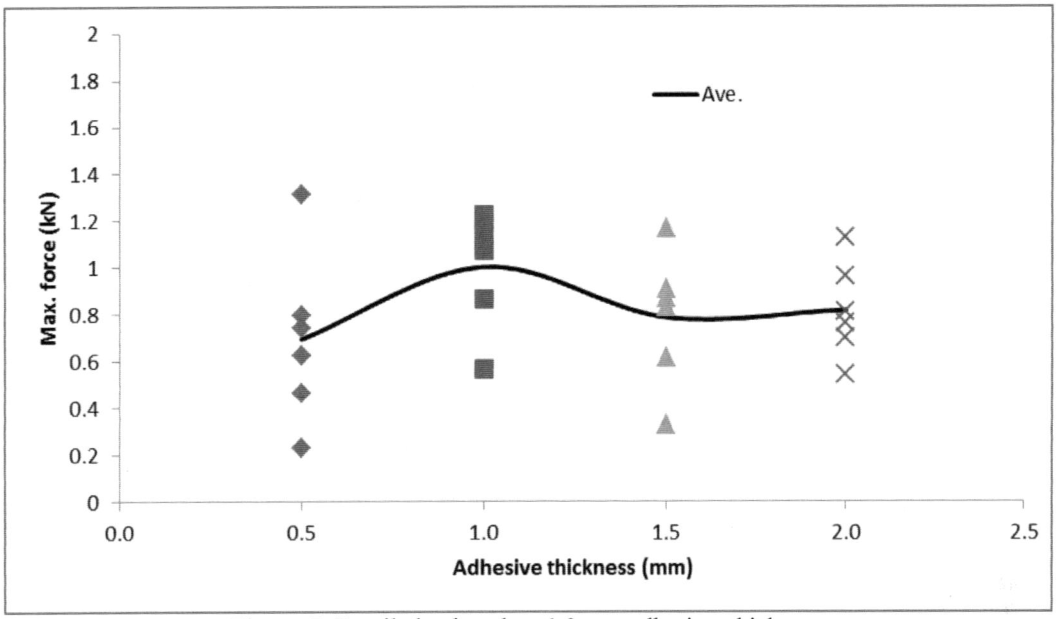

Figure 5. Tensile load peak at 1.0 mm adhesive thickness.

The experimental result indicates that failure load increases as adhesive thickness increases, until it reaches the optimum thickness. Once optimum thickness is reached, failure load will decrease slightly and consequently plateau. Kinloch [7] studied Goland and Reissner, Volkersen theory as well as finite-element analysis on thickness effect in adhesively bonded single lap joint. All theories in Kinloch's study indicate that thicker adhesive layer will result in increased failure load. In contrast, experimental result shows that fracture load does not increase in tandem with increasing adhesive thickness but rather fall slightly. The experimental result of this current work shows similar pattern as Kinloch's report.

4. Conclusion

In this paper, tensile tests were conducted on stainless steel T-joint specimen. The effect of adhesive thickness was analyzed. Optimum adhesive thickness was found at 1.0 mm where the failure load was the highest. Experimental result shows that failure load increases as adhesive thickness increase until it reaches 1.0 mm adhesive. Once it reaches 'optimum' thickness, failure load will reduce slightly and plateau.

5. References

[1] Ravi K and Narayanan R G *J. Testing and Evaluation* **40**
[2] Mattos H S C, Sampaio E M and Monteiro A H 2010 *Int. J. Adhes. Adhes.* **30** 774
[3] Khalili S M R and Ghaznavi A 2011 *Int. J. Adhes. Adhes.* **31** 347
[4] Guo S and Morishima R 2011 *Composite Structures* **94** 229
[5] Lucas F M S and Adams R D 2002 *Int. J. Adhes. Adhes.* **22** 311
[6] Ping H, Qi S, Weidong L and Xiao H 2012 *Int. J. Adhes. Adhes.* **38** 17
[7] Kinloch A J 1987 *Adhesion and Adhesives* (Chapman and Hall) 227

Optimization of quenching process in hot press forming of 22MnB5 steel for high strength properties for publication in

Nuraini Aziz[a] and S N Aqida[b]
Faculty of Mechanical Engineering, University Malaysia Pahang
26600 Pekan, Pahang, Malaysia
Phone : +609-424-2246 ; Fax : +609-424-2202

E-mail: [a]nurainibintiaziz@gmail.com, [b]aqida@ump.edu.my,

Abstract. This paper presents hot press forming of 22MnB5 steel blanks for high strength automotive components. The hot press forming was performed using Schenck press PEZ0673 machine with maximum press force of 1000 kN. Samples were square 22MnB5 blanks, of 50×60 mm dimension. A high temperature furnace was used to heat up the blanks to austenite temperature of 950°C. Samples were held at the austenite temperature prior to forming and quenching process. Three independent controlled parameters were cooling water temperature, press holding time and flow rate of water. Pressed samples were characterized for metallographic study, hardness properties and tensile properties. Metallographic study was conducted using Meiji optical microscope. Hardness was measured using Vickers indenter with load 1000gf. From metallographic study, the hot pressed 22MnB5 boron steel samples produced lath martensitic microstructure. Hardness of hot pressed samples increased with decreasing cooling time. The yield strength and the ultimate tensile strength of samples after hot forming were between 1546 and 1923 N/mm^2. These findings were important to design tailored ultra-high strength in automotive components at different process parameter settings.

1. Introduction

In recent years, weight reduction while has been strongly emphasized in the automotive industry while maintaining safety standards. The technology of hot stamping produced parts with extremely high strength, good ductility, high dimensional accuracy with (minimum spring-back effect) and possibility to produce complex geometries Boron steel is the most commonly used in hot stamping process with composition variant of 22MnB5, 8MnCrB3, 20MnB5, 27MnCrB5 and 37MnB4. Initial properties of boron steel were ferritic–pearlitic microstructure with tensile strength of 600 MPa. After being hot stamped, the strength increase to 1500MPa and experienced martensitic transformation, as shown in figure 1 (a) and (b) respectively. To achieve such properties, the boron steel blank was austenitized at 950 ◦C for at least 5 minutes. The blank was formed and quenched simultaneously by water-cooled die for 5–10 second. At cooling rate of 27 K/s, diffusionless martensitic transformation occurred thus exhibited high strength properties [3]. In figure 1 (b), the martensitic transformation begins at 425 ◦C and ends at 280 ◦C [4].

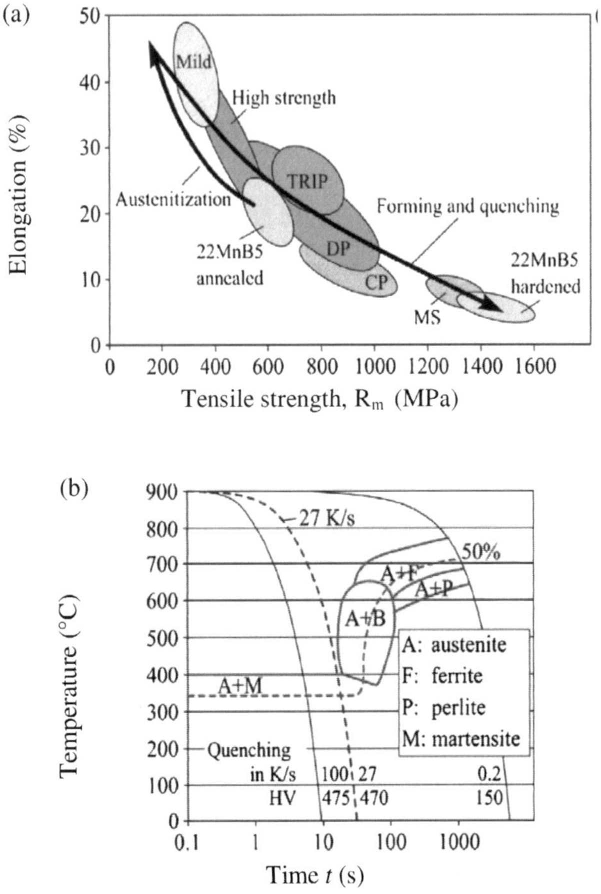

Figure 1. (a) Mechanical properties of 22MnB5; and (b) CCT diagram [5].

In hot stamping process, the tool surface temperature was maintained less than 200°C to produce high strength components [6]. Variations of cooling rate change mechanical properties of hot-stamping parts while dies are prone to failure due to rapid quenching during stamping. Thermal softening in die increased gradually with temperature where die hardness was reduced and eventually caused plastic deformation.

Cooling of hot-stamping dies was affected by several parameters like size of cooling channel, pressure holding time and cooling water velocity. Temperature changes of hot-stamping dies were studied for many times cycles by comparing thermal analysis and thermo-mechanical analysis, but the influences of processing parameters on temperature distribution have not been studied extensively [1]. The design method of hot-stamping dies and optimization schemes of cooling system were presented [12].In this paper three independent controlled parameters were investigated; cooling medium temperature, press holding time and flow rate of water. Pressed samples were characterized for metallographic study and hardness properties.

2. Experimental

As-received 22MnB5 steel plate of 1.0 mm thickness with the dimensions 60 mm x 50 mm was processed and analysed in this study. Chemical composition of 22MnB5 steel was analysed using (Software Vt Structure 5.2) spectroscopy in table 1. Hot press forming machine was used to press the sample surface with average pressure of 4bar. The hot press forming was performed using Schenck press PEZ0673 machine with maximum press force of 1000 kN. Sample was heated up to 950°C for 10 minutes and was transferred to press machine between 7 and 15 seconds. The processing was conducted in ambient temperature atmosphere.

Table 1. Chemical composition of alloying in 22MnB5 steel.

Element	C	Mn	Si	Ni	Cr	Cu	S	P	Al	V	Ti	B
wt%	0.221	1.29	0.28	0.013	0.193	0.01	0.001	0.018	0.032	0.005	0.039	0.0038

Sample surface was cleaned and processed at variation of cooling water temperature, press holding time and flow rate of water as given in table 2. Samples 1A, 2A and 3A were pressed and left for air cooling. Sample 1A and 1B were processed at same flow rate and quenching time but different cooling medium.

Table 2. Parameter settings of hot press forming for 22MnB5 boron steel blanks.

Sample	Medium	Flow rate (litre/min)	Quenching time (s)
1A		20	5
2A	Air	30	8
3A		40	11
1B			5
2B		20	8
3B			11
1C			5
2C	Water	30	8
3C			11
1D			5
2D		40	8
3D			11

Samples were prepared for metallographic study through grinding, polishing and chemical etching using 2% nital solution. Metallographic study was performed using IM7000 Series Inverted Optical microscopes with Progress Capture 28.8 Jenoptik Optical System image analyser software. Hardness properties of samples were measured using MMT Matsuzawa Vickers Hardness tester with 1000 gf load and dwell time of 10 second.

3. Result and Discussion

Micrographs of as-received and hot pressed 22MnB5 boron steel blank are shown in figure 2. The as-received sample contained mixture of pearlite phase and 73-77% ferrite and a small amount of carbide. Pearlite phase was located at ferrite grain boundaries and hardness varied between 160-200 $HV_{1.0}$. After hot pressed and water quenched at 20ℓ/min flow rate, sample contained martensitic structure as shown in figure 2(b).The martensitic content was more than 95%. During processing, sample was heated and water-cooled which allowed phase transformation to occur. Grain refinement occurred in the hot pressed blank as shown in figure 2(b) where smaller grain size analyzed.

Figure 2. Micrographs of 22MnB5 boron steel blank (a) before and (b) after water quenched at 20 ℓ/min.

Hardness properties of hot pressed samples at different quenching medium, water flow rate and quenching duration responding to micrograph are shown in figure 3. The highest hardness of 581 $Hv_{1.0}$ was measured in water quenched at 30 ℓ/min flow rate sample. Hardness properties of hot pressed samples were enhanced up to 3.5 times than the as-received boron steel sample. The lower quenching time of 5s produced a higher range of hardness. The hardness range of air quenched samples was between 490.1 and 530.4 $HV_{1.0}$.

At 500x magnification, samples show the martensite microstructure after hot press forming. Grain refinement occurred in the blank samples as shown in micrographs of figure 3(a). Isolated island of bainite can be easily distinguished in figure 3(a). Large martensite needles oriented in different angles are shown in micrograph of figure 3(b). A lath-shaped structure is clearly observed in figure 3(c) and 3(d), which consists of typical lath martensite (hereafter, this microstructure is referred to as fully lath-martensite).

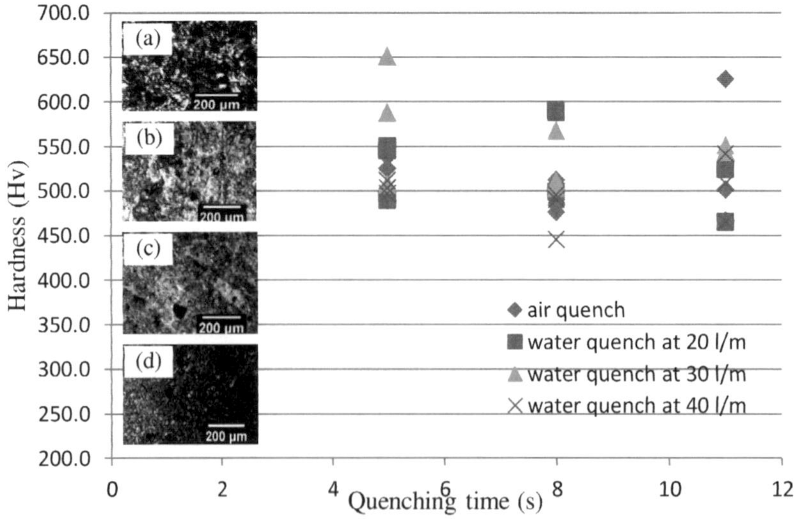

Figure 3. Hardness properties of blank samples surface area (a) air quench, (b) water quench at 20ℓ/m, (c) water quench at 30ℓ/m and (d) water quench at 20ℓ/m.

It is evident that air cooling is a weaker cooling media. It is seen that the cooling rate was not high enough for martensitic transformation. As a result, the microstructure converted to fully bainitic after air cooling in figure 3(a). In conclusion, using different cooling media influence the morphology of presented phases.

Calculated tensile properties of the samples are given in table 3 which corresponded to the hardness properties. It shows that after water-cooling quenching, the hardness of samples reached 528 $HV_{1.0}$ which was calculated as 1732 N/mm^2 tensile strength· The tensile strength of water quench hot pressed samples was higher than air quenched samples.

In summary, the following two factors cause the hardness changes observed in table 3: Bainite transformation occurring at a cooling rate at a temperature of more than 450 ◦C above the Ms point or auto-tempering at a cooling rate at a temperature less than 450 ◦C above the Ms point, or both factors. In addition, the major factor that causes the hardness variation in the hot-stamped specimens is auto-tempering below the Ms point. This is because the cooling rate at regions in direct contact with the die is assumed to exceed several hundreds of degrees Celsius per second.

Table 3. Hardness properties and tensile strength of as-received and hot pressed 22MnB5 boron steel blanks.

Properties	Flow rate (ℓ/m)	Hardness $(HV)_{0.1}$			Tensile strength, σ (N/mm^2)		
As-received sample		167			537		
Hot Pressed samples		Quenching time (s)					
		5	8	11	5	8	11
Air-quenched		517	490	530	1693	1595	1740
Water quenched	20	528	559	504	1732	1842	1643
	30	581	527	546	1923	1728	1797
	40	503	477	504	1640	1546	1643

4. Conclusion

The mechanical properties of hot press formed 22MnB5 steel varied with different parameter settings. Three different parameters; cooling water temperature, press holding time and flow rate of water from 22MnB5 steel grade were investigated. The die cooling medium was significant to the blanks properties. Hardness of 22MnB5 blank was enhanced to 3.9 times from the as-received samples. The maximum hardness of 650 $HV_{1.0}$ was obtained at water flow rate 30 ℓ/min and quenching time of 5 seconds. These findings were important to design hot press forming of 22MnB5 steel for high strength properties.

Acknowledgements

The authors would like to thank University Malaysia Pahang and Miyazu (M) Sdn Bhd for funding this research. This research was conducted for Industry Centre of Excellence (ICOE) of Mould and Die.

References

[1] Hoffmann H, So H, and Steinbeiss H, *Design of hot stamping tools with cooling system*", Annals of the CIRP, Vol **56/1**,2007

[2] Heinz Steinbeiss, Hyunwoo So, Thomas Michelitsch, Hartmut Hoffmann. *Method for optimizing the cooling design of hot stamping tools.* Prod. Eng. Res. Devel. 2007; **1**: 149-55

[3] Merklein M and Lechler J, 2008 *Determination of material and process characteristics for hot stamping processes of quenchable ultra high strength steels with respect to a FE-based process design. SAE World Congress*: Innovations in Steel and Applications of Advanced High Strength Steels for Automobile Structures, Paper No. 2008–0853.

[4] Somani M C, Karjalainen L P, Eriksson M, and Oldenburg M, 2001 *Dimensional changes and microstructural evolution in a B-bearing steel in the simulated forming and quenching process.* ISIJ International **4**, 361–67.

[5] Garcia Aranda L, Chastel Y, Fernandez Pascual J, Dal Negro T, 2002 *Experiments and simulation of hot stamping of quenchable steels.* Advanced Technology of Plasticity **2**, 1135-40

[6] Sikora S and Lenze F-J *Hot-forming important parameters for the production of high-strength BIW parts* IDDRG 2006: **295-01**

[7] Alexander Bardelcik, Christopher P Salisbury, Sooky Winkler, Mary A Wells, Michael J Worswick," *Effect of cooling rate on the high strain rate properties of boron steel*" International Journal of Impact Engineering **37** (2010) 694–02

[8] Akerstrom P,Wikman B and Oldenburg M,"*Material parameter estimation for boron steel from simultaneous cooling and compression experiments*"Modelling and simulation in material science and engineering **13**,1291-08,2005

[9] Berglund Amundsson K, Hellgren and L O,"*Hot stamped components with "soft zones" for improved crashworthiness simulation and validation of product performance*", Gestamp R & D, Luleå, Sweden, International Deep Drawing Research Group Proceedings, Olofström, Sweden, June 2008

[10] Kolleck R, Veit R, Merklein M, Lechler J and Geiger M,"*Investigation on induction heating for hot stamping of boron alloyed steels* ",CIRP Annals,**275-78**,2009

[11] Wilsius J, Hein P, and Kefferstein R "*Status and future trends of hot stamping of USIBOR 1500 P" Arcelor Research Automotive Applications*, **1**. Erlangener Workshop Warmblechumformung 2006, Bamberg, Meisenbach 2006

[12] Steinbeiss H, So H, Michelitsch T, Hoffmann H, "*Method for optimizing the cooling design of hot stamping tools*", Production Engineering Research and Development,Vol**1**,No.2,149-55,2007

[13] Naderi M, 2007 *Hot stamping of ultra high strength steels.* Doctoral Theses, RWTH Aachen.

IOP Conf. Series: Materials Science and Engineering **50** (2013) 012065 doi:10.1088/1757-899X/50/1/012065

Planning and design of a knowledge based system for green manufacturing management

Mohd Kamal Mohd Nawawi[1,a], **Nik Mohd Zuki Nik Mohamed**[2], **and Adam Shariff Adli Aminuddin**[3]
[1]School of Quantitative Sciences, Universiti Utara Malaysia, Malaysia
[2]Faculty of Mechanical Engineering, University Malaysia Pahang, Malaysia

E-mail: [a]mdkamal@uum.edu.my

Abstract. This paper presents a conceptual design approach to the development of a hybrid Knowledge Based (KB) system for Green Manufacturing Management (GMM) at the planning and design stages. The research concentrates on the GMM by using a hybrid KB system, which is a blend of KB system and Gauging Absences of Pre-requisites (GAP). The hybrid KB/GAP system identifies all potentials elements of green manufacturing management issues throughout the development of this system. The KB system used in the planning and design stages analyses the gap between the existing and the benchmark organizations for an effective implementation through the GAP analysis technique. The proposed KBGMM model at the design stage explores two components, namely Competitive Priority and Lean Environment modules. Through the simulated results, the KBGMM System has identified, for each modules and sub-module, the problem categories in a prioritized manner. The System finalized all the Bad Points (BP) that need to be improved to achieve benchmark implementation of GMM at the design stage. The System provides valuable decision making information for the planning and design a GMM in term of business organization.

1. Introduction

The environment has become a critical issue today. This is due to excessive and unjust use of natural resources. Since 40 years ago, several highly visible environmental disasters have demonstrated the importance of having a comprehensive environmental strategy in place [1-5]. Green manufacturing management (GMM) is a management system that contains only required resources and materials, manufactures only required quantity of quality products on time that meet customers' demands which driven the aim to reduce environmental impact. The Center for Green Manufacturing at the University of Alabama defines the goal of green manufacturing as:

"To prevent pollution and save energy through the discovery and development of new knowledge that reduces and/or eliminates the use or generation of hazardous substances in the design, manufacture, and application of chemical products or processes."

In the context of Malaysia, the government has proposed a fund of RM1.5 billion to promote the Green technology in 2010 through the National Green Technology Centre [6]. As are true of Total Quality Management (TQM) and other improvement initiative programmes, environmental strategies must be conceived and supported by top management, but deployed in every functional area of an organization to be meaningful [1]. With current competitive business environment and environment-friendly awareness, management should not only focus on the initiatives such as TQM, lean manufacturing, performance measurement, and supply chain but also the sustainability aspects of the initiatives.

2. Research Background

This paper introduces a new concept called Collaborative Green Manufacturing Management (CGMM) which can be implemented as an alternative for any manufacturer to improve their lean and green manufacturing processes. In the CGMM chain, all members must work together towards common objectives in order to make the lean and green manufacturing achievable in the collaborative environment. The framework presented consists of the conceptual design of the proposed CGMM system. The conceptual model is then converted into the structure of Knowledge-Based Collaborative Green Manufacturing Management System (KBCGMM) to enable the use of knowledge based system (KBS) which embed two powerful techniques; Gauging Absences of Pre-Requisites (GAP) and Analytical Hierarchy Process (AHP).

GAP analysis is a technique that is used to assess the gap between the organisation's actual environment and an ideal one, resulting in knowledge of the desirable prerequisites for an effective implementation [7-8]. On the other hand, AHP first developed and introduced by Saaty [9], is a powerful tool, which can be used to deal with multi-attribute and complex problems particularly in selecting and prioritising an alternative for improvement purposes. AHP has the capability to weigh the alternatives and make a comparison amongst the alternatives before the optimum solution can be suggested. However, in this paper, only the application of GAP technique will be shown and discussed.

2.1. Planning stage

The planning stage requires information that needs to be considered which focuses on two main aspects as shown in figure 1; the Collaborative Business and Green Manufacturing Chain perspectives. The function for the first part of planning stage, Collaborative Business is for gathering general information about the organisations environment, financial and market status. Organisation environment determines the particular environment the company is operating in. The information needed in this module are size of company, annual sales turnover, number of employees, age of company, position of company in automotive chain, competitors, suppliers, customers, and investment in green manufacturing activities. In CGMM, the position of a company in the supply chain is required to determine its suppliers and customers, since emphasis in not only within the organisation (internal), but also between organisations (external).

In the second part of planning stage, Green Manufacturing Chain component refers to connections between any two value-adding activities inside and across organisations. Activity in any process can be allocated as value-adding or non-value adding. In lean and green manufacturing, non-value adding activity is considered as a waste and must be eliminated. Green Manufacturing Chain can be divided into three subcomponents, Internal Chain, External Chain, and Product Design for Manufacture. In the Internal Green Chain, operators of the next process are the customers, and suppliers (current process) are committed to supply parts which are good in quality at the right time and right quantity. Customer satisfaction and supplier commitment are two major elements which contribute to the success of the internal green chain. In the External Green Chain, suppliers are considered as partners instead of outsiders. Suppliers are well informed about the demand and planning of the organisation and sometimes invited to involve in the product development and process design. The Product Design for Manufacture is developed with objectives of gathering product design information and analysing the product design process which covers from the conceptual design to the full launch of new products.

2.2. Design stage

The design stage requires information that needs to be considered which focuses on two main aspects as shown in figure 1; the Organization Competitive Priority and Lean Environment perspectives. The function of modules in Organization Competitive Priority is to discover the current organization capability towards CGMM in terms of these five competitive priorities i.e. quality, time, value, flexibility, and supply chain.

In the second part of design stage, Lean Environment component refers to connections between any two value-adding activities inside and across organizations. Activity in any process can be allocated as value-adding or non-value adding. In lean and green manufacturing, non-value adding activity is considered as a waste and must be eliminated. Lean Environment can be divided into three subcomponents, Employee Involvement, Waste Elimination, and Kaizen. The objective of this level is to identify and evaluate the current organisation CLMM alignment, which is based on these three identified processes to achieve customer satisfaction.

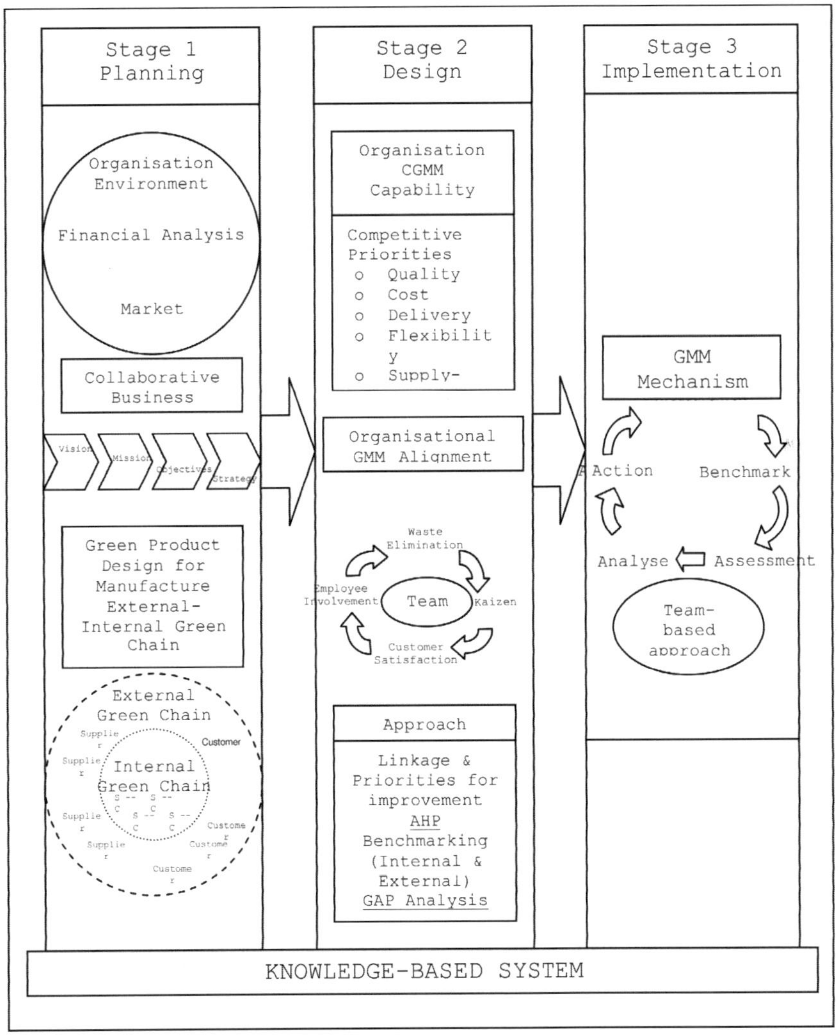

Figure 1. Planning and Design stages of KBCGMM conceptual model.

IOP Conf. Series: Materials Science and Engineering **50** (2013) 012065 doi:10.1088/1757-899X/50/1/012065

3. Example of Model Development

As an example, the Product Design for Manufacture Module (Level 2 of the KBCGMM System) is used to illustrate how the model was developed using KBS. Product design is one of the main activities of any manufacturing company, beside physical production and order taking process [10]. The Product Design for Manufacture module was developed with objectives of gathering product design information and analysing the product design process which covers from the conceptual design to the full launch of new products. Figure 2 shows two questions from this module which was developed using AM for Windows® software.

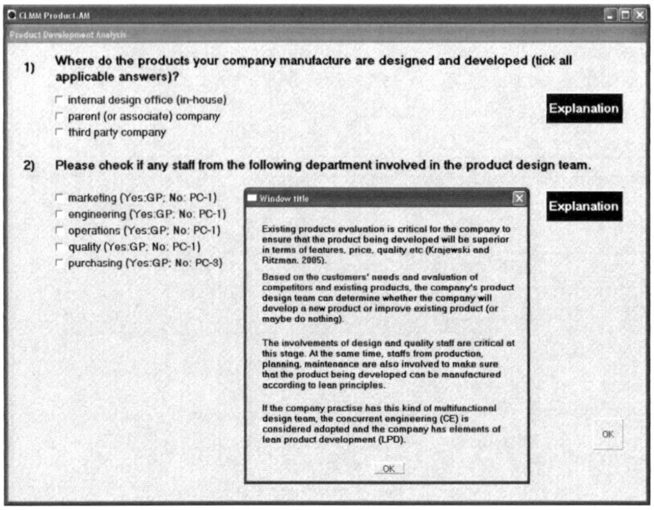

Figure 2. Example of questions in the Product Design for Manufacture Module.

A brief example of rules used in question number two is as follows:

IF the marketing team involved in the product design (Yes: GP; No: PC-1)
AND the engineering team involved in the product design (Yes: GP; No: PC-1)
AND the operations team involved in the product design (Yes: GP; No: PC-1)
AND the quality team involved in the product design (Yes: GP; No: PC-1)
AND the purchasing team involved in the product design (Yes: GP; No: PC-3)
THEN the product design team is multifunctional and the company design activity is good
ELSE the product design team is isolated and the company design activity needs improvement

An explanation facility is also provided in the system in order to assist the users in understanding the questions. Figure 3 shows two questions from this module which was developed using AM for Windows® software.

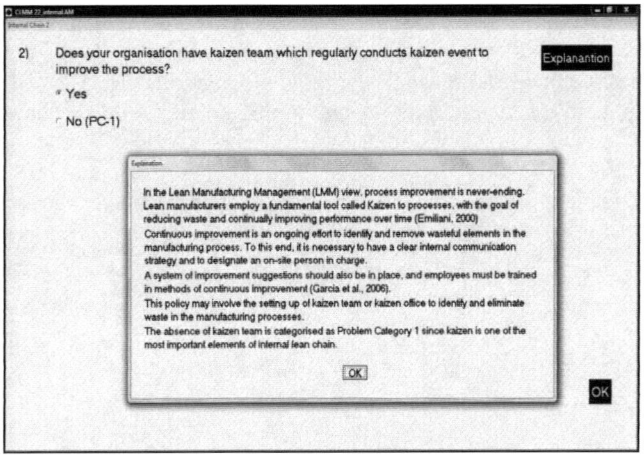

Figure 3. Example of explanation facility in the system.

Many of the questions are used with the GAP Analysis and are indicated by either Good Point (GP) code or Bad Point (BP) with problem categories code (PC-1 to PC-9). The description of the code is as described by Mohamed and Khan [11] and as shown in table 1.

Table 1. Problem categories and description of GAP analysis technique

Category	Description
PC-1	This indicates a **very serious problem**, which should and can be resolved in the short term and the result of the problem is quite likely to provide a real short-term benefit.
PC-2	This indicates a **serious problem**, which involves **pre-requisites to the system** and requires appropriate and logical improvement and implementation plan.
PC-3	This indicates a **major problem**, which is likely to have **pre-requisites to the system** and is better dealt with as part of an appropriate and logical improvement and implementation plan.
PC-4	This is **quite a major problem**, which is likely to have **pre-requisites to the sub-system** and is better dealt with as part of an appropriate and logical improvement and implementation plan.
PC-5	This indicates a **problem** and can be dealt with now. If resolved, it is likely to produce short-term benefits.
PC-6	This indicates a **minor problem** and can be dealt with now. If resolved, it is likely to produce short-term benefits.
PC-7	This is **not a serious problem**. Although it could be dealt with now, it is unlikely to produce short-term benefits. Therefore, it should only be dealt with if it is a pre-requisite for other things.
PC-8	This is **not really a problem**, However it is important to consider certain situations as future improvement.
PC-9	This is not **really a Good or Bad** point itself. The questions associated with this category are primarily asked to identify certain situations in the environment, which upon subsequent probing by succeeding questions may well reveal problems.

IOP Conf. Series: Materials Science and Engineering **50** (2013) 012065 doi:10.1088/1757-899X/50/1/012065

By answering the questions, the missing pre-requisites of the manufacturer position in relative to the benchmark can be identified through the number of Bad Points and its PC number. In order to evaluate the system performance and consistency, the prototype of CGMM model for the design stage has been tested by using artificial data. A simulated result for KBCGMM System – Stage 2 (design) is shown in Table 2.

Table 2. Example of summarized results of the GAP Analysis for the Design Stage

Module (and Sub-module)	**No of Questions**	**GAP Analysis**										
		GP	BP	PC 1	PC 2	PC 3	PC 4	PC 5	PC 6	PC 7	PC 8	PC 9
COMPETITIVE PRIORITY												
Quality												
Supply Quality Audit	18	10	8	2	1	0	2	1	0	1	1	0
Main Production Quality Audit	20	13	7	1	1	1	2	2	0	0	0	0
Customer Quality Audit	19	14	5	2	1	0	2	0	0	0	0	0
Sub-total	**57**	**37**	**20**	**5**	**3**	**1**	**6**	**3**	**0**	**1**	**1**	**0**
Cost												
Supply Cost	17	9	8	2	1	1	2	0	2	0	0	0
Main Production Cost	15	8	7	1	0	1	0	1	2	2	0	0
Resource Cost	12	6	6	1	1	0	3	0	0	1	0	0
Sub-total	**44**	**23**	**21**	**4**	**2**	**2**	**5**	**1**	**4**	**3**	**0**	**0**
Delivery												
Supply Timing	14	9	5	1	0	1	0	2	1	0	0	0
Main Production Timing	11	7	4	1	0	1	0	2	0	0	0	0
Delivery Timing	11	7	4	1	0	1	0	0	2	0	0	0
Sub-total	**36**	**23**	**13**	**3**	**0**	**3**	**0**	**4**	**3**	**0**	**0**	**0**
Flexibility												
Supply Flexibility	12	6	6	2	1	0	2	0	0	0	0	1
Main Prod Flexibility	12	8	4	1	1	0	0	0	0	1	1	0
Delivery Flexibility	11	6	5	1	1	0	1	0	0	1	0	1
Sub-total	**35**	**20**	**13**	**4**	**3**	**0**	**3**	**0**	**0**	**2**	**1**	**2**
Supply Chain												
Location	15	10	5	2	0	0	1	0	1	0	1	0
Logistics	17	12	5	0	2	0	1	1	1	0	0	0
Sub-total	**32**	**22**	**10**	**2**	**2**	**0**	**2**	**1**	**2**	**0**	**1**	**0**
LEAN ENVIRONMENT												
Employee Involvement												
Measurement	12	9	3	1	0	0	0	2	0	0	0	0
Benchmark	15	7	8	2	0	1	0	1	1	1	1	1
Assessment	10	4	6	0	1	1	1	2	1	0	0	0
Analyze	12	4	8	1	0	2	0	1	1	1	1	1
Action	12	8	4	1	1	0	1	1	0	0	0	0
Sub-total	**61**	**32**	**29**	**5**	**2**	**4**	**2**	**7**	**3**	**2**	**2**	**2**
Waste Elimination												
Measurement	13	7	6	1	0	2	0	1	0	1	0	1
Benchmark	14	6	8	1	1	0	2	0	2	1	1	0
Assessment	11	5	6	0	1	1	0	1	0	1	1	1
Analyze	10	5	5	1	0	1	0	1	1	1	0	0
Action	12	7	5	0	1	0	1	1	1	1	0	0
Sub-total	**60**	**30**	**30**	**3**	**3**	**4**	**3**	**4**	**4**	**5**	**2**	**2**
Kaizen												
Measurement	13	8	5	1	0	1	0	1	1	1	0	0

Benchmark	17	8	9	2	1	1	1	1	1	0	1	1
Assessment	16	9	7	1	0	1	1	2	0	2	0	0
Analyze	11	7	4	0	1	0	1	0	1	0	1	0
Action	14	10	4	1	0	1	0	1	1	0	0	0
Sub-total	**71**	**42**	**29**	**5**	**2**	**4**	**3**	**5**	**4**	**3**	**2**	**1**
GRAND TOTAL	**395**	**229**	**166**	**31**	**17**	**18**	**24**	**25**	**20**	**16**	**9**	**7**

A total number of 395 questions have been asked in this stage which also contains the number of Good Points (GP), the number of Bad Points (BP), together with the Problem Categories (PC) of the BP. The GAP analysis optimization technique suggests that only the BP are categorized into PC in order to identify the necessary pre-requisites that are required to achieve the CGMM. The KBGMM System has identified, for each modules and sub-module, the problem categories in a prioritized manner. Out of 395 questions, 229 have been categorized as GP whereas 166 have been considered as BP. The System finalized these 166 BP (31 PC-1, 17 PC-2, 18 PC-3, 24 PC-4, 25 PC-5, 20 PC-6, 16 PC-7, 9 PC-8, and 7 PC-9) need to be improved to achieve benchmark implementation of CGMM.

4. Conclusion

This paper has described the importance for automotive manufacturers to implement GMM in order to improve their lean and green manufacturing management system and compete in the globalize competition. A conceptual model for the design stage of GMM is developed and presented. The conceptual model then is converted into the structure of KBGMM which is supported by the knowledge based system (KBS). At the same time, Gauging Absences of Pre-Requisites (GAP) Analysis technique which is incorporated in the system assists users to understand the position of their organization in comparison to the ideal one. This would not only support in implementing GMM but also in benchmarking the strength of organizations in this area.

Acknowledgement

The authors would like to thank the School of Quantitative Sciences, Universiti Utara Malaysia (UUM) and Ministry of Higher Education for financial support under FRGS Grant SO11866.

References

[1] Walton S V, Handfield, R B and Melnyk, S A 2008 *J. Supply Chain Management.* **34** 2
[2] Garg H K and Singh R 2013 *Inter. J. Automot. Mech. Eng.* **7** 983
[3] Andrew-Muno M and Ibrahim R N 2013 *J Mech Eng Sci* **4** 488
[4] Mohamed N M Z N and Khan M K 2012 *Inter J Automot Mech Eng* **5** 545
[5] Rose M A N M, Deros B M and Rahman M N A 2013 *Inter J Automot Mech Eng* **7** 820.
[6] Razak M N 2009 *The 2010 Budget Speech* (Kuala Lumpur: Ministry of Finance Malaysia)
[7] Udin Z M 2004 *A Hybrid Knowledge-Based Approach for Planning and Designing a Collaborative Supply Chain Management System.* (Bradford: University of Bradford)
[8] Wibisono D 2003 *A Knowledge Based Approach to Assist in the Design of a Performance Measurement System for a Manufacturing Environment* (Bradford: University of Bradford)
[9] Saaty T L 1980 *The Analytic Hierarchy Process: Planning, Priority Setting, Resource Allocation* (New York: McGraw-Hill)
[10] Womack J P and Jones D T 2003 *Lean Thinking: Banish Waste and Create Wealth in Your Corporation* (New York: Simon and Schuster)
[11] Mohamed N M Z N and Khan M K 2011. *Proc. of 26th Int. Conf. on CAD/CAM, Robotics and Factories of the Future 2011* (Kuala Lumpur) p 310

Sustainable WEE management in Malaysia: present scenarios and future perspectives

Md Rezaul Hasan Shumon[1,a] **and S Ahmed**[2]
[1] Research Assistant, Department of Mechanical Engineering, University of Malaya, 50603 Kuala Lumpur, Malaysia.
[2] Associate Professor, Department of Mechanical Engineering, University of Malaya, 50603 Kuala Lumpur, Malaysia.

E-mail: [a]shumon330@gmail.com

Abstract: Technological advances have resulted development of a lot of electronic products for continuously increasing number of customers. As the customer taste and features of these products change rapidly, the life cycles have come down tremendously. Therefore, a large volume of e-wastes are now emanated every year. This scenario is very much predominant in Malaysia. On one hand e-wastes are becoming environmental hazards and affecting the ecological imbalance. On the other, these wastes are remaining still economically valuable. In Malaysia, e-waste management system is still in its nascent state. This paper describes the current status of e-waste generation and recycling and explores issues for future e-waste management system in Malaysia from sustainable point of view. As to draw some factual comparisons, this paper reviews the e-waste management system in European Union, USA, Japan, as a benchmark. Then it focuses on understanding the Malaysian culture, consumer discarding behavior, flow of the materials in recycling, e-waste management system, , and presents a comparative view with the Swiss e-waste system. Sustainable issues for e-waste management in Malaysia are also presented. The response adopted so far in collection and recovery activities are covered in later phases. Finally, it investigates the barriers and challenges of e-waste system in Malaysia.

1. Introduction

Since 1980's, massive volume of electric and electronic equipment has been produced, and used or consumed worldwide at a very high rate [11]. The rate of use is rather increasing over the time. However, most of such product's life cycles are known to have decreased significantly [8] Subsequently, the ever-increasing amount of waste electric and electronic equipment (WEEE) has become a common concern in many countries in the world[9], from the point of conservation of environment and reuse or recycling of such wastes. This E-waste has become a serious threat to the living environment as more and more appliances are thrown away after usage into bins. Obviously, this leads to toxic pollution, as electronic items break down and release of heavy metals, flame retardants, and other chemicals into soil and water [4]. Effective and efficient e-waste management, considering the social cost and environmental impact recovery seems to be a complex task [18]. However, it depends on the aspects of a particular country from e-waste management, despite the differences in the socio-economic and legal contexts [1].

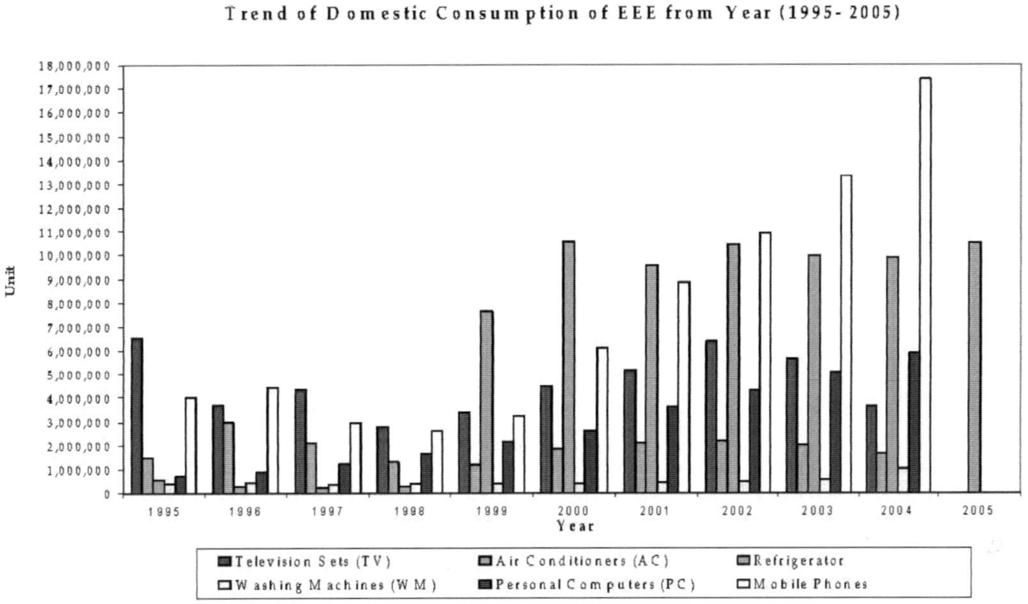

Figure 1. Trend of EEE product consumption in Malaysia: adapted from [3].

The methodology of this research consists of the several folds. Data on sales of E&E products and WEEE generation from household and industrial sources were collected. WEEE generation data were collected from the government sources like DOE, Ministry of Natural Resources and others. High ranking journal papers, literatures, e-books, and internet sources and field survey were done as the main sources of information collection on WEEE management activities to portray the future (expected) scenario. To find out the current position of WEEE management in Malaysia, field visit, interview, communication with Manufacturers, end users, DOE and other government agencies, and the relevant local municipalities were contacted. All these information were collected and analyzed to present a comparative picture of the current position in Malaysia to that of a few leading e-waste managing developed countries.

Interviews with the government officials and municipal authorities have given a thorough understanding of the management process and future perspective. People who are in charge of formulating the legislation coupled with a literature search obtained details of the recent progress comparable with other countries.

2. Review of e-waste regulations, laws and systems of various leading countries (Europe focusing on Switzerland, USA, Japan

2.1. Europe

E-waste management started in European countries under the WEEE directive (2002/96/EC), which was effective from February 13, 2003[15]. It imposes an obligation to the electronic and electronic product manufacturers to take back the end of life products and provide a framework of the various measures for manufacturers including collection, recovery and treatment of WEEE. Switzerland was the first country in the world to develop and implement a well-organized, formal e-waste management system for the collection, transportation, recycling/treatment and disposal of e-waste [17]. The country structured its legal and operational framework based on the Extended Producer Responsibility (EPR) model where manufacturers and exporters are liable for the physical and financial responsibilities for the environmentally sound handling, recycling and disposal of e-waste [7].

2.2. USA

In the USA, the fastest growing stream of Municipal Solid Waste (MSW) could be e-waste as because people now a days, like to buy, upgrade and discard electronic products [5]. According to U.S Environmental Protection Agency (EPA), about 80-85 percent of end of life electronic products ended up in U.S landfills [16]. US recycling systems vary according to the individual state perspectives. However, more than 20 states have enforced legislation to manage e-waste and most of them are based on extended producer responsibility policy [13] USA is staying behind other developed countries regarding e-waste management regarding structure development and collection. Some states are, however, advanced like in California; a number of companies are working on collection, recovery and reuse of e-waste [10].

2.3. Japan

In April 2001, Japan applied WEEE management regulation to recycle and treat four major electrical equipment wastes specifically air-conditioners, televisions, laundry machines, and refrigerators (including freezers) [5]. Consumers are required to work together with the collection of their WEEE to retailers. The "recycling fee" is introduced as pre-treatment fee and consumers need to agree with that. Additionally, consumers are often charged the transportation fees [6]. This, however, depends on retailers. Institutional and entrepreneurial consumers send their WEEE to retailers along with the pre-treatment fee or they can treat WEEE as industrial waste at their own expense as well. Retailers should have a commitment to accept designated WEEE on the request of consumers. Retailers also have an obligation to transfer the accepted WEEE to the producers' combined collection points. [12].

3. Malaysian E-waste scenario

3.1. Increasing trend of E-waste

Malaysia, as known to all, a fast developing country with a population of about 28 million [2]. Waste generations is learnt to have 0.8 kg waste per capita per day. According to the Department of Environment (DOE), Malaysia has generated 40,000 tons of e-waste in 2006 (NST Online, 2007). The figures below are showing the rapidly increasing trend of E-waste generation 40,000 tons in 2006 to 134,000 tons in 2009. These are the latest data available regarding e-waste at this moment in Malaysia

Table 1. Quantity of E-waste generated in Malaysia (Source: Department of Environment, Malaysia)

Year	Amount of Waste (Tons)
2006	40275.20
2007	52718.20
2008	102808.50
2009	134035.70

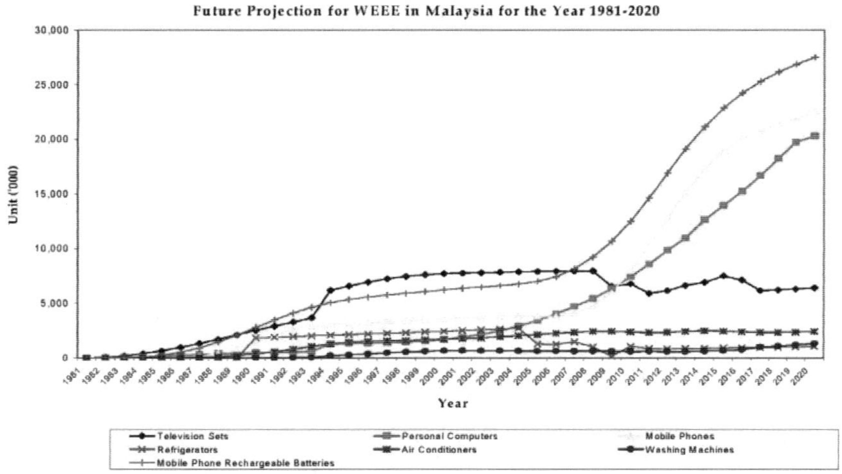

Figure 2. Projection of WEEE in Malaysia : adapted from [3].

The volume of e-waste is expected to rise up to 1.1 million metric tons in 2020, at a rate of 14 percent annually, according to the same report. The electrical and electronic items which have shared significantly to the volume of e-waste in Malaysia are television sets and mobile phones.

3.2. E-waste flow stream in Malaysia

Figure 3 shows the general flow of WEEE from the source of generation until the location of disposal is identified in Malaysia. It shows a significant disconnection between the largest sources of WEEE and the DOE-licensed e-waste collectors and processors.

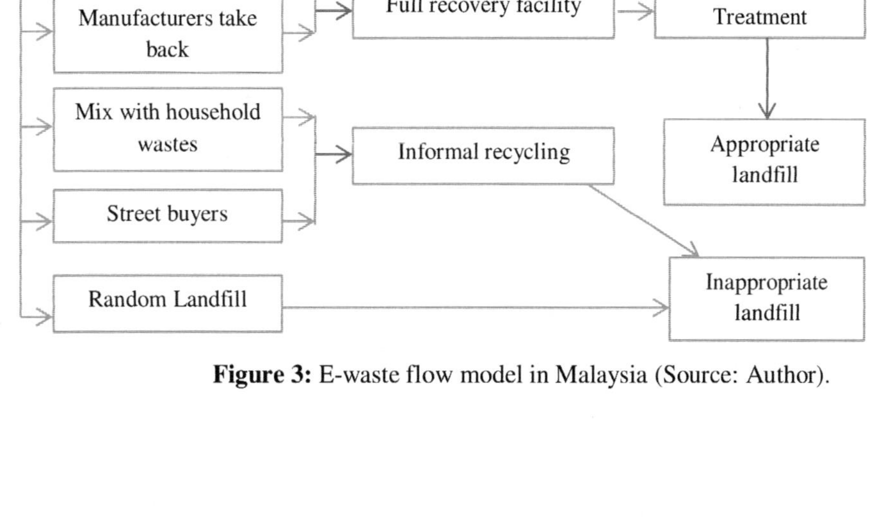

Figure 3: E-waste flow model in Malaysia (Source: Author).

3.3. Malaysian E-waste scenario in comparison with Switzerland

To understand the current position of Malaysia regarding e-waste management, it is worth to compare with an advanced context. Switzerland is the pioneer, and developed the most successful e-waste system in the world. Table 2 summarizes the characteristics of Swiss e-waste system and compares with the Malaysian one.

Table 2. Malaysian e-waste scenario in comparison with Switzerland

Stakeholders	E-waste system , responsibilities and Scenario	
	Swiss	Malaysia
Government	▪ The federal government plays the role of an overseer, framing the basic guidelines and legislation. ▪ Partly contribute to overall control and monitoring activities.	▪ Central government in association with state governments provides the legislative framework.
Manufacturers/ Importers	▪ Carry the economic and physical responsibilities of their products. ▪ PROs (e.g. SWICO, SENS)* have the role of managing the daily operations of the system, including setting the recycling fees, as well as licensing and auditing recyclers.	▪ Manufacturers and importers do not carry the physical responsibility of the product though some voluntary take back system is being seen as part of corporate social responsibility. ▪ No such effort like PRO's in Malaysia is seen yet.
Distributors & retailers	▪ Bear a part of the physical and informational responsibility of the product. ▪ Are obligated to take back products in categories they have on sale, irrespective of whether the product was sold by them, or whether the consumer purchases a similar product as replacement. ▪ Are responsible for clearly mentioning the amount of the ARF * in the customer invoice.	▪ Some retailers take back old items when delivers new one. On test basis this is running in several parts of the country, however, they are obliged to take back discarded products by law. ▪ No recycling fee's introduced to the customer yet.
Consumers	▪ Are responsible, and obligated by law, to return discarded appliances to retailers or designated collection points. ▪ Bear the final financial responsibility through the recycling fee on new product purchases.	▪ Are not responsible to return used products by law to the retailers or collection points. ▪ Do not bear financial responsibility in any form.
Collection points (specifically designated locations)	▪ Collect all kinds of WEEE free of charge and ensure the safety of the disposed products to prevent pilferage or illegal exports.	▪ Collects all of kinds of WEEE, but these points are not available everywhere.
Recyclers	▪ Must adhere to minimum standards on emissions and take adequate safety measures concerning employee health. ▪ Need authorization to operate a recycling facility from the cantonal government, as well as a license from the PROs.	▪ Recyclers need to take adequate safety measures for dismantling operations. ▪ Need license from Department of Environment (DOE).

3.4. Collection and recovery activities of e-waste in Malaysia

This study summarized the current e-waste collection activities found in Malaysia and their characteristics in table 3.

Table 3. Summary of the e-waste collection activities and characteristics of e-waste system in Malaysia(Compiled by Author)

E-waste collection activities	Characteristics of the collection system
DOE licensed contractors	▪ Collect industrial waste and send it to the licensed recovery facilities. ▪ They are also supposed to collect household waste but at present they are not working with e-waste from household and business entities.
Retailer's collection	▪ Retailers delivers new one and take back old one. ▪ Consumer can carry to the retailer and have a voucher relative to the type and condition of discarded items.
Voluntary collection organization ▪ Social organizations ▪ Environmental working groups ▪ Municipalities	Collects household e-waste by means of- ▪ Community based recycling programs. ▪ Drop box collection system ▪ Junk shop collection system ▪ Local government buy back center
Informal scrap collectors ▪ Street buyers ▪ Scavengers ▪ Traditional hawkers "Surat khabar lama"	▪ Collects e-waste from household consumer and less frequently business /institutional consumers. ▪ Door to door buying system. ▪ Exact information is available where they sell or treat even not sure whether inside or outside Malaysia. ▪ Constitutes a large portion of e-waste collection in Malaysia.
Manufacturer's initiative ▪ Panasonic Malaysia ▪ ECOMOTO Take-back ▪ Nokia Malaysia ▪ Dell Malaysia ▪ HP ▪ Pikom (National ICT	▪ Panasonic manages e-waste produces within their company but not from the external customer. They collect and send their e-waste to Alam Flora Sdn Bhd for proper treatment and disposal. ▪ ECOMOTO is a popular e-waste management program for Motorola employees and dealers. They initiated "20-to-1" trade-in program where the dealers or its employees will get one free battery for every 20 units of used or discarded rechargeable batteries returned. ▪ A `Take Back' scheme launched by Nokia Malaysia inspires consumers to join Nokia and take social responsibility to dispose used or old mobile phones, batteries and accessories by using designated recycling bins at particular locations in Malaysia ▪ Dell Malaysia introduces a system like, whenever a customer orders a new computer from Dell, an option is provided for the old computer to be collected when the new computer is delivered. Since April 2007, Dell Malaysia is collecting used monitors, desktop and notebook computers, printers and accessories. Visiting Penang Island (on 1st November, 2012), a platform is running under the State MPPP/ Dell Computer Recycling Program for used EEE collection from household. ▪ HP introduced a program "*HP Planet Partners Hardware Return and Recycling Programme*" for its corporate customer to easily dispose the used equipment in an environmentally conscious

	manner.
	▪ Pikom (National ICT Association of Malaysia) WEEE Collection Centers are collecting E-waste from households.

Electronic waste recovery systems have started in Malaysia recently. Today the electronic waste recycling business is in all areas in developed countries. It is becoming a large and rapidly consolidating business. Their recovery centers collect and treat mainly, computer and TV monitor, computer cover and casing, CPU, PCB board, IC tray, audiovisual components, VCRs, stereo equipment, mobile phones, other handheld devices, white goods like fridge, washing machine, computer components and others that contain valuable elements and substances suitable for reclamation, including lead, copper, and gold.

Currently there are 153 e-waste recovery facilities in Malaysia where 128 facilities are for partial recovery and 25 for fully recovery [14]. The majority of the e-waste management in Malaysia is centered in the Penang state. Figure 4 below shows the countrywide distribution of recovery facilities for E-waste.

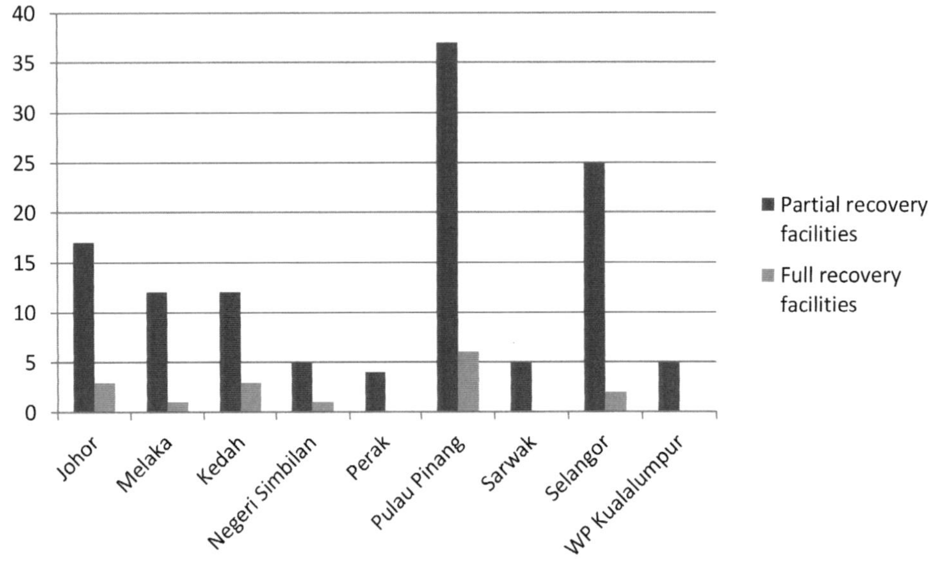

Figure 4. Distribution of recovery facilities for E-waste in Malaysia (Source: Department of Environment 2012).

4. Current problems and barriers for e-waste management in Malaysia

Current problems and barriers for e-waste management in Malaysia are summarized in table 4.

Table 4. Barriers for sustainable e-waste management in Malaysia (compiled by author)

	Barriers for sustainable e-waste management in Malaysia
Technological	▪ Recovery technology availability for precious metal is limited and obsolete. ▪ Inappropriate value addition from recycling reuse and remanufacturing. ▪ Eco friendly disposal of e-waste is almost not available. ▪ Information technology adaptation in take back system.
Engineering	▪ Efficient collection, dismantling, transportation system design and implementation. ▪ E-waste collection center's to set up at places convenient to consumers before being sent to recovery plants. ▪ Integrated reprocessing facilities are rarely seen. ▪ Probable technically and economically feasible solutions development for every step of e-waste management like recycling remanufacturing etc. ▪ Logistics and transportation difficulties of e-waste especially bulky home appliances such as refrigerators and washing machines.
Social and Policy	▪ Compulsory requirement of take back scheme through legislation is still to be implemented. ▪ Social organizations initiative for voluntary take back of e-waste is not prevalent. ▪ Consumers are in view to sell the e-waste for material value inside rather than to pay the recyclers for disposal. ▪ Revision on the law on import and export of e-waste.
Administrative	▪ Negotiation and settle the responsibility of the process with stakeholders. ▪ Protection of confidentiality of Intellectual Properties of e-waste ▪ Negotiating and fixing the responsibility for the costs incurred. ▪ Define and apply Individual Producer Responsibilities

5. Sustainable issues for E-waste management in Malaysia

We need to consider some issues to make the e-waste system sustainable namely, financial-economic, environment, socio-cultural, and technical aspects. Table 5 presents a synopsis of the sustainable issues in e-waste in Malaysia.

Table 5. Synopsis of sustainable issue for e-waste in Malaysia

Issues	Synopsis
Stakeholders engagement	To achieve a sustainable e-waste recycling system in Malaysia involvement of stakeholders is very much crucial. Stakeholders are formal and informal private sector, community, schools and social organizations, NGO's, Manufacturers, Importers, Retailers, Local council, Consumers, E-waste contractors, Scrap collectors, DOE, NSWMD, MHLG. Local authorities may foster cooperation among the stakeholders to develop a sustainable e-waste system.
Environmental	Illegal dumping and landfilling are the key sustainable issues in E-waste Malaysia. Another issues is, environmental toxicity from burning acid bath where materials are directly exposing to the environment from e-waste. So, managing e-waste successfully is vital and will lead to a sustainable environment in Malaysia.
Technological	Technology can play a pivotal role towards managing e-waste in Malaysia. The harmful effects from e-waste processing can be minimized by using latest technology while in Malaysia it is limited to wet chemical process and electrolysis. Effects on human and environment can be reduced by applying

	sustainable recycling production techniques. Also, continuous research and development is needed to search for new solutions for the problems associated to Malaysia.
Economic	Successful e-waste management can bring substantial economic benefits for Malaysia. Formal and informal sectors are collecting e-waste and adding value by sorting dismantling and other activities which ultimately hit the economy. Logistics and transportation of e-waste is one of the key sustainable economic issues in Malaysia. It is important to consider the income and economic activities of the informal recyclers and scavengers. It is unrealistic to eliminate informal sectors especially in the context of developing countries, so integrating them in the mainstream activities can lead to the economic sustainability of the e-waste system in Malaysia.
Social	Without considering social issues e-waste system can't be sustainable. The key social issues are Employment, Working conditions and Health risks. These risks are high because of manual handling and lack of protective clothing and equipment, resulting in direct contact with waste. Risks of manual handling waste may come from direct contact with broken glass, toxic materials like lead, mercury, cadmium etc. Sustainable e-waste system should comply with the need for this socially excluded people and improve their living conditions. Positively, in Malaysia investment is coming in this area which creates scope for employment of these people that certainly have a positive impact to the society.

6. Conclusion

Successful and productive e-waste management has strong economic and environmental significance for sustainable development of a country. In Malaysia, attention is being given for proper e-waste management. Although some recent initiatives have been taken, but a wide scope is left undone compared to several other countries. Collecting household, institutional and other e-wastes in a sustainable manner for recycling is crucial for maintaining its ever-green environment. Land filling of these economically useful materials and environmentally sensitive products cannot be an option any further. Therefore, an integrated system for collection, recovery and remanufacturing is needed to benefit everyone out of the process. It needs assessment, characterization of materials, investigation for recycling potential, optimization of recycling, shaping the behavior of stakeholders and so on. Researches are needed on these aspects. Enforcing legislation has much importance to frame all the stakeholders' role in a system and define individual responsibilities. On the other hand cost effective and environment friendly technologies are also needed to introduce in recycling processes. Finally, stakeholders have to be proactive and change their traditional viewpoint of making money from end of life products to save the environment they have to live in.

Acknowledgement

The authors thankfully acknowledge the financial support from University of Malaya through HIR PROJECT UM.C/HIR/MOHE/ENG/31.

References

[1] Balakrishnan R B, Anand K P and Chiya A B 2007 *Waste Management & Research* **25** 307-318
[2] Brandt T and Sue W C 2012 *Market Watch 2012. DE Int.* The German Chamber Network, Malaysian Electrical & Electronic Trade Fairs 2012

[3] Department of Environment M 2009 *The E-waste inventory project malaysia* Department of Environment, Malaysia
[4] Hendricks D 2012 *National Geographic*
[5] Kahhat R, Kim J, Xu M, Allenby B, Williams E and Zhang P 2008 *Resources, Conservation and Recycling* **52** 955-964
[6] Bi B and Yamamoto K 2010 *World Academy of Science, Engineering and Technology* **38** 2010
[7] Lee J C, Song H T and Yoo J M 2007 *Resources, Conservation and Recycling* **50** 380-397
[8] Mallawarachchi H and Karunasena G 2012 *Resources, Conservation and Recycling* **68** 44-53
[9] Nnorom I C and Osibanjo O 2008 *Resources, Conservation and Recycling* **52** 843-858
[10] Oliveira C R D, Bernardes A M and Gerbase A E 2012 *Waste Management* **32** 1592-1610
[11] Ongondo F O, Williams I D and Cherrett T J 2011*Waste Management* **31** 714-730
[12] Sasaki K 2004 *LUMES. Lund University*, Sweden p 36
[13] Silveira G T R and Chang S Y 2010 *Waste Management* **30**, 2278-2291
[14] Star O T 2012 *The Star online* Malaysia
[15] UNEP 2010 *United Nations Environment Protection (UNEP)*
[16] US EPA 2007 *Management of Electronic Waste in the United States: Approach One* EPA530-R-07-004b; 2007c.
[17] Wäger P A , Hischier R and Eugster M 2011 *Science of The Total Environment* **409** 1746-1756.
[18] Wath S B, Vaidya A N, Dutt P S and Chakrabarti T 2010 *Science of The Total Environment* **409** 19-32

IOP Conf. Series: Materials Science and Engineering **50** (2013) 012067 doi:10.1088/1757-899X/50/1/012067

Properties of polyurethane foam/coconut coir fiber as a core material and as a sandwich composites component

M. A. Azmi[a], H. Z. Abdullah and, M. I. Idris
Department of Material and Design Engineering,
Universiti Tun Hussein Onn Malaysia, Batu Pahat, Johor, Malaysia.

Email: [a]azham@uthm.edu.my

Abstract. This research focuses on the fabrication and characterization of sandwich composite panels using glass fiber composite skin and polyurethane foam reinforced coconut coir fiber core. The main objectives are to characterize the physical and mechanical properties and to elucidate the effect of coconut coir fibers in polyurethane foam cores and sandwich composite panels. Coconut coir fibers were used as reinforcement in polyurethane foams in which later were applied as the core in sandwich composites ranged from 5 wt% to 20 wt%. The physical and mechanical properties found to be significant at 5 wt% coconut coir fiber in polyurethane foam cores as well as in sandwich composites. It was found that composites properties serve better in sandwich composites construction.

1. Introduction

Sandwich panels consist of two outer skins and core in the middle. The combination of these parts offer sandwich panels a relatively high strength and stiffness at low densities. Skins can be made of composite laminate panels, aluminium alloys, titanium steel or plywood. Core is the constituent that requires low density materials such as polymer foams, balsa wood, synthetic rubbers or inorganic cements [1]. Commonly sandwich composites were used in aerospace, automotive, sporting goods, marine, construction and civil structures. Various materials and structures were used to design the sandwich composites to meet the application requirement. Composite material that formed with natural fibers constitutes a current area of interest in composites research. A great development in this field has been noticed and currently applied in automotive industries [2]. Natural fibers are low priced and sustainable natural resources and have good mechanical properties [3]. Therefore, the used of this fiber reduce the materials cost of sandwich composites and in the same time improve its properties [4] Furthermore the densities of natural fibers are close to the densities of thermoset polymer and glass fiber. On the other hand, polyurethane foam (PUF) resins are widely used in the engineering applications since exhibit its structural versatility as elastomer, thermoplastic, thermosetting, rigid and flexible foam. By combining the natural fiber with polyurethane foam (PUF) as a core, the sandwich construction development will enhance the properties of Polyurethane foam as well as sandwich composites panel [5].

2. Experimental

2.1. Panel Preparation

There are several stages in the construction of the sandwich composites panel of glass fiber and polyurethane/coconut coir foam core (GFRP - PUC):

(i) Coconut coir treatment: The coconut coir fibers were crushed using granulator machine with the fiber length ranging from 0.5-1cm. Later it has been treated by using alkaline treatment method. The treatment begins when the coir fibers treated with 5 wt.% Natrium Hydroxide (NaOH) and 95 wt.% water. The treatment with NaOH solutions were required to remove impurities from coconut fiber, such as hemi cellulose, lignin and aromatic acids, which could be reduced the adhesion between fiber and matrix. The duration for this treatment was 24 hours at room temperature. Lastly, the coir fibers were cleaned and dried up in oven for 12 hours at 80° [6].

(ii) Polyurethane foam: The polyurethane foam was produced by mixing the polyol and isocynates by using one shot process. When the mixing of polyol and isocynate were prepared, coir fibers ranging from 0 wt% to 20 wt% will be added and stirred to ensure the coir fiber was uniformly distributed. Lastly, these mixed components were poured into the mould and cured at a room temperature using rotational moulding method.

(iii) Skin preparation: Skins were prepared by using compression moulding technique with woven glass fiber mat and epoxy matrix with ratio of 2:1. Skin panels consist of three plies of fiber glass woven mat with panel dimension of 300 mm x 300 mm. These panels were then cured for 12 hours at room temperature.

(iv) Sandwich composite construction: Glass fiber skins were ground with 120 grid abrasive paper to provide rough surface with polyurethane foam core. Moreover epoxy adhesive was placed between skins and cores. Sandwich composites cured using compression moulding method at 100 KPa for 8 hours to ensure excellent bonding between them. Figure 1 shows the sandwich composites with two skins placed at the outer surface and foam core located between them. Figure 2 (a) and (b) show the polyurethane foam core and sandwich panels at 0 wt.% coir fibers.

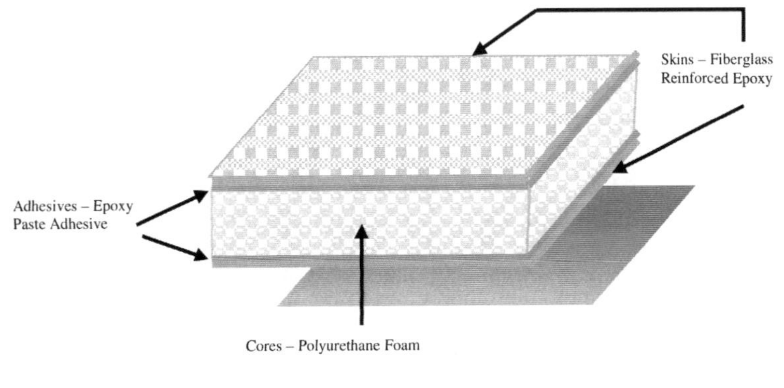

Figure 1. Sandwich composites panel of glass fiber and polyurethane/coconut coir foam core (GFRP - PUC).

(a) (b)

Figure 2. (a) Polyurethane foam core at 0 wt.% coir fibers
(b) Sandwich Panel at 0 wt.% coir fibers.

2.2 Test Methods

2.2.1. Density Test
The objective of this test is to determine the density of core materials. The density test was conducted under ASTM C 271. The samples dimensions were 300 mm x 300 mm x 18 mm. In this test, panels were tested to determine the effect of coir fiber to the panel's density. The density was measured as the mass of the panel divided by its volume [7].

2.2.2. Flexural Test
The flexural test conducted by using three point bending test [8] using universal testing machine. The objectives of this test are to determine the flexural strength and stiffness properties of sandwich composites panels. For this test, the specimen dimensions were 350mm x75mm and the test velocity was 2mm/min (ASTM C393, 2002).

3. Result and discussion

The physical and mechanical properties evaluated to analyze the comparison of between polyurethane foam cores and sandwich composites. The purpose of this analysis is to determine the ability of sandwich composites in enhancing the material properties. Sandwich composites offer excellent properties due to the way of skin and core deal with load applied.

3.1. Physical Properties
Physical properties of sandwich composites and polyurethane foam cores were determined by conducting density test. Figure 3 shows the differences between sandwich composites and cores. Polyurethane foam offered better density with lowest average value at 84.6 g/cm^3 compared to sandwich composite that have higher average density which is 382.45 g/cm^3. Sandwich composites density was increased to 352% compared to polyurethane foam cores. This increment is due to sandwich composite cores attachment to fiberglass composite skins. However the excellent properties contribution offered by the role of fiberglass composites skins in sandwich composites, make the density increment worth.

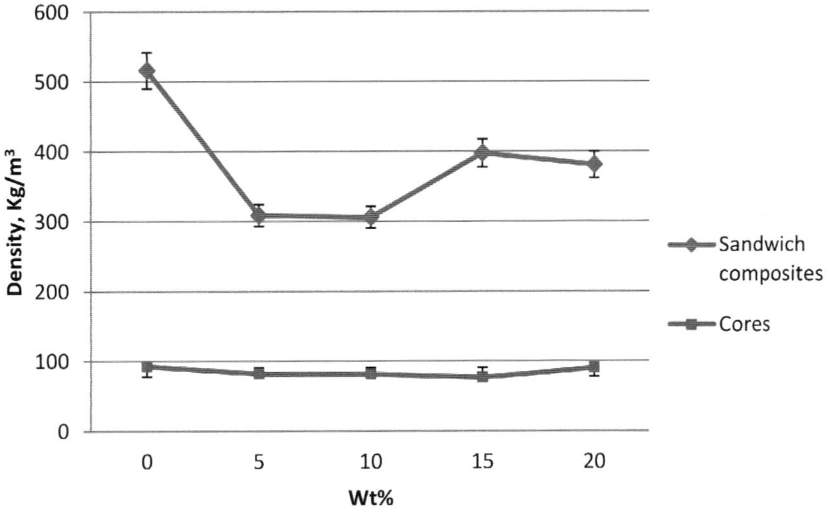

Figure 3. Density comparison of polyurethane foam cores and sandwich composites.

3.2. Mechanical Properties

Figure 4 shows the flexural shear stress comparison between cores and sandwich composites. The average flexural shear stress of sandwich composites is 190.95 kPa and cores flexural stress 31.46 kPa. The percent of shear stress increment for sandwich composites is 506.95% compared to cores. The maximum force of sandwich composites shows an increment 1145.42% compared to cores with the sandwich composites and cores maximum force are 587.04 N and 47.14 N respectively (refer figure 5). While for flexural modulus of sandwich composites and cores, the average flexural modulus are 362.76 MPa and 11.39 MPa respectively. Sandwich composites offer flexural modulus 3084% higher compared to cores. Figure 6 shows the flexural modulus comparison of polyurethane foam cores and sandwich composites.

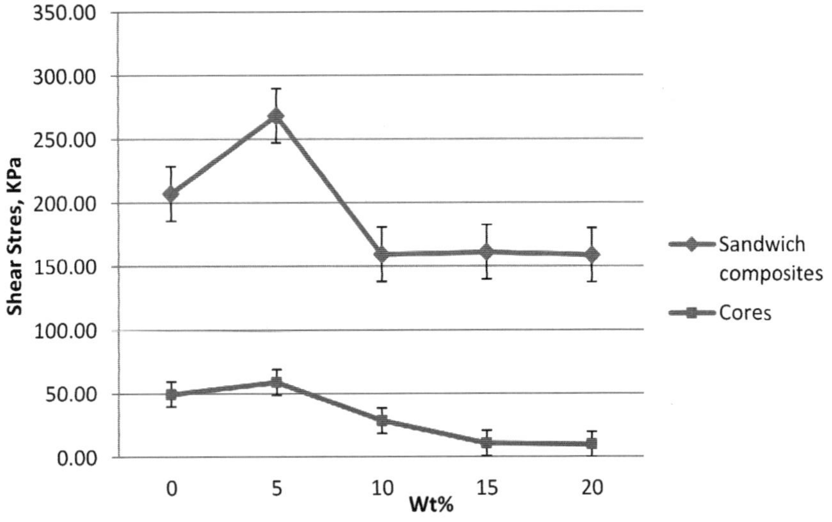

Figure 4. Flexural shear stress comparison of polyurethane foam cores and sandwich composites.

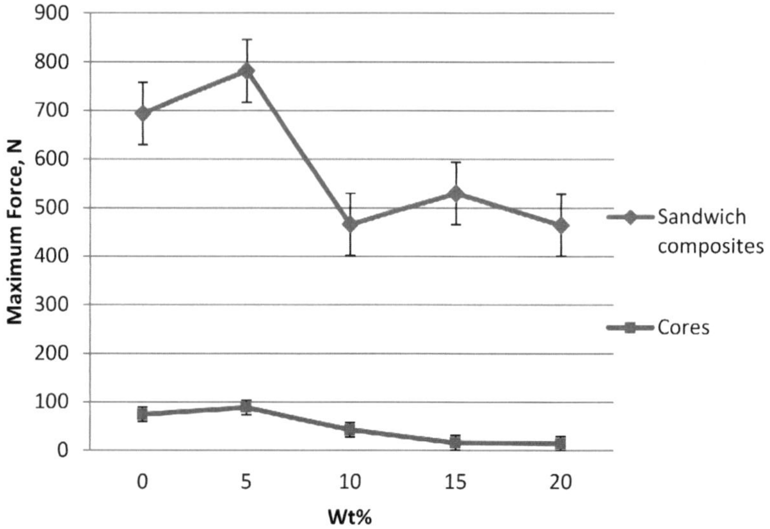

Figure 5. Maximum force comparison of polyurethane foam cores and sandwich composites.

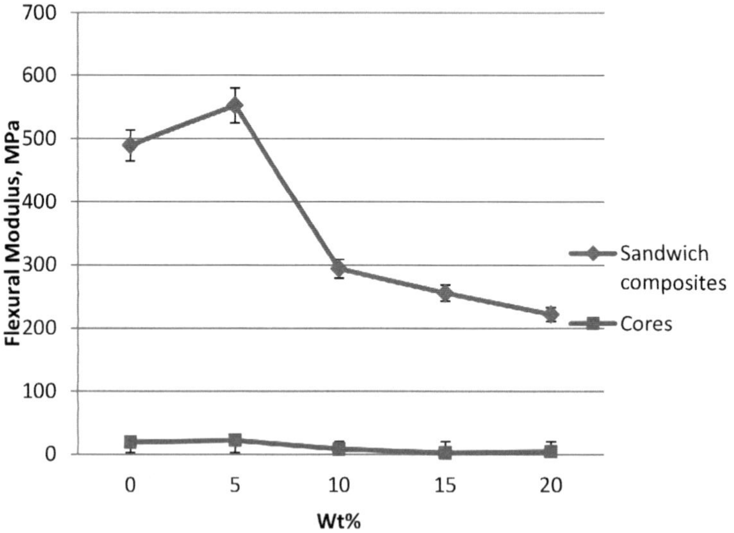

Figure 6. Flexural modulus comparison of polyurethane foam cores and sandwich composites.

The excellent increment in sandwich composites material properties compared to polyurethane foam cores is due to because of the sandwich composites structure itself. Sandwich composites structures contain skins and cores that resist each others from buckling during load application [9]. Cores of sandwich composite act as a web to stabilize the skins by carrying lateral forces applied. Most of in-plane loading and transverse bending stresses beared by the sandwich composite skins helped cores to be in compression and shear or tension and shear [10].The combinations of skins and cores properties lead sandwich composites to have better properties. Compared to polyurethane foam cores, they were found not to assist by any material in resisting the load applied. The polyurethane foam cores resist the load and buckling due to its own properties.

4. Conclusion

This study shows that by reinforcing polyurethane foam cores with coconut coir fibers, physical and mechanical properties of cores and sandwich composites had increased. The optimum performance was achieved by reinforced cores and sandwich composites with 5 wt% of coconut coir fibers. The polyurethane foam cores offered better properties as sandwich composites in resists bending stress to avoid panel buckling.

References

[1] Mallick P K 2008 Fibre reinforced composite material, manufacturing and design (Boca Raton : CRC Press)

[2] Pickering K L 2008 Properties and perfomances of natural-fibre composites (Cambridge : Woodhead Publishing)

[3] Chand N and Fahim M 2008 Tribology of natural fiber polymer composites (Cambridge : Woodhead Publishing)

[4] Bledzki A K Zhang W and Chate A 2001 Natural-fibre reinforced polyurethanemicrofoams *Composites Science and Technology* **61** 2405-11

[5] Silva R V 2005 Fracture toughness of natural fibers/castor oil polyurethane composites *Composites Science and Technology* **66** 1328-35

[6] Herrera P J and Valadez-González 2005 *A Fiber-matrix adhesion in natural fiber composites*: In A K Mohanty M Misra and L T Drzal 2005 Natural fiber (Biopolymers, and Biocomposites Florida:CRC Press) pp 187, 196-197

[7] American Society for Testing and Materials 2002 Standard test method for density of sandwich core materials (Philadelphia:ASTM C 271)

[8] American Society for Testing and Materials 2002 Standard test method for core shear properties of sandwich constructions by beam flexure (Philadelphia:ASTM C393)

[9] Papa E Corigliano A and Rizzi E 2001 Mechanical behaviour of a syntactic foam/glass fibre composite sandwich: experimental results *Structural Engineering and Mechanics* **12** 169 - 188

[10] Zheng X T Zhang J F Yang F Chai Y N and Li Y 2008 Experimental and analytical study on the mechanical behavior of stitched sandwich composite panel with a foam core Seventh *International Conference on Fracture and Strength of Solids* (FEOFS2007) (Xi'an : Northwestern Polytechnical University) 477-82

Biopolymer-based thermoplastic mixture for producing solid biodegradable shaped bodies and its photo degradation stability

Nurulsaidatulsyida Sulong and Anika Zafiah M. Rus[a]
[1] Sustainable Polymer Engineering, Advanced Manufacturing and Material Center (AMMC), Faculty of Mechanical and Manufacturing Engineering, Universiti Tun Hussein Onn Malaysia, 86400 Parit Raja, Batu Pahat, Johor

E-mail: [a]zafiah@uthm.edu.my

Abstract. In recent years, biopolymers with controllable lifetimes have become increasingly important for many applications in the areas of agriculture, biomedical implants and drug release, forestry, wild life conservation and waste management. Natural oils are considered to be the most important class of renewable sources. They can be obtained from naturally occurring plants, such as sunflower, cotton, linseed and palm oil. In Malaysia, palm oil is an inexpensive and commodity material. Biopolymer produced from palm oil (Bio-VOP) is a naturally occurring biodegradable polymer and readily available from agriculture. For packaging use however, Bio-VOP is not thermoplastic and its granular form is unsuitable for most uses in the plastics industry, mainly due to processing difficulties during extrusion or injection moulding. Thus, research workers have developed several methods to blend Bio-VOP appropriately for industrial uses. In particular, injections moulding processes, graft copolymerisation, and preparation of blends with thermoplastic polymers have been studied to produce solid biodegradable shaped bodies. HDPE was chosen as commercial thermoplastic materials and was added with 10% Bio-VOP for the preparation of solid biodegradable shaped bodies named as HD-VOP. The UV light exposure of HD-VOP at 12 minutes upon gives the highest strength of this material that is 17.6 MPa. The morphological structure of HD-VOP shows dwi structure surface fracture which is brittle and ductile properties.

1. Introduction

In recent years, eco-friendly biodegradable polymers and plastics have gained increasing attention because of growing recognition worldwide of the need to reduce global environmental pollution [1-7]. One alternative is renewable polymers made from plants. Sustainability, industrial ecology, eco-efficiency, and green chemistry are guiding the development of the next generation of materials, products, and processes. Biodegradable plastics and bio-based polymer products based on annually renewable agricultural and biomass feedstock can form the basis for sustainable, eco-efficient products that can compete and capture markets currently dominated by products based exclusively on petroleum feedstock.

Vegetable oils are becoming extremely important as renewable resources for the preparation of polyols required for the polyurethane industry. Polyols from natural oils, such as soybean, castor, and palm oils are increasingly being viewed by industry as a viable alternative to hydrocarbon based feedstocks. These oils are annually renewable, and are cost-competitive as well as environment friendly. According to a market summary published by the United Soybean Board in February 2000, vegetable oil-based polyurethanes are best suited to three markets namely: polyurethane foams, polyurethane binders and agricultural films [8]. The development of synthetic polymers using monomers from natural resources provides a new direction to develop biodegradable polymers from renewable resources. One of the examples in this regard is PLA, because it is made from agricultural products and is readily biodegradable [9].

One of the main disadvantages of biodegradable polymers obtained from renewable sources is their dominant hydrophilic character, fast degradation rate and, in some cases, unsatisfactory mechanical properties, particularly under wet environments. In principle, the properties of natural polymers can be significantly improved by blending with synthetic polymers. Polymer blending is a well-used technique whenever modification of properties is required, because it uses conventional technology at low cost. The usual objective for preparing a novel blend of two or more polymers is not to change the properties of the components drastically, but to capitalize on the maximum possible performance of the blend. In the 1970s and 1980s, numerous blends of starch with various polyolefins were developed. However, these blends were not biodegradable, and thus the advantage of using a biodegradable polysaccharide was lost [9]. In this research, thermoplastic polymer blends with renewable polymer from palm oil was discussed. An analysis of the mechanical properties and UV degradation was discussed in detail in a later section.

Normally all plastics products are manufactured using extrusion, injection molding, or extrusion blowing. The processing of polymers using heat and high shear into useful end products introduces impurities and reaction products that make them susceptible to photodegradation. Because of these complications, the extrapolation of research findings on UV-induced degradation of pure polymer resins to compounded and processed products of the same polymer, is often unreliable. Photo degradation data generated on the actual polymer formulations used in practice, processed in the conventional manner are the most useful for assessment of damage [10].

Polymer photo-degradation and stabilization is an extensive field of study. It is well established that photo-oxidation reactions play an important role in the degradation process of the UV-irradiated polymeric materials, and the controlling mechanisms were studied by some authors [11].

The polymers, used in the building and construction industry, undergo photolytic and photo-oxidative reactions during exposure to solar UV radiation. Microcracking and embrittlement of polymeric substrates are the most serious effects of photodegradation, these effects are often accompanied by extensive deterioration in the mechanical properties of the materials, such as tensile strength, impact strength and elongation, all of which are important parameters in the performance of a building product. It is therefore the prediction of the effect of UV radiation on plastics by natural weathering exposure, particularly where product development is concerned [10].

2. Methodology

2.1 Materials

Injection grade HDPE (HI1600) was supplied by Titan Petchem (M) Sdn. Bhd, having density of 0.94 to 0.965 g/cm^3 and the melting point of HDPE is >120°.

The origin bio-polymer (Bio-VOP), Vesawit Palm oil was used as original palm oil. The monomer conversion begins with the catalyst preparation to generate the epoxies from the unsaturated fatty compound, and second reaction is the acid-catalyst ring opening of the epoxies to form polyols or bio monomer [12- 17]. The bio-monomer mixed with cross linking agent; 4, 4'-Methylene Diphenyl Diisocyanate (MDI) and Toluene as the solvent. The proportion ratio is 1:0.5 of bio-monomer and MDI respectively. The mixture was stirred at 50°C for 15 minutes until the mixture become high viscous known as Bio-VOP and added to the HDPE in the Brabender Plastograph® EC machine.

For the preparation of polymer matrixs materials was started by mixing the HDPE with 10% bio-polymer (wt/wt ratio) by using Brabender Plastograph® EC machine from Germany at 130°C and 25 rpm speed rate for 20 minutes. This polymer/ biopolymer compound named as HD-VOP.

2.2 Injection Moulding

The compounding materials were then fed to injection moulding machine (Nissei Horizontal Screw Type Injection Moulding NP7 Real Mini from Japan). The mold was designed according to ISO 527 (5A) standards to produce specimens for tensile tests. Barrel temperature set points of nozzle zone, front zone, middle zone, rear 1 zone, rear 2 zone and feed zone were 150°C, 145°C, 140°C, 135°C,130°C and 70°C respectively.

2.3 Tensile Test

The tensile test sample was prepared according to ISO 527 (5A). The testing was consists of five samples for each different time exposure and Universal Tensile Machine AG-I, Shimadzu, 10 kN types with 5 mm/min crosshead speed was used.

2.4 UV Weathering Test Chamber

35 dumbbell shape samples produced from injection moulding was exposed to the UV light in UV Lamp Test Chamber Model HD-703 (Haida International Equipment Co., LTD) at different exposure time at 50° C. Different UV exposure time was conducted for each group of 5 sample to calculate the average value of mechanical testing. The sample were identified as A, B, C, D, E, F, G, and H for unexposed sample and exposed to 3 minutes, 6 minutes, 9 minutes, 12 minutes, 15 minutes, 18 minutes, and 21 minutes respectively. The UV- irradiation of the dumbbell sample was carried out using an array of UV fluorescent lamp emitting light in the region from 280 to 320 nm with a tail extending to 400 nm.

2.5 Scanning Electron Microscopy (SEM)

The morphological study based on the fracture surface of HD-VOP was examined using a SEM (JSM-6380LA Analytical Scanning Electron Microscope manufactured by JEOL Company Japan). The samples were cut into 1cm x 1cm x 1cm and was ultrathin coated with gold to produce electrically conductive specimens by JFC-1600 Auto Coater manufactured (JEOL Company Japan). The gold deposited on the sample either by low vacuum sputter coating or by high vacuum evaporation. Samples were examined after coated with ultrathin platinum, at 10kV voltage supplied by the SEM.

3. Discussions

Referring to figure 1, sample E was exposed in 12 minutes to UV light shows the highest mechanical property of tensile strength of 17.6 MPa while sample B exposed to 3 minutes of UV light shows the lowest mechanical property that is 17.2 MPa. According to figure 1, the strength of UV irradiated samples shows slight decrease of about 3 minutes and then increases up to exposure times of about 15 minutes then decrease again of about 21 minutes. The initial decrease in strength with exposure time may be due to occurrence of chain scission processes as a result of the photo-oxidation. However, the increase in strength with exposure time may be due to crosslink reaction of the polymer chains as a result of exposure to ultraviolet radiation [10-11].

The young's modulus of sample A of unexposed to UV irradiation gives the highest result of 1.23 MPa while sample B gives the lowest result of 1.07 MPa. The young's modulus exhibited small changes. However, these changes are not significant since all results fall within the same statistical range of newly manufactured material. Young's modulus is a measure of the stiffness of an elastic material and is a quantity used to characterize materials. Sample B shows the lowest result for both strength and young's modulus due to cross-linking of the material upon the photo degradation of UV irradiation.

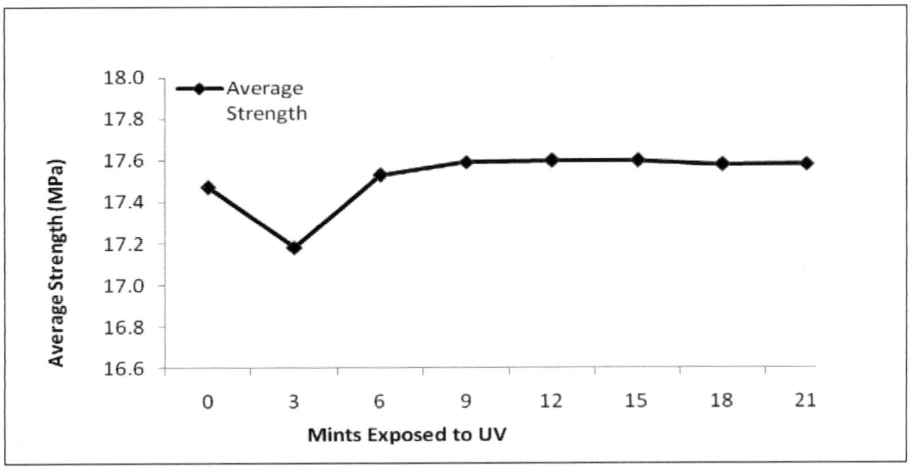

Figure 1. Average tensile strength of the HD-VOP sample.

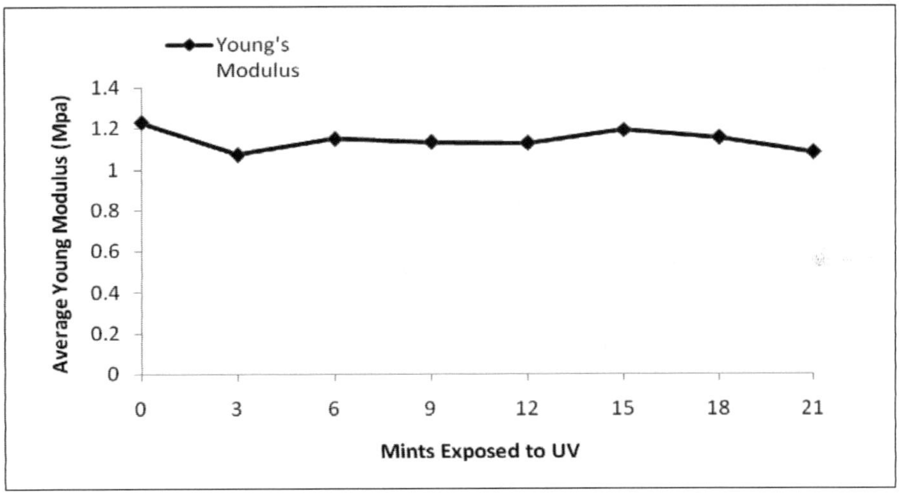

Figure 2. Young's Modulus of the HD-VOP sample.

Referring to figure 3 (a) (b) and (c), the fracture surface of sample A, B and C has two characteristic as shows in part A and part B. From figure 3, part A can be classified as a brittle fracture while for B is ductile fracture. SEM result for sample A, B and C shows the structure in part A have interconnected and twisted fibrils with large interfibrillar separation. Meanwhile, part B shows ductile fracture with numerous spherical 'dimples' resulting from uniaxial tensile failure. Each dimple is one half of a microvoid that formed and separated during the fracture process. It's also appears fibrous or dull and it have very rough surface compared to the matrix. As looking at the fracture surface of these samples, it's undergo a moderate amount of necking. The fracture process normally occurs in several stages. First, after necking begins, small cavities, or microvoids, form in the interior of the cross section. Next, as deformation continues, these microvoids enlarge, come together, and coalesce to form an elliptical crack, which has its long axis perpendicular to the stress direction. The

IOP Conf. Series: Materials Science and Engineering **50** (2013) 012068 doi:10.1088/1757-899X/50/1/012068

crack continues to grow in a direction parallel to its major axis by this microvoid coalescence process. Finally, fracture ensues by the rapid propagation of a crack around the outer perimeter of the neck, by shear deformation at an angle of about 45° with the tensile axis which is the angle at which the shear stress is at maximum. Sometimes a fracture having this characteristic surface contour is termed a *cup and-cone fracture* because one of the mating surfaces is in the form of a cup, the other like a cone [18].

For the sample D, E and F shows uneven fracture surface which is no void and dimples appear that cause higher in tensile strength. For sample G and H, there are numerous spherical 'dimples' resulting from uniaxial tensile failure that causes reduced in strength. The sample was exposed longer to UV light which makes these samples more brittle and the behaviour of the cracks propagation will be changed according to the weakness point in specimens.

a)

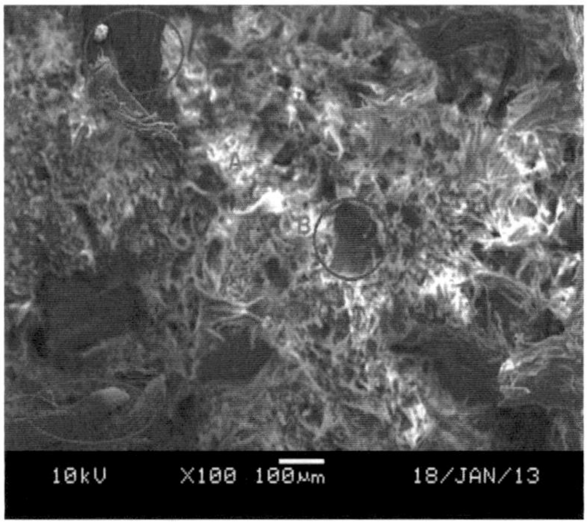

b)

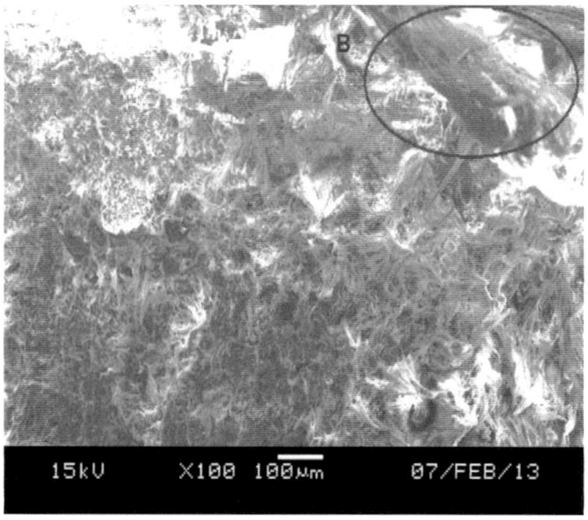

c)

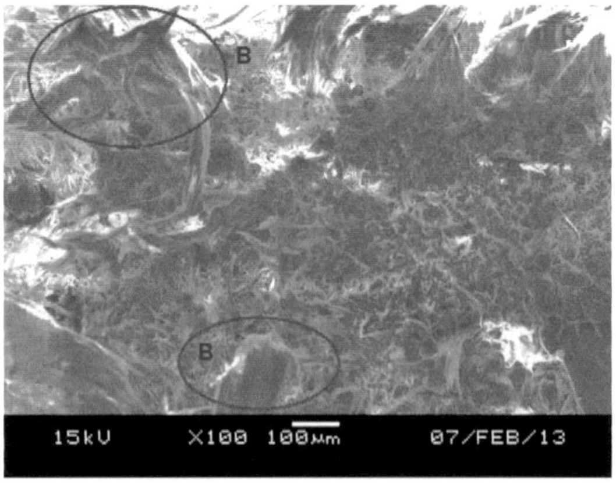

d)

e)

f)

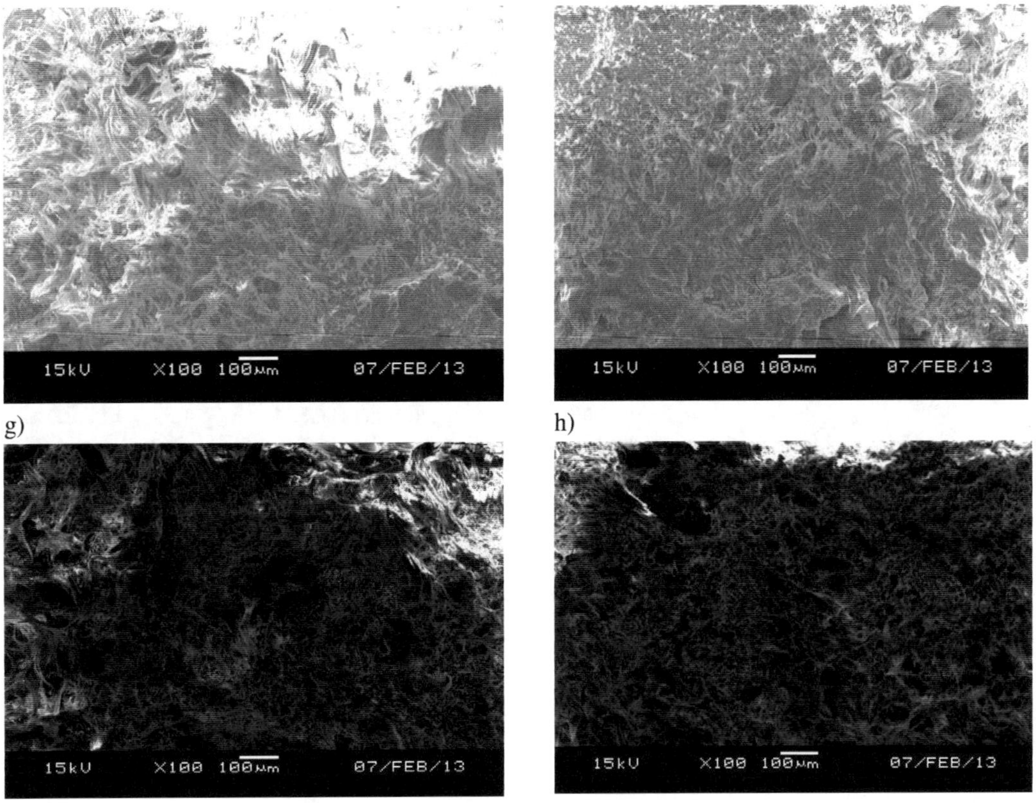

Figure 3. SEM morphology of fracture surface cross-section for HD-VOP.

4. Conclusions

As the conclusion, HD-VOP was successfully prepared by injection moulding process. The blending of thermosetting and thermoplastic polymer shows maximum tensile strength that is 17.6 MPa. The morphological structure of HD-VOP shows dwi structure surface fracture which is brittle and ductile properties. Therefore, this study revealed the potential of HD-VOP to be used in industrial applications based on mass production.

Acknowledgement

The author would like to thank the Malaysian Government and University Tun Hussein Onn Malaysia (UTHM), Johor and Malaysian Technical University Centre of Excellence (MTUN CoE) research grant Vot C014 and C015 for supporting this research.

References

[1] Kim H E 2006 Biodegradability of bio-flour filled biodegradable poly (butylenes succinate) bio-composites in natural and compost soil *Polymer Degradation and Stability.* **91** 1117-27
[2] Hadzreel M R A M and Aisha I S R 2013 *J Mech Eng Sci* **4** 548-553.
[3] Hardinnawirda K and Aisha I S.R 2012 *J Mech Eng Sci* **2** 181-186.
[4] Mohamed W A N W and Atan R 2012 *Inter J Automot Mech Eng* **5** 648-659.
[5] Umar A H, Zainudin E S and Sapuan S M 2012 *J Mech Eng Sci* **2** 198-205.

[6] Shan C W, Ghazali M I and Idris M I 2013 *Inter J Automot Mech Eng* **7** 1032-1042.
[7] Adebisi A A, Maleque M A and Rahman M M 2011 *Inter J Automot Mech Eng* **4** 471-480.
[8] Sharma V and Kundu P P 2008 Condensation polymers from natural oils *Progress in Polymer Science.* **33** 1199–1215
[9] Yu L., Dean K and Li L 2006 Polymer blends and composites from renewable resources *Progress in Polymer Science.* **31** 576–602
[10] AL-Shammary Z R 2011 The effect of inorganic pigment on the tensile properties of polycarbonate irradiated with UV – light *Iraqi Journal of Physics.* **9(16)** 52-56
[11] Salem M A 2001 Mechanical Properties of UV-Irradiated Low-Density Polyethylene Films Formulated With Carbon Black and Titanium Dioxide *Egypt. J. Sol.* **24(2)**141-150
[12] Andrew J C, Leornarde M and Mohd Rus A Z 2003 Private Comm *DEFRA*
[13] Mohd Rus A Z 2008 Degradation Studies of Polyurethanes Based On Vegetables Oils. Part I; Photodegradation *Prog in Reaction Kinetic And Mechanism.* **33** 363-391
[14] Mohd Rus A Z 2009 Degradation Studies of Polyurethanes Based On Vegetable Oils. Part 2; Thermal Degradation And Materials Properties *Prog React Kinet Mech.* **34** 1-43
[15] Mohd Rus A Z 2009 Effect of Titanium Dioxide on Material Properties for Renewable Rapeseed and Sunflower Polyurethane *International Journal of Integrated Engineering.* **1**(1)
[16] M. Rus A Z 2009 Material Properties of Novelty Polyurethane Based On Vegetable Oils *The 11th International Conference on QiR (Quality in Research)*, Depok, Indonesia, 3-6 August 2009
[17] M. Rus A Z 2010 Polymer from Renewable Materials, Science Progress available online; www.scilet.com/DB/SPROG
[18] Ahmad I and Mahanwar P A 2010 Mechanical Properties of Fly Ash Filled High Density Polyethylene *Journal of Minerals & Materials Characterization & Engineering.* **9(3)** 183-198

An experimental investigation of springback of AA6061 aluminum alloy strip via V- bending process

A B Abdullah[a] and Z Samad
School of Mechanical Engineering, Universiti Sains Malaysia Engineering Campus, Universiti Sains Malaysia, 14300, Nibong Tebal, Penang, Malaysia

E-mail: [a]mebaha@eng.usm.my

Abstract. Springback is one of the common geometrical defects found in the sheet metal forming process. Aluminium alloy with high content of magnesium such as AA6061 is preferred for their high formability limit, but commonly springback becomes a drawback. In this study, springback behaviour of the AA6061 will be observed. The effect of length, thickness and bend angle to the springback pattern was investigated and the result depicted that springback is more significant to thickness and bend angle, while the length gives less effect.

1. Introduction

Springback is one of considerable geometrical defect in sheet forming processes [1]. Springback can only be controlled and minimized but quite difficult to be eliminated. For that reason, understanding the mechanism and behavior of the springback is beneficial. In most literatures, the springback can be refereed as a condition that occurs when sheet metal or alloy formed, and the material has a tendency to return to its initial shape because of the elastic recovery of the material upon unloading [2]. It is not only influenced by the tensile and yield strengths, but also by thickness, bend radius and bend angle [3]. In most recent research, finite element is the main tool used to understand, predict and improved the component defect through springback [4-6]. In addition, there are few other tools that employed in predicting and measuring the springback such ANN [7], optical method [1] and optimization using the statistical method such as ANOVA and Taguchi method [8-10]. Recently the topic on springback measurement and compensation has devoted attention from researchers. The studies can be discussed into physical and numerical approaches. The physical parameters can be further explored into material properties and process characteristics as well as their sensitiveness. Typically within the former parameter, the issue of isotropic and kinematic hardening [11], Bauschinger effect [12], evolution of elastic properties [13] and elastic and plastic anisotropy [14] were raised. While the later, mostly the problem of sheet thickness, friction coefficient, blank holder force, tool geometry, contact pressure, temperature and unloading procedure discussed [15-17]. Springback measured on various types of sheet forming processes. For example, recently Zhan et al. [18] reviewed a creep age forming (CAF), and they found that to predict the amount of springback is the greatest challenge. Wang et al. [19] presented a practical incremental bending methodology to control punch displacement to achieve more accurate final bend angles. In this method, information collected from loading–unloading cycles were utilized.

The paper will present the experimental study on springback behavior of AA6061. The study focuses on the effect of length, thickness and bend angle to the springback. The paper starts with introduction and then follows by the experiment setup. The results are then discussed and, the paper ends with the conclusion.

2. Motivation

Twist springback is one the major geometrical defect in the manufacturing of the propeller blade. Conceptual the blade has complex design, which has hydrodynamic profile and twisted at varies twist angle as shown in figure 1 and summarized in table 1. The chord length also varies from the head to the tip. Preliminary works on the measurement of the twist springback on cold forged propeller blade have been conducted by Abdullah et al. [20]. In order to further understanding the problem of twist springback in the propeller blade, which is complex, pattern of springback in simple shape, need to be explored first. The output of this study will give the fundamental on how springback is happened and the parameters which may influence the occurrence of the springback.

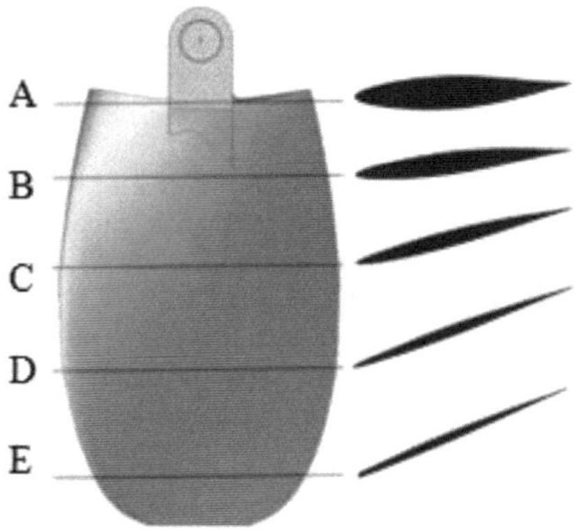

Figure 1. Nominal model of the propeller blade.

Table 1: Geometries of the produced blade.

Section	Chord Length, mm	Twist Angle, degree	Minimum Thickness, mm	Maximum Thickness, mm
A	17.79	48.80	0.410	2.768
B	19.87	51.76	0.182	2.177
C	21.72	58.90	0.154	1.560
D	21.13	65.48	0.120	1.005
E	18.96	67.35	0.010	0.703

3. Experimental setup

For v-bending experiment, the die set is prepared and as shown in figure 2. The die and puncher were made from D2 tool steel material and the experiment was performed using the same machine. The experiment will be performed and the springback pattern will be only observed based on the variation of bend angle, length and thickness of the workpiece. In the measurement of the springback, the Rax Vision Mitutoyo Profile Projector (PC 3000) machine was used. In the measurement, the 2 lines technique was employed to measure the angle as shown in figure 3. In this case, the difference between the angle after loading and after unloading is defined as springback. The die valley radius, punch radius and the die opening are constant at 90°, 2 mm and 32 mm respectively as illustrated in figure 4 (a) and (b). In this case, the die radius and punch radius has the same value.

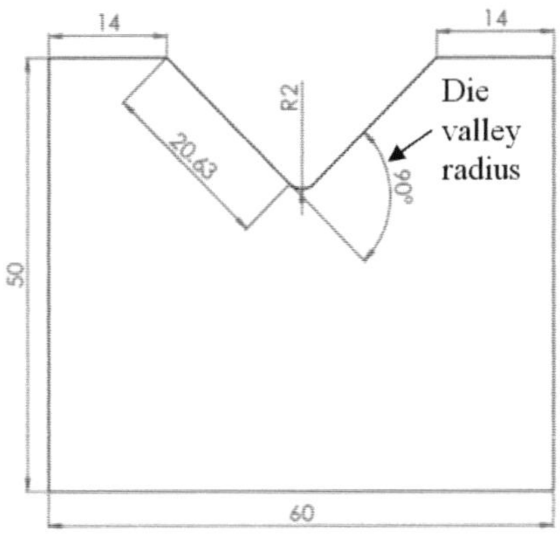

(a)

(b)

Figure 2. (a) V-bending bottom die; (b) V-bending setup.

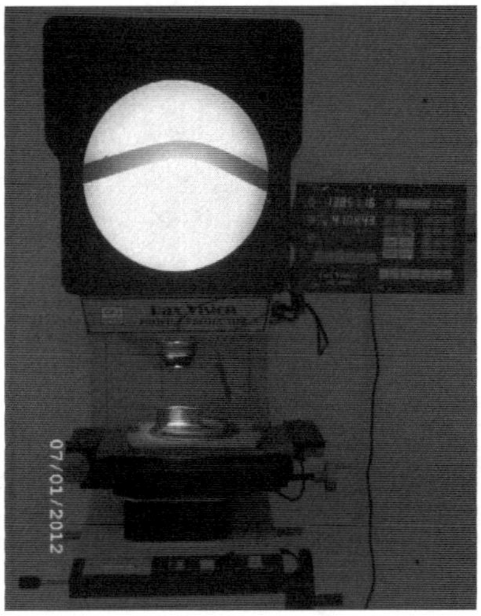

Figure 3. The profile projector used in measuring the springback.

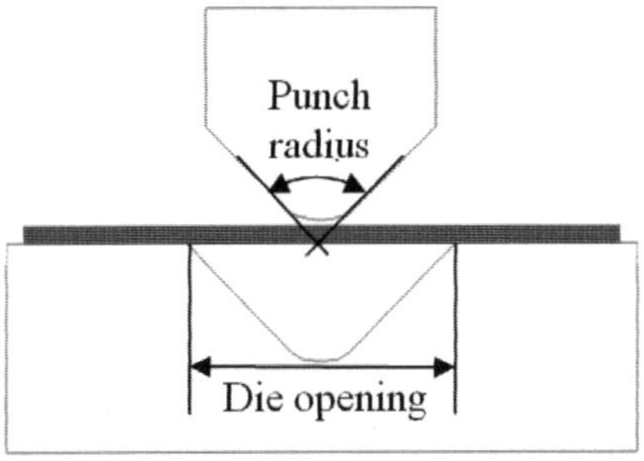

(a)

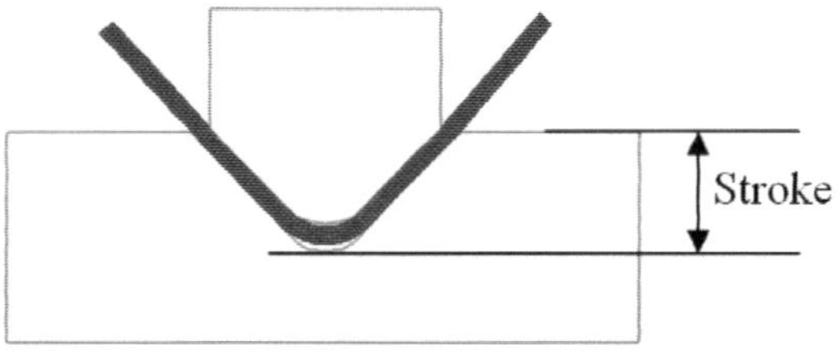

(b)

Figure 4. The definition of v-bending process, at (a) initial and (b) end.

4. Result and Discussion

For the v-bending experiment, observations were made at different stroke. The bend radius of 5, 7 and 10mm stroke is 17, 24 and 32 degree respectively. The maximum stress is approximately 90MPa and the pattern seems to be the same for each cases. For the 10 mm stroke, the workpiece tend to fail after strain is achieving 0.3. Failure can be observed from the crack mark on the bottom side of the workpiece as shown in figure 5.

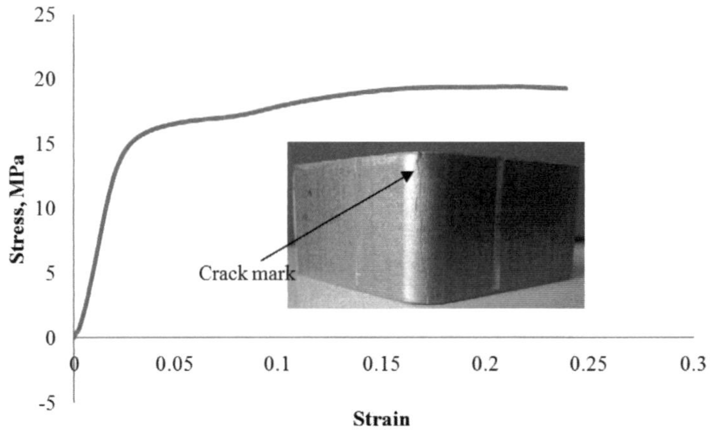

Figure 5. The stress strain behaviour of the material.

Figure 6 - 8 present the effect of the length and bend angle to the springback at different thickness. The result shows that the length of the workpiece has less effect to the springback compare to band angle. This can be seen in the results, even though there is negative and positive springback, but the pattern is about the same for each case. At higher thickness as shown in figure 6, the effect of bend angle can be clearly seen. In contrast, at lower thickness the amount of springback is higher but less effect of the bend angle as shown in figure 7 and 8. In other aspect, the effect of length to the springback is found to be insignificant. These results seem to be in line with the other studies conducted by Garcia-Romeu et al. [21] and Ouakdi et al. [22].

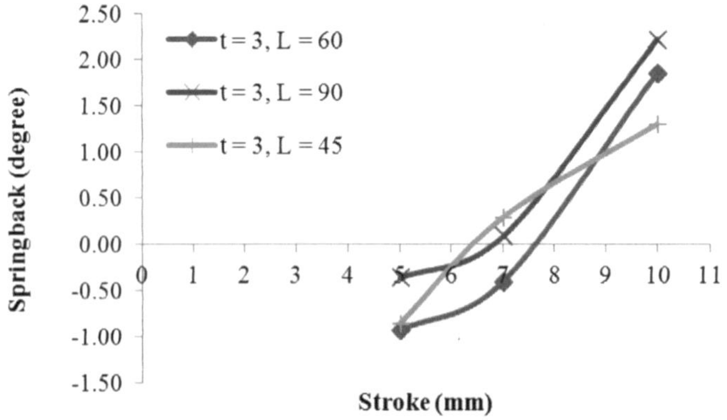

Figure 6. The springback behaviour at thickness of 3.0 mm.

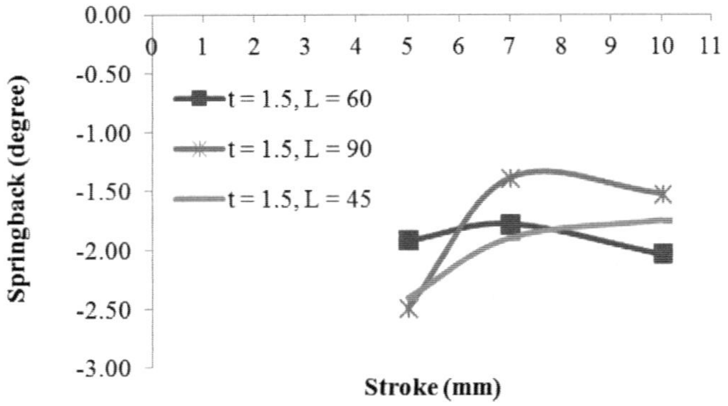

Figure 7. The springback behaviour at thickness of 1.5 mm.

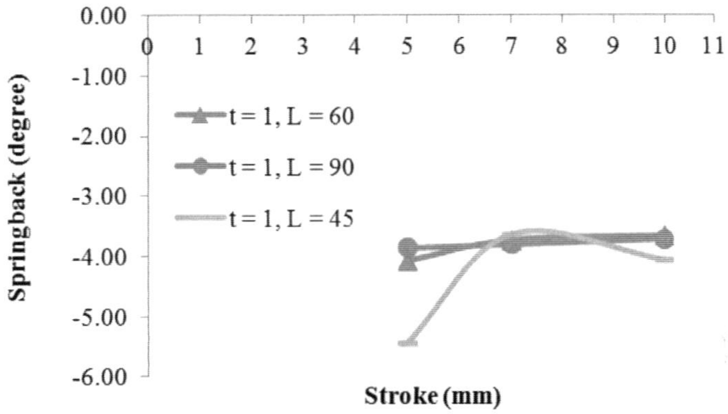

Figure 8. The springback behaviour at thickness of 1.0 mm.

5. Conclusion

For conclusion, the study has presented the experimental investigation on springback behavior of AA6061. The effects of length, thickness and bend angle have been studied and the result depicted that the springback is significant to the thickness and bend angle changes. In contrast, it has a very less effect to length of the workpiece. For future works, the springback behavior of the alloy with non-uniform thickness will be further explored. This data is very useful as the fundamental understanding and benchmark for further investigation of the twist springback in the cold forging of propeller blade made of aluminium alloy. Therefore, only length, bend angle and thickness were highlighted in the present study. These parameters may represent the chord length, twist angle and thickness of the blade and for the blade performance, those parameters are very critical and may affect the performance of the blade as well as the propeller.

Acknowledgements

Authors would like to acknowledge the Universiti Sains Malaysia for their sponsorship in this project through Short Term grant scheme (#60312014). Through this opportunity, authors would thank Mr. Fakruruzi and Mr. Zulhairi for their assistant in the experiment.

References

[1] Abdullah A B, Sapuan S M, Samad Z, Khaleed H M T and Aziz N A 2011 Aust. *J. Basic Appl. Sci.* **5(9)** 1756
[2] Marretta L, Ingarao G and Di Lorenzo R 2010 *Int. J. Mech. Sci.* **52(7)** 914
[3] Wang J F, Wagoner R H, Carden W D, Matlock D K and Barlat F 2004 *Int. J. Plasticity.* **20 (12)** 2209
[4] Schwarze M, Vladimirov I N and Reese S 2011 *Comput. Method. Appl.* M **200(5-8)** 454
[5] Shen H Q, Li SH and Chen GL 2010 *Mater. Design* **31(2)** 870
[6] Yu H Y 2009 *Mater. Design* 30(3) 846
[7] Bozdemir M and Golcu M 2008 *J Appl. Sci.* 8 3038
[8] Meinders T, Burchitz I A, Bonte M H A and Lingbeek RA 2008 *Int. J. Mach. Tool.* Manu. **48(5)** 499
[9] Khan M A R, Rahman M M, Kadirgama K, Maleque M A and Ishak M 2011 *J Mech Eng Sci.* 1 16
[10] Parate P R and Yarasu R B 2013 *J Mech Eng Sci.* **4** 479
[11] Vladimirov I N, Pietryga M P and Reese S 2009 *J. Mater. Process. Tech.* **209(8)** 4062
[12] Gau J T and Kinzel G L 2001 *Int. J. Mech. Sci.* **43(8)** 1813
[13] Sun L and Wagoner R H 2011 *Int. J. Plasticity.* **27(7)** 1126
[14] Li X, Yang Y, Wang Y, Bao J and Li S 2002 *J. Mater. Process. Tech.* **123** 209
[15] Li K P, Carden W P and Wagoner R H 2002 *Int. J. Mech. Sci.* **44(1)** 103
[16] Osman M A, Shazly M, El-Mokaddem A and Wifi A S 2010 *J. Achieve Mater. Manuf. Eng.* **38(2)** 179
[17] Grèze R, Manach P Y, Laurent H, Thuillier S and Menezes L F 2010 *Int. J. Mech. Sci.* **52(9)** 1094
[18] Zhan L, Lin J and Dean T A 2011 *Int. J. Mach. Tool. Manu.* **51(1)** 1
[19] Wang J, Verma S, Alexander R and Gau J T 2008 *J. Manuf. Processes.* **10(1)** 21
[20] Abdullah A B, Sapuan S M, Samad Z, Khaleed, H M T. and Aziz N A 2012 *Adv. Mater. Res.* 512-15 1840
[21] Garcia-Romeu M L, Ciurana J and Ferrer I 2007 *J. Mater. Process. Tech.* **191** 174
[22] Ouakdi E H, Louahdi R, Khirani D and Tabourot L 2012 *Mater. Design.* **35** 106

Impact strength on fiber-reinforced hybrid composite

S M Firdaus[1,a], **Nurdina**[1], **M Azmil Ariff** [1]
Faculty of Mechanical Engineering, Universiti Teknologi MARA,
13500 Permatang Pauh, Pulau Pinang.

E-mail: [a]sh.firdaus@ppinang.uitm.edu.my

Abstract. Acrylonitrile-Butadiene-Styrene (ABS) has been well known composite in automotive players to have light weight with high impact strength material compared to sheet metal material which has high impact strength but heavy in weight. In this project, the impact strength properties of fabricated pure ABS were compared to the eight samples of hybrid ABS composite with different weight percentages of short fibers and particle sizes of ground rubber. The objective was to improve the impact strength in addition of short fibers and ground rubber particles. These samples were then characterized using an un-notched Izod impact test. Results show that the increasing of filler percentage yielded an adverse effect on the impact strength of the hybrid composite. The effect of the ground rubber particulate sizes however are deemed to be marginal than the effect of varying filler percentage based on the collected impact strength data from all physically tested hybrid composites.

1. Introduction

Acrylonitrile-butadiene-styrene (ABS) has been a popular material in major automotive players and other sectors [1-6]. The plastic's composition flexibility makes it a suitable candidate for fabrication of composites where demands of engineering materials are steadily increasing over the past years [7-10]. Another aspect that drives the selection of ABS is its relative low cost to flexibility and overall durability compared to others engineering plastics or sheet metal [11]. The materials popularity has also disadvantage which they were non-environmentally-friendly wastage or disposal [7]. Various researchers have investigating the alternative to recycle the plastic or combine it with waste materials such as recycled PVC or ground rubber [1] to mitigate the problem. Ground rubber (GR) as ABS composite filler had garnered considerable interest, as it can potentially solve the problem of waste tire disposal and in the same time create a low-cost composite.

Another take to mitigate the impact of post-ABS usage is utilization of natural fibers to create fiber composites that uses renewable and environmentally-friendly sources. Natural fibers such as cotton, hemp and sisal had been investigated by other researchers as a potential replacement for synthetic fibers in composites and, they were found to be equivalent to synthetic fibers in certain aspects while being cheaper and renewable [12-13].

Addition of GR to ABS was touted to increase the plastic's dampening properties and they were manufactured to produce pipes but some studies had pointed it is more trouble than it is worth as addition of GR was found to significantly weaken the ABS. Several studies were done by other researchers to combat this problem via GR surface modification by pretreatment or vulcanization or addition of compatibiliser chemical [1]. Interesting prospects exist on adding fibers to particle-reinforced composites whereby it can potentially improve the composite to perform better in variety of areas to be used in various applications. Thus it is relevant to add fibers to ABS/GR composites to assess whether any improvement can be achieved, as numerous other researchers in other fields had found additions of fibers on their researched particle-reinforced composite are favorable [14-18]. Current literature on ABS/GR composites involves usage of novelty processes or complicated procedures and dangerous reagents while inclusion of natural fibers into such composites had not been investigated

Thus, a customized method had been investigated to create a similarly performing composite at vastly reduced costs and production complexity. It is also lucrative to incorporate short fibers into the composite to provide additional reinforcement as a mere ABS/GR composite is predicted to be significantly weaker than pure ABS [1] and the effects of the addition can be evaluated by conducting a simple impact strength test.

2. Experimental

2.1. Materials
Pure ABS pellets were used and mixed with ground rubber having sizes of 125, 250 and 500 microns. Short cotton fibers having an average fiber length of 5 mm to 8 mm were prepared and used as hybrid composite filler.

2.2. ABS Ratio Optimization
Approximately 100g of pure ABS pellets was liquefied in using ABS:Acetone ratios of 1:2, 1:4, 1:6 and 1:8 where each ratio sample is labeled as A, B, C and D and the liquefaction process was aided using an overhead mechanical stirrer. The process was conducted in a fume hood and left running for approximately 24 hours before retrieved for observation, preceding the sun drying process that utilizes halves of the samples respectively. The halves were subjected to five hour drying phase to direct sun with an observation for every one hour respectively. Based on the collected data, the optimum ratio was then determined for the preparation of hybrid ABS composites.

2.3. Hybrid ABS Composite Preparation
The weight of the thermoplastic was scaled down to 80g as the liquefied plastic was expected to overflow the carbon steel mould. The ground rubber particulates were prepared in sizes of 125, 250 and 500 microns. Short cotton fibers from bulk spinned yarn was then prepared into lengths of 5 mm to 8 mm. Three samples having filler percentage weight of 5%, 10% and 15% were prepared for each ground rubber particle size by initially liquefying the ABS for one hour, followed by addition of the filler materials and additional stirring time of fifteen minutes. Afterwards, the liquid mixtures were transferred to carbon steel moulds to be oven dried.

2.4. Unnotched Izod Impact Test
Three test samples were cut from each hybrid composite plate, bearing dimensions of 3.2 mm thickness, 12.7 mm width and 64 mm length. The Zwick swing arm impact test machine was then set up by attaching the Izod impact test arm and calibrated to deliver 7.5 joules of impact energy for each impact test conducted on all test samples respectively.

3. Results And Discussions

4. Analysis of ABS Ratio Optimization
Figure 1 show the undried sample C indicates intermediate elasticity between sample A and B while having uniform colour when compared to other samples. The undried sample D was ruled out due to presence of liquid mixture and this can be seen as potentially as a wasteful ratio. The sun-dried samples suggest the C sample is a best compromise in texture and colour. A uniform colour and median texture characteristics are approximated to indicate a good and complete liquefaction process thus the C sample ratio is deemed sufficient for ABS liquefaction.

IOP Conf. Series: Materials Science and Engineering **50** (2013) 012070 doi:10.1088/1757-899X/50/1/012070

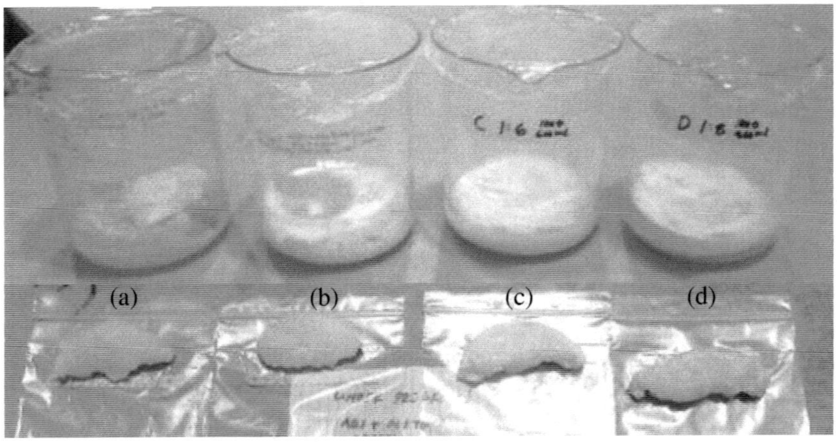

Figure 1. Sample ratio of ABS:Acetone (a) 1:2; (b) 1:4; (c) 1:6 and (d) 1:8.

5. Unnotched Izod Impact Test Analysis

Results from the impact test analysis conducted on the generated hybrid ABS composites suggest that sudden addition of the filler materials caused a drop on impact resistivity and there is a general decreasing trend of impact strength as the filler weight percentages were increased, as shown in figure 2, contrary to certain studies that indicated otherwise which may be explained due to lack of compatibiliser chemical and/or incompatible intermolecular interaction between the matrix and the fiber used in this research [19-22].

The increasing of GR particle sizes has smaller effect on impact strength in relative to filler weight percentages. The initial impact strength drop explained by the amount of additional GR to the ABS can significantly weaken the thermoplastic's physical properties [1].

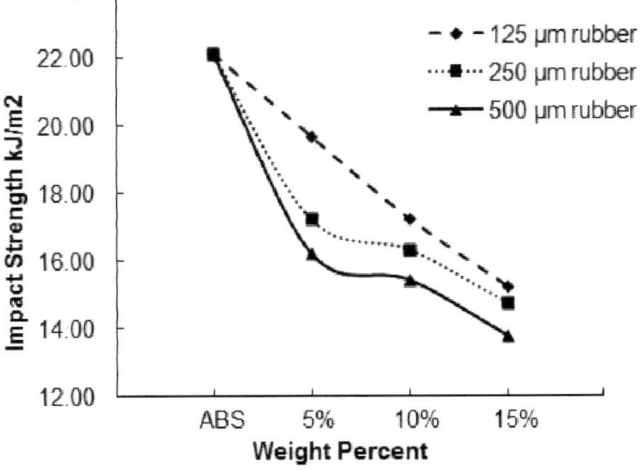

Figure 2. Impact strength versus weight percent, ABS denote reference point.

The decreasing of impact strength due to random orientation of the fibres figure 3(a), uneven distribution of the filler materials at figure 3(b) and presence of air bubbles at figure 3(c) shown.

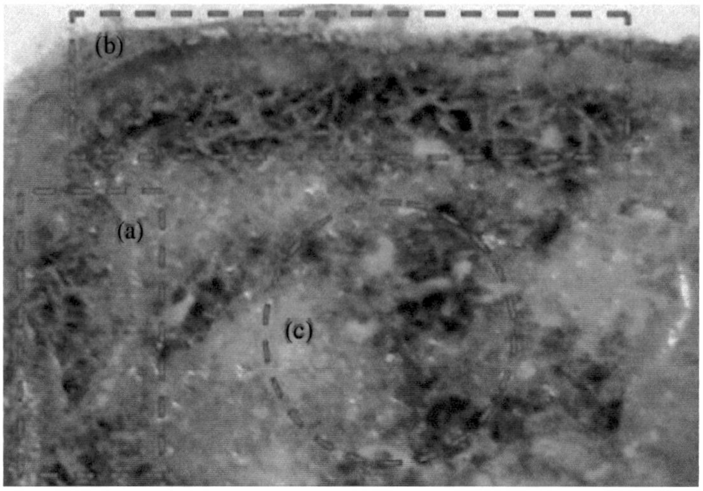

Figure 3. Defect occurred on the hybrid composite: (a) random orientation of fibers; (b) uneven distribution of filler material and (c) air bubble.

6. Conclusions

An optimum ratio for ABS liquefaction was found to be 1:6, for a customized fabrication method. As the filler increased, the impact strength shows a decreasing trend, it is due to the presence of air bubble and the non-uniformed filler distribution. The effect of the filler weight percentages is more profound than the effect of varying the GR particle sizes.

Acknowledgments

The authors wish to thank Universiti Teknologi MARA for the financial support Excellent Grant 2011 and also other supporting staff upon the completion of this project.

References

[1] Wu D Y, S Bateman and M Partlett 2007 *Composites Science and Technology* **67** 1909
[2] Ben D B, M Kharrat, M Dammak and G Monteil 2012 *Materials & Design* **34** 782
[3] Ibrahim M S, Sapuan S M and Faieza A A 2012 *J Mech Eng Sci.* **2** 133.
[4] Hardinnawirda K and SitiRabiatull Aisha I 2012 *J Mech Eng Sci.* **2** 181.
[5] Huzni S, Ilfan M, Sulaiman T, Fonna S, Ridha M and Arifin A K 2013 *Inter. J. Automot. Mech. Eng.* **7** 1023.
[6] Shan C W, Ghazali M I and Idris M I 2013 *Inter. J. Automot. Mech. Eng.* **7** 1032.
[7] Dobircau L, P A Sreekumar, R Saiah, N Leblanc, C Terrié, R Gattin and J M Saiter 2009. *Composites Part A: Applied Science and Manufacturing* **40** 329
[8] Adebisi A A, Maleque M A and Rahman M M 2011 *Inter. J. Automot. Mech. Eng.* **4** 471.
[9] Jeffrey K J T, Tarlochan F and Rahman M M 2011 *Inter. J. Automot. Mech. Eng.* **4** 504.
[10] Bachtiar D, Sapuan S M and Hamdan M M 2010 *Inter. J. Automot. Mech. Eng.* **1** 79.

[11] Scaffaro R, L Botta and G Di Benedetto 2012 *European Polymer Journal* **48**, 637
[12] Jarukumjorn, K. and N. Suppakarn 2009. *Composites Part B: Engineering* **40** 623
[13] Maadeed M A, R Kahraman, P Noorunnisa Khanam and N Madi 2012 *Materials & Design* **42** 289
[14] Lee D J, H Oh, Y S Song and J R Youn 2012 *Composites Science and Technology* **72** 278
[15] Umar A H, Zainudin, E S and Sapuan S M 2012 *J. Mech. Eng. Sci.* **2** 198.
[16] Bhaskar H B and Sharief A 2012 *J. Mech. Eng. Sci* .**3** 281.
[17] Hariprasad T, Dharmalingam G and Praveen Raj P 2013 *.J Mech. Eng. Sci.* **4** 518.
[18] Sevkat E, B Liaw, F Delale and B B Raju 2009 *Composites Part A: Applied Science and Manufacturing* **40** 1090
[19] Ravi Sankar H, Srikant R R, Vamsi Krishna P, Bhujanga Rao V and Bangaru Babu P 2013 *Inter. J. Automot. Mech .Eng* .**7** 968.
[20] Prachayawarakorn, J., P. Sangnitidej and P. Boonpasith 2010 *Carbohydrate Polymers* **81** 425
[21] Ibrahim, M. M, A Dufresne, W K El-Zawawy and F A Agblevor 2010 *Carbohydrate Polymers* **81** 811
[22] Shalwan A and B F Yousif 2012 *Materials & Design* **43** 283

IOP Conf. Series: Materials Science and Engineering **50** (2013) 012071 doi:10.1088/1757-899X/50/1/012071

A comparison of tensile properties of polyester composites reinforced with pineapple leaf fiber and pineapple peduncle fiber

J.M. Juraidi[1,a], **N. Shuhairul**[1], **S.A. Syed Azuan**[1] **and Noor Intan Saffinaz Anuar**[2]
[1]Universiti Kuala Lumpur Malaysia France Institute, 43650, Bangi, Selangor, Malaysia
[2]Malaysian Timber Industry Board, 50728, Kuala Lumpur, Malaysia

Email: [a]mohamadjuraidi@mfi.unikl.edu.my

Abstract. Pineapple fiber which is rich in cellulose, relatively inexpensive, and abundantly available has the potential for polymer reinforcement. This research presents a study of the tensile properties of pineapple leaf fiber and pineapple peduncle fiber reinforced polyester composites. Composites were fabricated using leaf fiber and peduncle fiber with varying fiber length and fiber loading. Both fibers were mixed with polyester composites the various fiber volume fractions of 4, 8 and 12% and with three different fiber lengths of 10, 20 and 30 mm. The composites panels were fabricated using hand lay-out technique. The tensile test was carried out in accordance to ASTM D638. The result showed that pineapple peduncle fiber with 4% fiber volume fraction and fiber length of 30 mm give highest tensile properties. From the overall results, pineapple peduncle fiber shown the higher tensile properties compared to pineapple leaf fiber. It is found that by increasing the fiber volume fraction the tensile properties has significantly decreased but by increasing the fiber length, the tensile properties will be increased proportionally. Minitab software is used to perform the two-way ANOVA analysis to measure the significant. From the analysis done, there is a significant effect of fiber volume fraction and fiber length on the tensile properties.

1. Introduction

Environmental awareness, new rules, and legislation are forcing industries to seek new materials that are environmental friendly. Over the past two decades, natural fibers have been receiving considerable attention as substitutes for synthetic fiber reinforcements [1-3]. Recently, natural plant fibers have been used in scientific researches as potential alternatives to glass fibers (GF) in fiber-reinforced plastics (FRP). Relative to glass fibers, these lignocelluloses fibers have lower densities, cost relatively lower, consume lesser energies during production, pose no abrasion to machines and have no health risk when inhaled [4-10].

From large selection of plant fibers, pineapple leaf fibers (PALF) have the highest cellulose contents which make the fibers mechanically sound [5]. According to this review, however, PALF are among the least studied fibers especially for reinforcing plastics. PALF come from the leaves of the pineapple plant, *Ananas comosus*, from Bromeliaceae family. Pineapple is one of the most important tropical fruits in Malaysia. Currently, the main focus of the pineapple industry in this country is the fruits and related foodstuffs leaving the leaves as agricultural wastes [4]. Even though PALF have been used traditionally as threads and textile in this country, works have only just started on studying them as commercial textile materials. Pineapple leaves are either composted or burned by farmers thus wasting potential source of good fibers.

There have been numerous studies carried out by researchers on various aspects of PALF. A number of authors analyzed the chemical composition of PALF [11-12]. PALF obtained from plants bearing inedible fruits were examined for textile purposes, and blends of PALF with silk and polyester fibers were studied [13]. Yu [14] studied chemically-treated PALF in terms of their properties and ability to be spun. Chemical treatments carried out on these technical fibers, while improving their fineness, resulted in degradation of their tensile properties [15].

Most of the recent studies, pineapple leaf fiber (PALF) has been used as a reinforcement instead of other part. This presented paper looks at the opportunity to utilise pineapple peduncle fibers (PAPF) as reinforcement in natural fiber composites. The aims of this paper is to compare the tensile properties between pineapple leaf fiber and pineapple peduncle fiber reinforced polyester composites by investigate the effects of fiber loading and fiber length on tensile properties.

2. Material and Methods

2.1. Preparation of Pineapple Leaf and Peduncle Fiber

Pineapple leaf and peduncle showed in figure 1, were obtained from Pekan Nanas, Johore. The pineapple leaf and peduncle fibers were extracted and were cut into 10, 20 and 30 mm of length. The pineapple leaf and peduncle were washed by using reverse osmosis water to remove the dirt and were dried for two days.

Figure 1: Pineapple leaf and peduncle fiber

2.2. Preparation of Composites

The specimens were manufactured by hand lay-up method, which was commonly used method for small scale composite manufacturing. All composite panels were prepared with various fiber volume fractions of 4, 8, and 12% and various fiber lengths of 10, 20 and 30mm. Polyester resin used as matrix with polyester / hardener ratio of 100: 1. The mixture was stirred until all the fiber was mixed with the polyester. The mixing was poured into mould and was pressed for 24 hours.

Figure 2: Fabrication by using hand layout

2.3. Test Specimen
The composite plates were cured in room temperature. After a week, the plates were cut into simple based on ASTM D638 Type 1 standard by using band saw machine.

2.4. Mechanical Testing
The tensile test was performed at the Fiber and Biocomposite Development Centre (FIDEC) at Olak Lempit, Banting. Tensile test of the specimens was carried out at room temperature with a constant cross-head rate of 10 mm/min using universal testing machine type Gotech model. The load cell that has been chosen is 5 tons. Five specimens were tested for each various fiber loading and fiber length, and the average value was obtained using Gotech software. Result of the tensile properties was analyzed by using analysis of variance (ANOVA). The purpose of this ANOVA analysis was investigated the design parameters significantly affect the quality characteristic of a product or a process.

3. Results and Discussion

3.1. Tensile Strength
Figure 3 shows a comparison of tensile strength between pineapple leaf and pineapple peduncle reinforced polyester composites. Both fiber showed the decreased trend in tensile strength when reached highest volume fraction. Result also shown by increasing the fiber length it also increased the tensile strength value. For overall result, fiber volume fraction at 4% and 30 mm of fiber length gave the highest value of tensile strength. Pineapple peduncle showed a higher tensile strength when compared to the pineapple leaf reinforced polyester composite. The highest value of the tensile strength value is 12.3 MPa which are from the pineapple peduncle while for the pineapple leaf fiber the highest is 10.2 MPa.

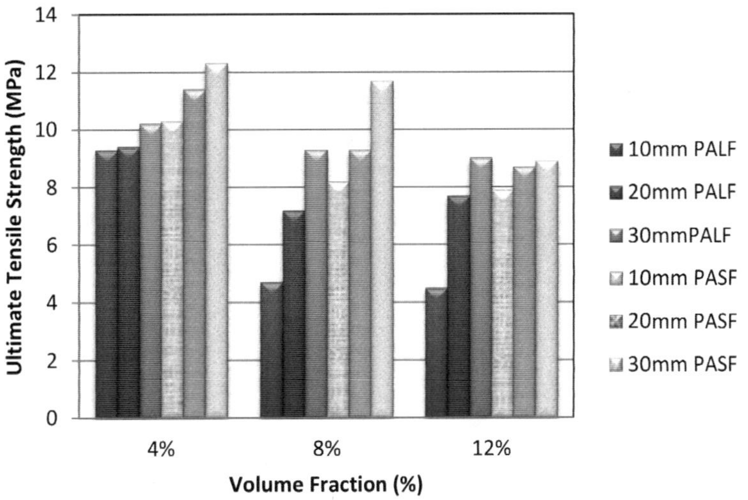

Figure 3: Comparison of tensile strength between pineapple leaf and peduncle fiber

The result was agreed with [16] that tensile strength increased with the volume fractions at the beginning and experienced slowly decreased when reach higher volume fraction. The reasons why tensile strength gave lower values for the volume fraction above 4% are possibly due to the fiber-to-fiber interaction, void and dispersion problems. In general, the composites showed an increasing trend in their mechanical properties over the fiber length. In this pineapple fiber, whether leaf or peduncle, increasing the fiber length showed the increasing trend in tensile strength. This is because the interlocked between the long fiber.

3.2. Analysis of Variance

An analysis of variance of tensile strength data in three different fiber volume fractions and three different fiber lengths was performed, with the objective of analyzing the influence of fiber loading and fiber length. For this study two away analysis of variance is performed because it involve two independent variable.

Table 1 and 2 show the results of the analysis of variance with the flexural strength in varies fiber volume fraction and fiber length.

In table 1 for pineapple leaf fiber reinforced polyester composites, the significant value of fiber volume fraction (0.00) and fiber length (0.00) are less than the threshold value (0.05), it can be concluded that the influence of fiber volume fraction and fiber length does effect the tensile strength. The interaction between the two factors fiber volume fraction and fiber length (0.084) are more than threshold value (0.05), leading to the conclusion that the combination of fiber volume fraction and fiber length does not significantly affect flexural properties of pineapple leaf fiber reinforced polyester composites.

In table 2, pineapple peduncle fiber reinforced polyester composites show a significant value of fiber volume fraction (0.00) and fiber length (0.00) are less than the threshold value (0.05). Thus, the influence of fiber volume fraction and fiber length did effect the tensile strength. The interaction between the two factors fiber volume fraction and fiber length (0.199) are more than threshold value (0.05), leading to the conclusion that the combination of fiber volume fraction and fiber length does not significantly affect flexural properties of pineapple peducle fiber reinforced polyester composites.

Table 1. Analysis of variance with the PALF tensile strength in varies volume fractions and fiber length.

Source	Type III Sum of Squares	df	Mean Square	F	P-value
Corrected Model	174.374[a]	8	21.797	7.517	.000
Intercept	2846.657	1	2846.657	981.729	.000
Volume	63.793	2	31.896	11.000	.000
Length	84.603	2	42.301	14.589	.000
Volume * Length	25.978	4	6.494	2.240	.084
Error	104.387	36	2.900		
Total	3125.418	45			
Corrected Total	278.761	44			

a. R Squared = .626 (Adjusted R Squared = .542)

Table 2. Analysis of variance with the PAPF tensile strength in varies volume fractions and fiber length.

Source	Type III Sum of Squares	df	Mean Square	F	P-value
Corrected Model	105.132[a]	8	13.141	8.541	.000
Intercept	4364.043	1	4364.043	2836.401	.000
Length	35.271	2	17.635	11.462	.000
Volume	60.100	2	30.050	19.531	.000
Length * Volume	9.762	4	2.440	1.586	.199
Error	55.389	36	1.539		
Total	4524.564	45			
Corrected Total	160.521	44			

a. R Squared = .655 (Adjusted R Squared = .578)

The values shown in this table are p-values of the two-way ANOVA tests. A p-value smaller than 0.05 indicates significant influence of the corresponding tensile on the corresponding property at the 5% significance level

4. Conclusion

In this paper, the tensile strength of pineapple leaf fiber and pineapple peduncle fiber reinforced polyester composites was measured. Based on the results, it was found that the tensile strength showed an increasing trend as the fiber length increased but decreased when increasing the fiber volume fractions. The optimum of fiber volume fractions and fiber length in polyester resin to obtain the highest tensile strength was found at 30 mm in fiber length and 4 % fiber volume fractions. In the analysis of variance, the interaction effect of fibre volume fraction and fiber length not significantly affect the tensile strength for both fibers. Pineapple peduncle fiber reinforced polyester composites showed a higher tensile strength compared to pineapple leaf fiber reinforced composites. The results confirmed that more studies are needed to determine the beneficial and cost effective applications of the pineapple peduncle fibre in composites.

Acknowledgment
The authors would like to thank the Department of Mechanical and Manufacturing, UniKL Malaysian France Institute and laboratory technician, Mr Mohd Radzi Bin Bujang of Fiber and Biocomposites Development Laboratory (FIDEC) for helping and guiding.

References

[1] Jeffrey K J T Tarlochan F and Rahman M M 2011 *Inter. J. Automot Mech Eng* **4** 504
[2] Adebisi A A Maleque M A and Rahman M M 2011 *Inter. J. Automot Mech Eng* **4** 471
[3] Shan C W Ghazali M I and Idris M I 2013 *Inter J Automot Mech Eng* **7** 1032
[4] Abdul Khalil H P S Siti Alwani M and Mohd Omar A K 2006 *Bioresources* **1** 220
[5] Mohanty A K Misra M and Drzal L T 2005 Natural Fibers Biopolymers and Biocomposites *Taylor & Francis* 890
[6] Ibrahim M S Sapuan S M and Faieza A A 2012 *J. Mech. Eng. Sci.* **2** 133
[7] Umar A H Zainudin E S and Sapuan S M 2012 *J. Mech. Eng. Sci.* **2** 198
[8] Wambua P Ivens J and Verpoest I 2003 *Composites Science and Technology* **23** 1259
[9] Bhaskar H B and Sharief A 2012 *J. Mech. Eng. Sci.* **3** 281
[10] Hadzreel M R A M and Aisha S R 2013 *J. Mech. Eng. Sci.* **4** 548
[11] Bhaduri S K and Sen S K 1983 *Carbohydrate Research* **121** 211
[12] Sharma U 1981 *Carbohydrate Research* **97** 323
[13] Anon 1992 *Textiles* **21** 21
[14] Yu C 1999 *Wool Record* **158** 49
[15] Kumar K B K Prabhakaran G Muruganandam R Raghu P Namasivayam N Rajendran K Durairaj V and Kannan G 1997 Study on pineapple fiber processing *Colourage* **44** 27-30
[16] Munirah M Lim K W Suriani A S Abdul R R Azman H 2005 *National Symposium of Polymeric Materials (NSPM)* UNITEN Bangi

IOP Conf. Series: Materials Science and Engineering **50** (2013) 012072 doi:10.1088/1757-899X/50/1/012072

Mechanistic model of stress corrosion cracking (scc) of carbon steel in acidic solution with the presence of H_2s

Y P Asmara[a], A Juliawati, A Sulaiman, and Jamiluddin
Department of Mechanical Engineering, Universiti Malaysia Pahang
26600, Pekan, Pahang, MALAYSIA.
Phone: +609-424-2246; Fax: +609-424-2202

E-mail:[a]YPanca@ump.edu.my

Abstract. In oil and gas industrial environments, H_2S gas is one of the corrosive species which should be a main concern in designing infrastructure made of carbon steel. Combination between the corrosive environment and stress condition will cause degradation of carbon steel increase unpredictably due to their simultaneous effects. This paper will design a model that involves electrochemical and mechanical theories to study crack growth rate under presence of H_2S gas. Combination crack and corrosion propagation of carbon steel, with different hydrogen concentration has been investigated. The results indicated that high concentration of hydrogen ions showed a higher crack propagation rate. The comparison between corrosion prediction models and corrosion model developed by researchers used to verify the model accuracy showed a good agreement.

1. Introduction

Problems on corrosion in the oil and gas industry are, frequently, caused by H_2S gas. Those gases widely exist in the oil fields which can increase corrosiveness of the environments. Its existence is indicated as a source of early failures of the pipeline. Degree of corrosiveness of H_2S gases are influenced by environmental conditions such as temperature, CO_2 partial pressure, corrosion film properties and mechanical properties of the materials [1.2].

For structures applications, materials subjected to static and exposure to the corrosive environment, there will be continuous degradation of the structural material due to mechanical and electrochemical effects. Interaction between those corrosion and mechanical process involve electrochemical and deformation of the crack tip region as the crack is developing. Crack growth characteristics of materials can be used to determine the retaining strength and durability of the materials [3].

Many studies have been conducted to deal with these gases and can be found in open literature [1,4]. Because it is very important, many experiments and field studies of H_2S corrosion has been done to provide a better understanding of the aspects that affect the whole process. To date, the use of research using pure empirical method to find corrosion model equations are still preferable. However, the problem is further complicated as the corrosion can be influenced not only by setting parameters during experimental works. It is influenced by crack geometry, film properties, tip region, atomic dissolution, and direction of slip dislocation. The complex interactions of various species and operating conditions are difficult to simulate in the laboratory. Thus, the technical accuracy of corrosion prediction will deviate with the real mechanism. An approach of corrosion prediction using combination of mechanistic and empirical model can simulate the real data conditions.

2. H_2S corrosion

Hydrogen sulfide, when dissolved in water, is involved in a series of chemical reactions in the pipeline [5]:

H_2S dissolution

$$H_2S_{(g)} \longleftrightarrow H_2S_{(aq)} \tag{1}$$

H_2S dissociation

$$H_2S_{(aq)} \longleftrightarrow HS^-_{(aq)} + H^+_{(aq)} \tag{2}$$

HS- dissociation

$$HS^-_{(aq)} \longleftrightarrow H^+_{(aq)} + S^{2-}_{(aq)} \tag{3}$$

H_2S Reduction

$$2H_2S_{(aq)} + 2e- \longleftrightarrow H_{2(g)} + 2HS^-_{(aq)} \tag{4}$$

FeS formation by precipitation

$$Fe_{(s)} + S^{2-}_{(aq)} \longleftrightarrow FeS_{(s)} \tag{5}$$

The role of H_2S in changing behavior of corrosion rate was studied by [6,7]. The following are three numerical models constructed by some researchers.

3. Mathematical Model

A mechanistic model of uniform hydrogen corrosion can be modeled using fundamental physicochemical laws [1,8]. It consists of convective diffusion equation (6), molecular diffusion equation (7) and diffusion via solid film equation (8).

(i) Convective diffusion reactions (*flux*) through boundary layer are caused by mass transfer processes (K_m), bulk concentration (C_{bi}) and outer concentration (C_{oi}).

$$Flux_i = k_{m,i}\left(c_{bi} - c_{o,i}\right) \tag{6}$$

(ii) Molecular diffusion (*flux*) through the liquid in the porous outer scale:

$$Flux_i = \frac{D_i \varepsilon \Psi}{\delta_{oc}}\left(c_{oi} - c_i\right) \tag{7}$$

where, D_i is diffusion coefficient for dissolved species i (m^2/s), ε is outer scale porosity, ψ is tortuosity factor, c_i is interfacial concentration of species *I*.

(iii) The very thin film and dense film at the metal surface forms a surface barrier. Then, solid state diffusion through the inner film can be formulated as:

$$Flux_i = A_i e^{\frac{B}{RT_K}} \ln\left(\frac{c_{i,i}}{c_{s,i}}\right) \tag{8}$$

Where, δf is film thickness, A is Arrhenius constant ,T_k is temperature (Kelvin),c_s is surface concentration, R is universal gas constant.

From the equations above, equation (1-8), expressed a mathematical corrosion theory *(CR)* as:

$$CR_i = A_i e^{\frac{B}{RT_K}} \ln \frac{c_{b,i} - CR_i\left(\frac{\delta_{oc}}{D_i \varepsilon \Psi} + \frac{1}{k_{m,i}}\right)}{c_{s,i}} \tag{9}$$

4. Electrochemical Theories

4.1. Anodic reaction

Based on the assumptions that dissolving metal ions into solution as a reaction is controlled by activation reaction, the equation reaction can be described as follows [8]:

$$i_{(Fe)} = i_{o,(Fe)} x 10^{\frac{\eta}{b_a}} \tag{10}$$

Where, b is tafel slope constant, i_o is current density, and η is polarisation potential.

4.2. Cathodic reaction

Cathode reactions occurring in the gas with some species undergo reduction reaction. Species that undergo reduction reactions are hydrogen, water, and acid bicarbonate (in the presence of CO_2 gas) as mentioned below.

4.3. Hydrogen ion reduction

To determine the effect of activation of the reaction and the mass flow of hydrogen reduction used the following equation (1) and equation (9):

$$\frac{1}{i_H} = \frac{1}{i_{H,a}} + \frac{1}{i_{H,\lim}} \tag{11}$$

Which i_a (H^+) is reaction activations (A/m^2), i_{lim} (H^+) is limiting current density (A/m^2).

4.4. Water reduction

The reaction of hydrogen ions (H^+) are also affected by the water. Reduction of water is formulated as follows:

$$i^o_{H2O} = i^*_{H2O} a_H^{-0.5} a_{H2O}^{2.2} \tag{12}$$

5. Mechanochemical Theories

The mechanochemical theories stated that a slip deformation is caused anodic dissolution at the bare surface which relates dislocation of slip plane, plastic shear and a constant stress [10] as expressed in the following equations.

$$\dot{\gamma} = \rho_d b \dot{X} \tag{13}$$

where ρ_d is the dislocation density, b is the Burgers vector and $\dot{X}$ is the average velocity of the dislocations. Then, crack tip strain rate at the tip, ε_{ct} , is formulated as:

$$\dot{\varepsilon}_{ct} = 2\dot{\gamma}\cos\theta \tag{14}$$

where γ is plastic shear deformation, θ is the angle between the direction of the slip plane and tensile stress.

5.1. Anodic dissolution current density

The average current density of anodic dissolution at the crack tip, i_a, can be related to the crack tip potential corrosion. It promotes slip dissolution and repassivation at the material surface. A slip step, t_{slip}, consist of passive film and anodic ions dissolutions due to slip formation [10].

$$i_a = \frac{1}{t_{slip}} \int_0^{tslip} i(t)dt \tag{15}$$

5.2. Crack growth rate

Mechanism of slip formation and dissolution at the crack tip can be related to crack growth rate, *da/dt*. Based on Faraday's law, average current density by anodic dissolution can be expressed as:

$$\frac{da}{dt} = \frac{M}{ZF\rho_m} ia \tag{16}$$

Where M is the atomic weight of the metal and ρ is the metal density. By substituting Eq. (13 - 16), the SCC crack growth rate is expressed by equation (17 - 19).

$$\frac{da}{dt} = A_o \left(\frac{\varepsilon_{ct}}{C_m} \right)^n \tag{17}$$

$$A_0 = \frac{M}{zF\rho_m} \frac{i_0 t_0^n}{(1-n)} \tag{18}$$

$$C_m = 2\rho_d \cos\theta N_{slip} n_d b^2 \tag{19}$$

Where A_0 is the rate coefficient, C_m is the material factor constant, and n is the numerical constant, water conductivity, and corrosion potential.

6. Implementation of the Model

6.1. Effects of H_2S concentration

Mathematical model calculates indications of increasing corrosion rate due to hydrogen ions concentration based on formula from equation (9). Figure 1 shows a tendency of relationship between corrosion rate and hydrogen ion concentration.

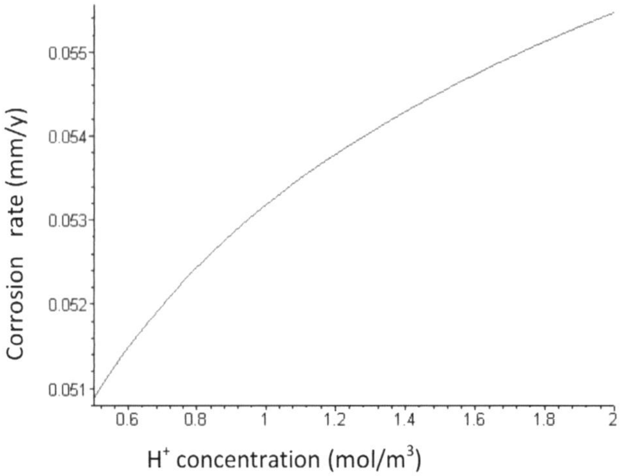

Figure 1. Simulated corrosion rate as a function of hydrogen ion concentration for a range of ions H^+ concentration from 0.4 mol/m ^{3} to 2 mol/m^3 at conditions: P_{total}= 1 bar and static.

It has been commonly accepted that the increase of time, corrosion current density will decrease exponentially. It becomes passive a certain period of time when film completely formed. But, at this condition, the films tend to dissolve which results corrosion rate start to grow. Figure 2 is a physical model used to describe corrosion and repassivation process. A mathematical relationship of current density and time used in this model is shown in figure 3.

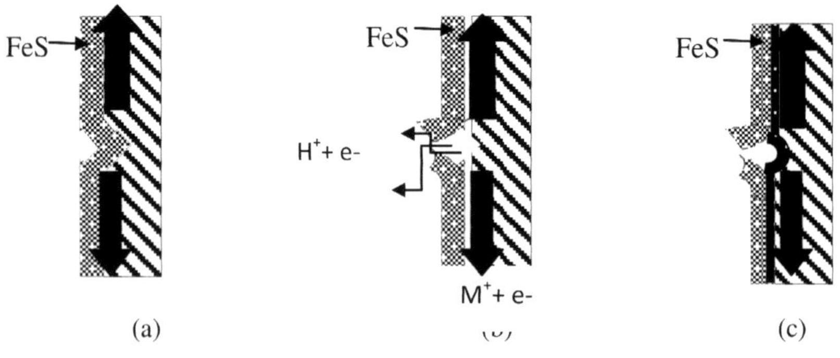

Figure 2. Models of film formation and film dissolution assumed at crack tip during corrosion process under stress condition. (a) Passive FeS film on crack tip (b) FeS film dissolution (c) FeS film formation.

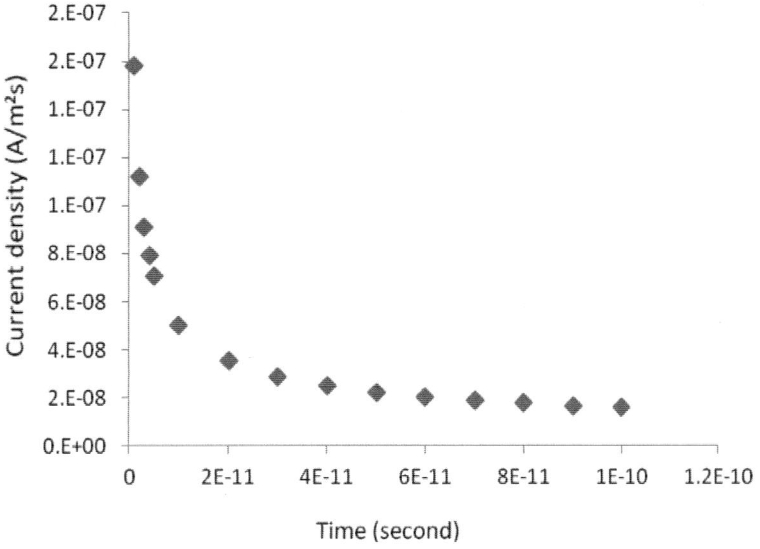

Figure 3. Current characteristic vs time of model at crack tip during corrosion.

Analytical observation from equation (17) indicates that crack growth has a decrease exponential function. This means that the corrosion rate decrease continuously within the SCC. The figures 4 shows a relationship between crack growth rate and material characteristics. Figure 5 present properties of dislocation densities related to crack growth rate.

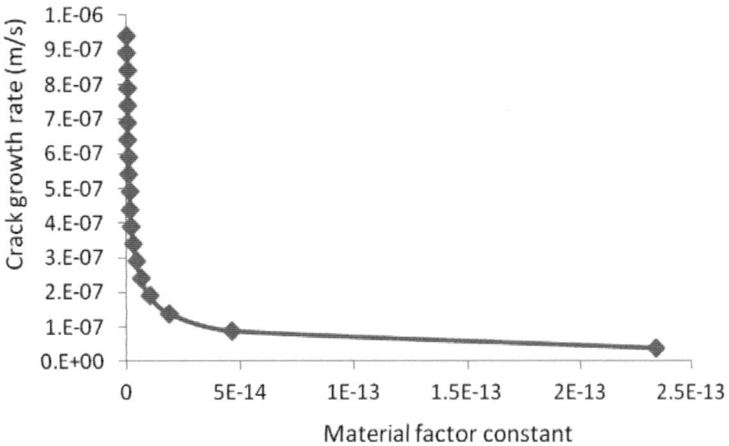

Figure 4. Observed relationship between crack growth and material factor constant (*Cm*).

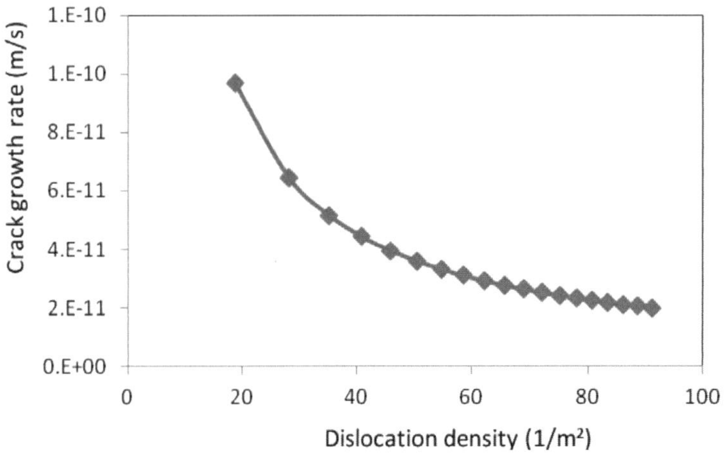

Figure 5. Observed relationship between crack growth and dislocation density.

7. Comparing models with published data experiments

Comparison experimental results are presented in this section to demonstrate the use of the model in determining crack extension during corrosion process. The data from calculations are validated using Irawan's experimental data [11] to assess the accuracy of the proposed model. The Irawan's experimental studies cover crack growth, corrosion current rate, and diffusion reaction rate. He used C ring samples materials with outer initial crack which were immersed in 1% - 3 % of H_2S solutions.

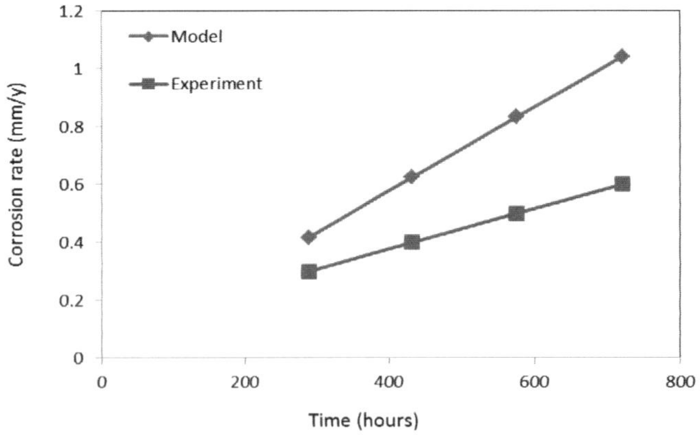

Figure 6. Comparison between the model to Irawan's experiment (in 1 % of H_2S saturated solutions at temperature 40°C, pH 4, total pressure 1bar).

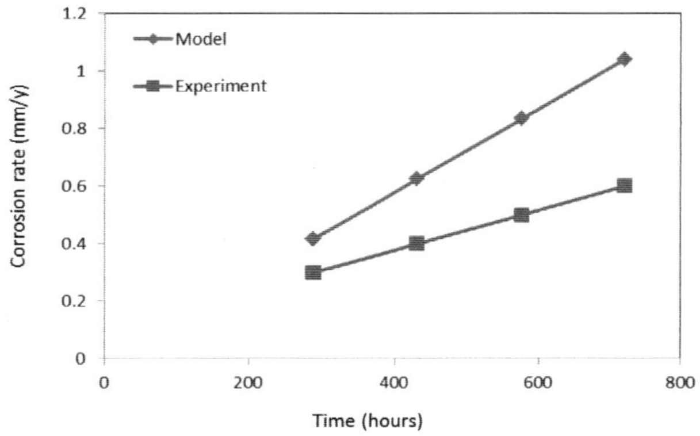

Figure 7. Comparison between the model to Irawan's experiment (in 2 % of H_2S saturated solutions at temperature 40°C, pH 4, total pressure 1bar).

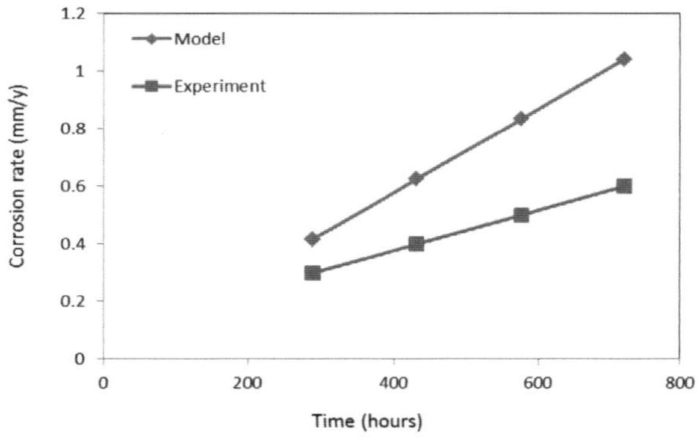

Figure 8. Comparison between the model to Irawan's experiment (in 3 % of H_2S saturated solutions at temperature 40°C, pH 4, total pressure 1bar).

Figure 6, 7 and 8 show crack growth (caused by corrosion and stress) as a function of time at 1%, 2% and 3% of H_2S gas concentration. The figures compare corrosion rate calculated mathematically based on equations developed by researchers [1,7,8] and experimental data reported by [11]. The graphs (figure 6, 7 and 8) indicated that the corrosion rate will increase when H_2S concentration increased. The increasing corrosion rate was proportional to the time. From the figures, based on experimental data and model calculation, corrosion rate increased sharply at the higher H_2S gas concentration and lower H_2S concentrations. It seemed there were interactions between H_2S gas concentration and exposure time. Therefore, it can be concluded that H_2S was a more dominant factor in controlling corrosion behavior then factor of time. It means that effect of scaling formation, as a result of corrosion decrease, was not considered by majority corrosion prediction models. Figures show good agreement between experimental and calculated data based on this corrosion prediction model. Those three comparisons show good correlations, coefficient determinations and significant errors estimations which have similarity more than 95%.

8. Conclusion

In general, the use of numerical model has a benefit to simulate real conditions. Thus, complicated experiments can be reduced. Using mechanistic theories, relationship among the variables tested will be more simple in the form of mathematical equations. Thus, effects of independent variables will be easier to identify and to develop. Furthermore, using mathematical operations, the certain conditions such as stationary conditions can be calculated analytically. In H_2S corrosion, stationary point conditions can be used to indicate scaling formation, limiting current density and independent flow conditions.

It must be known that each model is suitable for certain condition and several assumptions have been made. This model involving the most variables input such as mass flow rate, scale effects, H_2S gas, as inputs parameters. Thus, this model gives a realistic approach to calculate corrosion rate in H_2S environments. Is is a simple model which measures corrosion rate in the worst cases. So, the corrosion rate result given is always higher. In associated with determination of scaling formation, Future work on optimization should be started with complex variables. The complex variables can be selected using design experiments to determine the important variables that can be developed further.

Acknowledgments

The authors are thankful to Faculty Mechanical Engineering, Universiti Malaysia Pahang for providing grant and facilities for the research.

References

[1] Nesic S, Kun-Lin, John Lee 2002 A Mechanistic model of iron carbonate film growth and the effect on CO_2 corrosion of mild steel *NACE International, Houston*, 02237

[1] Vera J R Oil Characteristics 2006. Water /Oil wetting and flow influence on the metal loss, Corrosion. Part 1: Effect of Oil and Flow on CO_2/H_2S Corrosion *NACE International, Houston*. 6113

[2] Smith S N and Pacheco J L 2002 Prediction of corrosion in slightly sour environments *NACE International* 2241

[3] Mokhtar C I 2000 Predictiong CO_2 corrosion with the presence of acetic acid *PhD theses, UMIST*

[4] Keith G 2004 Electrochemical investigation and modeling of carbon dioxide corrosion of carbon steel in the presence of acetic acid *NACE International* 4379

[5] Brown B, Lee K, Nesic S 2003 Corrosion in multiphase flow containing small amount of H_2S *NACE International Corrosio*. 3341

[6] Kvarekval J 2007 Morphology of localised corrosion attacks in sour environments *NACE International* Houston. 7659

[7] Anderko, A. Robert, D Y 1999 Simulation of CO_2 / H_2S corrosion using thermodynamic and electrochemical models, Corrosion *NACE International* Houston, **31**

[8] Wei Sun S N 2006 Kinetics of iron sulfide and mixed iron sulfide/carbonate scale precitipation in CO_2/H_2S, *PhD Thesis, Department of Chemical Engineering, Russ College of Engineering and Technology, Ohio University*

[9] Dougherty and James A 2004 Aggressive agents as corrosion inhibitors in gas and oil production *NACE International, Houston*. 4423

[10] Koichi S and Jiro K 2001 Mechanochemical model to predict stress corrosion crack growth of stainless steel in high temperature water *Corrosion Science* **43** 1751-1766

[11] Irawan 2010 Analisa laju perambatan retak untuk jenis korosi scc pada pipeline akibat unsur H_2S, *Fyp Project Report*. FTK. ITS.

AUTHOR INDEX

AUTHOR INDEX